Handbook of Cereal Science and Technology

FOOD SCIENCE AND TECHNOLOGY

A Series of Monographs, Textbooks, and Reference Books

1. Flavor Research: Principles and Techniques, *R. Teranishi, I. Hornstein, P. Issenberg, and E. L. Wick (out of print)*
2. Principles of Enzymology for the Food Sciences, *John R. Whitaker*
3. Low-Temperature Preservation of Foods and Living Matter, *Owen R. Fennema, William D. Powrie, and Elmer H. Marth*
4. Principles of Food Science
   Part I: Food Chemistry, *edited by Owen R. Fennema*
   Part II: Physical Methods of Food Preservation, *Marcus Karel, Owen R. Fennema, and Daryl B. Lund*
5. Food Emulsions, *edited by Stig Friberg*
6. Nutritional and Safety Aspects of Food Processing, *edited by Steven R. Tannenbaum*
7. Flavor Research: Recent Advances, *edited by R. Teranishi, Robert A. Flath, and Hiroshi Sugisawa*
8. Computer-Aided Techniques in Food Technology, *edited by Israel Saguy*
9. Handbook of Tropical Foods, *edited by Harvey T. Chan*
10. Antimicrobials in Foods, *edited by Alfred Larry Branen and P. Michael Davidson*
11. Food Constituents and Food Residues: Their Chromatographic Determination, *edited by James F. Lawrence*
12. Aspartame: Physiology and Biochemistry, *edited by Lewis D. Stegink and L. J. Filer, Jr.*
13. Handbook of Vitamins: Nutritional, Biochemical, and Clinical Aspects, *edited by Lawrence J. Machlin*

14. Starch Conversion Technology, *edited by G. M. A. van Beynum and J. A. Roels*
15. Food Chemistry: Second Edition, Revised and Expanded, *edited by Owen R. Fennema*
16. Sensory Evaluation of Food: Statistical Methods and Procedures, *Michael O'Mahony*
17. Alternative Sweeteners, *edited by Lyn O'Brien Nabors and Robert C. Gelardi*
18. Citrus Fruits and Their Products: Analysis and Technology, *S. V. Ting and Russell L. Rouseff*
19. Engineering Properties of Foods, *edited by M. A. Rao and S. S. H. Rizvi*
20. Umami: A Basic Taste, *edited by Yojiro Kawamura and Morley R. Kare*
21. Food Biotechnology, *edited by Dietrich Knorr*
22. Food Texture: Instrumental and Sensory Measurement, *edited by Howard R. Moskowitz*
23. Seafoods and Fish Oils in Human Health and Disease, *John E. Kinsella*
24. Postharvest Physiology of Vegetables, *edited by J. Weichmann*
25. Handbook of Dietary Fiber: An Applied Approach, *Mark L. Dreher*
26. Food Toxicology, Parts A and B, *Jose M. Concon*
27. Modern Carbohydrate Chemistry, *Roger W. Binkley*
28. Trace Minerals in Foods, *edited by Kenneth T. Smith*
29. Protein Quality and the Effects of Processing, *edited by R. Dixon Phillips and John W. Finley*
30. Adulteration of Fruit Juice Beverages, *edited by Steven Nagy, John A. Attaway, and Martha E. Rhodes*
31. Foodborne Bacterial Pathogens, *edited by Michael P. Doyle*
32. Legumes: Chemistry, Technology, and Human Nutrition, *edited by Ruth H. Matthews*
33. Industrialization of Indigenous Fermented Foods, *edited by Keith H. Steinkraus*
34. International Food Regulation Handbook: Policy • Science • Law, *edited by Roger D. Middlekauff and Philippe Shubik*
35. Food Additives, *edited by A. Larry Branen, P. Michael Davidson, and Seppo Salminen*
36. Safety of Irradiated Foods, *J. F. Diehl*
37. Omega-3 Fatty Acids in Health and Disease, *edited by Robert S. Lees and Marcus Karel*
38. Food Emulsions, Second Edition, Revised and Expanded, *edited by Kåre Larsson and Stig E. Friberg*
39. Seafood: Effects of Technology on Nutrition, *George M. Pigott and Barbee W. Tucker*
40. Handbook of Vitamins, Second Edition, Revised and Expanded, *edited by Lawrence J. Machlin*

41. Handbook of Cereal Science and Technology, *Klaus J. Lorenz and Karel Kulp*
42. Food Processing Operations and Scale-up, *Kenneth J. Valentas, Leon Levine, and J. Peter Clark*

Other Volumes in Preparation

Fish Quality Control by Computer Vision, *edited by L. F. Pau and R. Olafsson*
Volatile Compounds in Foods and Beverages, *edited by H. Maarse*

Handbook of Cereal Science and Technology

edited by

KLAUS J. LORENZ
Colorado State University
Fort Collins, Colorado

KAREL KULP
American Institute of Baking
Manhattan, Kansas

MARCEL DEKKER, INC.
New York • Basel • Hong Kong

Library of Congress Cataloging-in-Publication Data

Handbook of cereal science and technology / edited by Klaus J. Lorenz, Karel Kulp.
p. cm. -- (Food science and technology ; 41)
Includes bibliographical references and index.
ISBN 0-8247-8358-1 (acid-free paper)
1. Cereal products. I. Lorenz, Klaus J. II. Kulp, Karel. III. Series: Food science and technology (Marcel Dekker) ; 41.
TP434.H36 1991 90-49154
664'.7--dc20 CIP

This book is printed on acid-free paper.

Marcel Dekker, Inc.
270 Madison Avenue, New York, New York 10016

Current printing (last digit):
10 9 8 7 6 5 4 3 2 1

PRINTED IN THE UNITED STATES OF AMERICA

PREFACE

This handbook brings together agronomists, cereal scientists and technologists, and chemists and engineering scientists with common interests in the raw materials, processes, and products of the cereals industries, to write a series of reviews on the chemistry and technology of cereals and also to provide an extensive and up-to-date bibliography. The chapters follow a logical sequence. Early chapters deal with the major cereals—wheat, corn, barley, oats, sorghum, millet, rice, rye, triticale, and wild rice. A series of reviews then covers the chemistry of components of these cereals—proteins, carbohydrates, lipids, vitamins, minerals, and enzymes—and the quality evaluation of grains and flours. Subsequent chapters cover the technology of baked products, pasta, breakfast cereals, malted cereals, and snack foods. The last three chapters discuss cereal enrichment, the nutritional quality of cereals, and the methods of ethanol production from cereal grains.

Knowledge of the production and utilization of cereal grains is fundamental to solving the food supply problem facing the majority of the world's population. In many developing areas of the world, as much as 75% of dietary energy comes from cereals, and in more affluent societies, the intake of cereals is seldom less than 25% of calories. Even in the latter case, the nutrient contribution of cereals is substantial. Cereals (including wheat, corn, barley, oats, sorghum, millet, rice, rye, triticale, and wild rice) are therefore not only important direct sources of food for humans, but also make a substantial contribution to the diet, indirectly as fodder, for farm livestock producing meat, milk, and eggs.

Optimal utilization of cereal grains requires knowledge of their structure and composition. The practical implications of kernel structure are numerous. They relate to the various stages of grain production, harvest, storage, marketing, and use. The chemical composition of different cereal grains varies widely, since it is influenced by genetics, soil, and cultural and climatic factors. Variations are encountered in the relative amounts of protein, carbohydrates, lipids, vitamins, minerals, and pigments. The composition, chemistry, and processing aspects of grains are therefore areas of scientific pursuit absolutely fundamental to producing more food for the world's population.

Although new varieties of cereal grains have been developed and there has been one cross of two grains (triticale), no new major cereal crops have been introduced in recent history. Genetic improvements are being made through increased worldwide efforts coordinated through international centers. New varieties, hopefully with increased yield, must continue to offset the adaptation of disease organisms and stabilize production. At the same time, efforts are directed towards increasing the nutritional value of each crop.

Increasing awareness of nutritional needs presents problems to those responsible for feeding our populations to plan wisely the strategies of food resources development. In this planning, both economically and technologically sound nutritional decisions have to be integrated. To assist in this effort, detailed knowledge of materials and processes for making breads, cakes, cookies, pasta, breakfast cereals, and snack foods—all of which represent highly complex and variable systems from a biochemical and physicochemical point of view—is essential. This text presents a comprehensive, up-to-date examination of these processes in light of current information derived from basic chemical and biochemical studies. In cereal processing, the modifications associated with these improvements have involved, in many instances, a drastic conversion of a time-developed craft into scientifically sound technology. This change has often required major changes in raw materials and processing procedures, occasionally resulting in finished products that are somewhat different from the traditional products.

Establishing enriched cereal food products for potential consumers requires research. There is a need to define the market, to delineate the varying types of consumers, and to interpret correctly their needs, habits, and attitudes so that their requirements can be met in a consistent manner. If consumers are to follow the eating patterns advised in national nutritional policies, it is necessary for them to partly replace caloric intake of alcohol, sugar, and fat with complex carbohydrate foods. The only convenient way of achieving this is to increase the intake of cereals and cereal products. This dietary change would also help to ensure a plentiful supply of dietary fiber, especially if the cereals were consumed as high extraction flour products. We have to communicate and interpret findings of cereal research to the public in concise, understandable terms. If we cannot get our message through to the public, it is the fault of the scientist and technologist and not that of the potential recipient of the information.

We extend credit and appreciation to the contributors to this book. Authorities in many fields, who hold responsible positions in universities, government, and industry, contributed chapters or reviewed the manuscripts. We are very grateful to them for giving much of their time and expert knowledge.

In editing this handbook, we have attempted to achieve some uniformity among the chapters and to eliminate obvious duplication. We appreciate the authors' cooperation, care, and patience in the preparation of their manuscripts.

This book contains the most up-to-date technical and scientific information. We hope that it will serve as a valuable text and reference book for both the student and the researcher involved in the cereal industry.

Sincere thanks go to Mrs. Inez Kwiatkowski and Ms. B. Martin for typing, and to Lisa A. Sisley of the Ruth Emerson Library, American Institute of Baking, for her expert proofreading.

Klaus J. Lorenz
Karel Kulp

CONTENTS

CONTRIBUTORS

Robert Becker, B.S. Agricultural Research Service, Western Regional Research Center, U.S. Department of Agriculture, Albany, California

Margaret Ann Bock, Ph.D Department of Home Economics, New Mexico State University, Las Cruces, New Mexico

James J. Boedicker, Ph.D North Central Experimental Station, University of Minnesota, Grand Rapids, Minnesota

Walter Bushuk, Ph.D Department of Food Science, University of Manitoba, Winnipeg, Manitoba, Canada

Okkyung Kim Chung, Ph.D Agricultural Research Service, Northern Plains Area, U.S. Grain Marketing Research Laboratory, U.S. Department of Agriculture, Manhattan, Kansas

Bruce E. Dale, Ph.D Department of Chemical Engineering, Texas A&M University, College Station, Texas

Anna R. Dera, Ph.D Department of Agronomy, University of Missouri, Columbia, Missouri

Brendan J. Donnelly, Ph.D Northern Crops Institute, North Dakota State University, Fargo, North Dakota

Hamed Faridi, Ph.D R. M. Schaeberle Technology Center, Nabisco Brands, Inc., East Hanover, New Jersey

J. Perry Gustafson, Ph.D Department of Agronomy, Agricultural Research Service, U.S. Department of Agriculture, and University of Missouri, Columbia, Missouri

Grace D. Hanners, B.S. Agricultural Research Service, Western Regional Research Center, U.S. Department of Agriculture, Albany, California

Eugene A. Hockett, Ph.D Agricultural Research Service, U.S. Department of Agriculture, and Plant and Soil Science Department, Montana State University, Bozeman, Montana

Lawrence A. Johnson, Ph.D Center for Crops Utilization Research, Iowa State University of Science and Technology, Ames, Iowa

Karel Kulp, M.S. Department of Research, American Institute of Baking, Manhattan, Kansas

George L. Lookhart, Ph.D Grain Quality and Structure Research Unit, U.S. Grain Marketing Research Laboratory, Agricultural Research Service, Northern Plains Area, U.S. Department of Agriculture, Manhattan, Kansas

Klaus J. Lorenz, Ph.D Department of Food Science and Human Nutrition, Colorado State University, Fort Collins, Colorado

Joseph A. Maga, Ph.D Department of Food Science and Human Nutrition, Colorado State University, Fort Collins, Colorado

Paul J. Mattern, M.S. Department of Agronomy, University of Nebraska—Lincoln, Lincoln, Nebraska

Cassandra M. McDonough, M.S. Cereal Quality of Laboratory, Department of Soil and Crop Sciences, Texas A&M University, College Station, Texas

Michael S. McMullen, Ph.D Crop and Weed Sciences Department, North Dakota State University, Fargo, North Dakota

Ervin A. Oelke, Ph.D Department of Agronomy and Plant Genetics, University of Minnesota, St. Paul, Minnesota

Richard E. Pyler, Ph.D Brewing Research and Development, Coors Brewing Company, Golden, Colorado

Gur S. Ranhotra, Ph.D Nutrition Research Group, American Institute of Baking, Manhattan, Kansas

Peter M. Ranum, M.S. Flour Service Department, Atochem North America, Buffalo, New York

Vladimir F. Rasper, Ph.D Department of Food Science, University of Guelph, Guelph, Ontario, Canada

Lloyd W. Rooney, Ph.D Cereal Quality Laboratory, Department of Soil and Crop Science, Texas A&M University, College Station, Texas

Sergio O. Serna-Saldivar, Ph.D Cereal Quality Laboratory, Department of Soil and Crop Sciences, Texas A&M University, College Station, Texas

Roy N. Sharp, Ph.D Department of Food Science, University of Arkansas, Fayetteville, Arkansas

David A. Thomas, M.S. Malting Research and Development, Coors Brewing Company, Golden, Colorado

Ronald E. Tribelhorn, M.S. Department of Agricultural and Chemical Engineering, Colorado State University, Fort Collins, Colorado

Handbook of Cereal Science and Technology

1

WHEAT

Paul J. Mattern
University of Nebraska–Lincoln, Lincoln, Nebraska

I. CEREAL GRAINS OF THE WORLD

A. Origin of Wheat

Wheat has been used as a food by man since prehistoric times, however, we do not have adequate records on its early history. It is generally believed that wheat evolved from wild grasses somewhere in the Near East. Humans probably first gathered wheat from the wild as long as 12,000–17,000 years ago. Types collected may have been *Triticum monococcum* (einkorn) or *T. dicoccum* (emmer) (8,48).

Triticum aestivum (common wheat), *T. compactum* (club wheat), and *T. sphaerococcum* (shot wheat) descended in several steps from emmer about 3000–4000 B.C. For the most recent information on the evolution of the cultivated wheats, see Ref. 35.

B. Classification of Wheat

Wheats belong to the grass family *Gramineae (Poaceae)* and to the tribe *Hordeae* (6), genus *Triticum*, species *aestivum*, and are given cultivar (variety) names by their plant breeders.

Linnaeus in 1753 made significant contributions to the classification of wheat, however, Sakamura (47) identified the chromosome numbers for commonly recognized wheat (Triticum) types. The diploids were found to contain 14 chromosomes ($n = 7$), the tetraploids had 28 chromosomes ($n = 14$), and the hexaploids had 42 chromosomes ($n = 21$). The three chromosome sets (or genomes), combined to form the cultivated wheats, have been the subject of intensive cytological research.

Common wheat (T. aestivum) consists of three genomes identified as A, B, and D. The source of the A and D genomes was clearly established as coming from *T. monococcum* and *T. tauschii (Aeqilops squarrosa)*. The source of the B genome has posed a difficult research problem in recent years. *Triticum searsii* is reported as the probable source of the B genome in common wheat (35).

Triticum durum, a tetraploid, is derived from a hybrid of diploid *T. boeticum* and an unknown wild grass. Each contributed a set of seven chromosomes (or a genome). Natural hybridization of a similar tetraploid with the diploid *T. tauschii* produced common wheat, *T. aestivum*, and club wheat, *T. compactum*.

Most of the wheat currently grown consists of three species of *Triticum*. *Triticum aestivum* L., a hexaploid, dominates world production. In the United States it is divided into four classes for marketing purposes. Hard Red Winter and Hard Red Spring classes are used primarily for leavened bread production. Soft Red Winter and Common White wheat classes are used primarily for pastries, crackers, and cookies. *Triticum compactum* Host, a club wheat, also a hexaploid type, is used for pastries in a similar manner as the Soft Red Winter wheats. Durum wheat, *Triticum durum* Dest., is a tetraploid species with extremely hard grain which is used for macaroni, spaghetti, and other pasta products. It is often used for arabic flat bread in North Africa and the Near East.

To facilitate marketing, plant breeders have kept distinctive combinations of grain protein and hardness. The common wheats can be produced with a wide combination of hardness and protein. However, to complement their normal end uses, bread wheats are selected to have a desired combination of hardness, protein level, and protein quality.

Soft wheat hexaploids are selected to have soft endosperm texture and low protein. Production in areas of high rainfall helps to promote the soft texture and low protein content.

The tetraploids are reported (68) to be associated with hard grain types. Diploids have soft grains and hexaploids can be variable. The club wheats, as hexaploids, could be selected as hard types, but have been maintained as a soft market class.

C. Morphology and Grain Characteristics

Wheat cultivars (varieties) are annual grasses which vary in habit, form, and structure. The wheat plant consists of roots, culms (stems), leaves, and spikes (heads). The spike is the inflorescence (flowering part). Plant parts may vary considerably among cultivars. A comprehensive review on the morphology of the wheat plant is given in Refs. 5 and 6.

Wheat is a self-pollinated crop. However, some cross-pollination can occur, but this is usually less than 1%. New wheat cultivars arise as rare combinations due to the hybridization between selected parents (49).

Plant breeders have made tremendous changes in wheat varieties. Increased production for this important food source has come about with shorter plant heights, which reduce lodging with fertilization and/or irrigation. Improved winter hardiness along with resistance to diseases and insects has stabilized production. In addition, the breeder has selected wheats for desired end use quality.

Red color of common hexaploid wheat grains is controlled by three independent genes. They are independently inherited and cumulative, but not equal, in effect. Total absence produces white wheat (38).

Durum grain also may be white or red. High-protein vitreous white durum is called amber durum. Red durum was preferred for producing puffed wheat at one time, but currently it is not an important U.S. crop.

Kernels of common wheats range from 5 to 8 mm in length and 2.5 to 4.5 mm in width; their average weight is reported at 37 mg, but it is not unusual in kernels of diverse genetic material to find a range of 20–60 mg. A longitudinal and cross section of a wheat grain along with an identification of components is shown is Figure 1. Proportions of the wheat grain components are given in Table 1.

D. Growing Conditions

Wheat is a cool weather plant and was considered for many years to require a low-humidity climate. Currently wheat is grown over a wide range of climates on diverse soil types. This adaptation contributes to its success as a major world crop (7).

Wheat has become important in crop rotations involving double-cropping with soybean or grain sorghum in the southeastern United States. Wheat production in this area requires

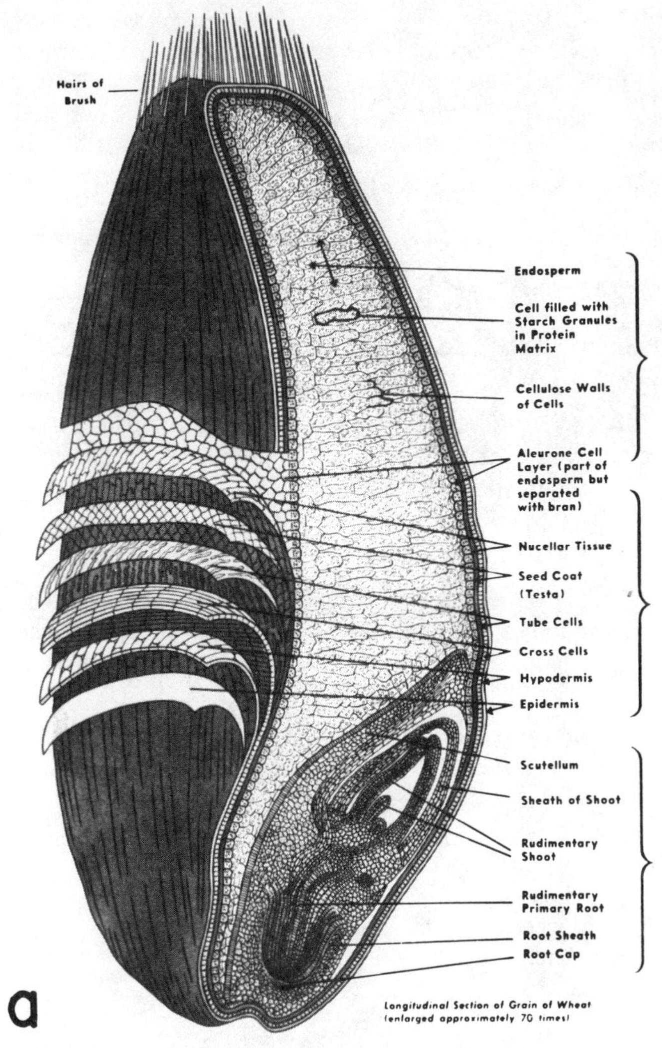

FIGURE 1 View of a wheat grain. In longitudinal section enlarged approximately 70 times (a), and in cross section (b). *Source*: Wheat Flour Institute, Washington, D.C.

proper cultivars, weed, insect, and disease control. In addition, adequate fertilization with macro–and micronutrients is essential.

Winter wheats must be fall planted and require a period of low temperature (vernalization). If spring planted, the growth usually remains prostrate on the ground and does not produce a flowering spike. Spring wheats do not require vernalization and are spring planted. They can be fall planted successfully in regions of mild winter temperatures. Wheat types range from true spring to true winter types. The expression of the trait depends on temperature, length of day, and time of seeding (8).

An in-depth report on all aspects of wheat improvement and associated topics can be found in Refs. 26, 29, and 45.

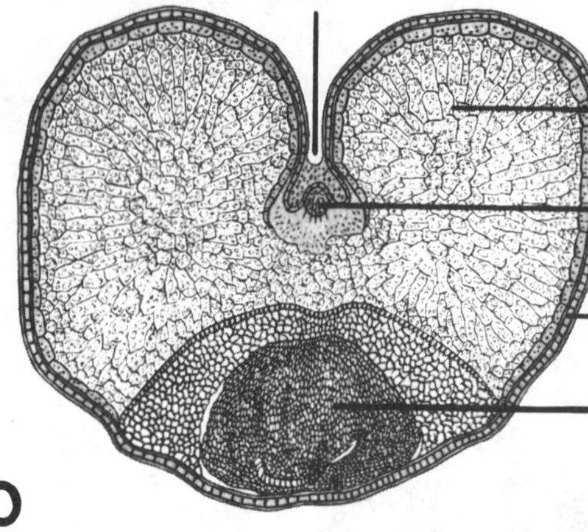

FIGURE 1 (Continued)

E. Production of Wheat

Areas of wheat harvest by major wheat producers are listed in Tables 2 and 3 by region and country, respectively. On a world basis, total area harvested increased only 6.6% from 1969 to 1985 (Table 2). Of interest is the 22.7% reduction in harvested acreage in the U.S.S.R. during this same period (65.230 million ha reduced to 50.384 ha).

World wheat production has grown significantly since 1965. Average world production during 1960–1964 was 232 million metric tons (Tg). During 1978–1980 this had grown to 432 Tg, an increase of 86%. The 1984–1985 average production of 516.8 Tg represented a 123% increase over the 1960–1964 average. Significantly improved management practices, especially greater use of nitrogen fertilizers, and improved new high-yielding cultivars contributed strongly to the increase.

Wheat production statistics are given in Tables 4 and 5 for world regions and countries, respectively.

Ten countries produced 76% of the worlds' wheat as based on the 1983–1985 average (Table 6). Three of these, the People's Republic of China (PRC), the U.S.S.R., and the United States accounted for 45.5% of the world production. Only the United States, France, Canada, Australia, and Argentina are consistent exporters. Among the 10 leading exporters, only Australia and Argentina are in the Southern Hemisphere.

Wheat yields (kg/ha) for major world regions and selected countries are shown in Tables 7 and 8, respectively. All regions have increased their wheat yields, but the highest increase is in Europe (Table 7). On a world average basis this has amounted to an increase of 45.5%. The highest yields in Europe were reported for the United Kingdom, West Germany, and France where ideal growing conditions were coupled with intensive agricultural practices. Most of the leading producers (Table 8) do not have impressive yields. Instead they compensate with larger planting areas. France has an impressive record with a combination of low harvested area coupled with high grain yields (Table 8).

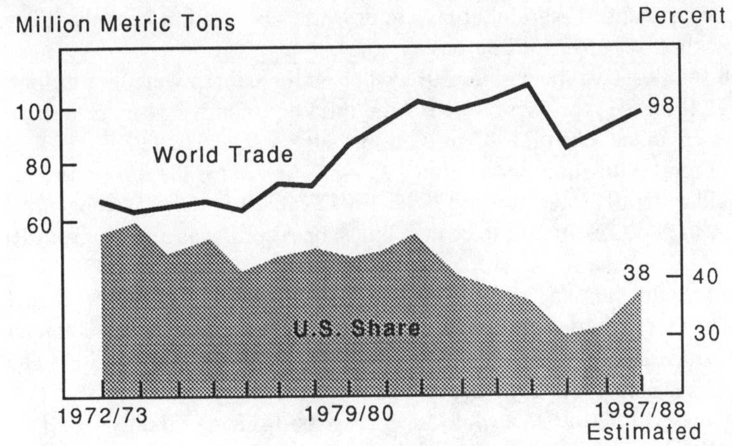

FIGURE 2 World wheat trade 1972/73 to 1987/88. (From Ref. 57, p.2.)

F. Wheat Exports and Imports

Major wheat and flour exporters are given in Tables 9 and 10 for world regions and selected countries, respectively.

Wheat production increased dramatically in certain countries that had historically been importers, thus lowering the demand for wheat from traditional exporting countries. In some instances they not only met their needs but became exporters with their overproduction. Ratios of exports between developed and developing countries were 17.7/1 in 1976 and 8.1/1 in 1985.

The leading wheat-exporting countries for 1983, 1985, and 1987, along with their percentages of world trade, are listed in Table 11. The United States and Canada had decreased exports from 1983 to 1985, whereas Australia and France experienced increases. During the same period Argentina remained a constant exporter. In 1987/88 the United States and Canada returned towards the 1983 totals, while Australian and Argentinian exports decreased. Data for France were not available for the 1987 crop year, but the EC–12, comprised mainly of French exports, also showed a decrease (in comparison to 1985 data for France).

Wheat imports for 1976–1978 and 1984–1986 are given in Tables 12 and 13 for regions and selected countries, respectively. Ratios of imports between developed and developing countries remained constant at 0.8/1 for both 1976 and 1985.

Dramatic developments took place in wheat production and trade during 1987/88 (25), when production was down, exports were up, stocks were reduced, and prices to farmers were 10% above \$2.28 loan value in December 1987 and prices had increased to 61.3% over \$2.17 loan value by June 1988 (59). Sources for grain market information are listed by Wellman (67).

U.S. exports as a share of total U.S. crop for selected years are given in Table 14. U.S. market year-ending wheat stocks peaked in 1985 (Table 15) at 1905 million bushels. The slightly reduced U.S. wheat stocks in 1986 reflected the slightly enlarged share of the world wheat trade (Fig. 2).

Table 15 was complied from May 1988 data (60) with an estimated carryover of 796 million bushels. Projections as of July 12, 1988 (71), issued by the Economic Research

Service of the U.S. Department of Agriculture, were down to 666 million bushels. [The August 11, 1988 projection was 597 million bushels (61).]

This would be the smallest carryover since 1975 when ending stocks were 435 million bushels. The July 12, 1988 (71), projection marks the third consecutive year for a draw down from the record ending stocks of 1905 million bushels on June 1, 1986.

Increased feed and food utilization of wheat in 1986–1987 resumed the long-term upward trend that had held generally steady during 1985–1986 (Table 16). Global wheat utilization in 1986–1987, which was estimated to be the highest on record, was up 34% from 10 years earlier.

World wheat stocks established a new record in 1986–1987 (Table 17). Based on annual utilization, the stocks were less burdensome (72). They were approximately 29% of annual utilization, the largest since 1969–1970 but below the 30–38% level of the late 1960s. The build-up of wheat stocks during 1986–1987 occurred outside the United States (Table 17). U.S. wheat exports increased 12% in 1986, while non-U.S. exports increased only 3%. This U.S. increase reflected a change in agricultural policies including lower price support levels, direct payments to farmers, and expanded export enhancement programs.

G. Grain Reserves and Disappearance

Yearly stocks of U.S. common and durum wheat for on-and off-farm storage since 1971 are shown in Tables 18 and 19, respectively. Total stocks for both wheat types were at their lowest point in 1973. The highest levels of durum and common wheat stocks were in 1982 and 1985, respectively.

A comparison of U.S. wheat production and total use, and the stocks-to-use ratios for marketing years 1972/73 to 1987/88 are given in Figure 3.

H. Wheat Storage

Mankind has a continuing challenge in maintaining an adequate food supply. Wheat and other cereal grains can be stored with confidence if certain precautions are observed, but too often we forget that the wheat grain is a living organism which was produced to maintain

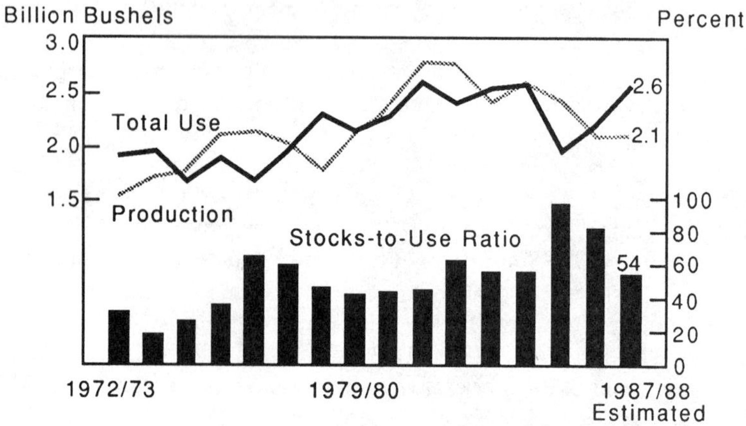

FIGURE 3 Total U.S. wheat production, use, and stocks-to-use ratio 1972/73 to 1987/88. (From Ref. 57, p.2)

another life cycle. Wheat life processes can be noted by respiration and germination of the grain. Respiration continues during storage, and it is essential to keep the germ alive. The level of respiration depends upon the amount of oxygen available and on temperature and moisture of the grain. Respiration is a normal process, but it should be kept at a minimal rate to reduce heating and to reduce the loss of dry matter of the grain. Wheat is considered to be dry if its moisture content is below 13%. Most commercial storage is kept below 13.5% moisture, although some may permit moisture levels to reach 14%.

The moisture content of stored grain, in a closed system, is dependent upon the relative humidity of the air surrounding the grain. Approximate moisture contents of three wheat classes stored under five different levels of relative humidity are given in Table 20.

During storage grain can be damaged by insects, microorganisms, rodents, and birds. For protection from the weather, grain is stored in flat warehouses or vertical silos. One successful long-term storage technique developed in Australia, and now used in the United States, is an air-tight bunker which can hold up to a million bushels per unit. Aeration fans reduce the moisture content to below 11%. Carbon dioxide is flushed through the grain and the bunker is hermetically sealed (28). One charge of carbon dioxide effectively inhibits insect activity and reduces the respiration rate. This process eliminates the use of toxic insecticides and can enhance consumer acceptance.

I. Wheat Standards and Grades

As the grain trade developed in the United States, there was need for a reliable commercial language so that buyers and sellers at separate locations could trade without an exchange of samples. Grain trade organizations took the lead in setting standards but could never agree on consistent standards. The demand for uniform grades and inspection resulted in 26 different bills between 1903 and 1916 in the United States, which provided for federal supervision of grain grading or for outright federal grain inspection (63). These bills and hearings produced the U.S. Grain Standards Act, which was passed August 11, 1916. The official standards for wheat were implemented on July 1, 1917, for certain classes and on August 1, 1917, for others (58). The act provided in part for (a) the establishment of official U.S. grain standards, (b) the federal licensing and supervision of the work of grain inspectors and, (c) establishment of an Appeals Board for complaints on grades assigned by the licensed inspectors. Currently, U.S. standards are in effect for wheat, corn, barley, oats, rye, triticale, sorghum, flaxseed, soybeans, and mixed grain (62).

According to the U.S. Standards for Wheat, wheat is defined as (19):

> grain that, before the removal of dockage, consists of 50 percent or more common wheat (*Triticum aestivum* L.), club wheat (*T. compactum* Host.), and durum wheat (*T. durum* Desf.) and not more than 10 percent of other grains for which standards have been established under the United State Grain Standards Act and that, after the removal of the dockage, contains 50 percent or more of whole kernels of one or more of these wheats.

Wheat is divided into seven classes based on color, kernel and varietal characteristics. The seven classes are: Hard Red Spring wheat, Hard Red Winter wheat, Soft Red Winter wheat, Durum wheat, White wheat, Unclassed wheat, and Mixed wheat. The classes Hard Red Spring wheat, Durum wheat, and White wheat are further divided into subclasses. Each class and subclass is divided into five U.S. numerical grades and U.S. Sample grades (Table 21). Special grades are provided to emphasize special qualities or conditions affecting the value of wheat and are added to and made a part of the grade designation. Special grades do not affect the numerical or sample grade designation.

1. Hard Red Spring wheat (HRSW) is divided into the following three subclasses: a) Dark Northern Spring wheat: HRSW with 75% or more of dark, hard, and vitreous kernels. b) Northern Spring wheat: HRSW with 25% or more, but less than 75% of dark, hard, and vitreous kernels. c) Red Spring wheat: HRSW with less than 25% of dark, hard, and vitreous kernels.

2. Durum wheat is divided into the following three subclasses: a) Hard Amber Durum wheat: Durum wheat with 75% or more hard and vitreous kernels of amber color. b) Amber Durum wheat: Durum wheat with 60% or more, but less than 75% of hard and vitreous kernels of amber color. c) Durum wheat: Durum wheat with less than 60% of hard and vitreous kernels of amber color.

3. White wheat. All varieties of white wheat are divided into the following four subclasses. a) Hard White wheat: White wheat with 75% or more of hard kernels. It may contain not more than 10% of White Club wheat. b) Soft White wheat: White wheat with less than 75% of hard kernels. It may contain not more than 10% of White Club wheat. c) White Club wheat: White Club wheat containing not more than 10% of other White wheat. d) Western wheat: White wheat containing more than 10% of White Club wheat and more than 10% of other White wheat.

Unclassed wheat is any variety not classed under the standards. It includes Red Durum wheat and wheats of other colors.

Mixed wheat consists of any mixture of wheat that consists of less than 90% of one class and more than 10% of one other class, or a combination of classes meeting the definition of wheat (40).

J. International Wheat Standards

Grain grading has developed differently in various countries, although the fundamental concepts are similar. Determinations used to decide wheat grade include: test weight, 1000 kernel weight, moisture, foreign material, dockage, shrunken and broken kernels, total kernel defects, hardness of grain, color of grain, insects and/or damage, protein or gluten, sedimentation value, falling number and Chopin "W" value (50).

Only test weight, moisture, dockage (sometimes), and protein are objective tests without judgment estimates. Other test factors require direct sensory judgments.

Buyers using wheat for specific end uses would prefer tests that identify samples for such uses. Cereal scientists generally agree that protein level, protein quality as determined by a mixograph or other test, and grain hardness are three minimal tests, on a "sound" wheat at 57 lb/bu test weight or above, which could identify a wheat for best end use (56).

Research has established near infrared reflectance (NIR) spectroscopy as a bulk sample method for wheat grain hardness (69). NIR has been routinely used in recent years for protein, moisture, and lipid determinations (among others) (42). A comprehensive review of NIR for grain applications is given in Ref. 70.

Instrumentation for determining grain hardness of individual wheat grains is under development and test. The Federal Grain Inspection Service could use such devices to determine the hardness class of a wheat and to detect blending of soft and hard classes. An example of such an instrument is reported in Ref. 44.

A comparison of wheat grading systems in major exporting countries is given in Table 22. Australia markets seven general classes. These classes are: Prime Hard, Hard, Standard White, Soft Durum, General Purpose, and Feed (Table 23). Canada markets seven general wheat classes (Table 24). Canadian Western Red Spring is marketed at three grades: No. 1 at three protein levels; No. 2 at two protein levels; and No. 3 at an unspecified protein level.

Complete information on world wheat grading systems is not readily available. Fortunately, the U.S. Congress Office of Technology Assessment was requested by Congress in the Food Security Act of 1985 to conduct a study on "Technology and Public Policy to Enhance Grain Quality in International Trade." The report is to be published in two volumes during 1988. It will analyze four subject areas: 1) grain quality attributes important to buyers, 2) the U.S. grain system's ability to produce and deliver high quality grain, 3) a comparison of the systems of major competitor countries with the U.S. system, and 4) policy options to enhance U.S. grain quality. The published work will be available through the U.S. Government Printing Office.

II. PROCESSING AND UTILIZATION OF WHEAT

In 1987 the United States processed 35.6% of its total wheat production. This was 750 million bushels (20.5 million metric tons) from 2105 million bushels (57.4 million metric tons) (64).

Wheat purchased under U.S. standards does not necessarily reflect end-use functional needs. Buyers rely on known regions for traditional uses which are maintained by state and private wheat breeding programs.

A. Wheat Milling

The miller selects and blends wheats which have the potential to produce the desired flour products. The most important criterion is the maintenance of a uniform quality flour product. Since wheats have a wide range of properties, wheat is binned according to certain critical quality factors, e.g., test weight, protein cultivar, etc. and is blended to maintain a desired flour product from milling.

B. Grain Cleaning

Grain received for milling contains field contamination, which includes plant parts, weed seeds, stones, lumps of soil, and dried or living insects. During shipping extraneous materials such as coal, metal fragments, and other grains are accumulated.

A typical wheat-cleaning facility for a 200 cwt (hundred weight) mill is shown in Figure 4. Mixing and weighing are followed by sieves, gravity table, destoner, and separators, which may be cylinder or disc types. The wheat is impacted with an entoleter and loosened dust is removed by aspiration. Final aspiration is combined with air blasts to improve separation efficiency. The cleaned wheat is ready for conditioning.

C. Conditioning

Prior to milling, water is added to wheat in a process known as "tempering." Hard wheats are normally brought to 15–16% moisture content, soft wheats to 13–14% moisture. Tempered wheat is held 18–24 hours at ambient temperature. This process toughens the bran and softens the starchy endosperm. Some millers may add 0.5–1.0% additional water 30–60 minutes before milling to improve bran separation.

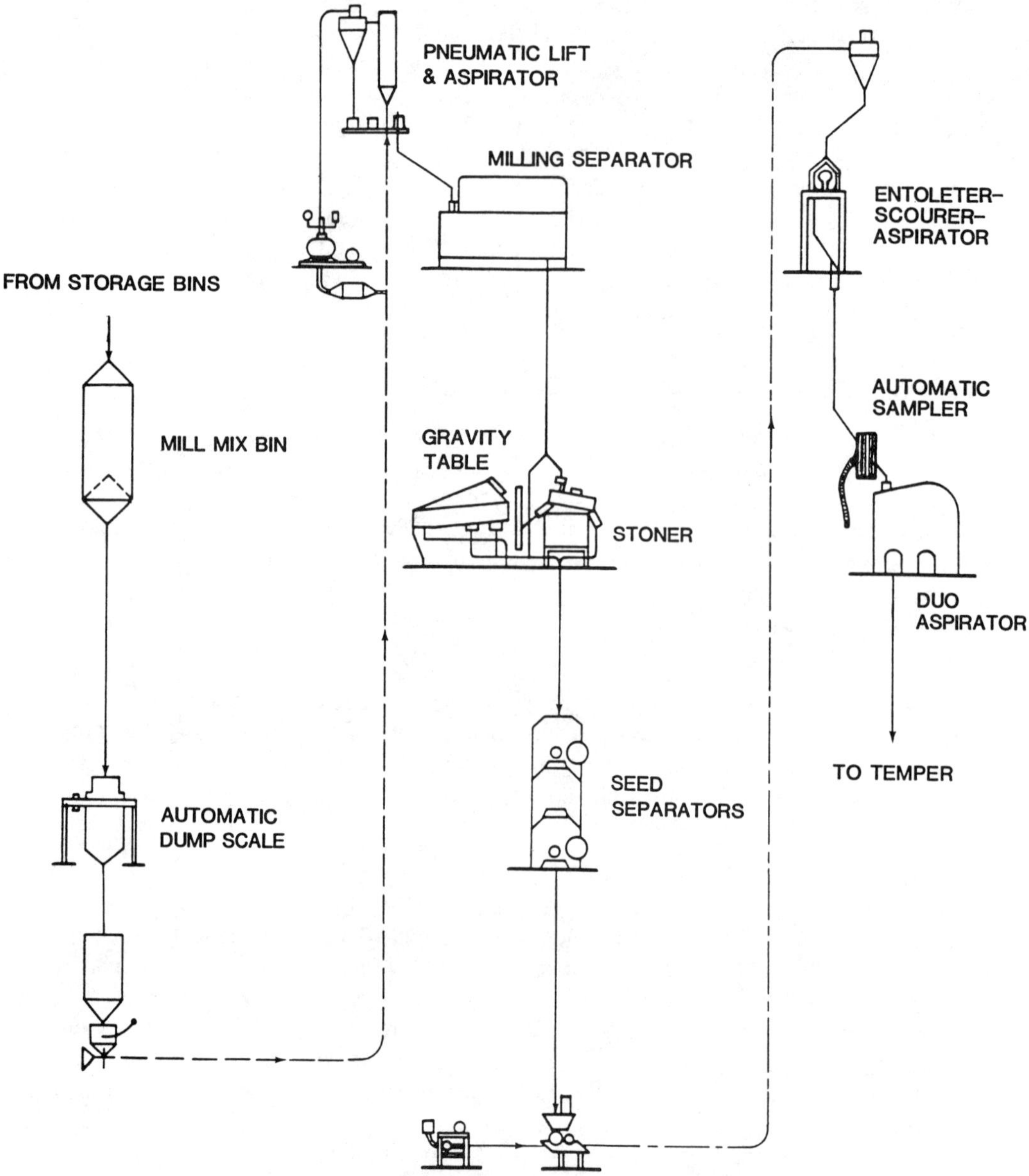

FIGURE 4 A typical grain-cleaning house used by Kansas State University for their 200 cwt pilot mill. (*Source*: Kansas State University.)

D. Milling for White Flour Production

The objective of flour milling is to separate the starchy endosperm portion of the grain from the bran and germ and reduce the endosperm to a desired granulation. Conventional flour milling uses a gradual reduction system. Traditionally, the Hard Red Winter and Hard Red Spring wheats have been used for bread production. At present, Kansas, Montana, and several private seed companies have active breeding and testing programs for Hard White Winter wheat, which would be selected for bread quality. Some Hard White wheat has been grown yearly in the Pacific Northwest.

Commercial mills can vary in design. A 200 cwt flour mill (capable of making 20,000 lbs of flour per day) illustrates a typical milling process (Fig. 5). For larger capacities more machines would be connected in parallel. The conditioned wheat is fed to prebreak rolls which are turning toward each other with a differential speed of 2.5:1. The faster roll speed in the United States would be approximately 500 rpm. The five break rolls (BK) have spiral corrugations the length of the roll. The slow roll tends to hold the wheat grain as the faster roll produces a crushing and shearing action. In Figure 5 the 1 BK (rolls) are 10 inches in

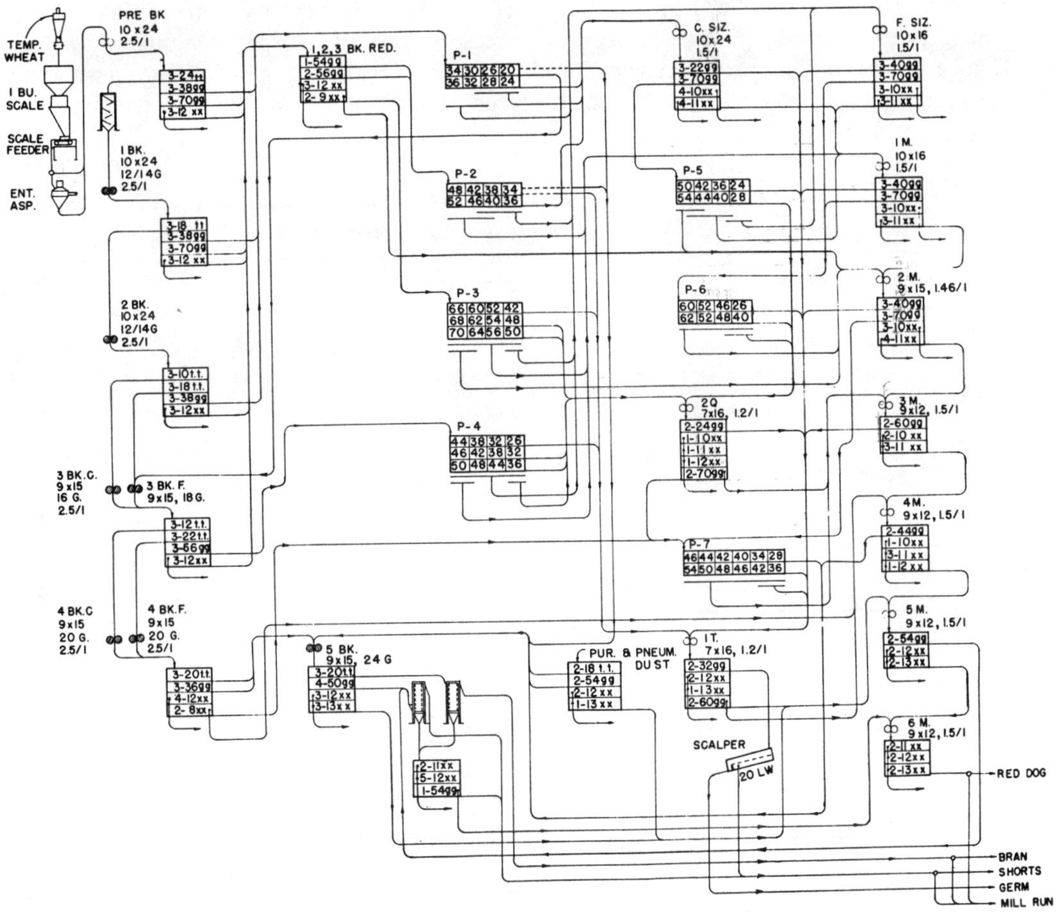

FIGURE 5 Milling diagram of Kansas State University 200 cwt flour mill, hard wheat flow. Each sifter box shown represents 1/4 2 × 14 sifter. (*Source*: Kansas State University.)

diameter and 24 inches long (10 × 24) and have 10 corrugations/inch on the slower roll (10/12 G). ("G" indicates a Getchell corrugation type spiral cwt.) Each miller has his own preference for types of corrugation, roll speed, etc.

Mill stocks are sifted after each successive set of break rolls, and the coarser material is conveyed to the next set of break rolls. Roll spacings are decreased at each successive break stage. Flour is removed at each break roll step. After the fifth break, the large bran flakes are conveyed to a bran duster for final cleanup. The larger endosperm chunks (middlings) are separated from small bran particles with the purifiers (P). The middlings are reduced to desired particle size with smooth rolls (M).

From a given sample of wheat being milled, the miller can produce a number of flour types or grades by combining different mill streams. The total number of mill streams will depend upon the size and sophistication of the mill. The weight of 100% straight flour which can be obtained from 100 lbs of wheat at an extraction rate of 72% would be 72 lb. Names of flour grades, or types, from various extraction rates and feed products are given in Figure 6. Milling extraction rates have increased during the past 30 years, and a typical mill today would achieve at least a 75% extraction rate.

Although flour milling was the first automated industrial system, the past 50 years have brought sophistication by application of computerization,, air-conveying systems, high-speed packaging, and rail and truck bulk delivery. One would expect variation in mills among and within countries. The trend to increase processing capacity continues. In Europe faster roll speeds have been employed, increasing from 300 rpm to 500 rpm. This has been accompanied by more efficient wheat feeding to the rolls by the Pratique feeding plate or auxiliary wind rolls (31).

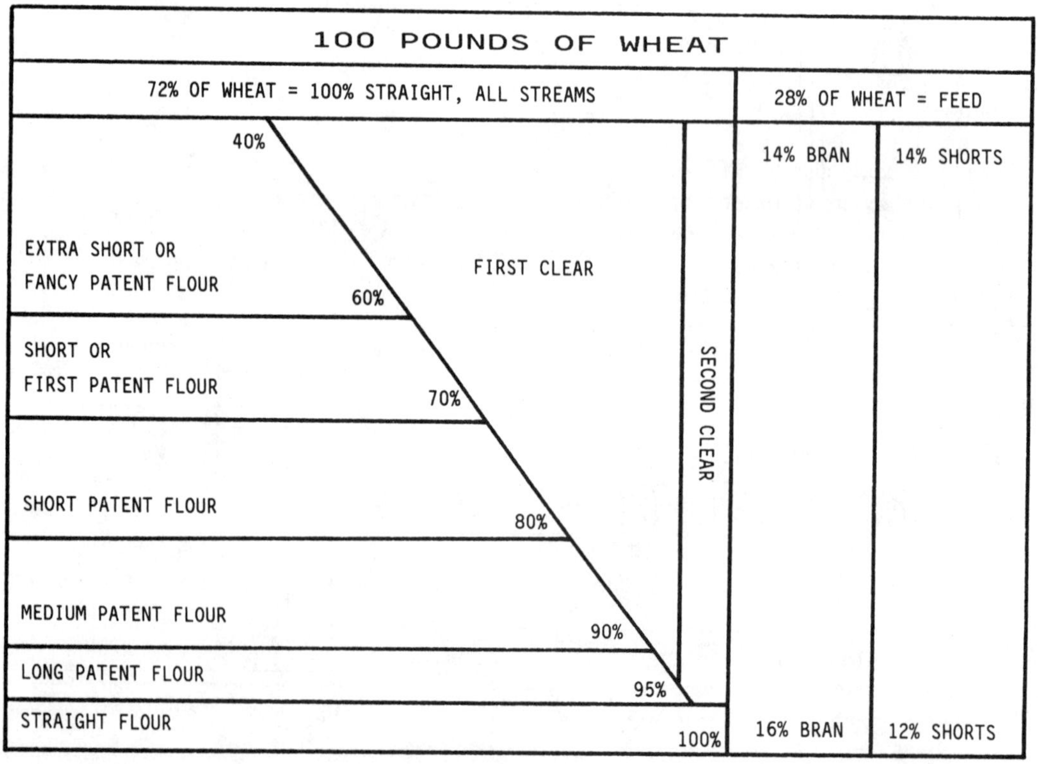

FIGURE 6 Flour grades for a typical milling system. (From Ref. 54.)

During milling there is separation of the endosperm from other kernel components. Tables 25 and 26 provide nitrogen and percent composition data from hand dissected kernels for embryo, scutellum, pericarp-testa, aleurone, and successive dissections of endosperm for a semi-hard and semi-soft wheat, respectively. Because starchy endosperm is separated from higher protein materials, there is a protein differential between flour and wheat. This loss of protein is reported in Table 27 for two wheats of varying hardness at two protein levels. This loss is generally 1.0%.

In an unpublished study the author found several lines of high protein germplasm in which the flour protein was higher than the wheat protein because bran proteins were low. If successful agronomic varieties could include these genetic factors, there would be a more efficient recovery of wheat protein in the milled flour.

A study (66) reports percent endosperm content along with flour and bran weight and flour remaining with the bran for diverse wheat germplasm (Table 28). The range for percent endosperm, 72.8–85.5, was large for these nonvariety samples. In this same study, means and the results of the analysis of variance of commercial check cultivars for kernel weight, milling yield, flour weight, bran flour weight, washed brand weight, sample recovery percent, and percent endosperm were compared and are listed in Table 29. The check cultivars differed significantly for milling yield, flour weight, bran flour weight, washed bran weight, and percent endosperm. Most of the variation was due to differences among the wheats analyzed. Sample recovery percentage was not affected by cultivar or field replication. Considering the number of steps in the modified milling procedure, the sample recovery percentages are satisfactory. The range in percent endosperm of sample was fairly large. Part of the variation for percent endosperm can be attributed to within-nursery environmental variation since replications had a slight effect on percent endosperm. Most of the variation for percent endosperm was probably genetic in origin because there were highly significant differences among the check cultivars.

Soft wheats such as Atlas 66 and Nap Hal had lower endosperm percentages than the hard wheats Centurk and Scout 66. Part of the variation among the wheats for percent endosperm could be due to some wheats having thicker bran and aleurone layers than others. Seed size could also be a factor.

There are large differences among the wheats for milling yield that are due almost entirely to kernel texture. The mean milling yield of Triumph 64, a Hard Red Winter wheat, is 8.4 percentage points greater than that of Atlas 66. Triumph 64, however, is only 2 percentage points higher in endosperm percent. Atlas 66 had almost twice as much endosperm adhering to the bran as Triumph 64.

In the evaluation of 40 Hard Red Winter wheats, regrown from varieties released during the past 70 years, no significant change was noted for flour yield (11).

E. Soft Wheat Milling

Soft wheat is milled by the process of gradual reduction similar to that used for hard bread wheats. Break flour yields from soft wheats are higher than those of hard wheats. Soft wheat flours agglomerate and require about double the sifter surface as needed for an equal capacity hard wheat mill.

Hardness in wheat is reversed by moisture. Therefore, soft wheats are milled at moisture levels of 13–14%. This keeps optimum milling properties and maintains the highest flour moisture for best economic return.

For comprehensive coverage of soft wheat, see Ref. 53.

F. Durum Wheat Milling

Cleaning and conditioning durum wheat prior to milling is similar to that of other wheat classes. An additional modification is the use of high speed beaters to remove as much of the germ as possible. This simplifies the purification of the semolina product during milling.

Durum wheat contains 1–3% of broken kernels due to breakage which occurs during threshing and transferring the extremely hard grain. These broken grains, which are separated during cleaning with seeds and other extraneous matter, are a valuable component which can be recovered with a specific gravity separator.

Durum wheat, which has the hardest starchy endosperm of all commercial wheats, must be tempered to a higher moisture level. Tempering times are similar to that of the bread wheats, but the tempering level is increased to 17–17.5% moisture. Tempering levels and holding times are dictated by climatic conditions. Colder temperatures require longer temper times for proper water penetration in the grain. Some millers use heat treatment to shorten temper times or to influence milling separation.

The main objective of durum milling is to produce the largest quantity of semolina possible. This goal is achieved by using more break rolls than for normal bread flour production, and by removing impurities from the semolina with more extensive cleaning and purifying systems. Traditionally pasta manufacturers prefer semolina in the 550–150 μm particle size range. Newer high-speed pasta manufacturers specify a finer granulation of 450–130 μm. Some producers have standardized on 350–130 μm size. A comprehensive review on the milling of durum wheat is given by Bizzarri and Morelli (4).

G. U.S. Durum Milling Production

U.S. production of straight semolina and durum flour reached its highest production point in 1987: 20,943,000 cwt (Table 30). This represents a 68.7% increase over the 1966 production value of 12,411,000 cwt. Yearly production from 1966 (when records were first kept) has been variable, though generally increasing.

Durum milling weighted extraction rates for years 1982–1987 are given in Table 31. Extractions ranged from 73.9 to 78.1% with a mean of 75.3%.

H. World Durum Outlook

World durum production was estimated at 953.2 million bushels, down 9% from the 1987/88 crop of 1052 million bu. and 1070 million bu. in 1986/87. Due to adverse weather in Canada and the United States, 1988 Canadian production was forecast at 99 million bushels compared to 146.6 million bu in 1987. U.S. production may be down to 58.7 million compared to 93.3 million in 1987. The U.S. imported 7.3 million bu of durum from Canada in 1987/88 because of poor quality.

World trade of durum for July, 1988 to June, 1989, was forecast at 159 million by compared to 185.5 million bu the previous year. U.S. exports were expected to be reduced by 25.7 million bushels.

Closing stocks of durum for Canada (73.3), the United States (51.3), and the E.E.C. (77.0) at the end of the 1988/89 season were projected at 201.6 million bushels. Supply should be adequate for the 1988/89 marketing season (61).

For a recent publication on durum wheat, see Ref. 14.

I. Feed Use of Wheat

A number of factors influence the use of wheat in animal rations. The most important are price, location, and nutritional value. It may prove more economical to buy wheat locally than to ship traditional feed grains greater distances.

Feed use can also vary in relation to quality. If wheat is unfit for milling for some reason, it may be fed to livestock. On a worldwide basis animal feed wheat use for 1980–1984 ranged from 19.00 to 20.93 million metric tons with a mean of 19.84 million metric tons (46).

Feeding of wheat to livestock in 1988/89 in the United States was estimated at 270 million bu (61). This is an increase of 20 million bu from the June 1988 projections and reflects shortages of traditional feed grains because of drought conditions and poor-quality market wheat because of disease and drought in the wheat-growing region.

III. CHEMICAL COMPOSITION

A. Protein

General information on wheat composition can be misleading. Various classes of wheat are intentionally bred and selected for a specific composition, usually to meet the end-use requirements for a product. For example, commercial soft wheats are maintained at low protein levels, although certain soft wheats are associated with genes for high protein and are used as germplasm in breeding programs to develop high protein hard wheats (30). Protein content in a single variety of wheat can vary from 7 to 20% depending upon growing environment and fertilizer use. Therefore, any simple data set or comparison for wheat should consist of average values.

Typical protein ranges for selected world wheats are given in Table 32. Protein content is negatively correlated with grain yield so that spring wheats are generally higher in protein content than winter grown types. Genetic potential is also important, as was noted earlier. Hard wheats, as they have been selected and maintained, contain more protein than other cereals except oats. Wheat contains less protein than other common foods, such as legumes (18–26% protein), meat (12–20%), fish (16–20%), eggs (13%), and cheese (19–25%) (2).

Constituents of wheat composition for a hard and a soft wheat are given in Table 33. The high-protein wheat is higher in protein in all constituents except the germ.

Constituents of wheat grains are not distributed uniformly. Composition of anatomical parts of the wheat grain along with caloric values are compared in table 34. The pericarp (bran) is high in pentosans, cellulose, and ash. The combined testa and hyaline are high in pentosans, cellulose, ash, and protein. The aleurone is botanically part of the endosperm but during milling is separated with the bran. It is high in protein, lipids, pentosans, and ash, thus contributing significantly to the nutritional quality of bran as a feedstuff. Starch is found in the endosperm. The outer endosperm (subaleurone) is higher in protein than the inner portion (32,33). The embryo and scutellum, which make up the germ, are high in protein, lipids, reducing sugars, and ash. Literature values for petroleum ether–extracted lipids of germ are about 30% of the weight. Commercial mill run germ has lipids in the 10–11.5% range because of contaminants separated with the germ.

Because of the structure of grain and its composition of various parts, as noted in Table 34, milling extraction rates affect flour composition. With an increase in extraction rates, protein, fat, and fiber increase, whereas carbohydrates decrease (Table 34). Specific data are influenced by kind of wheat, grain size, milling techniques, and others. Table 35 illustrates a typical situation.

In cereals only the bread wheat, and to some extent rye and tritcale, have storage proteins that form the gluten network in flour and water doughs which has the unique properties of elasticity and strength to produce yeast-leavened bread. Storage proteins comprise 85% of wheat endosperm proteins and consist of gliadin (alcohol-soluble) and glutenin (alkali- or acid-soluble) fractions.

Distribution of the cereal proteins according to solubility is found in Table 36. Part of the variation in the composition of cereal proteins based on solubility arises from differences in solvent composition and time and temperature of extraction. Growing environment and variety can affect the ratio of the storage proteins. As protein increases in a given cultivar, the gliadin fraction increases and the glutenin fraction decreases.

The amount of soluble protein fractions varies with different kernel components. Table 37 is a summary of a typical solvent fractionation for endosperm, germ, and bran. The germ has a lower amount of alcohol-soluble proteins, more salt and lactic acid solubles, and comparable alkali solubles. The bran is higher in salt solubles than the endosperm but lower in lactic acid solubles.

The amino acid contents of the four soluble fractions reported in Table 37 are shown in Table 38. Corresponding soluble fractions of germ and bran are similar in their amino acid compositions except for a larger amount of cystine in the salt-soluble fraction of the germ. All fractions of the endosperm have large amounts of glutamic acid and proline, whereas glycine and alanine contents of germ and bran are higher than that found in the endosperm. The serine and leucine contents of the salt-soluble fraction are distinctly lower than in other germ components.

Amino acid composition for four classes of wheat are listed in Table 39. The Hard Red Spring and Winter wheats have identical wheat proteins. Glycine is somewhat lower in the Hard Red Winter, otherwise the composition is similar. The two soft wheats are not typical since they are genetic sources for high protein. Atlas 66 (SRW) has a short mixing time and makes an average loaf of bread at higher protein content. NapHal with a wheat protein of 22.9% will not make a acceptable bread product. It has a mixograph mixing time of less than one minute. It has slightly less glutamic acid than the other three wheats and more proline. Otherwise the amino acid composition of the four wheats is similar.

The amino acid composition of a commercial hard red winter wheat mill mix and its flour, bran, break shorts, reduction shorts, and red dog are listed in Table 40.

Glutamic acid and proline are highest in the endosperm. Lysine, arginine, aspartic acid, and alanine are lowest in the wheat and flour.

Cereals have received much research attention because of their importance in the food supply. There have been attempts to increase quantitative levels of the protein, however, bioavailability is an important consideration. In rat-feeding trials a range in digestibility of the cereal proteins was found (Table 41). For rice, the percentage difference between total and digestible protein was less than 1%, while that for rye was more than 20%.

Along with variation in protein content, the digestibility of the amino acid can also fluctuate. Eggum (15) found that differences for total and digestible lysine are more critical than differences for protein (Table 42). Digestible lysine in wheat in this rat-feeding experiment was 20.8% less than the original level. These data are for raw wheat.

Lysine is the limiting essential amino acid in wheat and most cereals. Analyses of 12,613 common wheats for protein gave values that ranged from 6.9 to 22%, with a mean of 13%. Total variation for lysine ranged from 2.4 to 4.2% of the protein with a mean of 3.2%. Lysine percentage of the grain weight ranged from 0.25 to 0.65% with a mean of 0.40%. Lysine percentage of grain weight was positively correlated with protein (Fig. 7), whereas lysine percentage of protein was negatively correlated with protein (Fig. 8).

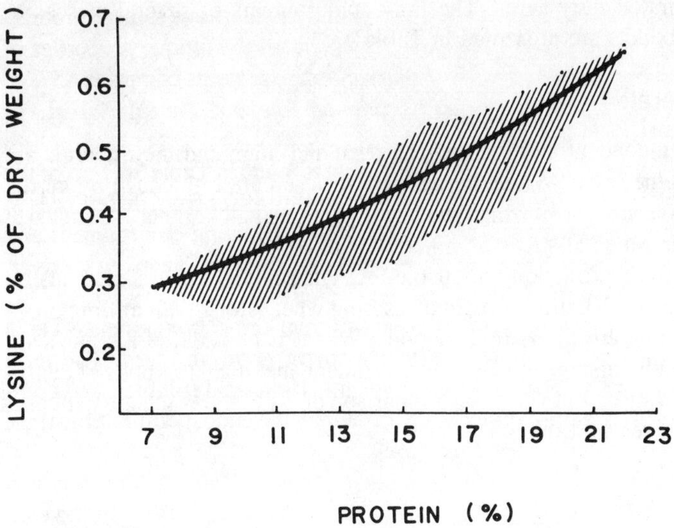

FIGURE 7 Relationship of grain protein level and lysine per unit weight of grain among 12,613 wheats in USDA world wheat collection. (From Ref. 65.)

B. Lipid

Lipid contents of wheat grain typically range from 2 to 4%. Lipid material is not dispersed evenly throughout the grain. The embryo (germ) contains 30% of its weight as oil. Commercial germ is in the 10–11% range. The endosperm is lowest in oil, and the outer layers have intermediate lipid levels between the germ and the endosperm. Wheat-germ oil includes a

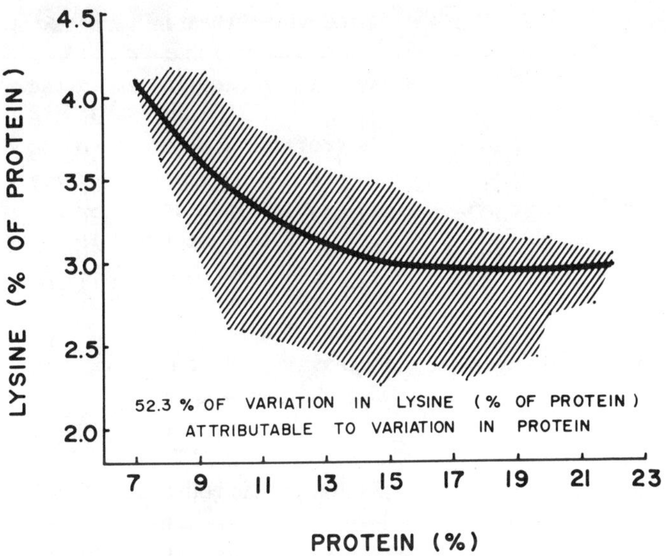

FIGURE 8 Relatioship of grain protein level and lysine per unit protein among 12,613 wheats in the USDA world wheat collection. (From Ref. 65.)

high proportion of unsaturated fatty acids. The fatty acid contents of several classes of wheat and their milled products are presented in Table 43.

C. Vitamins and Minerals

Vitamins are found in high concentrations in wheat germ and bran, and the minerals are especially concentrated in the bran. Whole kernel data for each are influenced by kernel size and the ratio of bran to endosperm, which may be higher in small kernels. Kernel size can be influenced by environmental stress or genetic factors.

Milling and degree of flour extraction will also effect vitamin and mineral analysis on flour and other milled products. Vitamin content of a spring wheat along with milling products of flour, shorts, and bran are given in Table 44. Wheat is considered a significant source of the vitamins thiamin, niacin, and B_6. More efficient analytical methods of analysis have been developed during recent years. A comparison of x-ray fluorescence with previously published data is given in Table 45.

TABLE 1 Proportions of Principal Parts of Mature Common and Durum Wheat Grain

	Embryo (%)	Scutellum (%)	Pericarp (%)	Aleurone (%)	Endosperm (%)
Common	1.2	1.54	7.9	6.7–7.0	81–84
Durum	I——1.6——I		12	I——86.4——I	

Source: Ref. 52.

TABLE 2 Area of Wheat Harvested—Regional Totals (1000 ha)

	1969–1971	1979–1981	1984	1985	1986
Africa	8,380	8,171	9,324	8,839	8,144
North America	27,131	41,054	41,369	41,159	40,016
South America	7,762	9,301	8,652	9,078	10,006
Asia	71,930	80,210	82,715	82,353	83,827
Europe	27,677	25,492	27,349	26,225	26,880
U.S.S.R.	65,230	59,463	51,061	50,384	48,803
Oceania	7,807	11,525	12,091	12,027	11,269
Developed Countries	128,720	138,818	132,984	130,733	127,978
Developing Countries	87,196	96,400	98,576	99,333	100,967
World	215,916	235,217	231,560	230,066	228,945

Source: Refs. 20, 21.

TABLE 3 Area of Wheat Harvested in Selected Countries (1000 ha)

	1969–1971	1979–1981	1984	1985	1986
North America					
United States	18,669	28,898	27,085	26,197	24,560
Canada	7,669	11,386	13,158	13,688	4,217
Mexico	761	723	1,089	1,242	1,206
South America					
Argentina	4,402	5,245	5,901	5,296	4,900
Brazil	1,857	2,958	1,742	2,658	3,908
Chile	737	513	471	506	569
Uruguay	376	281	226	300	188
Asia					
Afghanistan	2,199	2,220	2,310[a]	2,310[a]	2,100
China	28,336[a]	28,930[a]	29,577[a]	28,901[a]	29,697
India	16,941	22,364	24,672	23,614	23,074
Iran	5,370	5,894	5,800[a]	6,100[a]	6,458
Pakistan	6,122	6,865	7,343	7,301	7,403
Turkey	8,732	9,265	9,026	9,000[b]	10,000[a]
Europe					
Bulgaria	1,023	986	1,126	950[a]	1,000[b]
Czechoslovakia	1,076	1,121	1,210	1,209	1,205
France	3,892	4,473	5,105	4,832	4,877
East Germany	597	698	747	725[b]	749
West Germany	1,511	1,642	1,634	1,624	1,648
Greece	1,010	1,022	924	898	880
Hungary	1,289	1,187	1,361	1,358	1,318
Italy	4,089	3,373	3,280	3,032	3,133
Poland	2,004	1,525	1,706	1,885	2,025[b]
Romania	2,527	2,154	2,360	2,300[a]	2,450[a]
Spain	3,727	2,628	2,306	2,024	2,096
United Kingdom	980	1,434	1,939	1,902[b]	1,999
Yugoslavia	1,928	1,475	1,458	1,348	1,346
Oceania					
Australia	7,695	11,440	12,022	11,959	11,269
U.S.S.R.	65,230	59,463	51,061	50,384	48,803

[a]FAO estimate.
[b]Unofficial figure.
Source: Refs. 20, 21.

TABLE 4 Wheat Grain Production—Regional Totals (1000 metric tons)

	1969–1971	1979–1981	1984	1985	1986
Africa	7,945	8,979	9,350	10,535	11,547
North America	56,112	89,470	96,392	95,165	93,468
South America	9,648	12,235	16,855	14,714	16,754
Asia	80,243	136,328	176,899	175,835	187,998
Europe	72,910	91,855	129,537	113,847	115,996
U.S.S.R.	92,804	89,493	76,000[a]	83,000[a]	92,300
Oceania	9,371	14,777	18,609	16,933	17,780
Developed Countries	231,197	285,550	318,984	306,256	317,796
Developing Countries	97,835	157,617	204,659	203,773	218,046
World	329,033	443,167	523,642	510,029	535,842

[a]FAO estimate.
Source: Refs. 20, 21.

TABLE 5 Wheat Grain Production in Selected Countries (1000 metric tons)

	1969–1971	1979–1981	1984	1985	1986
North America					
United States	40,034	66,229	70,618	65,992	56,792
Canada	13,901	20,430	21,199	23,900	31,850
Mexico	2,141	2,754	4,542	5,228	4,772
South America					
Argentina	5,873	8,060	13,200	8,500[b]	8,900
Brazil	1,743	2,613	1,983	4,247	5,433
Chile	1,296	882	988	1,165	1,626
Uruguay	379	377	349	440	234
Asia					
Afghanistan	2,150	2,754	2,850[a]	2,850[a]	2,500[a]
China	31,005[a]	59,196[a]	87,817[a]	85,286[a]	89,002[a]
India	20,859	34,550	45,476	44,229	46,885
Iran	3,946	6,100	5,500[a]	6,000[a]	7,128
Pakistan	6,796	10,760	10,882	11,600[b]	13,923
Turkey	11,423	17,058	17,235	17,032	19,000
Europe					
Bulgaria	2,899	3,881	4,836	3,500[a]	1,265[b]
Czechoslovakia	3,436	4,482	6,170	6,023	5,305
France	14,112	22,362	32,977	29,030	26,587
East Germany	2,203	3,052	3,903	3,806[b]	4,193
West Germany	6,268	8,177	10,223	9,866	10,406
Greece	1,867	2,770	2,646	1,792	2,200
Hungary	3,410	4,800	7,392	6,573	5,803
Italy	9,756	8,989	10,047	8,516	9,070
Poland	4,925	4,189	6,010	6,461	7,390
Romania	4,433	5,471	7,580	6,800[a]	7,900[b]
Spain	4,734	4,516	6,052	5,326	4,292
United Kingdom	4,140	8,116	14,957	11,700	13,874
Yugoslavia	4,760	4,624	5,595	4,859	4,776
Oceania					
Australia	9,014	14,468	18,294	16,550	17,356
U.S.S.R.	92,804	89,493	76,000[a]	83,000[a]	92,300

[a]FAO estimate.
[b]Unofficial value.
Source: Refs. 20, 21.

TABLE 6 Average Wheat Area and Production 1983–1985 in the Leading Wheat-Producing Countries

	Harvested area (1000 ha)	Production (1000 metric tons)	World Production (%)
People's Republic of China (PRC)	29,176	84,832	16.7
U.S.S.R.	50,756	79,167	15.5
United States	26,042	67,489	13.2
India	23,951	44,166	8.7
France	4,921	28,917	5.7
Canada	13,514	23,868	4.7
Australia	12,304	18,987	3.7
Turkey	9,094	16,901	3.3
Pakistan	7,347	11,632	2.3
Argentina	6,026	11,333	2.2
10-Country total			76.0
World	230,739	509,469	100.0

Source: Ref. 21.

TABLE 7 Wheat Yields—Regional Totals (kg/ha)

	1969–1971	1979–1981	1984	1985	1986
Africa	948	1,099	1,123	1,192	1,418
North America	2,068	2,179	2,330	2,312	2,336
South America	1,243	1,315	1,948	1,621	1,674
Asia	1,116	1,700	2,139	2,135	2,243
Europe	2,634	3,604	4,737	4,341	4,315
U.S.S.R.	1,423	1,505	1,488	1,647	1,891
Oceania	1,200	1,282	1,539	1,408	1,578
Developed countries	1,796	2,057	2,399	2,343	2,483
Developing countries	1,122	1,635	2,076	2,051	2,160
World	1,524	1,884	2,261	2,217	2,340

Source: Refs. 20, 21.

TABLE 8 Wheat Yields in Selected Countries (kg/ha)

	1969–1971	1979–1981	1984	1985	1986
North America					
United States	2,144	2,292	2,607	2,519	2,312
Canada	1,813	1,794	1,611	1,419	2,240
Mexico	2,813	3,808	4,170	4,209	3,957
South America					
Argentina	1,334	1,537	2,237	1,605	1,816
Brazil	939	883	1,139	1,598	1,390
Chile	1,759	1,721	2,097	2,301	2,856
Uruguay	1,009	1,340	1,544	1,467	1,247
Asia					
Afghanistan	978	1,241	1,234	1,234	1,190
China	1,094	2,046	2,969	2,951	2,997
India	1,231	1,545	1,843	1,873	2,032
Iran	735	1,035	948	1,073	1,104
Pakistan	1,110	1,567	1,482	1,589	1,881
Turkey	1,308	1,841	1,909	1,892	1,900
Europe					
Bulgaria	2,835	3,937	4,293	3,684	4,000
Czechoslovakia	3,193	3,998	5,136	4,982	4,403
France	3,626	4,999	6,460	6,008	5,452
East Germany	3,689	4,373	5,227	5,250	5,601
West Germany	4,149	4,979	6,256	6,075	6,314
Greece	1,849	2,710	2,864	1,996	2,500
Hungary	2,645	4,043	5,433	4,842	4,403
Italy	2,386	2,565	3,063	2,808	2,895
Poland	2,458	2,746	3,522	3,428	3,649
Romania	1,754	2,540	3,213	2,957	3,224
Spain	1,270	1,718	2,625	2,631	2,048
United Kingdom	4,223	5,659	7,714	6,151	6,940
Yugoslavia	2,469	3,134	3,837	3,605	3,548
Oceania					
Australia	1,171	1,265	1,522	1,384	1,578
U.S.S.R.	1,423	1,505	1,488	1,647	1,891

Source: Refs. 20, 21.

TABLE 9 Major Exporters of Wheat Plus Flour (Wheat Equivalents) Regional Totals for 1976–1978 and 1984–1986 (1000 metric tons)

	1976	1977	1978	1984	1985	1986
Africa	52	187	147	113	121	87
North America	38,890	40,158	50,832	65,263	43,504	42,908
South America	3,300	6,042	1,835	7,445	9,716	4,142
Asia	148	1,062	3,305	1,761	1,706	879
Europe	16,940	14,264	15,419	29,260	32,926	30,092
Oceania	7,875	8,196	11,170	10,648	15,782	16,172
U.S.S.R.	1,691	2,966	2,218	1,952	1,558	1,450
Developed countries	61,331	61,429	75,914	107,569	94,151	91,043
Developing countries	3,466	7,113	5,099	8,920	11,070	4,687
World	68,938	72,929	84,963	116,489	105,221	95,730

Source: Refs. 22, 24.

TABLE 10 Major Exporters of Wheat for 1976–1978 and 1984–1986 (1000 metric tons)

	1976	1977	1978	1984	1985	1986
Africa						
South Africa	35	183	141	113	113	87
North America						
Canada	11,338	14,934	15,329	21,647	17,363	16,408
Mexico	13	23	14	—	—	—
United States	27,552	25,224	35,503	43,616	26,141	26,455
South America						
Argentina	3,264	5,970	1,835	7,370	9,716	4,141
Uruguay	35	71	—	74	—	—
Asia						
Japan	37	42	76	336	270	379
Singapore	66	154	206	42	13	39
Turkey	45	631	2,085	845	581	209
Europe						
Austria	87	—	235	554	660	532
Belgium–Lux	885	381	495	1,012	1,159	698
Bulgaria	260	281	208	332	677	300
Denmark	167	268	117	519	466	526
Finland	173	103	5	22	21	15
France	8,654	7,407	7,972	15,792	19,029	15,414
East Germany	50	50	75	87	95	95
West Germany	1,432	1,793	1,268	2,490	1,716	2,622
Greece	345	192	590[a]	1,469	721	681
Hungary	749	804	574	1,280	2,010	1,676
Italy	455	632	733	1,231	2,063	1,634
Netherlands	1,238	863	1,053	780	845	601
Romania	1,385	274	828	150	100	300
Spain	31	69	164	101	429	248
Sweden	879	903	603	809	700	526
United Kingdom	128	216	356	2,290	2,001	4,087
Yugoslavia	13	6	78	275	2	5
Oceania						
Australia	7,875	8,181	11,170	10,606	15,780	16,171

[a]FAO estimate

Source: Refs. 22, 24.

TABLE 11 Leading Wheat Exporting Countries 1983 to 1985 (1000 metric tons)

	1983[a]	%	1985[a]	%	1987/88[b]	%
United States	41,091	36.7	26,141	24.8	43,500	41.6
Canada	22,229	19.9	17,363	16.5	22,500	21.5
France	14,726	13.2	19,029	18.1	14,500[c]	13.9
Australia	8,311	7.4	15,780	15.0	12,500	11.9
Argentina	10,232	9.2	9,716	9.2	5,000	4.8
5–Country total	96,589	86.4	88,029	83.7	98,000	93.7
World	111,816		105,211		104,675	

[a]From Ref. 23.
[b]From Ref. 59.
[c]From EC–12 instead of France.

TABLE 12 Major Importers of Wheat Plus Flour (Wheat Equivalents) Regional Totals for 1976–1978 and 1984–1986 (1000 metric tons)

	1976	1977	1978	1984	1985	1986
Africa	9,756	11,401	13,425	19,052	20,466	19,018
North America	2,215	2,390	2,746	3,405	3,620	3,601
South America	6,803	5,457	8,147	8,794	7,932	6,132
Asia	26,898	27,201	32,458	41,111	31,224	33,967
Europe	19,070	17,455	17,115	16,685	18,520	19,056
Oceania	246	129	146	326	275	285
U.S.S.R.	7,214	7,142	7,206	27,279	20,573	14,396
Developed countries	32,689	30,820	30,403	51,110	45,727	40,488
Developing countries	39,513	40,356	50,841	65,543	56,883	55,966
World	72,202	71,175	81,244	116,653	102,610	96,454

Source: Refs. 22, 24.

TABLE 13 Major Importers of Wheat Plus Flour (Wheat Equivalents) for 1976–1978 and 1984–1986 (1000 metric tons)

	1976	1977	1978	1984	1985	1986
Africa						
Algeria	1,684	1,803	1,913	2,895	4,040	3,503
Egypt	3,888[a]	4,345	5,220	7,034[a]	6,996[a]	6,667
Libya	428	491	556	672[a]	498[a]	797[a]
Morocco	1,051	1,352	1,642	2,524	2,044	1,379
Mozambique	122	130	138	142[a]	135	126
Nigeria	735	770	1,109	1,071	1,679	1,321
Senegal	119	135	157	127	115	136
Sudan	186	109	288	494[a]	1,027[a]	596
Tunisia	352	544	551	778	487	951
Zaire	111	104	117[a]	158[a]	196[a]	253
North America						
Cuba	1,064	753	1,124	1,482	1,387	1,410[a]
Dominican Republic	163	111	160	126	256[a]	242[a]
El Salvador	101	105	117	148	151	175[a]
Haiti	114	122	100	177[a]	190[a]	171[a]
Trinidad & Tobago	116	104	106	103	83	128
United States	23	35	0.6	101	272	279
South America						
Bolivia	167	238	281	230	436[a]	528[a]
Brazil	3,435	2,626	4,334	4,869	4,048	2,250
Chile	1,135	473	1,074[a]	964	476	156
Colombia	398	216	449	661	688[a]	694
Ecuador	205	242	259	245	225	243
Peru	602	833	753[a]	763	837[a]	1,156[a]
Venezuela	740	705	764	968	1,106	952
Asia						
Bangladesh	1,173	623	2,805	1,872	1,360	1,161
China	2,599[b]	7,577[b]	8,304[b]	10,836	6,274	7,104
Hong Kong	167	185	183	239	226	254
India	6,298	855	169[a]	1,819	na[c]	3,820
Indonesia	981	810	843	1,448	1,333	1,623
Iran	443	1,238	1,539	3,919[a]	2,819[a]	2,401[a]
Iraq	616	723	997	3,317[a]	2,317[a]	2,463[a]
Israel	438	507	507	569	637	748
Japan	5,827	5,676	5,564	5,979	5,510	5,620
Jordan	253	289	292	468	389	285
Korea DPR	530[a]	500[a]	536[a]	200	200	200
Korea REP	1,826	2,026	1,664	2,649	2,986	3,449
Kuwait	149	58	154	284[a]	281	122

TABLE 13 (continued)

	1976	1977	1978	1984	1985	1986
Lebanon	486[a]	342[a]	358[a]	359[a]	439[a]	336[a]
Malaysia	433	503	566	593	572[a]	577
Pakistan	1,186	268	2,142	291	982	1,909
Philippines	722	671	766[a]	778	701	1,088
Saudia Arabia	494	428	650[a]	486	159	128
Singapore	163	341	268	202	156	223
Sri Lanka	549	877	920	660	724	699
Thailand	136	90	124	169	170	186
U.A. Emirates	120	87	138	187[a]	221[a]	236
Vietnam	532[a]	1,218[a]	1,295[a]	151[a]	89[a]	114
Yemen AR	203	275	446	566[a]	750	787
Yemen DEM	76	187	198	237	261	170
Europe						
Belgium-Lux	939	725	995	1,602	1,519	1,295
Czechoslovakia	803	492	379	206	248	93
France	289	183	629	520	319	413
East Germany	1,691	1,100	687	1,657	1,219	600
West Germany	1,983	1,449	1,394	1,695	3,190	2,517
Ireland	173	180	202	341	390	448
Italy	2,453	2,989	3,570	4,037	4,555	5,284
Netherlands	1,549	1,338	1,544	1,506	1,827	1,507
Norway	303	305	327	228	181	306
Poland	2,312	2,599	2,311	2,098	1,810	1,662
Portugal	267	417	706	794	564	649
Romania	989	540	300	100[a]	300[a]	—
Spain	68	204	227	144	232	800
Switzerland	370	321	414	320	242	260
United Kingdom	3,807	3,856	3,100	1,142	1,620	1,727
Yugoslavia	864	521	3	5	61	512
Oceania						
Fiji	49	45	51	58	61	61
New Zealand	112	na[c]	na[c]	126	66	73

[a]Unofficial figure.
[b]FAO estimate.
[c]Not available.
Source: Refs. 22, 24.

TABLE 14 U.S. Wheat Exports as Share of Total Crop (%)

1970/71	1979/80	1985/86	1987/88
49	64	47	53

Source:Ref. 57.

TABLE 15 U.S. Wheat Supply and Distribution (June 1–May 31) (million bushels)

	Supply			Distribution					
				Domestic					
Crop year	Production	Beginning stocks	Imports	Total	Food	Other	Exports	Total	Market year ending stocks
1970	1,352	885	1	2,238	517	255	741	1,513	823
1980	2,381	902	3	3,285	611	172	1,514	2,297	989
1984	2,595	1,399	10	4,004	651	502	1,425	2,578	1,425
1985	2,425	1,425	16	3,866	683	363	915	1,961	1,905
1986	2,092	1,905	20	4,018	724	469	1,004	2,197	1,821
1987	2,105	1,821	15	3,941	750	360	1,600	2,710	1,231
1988	2,170	1,231	15	3,416	775	345	1,500	2,620	796

Source: Ref. 60, p. 26.

TABLE 16 World Wheat and Wheat Flour Utilization[a], July/June, 1984/85–1988/89 (million metric tons)

	1984/85	1985/86	1986/87	1987/88	1988/89 (May 10)
United States	31.4	28.5	32.5	30.2	30.5
U.S.S.R.[b]	91.2	91.6	102.8	100.5	102.0
China	92.2	100.4	101.5	105.2	105.0
Others	278.0	275.2	284.9	297.9	298.8
Total non–United States	461.4	467.2	489.3	503.7	505.8
World total	492.8	495.7	521.7	533.9	536.3

[a]Utilization data are based on an aggregate of differing local marketing years. For countries for which stocks are not available (excluding the U.S.S.R.), utilization estimates represent apparent utilization.
[b]"Bunker weight" basis; not discounted for excess moisture and foreign material.
Source: Ref. 59, p. 11.

TABLE 17 World Wheat and Wheat Flour Market Year Ending Stocks[a], July/June 1984/85–1988/89 (million metric tons)

	1984/85	1985/86	1986/87	1987/88	1988/89 (May 10)
Total foreign[b]	125.4	116.3	126.6	113.7	109.8
U.S.S.R.: stock changes	5.0	1.7	5.0	4.3	4.0
United States	38.8	51.8	49.6	33.5	21.7
World totals	164.2	168.2	176.1	147.2	131.5

[a]Stock data are based on an aggregate of differing local marketing years and should not be construed as representing world stock levels at a fixed point in time. Stock data are not available for all countries and exclude parts of Eastern Europe and parts of Asia. Stock levels have been adjusted for estimated year-to-changes in U.S.S.R. grain stocks, but do not purport to include the entire absolute level of U.S.S.R. stocks.
[b]Inclusive of Soviet stock changes; see footnote[a].
Source: Ref. 59, p. 11.

TABLE 18 Grain Stocks (1000 bushels) of All Wheat in United States (Jan. 1 of each year)

Year	On farm	Off farm	Total
1971	694,549	853,100	1,547,649
1972	510,202	888,801	1,399,003
1973	364,382	563,952	928,334
1974	446,289	661,171	1,107,460
1975	548,600	838,028	1,386,628
1976	667,210	1,116,376	1,783,586
1977	834,577	1,162,459	1,997,036
1978	814,400	816,364	1,630,764
1979	773,920	942,247	1,716,167
1980	753,950	1,150,253	1,904,203
1981	948,675	1,223,397	2,172,072
1982	1,150,534	1,355,646	2,506,180
1983	1,015,409	1,310,964	2,326,373
1984	930,281	1,210,750	2,141,031
1985[a]	1,011,184	1,514,964	2,526,148
1986[b]	1,063,000	1,608,458	2,671,458

[a]Preliminary estimates.
[b]Year beginning December 1, 1986.
Source: Ref. 55.

TABLE 19 Grain Stocks (1000 bushels) of Durum Wheat in United States (Jan. 1 of each year)

Year	On farm	Off farm	Total
1971	94,340	24,928	119,268
1972	72,016	25,910	97,926
1973	49,251	18,433	67,684
1974	51,832	18,727	70,559
1975	65,412	20,528	85,940
1976	94,999	32,631	127,630
1977	81,891	26,524	108,415
1978	98,694	18,312	117,006
1979	87,384	19,776	107,160
1980	73,990	29,448	103,438
1981	119,151	31,034	150,185
1982	153,683	33,266	185,949
1983	97,776	35,267	133,043
1984	95,033	37,212	132,245
1985[a]	111,608	38,441	150,049
1986[b]	96,910	57,558	154,468

[a]Preliminary estimates.
[b]Year beginning December 1, 1986.
Source: Ref. 55.

TABLE 20 Approximate % Moisture Content of Wheat in Equilibrium with Air at Different Relative Humidities

Wheat class	Relative humidity (%)				
	15	30	45	60	75
Soft Red	6.3	8.6	10.6	11.9	14.6
Hard Red	6.4	8.5	10.5	12.5	14.6
White	—	8.6	9.9	11.6	15.0

Source: Permission from Ref. 13.

TABLE 21 Official U.S. Grades and Grade Requirement

	Minimum Limits of—		Maximum Limits of—						
	Test Weight per Bushel		Damaged Kernels					Wheat of other Classes[d]	
Grade	Hard Red Spring wheat or White Club wheat[a] (lb)	All other classes and subclasses (lb)	Heat–damaged kernels (%)	Total[b] (%)	Foreign material (%)	Shrunken and broken kernels (%)	Defects[c] (%)	Contrasting classes (%)	Total[e] (%)
U.S. No. 1	58.0	60.0	0.2	2.0	0.5	3.0	3.0	1.0	3.0
U.S. No. 2	57.0	58.0	0.2	4.0	1.0	5.0	5.0	2.0	5.0
U.S. No. 3	55.0	56.0	0.5	7.0	2.0	8.0	8.0	3.0	10.0
U.S. No. 4	53.0	54.0	1.0	10.0	3.0	12.0	12.0	10.0	10.0
U.S. No. 5	50.0	51.0	3.0	15.0	5.0	20.0	20.0	10.0	10.0

U.S. Sample grade is wheat that:

(1) Does not meet the requirements for the grades U.S. Nos. 1, 2, 3, 4, or 5; or

(2) Contains 32 or more insect-damaged kernels per 100 grams of wheat; or

(3) Contains 8 or more stones or any number of stones which have an aggregate weight in excess of 0.2 percent of the sample weight, 2 or more pieces of glass, 3 or more crotalaria seeds (*Crotalaria* spp.), 2 or more castor beans (*Ricinus communis* L.), 4 or more particles of an unknown foreign substance(s) or a commonly recognized harmful or toxic substance(s), 2 or more rodent pellets, bird droppings, or equivalent quantity of other animal filth per 1,000 grams of wheat; or

(4) Has a musty, sour, or commercially objectionable foreign odor (except smut or garlic odor); or

(5) Is heating or otherwise of distinctly low quality.

[a]These requirements also apply when Hard Red Spring or White Club wheat predominate in a sample of Mixed wheat.

[b]Includes heat-damaged kernels.

[c]Defects include damaged kernels (total), foreign material, and shrunken and broken kernels. The sum of these three factors may not exceed the limit for defects for each numerical grade.

[d]Unclassed wheat of any grade may contain not more than 10.0 percent of wheat of other classes.

[e]Includes contrasting classes.

Source: Ref. 19.

TABLE 22 Comparison of Wheat Grading Systems of the Major Exporting Countries

Australia:	7 Wheat types. 7 Area descriptions. Grade designations. Prime Hard, three protein levels. Hard, two grades. One grade for all other descriptions. Grade limitations for moisture, test weight, falling number, soundness, and foreign material.
Argentina:	2 Wheat types. 4 Zone or areas. 4 Grades. Numbered 1 through 4.
Canada:	7 Wheat classes. 19 Grade designations. Grades No. 1 and No. 2 Canadian Western segregated at three and two protein levels, respectively.
United States:	7 Classes. 13 Subclasses. 78 Grades. Numbered 1 through 5 plus sample grade.

Source: Refs. 10, 51.

TABLE 23 Australian Wheat Classes

Australian Prime Hard
Guaranteed minimum protein[a] at 13, 14, and 15%.

Australian Hard
No. 1 11.5–14% range.
No. 2 11.0–13% range.

Australian Standard White
An intermediate hardness with protein generally 9.5–12%.

Australian Soft
Low protein for production of cookies, cakes and pastry goods.

Australian Durum
No. 1 13% protein.
No. 2 11.5% protein.

Australian General Purpose
74–68 kg/hl (57.8–53.1 lb/bu) sold as No. 1 and No. 2 grades.

Australian Feed
Below 68 kg/hl (53.1 lb/bu).

[a]11 percent moisture basis.
Source: Refs. 1, 9.

TABLE 24 Canadian Seven General Wheat Classes

Canadian Western Red Spring	
No. 1	13.5% protein[a] and higher
	12.5% – 13.0% protein
	12.0% and lower
No. 2	13.5% protein and higher
	13.0% and lower
No. 3	unspecified protein
Canadian Western Amber Durum	
3 grades	
Canadian Western Utility[b]	
3 grades	
Canadian Prairie Spring	
Canadian Western Red Winter	
Candian Western Soft White Spring	
Canadian Eastern White Winter	

[a]N × 5.7; 13.5% moisture basis.
[b]New grade 1973–1974.
Source: Ref. 10.

TABLE 25 Nitrogen Content of Dissected Fractions of Maris Widgeon, a Semi-Hard Wheat

	Mealy grains (N 1.63%, 48 grains)				Vitreous grains (N 2.27%, 72 grains)			
Fraction	% of grain	% N content	N per 100 g of grain	N as % of total N in grain	% of grain	% N content	N per 100 g of grain	N as % of total N in grain
Embryo	1.2	5.83	0.070	4.2	1.2	6.07	0.073	3.3
Scutellum	1.5	4.80	0.072	4.3	1.7	4.97	0.085	3.8
Pericarp–testa	7.4	0.85	0.063	3.7	7.1	1.09	0.077	3.5
Aleurone layer	5.6	3.65	0.204	12.2	5.7	3.96	0.226	10.2
Endosperm 1	2.1	3.55	0.750	4.5	3.2	5.04	0.163	7.4
Endosperm 2	22.0	2.15	0.474	28.2	18.7	3.15	0.590	26.6
Endosperm 3	29.7	1.38	0.410	24.5	28.7	1.83	0.525	23.7
Endosperm 4	30.5	1.01	0.308	18.4	33.7	1.41	0.476	21.5
Endosperm total	84.3	1.52	1.267	75.6	84.3	2.08	1.754	79.2
Grain total	100		1.68	100	100		2.22	100

Source: Ref. 17.

TABLE 26 Nitrogen Content of Dissected Fractions of Cappelle-Desprez, a Semi-Soft Wheat

	Mealy grains (N 1.45%, 54 grains)				Vitreous grains (N 2.19%, 72 grains)			
Fraction	% of grain	% N content	N per 100 g of grain	N as % of total N in grain	% of grain	% N content	N per 100 g of grain	N as % of total N in grain
Embryo	1.1	6.05	0.066	4.6	1.1	6.11	0.067	3.0
Scutellum	1.4	4.82	0.067	4.7	1.6	5.05	0.081	3.7
Pericarp–testa	7.4	0.83	0.061	4.2	7.3	0.93	0.068	3.1
Aleurone layer	6.0	3.35	0.201	13.9	6.4	4.13	0.268	12.1
Endosperm 1	2.3	2.70	0.062	4.3	3.3	5.10	0.168	7.6
Endosperm 2	9.3	1.93	0.179	12.4	19.0	3.03	0.575	26.0
Endosperm 3	39.5	1.29	0.510	35.2	27.5	1.80	0.520	23.5
Endosperm 4	33.0	0.91	0.300	20.7	33.8	1.37	0.464	21.0
Endosperm total	84.1	1.25	1.051	72.6	83.6	2.04	1.723	78.1
Grain total	100		1.45	100	100		2.21	100

Source: Ref. 17.

TABLE 27 Comparison of Protein Loss Between Wheat and Flour from a Pilot Commercial Mill and a Laboratory Buhler Mill

Wheat	Mill Extraction, Commercial (%)	Mill Extraction, Buhler (%)	Protein Loss, Commercial (%)	Protein Loss, Buhler (%)	Estimated Protein Loss at 74% Extraction, Commercial	Estimated Protein Loss at 74% Extraction, Buhler
Maris Widgeon, high protein	74.6	76.6	1.1	1.2	1.1	1.3
Maris Widgeon, low protein	74.3	73.4	1.1	1.3	1.1	1.3
Cappelle–Desprez, high protein	77.0	70.6	1.3	1.7	1.4	1.6
Cappelle–Desprez, low protein	69.4	68.8	1.3	1.7	1.2	1.4

Source: Ref. 16.

TABLE 28 Mean, Range, and Standard Deviations for Kernel Components and Kernel Recovery Percentages from the Fractionation of 20g Samples for 153 Samples from the 1973 World Wheat Collection Nursery

Variables	Mean	Range	S.D.
Hectoliter wt. (kg/ hl)	77.89	70.95 to 83.85	2.279
1,000 kernel wt. g	40.63	24.95 to 57.94	7.351
Milling yield (%)	71.2	63.1 to 78.8	3.18
Flour wt. g[a]	12.70	11.28 to 14.09	0.58
Bran flour wt. g[a]	1.60	0.92 to 2.50	0.390
Washed bran wt. g[a]	3.15	2.56 to 4.77	0.309
% Endosperm	81.9	72.8 to 85.5	1.77
Sample recovery (%)	98.1	96.3 to 99.2	0.50

[a]Dry weight basis.
Source: Ref. 66.

TABLE 29 Means, Coefficients of Variation, F ratios, and L.S.D. Values for Kernel Components and Kernel Recovery Percentages for Check Cultivars in the 1973 World Wheat Collection Nursery

	Mean of check cultivars (r = 4)									
Variable	Atlas 66	Triumph 64	Scout 66	Inia 66	Centurk	Nap Hal	C.V.	F. ratio for cultivars	F. ratio for replications	L.S.D. 0.05
1000 kernel wt. g	34.7	40.1	42.2	50.5	36.0	26.4	3.56	140.0**	3.31*	2.06
Milling yield (%)	65.8	74.2	76.3	75.9	74.5	68.8	0.57	419.06**	4.47*	0.63
Flour wt. g	11.74	13.22	13.62	13.57	13.25	12.27	0.79	225.41**	3.21*	0.15
Bran flour wt. g	2.26	1.19	1.18	1.08	1.30	1.86	4.69	185.43**	11.20**	0.10
Washed bran wt. g	3.37	3.04	2.70	2.87	2.88	3.27	2.65	40.52**	3.20*	0.12
% Endosperm	80.59	82.56	84.56	83.63	83.44	81.23	0.55	44.42**	3.17*	0.68
Sample recovery (%)	97.67	98.11	98.43	98.50	98.02	97.82	0.54	1.53	1.31	0.80

*, ** Indicates significance at the 0.05 and 0.01 levels, respectively.
Source: Ref. 66.

TABLE 30 Annual Production of Straight Semolina and Durum Flour (million cwt)

Year	Production	Year	Production
1966	12.411	1977	16.363
1967	12.534	1978	16.292
1968	12.749	1979	17.404
1969	13.475	1980	14.720
1970	14.813	1981	13.455
1971	15.251	1982	14.124
1972	14.986	1983	14.776
1973	16.334	1984	16.922
1974	14.003	1985	18.362
1975	13.825	1986	20.099
1976	15.349	1987	20.943

Source: Ref. 64.

TABLE 31 U.S. Average Extraction Rates in Durum Milling

Year	% Weighted extraction
1982	73.9
1983	74.3
1984	78.1
1985	76.3
1986	75.2
1987	73.9
Mean	75.3

Source: Ref. 64.

TABLE 32 Protein Content Ranges of Wheat Types

Wheat type	Approximate protein range (%)
HRS (United States)	11.5–18
Durum	10–16.5
Plate (Argentina)	10–16
CWRS (Manitoba)	9–18
HRW (United States)	9–14.5
Russian	9–14.5
Australian	8–13.5
English	8–13
Other European	8–11.5
SRW (United States)	8–11
White (United States)	8–10.5

Source: Ref. 34.

TABLE 33 Protein[a] Composition of Wheat Fractions

Fraction	of Grain[a] %	Protein[b] %	Protein[c] %
Pericarp and testa	8.0	4.1	7.6
Testa and hyaline	—	—	15.7
Aleurone	7.0	18.4	24.3
Germ	1.0	31.1	26.3
Scutellum	1.5	24.9	—
Outer endosperm	12.5	12.8	16.2
Middle endosperm	12.5	8.2	—
Inner endosperm	57.5	5.8	8.0
Whole grain	100.0	8.2	12.1

Source: Refs. 2, 27.
[a]N × 5.83, 14% moisture.
[b]Soft Wheat.
[c]Hard Wheat.

TABLE 34 Wheat Constituents (%) and Calorie Values of Different Parts of Wheat Grain[a]

Part of grain	Crude protein	Lipids	Starch	Reducing sugars	Pentosans[b]	Cellulose	Ash	Calories
Whole grain	12.1	1.8	59.2	2.0	6.7	2.3	1.8	314
Pericarp	7.6	0.0	0.0	0.0	34.9	38.4	5.1	177
Testa and hyaline	15.7	0.0	0.0	0.0	51.1	11.1	8.1	177
Aleurone	24.3	8.1	0.0	0.0	39.0	3.5	11.1	247
Outer endosperm	16.2	2.2	63.4	1.6	1.4	0.3	0.8	349
Inner endosperm	8.0	1.6	72.6	1.6	1.4	0.3	0.5	348
Embryo and Scutellum	26.3	10.1	0.0	26.3	6.6	2.0	4.6	354

[a]14% moisture content.
[b]With other carbohydrates.
Source: Ref. 2.

TABLE 35 Composition[a] of Flours[b] at Various Extraction Rates.

Extraction rate	Protein	Crude fat	Carbohydrate	Crude fiber
100	12.2	2.4	64.1	2.00
85	12.1	1.6	69.8	0.40
80	11.7	1.4	70.2	0.21
70/72	11.3	1.1	72.0	0.10
near 40	10.0	0.8	74.5	trace

[a]Percentage.
[b]From wheat blend 70% Canadian Hard Red Spring and 30% English wheat.
Source: Ref. 2.

TABLE 36 Distribution of Wheat Endosperm Proteins According to Solubility (% total protein)

Cereal	Protein range (% dwb)	Albumins (water-soluble)	Globulins (salt-soluble)	Prolamin (alcohol-soluble)	Residue and glutelin (alkali-soluble)
Wheat (HRS)	10–15	5–10	5–10	40–50	30–40
Wheat (Durum)	12–16	10–15	5–10	40–50	30–40

Source: Ref. 52.

TABLE 37 Relative Concentrations of Protein Fractions of Endosperm, Germ, and Bran

Grain part	Isopropyl alcohol soluble[a] (%)	NaCl soluble[b] (%)	Lactic acid soluble[c] (%)	KOH soluble[d] (%)	Total (%)
Endosperm	47.1	2.4	10.1	36.5	96.1
Germ	22.9	14.5	20.1	35.7	93.1
Bran	43.3	9.2	3.0	37.4	92.9

[a]40% isopropyl alcohol.
[b]2% NaCl.
[c]3.85% lactic acid.
[d]0.1% KOH.
Source: Ref. 41.

TABLE 38 The Amino Acid Compositions (mol%) of the Soluble Protein Fractions of Endosperm, Germ, and Bran

	Endosperm				Germ				Bran			
Amino acid	IPA[a]	NaCl	Lactic	KOH	IPA[a]	NaCl	Lactic	KOH	IPA[a]	NaCl	Lactic	KOH
Lysine	1.16	3.61	4.68	3.30	4.80	7.68	8.10	6.15	2.08	4.60	6.89	4.30
Histidine	1.87	2.86	2.44	1.75	2.35	2.75	2.67	2.19	1.70	3.90	3.81	2.20
Arginine	2.23	5.33	4.71	2.59	5.00	8.83	8.37	6.12	1.83	9.54	6.49	5.40
Aspartic acid	2.19	5.92	0.80	4.25	7.07	6.97	8.02	8.85	3.79	8.30	7.59	5.76
Threonine	2.10	4.15	4.32	3.55	4.79	5.08	4.74	4.76	3.02	3.86	4.79	3.85
Serine	4.44	6.71	6.97	5.50	5.45	0.40	6.13	6.00	5.47	5.80	7.16	6.34
Glutamic acid	39.18	24.67	26.68	32.00	21.96	13.83	13.98	13.12	36.04	16.07	14.17	22.28
Proline	16.20	7.05	9.68	11.36	7.32	2.68	2.79	3.81	15.55	3.55	6.04	10.42
Glycine	3.88	9.45	6.28	6.80	10.90	14.69	10.81	11.22	4.92	13.37	14.46	10.85
Alanine	2.77	3.24	5.92	5.22	8.62	9.74	9.57	9.32	4.70	7.74	9.05	6.62
Cystine	2.38	4.81	6.44	1.79	1.54	11.89	trace	trace	1.00	3.40	4.15	0.51
Valine	3.64	5.54	5.30	3.48	5.27	5.24	7.20	7.31	3.87	5.49	4.79	4.83
Methionine	1.01	1.63	1.68	1.58	1.41	2.68	1.13	2.00	0.47	0.88	0.46	1.78
Isoleucine	3.64	3.47	4.37	3.42	3.19	2.95	4.12	4.61	3.68	3.35	2.93	3.41
Leucine	5.30	6.14	3.08	7.16	5.28	0.16	6.93	7.72	6.15	4.90	4.63	6.25
Tyrosine	1.77	2.09	2.70	2.86	2.37	2.00	2.24	3.01	1.49	2.44	1.13	1.96
Phenylalanine	6.24	3.31	3.95	3.39	2.68	2.43	3.20	3.81	4.24	2.81	1.46	3.24

[a]Isopropyl alcohol.
Source: Ref. 41.

TABLE 39 Amino Acid Composition of Wheats (% by weight)

Amino acid	HRS wheat[a]	HRW wheat[b]	SRW wheat[c]	SRS wheat[d]
Tryptophan	1.24	—	—	—
Threonine	2.88	3.1	3.2	3.01
Isoleucine	4.34	3.9	4.3	4.10
Leucine	6.71	7.2	7.3	7.12
Lysine	2.82	2.9	2.9	2.88
Methionine	1.29	1.5	1.4	1.35
Cystine	2.19	1.8	—	2.18
Phenylalanine	4.94	4.7	4.8	5.41
Tyrosine	3.74	2.7	2.0	1.79
Valine	4.63	4.5	4.7	4.76
Arginine	4.79	5.0	4.5	4.85
Histidine	2.04	2.5	2.4	2.58
Alanine	3.50	3.6	3.7	3.54
Aspartic acid	5.46	5.3	5.4	5.63
Glutamic acid	31.25	31.9	34.8	30.48
Glycine	6.11	4.3	4.3	3.80
Proline	10.44	11.0	10.4	11.57
Serine	4.61	4.8	4.9	4.63
Wheat protein[e]	14.0	14.0	17.5	22.9

[a]Hard red spring.
[b]Hard red winter (NE701132), unpublished data.
[c]Soft red winter (Atlas 66), unpublished data.
[d]Soft red spring (NapHal), unpublished data.
[e]($N \times 5.7$) dry weight basis.
Source: Ref. 52.

TABLE 40 Amino Acid Composition of Hard Red Winter Commercial Mill Wheat Mix and Its Milling Fractions[a]

	Wheat	Flour	Bran	Break shorts	Red[b] shorts	Red dog
Lysine	3.0	2.2	4.5	4.5	4.5	3.7
Histidine	2.3	2.6	2.8	2.8	2.6	2.4
Arginine	4.2	3.5	6.4	6.5	6.2	5.2
Aspartic acid	5.5	4.2	7.3	7.6	7.9	6.5
Threonine	3.4	3.0	3.5	3.6	3.8	3.4
Serine	5.1	4.8	4.6	4.8	4.7	4.9
Glutamic acid	32.8	36.9	20.8	20.7	19.1	26.8
Proline	10.1	11.5	6.9	6.9	6.2	8.1
Glycine	4.3	3.5	5.5	5.5	5.7	4.8
Alanine	3.7	2.9	4.9	5.2	5.3	4.5
Cystine	1.5	1.6	—	0.7	0.7	0.9
Valine	4.6	4.2	5.1	5.2	5.2	5.0
Methionine	1.4	1.5	1.4	1.3	1.1	1.5
Isoleucine	3.9	3.9	3.8	3.8	3.7	3.9
Leucine	7.2	7.2	6.7	6.8	6.8	7.1
Tyrosine	2.0	2.3	2.1	2.0	2.1	2.2
Phenylalanine	4.5	4.7	4.0	4.0	4.0	4.3
Protein[c]	13.3	12.2	17.7	16.8	14.4	14.1
Milling yield(%)[a]	100	72.8	21.4	3.2	2.2	0.23

[a]Unpublished data.
[b]Reduction shorts.
[c](N x 5.7) dry weight basis.
[d]Milled on Kansas State University 200 cwt flour mill.

TABLE 41 Total Protein and *in vivo* Protein Digestibility (D) in Eight Cereal Grains

Cereal grain	Total protein in dry matter (%)	D protein in dry matter (%)	Difference[a] (%)
Barley	10.1	8.3	18.0
Oat	10.8	9.0	15.9
Wheat	12.6	11.3	10.4
Rye	9.1	7.0	23.0
Maize	10.1	8.8	12.4
Sorghum	12.5	10.6	15.2
Rice	9.0	8.9	0.7
Triticale	13.1	12.1	7.3
Soya bean meal	53.1	50.4	5.2

[a]Difference between total and digestible protein. The smaller the value the more digestible the protein.
Source: Ref. 15.

TABLE 42 Total and Digestible Lysine in Cereal Grains

Lysine source	Total lysine in dry matter (g/16g N)	Digestible lysine in dry matter (g/16g N)	Difference (%)
Barley	3.69	2.80	24.1
Oat	4.03	3.21	20.3
Wheat	2.55	2.02	20.8
Rye	3.67	2.90	21.0
Maize	2.73	2.31	15.4
Sorghum	1.83	1.33	27.3
Rice	3.54	3.51	0.9
Soya bean meal	5.98	5.48	8.3

Source: Ref. 15.

TABLE 43 Fatty Acid Compostion of Wheat and Kernel Parts

Grain	Water	Total lipid	Saturated: Sum	14:0	16:0	18:0	20:0	Unsaturated: Sum	16:1	18:1	18:2	18:3
			Fatty acid[a]									
	g/100 g food, edible portion											
Wheat												
Whole grain												
Hard Red Spring	14.0	2.7	0.37	—[b]	0.36	0.01	—	1.56	0.01	0.25	1.20	0.10
Hard Red Winter	14.0	2.5	0.35	—	0.33	0.02	—	1.47	0.01	0.28	1.08	0.10
Soft Red Winter	14.0	2.4	0.35	—	0.33	0.02	0	1.40	0.01	0.25	1.07	0.07
White	14.0	2.0	0.30	—	0.28	0.02	0	1.14	0.01	0.18	0.88	0.07
Flours												
Hard Red Spring	14.0	1.5	0.23	—	0.21	0.01	0.01	0.84	—	0.12	0.69	0.03
Hard Red Winter	14.0	1.5	0.20	—	0.19	0.01	—	0.74	—	0.10	0.64	0.03
Soft Red Winter	14.0	1.4	0.22	—	0.20	0.01	0.01	0.76	—	0.08	0.65	0.03
All purpose	14.0	1.4	0.23	—	0.22	0.01	—	0.72	—	0.11	0.58	0.03
Bran	14.0	4.6	0.74	0.01	0.69	0.04	—	3.09	0.02	0.71	2.20	0.16
Germ	14.0	10.9	1.88	—	0.01	1.81	0.06	8.18	0.04	1.54	5.86	0.74
Wheat durum												
Whole grain	14.0	3.3	0.54	—	0.51	0.03	—	1.88	0.01	0.40	1.36	0.11
Semolina	14.0	1.8	0.33	—	0.31	0.02	0	0.90	0.01	0.17	0.68	0.04

[a]14:0 = myristic; 16:0 = palmitic; 18:0 = stearic; 20:0 = arachidonic; 16:1 = palmitoleic; 18:1 = oleic; 18:2 = linoleic; 18:3 = linolenic.
[b]Dashes denote <0.005 g.
Source: Ref. 36.

TABLE 44 Vitamin Contents of Chris Hard Red Spring Wheat and Its Milling Fractions (μg/g. dry basis)

Vitamin	Grain	Flour	Shorts	Bran
Thiamin	9.9	0.7	10.1	13.2
Riboflavin	3.1	1.5	1.8	5.5
Niacin	48.3	9.5	23.5	171.4
Biotin	0.056	0.013	0.055	0.162
Folacin	0.56	0.09	0.59	1.59
Pantothenic acid	9.1	2.5	7.0	31.7
Vitamin B_6	4.7	0.48	5.3	13.0

Source: Ref. 39.

TABLE 45 Comparison of X–Ray Fluorescence (XRF) Bran Mineral Values with Literature Mineral Values for Wheat Whole Grain and Bran in ppm Dry Weight Basis

Constituent	Mg	P	S	Cl	K	Ca	Mn	Fe	Cu	Zn
Grain[a]	900–2900	1500–5400	1000–2900	200–1900	2900–6200	50–1220	5–260	28–420	4–24	19–100
Grain[b]	1800 ± 140						37.9 ± 3.4		5.1 ± 0.5	24 ± 4.5
Hard wheat										
grain[c]	1326				3580	290	47.7	40.1	4.9	30.9
Soft wheat										
grain[c]	1300				4035	300	47.2	36.4	4.7	29.2
Bran[d]	6880	15700			16710	1280	136.5	141.3	15.2	99.4
Bran[e]	4500–7400	10500–17440			13950–18600	660–1510	100–167	86–120	9.8–18.6	65–164
Bran[f]—Mean	3508	6805	2448	779	10359	977	127.3	94.0	9.6	68.2
XRF										
Range	2691–5212	4711–10295	1654–3205	435–1245	7014–14501	443–1680	80.4–237.3	54.1–146.1	5.3–16.1	42.0–120.9
C.V.	6.7	6.4	3.4	7.7	6.4	7.9	9.7	9.8	10.2	8.9

[a]From Ref. 3.
[b]From Ref. 73.
[c]From Ref. 37.
[d]From Ref. 12.
[e]From Ref. 18.
[f]From Ref. 43.

REFERENCES

1. Australian Wheat Board, Annual Report 1986/87, GPO Box 4562, Melbourne, 3001, Australia.
2. Aykroyd, W. R., and Doughty, J., Wheat in Human nutrition, FAO Nutritional Studies No. 23, FAO, Rome, 1970.
3. Beeson, K. C., The mineral composition of crops with particular reference to the soils in which they were grown, U.S. Dept. of Agr., Misc. Publ. 369, 1941.
4. Bizzarri, O., and Morelli, A., Milling of durum wheat, in: *Durum Wheat: Chemistry and Technology*, G. Fabriani and C. Lintas (Eds.), American Association of Cereal Chemists, Inc., St. Paul, MN, 1988.
5. Briggle, L. W., Morphology of the wheat plant, in: *Wheat and Wheat Improvement*, K. S. Quinsenberry and L. P. Reitz (Eds.), American Society of Agronomy, Madison, WI, 1967, pp. 89–116.
6. Briggle, L. W., and Reitz, L. P., Classification of *Triticum* species and of wheat varieties grown in the United States, *U.S. Dept. of Agr. Bull. 1278* (1963).
7. Briggle, L. W., and Curtis, B. C., Wheat worldwide, in: *Wheat and Wheat Improvement*, E. G. Heyne (Ed.), 2nd ed. American Society of Agronomy, Inc., Madison, WI, 1987, pp. 1–32.
8. Briggle, L. W., Origin and botany of wheat, in: *Wheat*, Documenta CIBA-GEIGY, E. Hafliger (Ed.), CIBA-GEIGY, Basel, Switzerland, 1980, pp. 6–13.
9. *Bulk Wheat Yearbook*, Grain Handling Authority of N.S.W., Vol. 17, 1983.
10. Canadian Grain Research Laboratory, *Quality of Canadian Grain 1987*, 1404-303 Main Street, Winnipeg, Manitoba, R3C 3G8, 1987.
11. Cox, T. S., Shogren, M. D., Sears, R. G., Martin, T. J., and Bolte, L. C., Genetic improvement in milling and baking of hard red winter wheat cultivars, 1919–1988, *Crop Sci. 28*:750–755 (1988).
12. Czerniejewski, C. P., Shank, C. W., Bechtel, W. G., and Bradley, W. B., The minerals of wheat, flour, and bread, *Cereal Chem., 41*:65–72 (1964).
13. *Dehumidification Engineering Manual*, Airflow Co., Frederick, MD 21701.
14. G. Fabriani and C. Lintas (Eds.), *Durum Wheat: Chemistry and Technology*, American Association of Cereal Chemists, St. Paul, MN, 1988.
15. Eggum, B. O., Nutritional aspects of cereal proteins, in: A. Muhammed, et al. (Ed.), *Genetic Diversity in Plants*, Plenum Press, New York, 1977, pp. 349–369.
16. Farrand, E. A., Study of the relationships between wheat protein contents of two U.K. varieties and derived flour protein contents at varying extraction rates. I. Studies on an experimental commercial mill and a laboratory Buhler mill, *Cereal Chem., 51*:56–66 (1974).
17. Farrand, E. A. Study of the relationships between wheat protein contents of two U.K. varieties and derived flour protein contents at varying extraction rates. II. Studies by hand-dissection of individual grains, *Cereal Chem., 51*:66–74 (1974).
18. Farrell, E. P., Ward, A., Miller, G. D., and Lovett, L. A., Extensive analyses of flours and millfeeds made from nine different wheat mixes, *Cereal Chem., 44*:39–59 (1967).
19. *Federal Register*, Volume 52, No. 125, June 30, 1987, pp. 24430–24431.
20. Food and Agricultural Organization (FAO), *Production Yearbook*, Volume 32, FAO, Rome, 1978.
21. Food and Agricultural Organization (FAO), *Production Yearbook*, Volume 40, FAO, Rome, 1986.
22. Food Agricultural Organization (FAO), *Trade Yearbook*, Volume 32, FAO, Rome, 1978.
23. Food and Agricultural Organization (FAO), *Trade Yearbook*, Volume 39, FAO, Rome, 1985.
24. Food and Agricultural Organization (FAO), *Trade Yearbook*, Volume 40, FAO, Rome, 1986.
25. Gomme, Frank R., Review of the world foods grains situation, in: *64th Annual Agricultural Outlook Conference*, USDA, Washington, D.C., Dec. 1–3, 1987, pp. 101–108.
26. Heyne, E. G. (Ed.), *Wheat and Wheat Improvement*, 2nd ed., American Society of Agronomy, Inc., Madison, WI, 1987.

27. Hinton, J. J. C., The distribution of protein in the maize kernel in comparison with that in wheat, *Cereal Chem., 30*:441–445 (1953).
28. Holmay, K., Storage system combines aeration. Fumigation technologies, *The Wheat Grower, 11*:15–17 (1988).
29. Inglett, G. E. (Ed.), *Wheat: Production and Utilization*, AVI Publishing Co., Westport, CT. 1974.
30. Johnson, V. A., and Mattern, P. J., Wheat, rye and triticale. In: Olson, R. A. and Frey, K. J. (Eds.), *Nutritional Quality of Cereal Grains: Genetic and Agronomic Improvement*, Agronomy Series 28. American Society of Agronomy, Madison, WI., 1987.
31. Jones, C. R., Recent developments in flour milling technology, *Milling, 142*(13):324–327 (1964).
32. Kent, N. L., Subaleurone endosperm cells of high protein content, *Cereal Chem., 43*:585–601 (1966).
33. Kent, N. L., *Technology of Cereals*, 3rd ed., Pergamon Press, Inc., Elmsford, NY, 1983.
34. Kent-Jones, D. W., and Amos, A. J., *Modern Cereal Chemistry*, 4th ed., Northern Pub. Co. Ltd., Liverpool, 1947.
35. Kimber, G., and Sears, E. R., Evolution in the genus *Triticum* and the origin of cultivated wheat, in: E. G. Heyne (Ed.), *Wheat and Wheat Improvement*, 2nd ed., American Society of Agronomy, Inc., Madison, WI, 1987, pp. 154–164.
36. Lockhart, H. B., and Nesheim, R. O., Nutritional quality of cereal grains, in: Y. Pomeranz, (Ed.), *Cereals '78: Better Nutrition for the World's Millions*. American Association of Cereal Chemists, St. Paul, MN, 1978, pp. 201–221.
37. Lorenz, K., and Loewe, R., Mineral composition of U.S. and Canadian wheats and blends, *J. Agric. Food Chem., 25*:806–809 (1977).
38. Metzger, R. J., and Silbaugh, B. A., Location of genes for seed coat color in hexaploid wheat, *Triticum aestivum* L, *Crop Sci., 10*:495–496 (1970).
39. Michela, P., and Lorenz, K., The vitamins of triticale, wheat, and rye, *Cereal Chem., 53*:853–861 (1976).
40. Miller, W. Kirk, "Grain Standards; Official U.S. Standards, Handling Practices and Insect Infestations; Final Rules," *Federal Register, 52*(125):24414–24442. Federal Grain Inspection Service, USDA, June 30, 1987.
41. Mitsunaga, T., and Mitsuda, H., Comprehensive examination of protein distribution profile in wheat grain. *J. Nutr. Sci. Vitaminol., 21*:363–372 (1975).
42. Panford, J. A., Application of near-infrared reflectance spectroscopy in North America, in: Williams, P. and Norris, K. (Eds.), *Near-Infrared Technology in the Agricultural and Food Industries*, American Association of Cereal Chemists, St. Paul, MN, 1987, pp. 201–212.
43. Peterson, C. J., Evaluation of variation in trace mineral levels of wheat grain using x-ray fluorescence spectrometry, M. S. dissertation, University of Nebraska-Lincoln, 1982.
44. Pomeranz, Y., Martin, C. R., Rousser, R., Brabec, D., and Lai, F. S., Wheat hardness determined by a single kernel compression instrument with semiautomated feeder, *Cereal Chem., 65*:86–94 (1988).
45. Quinsenberry, K. S., and Retiz, L. P. (Eds.), *Wheat and Wheat Improvement*, American Society of Agronomy, Inc., Madison, WI, 1967.
46. *Review of World Wheat Situation 1984/85*, International Wheat Council, 28 Haymarket, London, SW1Y 4S5, p. 33.
47. Sakamura, T. Kurze mitteilung über die Chromosomenzahlen und die Verwandtschaftsverhältnisse der Triticum-arten. *Bot. Mag.* (Tokyo), *32*:150–153 (1918).
48. Schmidt, J. W., Breeding and genetics, chapter 2, in: *Wheat: Production and Utilization*, G. E. Inglett (Ed.), AVI Publishing Co., Inc., Westport, CT, 1974.
49. Schmidt, J. W., Genetics and breeding in wheat. In: *Wheat Documenta CIBA-GEIGY*, E. Hafliger, (Ed.), CIBA-GEIGY, Basel, Switzerland, 1980, pp. 14–18.
50. Shellenberger, J. A., World wheat, standards compared, *Cereal Foods World, 20*:345–347, 368–370 (1975).
51. Shellenberger, J. A., International standards for grains, *Cereal Foods World, 22*:73–74, 1977.

52. Simmonds, D. H., Structure, Composition and Biochemistry of Cereal Grains. In: *Cereals '78: Better Nutrition for the World's Millions*, Y. Pomeranz (Ed.), American Association of Cereal Chemists, St. Paul, MN, 1978, pp. 105–137.
53. W. T. Yamazaki and C. T. Greenwood (Eds.), *Soft Wheat: Production, Breeding, Milling, and Uses*, American Association of Cereal Chemists, Inc., St. Paul, MN, 1981.
54. Swanson, C. O., *Wheat and Flour Quality*, Burgess Publishing Co., Minneapolis, MN, 1938.
55. U.S. Department of Agriculture, *Agricultural Statistics 1986*, U.S. Government Printing Office, Washington, D.C., 1987.
56. USDA, ARS, *Report of Workshop on Wheat Classing, March 8–9, 1984*, Beltsville, MD, 1984.
57. USDA, ERS, *Outlook '88 Charts*, 64th Annual Agricultural Outlook Conference. Dec. 1–3, 1987. Published Jan. 1988, USDA, Washington, D.C.
58. U.S. Department of Agriculture, Agricultural Marketing Service. Data on proposed changes in wheat standards, p. 4, AMS Grain Division Publication AMS-514, Sept. 1963.
59. USDA, FAS, *Export Market for U.S. Grain and Products*, Circular EMG, 6-88, p. 22.
60. USDA, FAS, *World Grain Situation and Outlook*, Circular Series FG 5-88, May 1988, p. 26.
61. U.S. Department of Agriculture, ERS, FAS, *World Agricultural Supply and Demand Estimate*, WASDE-221, August 11, 1988.
62. U.S. Department of Agriculture, *FGIS Grain Inspection Handbook*, Book II, Wheat, May 1, 1988.
63. USDA, *Grain Grading Primer*, Miscellaneous Publication No. 740. September 1957.
64. U.S. Department of Commerce, Bureau of Census, Current Industrial Reports, Flour Milling Products, M20A(87)12, Washington D.C., issued February, 1988.
65. Vogel, K. P., Johnson, V. A., and Mattern, P. J., Results of systematic analyses for protein and lysine composition of common wheats (*Triticum aestivum* L.) in the USDA World Collection, *Nebr. Res. Bull., 258* (1973).
66. Vogel, K. P., Mattern, P. J., and Lenser, G. W., Modified milling procedure for separating endosperm and nonendosperm portions of the wheat kernel for protein and lysine analysis, *Agron. J. 68*:686–690 (1976).
67. Wellman, A. *Sources of Grain Market Information*, NebGuide G82-622 (revised April 1988), Cooperative Extension Service, University of Nebraska-Lincoln, Lincoln, NE, 1988.
68. Williams, P. C., The influence of chromosome number and species on wheat hardness. *Cereal Chem. 63*(1):56–57 (1986).
69. Williams, P. C., and Sobering, D. C. Attempts at standardization of hardness testing of wheat. II. The near-infrared reflectance method, *Cereal Foods World, 31*:417–420 (1986).
70. Williams, P., and Norris, K. (Eds.), *Near-Infrared Technology in the Agricultural and Food Industries*, American Association of Cereal Chemists, St. Paul, MN, 1987.
71. *World Agricultural Supply and Demand Estimates*, Economic Research Service, U.S. Dept. of Agric., July 12, 1988.
72. *World Food Trade and U.S. Agriculture 1960–1986*, Trade Tensions Intensify, 7th Annual Edition. The World Food Institute, Iowa State University, Ames, IA, Oct. 1987.
73. Zook, E. G., Greene, F. E., and Morris, E. R., Nutrient composition of selected wheats and wheat products. VI. Distribution of manganese, copper, nickel, zinc, magnesium, lead, tin, cadmium, chromium, and selenium as determined by atomic adsorption and colorimetry. *Cereal Chem., 47*:720–731 (1970).

2

Corn: Production, Processing, and Utilization

Lawrence A. Johnson
Iowa State University of Science and Technology, Ames, Iowa

I. HISTORY OF CORN

Corn (*Zea mays* L.) is native to the Americas and is their most important cereal crop. Corn originated in Mexico, evolving from the wild grass Teosinte (*Zea mays* sp. *mexicana*). Archeological evidence indicates that corn was domesticated and grown as a crop by native Americans as early as 5000 B.C. in Mexico's valley of Tehuacan (14). Corn became a symbol of religion and prosperity in the Mayan civilization. From this center corn spread northward to Canada and southward to Argentina.

Among the discoveries made by Christopher Columbus during his visit to the New World in 1492 was this staple food of the native Americans, known to them as "mahyz." Corn is still universally known as "maize" from the early Spanish translation. Maize was not popularly known as "corn" until the American colonists applied this British term for grain to maize. Following Columbus's discovery, corn was transplanted to Spain from where it quickly spread across Europe and to Africa and Asia.

Corn played an important role in the colonialization of the United States. The native Americans taught the colonists how to convert the grain into food and livestock feed. Breads made from corn quickly became staple foods in colonial America, and other foods, such as mush, porridges, and breakfast grits, were important foods to pioneers migrating west. Sites suitable to providing water power for stone mills to process corn and other cereals were usually critical in decisions about where to locate frontier towns during the the settling of the midwestern United States.

Early improvement in corn varieties was accomplished by farmers selecting seed from the hardiest and most productive plants in their open pollinated fields. However, early in the 1800s two predominant races of corn—Virginia Gourdseed and Northeastern Flints—grown in the Eastern Seaboard states were crossed, demonstrating the superiority of hybrid corn. The cross was repeated many times as settlers migrated west, and from these lines the Corn Belt Dents emerged. Corn Belt Dents have been steadily improved to become the most productive race of corn in the world—indeed, the most productive cereal in the world.

II. HIGH PRODUCTIVITY OF CORN

Every continent except Antarctica now produces corn (Table 1). Corn ranks as the second most widely produced cereal crop worldwide (78). Only wheat is produced in greater quantity. The U.S. Corn Belt is by far the largest producer, producing almost one-half of the world's corn production. China and Brazil are also major producers.

Corn yields in the United States are often more than double those elsewhere in the world. The 1987 average yield in the U.S. Corn Belt approached 120 bu/acre (6720 lb/acre) (77) and yields over 240 bu/acre have been reported. Only 35 years ago the average yield in this area was 38 bu/acre. During this period, yields have increased yearly at an average pace 2.2 bu/acre and there is little indication that this level of annual increase cannot be sustained for quite some time to come. Because of this high productivity, corn is by far the most economical cereal to produce and, therefore, is also one of the most economical sources of metabolizable energy for feeds and of starch and sugar for food and industrial products. Due to the recent world wide glut of grains, the price paid for corn in the United States has fallen over the past 5 years. The weighted average price paid to U.S. farmers in 1983 was $3.25/bu but by 1986 had fallen to $1.49/bu (82).

III. ANATOMICAL STRUCTURE, COMPOSITION, AND PROPERTIES

Table 2 tabulates the physical properties of corn, which are important in designing handling equipment and storage facilities. These properties, of course, are somewhat affected by moisture content and are reported in Table 2 on a 15% moisture basis. Compared to other grains, corn has a unique shape and specific gravity, but many other properties are similar to other cereals.

Corn kernels are the largest cereal seed, weighing 250–300 mg each. They are flat seeds due to pressure during growth from adjacent kernels on the cob. The kernels have a blunt crown and a pointed conical tip cap (Fig. 1). Corn kernels are botanically classified as a caryopsis (dry, indehescent, single-seeded fruit) and are attached to the cob by the pedicle. The kernel contains a complete embryo and all the structural, nutritional, and enzymatic functions required for growth and development into a plant. The kernel is composed of four anatomical parts: the tip cap, which provides the point of attachment between the cob and the kernel; the bran or protective outer covering; the germ and embryo; and the endosperm, which is the reservoir of nutrients to support germination.

Each anatomical part of the corn kernel has a different composition (Fig. 2) The largest fraction of the kernel is the endosperm, which is largely composed of starch, the reserve energy supply. Endosperm cells are packed with starch granules embedded in a continuous matrix of amorphous protein. Also embedded in this matrix are protein bodies composed almost entirely of the storage protein zein.

Corn is often harvested as soon as the moisture content drops below 28%. Unless quickly dried, high-moisture corn is subject to rapid deterioration, especially be mold infestation. Any hygroscopic material, including corn, loses (desorbs) or gains (adsorbs) moisture depending on whether its water vapor pressure (water activity) is more or less than its environment. Equilibrium moisture content for corn is complicated by the hysteresis effect, in which the equilibrium moisture content depends on whether the grain is desorbing or adsorbing (Fig. 3) Also, each component, be it protein, starch, fat, or fiber, has a different affinity for moisture, so it is not surprising that different corn fractions have different moisture sorption isotherms. Because different types of corn have slightly different compositions, their moisture sorption isotherms may vary accordingly. A general model for describing moisture sorption isotherms was developed by Chung and Pfost (16). The critical moisture content for "safe" storage of corn is generally regarded as 15%, but as Figure 4 indicates, lower moisture contents are required for corn to be stored for long periods at warm temperatures as occur in the tropics.

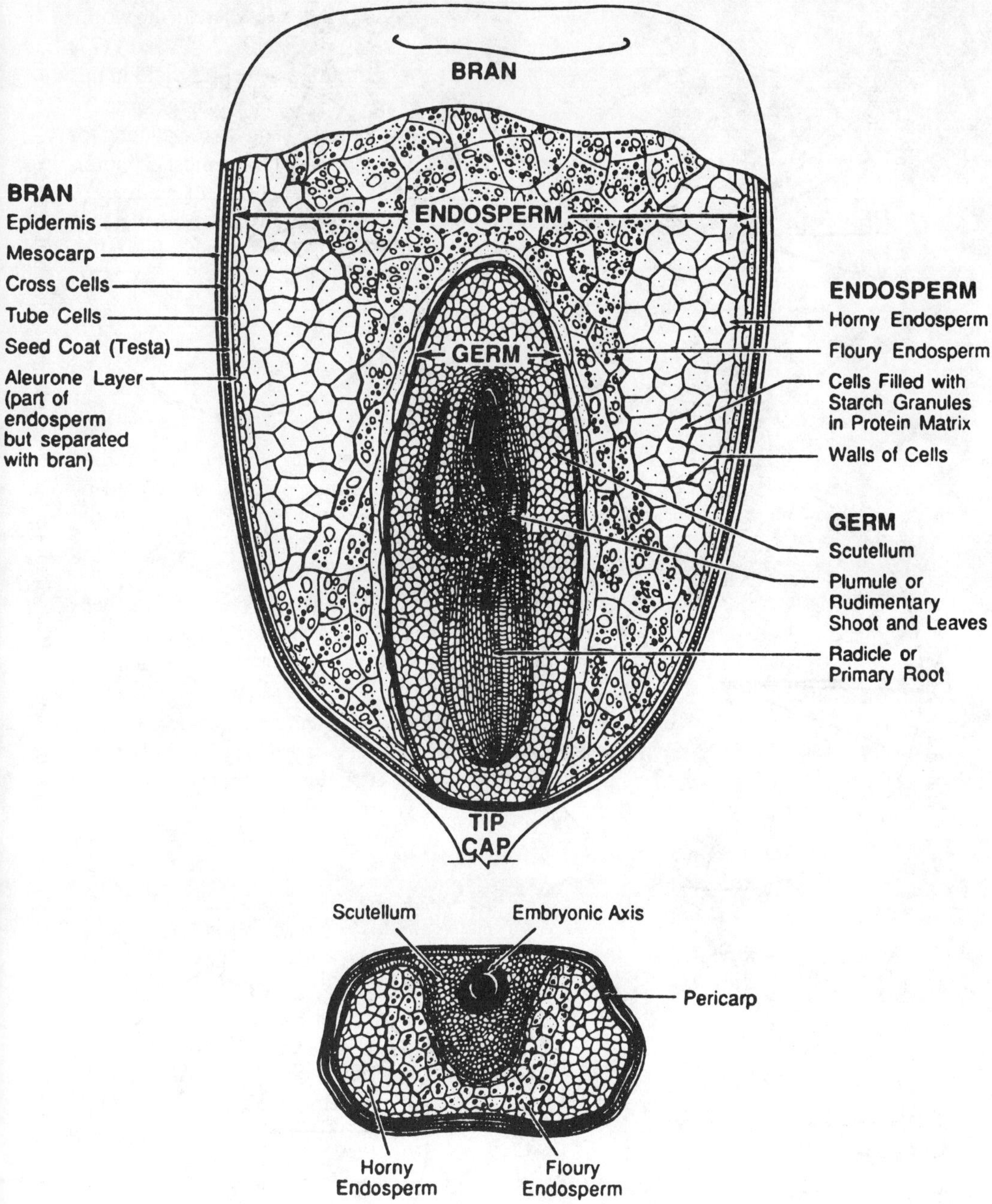

FIGURE 1 Structure of the corn kernel. (Modified from figures provided by the Corn Refiners Association.

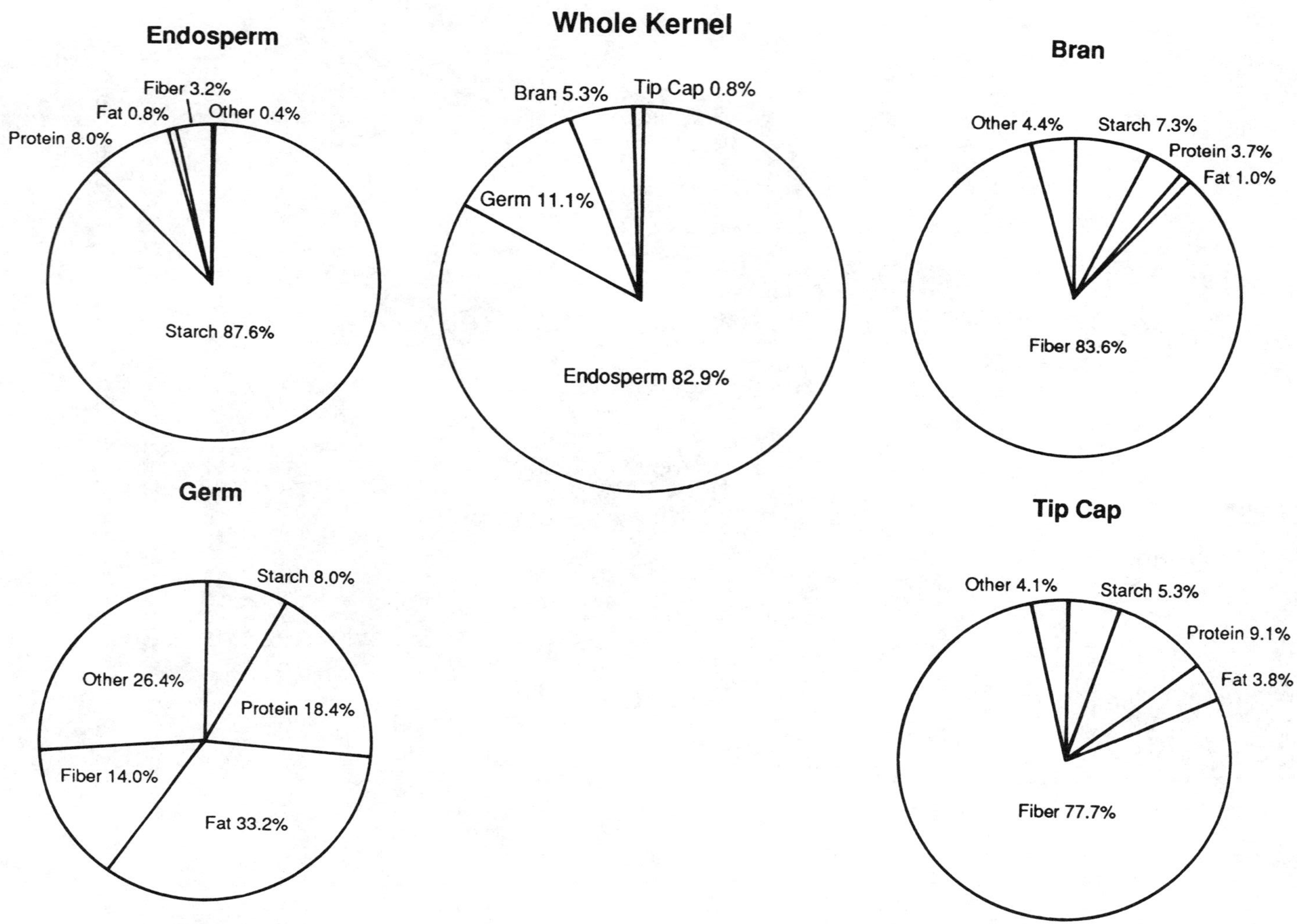

FIGURE 2 Compositions of normal dent corn and its anatomical parts. (Data from Earl et al. (28) (samples 1–6 and 8) Watson (88) and estimates of author.)

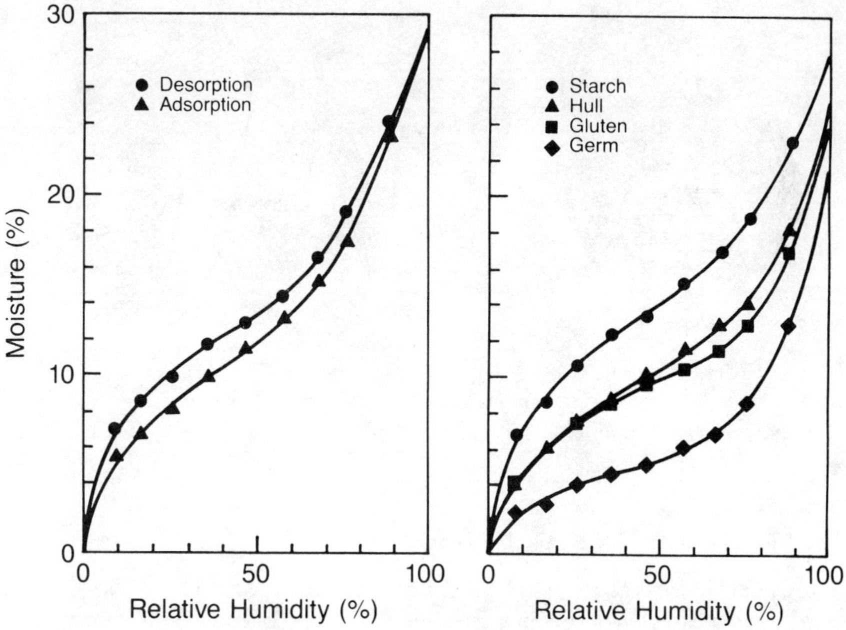

FIGURE 3 Equilibrium moisture isotherms for corn and corn products. (Modified from Ref. 15.)

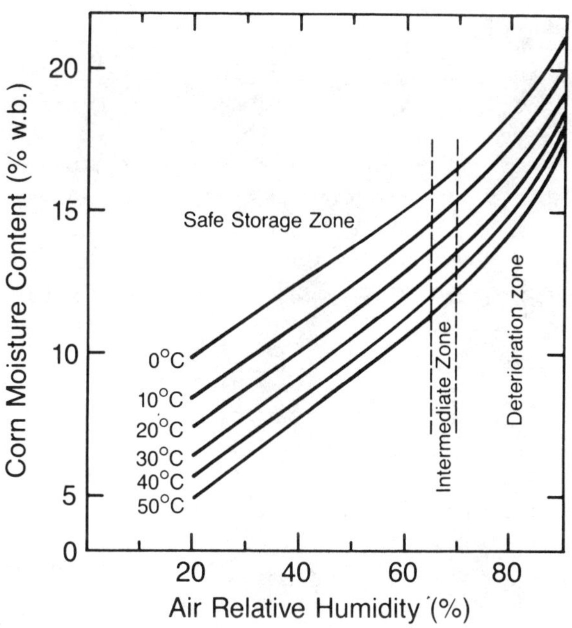

FIGURE 4 Desorption equilibrium moisture isotherms for corn at different temperatures. (From Ref. 37, originally Ref. 16. Used with permission.)

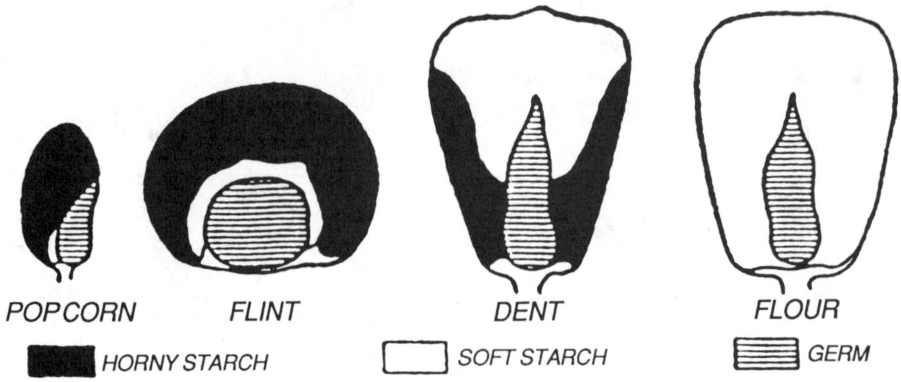

FIGURE 5 Types of corn. (From Ref. 75.)

IV. CORN TYPES AND THEIR COMPOSITIONS

There are six major types of corn kernels: dent, flint, flour, sweet, pop, and pod corns. Major differences are largely based on the quality, quantity, and pattern of endosperm composition (Fig. 5 and Table 3). Dent corn is characterized by the presence of corneous, horny endosperm at the sides and back of the kernels, while the central core is soft and floury. The soft endosperm extends to the crown, where it collapses on drying to form an indentation. The indented crown is peculiar to dent types and is the basis for the term "dent corn."

There are white as well as yellow dent corns. Higher prices are usually paid for white dents, which are often used by the dry corn milling industry to produce light-colored food products. In addition to color, dent corns vary in their proximate compositions and starch properties. The University of Illinois has bred varieties that are high in protein and others that are high in oil. Some lines have over 10% fat compared to more typical values of 4–4.5%. Other lines have as much as 26% protein compared to typical levels of 9–10%. High oil and high protein lines have little commercial value at present, but may have competitive advantages in livestock feed if yields can be improved. Other dent lines differ in their starch properties. Corn starch is normally composed of 74–76% amylopectin and 24–26% amylose. Some corn lines, called "waxy," contain over 99% amylopectin (48); others contain as little as 20% (80% amylose) and are known as "amylomaize." Both waxy and amylomaize are grown on contracts with growers for wet corn mills to process them into specialty starches. Waxy and modified waxy starches are widely produced because of their viscosity, thermal and pH stability, and other properties after hydration. Waxy grain is occasionally grown for livestock feed. Amylomaize starch is used in textiles, gum candies, and adhesives for manufacturing corrugated cardboard.

Flint corns have thick, hard, vitreous endosperm layer surrounding a small, soft, granular center. The relative amounts of corneous (horny) and soft (floury) endosperm may vary. The ears are long and slender with fewer rows of kernels than dent. The kernels are smooth and rounded. Very little flint corn is grown in the United States today even though it was grown extensively during colonial times. Yields of flint corn are generally 10% less than those of dent corns. Flints are extensively grown in South America, Africa, and southern Europe.

The most primitive races of corn are the popcorns. Popcorn is characterized by a very hard, corneous endosperm and is essentially a small-kerneled flint. Popcorn is usually grown for use as a snack food.

Flour corn is one of the oldest corns, having been raised by the Aztecs and Incas. Flour types are characterized by soft endosperm throughout the kernel. They are easy to grind but are susceptible to mold in wet areas. small amounts of flour corns are grown in dry areas of the United States, but large amounts are grown in Andean region of South America.

High lysine or opaque-2 corn has been bred to improve the amino acid balance in corn protein. Compared to dent corn, opaque-2 corn is high in lysine, an essential amino acid. Opaque-2 corn is similar to flour types having soft chalky endosperm. Opaque-2 corn contains relatively few and small protein bodies in its endosperm. Despite considerable promise for improving the nutrition of populations with diets high in corn, high lysine corns have not attained a widespread use due to 7–10% lower yields than dents.

Sweet corns are believed to have originated from a mutation in Chullpi, a Peruvian race of corn. The sugary gene of sweet corn retards normal conversion of sugar to starch and the kernel accumulates phytoglycogen, a water-soluble polysaccharide which alters texture and increase sweetness. Soluble saccharides comprise about 12% of the dry weight of sweet corn versus 2–3% in other types. Sweet corn is normally consumed as a food vegetable in the immature milk stage.

Pod corn is an ornamental type with long glumes enclosing each kernel in addition to husks covering the ear. There are dent, sweet, waxy, pop, and flint types within pod corn. Pod corn is not grown commercially and is merely regarded as a curiosity.

V. CORN QUALITY AND GRADING STANDARDS

Grading standards are used to facilitate marketing of most grains, and corn is exception. U.S. grading standards for corn have been in effect with few changes since the U.S. Grain Standards Act was enacted in 1916. Current U.S. grading standards are shown in Table 4. There are five numerical grades and a lower grade known as "sample grade." Currently, four factors determine grade: test weight, a measure of bulk density; amount of broken kernels and foreign material; amount of total damaged kernels; and amount of heat-damaged kernels. Moisture was also a grading factor prior to 1985, but now is only a standard, the value of which must be shown on invoices (32). These standards specify three classes of corn: Yellow Corn (that which contains less than 5% white corn); White corn (that which contains no more than 2% yellow corn); and Mixed Corn (that which contains more than 10% of other grains).

VI. CORN UTILIZATION

Depending on the provisions of the annual farm bill and weather conditions during the growing season, the United States typically produces over 8 billion bushels of corn (Table 5). In 1986, the United States produced 8.25 billion bushels, and 7.41 billion bushels were utilized, leaving a surplus, accumulated over several years, of 4.9 billion bushels in storage. During that year, 63% was consumed as feed by livestock; 21%, exported; 16%, for food, alcohol, and industrial purposes; and 0.2%, as planting seed. The market for corn in export channels has greatly expanded since 1960 (Fig. 6), and the amount utilized by processing industries to produce starches, sweeteners, and ethanol has expanded considerably since 1975. Corn, once regarded almost exclusively as a feed grain and largely consumed right on the farm where it was produced, now has other important market outlets.

Utilizations of corn can be divided among those involving direct consumption with minimal preparation and those involving substantial value-added processing. For instance, corn can be used directly by livestock to produce meat and dairy products. Direct uses of corn include sweet corn, popcorn, alkali-cooked corn, and other foods made from whole

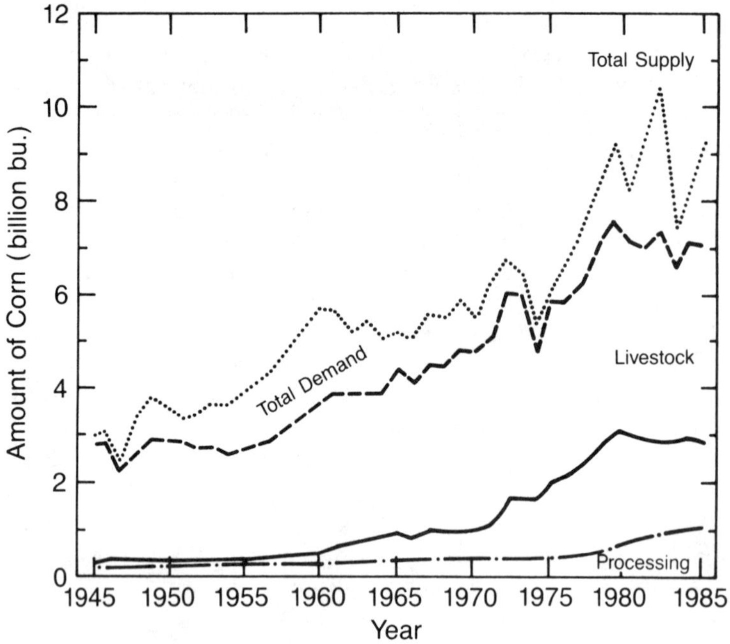

FIGURE 6 U.S. corn supply and demand estimates. (Modified from Ref. 44.)

corn or by traditional stone grinding. Other ways to utilize corn involve one or more levels of value-added processing. These processing industries include: fractionation processes which take corn apart and separate it into its component fractions for use as food ingredients or industrial products, such as wet and dry milling; conversion processes which convert a fraction into more valuable ingredients, such as the enzymatic conversion of starch into sugar or the fermentation of the sugar into ethanol; and refabrication processes which recombine corn or corn ingredients with other ingredients to produce "engineered" or "refabricated" foods or industrial products. Of course, value must be added at each stage in order to justify processing costs.

A. Corn as Livestock Feed

In 1986, about 4.65 billion bushels of corn grain, 125 million tons of corn silage, 4.35 million tons of corn gluten feed, 0.83 million tons of corn gluten meal, and 1.5 million tons of corn hominy feed were used to feed livestock. Corn and corn products are generally the most cost-effective feeds or feed supplements available. The high yield, abundant predictable supply, low cost, close proximity of large animal populations to corn-growing areas, and high digestibility make corn the preferred feed. In addition, corn is highly palatable to livestock and poultry, is easily ground and processed, does not contain any antinutritional factors, and only rarely contains toxigenic compounds such as aflatoxins. Tables 6–8 provide nutritional information that is important when feeding corn and corn products.

Typical corn-based cattle rations using shelled corn and cracked corn are shown in Table 9. Corn often comprises 85–94% of cattle finishing rations. Typical daily gains range from 3.5 to 4.0 lb per steer with feed efficiencies around 5 lb of feed per lb of weight gain.

Corn silage is an important feed to dairy farmers. Whole green plants are harvested and chopped. The chopped corn is placed in a silo and allowed to undergo natural lactic acid fermentation where sugars are converted by *Streptococcus lactis* and *Lactobacillus plantarum* to lactic acid, which acts as a preservative. About 10% of the corn acreage in the United States is harvested for silage (Watson, 1982).

Often corn is processed to improve palatability, feed conversion, and handling characteristics or to reduce wastage by animals. These processes include: mechanical processes such as dry rolling (cracking and crushing) and pelleting; dry heat treatments, such as popping, roasting, micronizing (heated by infrared heaters and rolling), and extruding; and moist heat treatments such as steam flaking and exploding.

B. Direct Utilization of Corn as Food

Despite widespread acceptance and importance of corn-based foods early in American history, direct use of corn as food in the United States has decreased over the past century in favor of increased consumption of more refined and processed foods. This is due to increased demand for more convenience and longer shelf life. However, in many other countries, especially, in Latin America, Africa, and Asia, corn remains a staple food. Table 10 lists many of these foods consumed throughout the world. Alkali-cooked products, sweet corn, and popcorn are probably the most important of these in Western diets.

1. Alkali-Cooked Corn-Based Foods

The ancient Aztecs discovered that cooking corn in the presence of the alkali from wood ash or quicklime transformed corn into a variety of highly palatable foods, now referred to as "Mexican" foods (Fig. 7). These foods are become increasingly popular, even in fast food

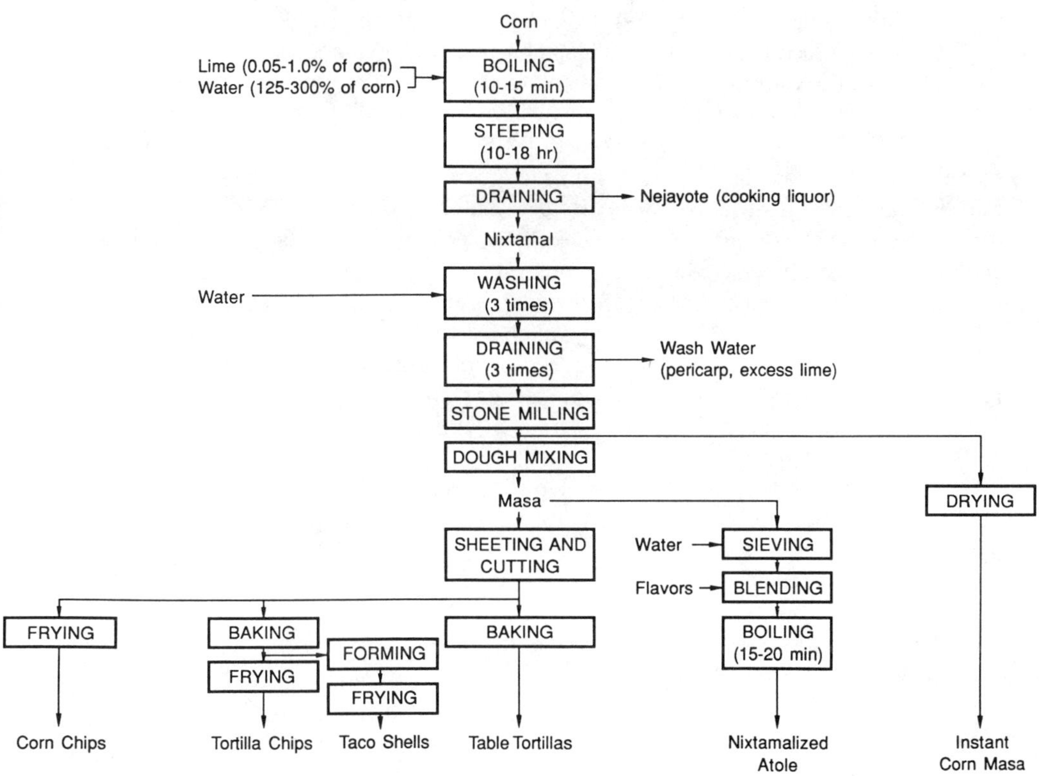

FIGURE 7 Flow sheet for producing Mexican-type foods from corn.

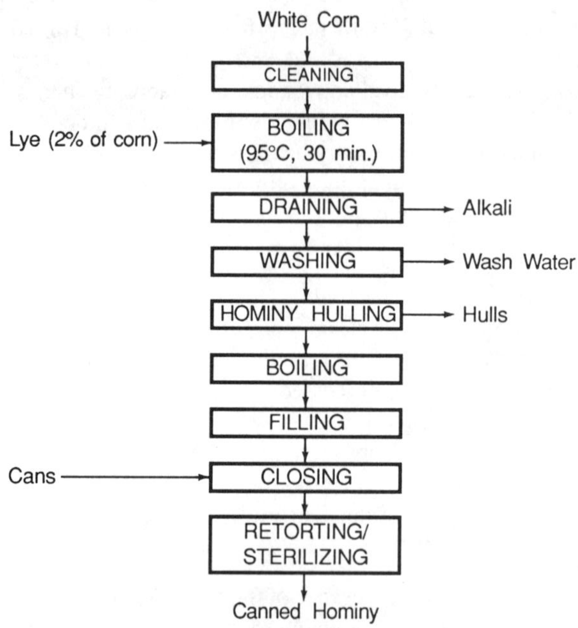

FIGURE 8 Flow sheet for processing corn hominy.

establishments in the United States, where reached $4 billion in 1986 (69). There is a growing trend in most developed countries toward foreign and ethnic foods.

This type of food includes enchiladas and burritos (made with tortillas); tacos, tostadas, and nachos (made with fried tortillas); and nixtamalized atole and tamales (made with corn masa) (33). These foods have been extensively discussed by Rooney and Serna-Saldivar (63). Traditional methods involve cooking (90–100°C for 10–15 min) and steeping the corn overnight in water containing about 1% lime to produce nixtamal. This process loosens the hull and some of the germ, allowing them to be removed by simple washing, gives the corn a desirable unique flavor and texture, and partially gelatinizes the starch. The amount of cooking is important to achieve the optimum texture, as excessive starch gelatinization leads to stickiness and makes the dough difficult to handle, especially if automated sheeting equipment is used. The cooking/steeping water, called "nejayote," contains up to 6% dissolved and suspended solids and is discarded. The nixtamal is washed to remove the hulls and residual lime and stone-ground to a dough called "masa." The masa is kneaded, molded into disks, and baked into tortillas on a hot griddle. Tortilla chips or tostadas are traditionally the leftover, stale tortillas which are fried.

Traditionally, white corns are preferred in alkali-cooked foods. Opaque-2, a potentially desirable corn for corn-based foods due to its improved lysine content, is largely unacceptable in alkali-cooked foods because its starch undergoes rapid gelatinization, yielding masa that is excessively sticky. Floury corns do not produce masa doughs with acceptable rheological properties; corns with a high degree of corneous endosperm are desired. Corn lots containing more than 4% cracked kernels will produce sticky masa doughs unless cooking time is significantly reduced (10).

Early American settlers produced another type of alkali-cooked food from white corn varieties with unusually large kernels—corn hominy (Fig. 8). The corn is cooked in lye rather than lime and washed to remove hulls. Then the cooked corn is boiled in fresh water, salted, and canned. Corn hominy is consumed as a vegetable food.

2. Sweet Corn

Although field corn can be harvested in the milk stage for consumption as a vegetable food, its palatability is much inferior to that of sweet corn lines. Sweet corn differs genetically from field corn in that standard sweet corn is homozygous recessive for sugary endosperm. This homozygous recessive gene causes sweet corn to accumulate as much as 25% (db) phytoglycogen, a polysaccharide consisting of glucose that gives the desired creamy texture not possible with only starch. Optimum quality of standard sweet corn is obtained if harvested before most sugars are converted to phytoglycogen. This period is very short, often one or two days. Breeders have selected for high sugar content within standard sweet corn lines. Therefore, in order to assure a steady supply of fresh, high quality sweet corn during the growing season, planting must be staggered and closely monitored for maturity.

Since the mid-1970s, augmented sugary sweet corns have been developed that have higher sugar levels and are sweeter than standard sweet corns. These are mutants within sugary types involving the sugary enhancer gene. These lines convert sugar to phytoglycogen at the same rate as standard sweet corns but have a higher initial sugar content. Consequently, the harvesting period can be prolonged to 4–5 days and these lines maintain optimal sensory qualities longer after harvest.

Sweet corn can be consumed in a variety of ways: as corn-on-the-cob, cut kernels, and cream-style (43). Corn products are commonly preserved by both canning and freezing. Methods of processing are detailed in Figure 9. Production and consumption have been steady for the past 10 years (Table 11). Per capita consumption of sweet corn is about 28 lb/year with 27% consumed fresh; 41% canned: and 32% frozen. Over the past 10 years the trend is towards consumption of frozen sweet corn.

3. Popcorn, the Original Snack Food

Popcorn is undoubtedly the oldest snack food and has been consumed for centuries. Total U.S. popcorn sales have been steadily increasing for the past 10 years, growing more than 72% during that period (Fig. 10). The recent surge in popcorn sales is due to the proliferation of microwavable popcorn products appealing to consumers who demand improved convenience. Also, there has been a proliferation in recent years of flavored ready-popped popcorns in addition to the usual salted, buttered, and caramel-coated popcorns. Besides home consumption, popcorn is widely consumed at theaters, amusement parks, and sporting events. Popcorn is now a $2 billion industry (Popcorn Institute, Chicago, IL).

Almost all the popcorn consumed is produced in the United States. Americans are the biggest consumers, at 46 quarts of popped popcorn per capita per year. Popcorn is widely grown throughout the corn belt states; Indiana and Nebraska are the largest producers (Table 12). Total popcorn production in 1986 was about 475 million lb.

Not all cereals will pop. Popcorn appears to pop exceptionally well because the failure of the pericarp occurs at an exceptionally high temperature, 177°C. This temperature is well above the point where the moisture in the kernel vaporizes into voids surrounding starch granules creating steam pressure 135 psi and having the potential to cause an explosive expansion when the pericarp finally ruptures (42). Popcorn is a special type of flint corn selected for its popping potential. Compared to other flint corns, popcorn kernels are smaller and have essentially no soft endosperm (see Fig. 5). Commercial popcorns are usually either yellow or white, although varieties with other pericarp colors have acceptable eating qualities. High quality popcorn will have an expansion ratio (vol. popped/vol. unpopped) in excess of 40:1, yield at least 98% popped kernels, and have test weights from 63 to 68 lb/bu. Kernels with high specific gravity have been observed to have high expansion ratios (36).

Eating quality of popcorn is affected by popped volume, shape of popped kernels, tenderness, and flavor (41), all affected by moisture content before and after popping. Opti-

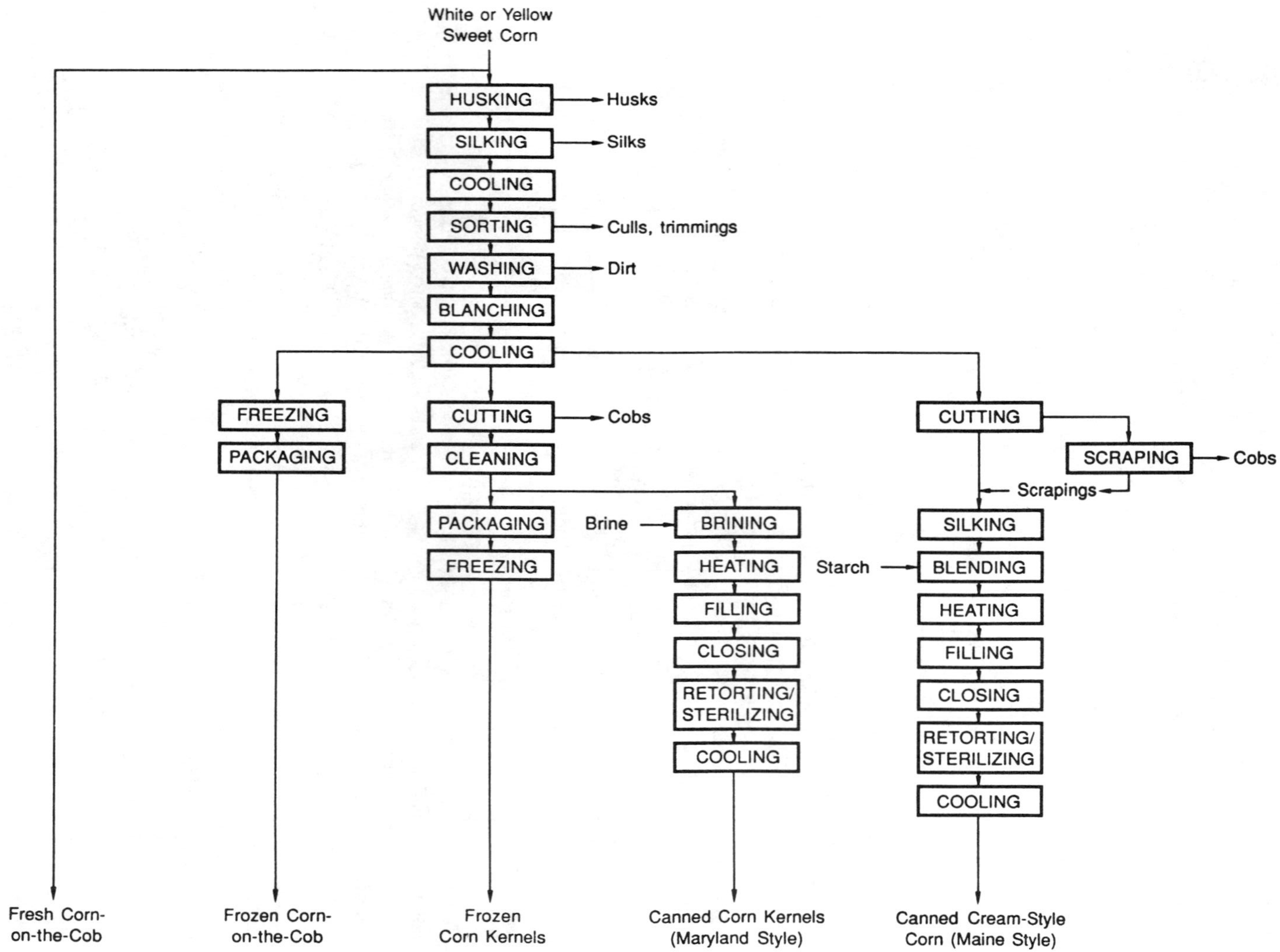

FIGURE 9 Flow sheet for processing sweet corn products.

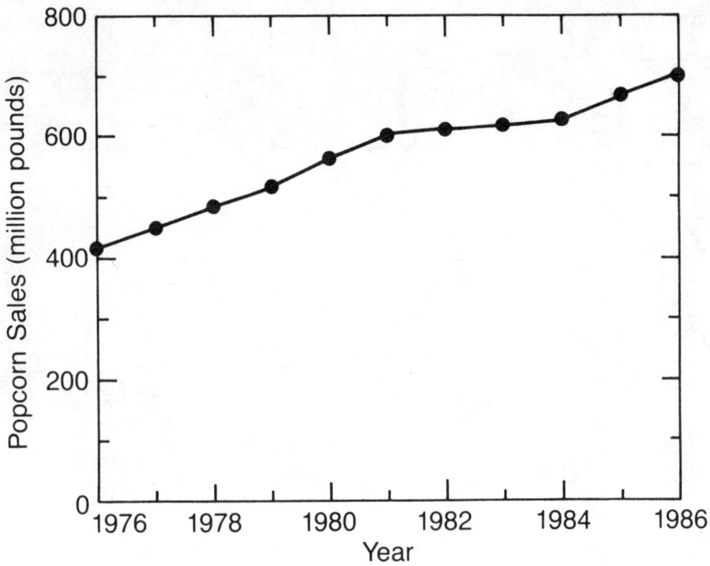

FIGURE 10 Ten year history of popcorn sales. (Data provided by the Popcorn Institute, Chicago, IL.)

mum moisture content for popping depends upon whether oil or air is used to heat the kernels (Fig. 11). When air popping, kernels with 14.0% moisture content give the highest expansion; when oil popping, kernels with 13.5% moisture give the greatest expansion. Popcorn should be packaged in containers that provide an excellent moisture barrier so that the moisture content does not deviate far from these optimum values. Special packages have been developed and patented for use with microwavable popcorns. Popped popcorn adsorbs moisture rapidly from the atmosphere unless adequately protected because of the large surface area and because of the gelatinized starch that is more hygroscopic than native starch.

VII. SEPARATION OF CORN INTO ITS COMPONENT FRACTIONS

Corn is subjected to wet-and dry-milling processes to separate the grain into its components; however, fractionation by the two methods differ in the types of component fractions sought. The objective of dry milling is to separate the kernel into its anatomical parts (endosperm, bran, and germ); the wet corn-milling objective is separation of corn into its chemical constituents (starch, protein, fiber, and oil).

The annual amount of corn ground and fractionated in the United States is about 1.1 billion bushels of which over 75% is wet milled (Fig. 12). While the amount of corn processed by the dry-milling industry has remained relatively constant over the past 8 years, the amount of corn processed by the wet-milling industry has doubled during the same period. The growth is the wet-milling industry is the result of increased demand for high fructose corn syrup and ethanol. Table 13 shows changes in production levels for wet-milled and dry-milled corn products over the past decade.

Both dry and wet corn milling are mature industries in the United States that are rapidly becoming consolidated into a few large companies. The major U.S. corn-milling companies, both wet and dry, are given in Table 14. These companies account for over 90% of the corn-processing capacity.

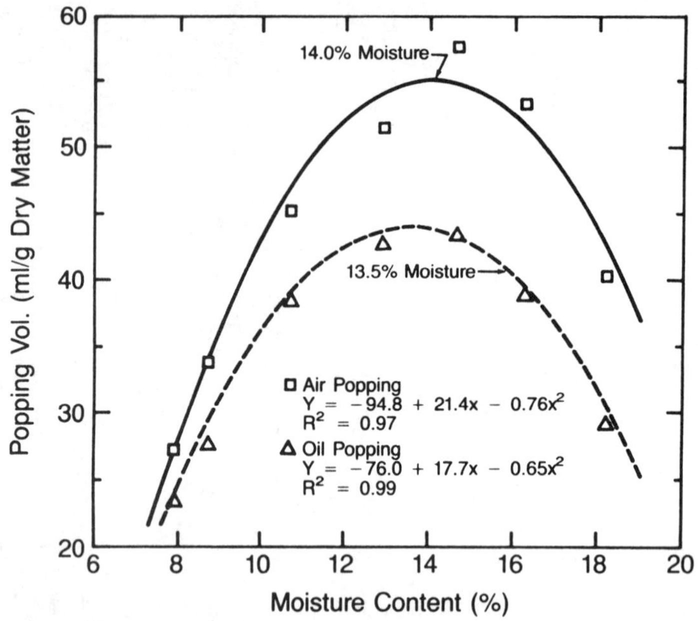

FIGURE 11 Moisture effect on popping volume of popcorn. (Data provided by K. Hsu, Iowa State University, Ames, IA.)

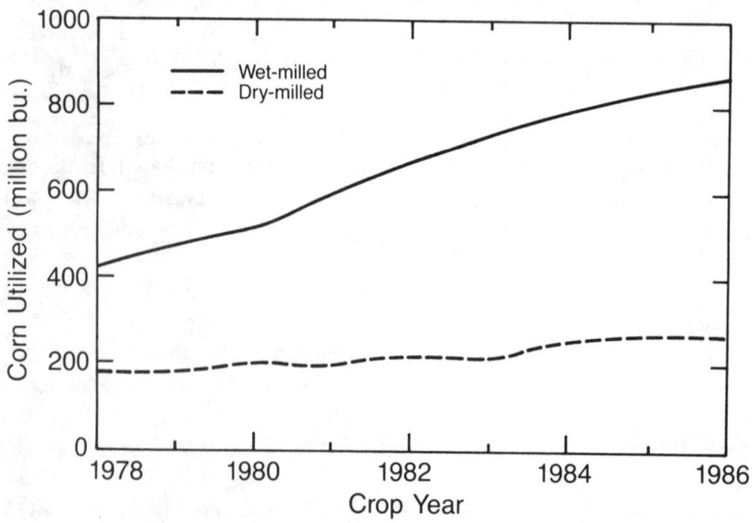

FIGURE 12 The U.S. annual corn grind. (Data provided by the Iowa Corn Promotion Board.)

A. Dry Corn Milling

Two different systems are used to dry mill corn: degerming and nondegerming. The nondegerming system grinds corn, preferably white dent corn, into meal with little, if any, separation of germ (12). Traditional stone-grinding mills continued to be used in some nondegerming systems. Occasionally stone-ground meal is bolted or sieved to remove the coarsest particles of hull and germ. Traditional or bolted undegermed corn meal has a comparatively short shelf life compared to products from degerming systems. The germ, which contains 32–35% (db) oil, is retained. The oil is spread over the surface of many particles increasing the area of exposure to oxygen and lipolytic enzymes, making it prone to oxidative and hydrolytic rancidity. No more than 20% of the corn subjected to dry corn milling is processed by nondegerming systems.

1. The Tempering-Degerming Milling Process

In 1906 the Beal degerminator was introduced, giving birth to tempering-degerming (TD) systems, well described by Stiver (70) and Easter (27). This technology facilitated the production of corn products with much lower fat contents, greater shelf life, and servicing more distant markets. Therefore, larger mills became more cost effective than older, local, small grist mills. A general flow sheet conceptualizing TD systems is shown in Figure 13; more detailed flow sheets are shown in Figures 14 and 15.

The objectives of the TD systems are: to remove essentially all of the germ and hull and leave the endosperm as free of oil and fiber as possible; to recover maximum yield of endosperm as large clean particles; and recover maximum yield of germ as large clean parti-

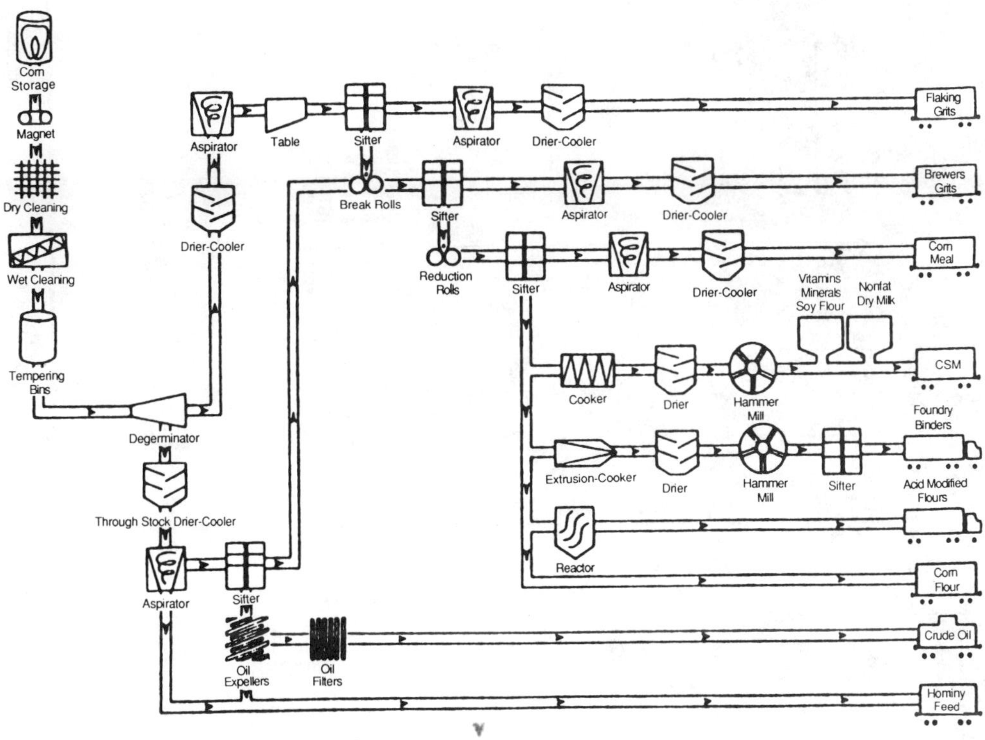

FIGURE 13 Production flow chart for dry corn milling (tempering-degerming). (From Ref. 91 with permission.)

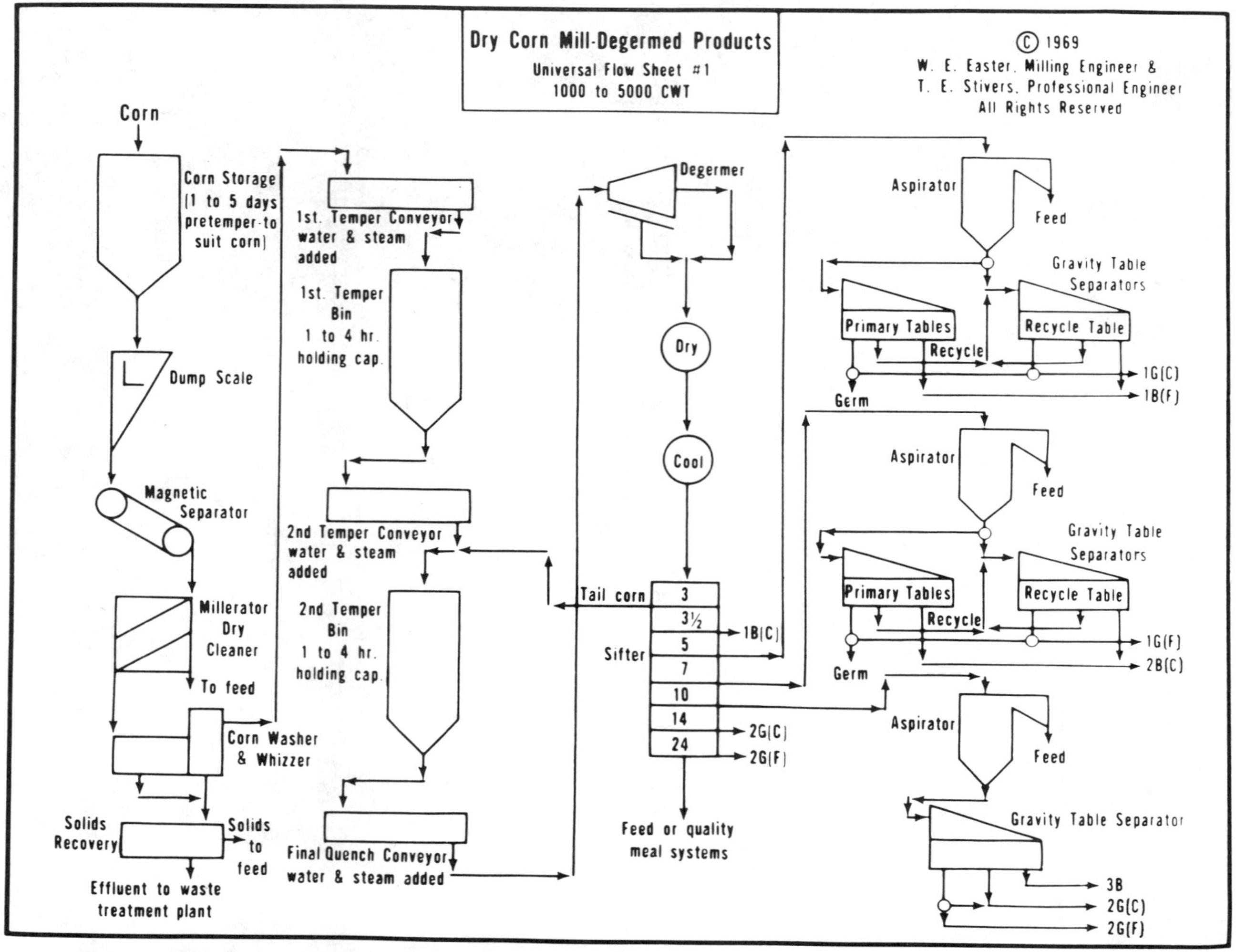

FIGURE 14 Detailed flow sheet for cleaning and degerming operations in dry corn milling (tempering-degerming). (From Ref. 70 with permission.)

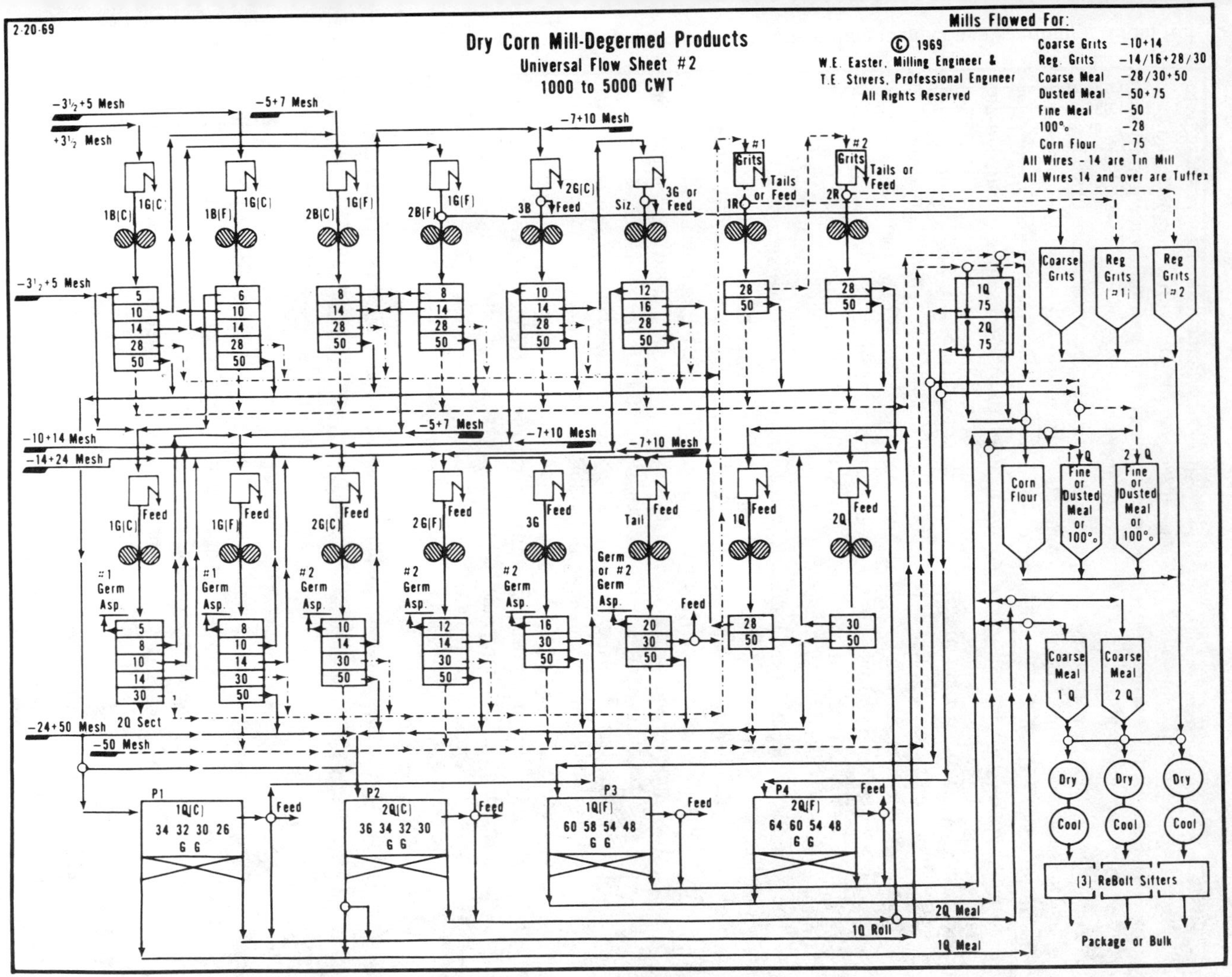

FIGURE 15 Detailed flow sheet for milling operations in dry corn milling (tempering-degerming). (From Ref. 70 with permission.)

cles (5). The TD system uses physical and mechanical treatments to achieve these objectives. Corn is cleaned to remove dirt, stones, insects, tramp iron, broken kernels, and extraneous plant material. Water is added to the corn to increase the moisture content to 20%, and the moistened corn is allowed to equilibrate (tempering) for 1–3 hr. This toughens the germ and bran so that their particle sizes remain large, making separation easier. Once the germ and hull are removed, the endosperm is reduced in size to grits with roller mills. A complex array of additional roller mills and particle size-separating equipment is used to purify and size endosperm particles. All products must be dried prior to packaging or bulk storage.

2. Products from the Tempering-Degerming Process

All dry corn mills produce a line of basic products that include grits, meals, flour, and germ. Hominy feed is produced as a by-product and sold for feed. Table 15 tabulates typical yields and particle size ranges. Grits comprise the largest fraction. Table 16 tabulates typical compositions for dry corn-milled products. Grits usually contain less than 1% (db) fat; finer meals, 1.0–1.5%; and flours, 2.0%. Protein content is inversely proportional to particle size. The germ fraction contains the highest levels of fat and protein. The germ is subjected to either hard screw pressing or prepress solvent extraction to recover crude edible oils. The high protein, defatted germ meal is used for feed or occasionally finds its way into foods.

Generally, larger endosperm particles, flaking grits, are the more valuable products from dry corn milling, and the process and its machinery are designed to maximize production without exceeding quality specifications. Grits are used in high-value consumer products such as breakfast cereals, extruded snacks, and fermented beverages. It is important that grits be bright yellow or clear white, depending on the type of corn being processed, and free of dust and bran. Grits should be relatively pure horny endosperm, and their yields are affected by the relative proportions of horny endosperm to soft endosperm in the corn being milled. Incoming corn is often graded on the basis of this ratio, which is often determined subjectively by hand dissection. Other dry-milled corn products have a variety of food and industrial uses (Table 17). Approximately 36% of dry corn mill production is used in livestock feeds and pet foods; 54% in food applications (breweries and breakfast cereal manufacturers being the largest customers); and 10% in nonfood industrial applications (Table 18).

B. Wet Corn Milling

By comparison to other processing methods for corn, wet milling is a relatively recent innovation, although wet milling of wheat has roots in ancient Egypt where archeological evidence indicates starch was used over 3000 years ago as paste and adhesives in prototypes of paper (66). It was not until 1949 that corn was subjected to wet milling on a commercial scale when a wheat starch plant in Utica, New York, was converted to corn (92). Today the wet corn-milling industry consumes over 855 million bushels of corn and adding over $3.2 billion of value (Table 19). The wet corn-milling industry no longer merely fractionates corn into its constituents but also converts some of these fractions into syrups and ethanol, and therefore is known as the corn-refining industry. Table 20 summarizes product shipments for the corn-refining industry for the past 5 years.

1. The Wet-Milling Process

Whereas the separation achieved in dry corn milling is purely mechanical, the separation of constituents in wet milling is achieved by a combination of chemical and mechanical means. Figure 16 shows the sequence of major operations involved.

Wet corn milling begins with steeping cleaned corn for 36–48 hr with process water recycled from downstream operations. Sulfur dioxide is added to the water to achieve a level

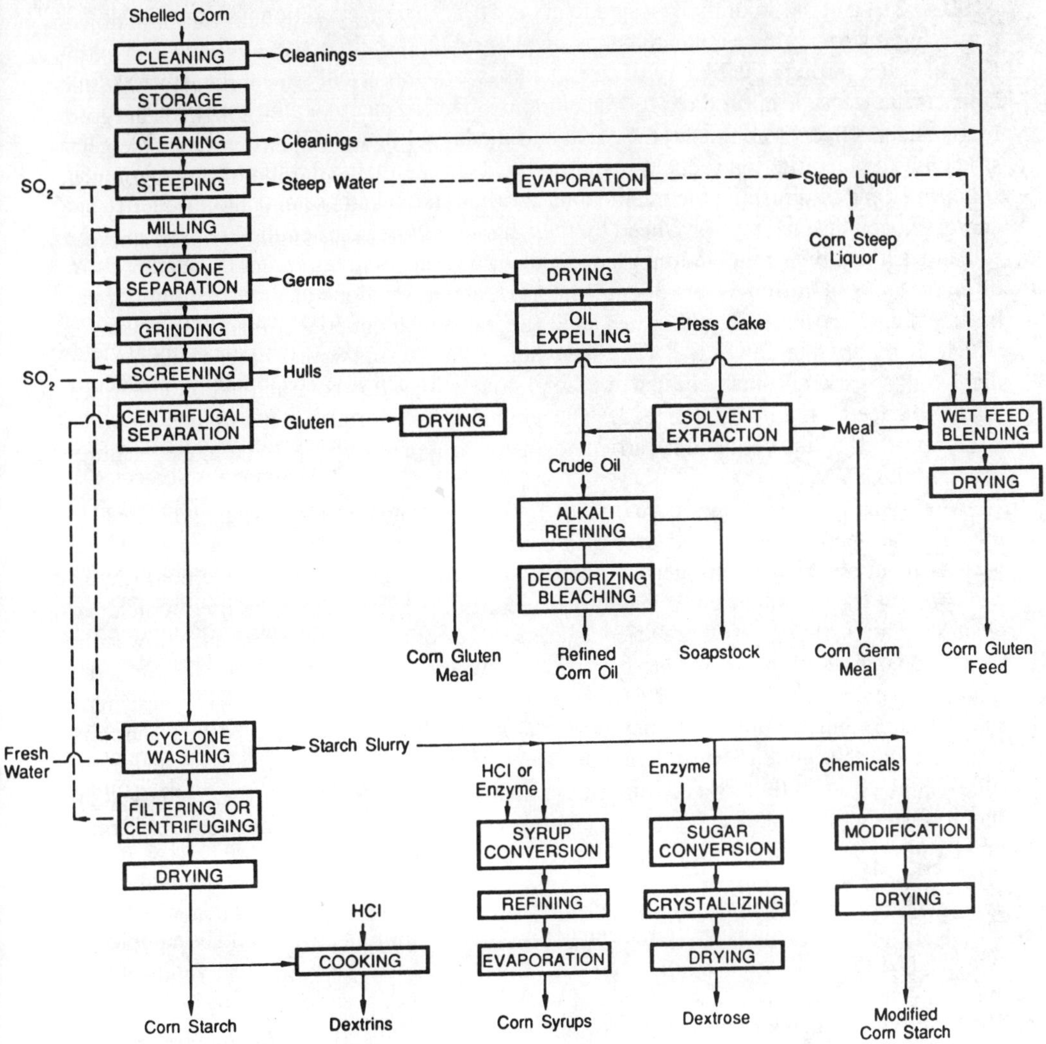

FIGURE 16 Process flow chart for wet corn milling. (Modified from Ref. 5.)

of 0.1–0.2%, and the solution is heated to about 50°C. These conditions are unsuitable for putrifying microorganisms but are conducive to growth of indigenous *Lactobacillus*. The flow of water through the battery of steep tanks is countercurrent so that the freshest corn contacts the weakest sulfur dioxide concentration. During steeping, the kernels absorb solution and swell, activating enzymes native to the kernel that assist in breaking down the structure; the bisulfite ion reduces disulfide bonds in the matrix proteins increasing their solubilities and diminishing the interactions between starch and protein; and the lactic acid and/or exogenous enzymes produced by the *Lactobacilli* help soften the endosperm. The role of each and their relative importance are not well understood. After contacting 8–16 different tanks of corn over the 36–48-hr period, about 5 gallons of light steep liquor per bushel of corn processed is withdrawn and evaporated to about 50% solids, the contents of which are reported in Tables 6–8. The steep liquor may be sold for use in fermentation media, but this market is small compared to total production. It is common to combine steep liquor with feed fractions (gluten feed). The remaining 3–4 gal/bu of water used in wet milling is absorbed by the corn and is carried downstream, and must be dried from the resulting corn fractions.

After steeping, the softened corn is ready for grinding and fractionating as shown in Figure 17. The steeped corn is sluiced to disk attrition mills where the objective is to free the germ without breaking it into pieces. The ground slurry is pumped to hydroclones (liquid cyclones) to separate the lighter weight germs. Multiple passes may be used to achieve clean separations. The germs are dried and processed for oil. The heavier underflow from the hydroclones is screened, and larger particles are reground with an impact mill. Next, the fiber is separated, washed in a series of screens, and dewatered. The remaining stream of starch and protein is pumped to disk-nozzle type centrifuges where the heavier starch is separated from the gluten. The gluten is dewatered using additional centrifuges and vacuum filters. Some protein (3–5%) and impurities remain in the starch slurry that are removed by starch washing hydroclones. It is at this point where fresh water enters the process. Centrifuges and/or vacuum filters dewater the purified starch.

2. Wet Corn Mill Products

The most valuable fractions are starch and germ. The wet-milling process and its equipment are designed to maximize the yields and qualities of these fractions. Typical yields of wet mill products are shown in Table 21 along with their respective compositions in Table 22. The by-products account for one-third of the mill output. Most of the by-products, except the oil recovered from the germ, are utilized as livestock feed (87).

a. Feed Products Corn gluten feed, corn gluten meal, and occasionally spent corn germ meal and steep liquor are sold to the formulated feed industry or directly to farmers and feedlot operators. Usually the fiber, spent germ meal after extraction of oil, and steep liquor are combined and sold either wet or dry as corn gluten feed. Despite the additional expense in transporting water, wet gluten feed is increasing in popularity with local cattle feeders due to greater protein digestibility and lower sales prices, which result from reduced drying. Lower prices for wet gluten are due to savings of drying costs. Corn gluten feed contains a minimum of 21% protein at 11% moisture. Because of the high fiber content of corn gluten feed, very little is used in poultry or swine rations.

Corn gluten meal is considerable more valuable as a feed ingredient because of its high protein, vitamin A, and xanthophyll contents. Most corn gluten meal contains 60% protein at 11% moisture but occasionally small amounts of corn gluten meal are sold at 41% protein. This lower protein level is achieved by diluting corn gluten meal with corn gluten feed. The poultry-feeding industry utilizes almost all of the corn gluten meal because the xanthophylls are important for pigmenting poultry skin and egg yolks.

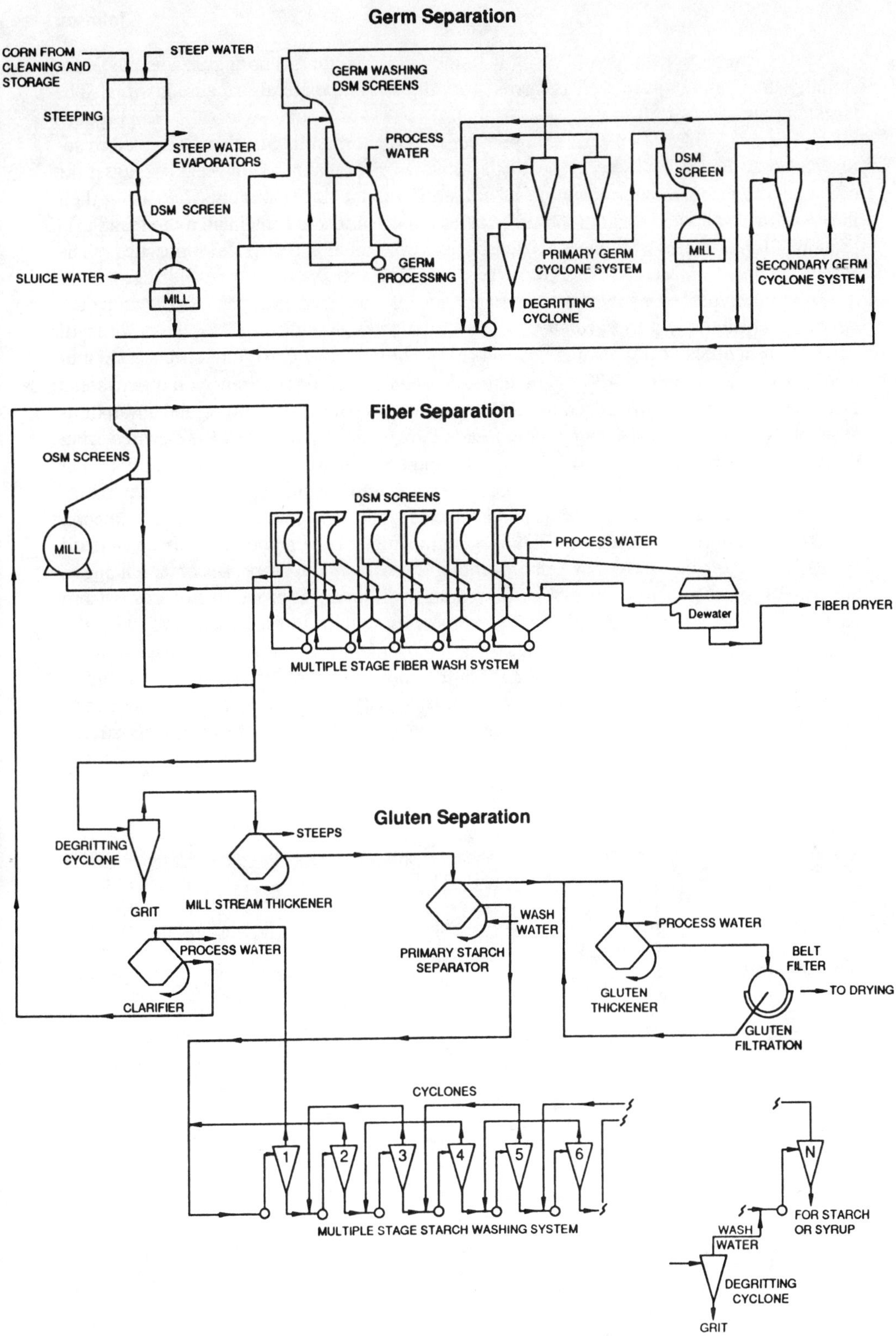

FIGURE 17 Process flow for starch separation in wet corn milling. (Modified from Ref. 11.)

Most spent germ meal is sold along with the other fractions as corn gluten feed. Occasionally small amounts are used as carriers for vitamins and minerals because it will absorb large amounts of water and oil.

b. Starch Starch is the primary product of a wet corn mill. Starch may be sold as native and after modifying by heat, chemicals, or enzymes or it may be converted to sugars for use as sweeteners or as a fermentation substrate. Starch yields in excess of 67% (db) or 32.7 lb/bu are strived for. The starch should be at least 99% pure and contain no more than 0.3% protein. Higher levels of impurities alter critical functional and physical properties of the starch.

Different types of native corn starches can be produced: normal, waxy, and high amylose. The latter two are specialty starches commanding premium prices. Table 23 compares the properties of normal and waxy corn starches. Native corn starch is composed of nearly spherical granules, 5–30 μm in diameter, and exhibits birefringence on microscopic examination with plane-polarized light. Starch granules contain crystalline and amorphous regions. Normal dent corn starch is composed of 26–28% amylose and 74–72% amylopectin; waxy, 99% amylopectin and 1% amylose; and amylomaize, 50–20% amylopectin and 50–80% amylose. On heating, the crystalline regions are disrupted with consequent loss of birefringence, the granule swells, and amylose exudes from the granule. On continued heating the granule disintegrates, completing gelatinization. The solution of starch, termed "paste," increases in viscosity as gelatinization progresses. The properties of starch pastes are usually measured using a viscoamylograph. Viscometric properties of corn starch pastes are compared to pastes of starches from other sources in Figure 18.

Table 24 estimates the utilization of corn starch in various food and industrial applications. Food applications utilize about 35% of the corn starch production (53). Major food uses for starch include brewing adjuncts; chemicals, drugs, and pharmaceuticals; viscosity control agents in canning; starch-based confectionery products; and bakery applications.

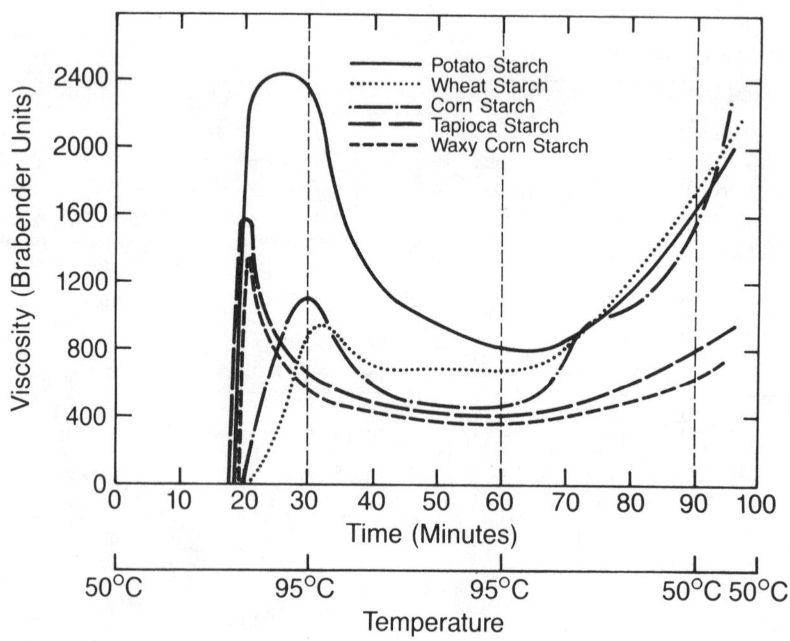

FIGURE 18 Visco-amylograph curves for important unmodified starches. (Modified from Ref. 49.)

Starches are used in foods as stabilizers, thickeners, and gelling agents. Additional food applications for corn starches, native and modified, are described by Moore et al. (49). The other 65% of the starch production is utilized in nonfood industrial applications including adhesives in paper and building materials and coatings and sizings in textiles and paper products. More recently, degradable plastics are being manufactured using about 6% starch to enhance degradation of petroleum-derived polymers.

c. Corn Bran as a Source of Dietary Fiber Recent increased consumer awareness of the importance of fiber in the human diet has increased demand for high fiber ingredients, such as those derived from corn bran. Corn fiber from wet mills can be refined into products having compositions and properties similar to those in Table 25. Refined corn bran has at least 92% dietary fiber, which places it among the most concentrated sources. Refined corn bran is used as a fiber supplement in dietary beverages, extruded breakfast cereals and snack foods, and breads and other bakery products.

d. Edible Oil Recovery from Corn Germ Edible oils are routinely recovered from both dry-milled and wet-milled corn germs. However, due to wide differences in their oil contents, the two sources are processed differently (Fig. 19). Dry-milled germs contain about 18% (db) oil and can be rolled into thin flakes and directly extracted with hexane. Wet-milled germs, on the other hand, contain about 52% (db) oil and cannot be rolled into flakes that will retain their integrity during extraction. Consequently, wet-milled germs are processed by prepress solvent extraction, where the germs are heated and screw pressed to reduce the oil content to 18–20% prior to extraction with hexane. In both processes the desolventized, defatted germ meal contains less than 1.2% (db) residual oil.

The crude oil is subjected to refining operations to remove undesirable compounds affecting appearance and sensory properties (Fig. 20). These compounds include phosphatides, free fatty acids, pigments, waxes, and trace amounts of flavors and odors. Properly refined corn oil is bland and pale yellow and does not develop a cloudy haze at refrigerator temperatures. Often corn oil is partially hydrogenated to convert the liquid oil into a semi-plastic fat suitable for use in margarines. Although some sites of unsaturation are reduced, corn oil margarines usually have higher amounts of polyunsaturated fatty acids than many competing margarine fats.

Compositions and important properties of crude and refined corn oils are compared in Tables 26 and 27. The low level of linolenic acid and high level of the natural antioxidant tocopherols give corn oils unusual stability towards oxidation. The high smoke point and low solidification point make corn oils ideal cooking and salad oils, respectively.

Corn oil is used primarily as a cooking and salad oil, and as a base fat for margarine after partial hydrogenation (Table 28). Recent expansion of the wet corn-milling industry has catapulted corn oil to the second most utilized vegetable oil in the United States, and it is widely available at competitive prices.

e. Recovery of Zein from Corn Gluten Meal There are four major classes of proteins in corn distinguishable on the basis of their solubilities: albumins, globulins, glutelins, and zein (prolamines) (Table 29). Zein is the major protein in corn gluten meal, comprising about 47%. Zein, as a sole source of protein, will not support the growth of animals because it is deficient in lysine and tryptophan. Zein has a low content of ionizable amino acids and high content of nonpolar amino acids which confer unique solubility properties. Zein is poorly soluble in water but highly soluble in 70% ethanol and is the basis for its industrial separation.

Zein is extracted from corn gluten meal with hot aqueous ethanol (Fig. 21). The extract is treated with sodium hydroxide, cooled, acidified, and filtered. The extract is then subjected to liquid-liquid extraction to remove oil, carotenoids, and some of the ethanol. The protein is precipitated by adding cold water, recovered by filtering, and dried.

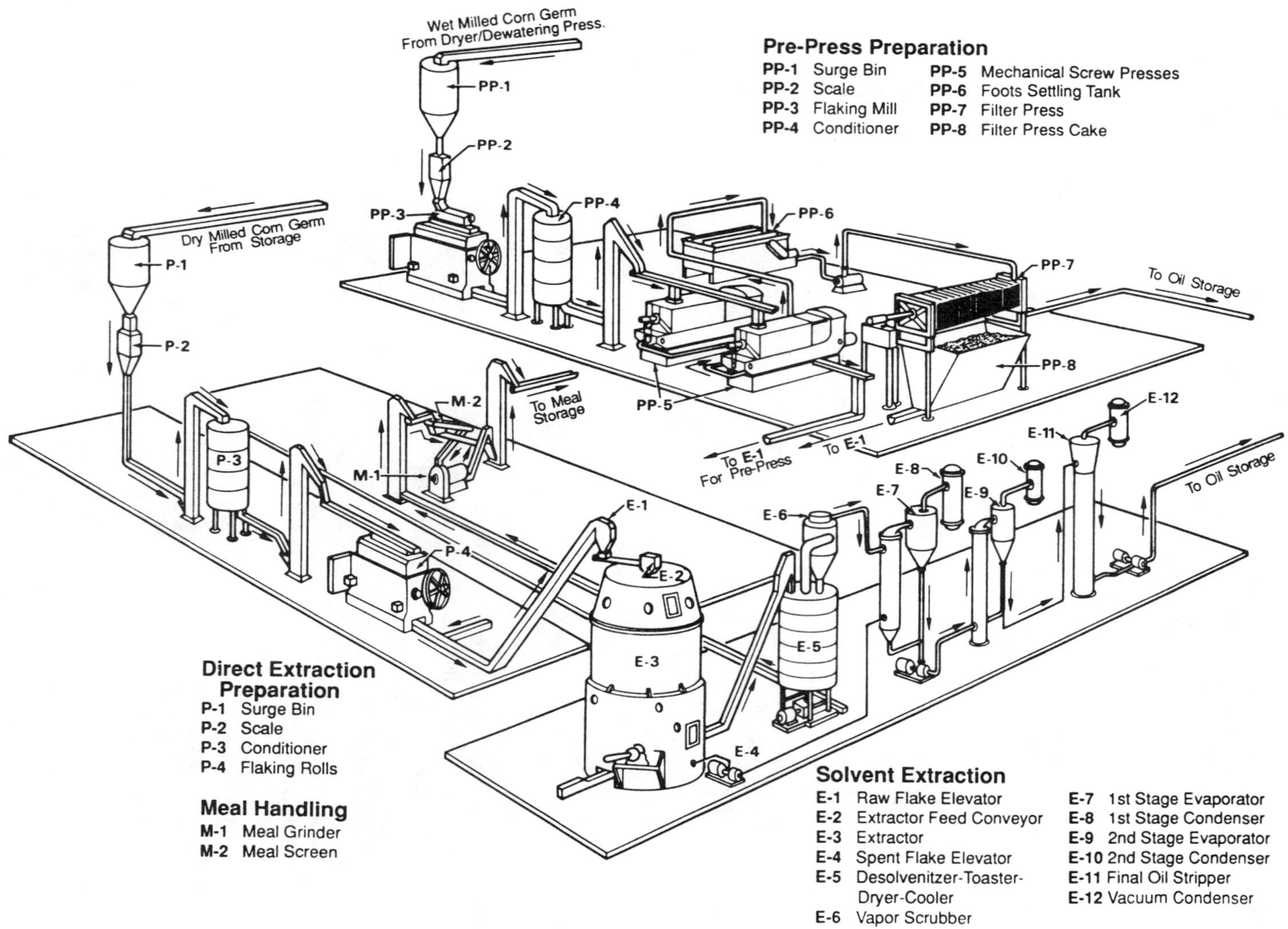

FIGURE 19 Process flow for recovery of oil from corn germ. (Modified from figures provided by French Oil Mill Machinery Co., Piqua, OH.)

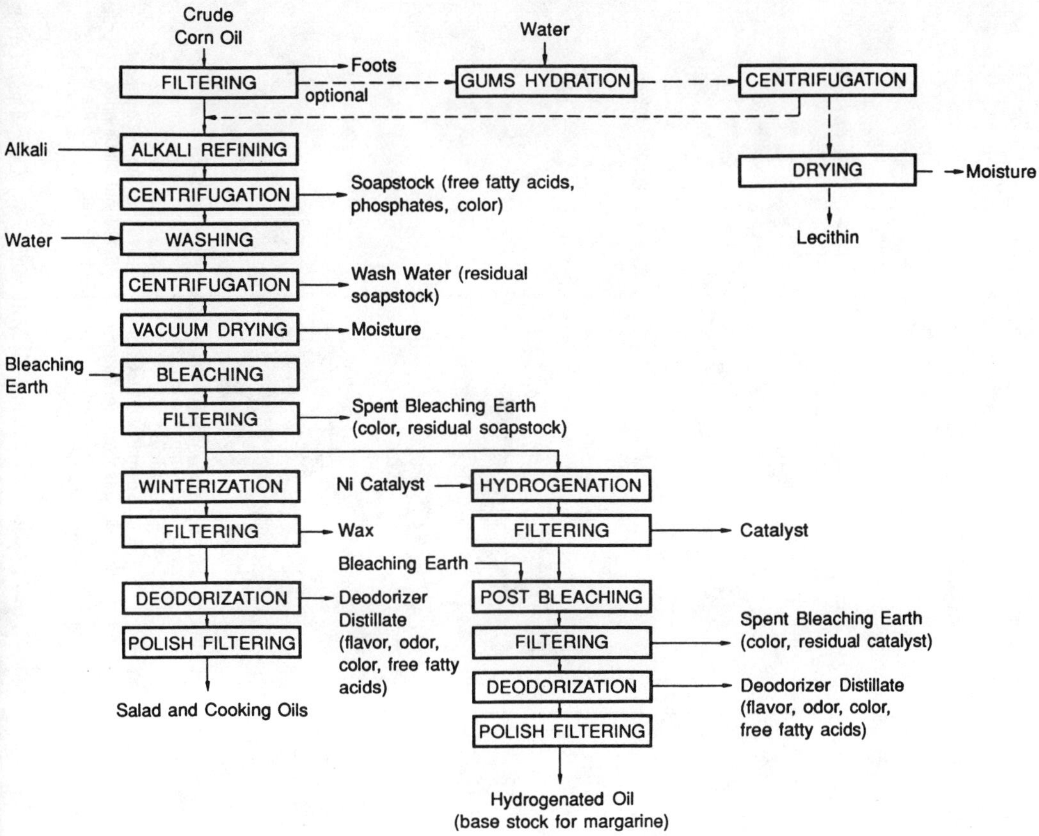

FIGURE 20 Process flow for refined corn oil.

Unlike most other commercially available proteins, zein is highly resistant to water and forms tough films and fibers. As a result, zein has a number of industrial applications (59), which are tabulated in Table 30. Zein has been commercially produced since 1938 (23). Peak production occurred in 1954 when nearly 6 million pounds were produced (62). Around 1960, nonfood markets for zein eroded almost overnight as newer synthetic plastics became available (77). Today's market is about 0.5 million lb for food coatings and binding and coating agents in pharmaceutical tablets.

VIII. CONVERSION OF RAW FRACTIONS INTO VALUE-ADDED INGREDIENTS AND CHEMICALS

Once separated by wet or dry milling, corn fractions can be converted to a large number of food ingredients and industrial chemicals of increased value. These conversions can be either chemical (Table 31) or biochemical (enzymatic or microbial fermentation) (Table 32). The largest conversion industries using corn products are nutritive sweeteners, modified starches, and ethanol. In the past, conversion of corncobs into furfural has also been important.

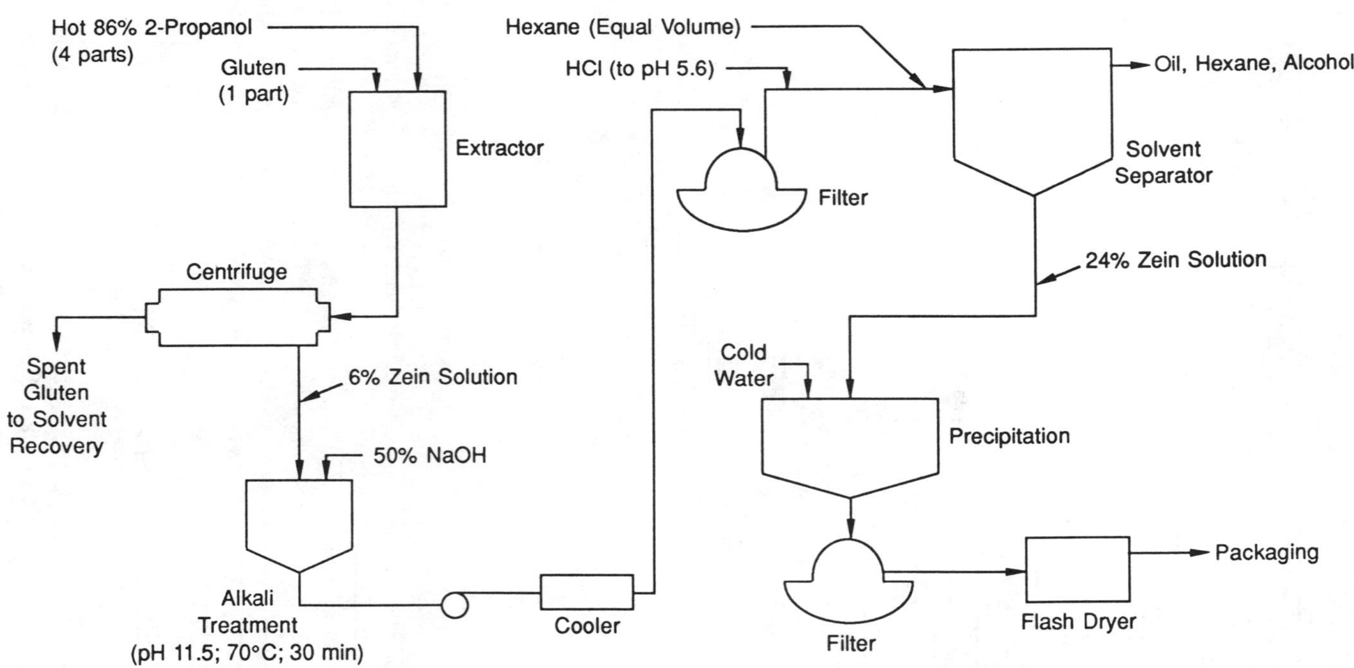

FIGURE 21 Process flow scheme for production of zein from corn gluten. (Modified from Ref. 50.)

A. Modified Starches

Starch modification operations are usually integrated with wet-milling operations because most reactions are aqueous and take place in the wet slurry stage after the starch-washing step in wet milling. This saves one drying step. Modifications are carried out in batch or continuous reactors (Fig. 22). The major types of modifications to starch include acid thinning, dextrinizing, pregelatinizing, oxidizing, bleaching, and derivatizing (29). Reactions, properties, and applications for major modified starches are described in Table 33. Whistler et al. (93) and Wurzburg (95) have published detailed treatises on starch and its modification.

In 1899, the first patent was granted for a process to modify starch—acid modification (20). Treatment with acid is done in a controlled manner to hydrolyze a small number of glycosidic bonds between glucose units and weaken the structure of the starch granule. This reduces the viscosity of the warm paste but allows the starch to retain a strong tendency to gel on cooling. These properties are important in starch-based gum candies, paper coatings, and textile sizings.

Oxidation is another means of reducing the viscosity and altering other functional properties of native starch. A number of agents are capable of oxidizing starch, but sodium hypochlorite is most extensively used today. Sodium hypochlorite randomly oxidizes hy-

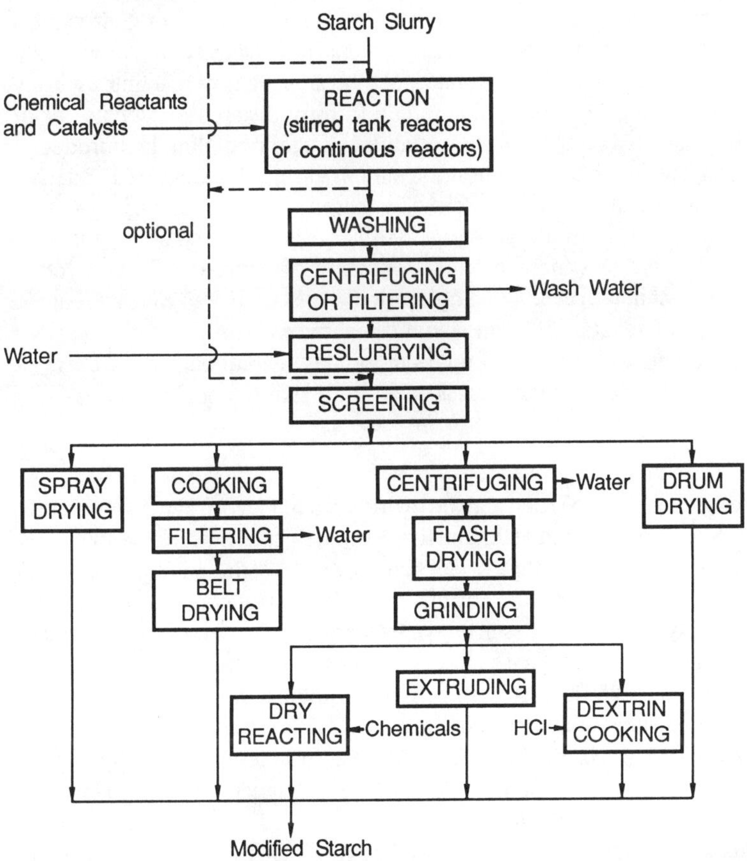

FIGURE 22 Process for modifying corn starches. (Modified from Ref. 68.)

droxyl groups to carbonyl and carboxyl groups with consequential breakage of the glycosidic bond. Once the desired degree of oxidation is achieved, the reaction is stopped by adding sodium bisulfite. Oxidized starch retains its granular structure, is insoluble in water, produces clear pastes with reduced thickening tendency on cooling, and produces clear tough films. The major use for oxidized starches is in laundry starches and paper manufacture.

Dextrinization is the process of dry heating or roasting starch with or without an acid or alkaline catalyst. The granule integrity is disrupted and weakened without destroying the granule. Dextrinized starch granules swell when suspended in water and heated, and layers of the granule separate and disperse. Dextrins have reduced viscosities and tendencies to gel, have appreciable cold water solubilities, and produce tacky films. Dextrins are widely used as quick-setting pastes in paper products.

Starches are often pregelatinized by cooking the raw starch before drying. When slurried in water these starches have properties comparable to gelatinized starch without the need for additional cooking. This facilitates "instantized" products, such as instant puddings.

Occasionally very pure white-colored products are required. Stark white color is achieved by treating the starch with small amounts of hydrogen peroxide or other bleaching agents. The functionality of bleached starches is substantially the same as that of the parent starch.

Starch can be derivatized by a number of agents reacting with hydroxyl groups. There are two types of derivatives, cross-linking derivatives and stabilization derivatives. Starches are cross-linked to impart greater stability so that the viscosity of starch sols to shear, heat, and acid is maintained. Cross-linking can be achieved by reacting hydroxyl groups on two different molecules within the granule to form distarch phosphates or distarch adipates. Starch can also be derivatized to stabilize it against gelling by introducing substituents which interfere with inter-and intramolecular hydrogen bonding. Other derivatives achieve specific functional properties, such as increased water-combining capacity or viscosity, impeded retrogradation, imparting a positive charge or improved freeze-thaw stability. These types of derivatives include the introduction of phosphate, hydroxyethyl, hydroxypropyl, acetate, succinate, carboxymethyl, cationic, and xanthate groups. Figure 23 shows pasting properties of several different starch derivatives.

Modified starches comprise about 40% of total starch sales in nonfood applications (Table 34). Acid treating is the most widely practiced form of starch modification.

B. Corn Sweeteners

Starch is a polymer of glucose and, as such, can be hydrolyzed to glucose syrups and further converted to crystalline sugars. Around 1820, the Russian chemist G. S. C. Kirchoff discovered that starch could be transformed into a sweet substance by heating in dilute acid (21). However, it was not until 1857 at a plant in Buffalo, New York, that corn starch was commercially converted into syrups. Advances in enzyme technology in the early 1950s greatly advanced the commercialization of glucose syrups. Later advances in the use of glucose isomerase in the 1970s made high fructose corn syrups possible. Now the industry is capable of making both simple sugars that comprise sucrose.

Both the α-1,4 and α-1,6 glycosidic bonds of starch are susceptible to acid hydrolysis, and acid hydrolysis is the first step in the production of corn sweeteners (Fig. 24). However, acid hydrolysis alone does not yield satisfactory sweeteners because of the potential for numerous side reactions which may darken the color of the syrup. Therefore, acid is used only to thin the starch and increase the susceptibility of starch to enzymic attack. Thinning can

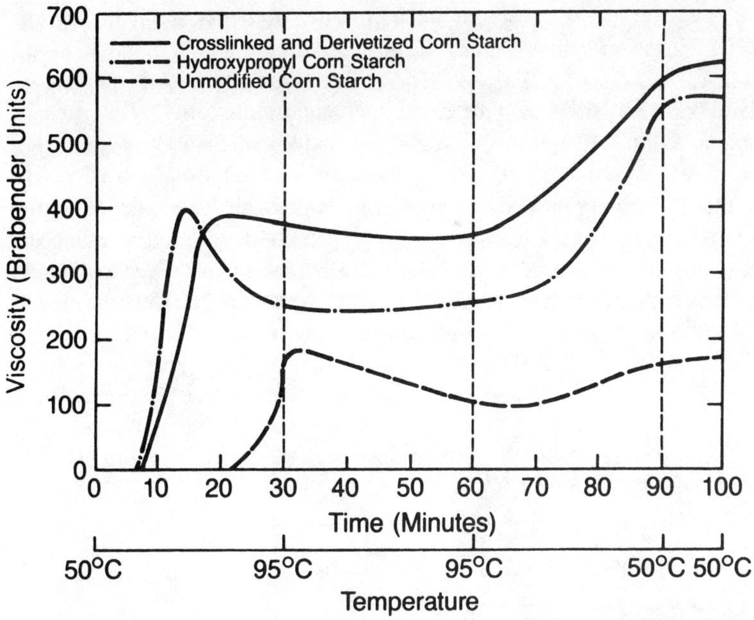

FIGURE 23 Visco-amylograph curves for important modified corn starches. (Modified from Ref. 49.)

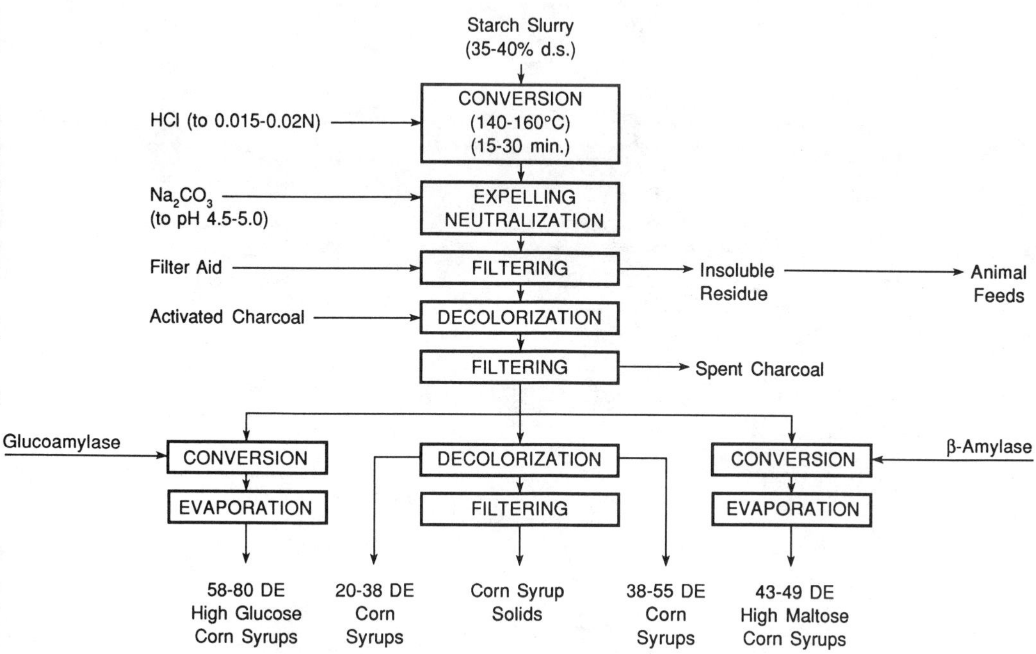

FIGURE 24 Process flow for acid-converted dextrose corn sweeteners. (Modified from Ref. 5.)

also be accomplished by gelatinizing the starch in the presence of thermostable β-amylase. α Amylase randomly attacks starch in an "endo" fashion producing high molecular weight polysaccharides, such as maltodextrins (Fig. 25). The acid-thinned starch is neutralized, filtered, and decolorized before treating with the enzyme glucoamylase (amyloglucosidase), which hydrolyzes starch to glucose unit by unit along the chain. Glucoamylase can hydrolyze α-1,6 linkages as well as α-1,4. Alternatively, the thinned starch may be treated with β-amylase, which hydrolyzes every other α-1,4 glycosidic bond in an "exo" fashion to produce maltose and a few higher saccharides where α-1,6 branch points on amylopectin occur and stop the reaction. Depending on how far towards complete hydrolysis the reaction is allowed to proceed, syrups with different sweetness intensities are produced. Corn syrups can be concentrated and dried to produce corn syrup solids.

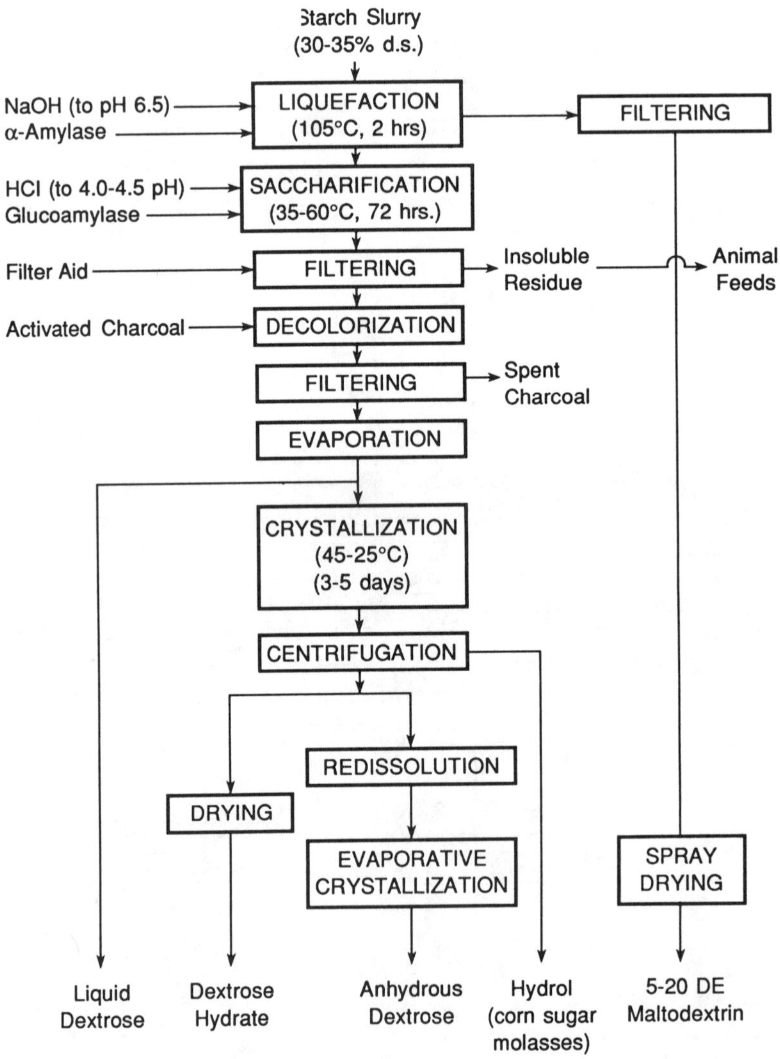

FIGURE 25 Process flow for enzyme-converted dextrose corn sweeteners. (Modified from Ref. 5.)

Dextrose equivalent (DE) is the term used to describe the extent of hydrolysis and reducing power of a syrup relative to dextrose (glucose). The higher the DE, the greater the sweetness intensity. Table 35 summarizes the saccharide composition and properties of different maltose and glucose syrups. More detailed data on properties of corn sweeteners have been published by Pancoast and Junk (54).

Within the past 10 years, the technology to convert glucose syrups to fructose-containing syrups has been substantially improved (84). Fructose is much sweeter than glucose and somewhat sweeter than sucrose, and consequently less sugar and therefore calories are required to achieve the same sweetness intensities. The first steps in producing high fructose corn syrups (HFCS) are the same as those used to produce high DE glucose syrups (Fig. 26). Glucose syrups are converted to an equilibrium mixture containing 42% fructose, 52% glucose, and 6% higher saccharides after treating with glucose isomerase. Glucose isomerase has been immobilized on a solid support phase so that the reaction can be carried out continuously in column reactors and thus prevent the loss of expensive enzyme. The mixture of sugars can be fractionated using fructose adsorbent columns. As high as 98% pure fructose can be recovered. This level of purity is required in order to crystallize fructose. Usually 90% fructose is blended with the lower fructose-containing syrup exiting the isomerization columns to produce 55% fructose syrup, which is the HFCS product having the highest market volume. The 55% HFCS has the same sweetness intensity as crystalline sucrose (invert sugar) on an equivalent weight basis. Other properties and the compositions of common HFCS products are shown in Table 36.

Corn sweeteners have become widely accepted food ingredients for imparting sweetness and various functional properties (40,65). However, major shifts in types of sweeteners have occurred over the past 10 years as a result of HFCS development (Table 37). Total consumption of sweeteners has remained constant during this period at about 130 lb/year/person but the market share supplied by corn sweeteners has steadily grown at the expense of refined sugar. The consumption of HFCS has doubled in this period and now stands at about 46 lb/year/capita, or 35% of the sweetener market. This is largely the result of the HFCS becoming the preferred sweetener in soft drinks, which has now obtained 91% of this market (Table 38). The potential market for HFCS is approaching maturity, and fewer dramatic shifts in future consumption of sweeteners are anticipated.

C. Furfural Production from Corncobs

Corncobs are composed of inner and outer glumes which subtend the kernels a woody ring of lignified conducting tissue and an inner pith (5). Corncobs can be ground and fractionated into three fractions: grits, pith/chaff, and mixed. The properties of these fractions are shown in Table 39. Each fraction has a number of industrial uses.

The largest single use for corncobs is the furfural industry. Furfural has been produced from corncobs since World War II and has been the largest single use for corncobs. Furfural is produced by destructive distillation over sulfuric acid. The hemicellulose fraction (xylan) of corncob cell walls is hydrolyzed to pentose sugars (Fig. 27). The pentoses are dehydrated with heat to furfural, which is then recovered by steam distillation. Furfural is marketed directly or converted to a number of other useful chemical derivatives and products. Although the theoretical yield of furfural from corncobs is 21–23%, actual yields are closer to 10% (26,30).

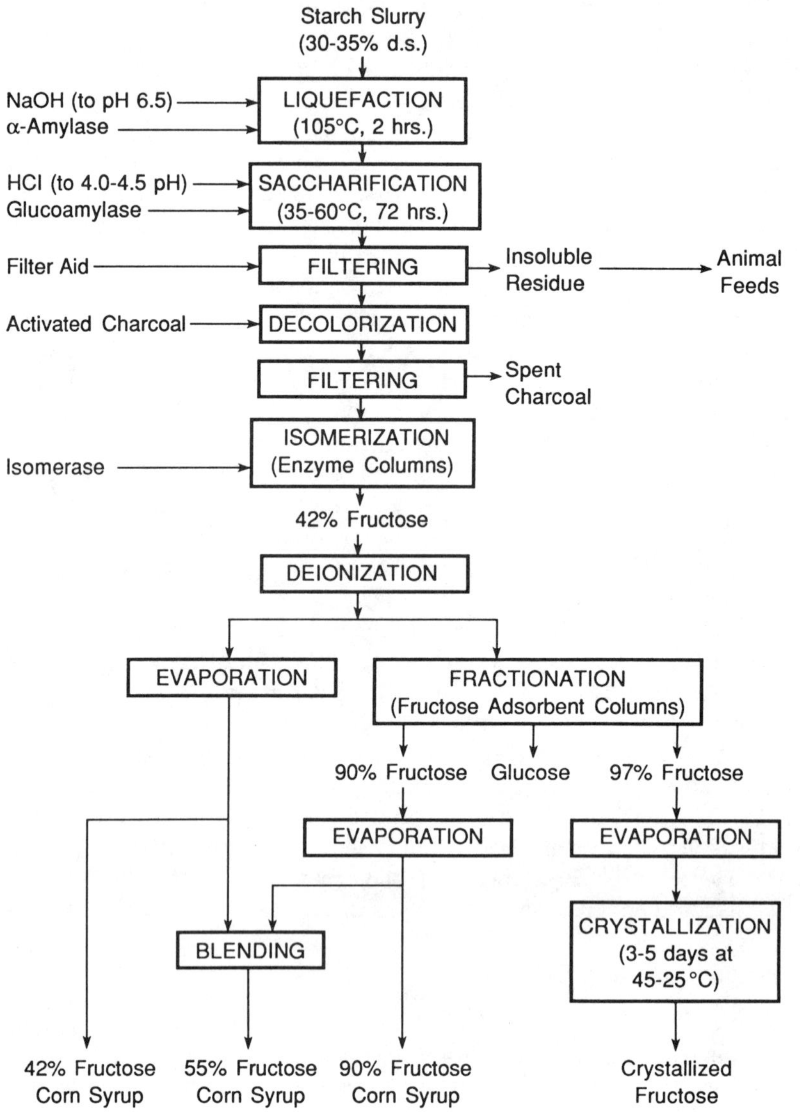

FIGURE 26 Process flow for producing high fructose corn sweeteners. (Modified from Ref. 5.)

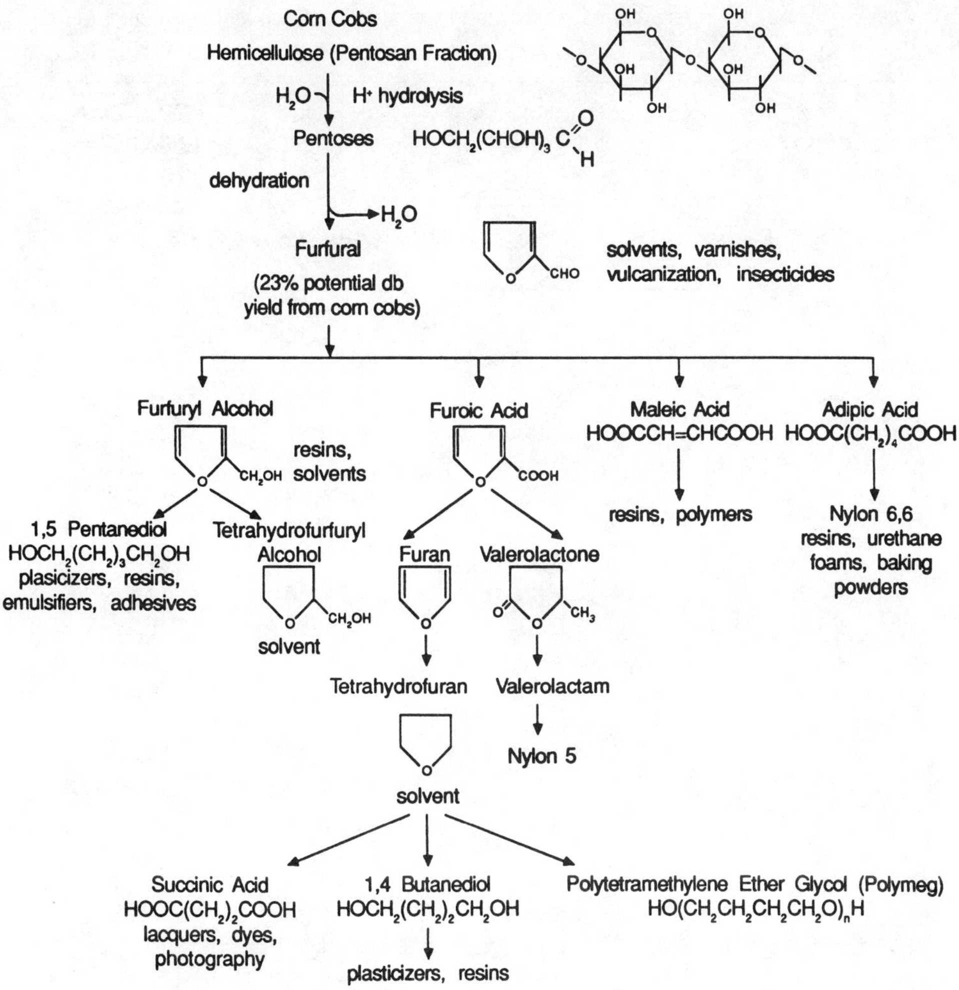

FIGURE 27 Process scheme for producing furfural and furfural derivatives.

TABLE 1 Annual Harvested Area, Yield, and Production of Corn by Continent and Major Producing Country During 1985–1986[a]

Continent/Country	Harvested area[b] (1,000,000 ha)	Yield[b] (t/ha)	Production[b] (1,000,000 t)
World total	129.07	3.73	481.02
North and Central America	39.62[c]	6.19[c]	245.37[c]
United States	30.44	7.41	225.48
Mexico	6.20	1.69	10.50
Canada	1.20	6.24	7.47
Guatemala	0.69	1.61	1.11
South America	18.02[c]	2.02[c]	36.36[c]
Brazil	12.00	1.83	22.00
Argentina	3.35	3.43	11.50
Venezuela	0.48	1.88	0.90
Colombia	0.55	1.43	0.78
Europe	11.57	5.29	61.17
France	1.86	6.66	12.37
Italy	0.92	6.88	6.31
Romania	3.10	4.52	14.00
Yugoslavia	2.40	4.12	9.89
EEC Total	4.02	6.40	25.77
Soviet Union	4.48	3.21	14.40
Asia	34.55[c]	2.69[c]	92.90[c]
China	17.69	3.61	63.83
India	5.88	1.17	6.89
Indonesia	2.22	2.05	4.55
Thailand	1.75	1.20	2.10
Africa	19.91	1.52	30.22
South Africa	3.89	2.00	7.76
Zimbabwe	1.43	2.07	2.95
Kenya	1.65	1.61	2.65
Nigeria	2.00	1.00	2.00
Oceana	0.10	4.60	0.46
Australia	0.08	3.04	0.23
New Zealand	0.02	10.00	0.22

[a]Data compiled by USDA (78) unless otherwise noted.
[b]Conversions: 1 hectare (ha) = 2.47 acres; 1 metric ton (t) = 2,205 lb = 39.38 bu; 1 t/ha = 892 lb/acre = 15.93 bu/acre.
[c]USDC (81).

TABLE 2 Physical Properties of Corn[a]

Property	Typical value	Range
Kernels per ear[b]	800	500–1,200
Kernels per lb[b]	1,300	900–1,800
Test weight (lb/bu)		
Sweet corn in husk[c]	35	NA
Shelled dent[c]	58.3	52–60
Shelled popcorn[c]	56	NA
Bulk density (lb/ft^3)		
Sweet corn in husk[c]	28	NA
Husked corn[c]	28	NA
Shelled dent[d]	45.4	40–46
Shelled popcorn[c]	44.8	NA
Specific gravity[e] (g/cc)	1.26	NA
Angle of repose[d] (degrees)	35	34–43.5
Void volume[f] (%)	42.3	NA
Thermal conductivity[g] (BTU/hr.ft.°F)	0.0945	0.0812–0.0996
Specific heat[g] (BTU/lb.°F)	0.484	0.350–0.588
Heat energy content[h] (BTU/lb)	8,075	NA
Diffusivity[g] (ft^2/hr)	0.00351	0.00336–0.00395

NA denotes data not available.
[a]Dent corn at 15% moisture.
[b]From Ref. 88.
[c]From Ref. 9.
[d]From Ref. 13.
[e]From Ref. 52.
[f]From Ref. 74.
[g]From Ref. 45.
[h]From Ref. 46.

TABLE 3 Typical Compositions (% db) and Characteristics of Different Types of Corn

Component/ Characteristic	Dent: Normal	Dent: Waxy	Dent: Amylomaize	Dent: High-oil	Dent: High-protein	Flint	Popcorn	Flour	Opaque-2	Sweet corn
Starch (%)	71.7[b]	NA	57.9[g]	NA	NA	NA	62.3[m]	NA	64.4[n]	54.1[m]
Protein, N x 6.25 (%)	9.5[b]	NA	13.7[g]	NA	26.6[j]	11.1[k]	11.9[m]	NA	13.6[n]	12.7[m]
Fat (%)	4.3[b]	NA	5.9[g]	19.6[h]	NA	4.7[k]	5.3[m]	NA	5.5[n]	8.4[m]
Fiber (%)	9.5[b]	NA	NA	NA	NA	NA	2.6[m]	NA	NA	3.5[m]
Sugars[a] (%)	2.6[b]	NA	NA	NA	NA	NA	9.3[m]	NA	2.8[n]	12.0[m]
Ash (%)	1.4[b]	NA	NA	NA	NA	NA	1.9[m]	NA	NA	2.1[m]
Amylose (g/100 g starch)	24[c]	1[c]	50–80[h]	24[f]	24[f]	24[f]	24[f]	24[f]	24[f]	24[f]
Amylopectin (g/100 g starch)	76[c]	99[c]	50–20[h]	76[f]	76[f]	76[f]	76[f]	76[f]	76[f]	76[f]
Lysine (g/100 g protein)	2.7[d]	2.7[f]	2.7[e]	2.4[i]	2.3[i]	2.0[l]	2.0[f]	2.7[f]	3.8[d]	2.5[i]
Endosperm variation	Horny at sides, soft floury in central core, top indents on drying					Almost entirely horny endosperm, no indent on drying	Small flint type	Entirely soft endosperm little or no indent on drying	No horny endosperm, entirely soft	Shrunken and wrinkled when dry
Preferred uses	Feeds, wet milling, dry milling	Wet milling and feeds	Wet milling	Limited use in feeds	Limited use in feeds	Feeds and dry milling in South America and Africa	Snack foods	Food in South America	Food	Food
Typical yields (bu/A)	120[e]	NA	NA	107[f]	NA	92[f]	54[l]	91[f]	110[f]	94[f]

NA denotes data not available.
[a]Calculated by difference after subtracting starch, protein, fat, fiber and ash.
[b]Distillers Feed Research Council (25).
[c]Medcalf (48).
[d]Dimler (24).
[e]USDC (82).
[f]Estimates of author.
[g]Anderson (4).
[h]Zuber and Farrah (97).
[i]Paulis and Wall (56).
[j]Alexander and Creech (3).
[k]Tello et al. (73).
[l]Popcorn Institute, Chicago, IL.
[m]Tawfik (72).
[n]Watson and Yahl (89).

TABLE 4 United States Grading Standards for Corn[a]

Grade	Minimum test weight (lb/bu	Maximum limits: Broken corn and foreign material (%)	Maximum limits: Damaged kernels: Total (%)	Maximum limits: Damaged kernels: Heat damaged (%)
U.S. No. 1	56	2.0	3.0	0.1
U.S. No. 2	54	3.0	5.0	0.2
U.S. No. 3	52	4.0	7.0	0.5
U.S. No. 4	49	5.0	10.0	1.0
U.S. No. 5	46	7.0	15.0	3.0
Sample grade[b]	—	—	—	—

[a]Grades and grade requirements for yellow corn, white corn and mixed corn (31).
[b]U.S. Sample Grade is corn that does not meet the requirements for any of the grades from U.S. No. 1 to U.S. No. 5, inclusive; or which, in a 1000-g sample, contains eight or more stones with an aggregate weight in excess of 0.20% of the sample weight, two or more pieces of glass, three or more crotalaria seeds, two or more castor beans, eight or more cocklebury, four or more particles of an unknown substance(s) or a commonly recognized harmful or toxic substance(s) or animal filth in excess of 0.20%; or has a musty, sour, or commercially objectionable foreign odor; or is heating or otherwise of distinctly low quality.

TABLE 5 Supply and Utilization of Corn (million bushels)

Year, begin–ning Sept. 1	Supply				Utilization						Ending stocks		
					Domestic Utilization								
	Beginning stocks	Production	Imports	Total	Food, alcohol and industrial	Seed	Feed and residual	Total	Exports	Total utilization	Government owned	Privately owned	Total
1980	2,034	6,639	1	8,675	698	20	4,157	4,875	2,408	7,283	242	1,150	1,392
1981	1,392	8,119	1	9,512	778	19	4,169	4,966	2,010	6,975	280	2,257	2,537
1982	2,537	8,235	1	10,772	880	15	4,521	5,416	1,834	7,249	1,143	2,380	3,523
1983	3,523	4,175	3	7,701	956	19	3,818	4,793	1,902	6,694	202	805	1,006
1984	1,006	7,674	4	8,684	1,070	21	4,079	5,170	1,865	7,036	225	1,423	1,648
1985	1,648	8,877	11	10,536	1,140	20	4,095	5,255	1,241	6,496	546	3,494	4,040
1986	4,040	8,250	2	12,292	1,175	16	4,717	5,908	1,504	7,412	1,443	3,439	4,882
1987[a]	4,882	7,064	2	11,948	1,219	17	4,900	6,136	1,700	7,836	1,350	2,762	4,112

[a]Projected values.
Source: Ref. 79.

TABLE 6 Gross Nutrient Analysis of Corn and Corn Products (dry basis) for Livestock Feeds[a]

Nutrient	Dent yellow corn	Corn hominy feed	Corn gluten feed	Corn gluten meal (41% protein)	Corn gluten meal (60% protein)	Corn germ meal (sol. ext.)	Corn steep liquor
Proximate analysis (%)							
Protein (N x 6.25)	10.9	11.5	25.6	46.8	67.2	22.3	47.2[b]
Fat (ether extractable)	4.3	7.7	2.4	2.4	2.4	4.1	0.0[b]
Saturated	0.9	1.2	0.5[c]	0.5[c]	0.5[c]	0.8[c]	0.0[c]
Unsaturated	3.7	6.1	1.9[c]	1.9[c]	1.9[c]	3.3[c]	0.0[c]
Linoleic acid	2.05	3.71	1.15[c]	1.15[c]	1.15[c]	1.95[c]	0.0[c]
Starch	71.7[d]	NA	15.6[b]	NA	15.6[c]	26.7[b]	NA
Crude fiber	2.9	6.7	9.7	4.8	2.2	13.1	0.0
Acid detergent fiber	3.0	NA	13.0	9.0	5.0	NA	0.0[e]
Ash	1.5	3.1	7.5	3.4	1.8	4.2	14.7[b]
Nitrogen-free extract	81.8[f]	72.7[f]	53.4[f]	43.4[f]	25.0	58.0	38.4[b]
Energy (kcal/kg)							
Gross	4,402	4,702	4,490	NA	NA	NA	NA
Digestible							
Ruminants	3,840	4,140	3,660	3,790	3,920	3,260	NA
Swine	3,837	4,017	3,562	3,876	4,409	3,061	NA
Metabolizable							
Ruminants	3,420	4,740	3,250	3,380	3,510	2,850	NA
Chickens	3,818	3,208	1,924	3,278	4,086	1,854	3,110[g]
Swine	3,724	3,748	2,751	3,436	3,907	3,295	NA
Total digestible nutrients (%)							
Ruminants	87	94	836	86	89	74	80
Swine	90	91	84	88	89	NA	NA

NA denotes data not available.
[a]Data compiled by National Academy of Sciences (51) unless otherwise noted.
[b]From Ref. 67.
[c]From Ref. 25.
[d]From Ref. 5.
[e]From estimates of author.
[f]From Ref. 50.
[g]From Ref. 96.

TABLE 7 Protein Digestibilities and Amino Acid Composition of Corn and Corn Products (dry-basis) for Livestock Feeds[a]

	Dent yellow corn	Corn hominy feed	Corn gluten feed	Corn gluten meal (41% protein)	Corn gluten meal (60% protein)	Corn germ meal (sol. ext.)	Corn steep liquor
Protein (%)	10.9	11.5	25.6	46.8	67.2	22.3	47.2[b]
Digestible protein (%)							
Cattle	7.6[c]	7.9[c]	24.2[c]	39.2[c]	NA	.NA	NA
Swine	8.2[c]	9.4[c]	NA	40.5[c]	NA	NA	NA
Amino acid composition (% of sample)							
Alanine	0.78[d]	NA	1.7[b]	3.7[b]	5.8[b]	1.6[b]	3.6[e]
Arginine	0.48	0.52	0.87	1.53	2.31	1.43	2.2[e]
Aspartic acid	0.68[d]	.NA	1.3[b]	2.7[b]	4.0[b]	1.6[b]	2.8[e]
Cystine	0.25	0.16	0.49	0.73	1.10	0.44	1.6[e]
Glutamic acid	1.77[d]	NA	3.7[b]	9.6[b]	15.3[b]	3.6[b]	7.0[e]
Glycine	0.42	0.38	0.94	1.65	2.33	1.20	2.2[e]
Histidine	0.29	0.22	0.68	1.06	1.55	0.76	1.4[e]
Isoleucine	0.39	0.43	0.98	2.46	2.82	0.76	1.4[e]
Leucine	1.37	0.94	2.44	7.92	11.33	1.97	4.0[e]
Lysine	0.28	0.42	0.71	0.87	1.12	0.98	1.6[e]
Methionine	0.19	0.18	0.41	1.14	1.98	0.64	1.0[e]
Phenylalanine	0.54	0.36	0.90	3.05	4.45	0.98	1.6[e]
Proline	0.84[d]	NA	1.9[b]	4.0[b]	6.1[b]	1.4[b]	4.0[e]
Serine	0.57	NA	0.94	1.97	3.71	1.09	2.0[e]
Threonine	0.40	0.44	0.87	1.56	2.46	1.19	1.8[e]
Tryptophan	0.09	0.12	0.17	0.23	0.33	0.21	0.1[e]
Tyrosine	0.34	0.55	0.81	1.11	3.54	0.76	1.0[e]
Valine	0.50	0.55	1.22	2.40	3.43	1.31	2.4[e]

NA denotes data not available.
[a]Data compiled by National Academy of Sciences (51) unless otherwise noted.
[b]From Ref. 67.
[c]From Ref. 50.
[d]From Ref. 96.
[e]From Ref. 18.

TABLE 8 Vitamin and Mineral Contents of Corn and Corn Products (dry-basis) for Livestock Feeds[a]

Nutrient	Dent yellow corn	Corn hominy feed	Corn gluten feed	Corn gluten meal (41% protein)	Corn gluten meal (60% protein)	Corn germ meal (sol. ext.)	Corn steep liquor
Vitamins							
Biotin (mg/kg)	0.08	0.15	0.36	0.20	0.21	0.24	0.66[b]
Carotene (mg/kg)	3	10	7	18	34	2	0[b]
Choline (mg/kg)	567	1,280	1,684	391	391	1,785	6,996[b]
Folic acid (mg/kg)	0.3	0.3	0.3	0.3	0.3	0.2	0.2[b]
Inositol (mg/kg)	NA	NA	5,923[b]	NA	2,102[b]	NA	12,012[b]
Niacin (mg/kg)	28	52	79	55	66	33	167[b]
Pantothenic acid (mg/kg)	6.6	9.1	15.1	11.2	3.9	4.6	30[b]
Pyridoxine (mg/kg)	5.3	12.1	14.8	8.8	7.6	6.8	18[b]
Riboflavin (mg/kg)	1.4	2.3	2.5	1.8	2.2	4.2	12[b]
Thiamine (mg/kg)	3.8	8.9	2.2	0.2	0.3	4.9	6[b]
Vit A equiv. (IU/g)	8.0[c]	16.8[c]	NA	24.5[c]	48.9[c]	NA	NA
Vit E (mg/kg)	25	NA	14	34	26	94	NA
Xanthophylls (mg/kg)	19	4	42	191	326	NA	NA
Minerals							
Calcium (%)	0.03	0.01	0.36	0.16	0.08	0.04	0.28[b]
Chlorine (%)	0.05	0.06	0.25	0.07	0.10	0.04	0.86[b]
Chromium (mg/kg)	NA	NA	<1.5[b]	NA	<1.5[b]	<1.5[b]	<2.0[b]
Cobalt (mg/kg)	0.05	0.06[d]	0.10	0.08	0.05	NA	0.28[b]
Copper (mg/kg)	4	15[d]	52	30	29	5	31[b]
Iodine (mg/kg)	NA	NA	0.07	NA	0.02	NA	NA
Iron (mg/kg)	30	17	471	423	313	370	220[b]
Magnesium (%)	0.14	0.02	0.36	0.06	0.09	0.34	1.42[b]
Manganese (mg/kg)	5	16	26	8	7	4	58[b]
Molybdenum (mg/kg)	NA	NA	0.9[b]	NA	0.67[b]	0.56[b]	2.0[b]
Phosphorus (%)	0.29	0.10	0.82	0.50	0.54	0.47	3.6[b]
Potassium (%)	0.37	0.09	0.64	0.03	0.21	0.31	4.8[b]
Selenium (mg/kg)	0.08	0.11[d]	0.30	1.11	0.92	0.37	0.7[b]
Sodium (%)	0.03	0.01	1.05	0.10	0.06	0.08	0.22[b]
Sulfur (%)	0.12	0.19	0.23	0.39	0.72	0.33	1.18[b]
Zinc (mg/kg)	14	3[d]	72	190	35	114	132[b]

NA denotes data not available.
[a]Data compiled by National Academy of Sciences (51) unless otherwise noted.
[b]From Ref. 18.
[c]From Ref. 50.
[d]From Ref. 25.

TABLE 9 Typical Corn-Based Cattle Finishing Diets and Performance

	Ration 1: whole shelled corn	Ration 2: cracked corn
Ration composition (%)		
Corn, whole shelled (air dry)	88.6	—
Corn, cracked (air dry)	—	88.6
Soybean meal, 44% protein	10.0	10.0
Dicalcium phosphate	0.55	0.55
Limestone, feeding grade	0.35	0.35
Trace mineralized salt	0.50	0.50
Aurofac-10	0.10	0.10
Quadrex-11	0.025	0.025
Comparative performance		
Number of steers	12	12
Number of days on test	82	82
Average initial weight of steers (lb)	717	728
Average final weight of steers (lb)	1,072	1,018
Weight gain/steer (lb)	355	290
Average daily gain/steer (lb)	3.99	3.44
Feed efficiency (feed/gain)	4.84	5.29

Source: Ref. 38.

TABLE 10 Corn-Based Foods Made with Whole Corn or by Traditional Stone Grinding

Type	Specific Products
Raw corn	Corn salad
Boiled sweet corn	Fresh and frozen corn-on cob; fresh, frozen and canned corn kernels; cream-style corn
Fermented corn	Pickled miniature corn
Fried products	Corn mush, corn fritters, hush puppies
Alkali-cooked products	Hominy corn, corn chips, tortilla chips, table tortillas, taco shells, sopapillas, atoles, pazole, menudo, tostados
Steamed products	Tamales, couscous, Chinese breads, ricelike products, dumplings, chengu
Roasted products	Pinole, roasted whole kernels (Corn Nuts®), greuls, porridges
Unleavened, unfermented bread	Corn muffins, hoecake, tortillas, arepa
Fermented and/or leavened bread	Boston brown bread, Johnny cakes, pancakes, corn bread, blintzes
Porridges	Indian pudding, atole, ogi, kenkey, ugali, agi, edo pap, maizena, posho, asidah
Breakfast cereal	Southern corn grits, corn flakes
Snacks	Popcorn, tostados
Nonalcoholic beverages	Corn coffee, magou, chica dulce
Alcoholic beverages	Corn liquor, maize beer, koda, chicha, Kaffir beer

Sources: Refs. 63, 94, and author.

TABLE 11 Production, Utilization, and Consumption of Sweet Corn

	1977	1978	1979	1980	1981	1982	1983	1984	1985	1986
Production (1000 tons in hush)[a]	3,093	3,138	3,166	2,872	3,093	3,462	2,971	3,332	3,426	3,271
Processing for freezing[a]	796	860	832	769	832	1,127	901	969	1,111	1,055
Processing for canning[a]	1,580	1,575	1,632	1,389	1,546	1,614	1,326	1,583	1,529	1,476
Fresh market[a]	717	703	702	714	714	722	743	780	786	740
Canned corn pack (1000 cs)[b]	55,956	56,913	58,836	50,256	57,561	59,456	50,818	56,201	55,190	54,707
Golden, whole kernel[b]	40,571	41,838	42,110	37,103	41,886	44,460	38,454	41,549	41,529	40,991
Golden, cream style[b]	15,385	15,075	16,726	13,154	15,675	14,996	12,364	14,652	13,661	13,716
White, whole kernel[b]	859	600	873	NA	NA	679	385	674	726	731
White, whole kernel[b]	265	554	470	NA	NA	348	156	405	335	157
Canned hominy pack (1000 cs)[c]	2,300	NA	NA	NA	NA	1,600	NA	NA	NA	NA
Frozen corn-on-cob (million lb)[d]	265	307	264	259	305	379	294	350	423	392
Frozen cut corn packs (million lb)[d]	261	303	311	271	314	456	367	324	360	365
Per capita consumption (lb fresh on-cob basis)										
Fresh[e]	7.76	7.41	7.32	7.32	7.24	7.24	7.39	7.70	7.70	NA
Canned[e]	13.73	13.23	13.07	10.18	11.89	12.41	9.78	11.97	11.54	NA
Frozen[e]	7.21	7.75	7.31	6.75	7.23	9.64	7.66	8.17	9.23	NA
Total[e]	28.70	28.39	27.70	24.25	26.36	29.29	24.88	27.84	28.47	NA

[a]Compiled by Crop Reporting Board, Statistical Reporting Service, USDA (8).
[b]Compiled by Division of Economics and Statistics, National Canners Association (8).
[c]Compiled by Census Reports (8).
[d]Compiled by American Frozen Food Institute (8).
[e]Compiled by Economics Research Service, USDA (8).

TABLE 12 U.S. Popcorn Production (1000 pounds)

State	Yellow			White			Total		
	1984	1985	1986	1984	1985	1986	1984	1985	1986
Illinois	64,591	72,732	59,464	399	786	61	64,990	73,518	59,525
Indiana	146,560	186,475	128,072	6,362	4,825	4,990	152,922	191,300	133,062
Iowa	30,904	37,050	32,761	5,650	6,514	3,163	36,554	43,564	35,924
Kansas	12,540	14,023	17,997	0	0	0	12,540	14,023	17,997
Kentucky	17,488	25,044	2,358	0	0	0	17,488	25,044	2,358
Michigan	10,975	11,340	11,019	2,150	1,187	719	13,125	12,527	11,738
Missouri	33,570	42,536	41,120	770	1,030	870	34,340	43,566	41,990
Nebraska	104,435	104,300	65,603	14,931	14,050	11,820	119,366	118,350	77,423
Ohio	62,148	70,016	72,297	587	1,417	2,046	62,735	71,433	74,343
Other states	52,390	38,579	20,347	2,378	1,865	590	54,768	40,444	20,937
Total	535,601	602,095	451,038	33,227	31,674	24,259	568,828	633,769	475,297

Source: Popcorn Institute, Chicago, IL.

TABLE 13 U.S. Corn Utilization as Food and Industrial Products (million bushels)

	Wet-milled products					
Year beginning Sept. 1	Starch	Glucose and dextrose	HFCS	Alcohol	Other alcohol processes	Dry-milled and alkali-cooked products
1975	115	162	45	5	20	154
1976	116	164	62	10	15	155
1977	124	170	80	10	20	158
1978	124	170	105	15	20	155
1979	120	170	127	25	20	158
1980	120	183	165	35	35	160
1981	130	183	185	83	35	162
1982	127	188	215	130	50	170
1983	145	191	256	150	50	164
1984	142	188	310	170	100	160
1985	152	188	328	185	127	161
1986	155	185	339	200	135	161
1987[a]	160	186	348	207	145	163

[a]Projected utilization.
Source: Ref. 79.

TABLE 14 Major U.S. Corn Milling Companies (1988)

Company	Headquarters	Plants	Products[ab]
Wet Milling			
Archer Daniels Midland	Decatur, IL	Cedar Rapids, IA Clinton, IA Decatur, IL	FS, IS, CS, HFCS, CB E, CO, GM, GF, CC
American Maize Products Co.	Stamford, CT	Decatur, AL Dimmit, TX Hammond, IN	FS, IS, SS, MD, MS, CS, CD, GM, GF
Cargill Co.	Minneapolis, MN	Cedar Rapids, IA Dayton, IA Eddyville, IA Memphis, TN	FS, IS, MS, CS, HFCS, E, CO, GM, GF
Coors	Golden, CO	Johnstown, CA	CS, HFCS, GM, GF,
Corn Processing Co. International, Inc.	Englewood, NJ	Argo, IL Stockton, CA Winston–Salem, NC	FS, IS, CS, HFCS, CD, CO, GM, GF, SW, H
Grain Processing Corp.	Muscatine, IA	Muscatine, IA	FS, IS, MD, MS, CS, GM, GF
Hubinger	Keokuk, IA	Keokuk, IA	FS, IS, MS, CS, GM, GF
Minnesota Corn Processors	Marshall, MN	Marshall, MN	FS, IS, MD, CS, MS, E, GM, GF, SW
National Starch & Chemical	Bridgewater, NJ	Indianapolis, IN N. Kansas City, MO	FS, IS, SS, GM GF
Penford	Cedar Rapids, IA	Cedar Rapids, IA	FS, IS, SS, CS, GM, GF
A. E. Staley Mngf. Co.	Decatur, IL	Decatur, IL Lafayette, IN Loudon, TN Morrisville, PA	FS, IS, SS, MD CS, HFCS, CD, CF E, CB, CO, GM GF

TABLE 14 (Continued)

Company	Headquarters	Plants	Products[ab]
Dry Milling			
Archer Daniels Midland	Decatur, IL	Lincoln, NE (Gooch Mills)	FG, F, M, C
		Shawnee Mission, KS (ADM Milling)	F, SFP, SIP, CSM
		Milwaukee, WI (Krause Milling)	HF
Illinois Cereal Mills	Paris, IL	Paris, IL	FG, G, M, C
		Indianapolis, IN (Evans Milling)	F, SFP, SIP, HF
Lauhoff Grain Co.	Danville, IL	Danville, IL	FG, G, M, C
		Crete, NE (Crete Mills)	F, CB, HF
Lincoln Grain, Inc.	Atchinson, KS	Atchinson, KS	FG, G, M, C, F, CM, CO, HF
Martha White Foods	Nashville, TN	Jackson, TN	FG, G, M, C,
		Lebanon, TN	F, HF
		Tifton, GA	
Midstate Mills, Inc.	Newton, NC	Newton, NC	FG, G, M, C, F, HF
The Quaker Oats Co.	Chicago, IL	Cedar Rapids, IA	FG, G, M, C
		Chattanooga, TN	F, HF
		St. Joseph, MO	
J. R. Short Milling Co.	Kankakee, IL	Kankakee, IL	FG, G, M, C, F, SFP, GM, HF

[a]Wet milling product symbol FS denotes food starch: IS, Industrial starch; SS, specialty starches (modified); MD, maltodextrins; MS, maltose syrups; CS, corn syrups; HFCS, high fructose corn syrup; CD, crystalized dextrose; CF, crystalized fructose; E, ethanol; CB, corn bran; CO, corn oil; GM, gluten meal; GF, gluten feed; SW, steep water; H, hydrol; and CC, carmel color.

[b]Dry milling product symbol FG denotes flaking grits; G, grits; M, meal; C, cones; F, flour; CM, corn masa; CSM, corn:soy:milk blends; SFP, specialty food products; SP, specialty industrial products; GM, germ meal; CB, corn bran; and HF, hominy feed.

TABLE 15 Products Produced from Dry Corn Milling (Tempering-Degerming)

Product	Particle size range: U.S. standard sieve[a] Less than	More than	Diameter (μm) Less than	More than	Yield[a] (%)
Usual products					
Cereal flaking (hominy) grits	3.5	6	5,660	3,360	12
Coarse grits	10	14	2,000	1,410	15
Regular grits	14	28	1,410	638	23
Coarse meal	28	50	638	297	3
Dusted meal	50	75	297	194	3
Flour	75	pan	194	pan	4
Oil	—	—	—	—	—
Hominy feed	—	—	—	—	35
Shrinkage	—	—	—	—	4
Alternative products					
Brewers grits	12	30	1,680	590	30
100% meal	28	pan	638	pan	10
Fine meal (coarse cones)	50	80	297	177	7
Germ fraction	3.5	20	5,660	840	10

NA denotes data not available.
[a]From Ref. 12.

TABLE 16 Compositions of Dry-Milled Corn Products (Tempering-Degerming)

Product	Moisture (%)	Protein (N × 6.25)	Composition (% db)		
			Fat	Crude fiber	Ash
Cereal flaking (hominy grits)	14.0	8.4	0.7	0.4	0.4
Coarse grits	13.0	8.4	0.7	0.5	0.4
Regular grits	13.0	8.0	0.8	0.5	0.5
Brewers grits	13.0	8.3	0.7	0.5	0.5
Coarse meal	12.0	7.6	1.2	0.5	0.6
Dusted meal	12.0	7.5	1.0	0.5	0.6
100% meal	12.0	7.2	1.5	0.6	0.6
Fine meal	12.0	7.0	1.6	0.6	0.7
Flour	12.0	6.6	2.0	0.7	0.7
Germ fraction	15.0	14.9	18.0	4.6	4.7
Hominy feed	13.0	12.5	6.3	5.4	3.3

Source: Ref. 12.

TABLE 17 Uses for Dry-Milled Corn Products

Product	Food use	Industrial use
Flaking grits	Breakfast cereals	
Coarse grits	Breakfast cereals	
	Snacks	
Medium grits	Breakfast cereals	
	Snacks	
Fine grits	Brewing	Wallpaper paste
	Maize porridge	
Meal	Brewing	Floor wax
	Snacks	Handsoap
	Pancake, waffle, muffin mixes	
Cones	Snacks	Dusting agents
	Cornbread	Handsoap
	Breakfast cereals	
Flour	Baby foods	Fermentation media
	Frozen foods	Explosives
	Biscuits	Wallboard
	Wafers	Briquetting
	Pancake mixes	
	Meat extenders	
	Breading, batters	
Speciality products	Confectionary toppings	Foundry binders
	Jam and jelly extenders	Label adhesives
	Fig paste extenders	Edge paste
	Coconut extenders	
	Thickening agents	
	Dough conditioners	
	Brewers flakes	
	Meat extenders	

TABLE 18 Estimated Utilization of Corn Dry-Milling Products

Application	Consumption of corn products	
	Million pounds	% of Total
Animal feeds	2,200[a]	36.2
Food uses	3,300	54.3
Brewing	1,850[b]	30.4
Breakfast cereals	800[b]	13.2
Fortified foods (PL 480)	325[b]	5.4
Mixes (pancake, cookie, muffin, etc.)	100[b]	1.6
Snackfoods	100[b]	1.6
Baking	50[b]	0.8
Other foods (breadings, batters, baby foods, etc.)	75[b]	1.2
Industrial (nonfood) uses	580	9.5
Gypsum board	100[b]	1.6
Pharmaceuticals/fermentation	200[b]	3.3
Foundary binders	90[b]	1.5
Ore refining	65[a]	1.1
Charcoal binders	75[b]	1.2
Drilling fluids	25[a]	0.4
Paper, paperboard, corrugating	15[c]	0.3
Other	10[c]	0.2

[a]From Ref. 1.
[b]From Ref. 2.
[c]Estimates of author.

TABLE 19 Size and Significance of Corn Wet-Milling Industry (1985–1986)[a]

Number of plants	27
Location	15 states
Annual corn grind	855 million bushels
Value of corn purchased	$2 billion
Employment	10,000 persons
Capital investment	$5.3 billion
Major products	
High fructose corn syrups	10.8 billion pounds
Dextrose corn syrups	4.3 billion pounds
Dextrose	1.0 billion pounds
Corn starches	4.4 billion pounds
Industrial starches[b]	2.9 billion pounds
Food starches [b]	1.5 billion pounds
Other products	11.7 billion pounds
Value addition	$3.2 billion

[a]Data provided by the Corn Refiners Association, Washington, DC, unless otherwise noted
[b]Based on estimates that nonfood uses for starch account for 65% of total starch usage (53).

TABLE 20 Shipments of Products of the Corn-Refining Industry (million pounds)

Products	1982	1983	1984	1985	1986	1987
Corn starch, modified starch, dextrin	3,678	4,019	4,183	4,225	4,338	4,401
Maltodextrin, glucose syrup, dextrose, corn syrup solids	6,282	6,298	6,419	6,422	6,400	6,245
High fructose corn syrup	8,400	9,707	11,502	13,920	14,136	14,328
Corn oil, crude	82	73	116	164	NA	NA
Corn oil, refined	371	400	407	382	NA	NA
Corn gluten feed	6,441	7,391	8,740	8,811	8,710	8,748
Corn gluten meal, 41% protein	14	19	20	19	19	NA
Corn gluten meal, 60% protein	1,272	1,383	1,635	1,609	1,600	1,604
Corn germ meal	19	29	29	49	NA	NA
Steepwater	245	212	301	282	299	183
Hydrol	219	209	217	229	NA	NA

NA denotes data not available.
Source: Corn Refiners Association, Washington, DC.

TABLE 21 Typical Yields of Wet-Milled Products

Product	Yield (% db)	Yield (lb/bu db)
Starch	67.5	32.7
Fiber	11.5	5.6
Germ	7.5	3.6
Oil	3.9	1.9
Cake	3.6	1.7
Gluten Meal (60% protein)	5.8	2.8
Steep liquor solubles	7.5	3.6
Loss	0.2	0.1
Gluten Feed (fiber, germ cake, steep liquor solubles to yield 21% protein)	22.6	11.0

Source: Ref. 5.

TABLE 22 Compositions of Wet-Milled Products (dry-basis)

Product	Starch (%)	Protein (%) (N x 6.25)	Fat (%)	Crude Fiber (%)	Pentosans (%)
Starch	99.0	0.3	0.02	0.03	NA
Fiber	23.0	12.0	1.0	14.0	30.0
Germ	10.0	12.0	52.0	10.0	12.0
Oil	0	0	100	0	0
Solvent-extracted meal	20.0	25.0	1.0	10.0	25.0
Gluten	18.0	70.0	7.0	2.0	NA
Steep liquor solubles	NA	46.0	0	0	0

Source: Ref. 5.

TABLE 23 Compositions and Properties of Commercial Corn Starches[a]

	Normal	Waxy
Composition (% as is)		
Moisture	11	11
Starch	88	88
Portion as amylose	28	0
Portion as amylopectin	72	100
Protein (N × 6.25)	0.35	0.28
Fat (ether extractables)	0.04	0.04
Total lipids	0.87	0.23
Fiber	0.1	0.1
Ash	0.1	0.1
SO_2	0.0004	NA
Properties		
Granule size, average (μm)	9.2	NA
Granule size, range (μm)	5–30	NA
Gelatinization temperature range (°C)	62–72	63–72
Swelling power at 95°C[b] (%)	24	64
Solubility at 95°C (%)	25	23
Critical concentration at 95°C[b] (%)	4.4	1.6
Brabender pasting temperature (8%,°C)	75–80	65–70
Brabender peak viscosity (8%, BU)	700	1,100
Paste texture[b]	short	long
Paste clarity[b]	opaque	translucent
Resistance to shear[b]	medium	low
Rate of retrogradation[b]	high	very high
Specific gravity	1.5	1.5
Bulk density (lb/ft^3)	44–45	44–45

NA denotes data not available.
[a]Data compiled by Watson (86) unless otherwise noted.
[b]From Ref. 71.

TABLE 24 1986 Estimated Utilization of Corn Starch

Application	Starch consumption	
	Million pounds	% of Total
Food Applications[a]	1,536	35
Brewing	307	20
Chemicals, drugs, pharmaceuticals	307	20
Canning	115	7.5
Confectionary	100	6.5
Baking	92	6.0
Other (prepared mixes, condiments, etc.)	614	40
Nonfood[b]	2,852	65
Paper and paperboard	1,231	43
Coatings	406	33
External surface sizing	719	58
Internal sizing at wet end	106	9
Adhesives	806	30
Corrugating and laminating	387	45
Envelopes, stamps, labels	472	55
Textiles (finish sizing, warp sizing)	254	9
Building materials (ceiling tiles, gypsum board, plywood, rigid foams, etc.)	171	6
Other (cosmetics, briquettes, crayons, dyes, explosives, plastics, laundry, oil drilling, water treatment, etc.)	342	12

[a]Market shares based on data published by Orthoefer (53) and total starch production estimated by the Corn Refiners Association (22).
[b]Market shares based on data compiled by Russell (64) and total starch production estimated by the Corn Refiners Association (22).

TABLE 25 Typical Composition and Properties of Refined Corn Bran (as is basis)[a]

Moisture (%)	3
Dietary fiber (%)	92
Cellulose[b] (%)	22
Xylan[b] (%)	41
Ligin[b] (%)	5
Fat (%)	3
Protein (%)	5
Starch (%)	5
Ash (%)	0.04
Energy (calories/g)	0.5
Water holding capacity (g H_2O/g)	3.5
Total titratable acidity	4
SO_2 (ppm)	less than 25

[a]Data provided by A. E. Staley Mfg Co., Decatur, IL, unless otherwise noted.
[b]From Ref. 39.

TABLE 26 Compositions of Crude and Refined Corn Oils

Component	Crude	Refined
Triglycerides (%)	95.6[a]	98.8[b]
Saturation		
Saturates (%)	—	12.9[b]
Mono-unsaturates (%)	—	24.8[b]
Polyunsaturation (%)	—	61.1[b]
Ratio of polyunsaturation to saturated	—	4.8[b]
Triglyceride fatty acid profile (%)		
Palmitic (16:0)	11.1–12.8[c]	11.1–12.8[c]
Stearic (18:0)	1.4–2.2[c]	1.4–2.2[c]
Oleic (18:1)	22.6–36.1[c]	22.6–36.1[c]
Linoleic (18:2)	49.0–61.9[c]	49.0–61.9[c]
Linolenic (18:3)	0.4–1.6[c]	0.4–1.6[c]
Arachidic (20:0)	0.0–0.2[c]	0.0–0.2[c]
Phospholipids (%)	1.5[d]	0.04[a]
Free fatty acid (% as oleic)	1.5–4.0[d]	0.02–0.03[a]
Waxes (%)	0.05[a]	0[a]
Cholesterol (%)	0[a]	0[a]
Phytosterols (%)	1.2[a]	1.1[a]
Tocopherols (%)	0.12[d]	0.09[a]
Carotenoids (%)	0.0008[d]	NA

NA denotes data not available.
[a]From Ref. 60.
[b]From Ref. 19.
[c]From Ref. 90.
[d]From Ref. 61.

TABLE 27 Chemical and Physical Properties of Crude and Refined Corn Oils

Property	Crude	Refined
Refractive index	1.470–1.478[a]	1.470–1.474[b]
Iodine value	110–128[a]	125–128[b]
Thiocyanogen value	77[a]	NA
Saponification value	187–196[a]	NA
Unsaponifiable matter (%)	1.0–2.2[a]	1.2[b]
Titer (°C)	16–20[a]	NA
Acetyl value	8–12[a]	NA
Reichert-Meissl value	<0.5[a]	<0.5[b]
Polenske value	<0.5[a]	<0.5[b]
Hehner value	—	93–96[b]
Cold test (hr)	—	24[b]
Solidifying point (°C)	—	–20 to –10[b]
Melting point (°C)	—	–16 to –11[b]
Smoke point (°C)	—	221 to 260[b]
Flash point (°C)	—	302 to 338[b]
Fire point (°C)	—	310 to 371[b]
Specific gravity		
at 15/15°C	0.920–0.928[a]	NA
at 25/25°C	0.914–0.921[a]	0.918–0.925[b]
Bulk density at 20°C		
(kg/liter)	0.92c	0.92[c]
(lb/gal)	7.672[b]	7.672[b]
Viscosity at 20°C (cp)	NA	15.6[d]
Flavor and odor	trace	bland[b]
Color		
yellow (Lovibond)	—	20–35[b]
red (Lovibond)	—	2.5–5.0[b]
Heat of combustion (cal/g)	—	9,420[d]

NA denotes data not available.
[a]From Ref. 6.
[b]From Ref. 60.
[c]Estimates of author.
[d]From Ref. 17.

TABLE 28 Utilization of Corn Oil Compared to Other Vegetable Oils (million pounds, 1985 data)[a]

Oil	Application				Total utilization
	Shortening	Margarine	Salad and cooking oils	Other	
Corn	—	208	514	12[b]	734
Soybean	3,609	1,625	4,726	126[b]	10,086
Cottonseed	180	30[b]	387	15[b]	612
Palm	230	—	—	50[b]	280
Coconut	120[b]	—	—	155[b]	275
Peanut	—	9	107[b]	—	116
Sunflower	—	—	94[b]	—	94

[a]Data compiled by U.S. Department of Commerce, Census Bereau: Current Industrial Reports, monthly reports, 1985 (published in Corn Refiners Association (19)) unless otherwise noted.
[b]Estimates of author.

TABLE 29 Compositions and Properties of Major Proteins in Corn

	Zein	Albumins	Globulins	Glutelins
Portion of whole grain protein (%)	39[a]	8[a]	9[a]	40[a]
Portion of endosperm protein (%)	47[b]	4[c]	4[c]	39[b]
Portion of germ protein (%)	5[a]	30[c]	30[c]	25[a]
Soluble in	60–90% ethanol, 70–90% isopropanol, aqueous acetone, ethanol: benzene, ethanol: chloroform, ethanol: nitromethane ethylene glycol	Water	0.5 M NaCl	Dilute alkali (pH 10) + 0.6% ME Dilute alkali (pH 10) + 0.05% SDS
Amino acid composition (mg/100 g protein)				
Alanine	110[b]	85[c]	60[c]	79[b]
Arginine	10[b]	43[c]	72[c]	30[b]
Aspartic acid	41[b]	73[c]	58[c]	49[b]
Glutamic acid	166[b]	86[c]	114[c]	131[b]
Glycine	17[b]	93[c]	73[c]	62[b]
Histidine	8[b]	16[c]	25[c]	18[b]
Isoleucine	31[b]	28[c]	23[c]	29[b]
Leucine	151[b]	49[c]	45[c]	85[b]
Lysine	1[b]	44[c]	41[c]	18[b]
Methionine	10[b]	10[c]	7[c]	25[b]
Phenylalanine	43[b]	21[c]	28[c]	27[b]
Proline	94[b]	45[c]	33[c]	73[b]
Serine	52[b]	48[c]	53[c]	47[b]
Threonine	24[b]	45[c]	28[c]	34[b]
Tryptophan	0[b]	6[c]	4[c]	3[b]
Tyrosine	31[b]	49[c]	17[c]	31[b]
Valine	31[b]	50[c]	49[c]	45[b]
Ammonia	148[b]	84[c]	80[c]	202[b]

[a]From Ref. 62.
[b]From Ref. 55.
[c]From Ref. 57.

TABLE 30 Properties and Uses for Zein

Normal industrial yield[a]	20–24% db gluten meal
Peak market[a]	6 million pounds in 1954
Current market	Less than 0.5 million pounds
Properties	Forms tough, glossy, scuff-proof, and grease-proof coatings Unique film and fiber forming properties Thermoplastic Films, fibers, etc. are resistant to microbial attack Films, fibers, etc. can be cured with formaldehyde to an essentially inert product
Uses	Paper sizings Paper and paperboard adhesive Additives in oil cloth and linoleum Felt stiffener Moisture and oxygen barrier in food coatings (nuts and confections) Pharmaceuticals (binding and seal coating tablets) Varnish substitutes Printing inks Cork binding Spun textile fibers Floor coatings Paper coatings
Suppliers	Freeman Industries Tuckahoe, NY

[a]From Ref. 62.

TABLE 31 Chemical Conversions of Corn Products

Corn product	Product class	Product
Starch	Modified starches	Pregelatinized starch
		Oxidized starch
		Esterified starch (starch acetate, maleate, succinate, phosphates)
		Starch ethers (hydroxypropyl, hydroxyethlyl, carboxymethyl, cationic derivatives)
		Cross-linked starch
		Dialdehyde starch
		Grafted starch
	Starch hydrolysates	Dextrins
		Acid-modified starch
		Glucose syrups
		Maltose syrups
		Maltodextrins
		Dextrose
		Maltose
		Massee
Saccharides	Organic acids	Gluconic acid
		Lactic acid
		Oxalic acid
		Glucuronic acid
		Glucaric acid
		Fumaric acid
		Saccharinic acid
		Arabonic acid
		Maltobionic acid
		Levulinic acid
	Polyols	Sorbitol
		Mannitol
		Maltitol
		Glycerol
		Ethylene glycol
		Methylglycerol
		Sucropolyols
		Erythritol
	Glucosides	Glycerol glucoside
		Methylglucoside
		Alkylglucoside
		Hydroxyalkylglucoside
	Glucose esters	Glucose phosphate
		Acetone glucose
		Polyurethanes
	Polyol ether	
	Levoglucosan	
	Hydroxymethylfurfural	
Corn cobs (pentosans)	Furfural derivatives	Furfural
		Furfuryl alcohol
		Tetrahydrofurfuryl alcohol
		Furan
		Tetrahydrofuran
		Polytetramethylene ether glycol

Source: Adapted from Ref. 83.

TABLE 32 Biochemical Conversions of Corn Products

Conversion	Product class	Product
Enzyme modification of starch	Starch hydrolysates	Maltodextrins
		Cyclodextrins
		Maltose syrups
		Glucose syrups
Enzyme modification of glucose	Sugar isomerization	Fructose syrups
Fermentation of cellulose (corn cobs or stover)	Alkanes	Methane
		Ethane
	Silage	Corn silage feeds
Fermentation of starch	Organic acids	Acetic acid
		Citric acid
		Gluconic acid
		Malic acid
		Succinic acid
		Fumaric acid
		Lactic acid
		Propionic acid
		Butyric acid
		Ketogluconic acid
		Itaconic acid
		Kojic acid
	Alcohols	Ethanol
		Isopropanol
		Butanol
	Ketones	Acetone
		Glucosone
	Polyols	2,3 Butanediol
		Glycerol
		Mannitol
		Arabinitol
	Amino acids	L-Glutamic acid
		L-Methionine
		L-Lysine
	Nucleotides	Guanyl acid
		Inosinic acid
		Xanthyl acid
	Biopolymers	Xanthan
		Pullulan
		Alginate
		Scleroglucane
		Curdlane
		Erwinia polysaccharide
		Polyhydroxybutyric acid
		Glycane
		Polysaccharide P.S.7
		Beyerinckia polysaccharide
	Fats and lipids	Single-cell oil
	Proteins	Single-cell protein
	Enzymes	
	Vitamins	
	Antibiotics	
	Hormones	
	Phosphorylated sugars	

Source: Adapted from Ref. 83.

TABLE 33 Important Modifications to Corn Starch.

Modification	Reactants	Mechanism	Reaction	Result	Application
Acid modified	Mineral acid	Hydrolysis Weakened granule	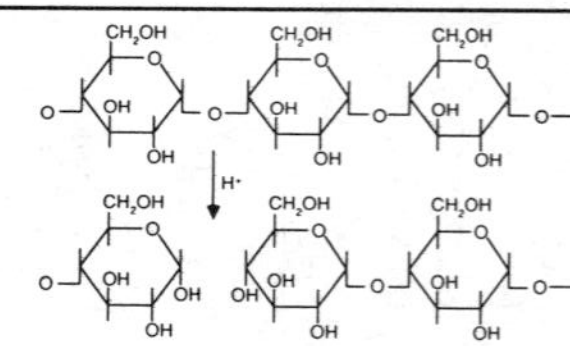	Reduces viscosity. Firmer gels.	Gum candies. Textile warp sizes. Paper coatings
Oxidation	Sodium hypochlorite	Random oxidation of hydroxyl groups to carboxyl or carbonyl groups, with rupture of adjacent glycosidic bond	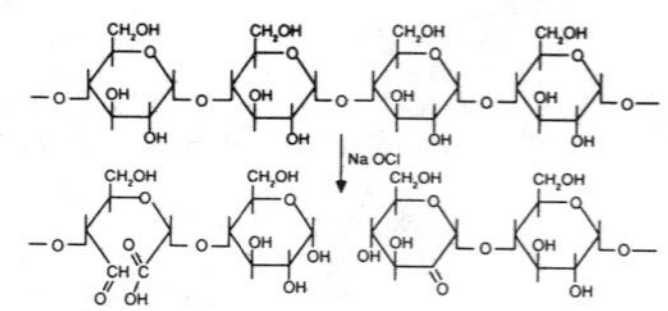	Reduces viscosity. Gives clear pastes. Reduces gelling and retrogradation. Retains original granule structure.	Paper and textile sizings. Adhesives. Food applications for high solids, low viscosity.
Dextrinization	Dry heat with acid or alkaline catalyst	Disprupts granule integrity		Reduces viscosity. Increases water solubility.	Adhesives in paper products.
Pregelatinization	Heat and water	Gelatinizes starch granule		Cold water dispersible.	Instant adhesives and puddings.
Bleaching	Hydrogen peroxide Peracetic acid Ammonium persulfate Potassium permaganate Sodium chlorite Sodium hypochlorite	No reaction with starch Destroys chromophore of trace contaminating pigments such as xanthophyl		Produces stark white color.	Fluidizing agents for dry powders such as confectioner's sugar.

Phosphated derivatives	Monosodium orthophosphate or sodium tripolyphosphate	Introduction of phosphate group	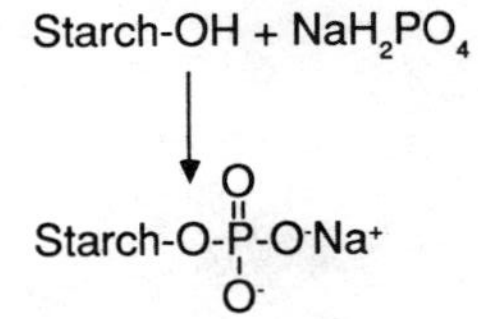	Produces stable gels. Improves clarity. Increased viscosity. Longer, more cohesive textures. Improved paper strength.	Foods, paper textile sizings, flocculants, emulsifiers, and adhesives.
Hydroxyethyl derivatives	Ethylene oxide	Introduction of hydroxyethyl group	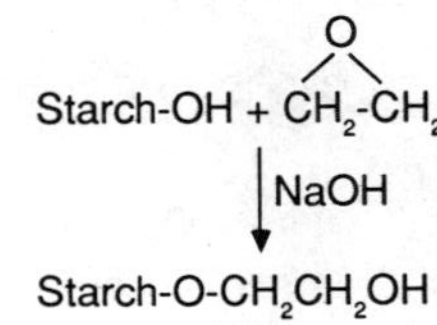	Reduces gelatinization temperature. Produces more clear stable pastes. Reduces gelling on cooling. Improves strength and stiffness of paper.	Non-food uses such as surface sizings, paper coatings and adhesives.
Hydroxypropyl derivatives	Propylene oxide	Introduction of hydroxypropyl group	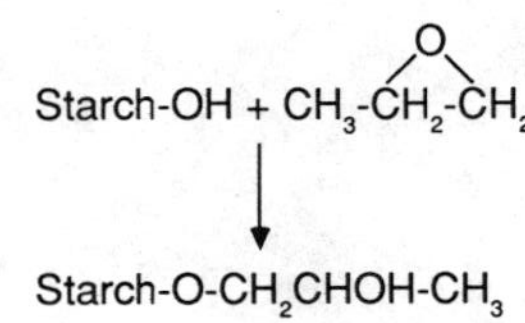	Improved viscosity, texture and stability.	Food uses in gravies, sauces, pie fillings, and salad dressings. Also used in adhesives and wallboard binders.

TABLE 33 (Continued)

Modification	Reactants	Mechanism	Reaction	Result	Application
Acetate derivatives	Acetic anhydride or vinyl acetate	Introduction of acetyl groups	$\text{Starch-OH} + CH_3\text{-}C(=O)\text{-}O\text{-}C(=O)\text{-}CH_3 \xrightarrow{NaOH} \text{Starch-O-}C(=O)\text{-}CH_3 + CH_3COOH$	Prevents retrogradation. Reduces tendency to congeal. Reduces gelatinization temperature. Increases hot-peak viscosity. Reduces cold paste viscosity.	Food thickeners and stabilizers in fruit pies, gravies, salad dressings and filled cakes. Warp sizings for textiles. Gummed tape adhesives.
Succinate derivatives	Succinic anhydride	Introduction of succinate groups	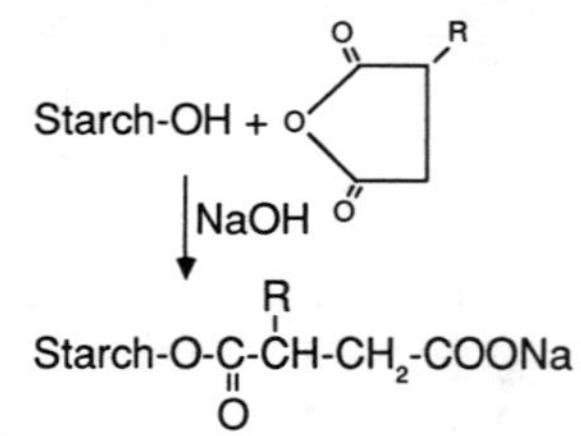	Imparts surface activity.	Food thickeners. Encapsulating media. Paper and textile sizings. Adhesives.
Carboxymethylation	Sodium monochloroacetate	Introduction of anionic groups	$\text{Starch-OH} + Cl\text{-}CH_2\text{-}COONa \xrightarrow{OH^-} \text{Starch-O-}CH_2\text{-}COONa$	Improves cold water solubility.	Paints, oil drilling muds, wallpaper adhesives, and detergents.
Cationic derivatives	Tertiary or quarternary amines	Introduction of cationic groups	$\text{Starch-OH} + Cl\text{-}CH_2\text{-}(CH_2)_n\text{-}NH^+(R_1)(R_2) \longrightarrow \text{Starch-O-CH-}(CH_2)_n\text{-}NH^+(R_1)(R_2) + HCl$	Improves adsorption by cellulose. Improved pigment retention of paper.	Paper strengtheners. Surface sizing, adhesives and flocculants.

Starch xanthates	Carbon disulfide	Introduction of xanthate groups	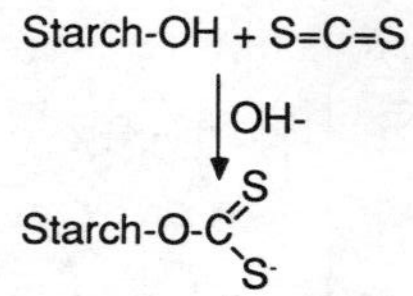	Improves wet strength in paper. Bonds cellulose fibers. Accelerates vulcanization.	Paper manufacturing. Rubber manufacturing.
Cross-linking 1. Distarch phosphates	Phosphorus oxychloride or trimetaphosphate	Cross linking glycosidic chains	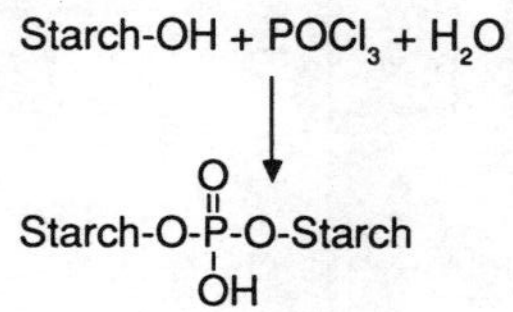	Reduces swelling. Reduces loss of viscosity due to temperature, shear, and acid.	Retortable thickeners. Salad dressings. Baby foods and fruit pie fillings.
2. Distarch adipate	Adipic acid and acetic anhydride		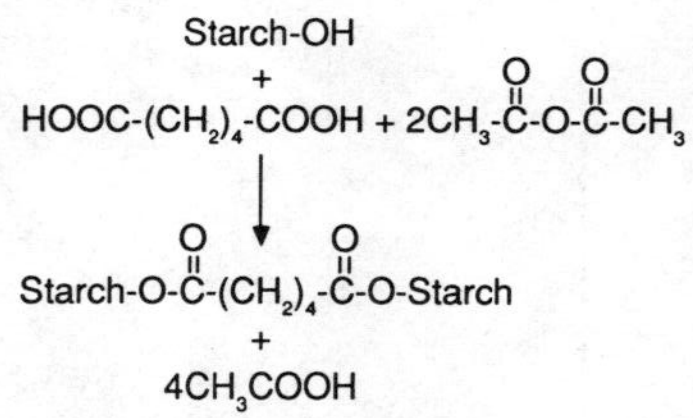		

TABLE 34 1986 Estimated Production of Different Types of Starches for Industrial (nonfood) Uses

Type of starch	Production (million pounds)	
	1972[a]	1986[b]
Unmodified starch	1,817	1,675
Acid-modified starches	340	314
Cationic starches	100	92
Dextrins	150	138
Oxidized starches	166	153
Pregelatinized starches	100	92
Others	420	387
Total	3,093	2,852

[a]Data from Ref. 64.
[b]Estimated by author. Total nonfoods usage of starch estimated as 65% (53) of 4,388 million pounds of total starch (estimate provided by Corn Refiners' Association, Inc.). Market share of individual modified starches assumed to be the same in 1986 as in 1972.

TABLE 35 Compositions and Properties of Maltose and Dextrose Corn Sweeteners

Product	Dextrose equivalent	Process	Carbohydrate composition[a] (%db) DP1	DP2	DP3	DP4	DP5	DP6	DP7 & higher	Solids[c] (%)	Ash[b] (%)	RI at 20°C[b]	Viscosity at 38°C (centipoise)	Bulk density at 38°C[b]	Relative sweeteness to sucrose[c] (%)
Maltodextrin	5	Enzyme	t[b]	t[b]	1[b]	1[b]	1[b]	1[b]	96[b]	97	0.5	NA	NA	35 lb/ft³	0–5
	12	Enzyme	1	3	4	3	3	6	80	97	0.5	1.42577	NA	38 lb/ft³	10–15
High maltose corn syrup	43	Dual	8	40	15	7	2	2	26	81	0.05	1.49709	21,000[c]	NA	35
	49	Dual	9	52	15	1	2	2	19	81	0.05	1.49573	17,000[b]	11.9 lb/gal	40
Dextrose corn syrup	27	Acid	9	9	8	7	7	5	54	78	0.3	NA	22,000[c]	11.7 lb/gal	30–35
	36	Acid	14	12	10	9	8	7	40	80	0.3	NA	20,000[c]	11.8 lb/gal	35–40
	42	Acid	20	14	12	9	8	7	30	80	0.3	1.49670	12,500[c]	11.9 lb/gal	45–50
	55	Acid	31	18	12	10	7	5	17	80	0.3	1.49434	7,500[c]	11.8 lb/gal	50–55
	65	Dual	39	31	7	5	4	3	11	82	0.3	1.49263	5,500[c]	11.9 lb/gal	60–70
	70	Dual	47	27	5	5	4	3	9	82	0.3	1.49135	5,000[b]	11.8 lb/gal	65–75
	95	Dual	92	4	1	1	t	t	t	71	0.3	NA	NA	11.2 lb/gal	90–100
Crystalline maltose	NA	NA	99[b]	t[b]	0[b]	0[b]	0[b]	0[b]	0[b]	100	0.01	NA	NA	NA	16–32
Crystalline dextrose	NA	Dual	99[b]	t[b]	0[b]	0[b]	0[b]	0[b]	0[b]	100	0.01	NA	NA	NA	65–70

NA denotes data not available; t denotes trace amounts present
Acid denotes acid conversion, enzyme denotes enzyme conversion, and dual denotes acid-enzyme conversion
[a]DP denotes degree of polymerization. Composition data compiled by Corn Refiners Association (21) unless noted otherwise.
[b]Estimates of author based on product specifications of suppliers and data compiled by Lloyd and Nelson (47).
[c]From Ref. 34.

TABLE 36 Compositions and Properties of Fructose Sweeteners[a]

	42% HFCS	55% HFCS	90% HFCS	Crystalline fructose
Dry matter (%)	71	77	80	100
Ash[b] (%)	0.05[b]	0.05[b]	0.03[b]	0.03[b]
Sugar composition (%)				
Fructose	42	55	90	97
Dextrose	52	41	8	3
Maltose	4	3	1	0
Higher saccharides	2	1	1	0
pH	4.0–4.4	3.5–4.0	3.5	—
Viscosity at 28°C (centipoise)	160	670	1,100	—
Bulk density (lb/gal)	11.23	11.55	11.75	—
Refractive index at 20°C[c]	1.46441	1.47869	—	—
Relative sweetness to sucrose[b] (%)	90–100	100–110	120–160	180

[a]Data compiled by Hanover (34) unless otherwise noted.
[b]Estimates of author based on product specifications provided by suppliers and data compiled by Lloyd and Nelson (47).
[c]At 50% solids, from Corn Refiners Association (21).

TABLE 37 Total and Per Capita Consumption of Various Sweeteners

	1978	1979	1980	1981	1982	1983	1984	1985	1986	1987
Total U.S. consumption (million tons)	14.87	15.12	15.12	15.33	15.41	16.13	16.87	17.82	17.78	18.59
Total caloric sweeteners (million tons)	14.10	14.30	14.24	14.39	14.31	14.61	15.00	15.66	15.55	16.15
Refined sugar (million tons)	10.18	10.05	9.52	9.13	8.56	8.33	7.98	7.54	7.27	7.59
Total corn sweeteners (million tons)	3.75	4.09	4.58	5.12	5.60	6.12	6.84	7.96	8.10	8.39
HFCS (million tons)	1.35	1.67	2.18	2.67	3.10	3.60	4.30	5.38	5.50	5.77
Glucose (million tons)	1.98	2.02	2.00	2.05	2.09	2.11	2.13	2.16	2.17	2.19
Dextrose (million tons)	0.42	0.40	0.40	0.40	0.41	0.41	0.41	0.42	0.43	0.43
Total low-caloric sweeteners (million tons)	0.77	0.82	0.88	0.94	1.10	1.52	1.87	2.16	2.23	2.44
Saccharin (million tons)	0.77	0.82	0.88	0.92	0.98	1.11	1.18	0.72	0.66	0.73
Aspartame (million tons)	0.00	0.00	0.00	0.02	0.12	0.41	0.69	1.44	1.57	1.71
Per capita consumption, total sweeteners (lb)	133.5	134.4	132.8	133.3	132.6	137.6	142.5	142.4	148.9	147.2
Total caloric sweeteners (lb)	126.6	127.1	125.1	125.1	123.2	124.6	126.6	130.9	128.7	132.4
Refined sugar (lb)	91.4	89.3	83.6	79.4	73.7	71.1	67.4	63.0	60.2	62.2
Total corn sweeteners (lb)	33.7	36.3	40.2	44.5	48.2	52.2	57.8	66.5	67.1	68.8
HFCS (lb)	12.1	14.9	19.1	23.2	26.7	30.7	36.3	45.0	45.6	47.3
Glucose (lb)	17.8	17.9	17.6	17.8	18.0	18.0	18.0	18.0	18.0	18.0
Dextrose (lb)	3.8	3.6	3.5	3.5	3.5	3.5	3.5	3.5	3.5	3.5
Total low-caloric sweeteners (lb)	6.9	7.3	7.7	8.2	9.4	13.0	15.8	18.0	18.5	19.0
Saccharin (lb)	6.9	7.3	7.7	8.0	8.4	9.5	10.0	6.0	5.5	5.5
Aspartame (lb)	0.0	0.0	0.0	0.2	1.0	3.5	5.8	12.0	13.0	13.5

Source: Ref. 80.

TABLE 38 1985 Penetration of HFCS in U.S. Sweetener Markets

Application	Sweetener consumption (1000 mt, dry)			HFCS market share (%)	Max. HFCS market share (%)	Max. possible HFCS demand (mt)
	Total	Sugar	HFCS			
Beverages	3,360	307	3,053	91	90–100	3,360
Baking	1,818	1,360	458	25	25	455
Canning	1,018	383	635	62	60–70	710
Dairy	595	400	195	33	35	208
Confectionary	985	962	23	2	5	50
Other foods	469	397	72	15	40	188
Other nonfoods	182	—	182	—	—	550
Total	8,323	3,809	4,514	54	66	5,521

Source: Data compiled by USDA and McKeany Flavell Company (22).

TABLE 39 Composition and Properties of Corncobs

	Corncobs	Grit fraction	Pith/chaff fraction	Mixed fraction
Component parts (%)				
Woody ring	60	100	0	75
Coarse chaff	34	0	85	21
Fine chaff	4	0	10	3
Pith	2	0	5	1
Constituents (% as is basis)				
Moisture	9.6	7	6	8
Cellulose	41	47	36	44
Hemicellulose	36	37	37	37
Xylan	30	32	30	31
Lignin	6	7	5	6
Pectins	3	3	3	3
Starch	0.014	0.014	0.015	0.014
Nutritional values				
Protein, N × 6.25 (%)	2.5	1.4	4.4	2.0
Fat, ether ext. (%)	0.5	0.2	0.9	0.4
Crude fiber (%)	32	37	29	36
Ash (%)	1.5	1.2	1.6	1.3
Nitrogen-free extract (%)	53.5	53	58	52.5
Neutral detergent fiber (%)	83	92	78	87
Total digestible nutrients (beef) (%)	42	44	43	43
Metabolizable energy (beef) (kcal/kg)	1,537	1,604	1,536	1,573
Industrial values				
Constituent sugars (%)				
Xylose	22	24	21.5	23
Arabinose	3.1	2.2	4.7	2.7
Glucose	28	32	24	30
Galactose	1	0.5	1.8	0.8
Mannose	0.6	0.6	0.6	0.6
Hexuronic acid	7	7	7	7
Furfural potential (%)	20	21	20	20
Acetic acid yield (%)	19	19	20	19
Gross energy (kcal/kg)	4,000	4,110	4,160	4,070
(BTU/lb)	7,200	7,400	7,480	7,320
Ignition temp. (°C)	205	205	205	205
Bulk density (lbs/ft^3)	15.5	28	8.5	14
Specific gravity	1.4	1.3	1.6	1.4
Water absorption	369	133	727	279
Oil absorption	259	100	500	198
Mohs' hardness	—	4.5	—	—

Source: Ref. 5.

REFERENCES

1. Alexander, R. J., Industrial uses of dry-milled corn products, in *Industrial Uses of Cereals*, Y. Pomeranz (Ed.), American Association of Cereal Chemists, Inc., St. Paul, MN, 1973.
2. Alexander, R. J., Corn dry milling. In *Corn: Chemistry and Technology*, S. A. Watson and P. E. Ramstad (Eds.), American Association of Cereal Chemists, Inc., St. Paul, MN, 1987.
3. Alexander, D. E., and Creech, R. G., Breeding special industrial and nutritional types. In *Corn and Corn Improvement*, G. F. Sprague (Ed.), American Society of Agronomy, Inc. Madison, WI, 1977.
4. Anderson, R. A., Wet-Milling characteristics of class 8 amylomaize, *Cereal Chem., 42*:580 (1965).
5. Anderson, R. A., and Watson, S. A., The corn milling industry. In *CRC Handbook Series in Agriculture Processing and Utilization*, Vol 2, I. A. Wolf (Ed.), CRC Press, Inc., Boca Raton, FL, 1982.
6. AOCS, *Official and Tentative Methods of the AOCS*, American Oil Chemists Society, Champaign, IL, 1962.
7. Anonymous, Zein: a natural moisture barrier, *Food Engineering* (May): 104 (1978).
8. Anonymous, *The Almanac of Canning, Freezing, Preserving Industries*, 72nd edition, Edward E. Judge & Sons, Inc., Westminster, MD, 1987.
9. ASAE, *Agricultural Engineers Yearbook*, American Society of Agricultural Engineers, St. Joseph, MI, 1987.
10. Bedolla, S., and Rooney, L. W., Cooking maize for masa production, *Cereal Foods World, 27*:219 (1982).
11. Bier, T. H., *Seminar notes: corn wet milling*, Dorr-Oliver, Inc., Stamford, CT, 1978.
12. Brekke, O. L., Corn dry milling industry. In *Corn: Culture, Processing, and Products*, G. E. Inglett (Ed.), AVI Publishing Co., Westport, CT, 1970.
13. Brooker, D. H., Bakker-Arkema, F. W., and Hall, C. W., *Drying Cereal Grains*, AVI Publishing Co., Westport, CT, 1974.
14. Brown, W. L., Zuber, M. S., Darrah, L. L., and Glover, D. V., Origin, adaptation and types of corn, *National Corn Handbook*, Cooperative Extension Service, Iowa State University, Ames, IA, 1985.
15. Chung, D. S., and Pfost, H. B., Adsorption and desorption of water vapor by cereal grains and their products. Part I. Heat and free energy changes of adsorption and desorption, *Trans ASAE,10*:549 (1967).
16. Chung, D. S., and Pfost, H. B., Absorption and desorption of water vapor by cereal grains and their products. Part II. Development of the general isotherm equation, *Trans. ASAE,10*:552 (1967).
17. Corn Refiners Association, Inc., *Critical Data Tables*, Washington,DC, 1975.
18. Corn Refiners Association, Inc., *Corn Wet-Milled Feed Products*, 2nd ed., Washington, DC, 1982.
19. Corn Refiners Association, Inc., *Corn Oil*, Washington, DC, 1986.
20. Corn Refiners Association, Inc., *Corn Starch*, Washington, DC, 1986.
21. Corn Refiners Association, Inc., *Nutritive Sweeteners from Corn*, Washington, DC, 1986.
22. Corn Refiners Association, Inc., *Corn Annual*, Washington, DC, 1987.
23. Croston, C. B., and Evans, C. D., The industrial uses of corn proteins. Crops in Peace and War, *The 1950–51 Yearbook of Agriculture*, U.S. Department of Agriculture, Washington, DC., 1951.
24. Dimler, R. J., Kernel structure and wet milling of high-lysine corn, *Am. Miller Processor, 94*:7 (1966).
25. Distillers Feed Research Council, *Distillers Feed Research*, Cincinnati, OH, 1982.
26. Dunlop, A. P., The furfural industry. In *Industrial Uses of Cereals*, Y. Pomeranz (Ed.), American Association of Cereal Chemists, Inc., St. Paul, MN, 1973.
27. Easter, W. E., The dry corn milling industry, *Assoc. Oper. Mill. Bull.*, 3112 (1969).
28. Earle, F. E., Curtis, J. J., and Hubbard, J. E., Composition of the component parts of the corn kernel, *Cereal Chem., 23*:504 (1946).

29. Felche, G., Chemical modification and degradation of starch. In *Starch Conversion Technology*, G. M. van Beynum and J. A. Roels (Eds.), Marcel Dekker, Inc., New York, 1985.
30. Foley, K. M., and Vander Hooven, I. B., Properties and industrial uses for corncobs. In *Cereals: A Renewable Resource*, Y. Pomeranz and L. Munck (Eds.), American Association of Cereal Chemists, Inc., St. Paul, MN, 1981.
31. FGIS, *The Official United States Standards for Grain*, Federal Grain Inspection Service, U.S. Department of Agriculture, Washington, DC, 1984.
32. FGIS, Revision to the U.S. Grading Standards for corn, U.S. standards for sorghum, U.S. standards for soybeans, *Fed. Regist., 49*:35743 (1984).
33. Gomez, M. H., Rooney, L. W., Waniska, R. D., and Pflugfelder, R. L., Dry corn masa flours for tortilla and snack food production, *Cereal Foods World, 32*:372 (1987).
34. Hanover, L. M., Functionality of corn-derived sweeteners in formulated foods. In *Chemistry of Foods and Beverages: Recent Developments*, G. Charalambous and G. Inglett (Eds.), Academic Press, New York, 1982.
35. Hardeman, N. P., *Shucks, Shocks, and Hominy Blocks*, Louisiana State University Press, Baton Rouge, LA, 1981.
36. Haugh, C. G., Lien, R. M., Hanes, R. E., and Ashman, A. H., Physical properties of popcorn, *Trans. ASAE,19*:168 (1976).
37. Herum, F. L., Harvesting and postharvest management. In *Corn: Chemistry and Technology*, S. A. Watson and P. E. Ramstad (Eds.), American Association of Cereal Chemists, Inc., St. Paul, MN, 1987.
38. Hixon, D. L., Hatfield, E. E., and Lamb, P. E., Whole shelled corn versus cracked corn in cattle finishing diets, University of Illinois 1969 Beef Day Report, 1969.
39. Hooper, F. E., Chemical examination of corn bran, *Ind. Eng. Chem., 34*:728 (1942).
40. Horn, H. E., Corn sweeteners: functional properties, *Cereal Foods World, 26*:219 (1981).
41. Hoseney, R. C., *Principles of Cereal Science and Technology*, American Association of Cereal Chemists, Inc., St. Paul, MN, 1986.
42. Hoseney, R. C., Zeleznak, K., and Abdelrahman, A., Mechanism of popcorn popping, *J. Cereal Sci., 1*:43 (1983).
43. Huelsen, W. A., *Sweet Corn*, Interscience Publishers, Inc., New York, 1954.
44. Iowa Corn Promotion Board, *Corn: Its Your Business*, Des Moines, IA, 1986.
45. Kazarian, E. A., and Hall, C. W., Thermal properties of grains, *Trans. ASAE, 8*:33 (1965).
46. Keener, H. M., Henery, J. E., Schonauer, S. L., and Anderson, R. J., Burning shelled corn as an alternative fuel, *Ohio Report* (July-August) (1985).
47. Lloyd, N. E., and Nelson, W. J., Glucose-fructose containing sweeteners from starch. In *Starch: Chemistry and Technology*, R. L. Whistler, J. N. Bemiller, and E. F. Paschall (Eds.), Academic Press, Inc., New York, 1984.
48. Medcalf, D. G., Structure and composition of cereal components as related to their potential industrial utilization: starch. In *Industrial Uses of Cereals*, Y. Pomeranz (Ed.), American Association of Cereal Chemists, Inc., St. Paul, MN, 1973.
49. Moore, C. O., Tuschhoff, J. V., Hastings, C. W., and Schanefelt, R. V., Applications of starches in foods. In *Starch: Chemistry and Technology*, 2nd ed., R. L. Whistler, J. N. Bemiller, and E. F. Paschall (Eds.), Academic Press, Inc., New York, 1984.
50. National Academy of Sciences, *United States-Canadian Tables of Feed Composition*, 2nd ed., Washington, DC, 1969.
51. National Academy of Sciences, *United States-Canadian Tables of Feed Composition*, 3rd ed., National Academy Press, Washington, DC, 1982.
52. Nelson, S. I., Moisture-dependent kernel- and bulk density for wheat and corn, *Trans. ASAE, 23*:139 (1980).
53. Orthoefer, F. T., Corn starch modification and uses. In *Corn: Chemistry and Technology*, S. A. Watson and P. E. Ramstad (Eds.), American Association of Cereal Chemists, Inc., St. Paul, MN, 1987.
54. Pancoast, H. M., and Junk, W. R., *Handbook of Sugars*, AVI Publishing Co., Inc., Westport, CT, 1980.

55. Paulis, J. W., and Wall, J. S., Albumins and globulins in extracts of corn grain parts, *Cereal Chem., 46*:263 (1969).
56. Paulis, J. W., and Wall, J. S., Comparison of the protein composition of selected corns and their wild relatives, Teosinte and Trypsacum, *J. Agric. Food Chem., 25*:265 (1977).
57. Paulis, J. W., James, C., and Wall, J. S., Comparison of glutelin proteins in normal and high-lysine corn endosperm, *J. Agr. Food Chem., 17*:1301 (1969).
58. Pomes, A. F., Zein, *Encycl. Polym. Sci. Technol., 15*:125 (1971).
59. Rathmann, D. M., *Zein: An Annotated Bibliography 1891–1953*, Mellon Institute Biliographic Series Bulletin, No. 7, Mellon Institute, Pittsburgh, PA, 1954.
60. *Reiners, R. A.*, Seminar Proceedings: Products of the Wet Corn Milling Industry in Food, *1978*.
61. Reiners, R. A., and Gooding, C. M., Corn Oil. In *Corn: Culture, Processing, and Products*, edited by G. E. Inglett (Ed.), AVI Publishing Co., Westport, CT, 1970.
62. Reiners, R. A., Wall, J. S., and Inglett, G. E., Corn proteins: potential for their industrial use. In *Industrial Uses of Cereals*, Y. Pomeranz (Ed.), American Association of Cereal Chemists, Inc., St. Paul, MN, 1973.
63. Rooney, L. W., and Serna-Saldivar, S. O., Food uses of whole corn and dry-milled fractions. In *Corn: Chemistry and Technology*, S. A. Watson and P. E. Ramstad (Eds.), American Association of Cereal Chemists, Inc., St. Paul, MN, 1986.
64. Russell, C. R., Industrial use of corn starch. In *Industrial Uses of Cereals*, Y. Pomeranz (Ed.), American Association of Cereal Chemists, Inc., St. Paul, MN, 1973.
65. Schanefelt, R. V., Corn sweeteners: products and properties, *Cereal Foods World, 22*:44 (1977).
66. Shreve, R. N., and Brink, Jr., J. A., *Chemical Process Industries*, McGraw-Hill Book Co., New York, 1984.
67. Shroder, J. D., and Heiman, V., Feed Products from corn processing. In *Corn: Culture, Processing, and Products*, G. E. Inglett (Ed.), AVI Publishing Co., Inc., Westport, CT, 1970.
68. Sims, R. L., The technology of wet corn milling. In *Starch Conversion Technology*, G. M. van Beynum and J. A. Roels (Eds.), Marcel Dekker, Inc., New York, 1985.
69. Spencer, T., Current status and future changes in the Mexican food industry. Presented at the second annual Mexican Food Short Course, Texas A&M University, College Station, TX, Nov. 19–21, 1986.
70. Stiver, T. E., American corn milling systems for de-germed products, *Assoc. Oper. Mill. Bull.*, 2168 (1955).
71. Swinkels, J. J., Sources of starch, its chemistry and physics, In *Starch Conversion Technology*, G. M. van Beynum and J. A. Roels (Eds.), Marcel Dekker, Inc., New York, 1985.
72. Tawfik, M. E., Association among biochemical constituents of corn, *Agr. Res. Rev., 54*:75 (1976).
73. Tello, F., Alvarez-Tostado, M. A., and Alvarado, G., A study on the improvement of the essential amino acid balance of corn protein, *Cereal Chem., 42*:368 (1965).
74. Thompson, R. A., and Isaacs, G. W., Porosity determination in grains and seeds with an air-compression pycnometer, *Trans. ASAE 10*:693 (1967).
75. USDA, *Pop corn*, Farmers' Bulletin No. 1679, U.S. Department of Agriculture, Washington, DC, 1933.
76. USDA, *Feed Situation and Outlook Report*, Economic Research Service, U.S. Department of Agriculture, Washington, DC, September, 1987.
77. USDA, *Crop Production 1987 Summary*, National Agricultural Statistics Service, U.S. Department of Agriculture, Washington, DC, January, 1988.
78. USDA, *World Grain Situation and Outlook*, Foreign Agricultural Service, U.S. Department of Agriculture, Washington, DC, January, 1988.
79. USDA, *Feed-Situation and Outlook Report*, Economic Research Service, U.S. Department of Agriculture, Washington, DC, May, 1988.
80. USDA, *Sugar and Sweetener-Situation and Outlook Yearbook*, Economic Research Service, U.S. Department of Agriculture, Washington, DC, June, 1988.

81. USDC, *Statistical Abstract of the United States 1987*, 107th ed., Bureau of Census, U.S. Department of Commerce, Washington, DC, 1987.
82. USDC, *Statistical Abstract of the United States 1988*, 108th ed., Bureau of Census, U.S. Department of Commerce, Washington, DC, 1988.
83. van Beynum, G. M., and Roels, J. A., The multidimensionality of starch conversion technology. In *Starch Conversion Technology*, G. M. van Beynum and J. A. Roels (Eds.), Marcel Dekker, Inc. New York, 1985.
84. van Tilburg, R., Enzymatic isomerization of starch-based glucose syrups. In *Starch Conversion Technology*, G. M. van Beynum and J. A. Roels (Eds.), Marcel Dekker, Inc., New York, 1985.
85. Watson, S. A., Industrial utilization of corn. In *Corn and Corn Improvement*, G. F. Spraque (Ed.), American Society of Agronomy, Inc., Madison, WI, 1977.
86. Watson, S. A., Corn and sorghum starches: production. In *Starch: Chemistry and Technology*, 2nd ed., R. L. Whistler, J. N. Bemiller, and E. F. Paschall (Eds.), Academic Press, Inc., New York, 1984.
87. Watson, S. A., Coproducts overview. Corn Refiners'' 1986 Scientific Conference, Corn Refiners'' Association, Washington, DC, 1986.
88. Watson, S. A., Structure and composition. In *Corn: Chemistry and Technology*, S. A. Watson and P. E. Ramstad (Eds.), American Association of Cereal Chemists, Inc., St. Paul, MN, 1987.
89. Watson, S. A., and Yahl, K. R., Comparison of the wet milling properties of opaque-2 high lysine corn and normal corn, *Cereal Chem., 44*:488 (1967).
90. Weber, E. J., Structure and composition of cereal components as related to their potential industrial utilization. IV. Lipids. In *Industrial Uses of Cereals*, Y. Pomeranz (Ed.), American Association of Cereal Chemists, Inc., St. Paul, MN, 1973.
91. Wells, G. H., The dry side of corn milling, *Cereal Foods World, 24*:443 (1979).
92. Whistler, R. L., History and future expectations of starch use. In *Starch: Chemistry and Technology*, 2nd ed., R. L. Whistler, J. N. Bemiller, and E. F. Paschall (Eds.), Academic Press, Inc., New York, NY, 1984.
93. Whistler, R. L., Bemiller, J. N., and Paschall, E. F., *Starch: Chemistry and Technology*, Academic Press, Inc, New York, 1984.
94. Woods, C. D., *Food Value of Corn and Corn Products*, Farmers' Bulletin No. 298, U.S. Department of Agriculture, Washington, DC, 1907.
95. Wurzburg, O. B., *Modified Starches: Properties and Uses*, CRC Press, Inc., Boca Raton, FL, 1986.
96. Wright, K. N., Nutritional properties and feeding value of corn and its by-products. In *Corn: Chemistry and Technology*, S. A. Watson and P. E. Ramstad (Eds.), American Association of Cereal Chemists, Inc., St. Paul, MN, 1986.
97. Zuber, M. S., and Darrah, L. L., Breeding, genetics, and seed corn production. In *Corn: Chemistry and Technology*, S. A. Watson and P. E. Ramstad (Eds.), American Association of Cereal Chemists, Inc., St. Paul, MN, 1987.

3

Barley

Eugene A. Hockett
U.S. Department of Agriculture and Montana State University, Bozeman, Montana

I. HISTORY

A. Origin

Barley has been associated with the earliest beginnings of agriculture. Reviews of the origin of barley have been made by several authors (14,15,28,59,82,89,90). The most likely origin of cultivated barley is from a wild progenitor growing in the fertile crescent 35,000–40,000 years ago (59). Probably this progenitor was a two-rowed type from which evolved the present wild-type weed barley *Hordeum vulgare* sp. *spontaneum* and the *Hordeum vulgare* cultivated two– and six-rowed types (28). The present distribution of the *H. vulgare* sp. *spontaneum* is shown in Figure 1, with additional sites found in northwestern India and Morocco. The archeological sites containing the oldest samples of barley are 9,000–10,000 years old (90) in the Near East (Fig. 2). From this area barley culture spread to Europe, west Asia, and the Nile valley (Fig. 3) (90).

Recorded history began on Sumerian clay tablets (28), which described and recommended cultural practices for raising barley ca. 1700 B.C. Barley was an important crop throughout the Neolithic period and the Bronze Age (90) and continues to be an important worldwide crop up to the present time. Barley has been associated throughout its history with both the production of beer and as a food and feed crop.

B. Distribution

Barley was first grown in North America in 1492 or 1494 by a colony founded by Columbus (86) (see Fig. 4). The Spaniards introduced barley into Mexico during the conquest, and from there it was introduced by missionaries into California. The second point of introduction was in 1602 on the south coast of Massachusetts. Barley spread from these points of introduction westward and northward as shown in Figure 4. By 1958 the crop was widely grown in North America (Fig. 5). Since 1958 the distribution of barley has changed somewhat to the present areas (Table 1). Winter barley is grown mainly in the southern United States except for some acreage in the eastern and northwestern United States. Spring barley makes up the bulk of the barley acreage in both the United States and Canada, with the leading states being North Dakota, Idaho, Minnesota, Montana and Washington.

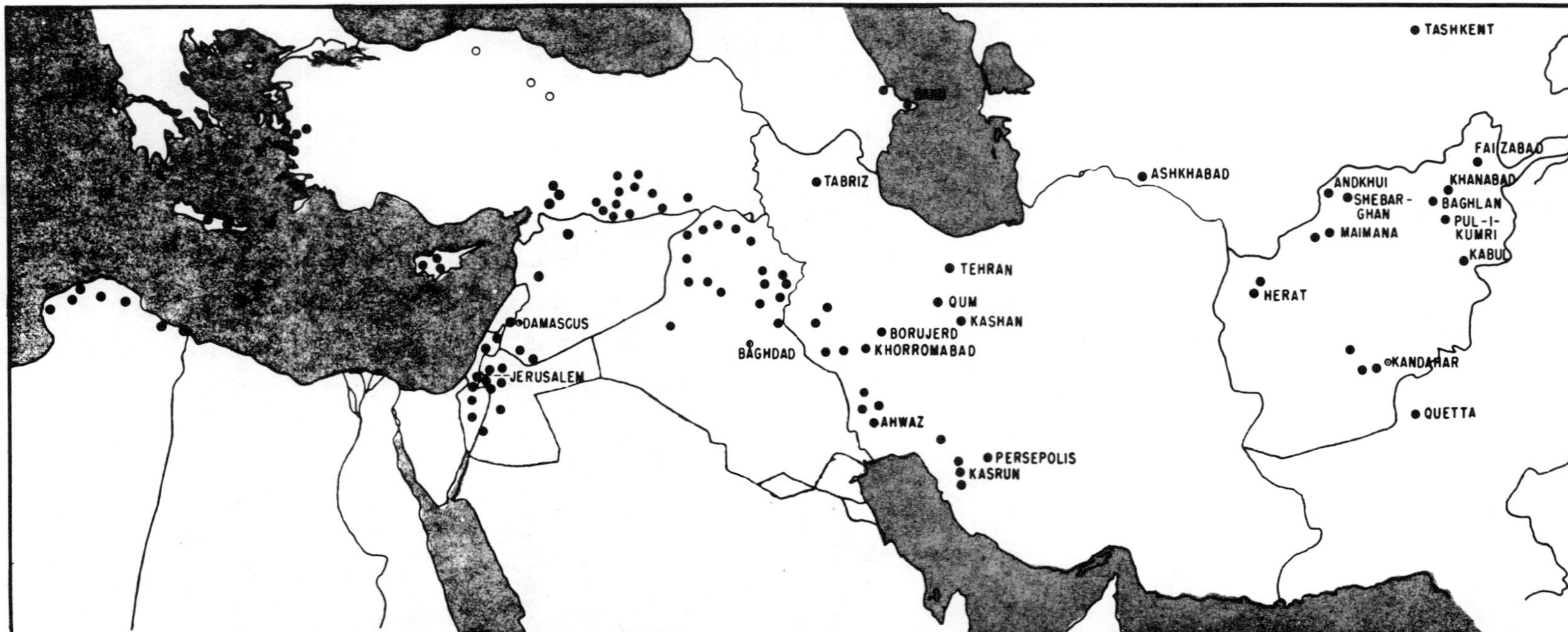

FIGURE 1 Distribution of *Hordeum spontaneum*. Each dot represents a collection site that has been verified by J. R. Harlan. (From Ref. 28.)

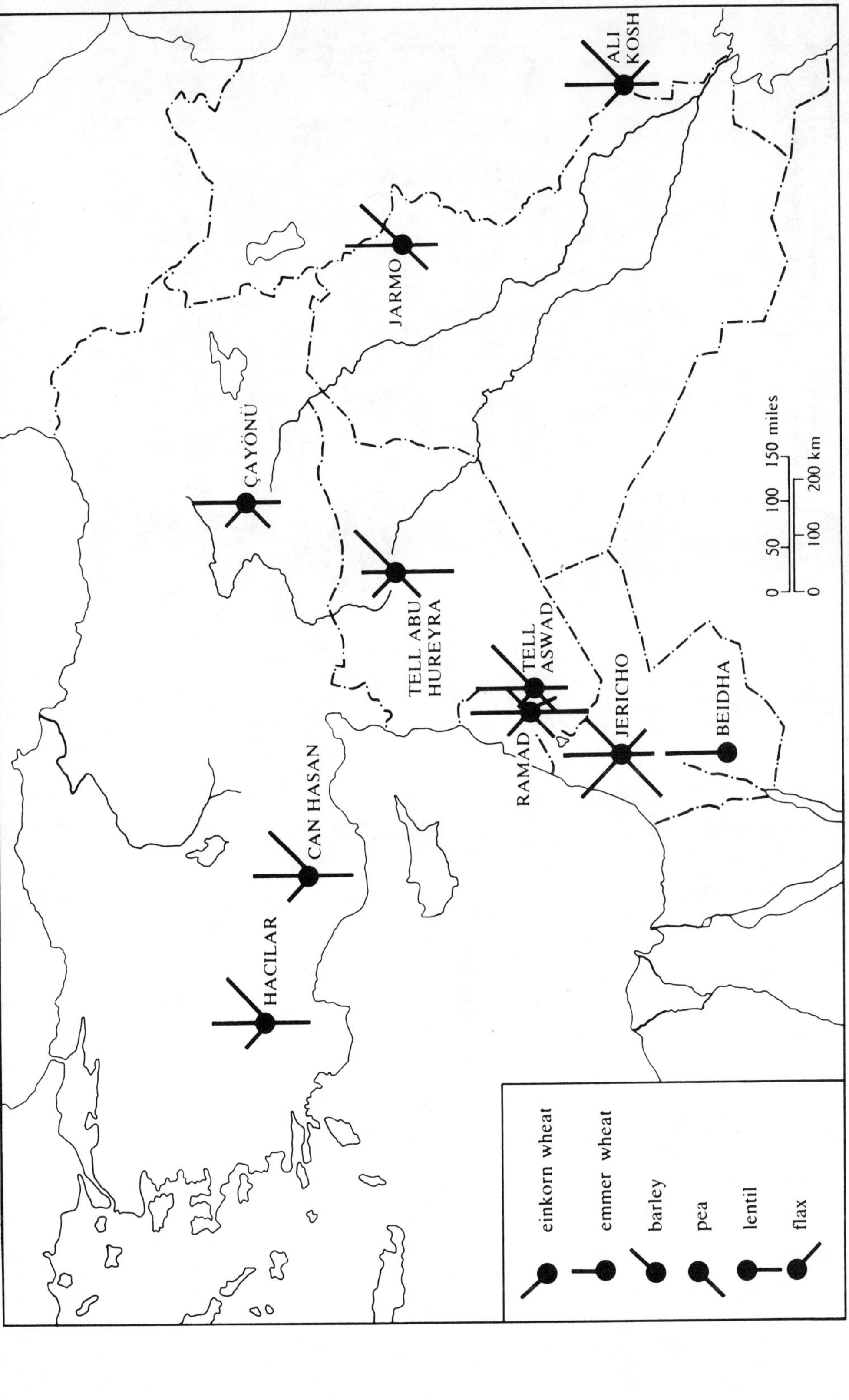

FIGURE 2 Crop assemblage in the early Neolithic farming villages in the Near East (before 6000 B.C.) A short whisker indicates that the crop is relatively rare, a long whisker that it is relatively common among the excavated plant remains. (From Ref. 90.)

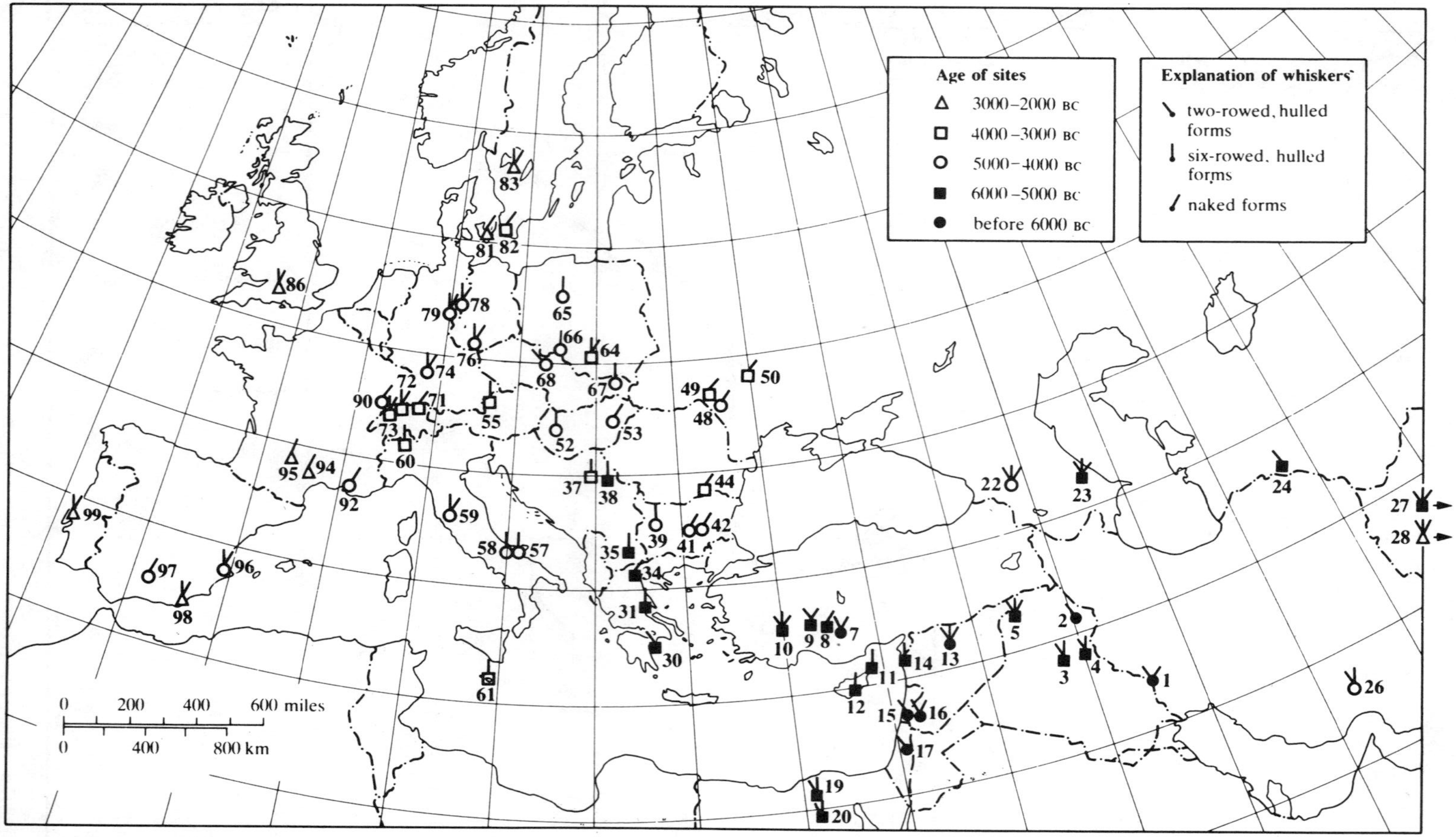

FIGURE 3 The spread of barley to Europe, western Asia, and the Nile valley. (For details of numbered sites, see Ref. 90.)

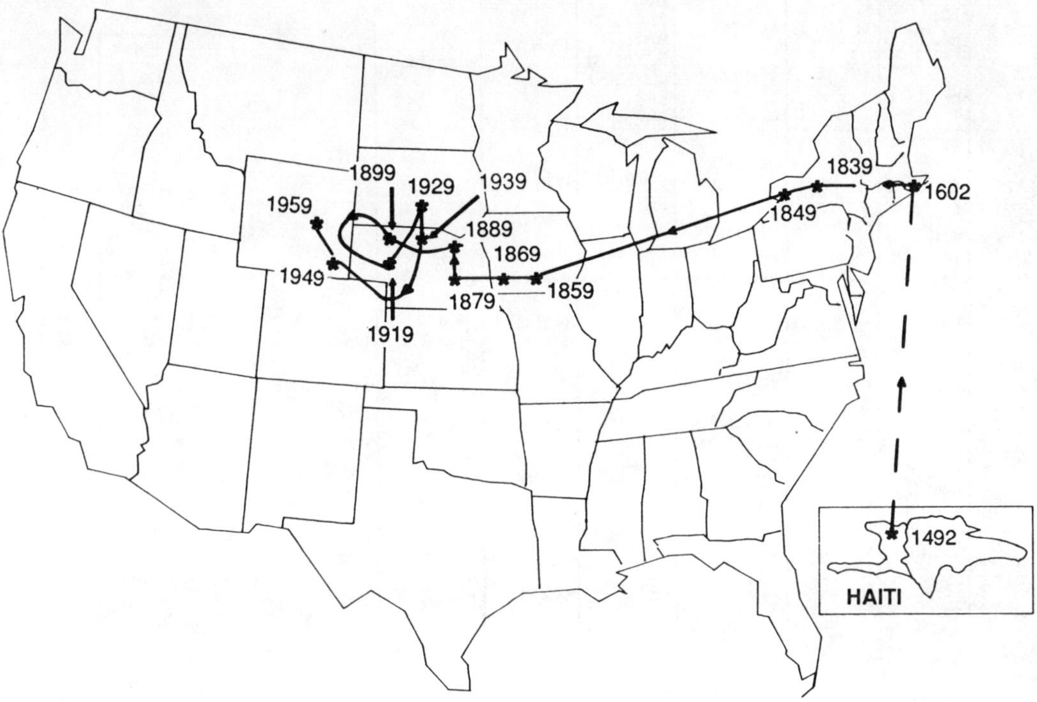

FIGURE 4 National centers of barley production, 1839–1959, and two historical sites, 1492 and 1602. (From Ref. 86.)

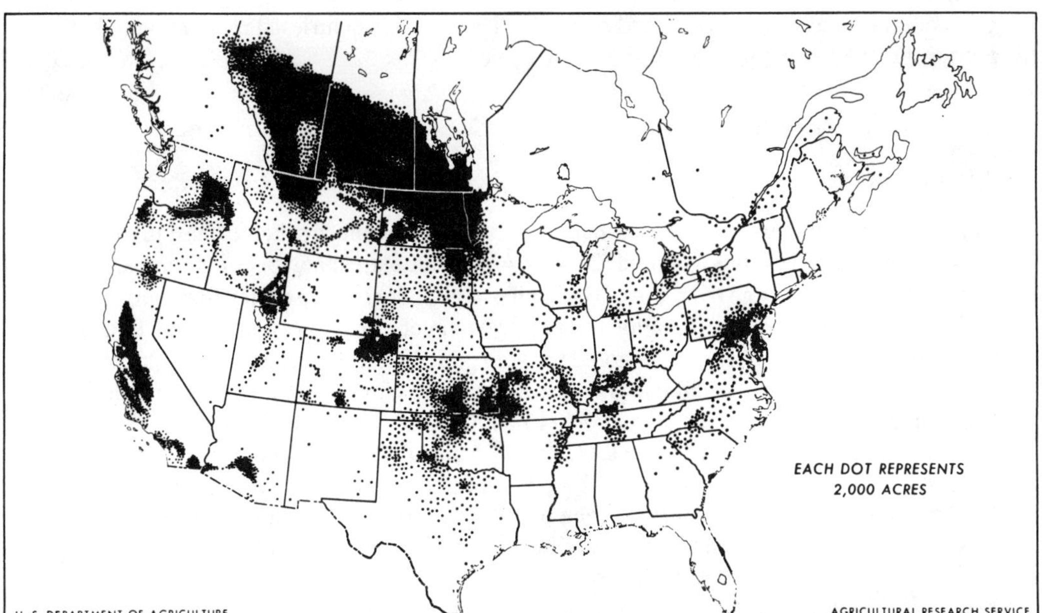

FIGURE 5 Barley-growing areas in the United States and Canada, 1955–1958. (From Ref. 87.)

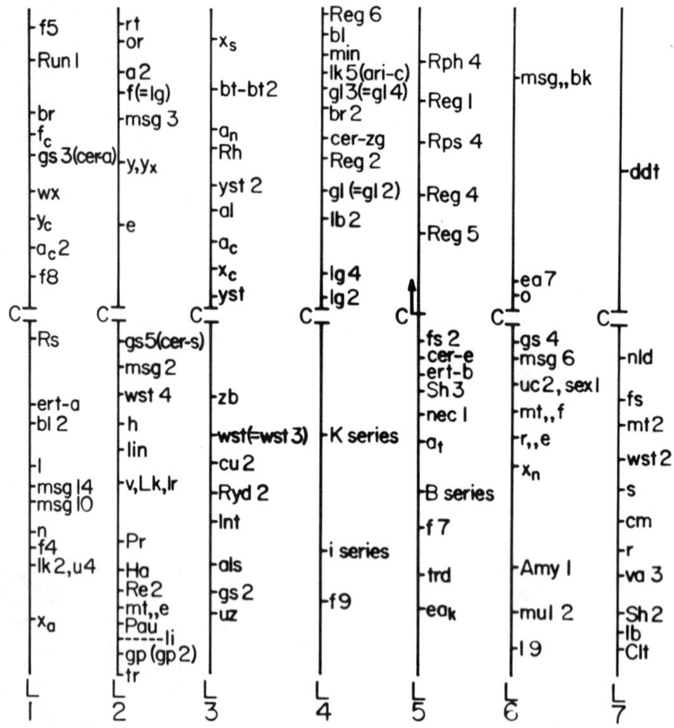

FIGURE 6 Linkage maps of barley. C = centromere, L = long arm, 1 through 7 = chromosome number. Refer to Table 3 for gene list. (From Ref. 78.)

Barley has always been an important crop worldwide, and at this time is fourth in cereal production behind wheat, maize, and rice (54). The leading countries in production of barley for the 1983–1985 period are shown in Table 2.

II. GENETICS AND BREEDING

A. Inheritance and Heritability

Barley is a crop with worldwide distribution, and it is a preeminent plant for use in experimental genetic studies. It is a diploid (2n = 2x = 14) with a small number of chromosomes, which are relatively large (6–8 μm). Barley has a high degree of self-fertilization, but is easily hybridized. Several generations can be grown in one year over a wide array of environmental conditions. Seed can be stored for long periods and numerous genetic stocks are available for physiological and morphological characteristics. Reviews of the genetics of barley have been made by several authors (15,34,52,70,80,85). Lists of the genetic stocks have been published by Sogaard et al. (71) and Tsuchiya (79), with over 900 genes listed.

Figure 6 is a current linkage map of the genes located on the seven chromosomes of barley. A list describing these genes is given in Table 3. Special collections are available of some genes, such as the genetic male sterile collection with 48 separate loci and approximately 450 stocks available. Heritability of agronomic and quality traits in barley have been extensively studied (Table 4). The genetics of breeding for disease and pest resistance have been reviewed by Sharp (67).

B. Biotechnology

The barley nuclear genome contains nearly 12 billion base pairs of DNA among its seven pairs of chromosomes. In spite of its complexity, a series of molecular markers have been identified (Fig. 7) which span half its recombinational distance (8,42,68). Linkage maps consisting largely of DNA markers are being constructed in the United Kingdom (21), Germany (30), the United States, and Canada. The application of these maps ranges from identifying the location of genes which modify characters including disease resistance, grain protein content, enzymatic characters, and extract, to providing excellent markers for varietal identification.

The barley chloroplast genome is small (@134,000 base pairs of DNA), and its structure was first elucidated at Carlsberg Laboratory (56). The undetermined structure of the mitochondrial genome is much larger than the chloroplast, and it exhibits more variability than the chloroplast among genotypes of barley (36).

Transformation, the direct insertion of specific cloned genes, is now being attempted in barley. It has been reported to be successful in rye, soybeans, and maize. Published examples of stable transformants in barley are not yet available (10).

The molecular tools currently being developed have great promise to improve the efficiency with which we manipulate quality and other production-related quantitative characters and move the genes controlling their expression into varieties with better local or regional adaptation. Transformation has the potential to modify the resistance of barley to diseases such as Barley Yellow Dwarf Virus, to produce industrially proteins, and to modify the nutritional quality of the grain (10). Barley can be regenerated into fertile plants from various cells and tissues, but regeneration from protoplasts is not yet possible. These techniques would allow the use of embryo rescue, double haploid production, creating new variation, providing new cytoplasmic or chromosomal combinations, selection of mutants, in vitro fertilization, and attempting transformation (39).

C. Cytogenetics

The cytogenetics of barley has been reviewed recently by Ramage (58) and Tsuchiya (77). Collections of inversions and translocations are maintained in barley by international coordinators. Deficiencies and duplications have also been studied in barley (58).

Aneuploids have been extensively used in barley (58,77) to associate genes with chromosomes (primary trisomics); produce hybrid barley (tertiary trisomics); associate genes with chromosome arms and localize the centromere on the linkage maps (telotrisomics); locate genes in a linkage group (acrotrisomics); and establish eight-paired barley (partial tetrasomics). Euploids are found in barley (58) and have been used to produce homozygous lines and study segregation ratios and linkage values in gametes produced by F_1 plants (haploids); produce aneuploids (triploids); and attempt to produce commercial, $2n = 4x = 28$, chromosome barleys (autotetraploids). Individual pairs of barley chromosomes have been added to the chromosome complement of wheat (58) and used to make genetic and evolutionary studies of barley. Figure 8 shows a micrograph of barley chromosomes.

D. Germplasm Resources

1. Collections

About 25,000 barley landraces plus 25,000 breeder lines and cultivars are preserved in collections of barley throughout the world (20). In addition, the wild *H. vulgare* sp. *spontaneum* and *bulbosum* have about 3000 and 600 ascensions, respectively (13,20). The location and addresses of the major base germplasm collections are shown in Table 5. Working

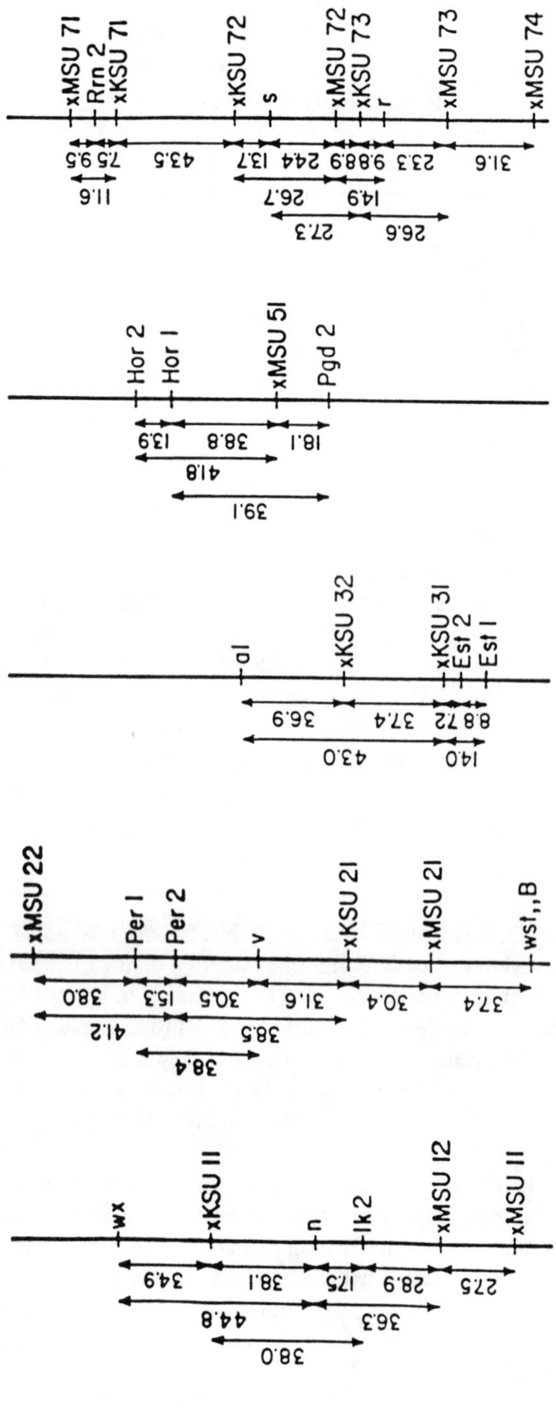

FIGURE 7 Linkage maps of barley showing molecular markers; 1, 2, 3, 5, and 7 represent chromosome number. (From Ref. 68.)

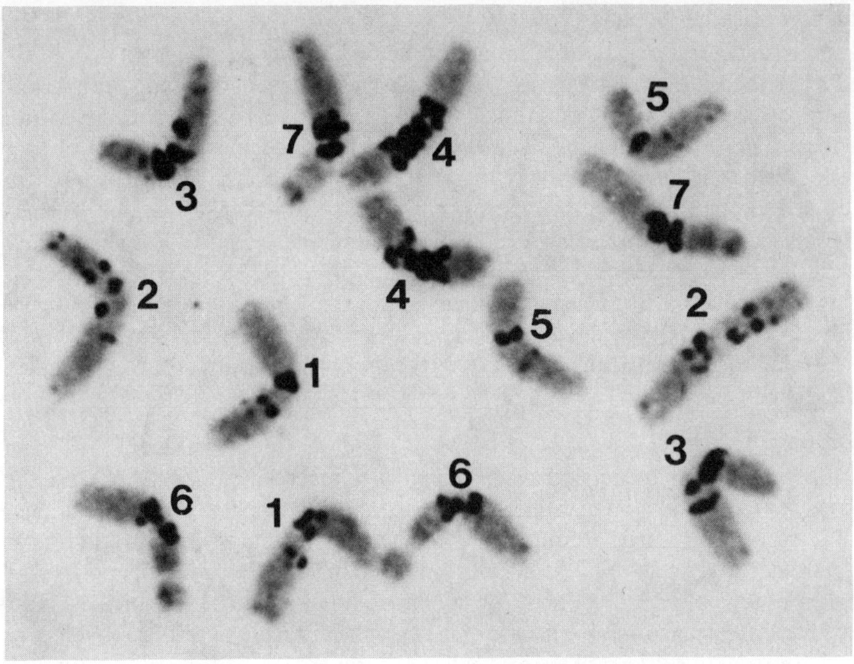

FIGURE 8 Giemsa-banded somatic chromosomes of barley (x3000). (From Ref. 69.)

gemplasm collections are found in Brazil, Bulgaria, Czechoslovakia, England, the Federal Republic of Germany, the German Democratic Republic, Syria, The Netherlands, the United States, and the U.S.S.R. (48). Many composite crosses of barley are maintained in the United States with CCXLIII being the last one assigned a number by the USDA-ARS collection (48).

2. Wide Crosses

Reviews describing the wide crosses made in barley are given by von Bothmer (13,14) and Fedak (24,25). The origin, taxonomy, and related species of barley are described (14), as are the incompatibility, mechanisms, and cytogenetics of wild barley crosses (24). There is a general lack of knowledge of the agronomic potential of wild species of *Hordeum* (13), but Table 6 shows some uses that have been made of wide crosses to produce desirable agronomic traits. Intergeneric crosses of barley and *Triticum*, *Secale*, *Elymus*, *Sitanion*, *Hystrix*, *Agropyron*, *Horedelymus*, and *Psathyrostachys* species have been made (24,25).

E. Breeding

Reviews of barley breeding dating from 1936 to 1987 are given by several authors (6,15,27,59,67,84). Barley is the most widely grown cereal crop, and with this diverse mix of environmental conditions the breeding objectives become quite diverse. The major objectives of barley-breeding programs in the United States are (priority depends on the location of the breeding program): yield, quality (malting and feed), disease resistance, drought resistance, winterhardiness, straw strength, maturity, insect resistance, plant height, seed size, salinity tolerance, and shattering resistance.

1. Population-Breeding Methods

Barley was the first cereal crop used in an innovative way for the development of population-breeding methods. H. V. Harlan was the first to advocate the mixing and later hybridization of cultivars as composite crosses to be grown on a long-term basis (59). Suneson and Wiebe continued the development of the theory and release of composite crosses utilizing genetic male sterility and the concept of evolutionary plant breeding. Composite crosses were developed for many different specific uses by other scientists (59). Recurrent selection has been used in many crops, but a refinement was suggested by R. F. Eslick called "male sterile facilitated recurrent selection" (MSFRS) (59). MSFRS has been used by Ramage, Eslick, Hockett, D. R. Clark, and others to develop populations for specific purposes. A disadvantage of the population method of breeding can be an inability to devise a suitable screening method on a single plant basis for the characteristics desired (examples are yield and quality).

2. Hybrid Barley

Methods of hybrid barley production have been reviewed recently (6,31,57,59,66). Methods proposed for the production of hybrid barley are use of synthetic hybrid populations, DDT resistance linked to male sterility, balanced tertiary trisomics, preflowering gene linked to male sterility, haplo-viables, chemical restoration of fertility, photoperiod sensitive male steriles, cytoplasmic male sterility, apomictic lines, and male gameticides. The most promising method appears to be the use of the cytoplasmic male sterility discovered by Åhokas (2). Apparently sufficient heterosis exists in barley for it to be commercially successful (29,57) if a suitable method can be found. Other problems in hybrid barley production are sufficient female receptivity and male pollination ability. Composite crosses have been developed which greatly enhance outcrossing potential (32), but these genotypes need to be integrated with appropriate parental stocks.

III. STRUCTURE

The structure of barley is of interest from both practical and theoretical aspects and has been reviewed by several authors (15,60,62,87). The taxonomic value of the morphological characteristics of barley was studied by Åberg and Wiebe (1). Ward (81) considered the evolutionary behavior of morphological markers. He determined their frequency of occurrence and geographical distribution, and how these markers are associated with each other.

A. Plant

A covered barley kernel is shown germinating in Figure 9 at from 3 to 7 days (depending on the cultivar) after conditions are favorable for germination. Figure 10 shows a later stage in plant development, when tillering occurs. Barley has large overlapping auricles (Fig. 11) compared to wheat (smaller nonoverlapping auricle) and oats (rudimentary auricles), which are readily distinguished at the seedling stage. The beginning of root development is also shown in Figure 10. Figure 12 shows root development in different types of barley depending on whether the area from which they come are dry (xerophytic) or wet (mesophytic). The apex of a barley root is shown in Figure 13.

B. Spike

The most common types of barley spikes are shown in Figure 14. The spike is made up of spikelets with three florets attached to each of the nodes of the rachis. Two-rowed barleys have on one fertile floret, whereas six-rowed barleys have all three fertile. Samples of two-

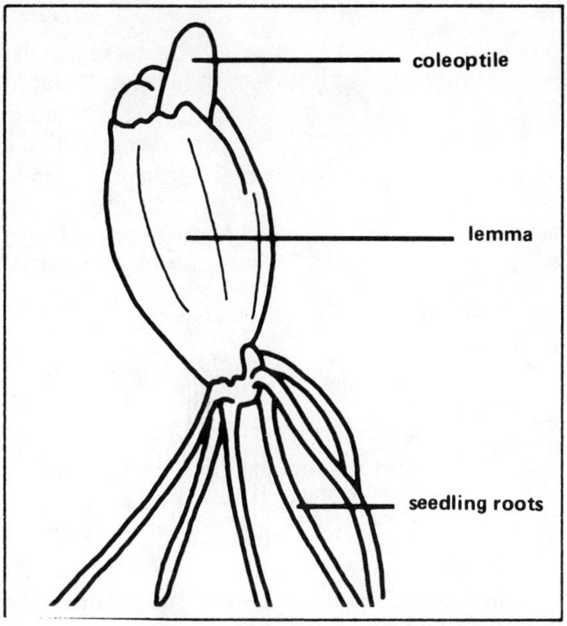

FIGURE 9 Germination of barley grain. (From Ref. 41.)

and six-rowed barley kernels can be distinguished from each other because two thirds of the kernels in the six-rowed sample are twisted.

C. Kernel

The barley floret at anthesis is shown in Figure 15. Double fertilization of the barley egg cell is shown in Figure 16. A mature barley caryopsis or grain is shown in Figure 17. The kernel dry weight is made up of approximately the following percentages of tissue: husk and pericarp, 10%; aleurone and associated testa, pigment strands, and some nuclear tissue, 14%; starchy endosperm and nuclear remains in sheaf cells region, 73%; embryo, 3% (15).

IV. PRODUCTION

Cultural practices of barley have been reviewed by several authors (7,15,61,65), while Poehlman (54) discussed adaptation of cultivated barley.

A. Soil and Climatic Requirements

Barley grows best under cool dry conditions and can stand heat under low humidity or high humidity under cool conditions. It does not grow well under a hot humid environment primarily because of diseases. When both wheat and barley grow well in an area, wheat is the cereal of choice for food and barley for feed and malting. Barley matures more quickly than other cereals and thus can be grown at higher latitudes and altitudes. The winterhardiness of barley is not as good as that of rye or wheat, but is better than that of oats. Barley grows under a wide range of photoperiod, temperature, and moisture conditions, but it grown best in the area where the cultivar was bred. Photoperiod was more important than temperature

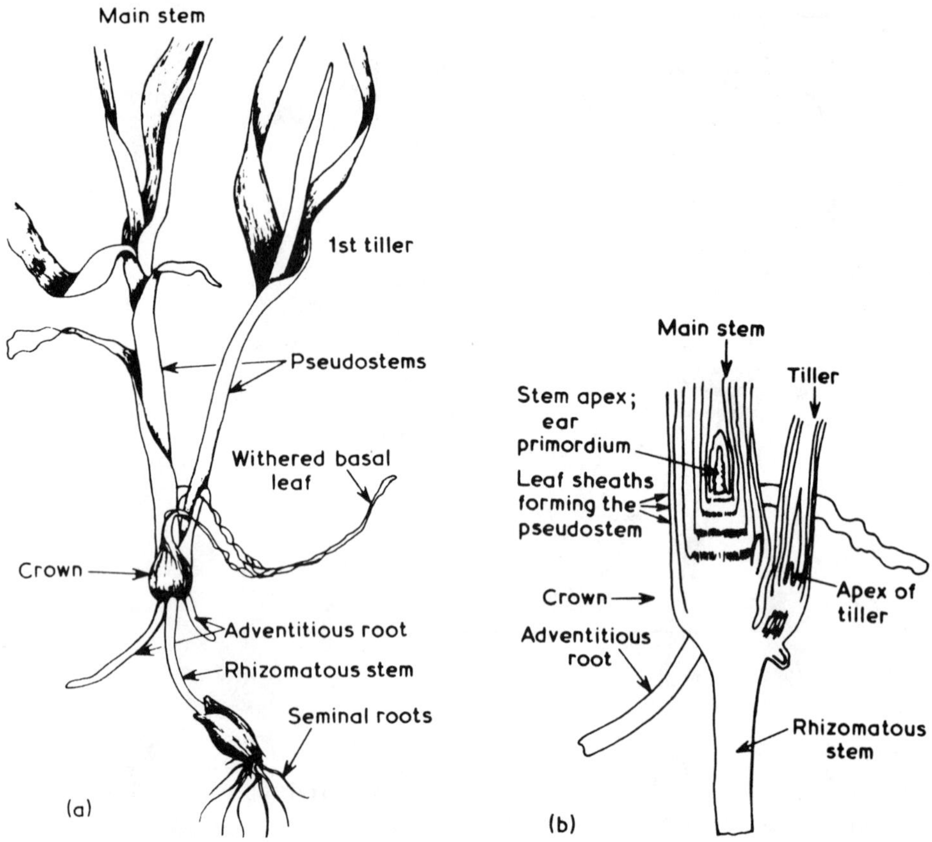

FIGURE 10 A seedling in which the pseudostems of the main axis and the first tiller are visible and erect. (From Ref. 15, after Ref. 53.)

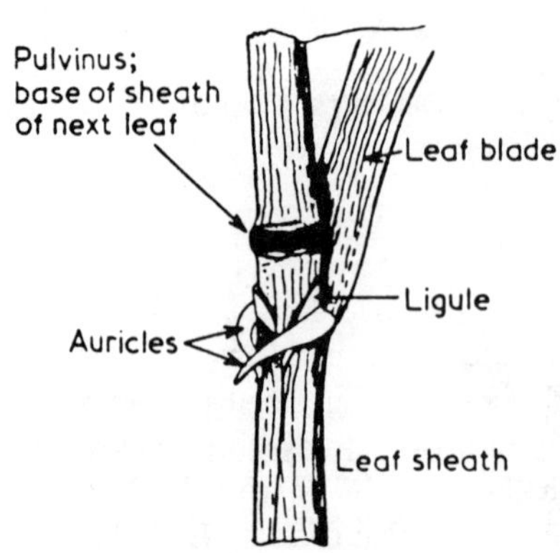

FIGURE 11 Schematic view of the manner in which the auricles and ligule are placed at the junction between the leaf sheath and the leaf blade. (From Ref. 15.)

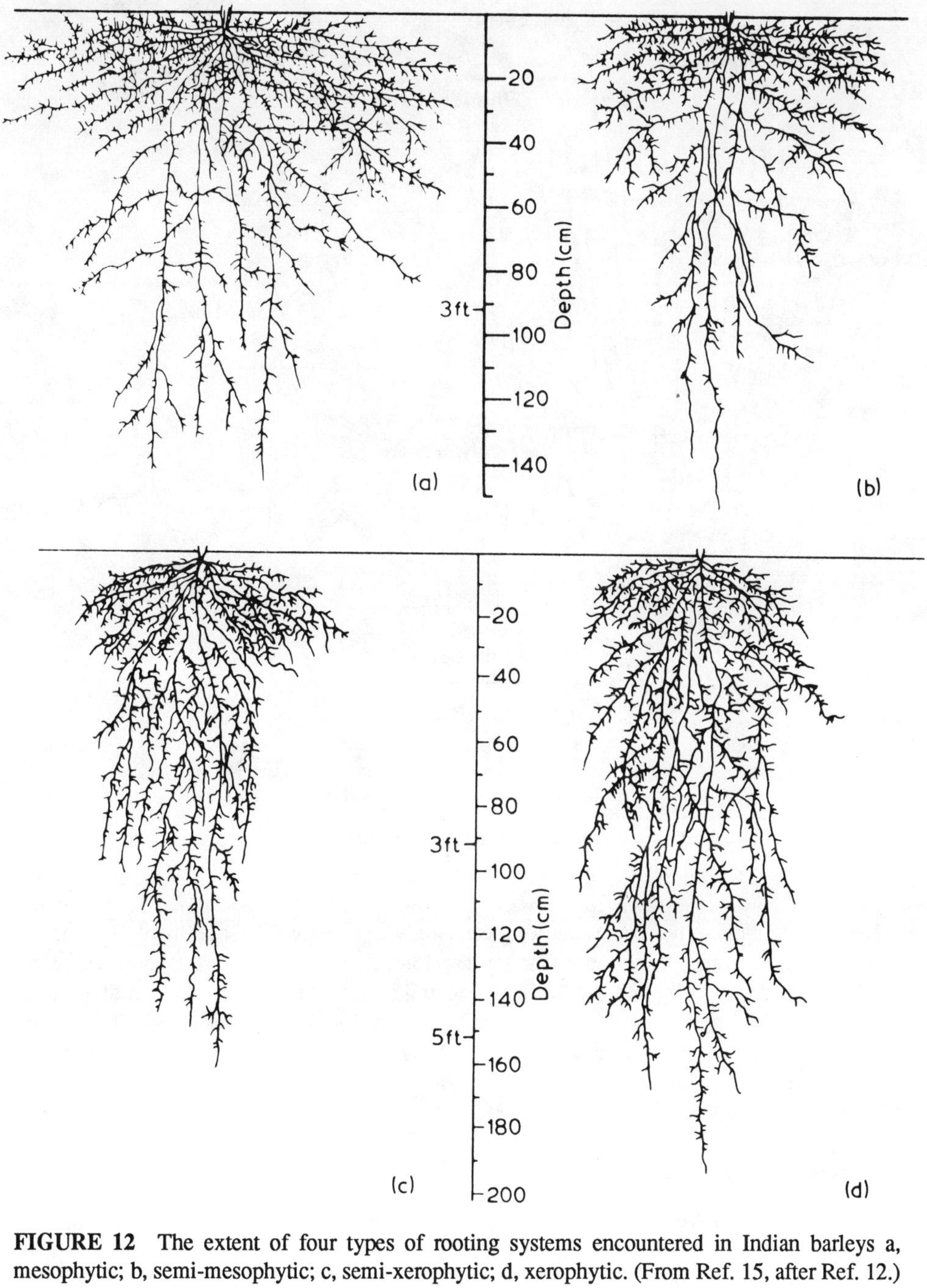

FIGURE 12 The extent of four types of rooting systems encountered in Indian barleys a, mesophytic; b, semi-mesophytic; c, semi-xerophytic; d, xerophytic. (From Ref. 15, after Ref. 12.)

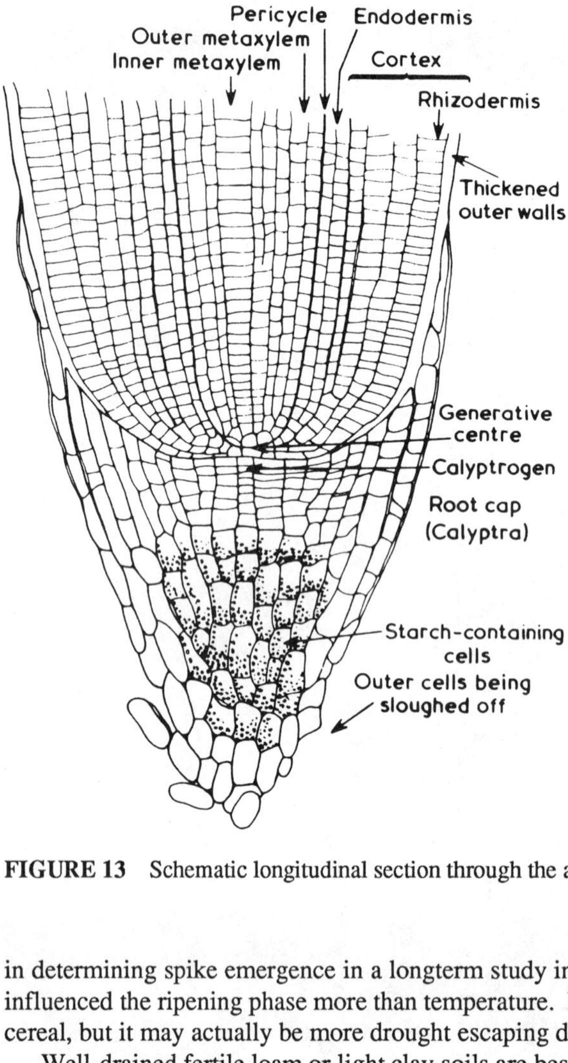

FIGURE 13 Schematic longitudinal section through the apex of a barley root. (From Ref 15.)

in determining spike emergence in a longterm study in Europe (37), and available energy influenced the ripening phase more than temperature. Barley is the most drought-resistant cereal, but it may actually be more drought escaping due to the shorter maturation period.

Well-drained fertile loam or light clay soils are best for barley production, since sandy soils do not hold enough water and heavy clay soils may become waterlogged. Of the three cereals, barley, wheat, and oats, barley is more tolerant than the others of alkaline conditions and more sensitive to acid soils (54).

B. Cultural Practices

Production costs of several cropping regions of Montana are shown in Table 7. The break-even price for barley production changes drastically as farm size and other physiographic characteristics change between these areas.

1. Rotations

The use of barley in a rotation depends on the length of the growing season, moisture available in the soil profile and expected rainfall, availability of other crops adapted to the region, disease problems, soil type, and soil fertility (61). Barley is a desirable companion crop, since its early maturity and early harvest aids in the establishment of the interseeded crop.

FIGURE 14 Barley spikes: A, six-rowed; B, two-rowed.

The inclusion of barley in the farm rotation in the typical wheat-fallow system of the northern Great Plains helps control winter annual weeds and diseases of wheat as well as balance the workload (7). Barley is an important component in flexible cropping systems presently used in the major spring barley areas, since its early maturity and drought tolerance make it the crop of choice when stored water is low.

Barley also fits into southern and southeastern cropping rotations with corn and double cropped soybeans (7). Double cropping systems in the southwestern United States benefit by using barley in the rotation because of its early maturity or suppression of cotton diseases.

2. Planting

In general, spring barley should be seeded as early in the spring as possible for the highest yields, unless early spring frosts are a problem. The best seeding time for fall-planted spring barley in the southwest is usually in November unless the area is subject to frost at heading time. Winter barley should not be planted too early in the fall, because of increased aphid and disease problems (7). In the southeast barley is generally planted 1 or 2 weeks before winter wheat (61). Barley should be seeded 2–6 cm in depth and at a rate of 54–108 kg/ha (7).

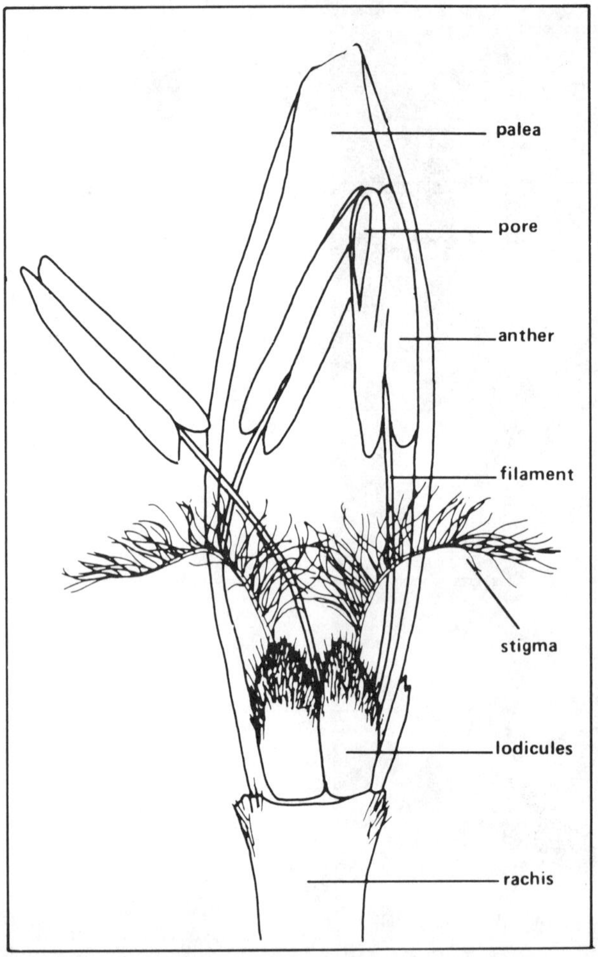

FIGURE 15 Barley flower with anthesis beginning. (From Ref. 41.)

3. Fertilizing and Water Use

Barley responds well to applied fertilizers and increased water, but a good management plan must be in place to utilize these inputs. Soil tests are absolutely essential to apply the correct amount of fertilizer. Barley is not as lodging resistant as wheat and oats, so it may not tolerate such high fertility levels.

Nitrogen balance is the most critical nutritional factor determining the yield and quality of the barley crop and is especially critical when recropping after a nonlegume crop. Barley production is affected by the nitrogen source, type of growing season, the level of carryover in the soil, time and method of application, and cultivar grown (7). In subhumid and humid areas measurement of the residual nitrate, N, in the soil is not as effective in determining a correct recommendation as it is in the semi-arid areas. For winter barley a split application of nitrogen in the fall and spring usually gives the best results. The interaction of nitrogen and water is the major factor determining yield (Fig. 18). Much more N can be applied if adequate water is available through irrigation. With irrigation yield can be increased compared to dry land where yields stabilize at 112 kg/ha of N. The protein content of the grain (Fig. 19) is also strongly influenced by N application. Excess N can decrease the value of

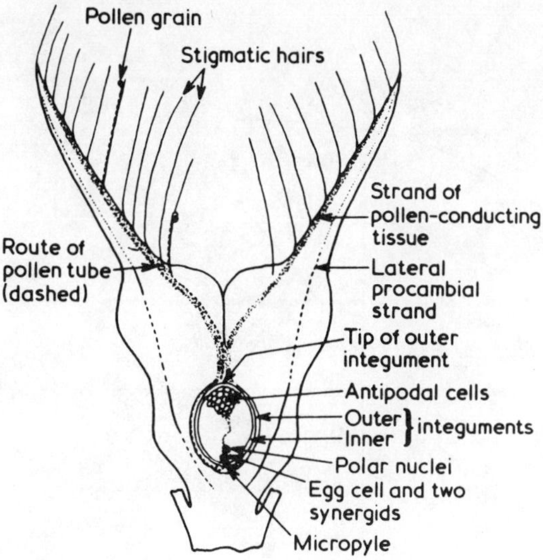

FIGURE 16 The route of the pollen tube from the stigmatic hair to the egg cell. (From Ref 15, after Ref. 55.)

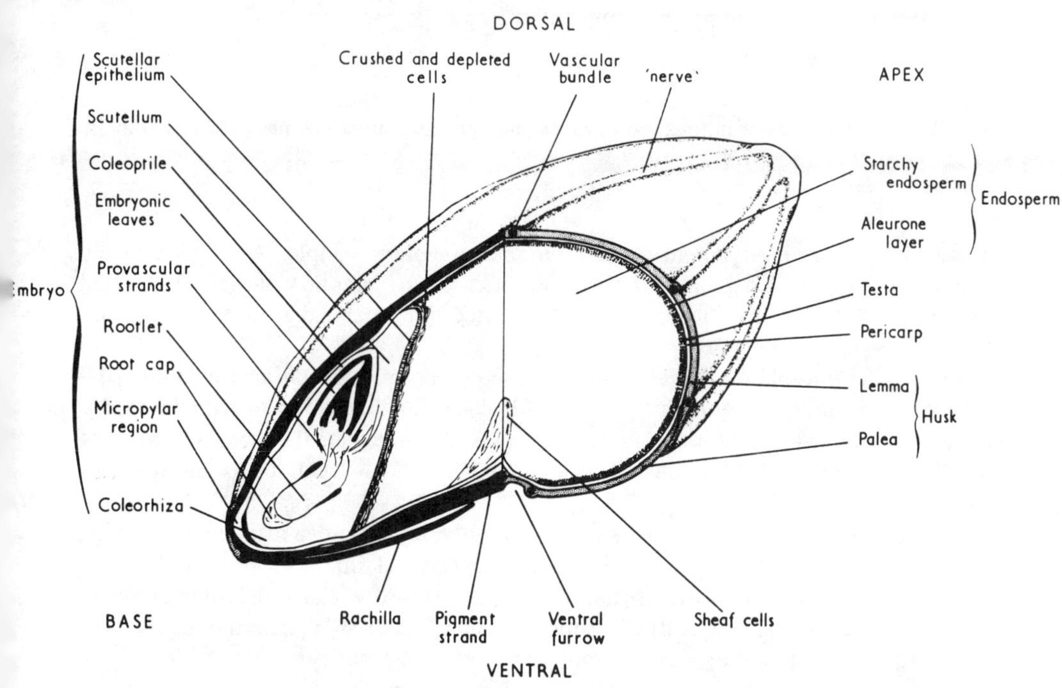

FIGURE 17 Idealized diagram of a barley grain, with a sector removed, to show the disposition of the tissues. (From Ref. 15.)

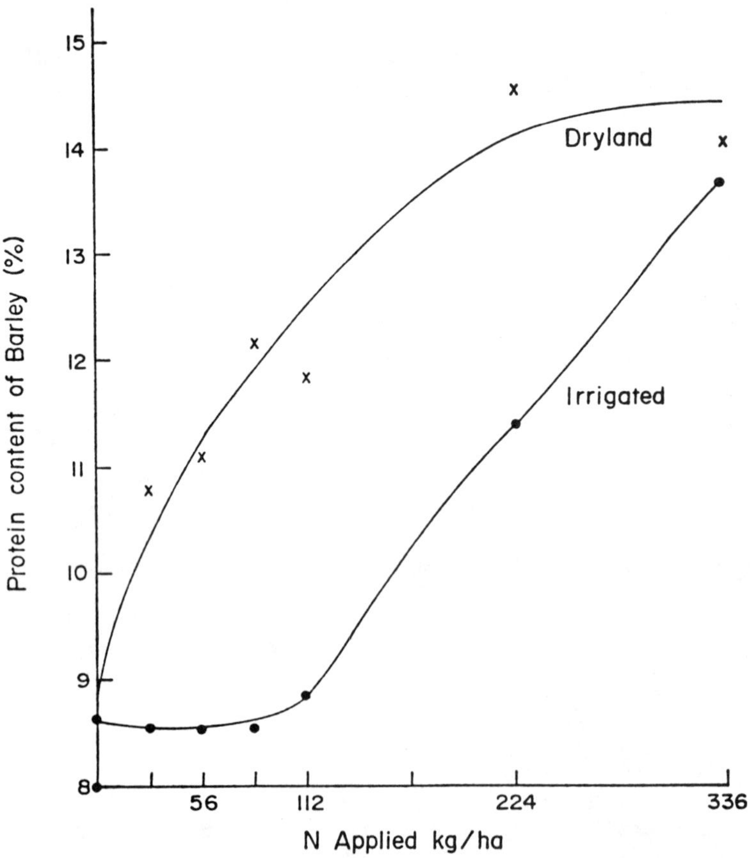

FIGURE 18 The effect of nitrogen fertilization and moisture supply on the yield of barley in Saskatchewan. (From Ref. 7.)

malting barley, since high protein can result in rejection of the sample. An expected yield curve shown in Figure 20 indicates that in the northern Great Plains no yield can be expected with less than 160 mm of plant-available water (stored soil water and growing season rainfall).

Soil tests for phosphorus, P, are reliable and should be used for fertilizer recommendations for barley. In the Great Plains banding of P at seeding time below or with the seed is the preferred method of application in deficient soils (7). Adequate P increases survival of winter barley and increases the percentage of plump grain and malt extract in malting barley (61).

Correct timing of irrigation water is important for barley culture in maintenance of high yields and malting quality. If water is limited, the most critical time of application is the late boot-heading stage (7). Temperature and water use are closely related in barley grown in the northern Great Plains (Table 8). This relationship is important in determining the time of application and amount of water to apply when irrigating barley.

Many studies have been made on drought tolerance or resistance in barley. There are several recent reviews of the morphological and physiological characteristics associated with barley yield (11,19).

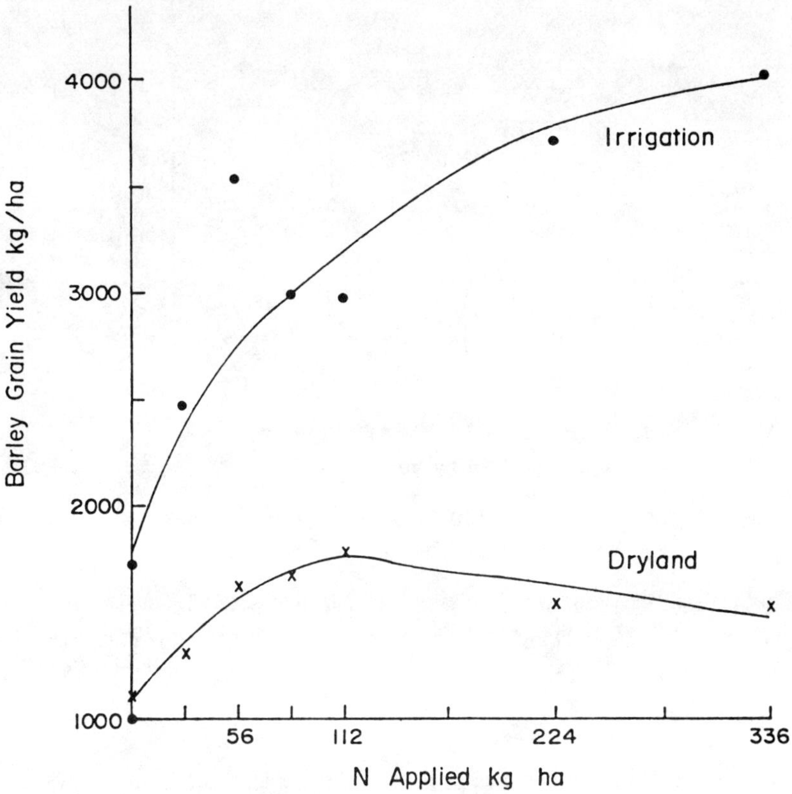

FIGURE 19 The effect of nitrogen fertilization and moisture supply on the protein content of barley. (From Ref. 7.)

4. Harvesting

Barley harvested for grain can be cut and threshed directly with a combine at <14% moisture, but wheat and barley exhibit different threshing characteristics, so close attention is needed to cylinder speeds. Operators need to be especially careful with malting barley, because skinning and breaking of the kernels results in discounts or rejection by the maltster. Barley reaches physiological maturity at about 40% moisture and can be swathed at this moisture content or below with no loss of yield. Swathing can reduce kernel shattering, reduce losses for hail, insects, and frost and facilitate harvest of unevenly ripening fields.

C. Pest Control

1. Diseases

Although a strict definition of disease includes any abnormality of the plant, only infectious pathogens are considered in this section. Diseases which most commonly infect barley are shown in Table 9. Most barley diseases occur under irrigated or high-moisture conditions, but some diseases thrive when plants are stressed (65). Losses due to disease for 1951–1960 were estimated to be 13.5% of the total barley production in North America (47) without

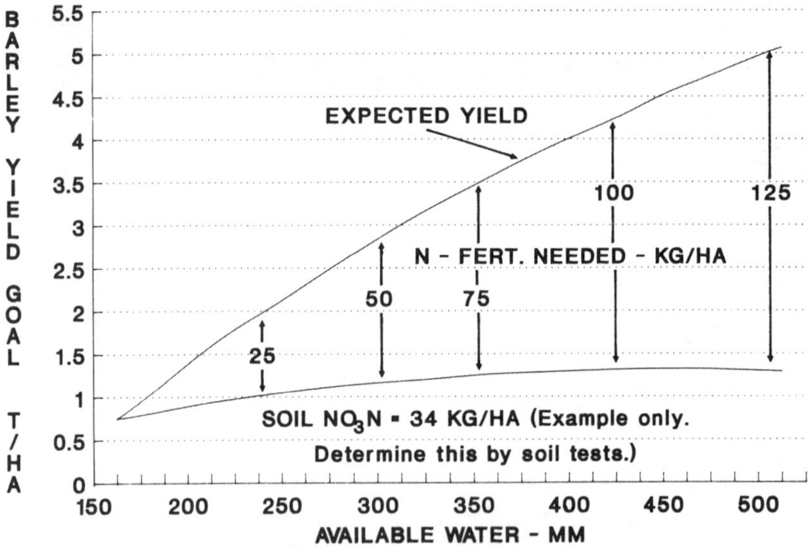

FIGURE 20 Expected yield curve for spring barley on recrop land in eastern Montana as related to N requirements (soil + fertilizer) and plant available water (stored soil water + growing season rainfall). (From Ref. 16.)

including those areas where disease has made barley unprofitable. North America is affected by about 15 economically important barley diseases. The region and climatic conditions of that area determine which ones are serious in a growing area. Diseases reduce quality as well as yield, causing shriveled kernels, high protein, reduced seed viability, and poor quality kernels. Disease organisms present may be toxic to animals and man (40,47). Diseases are controlled best by resistant cultivars, fungicides, or cultural practices. Breeding of disease-resistant cultivars is difficult since new pathogen strains often evolve as the new resistant cultivar is grown in monoculture (15,67). Attempts have been made to overcome some of the pathogen's ability to change through use of multilines (67,75) or polygenic resistance (67).

2. Weeds

Weeds may be defined as "plants out of place." Losses due to weeds in barley in the United States are estimated at $5 billion each year (38). A wide spectrum of weeds are found in barley because it is grown in such a wide range of environments. Probably the most troublesome weed in barley is wild oats (*Avena fatua* L.). Weeds compete for light, nutrients, and moisture and reduce yield, grain, and forage quality, increase labor and equipment costs, and reduce land values. Reduction in weeds can be accomplished through prevention, suppression, control, and eradication (38). Sanitation and prevention of weed establishment through cultural and chemical means is the best way of accomplishing the above objectives. Chemical control is only part of an integrated management program. Operators should use specific herbicides for specific weeds or herbicide combinations best suited to a particular situation. Table 10 gives recommendations current at this time for the Great Plains area of the United States. Effectiveness of the different chemicals vary, and tolerance of barley to the chemicals also should be taken into account (Table 11). Barley cultivars also vary in sensitivity to a particular chemical (7).

3. Insects

Over 100 species of arthropod pests attack barley in the United States and Canada, and they are estimated to cause 5–7% loss to the annual grain crop (74). Other losses are to the forage value of barley, reduction of quality of malting and food barley, and indirectly as vectors of virus diseases. The presence of insects does not automatically mean damage will occur, so an "economic injury level" must be reached before control is advisable (74). This injury level has been established for only a few pests of barley, and only a few pests such as the greenbug and Russian wheat aphid cause damage every year. The Russian wheat aphid spread from Mexico in 1980 as far north as Montana in 1987 and is reported more likely to damage barley than wheat, with oats least susceptible of the small grains.

Crop damage due to insects usually occurs during the larval stage, with most insects having one generation per year (an exception is aphids, which have many generations per year) (46). Some insects can be controlled with contact and some with systemic insecticides depending on their life cycle and method of plant infestation. Beneficial insects are an important agent in controlling many pest species, but they usually work slowly, and often chemical control is necessary to prevent crop loss (46). Care must be taken that the chemicals to control the pests do not also kill the predators and parasites. Insects of economic importance in North America and suggested control measures are shown in Table 12.

Postharvest pests of barley are also very important, but they will be discussed in Section VI.

V. CHEMICAL COMPOSITION

The chemical composition of the growing plant is described by Briggs (15) and is best known in regard to its value as hay or silage. Barley grain is rich in starch and sugars, relatively poor in protein, and very low in fat. The husk (lemma and palea) is mostly composed of lignin, pentosans, mannan, uronic acids, hemicelluloses, and cellulose fibers. Silica is present in the outer walls of the husk, and the awns contain large amounts of silica (15). The pericarp lacks lignin but otherwise resembles the husk in chemical composition (49). The testa contains crude cellulose and pigment strands of alkane waxes, which are a barrier for chemical substances and microbes. Polyphenols, which may complex with proteins, are abundant in the pericarp, testa, and aleurone layer. The aleurone has thick cell walls comprised of arabinoxylans and has aleurone grains of protein and phytic acid, spherosomes rich in fat, and abundant minerals. The subaleurone layer of the endosperm has relatively low starch content, is rich in proteins and β-amylase. The starchy endosperm is composed of about 85–89% starch enclosed in cell walls. β-Glucans make up 75% of the cell wall, and the rest is arabinoxylans (49). The embryo consists of about 7% cellulose, 14–17% lipids, 14–15% sucrose, 5–10% raffinose, 5–10% ash, and 34% protein (15). The cell walls of the embryo contains uronic acids, pectin, and hemicellulose (49). The chemical composition of parts of the barley grain are shown in Table 13.

A. Carbohydrates

1. Starch

Starch is a polysaccharide, α-glucan, and can be divided into straight-chain amylose and branched-chain amylopectin. The starch granules in the barley endosperm are laid down within amyloplasts and fall into two size groups: 1.7–2.5 μm and 22.5–47.5 μm. These starch grains contain traces of lipids, minerals, proteins, and nucleotides (15). The starch is synthesized in plastids, but the method is uncertain. The "diastatic" enzymes responsible for degrading the starch in germination are phosphorylase, α-glucosidase, α-amylase, β-amylase, debranching enzymes, and transglucosylase (15). The starch in the endosperm is 95% deposited in 11–28 days after ear emergence, and the ratio of amylose to amylopectin increases to the final ratio. In normal barleys the ratio of amylose to amylopectin is about 1:3, in high amylose Glacier about 1:1, and waxy barley is 97–100% amylopectin (50). Some workers have reported altered starch and free sugars in Riso 1508 barley, but others found no changes in the major carbohydrate constituents of the high-lysine barleys (50). The starch composition of several normal, high-lysine (Hiproly and Riso 13) and high-sugar (MT 1337–1) barleys are shown in Table 14 and vary from 21.2 to 64.4% in starch.

2. Soluble Sugars

At least nine monosaccharides and seven closely related compounds occur in barley (15). Glucose and fructose are found free and in combination, whereas other monosaccharides are polymerized as oligosaccharides, polysaccharides, glycosides, glycolipids, and glycoproteins (15). The composition of some barleys grown in Montana are shown in Table 15. Soluble sugar content of normal barley is about 2–3%, hulless barleys 2–4%, high lysine barleys 2– 6%, and high sugar barleys 7–13%. The total sugars and reducing sugars decline from anthesis to maturity, whereas nonreducing sugars are constant in amount during plant growth (15). The major sugar of living tissues of barley is sucrose.

3. Nonstarch Polysaccharides

Briggs (15) defines these polysaccharides in barley as "pectin fractions" or Hemicellulosic materials, which are then divided into gums if soluble in hot water and hemicellulose if soluble in alkali. The material remaining after delignification and removal of gums and hemicelluloses is termed "holocellulose." Endosperm gums consists of β-glucan and an arabinoxylan. Endosperm cell walls in barley are unique in cereals, since they completely enclose the cell-forming barriers to proteolytic and amylolytic enzymes. Some of the polysaccharides are linked to phenolic acid and some to lignin (15). The dietary fiber of barley is comprised of lignin plus the nonstarch polysaccharides. It was determined by Åman and Newman for the barleys shown in Tables 14 and 15.

B. Protein

"Crude protein" (N x 6.25) contains about 80% "true protein" with the rest nonprotein N and amides (15). The growing plant contains 6–7% of the dry matter as protein, but this declines to 0.6–0.8% of the dry matter in latter growth stages. The crude protein levels of barley grains for normal and exotic barleys are shown in Tables 14 and 15. The limiting amino acid in barley protein is lysine followed by methionine, threonine, and tryptophane. Amino acid contents of normal two-and six-rowed barley are shown in Table 16. When barley is grown at high levels of N fertilizer, the crude protein in the grain increases, but the proportion of lysine in the protein declines (15,50). The cereal proteins are classed as albumins, globulins, prolamines, and glutelins. In barley the albumins and globulins (salt soluble), which are rich in lysine (5– 7%) and threonine, are mainly metabolic proteins. The prolamines are the major storage proteins of the endosperm and are low in lysine (<2%). The glutelins, also found in the endosperm, are associated with bound structural proteins of the membranes and are about 4% lysine (50). High-lysine barleys (Table 17) such as Riso 1508 have increased lysine, lower seed weight, and decreased yield when compared to the mother line Bomi. Table 18 shows maximum and minimum values for the limiting amino acids of barley. These samples include high-lysine and high-protein lines, none of which are commercially available. *Hordeum spontaneum* or wild barley lines were found to have protein from 15 to 28%, and lysine from 2.2 to 3.6%, indicating about 5–10% of the strains could be considered high-lysine, high-protein lines (3).

C. Fats

Barley lipid content is low (2–3%) compared to maize and oats (50). Triglycerides comprising 77.9% of barley lipids contain palmitic acid and the unsaturated fatty acids oleic, linoleic (principal one in barley kernel), and linolenic. The barley kernel also contains diglycerides, free sterols, free fatty acids, and sterol esters and hydrocarbons (50). Most of the lipids are in the endosperm (77%), embryo (18%), and hull (5%). The fat content of a developing plant increases and then may decline slightly (15). The fat content of normal covered, normal hulless, high-sugar, and high-lysine barleys range in percentage from 1.9 to 2.4, 2.7 to 3.9, 4.4 to 7.3, and 2.9 to 5.8, respectively (Table 15).

D. Minerals

The ash content of barley (2–3%) is influenced by the growing season, soil zone, soil type, and soil fertility (15,50). The distribution of minerals is uneven throughout the kernel (Table 19), and the rachis (5–14%) and awns (17–38%) contain a large portion of the minerals (15). The ash content of barleys in Table 14 and 15 range from 2.2 to 3.9%, with MT 1337–2 the lowest and Riso 1508 the highest in content.

E. Vitamins

Barley is an excellent source of B-complex thiamine (B1), pyridoxine (B6), riboflavin (B2), pantothenic acid and is also quite high in niacin, but only 10% of the niacin is available to monogastric animals (50). A small amount of vitamin E is found in the germ, and some biotin and folacin, but no carotene, or vitamin A, B12, or D are found in ungerminated grain (15,50).

F. Phenolic Compounds

In plants the phenolic compounds can be divided into benzoic acids, cinnamic acids, terpenoids, and flavonoids (9). Barley contains a wide range of phenolic compounds in one form or another ranging from free and combined tyrosine, tyramine and derivatives, phenolic acids, esters and glycosides, and numerous other phenols through lignans and substances related to lignans (15). The anthocyanidines have been of interest recently since proanthocyanidin free barley used in brewing may prevent the formation of chill haze in beer (83).

VI. PROCESSING AND UTILIZATION

Barley is used worldwide for animal feed, brewing, and human food. In the United States the amount used for animal feed fluctuates with the supply available, since other feeds can be substituted if the supply is short and the price high (Fig. 21). Malting barley demand is relatively inflexible from year to year since substitution by other crops is very limited. The use of barley for animal feed was slightly higher than the nonfeed uses (malting, seed, and other uses) in 1987 and 1988. The supply has decreased drastically in the last 2 years because of increased exports in 1986 and 1987 and greatly decreased production in 1988 (Fig. 21).

A. Feed and Food Barley

1. Animal

a. Grain Barley nutritional quality refers to relative merit in providing essential total nutrients biologically balanced and available to humans and other animals. Payment to farmers is on the basis of quantity of feed barley produced, not feed quality produced, so breeders have concentrated on yield and malting quality. Hockett and White (35) found that the best malting cultivars have the highest feed value. Barley is high in starch and sugars, low in fat, and relatively poor in protein (15,50).

Analysis of hulled and hulless barleys and brewing by-products, which are fed to animals, are shown in Table 20. The barley fiber is of little or no value for nonruminant animals, and the water-soluble mixed linked β-glucans may cause digestive problems in poultry (50). When barley grain is fed to swine, grinding, pelleting, cubing, rolling, and micronizing improves feed value. Pigs fed high-lysine barleys showed improved growth rate and increased biological value of the barley. Removal of hull genetically improved feed value of barley in most pig-feeding trials (50). Numerous reports show beneficial effects of feeding brewers dried grains to pigs, but the reasons for this improvement are not clear.

Barley fed to poultry in combination with corn improved egg production and feed efficiency compared to either fed alone (50). Although barley has a higher protein content and a better balance of amino acids than corn, its commercial use in feeding poultry is restricted because of its lower productive energy and the sticky wet droppings that create sanitation problems. Supplementing barley with various enzyme preparations such as β-glucanase

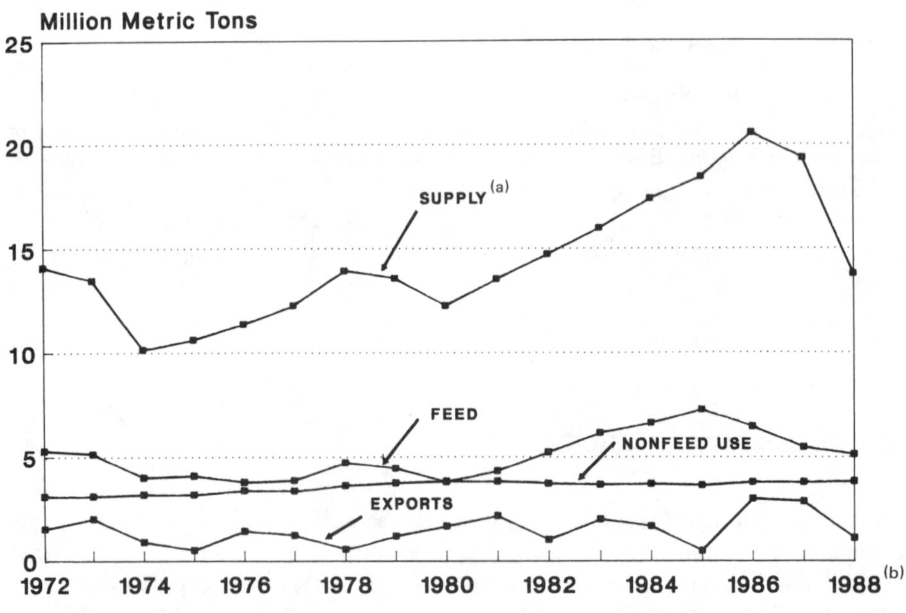

FIGURE 21 Supply and disposition of barley in the United States, 1972–1988. (1972–85 data from Agricultural Statistics 1987, USDA; 1986–1988 data from Barley Bulletin, North Dakota Barley Council, August 1988.)

(a) Supply includes beginning stocks, production, and imports.

(b) 1988 values are projected.

improved feed consumption, weight gain, feed efficiency, and cage cleanliness of chickens (15,50).

In the western, northwestern, and northern Great Plains of the United States, barley is a major energy source and major cereal grain of feedlots and dairy diets. When it is processed properly to obtain its maximum feeding value, it is equal to maize in feed value for ruminants (50). Cattle tend to bloat more and tire more quickly of barley than other cereal grains, but less roughage and protein supplementation is required because of the high-fiber and high-protein content of barley.

b. Hay, Silage, and Straw Barley hay should be cut when the nutritive value is suitable and yield is high. If harvested at the milk to soft dough stage and promptly dried, this objective can be reached (Table 21). If awned barleys are harvested too late, the awns may damage animals' eyes, mouths, and gut linings (15,50).

If barley is ensiled it should contain more than 30% dry matter and enough carbohydrates to support adequate lactic fermentation (at early dough stage) (15) (Tables 21,22). Consistent differences have not been found in feed values of barley, wheat, and oat silage, but all are inferior to maize silage for beef and dairy cattle (50). Barley straw is a useful roughage which can provide significant energy to ruminants (15,50). Cultivar differences in straw feeding value have been found. The barley straw has a low protein content, low digestibility, and low energy value (Table 23), but chopping and grinding the straw increases its value. The digestibility of straw can be significantly increased by treatment with dilute solutions of NaOH (15,76) (Table 23). Proper treatment of straw with ammonia is more convenient than with NaOH and improves digestibility, energy value, protein, and intake (76).

2. Human

In the last century barley has been used less for food than has wheat because of lesser palatability, baking quality, and milling characteristics. From the nutritional point of view, barley is equal to or better than wheat (49). Barley is eaten as popped barley, barley flakes, sprouts, barley starch and as sweeteners, barley malt flour supplements, malted milk, infant food syrups, barley tea or coffee substitute, and rice extender (50). After Riso 1508 barley is steamed, milled, and supplemented with vitamins and minerals, it will produce an infant baby food where all the protein and carbohydrates come from barley. This infant food would be equal to presently produced commercial baby foods marketed in developing countries at a far lower cost (49). The β-glucan content of barley is of great interest to the chicken feed industry as well as from the human nutrition aspect. The β-glucans of barley will lower blood cholesterol levels in chickens, rats, and swine (15,50,51) and in humans (51). Barley cultivars differ in β-glucan, content, which can be estimated by viscosity measurement (Table 24). Soluble dietary fiber (relatively high in barley) may have a significant effect on human health and nutrition. Increased amounts of dietary fiber have been shown to help in controlling diabetes, hyperlipidemia, obesity, hypertension, coronary heart disease, and gastrointestinal disorders such as hiatus hernia, gallstones, appendicitis, diverticular disease of colon, bowel polyps and colorectal cancer, and hemorrhoids (5,73).

B. Malting Barley

1. Uses

Barley is used in preference to wheat and rye for malting because of the cemented lemma and palea, which protects the acrospire during germination, serves as an aid for filtering during brewing, and produces a firmer kernel at the high moisture content required in steeping and malting (17). Although 85 breweries operated in the United States in 1987, the top six companies sold 96.7% of the beer (26) (Table 25). Most of these 85 breweries are "boutique" or local companies and sell as little as 108 barrels per year (26). Total malt utilization has almost doubled from 1959 to 1986, while malt used per barrel has decreased (Table 26). In addition to brewing beer, the remaining 5% of barley malt is used in the United States for baking, malt extracts (powders and syrups), diastase (energy rich powder), whisky, gin, vodka, alcohol, and malt vinegar (15). Barley syrups and malt extracts, which are variable in composition (Table 27), are prepared by concentrating worts by evaporation (15).

2. Malting Cultivars

The spring barley cultivars are concentrated in the states shown in Table 28. Minnesota, North Dakota, South Dakota, and Wyoming all have more than 65% of their acreage planted to malting cultivars. The industry requirements for acceptable malting barley cultivars are shown in Table 29, and the composition of the six-and two-rowed standards in Table 30. Cultivars for breeding programs are evaluated by micro malting (20–100 g), pilot scale tests (25–36 kg), and commercial plant scale tests (≥45 t amounts) (17). Near infrared reflectance spectroscopy shows promise in separating poor malting lines from good lines (17). Modern malting barley is subject to more quality requirements than any other crop and is the only grain crop sold on the basis of a specific cultivar (see Chapter 21).

C. Marketing

1. Classification and Prices Received

The classes and subclasses of barley are shown in Table 31. Further details of the United States market classes of barley can be found in Wilson (88). Malting barley is usually purchased on the basis of samples drawn from rail cars loaded at country elevators and sold directly to the maltster as consigned by a commission firm, or sometimes directly to the maltster by farmer contract. No futures market exists for barley, and the regional markets are located in Minneapolis, Portland, Stockton, Los Angles, and Duluth (88). Acceptable malting barley lots command a higher price than feed barley (Fig. 22).

2. Storage

Most barley in the United States is harvested directly or from the swath by combining. Malting samples are binned separately by the farmer, country or regional elevator, and the maltster. Some of the stored grain insects and their description are shown in Table 32. Current recommendations in local areas should be consulted for control of storage pests.

3. Exports

In some years barley exports from the United States are low (see Fig. 21), but in other years they are quite significant. World price, world production, and political considerations, such as export subsidies, largely determine the amount of barley exported. Tables 33 and 34 show exports of barley and malt, respectively.

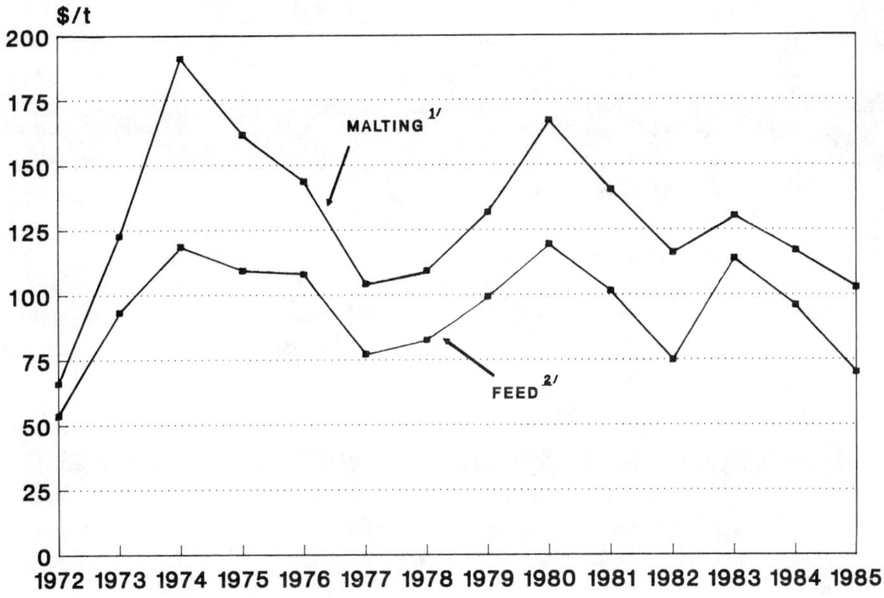

FIGURE 22 Average annual malting and feed barley prices at Minneapolis, 1972–1985. (Data from Agricultural Statistics 1987, USDA.)
[a]No. 3 or better malting barley.
[b]No. 2 feed barley (No. 3 or better for 1972).

TABLE 1 U.S. Production of Barley: Harvested Area, Yield, and Production by State, Three-Year Average, 1978–1980 and 1985–1987[a]

	Harvested area (1000 ha)		Yield (kg/ha)		Production (1000 t)	
State	1978–80	1985–87	1978–80	1985–87	1978–80	1985–87
Arizona	17.3	14.2	4036	5499	70.2	75.9
California	331.0	148.5	3049	3279	996.2	483.1
Colorado	101.3	118.8	3498	3370	355.1	398.5
Delaware	10.1	18.5	2475	3241	24.7	59.6
Idaho	382.1	432.0	3372	3370	1286.2	1446.3
Kansas	22.4	80.3	2278	2026	51.0	166.3
Kentucky	10.4	7.7	2655	2311	27.9	16.8
Maryland	30.5	33.7	2637	3316	80.3	110.9
Michigan	8.5	18.6	2637	2994	22.4	55.4
Minnesota	355.7	376.0	2601	2618	925.6	1042.7
Montana	457.7	665.5	2224	1451	1021.0	1034.2
Nebraska	11.1	42.5	2134	1919	23.7	82.8
Nevada	11.1	13.6	3408	4660	37.8	63.3
New Jersey	5.9	6.2	2690	3349	16.0	20.8
New Mexico	11.7	5.7	3085	3402	36.2	19.6
North Carolina	25.0	20.1	2708	2526	67.7	48.8
North Dakota	756.0	1174.5	2224	2311	1718.7	2930.8
Oklahoma	25.0	14.2	2027	2059	50.1	28.9
Oregon	69.5	123.5	3067	3349	212.1	399.5
Pennsylvania	34.4	25.6	2583	3370	88.7	86.1
South Carolina	9.4	9.2	2493	2112	23.6	18.8
South Dakota	203.2	280.1	1973	1919	404.2	568.2
Texas	17.0	13.9	1847	2241	32.4	32.9
Utah	59.8	59.3	3767	4085	225.1	241.6
Virginia	39.3	34.3	2744	3118	107.8	105.8
Washington	150.5	356.4	3408	2757	520.8	965.1
Wisconsin	10.4	30.8	2816	2779	29.3	83.6
Wyoming	57.8	62.7	3372	3548	194.6	222.3
Total U.S.	3240.4	4186.5	2685	2521	8688.5	10814.3

[a]1978–80 barley statistics from Poehlman (54); 1985 and 1986 barley statistics from United States Department of Agriculture, Agricultural Statistics, 1987; and 1987 barley statistics from American Malting Barley Association, Inc., Gleanings, August, 1988.

TABLE 2 Harvested Area, Yield and Production in 10 Leading Barley-Producing Countries, Three-Year Average, 1983–1985[a]

Country	Harvested area (1000 ha)	Yield (t/ha)	Production (1000 t)	Percent of world production
U.S.S.R.	29,816	1.59	47,396	26.6
United States	4,700	2.78	13,071	7.4
Canada	4,764	2.63	12,588	7.1
France	2,143	5.14	11,023	6.2
United Kingdom	1,955	5.23	10,265	5.8
Federal Republic of Germany	1,967	4.97	9,784	5.5
Spain	4,144	2.21	9,104	5.1
China	3,518	1.86	6,548	3.7
Turkey	3,100	1.95	6,033	3.4
Denmark	1,124	4.88	5,486	3.1

[a]Data from Agricultural Statistics 1987, USDA.

TABLE 3 Genes Located on the Barley Chromosomes and Their Symbols and Description

Symbol	Characteristic	Symbol	Characteristic
	Chromosome 1		*Chromosome 2*
f5	Chlorina 5	rt	Rattail spike
Run1	Resistance to *U. nuda*	or	Orange seedling
br	Brachytic plant	a2	Albino seedling 2
f_c	Chlorina seedling	f(=lg)	Chlorina seedling (viridis)
gs3(cera)	Glossy sheath	msg3	Male sterile 3
wx	Waxy endosperm	y	Virescent seedling
yc	Virescent seedling	yx	Xantha (yellow) seedling
ac2	Albino seedling	e	Wide outer glume
F8	Chlorina 8	gs5(cera)	Glossy sheath
Rs	Red Stem	msg2	Male sterile 2
ert-a	Erectoides	wst4	White streak 4
bl_2	Non-blue (white aleurone)	h	Short culm
l	Lax (long spike)	lin	Low internode number
msg14	Male sterile 14	v	Six-rowed
msg10	Male sterile 10	Lk	Long awn
n	Naked caryopsis	lr	Lateral spikelet lemma appendage reduced
f4	Clorina 4		
lk2	Short (fine) awn	Pr	Purple straw
u4	Unbranched style	Ha	Reaction to *Heterodera* (cereal cystnematode)
x_a	Xantha seedling	Re2	Purple lemma and pericarp
		mt,,e	Mottled leaf
		Pau	Purple auricle
		li	Liguleless
		gp(gp2)	Grandpa
		Tr	Triple awned lemma
	Chromosome 3		*Chromosome 4*
x_s	Xantha seedling	Reg6	Resistance to *E. gramini*
bt-bt2	Non-brittle rachis	bl	Non-blue (white) aleurone
an	Albino seedling	min	Minute (dwarf)
Rh	Reaction to *Rynchosporium secalis*_(scald)	lk5(ari–c)	Short awn
		gl3(=gl2)	Glossy seedling (leaf) 3
yst2	Yellow streak	br2	Brachytic 2
zb	Zebra stripe	cer-zg	Eceriferum-zg214

TABLE 3 (Continued)

Symbol	Characteristic	Symbol	Characteristic
wst(=wst3)	White streak 3	Reg2	Resistance to *E. gramini*
cu2	Curly	gl(=gl 2)	Glossy seedling (leaf)
Ryd2	Resistance to barley yellow dwarf virus	lb2	Long basal internode 2
		lg4	Light green 4
lnt	Low number of tillers	lg2	Light green 2
als	Absent lower laterals	K series	Hooded lemma
gs2	Glossy sheath	i series	Deficiens
uz	Uzu or semi-brachytic	f9	Chlorina 9
	Chromosome 5		*Chromosome 6*
Rph4	Reaction to *P. hordei*	msg 36bk	Male sterile
Regl	Reaction to *E. graminis*	ea7	Early heading 7
Rsp4	Reaction to *P. hordei*	o	Orange lemma
Reg4	Reaction to *E. graminis*	gs4	Glossy sheath
Reg5	Reaction to *E. graminis*	msg6	Male sterile 6
fs2	Fragile stem	uc2	Uniculm 2
cer-e	Ecerferium-e8	sex1	Shrunken endosperm, xenia
ert-b	Erectoides	mt,,f	Mottled leaf
Sh3	Spring habit of growth	r,,e	Smooth awn
nec1	Necrotic leaf	xn	Xantha seedling, Nepal
at	Albino seedling, Trebi	Amy1	Alpha amylase
B series	Black lemma and pericarp	mul2	Multiflorous 2
f7	Chlorina 7	l9	Dense spike
trd	Third outer glume		
eak	Early heading		
	Chromosome 7		
ddt	DDT resistance		
nld	Narrow leafed dwarf		
fs	Fragile stem		
mt2	Mottled 2		
wst2	White streak 2		
s	Short-haired rachilla		
cm	Cream seedling, Black Hulless III		
r	Smooth awn		
va3	Variegated 3		
Sh2	Spring habit 2		
lb	Long basal rachis internode		
Clt	Purple color at leaf tips		

Source: Ref. 34.

TABLE 4 Broad and Narrow Sense Heritability and Genetic Advance Estimates for Grain Yield Components and Other Agronomic Characteristics Since 1964

	Heritability[a]								
	Broad sense			Narrow sense			Genetic advance[b]		
Characteristic	Average (%)	Range (%)	Number references reviewed	Average (%)	Range (%)	Number references reviewed	Average (%)	Range (%)	Number references reviewed
Grain yield	44	5–93	26	27	0–54	11	23	3–46	10
Spike number	49	3–98	24	34	14–66	9	33	4–113	12
Kernels/spike	64	15–99	23	39	2–91	12	28	3–71	8
Kernel weight	63	24–99	22	43	13–78	10	12	2–22	9
Heading date	74	19–100	17	60	34–92	6	10	1–23	9
Lodging score	66	41–88	5	27	6–38	3	123	—	1
Plant height	66	4–99	30	41	8–73	13	15	1–44	11
Grain protein	53	5–98	14	32	8–76	4	16	5–22	3
Grain plumpness	62	34–90	5	43	24–58	3	18	11–24	2
Diastatic power									
Barley	82	55–94	5	58	23–94	3	20	—	1
Malt	68	50–86	2	—	—	—	—	—	—
Extract									
Barley	59	43–71	3	12	8–16	2	1	—	1
Malt	57	46–69	3	—	—	—	—	—	—
Spike length	66	3–98	17	50	44–56	5	20	4–34	8

[a]Computations were most often on the plot basis, but some were on a plant or trial mean basis.
[b]Given as percentage of the mean.
Source: Ref. 34.

TABLE 5 Information on Five Base Germplasm Collections

Collection and address	Accessions	Type of storage
National Seed Storage Laboratory, U.S. Department of Agriculture, Colorado State University, Fort Collins, Colorado, 80523	25,284 landraces, cultivars, and wild species. Includes duplicates of USDA Barley and Barley Genetic Stock Collections. Collected worldwide.	Long-term. Seeds with 4–7% moisture content stored in sealed metal cans at –18° to 20°C.
Canadian Barley Collection, Plant Gene Resources of Canada (PGRC), Agriculture Canada, Ottawa Res. Station Bldg. 75, Ottawa, Ontario K1A 0C6 Canada	21,000 landraces, cultivars, and wild species. About 16,000 accessions from USDA Barley Collection. Collected worldwide.	Long-term. Seeds with 5–6% moisture content stored in sealed foil pouches at –20°C.
National Institute of Agricultural Sciences (NIAS), Division of Genetics, Seed Storage Laboratory Kannondai 3.1.1, Yatabe-machi, Ibaraki-ken 300 21 Tsukuba–gun, Japan	5263 landraces and cultivars. Collected worldwide.	Seeds with 6–7% moisture content stored in vacuum sealed cans at –10°C for long-term storage; at –1°C and 35% relative humidity in screw-lid cans for medium term storage.
Plant Genetic Resources Center, P.O. Box 30726, Addis Ababa, Ethiopia	5000 landraces collected from Ethiopia and North Africa.	Seeds with 5–6% moisture content stored in cans and foil packets at 4°C and 30% relative humidity for medium-term storage and at –10°C for long-term storage.
Nordic Gene Bank, Box 1543 S–221 01 Lund, Sweden	Landraces, culrivars, and wild species collected primarily from Denmark, Finland, Iceland, Norway, and Sweden.	Long-term storage under construction (–20°C capacity 50,000 samples of ca. 15,000–20,000 seeds).

Source: Ref. 48.

TABLE 6 Potential Agronomic Traits as Identified in Wild Species and Interspecific Hybrids

Species or hybrid	Trait
H. vulgare cv. (Ono) 4x X *H. bulbosum* 4x X *E. Mollis* 4x	Resistant to leaf rust Partially resistant to net blocth Partially resistant to spot blotch Tolerant to Barley Yellow Dwarf Virus
(*H. brachyantherum* 4x X *H. bogdannii* 2x) 2 X *H. vulgare* cv. Traill 4x	Tolerant to Barley Yellow Dwarf Virus Partially resistant to net blotch Partially resistant to spot blotch Leaf Pubescence Winterhardiness
H. vulgare X [*H. jubatum* X *H. compressum*[a] X *H. vulgare* (2n = 28)]	Cytoplasmic male sterility
H. compressum x *H. pusillum*	Resistance to spot blotch
H. vulgare cv. Traill 4x X *H. bulbosum* 4x	Resistance to spot blotch
H. jubatum X *S. cereale*	Winterhardiness
H. arizonicum X *A. spicatum*	Forage grass Perennial Postharvest dormancy
H. vulgare ssp. *spontaneum*	Mildew resistance Salt tolerance Winterhardiness Flooding tolerance Epidermal trichomes Cytoplasmic male sterility[b]
H. bulbosum	Mildew resistance Winterhardiness Outcrossing habit Anther extrusion
Elymus spp.	Rust resistance Leaf spotting resistance Strong straw Drought tolerance

[a]Synonym for *H. stenostachys*.
[b]See Ref. 2.
Source: Ref. 24.

TABLE 7 Estimated Production Costs for Feed Barley, Based on 50–50 Crop Fallow Cropping System and 1982–1983 Input Prices

	Montana Cropping Areas		
	North central	East central	North western
Average hectare cropland/farm	810	405	122
Expected yield, in metric tons/hectare	2.04	1.83	2.58
Variable costs/hectare		$	
Pre-harvest:			
Seed	11.86	14.82	11.86
Chemicals & application	9.26	9.26	5.71
Fertilizer & application	21.00	18.77	46.31
Crop insurance	12.35	12.35	12.35
Tractor operating costs	3.56	3.95	12.70
Machine operating costs	5.11	7.14	31.52
Hired labor	7.58	9.39	19.64
Miscellaneous	8.25	4.62	3.93
Interest on operating capital	4.92	5.85	9.34
Subtotal, pre-harvest costs	83.89	86.15	153.36
Harvest:	30.11	49.77	68.96
Total variable costs	114.00	135.92	222.32
Break-even price/metric ton	55.60	74.44	85.47
Fixed costs/hectare			
Fallow costs	113.87	119.00	149.58
Fixed costs for growing crop			
Farm machinery & building depreciation	40.63	65.53	114.53
Insurance on machinery & buildings	1.14	1.83	3.19
Personal property taxes	3.38	5.46	9.53
Real estate taxes	6.18	6.18	6.18
Opportunity cost on inventory	57.60	64.52	78.13
Family labor opportunity cost	0.00	0.00	0.00
Management opportunity cost	5.31	6.32	10.13
Subtotal, fixed costs for growing crop	114.24	149.84	221.69
Total fixed costs	228.11	268.84	971.27
Total costs/hectare	342.11	404.76	593.59

Source: Ref. 65.

TABLE 8 Average Barley Water Use (mm/day), Based on Week During Crop Growth Period and Average Air Temperature

	Week after crop emergence							
	1	3	5	7	9	11	13	
	Average daily water use (mm/day)							Average total (mm/season)
Air temperature								
10–15°C (50–59°F)	1.02	1.02	2.29	2.29	2.29	2.03	1.02	1370
15.5–20.5 (60–69)	1.02	2.03	3.81	3.81	3.81	2.54	1.02	2490
21–26 (70–79)	1.27	3.05	5.33	5.33	5.59	3.81	1.02	3540
26.5–31.5 (80–89)	1.27	3.56	6.35	6.38	6.60	4.32	1.02	4162
32–37 (90–99)	2.03	4.06	7.37	7.37	7.62	5.08	1.27	4856
Approximate growth stage		4–5 leaf		Heading			Milk	

Source: Ref. 65.

TABLE 9 Diseases of Barley, Their Casual Organisms, and Suggested Control Methods

Disease	Casual organism	Control
Bacterial		
Bacterial leaf blight	*Pseudomonas syringae*	None
Bacterial leaf streak and black chaff	*Xanthomonas campestris* pv. *translucens*	Clean seed, crop rotation, seed treatment
Fungal		
Common root rot and seedling blight	*Cochliobolus sativus, Fusarium culmorum, F. graminearum*	Clean seed, seed treatment, tolerant cultivars, optimum fertilization
Pythium root rot	*Pythium* species	Cultural practices
Take-all	*Gaeumannomyces graminis* var. *tritici*	Crop rotation, ammonium nitrogen fertilizer
Barley stripe	*Pyrenophora graminea*	Clean seed, seed treatment, resistant cultivars
Net blotch	*Phrenophora teres*	Resistant cultivars, clean seed, seed treatment, crop rotation, foliar fungicides
Powdery mildew	*Erysiphe graminis* f. sp.*hordei*	Resistant cultivars, foliar fungicides
Scald	*Rhynchosporium secalis*	Resistant cultivars, sanitation, crop rotation, foliar fungicides
Spot blotch	*Cochliobolus sativus*	Resistant cultivars, clean seed, seed and foliar fungicides, crop rotation
Septoria leaf blotch and glume blotch	*Septoria avenae* f.sp. *triticea, S. passerinii, S. nodorum*	Clean seed, crop rotation, seed treatment, sanitation, resistant cultivars, foliar fungicides
Leaf rust	*Puccinia hordei*	Resistant cultivars, foliar fungicides
Stem rust	*Puccinia graminis*	Resistant cultivars
Stripe rust	*Puccinia striiforms*	Resistant cultivars, foliar fungicides
Covered smut	*Ustilago hordei*	Resistant cultivars, seed treatment

TABLE 9 (Continued)

Disease	Casual organism	Control
True loose smut	*Ustilago nuda*	Resistant cultivars, seed treatment
Semi loose smut	*Ustilago nigra*	Resistant cultivars, seed treatment
Ergot	*Claviceps purpurea*	Crop rotation, clean seed, deep planting, grass weed control, closed flowering cultivars
Scab	*Fusarium* species	Crop rotation, deep plowing, seed treatment
Storage molds	*Aspergillus* and *Penicillium* species	Proper harvest and storage conditions
Black point	*Alternaria* & *Gusarium* species & *Cochliobolus sativus*	Tolerant cultivars, seed treatment
Viral		
Barley stripe mosaic	Barley stripe mosaic virus	Virus-free seed, resistant cultivars
Barley yellow dwarf	Barley yellow dwarf virus	Resistant cultivars, aphid control, cultural practices
Oat blue dwarf	Oat blue dwarf virus	None
Other		
Aster yellows	Aster yellows	None
Cereal cyst nematode	*Heterodera avenae*	Resistant cultivars, crop rotation
New virus-like disease of barley[a]	New virus-like agent	Crop rotation, prevent drought stress

[a]Ref. 64.
Source: Ref. 44.

TABLE 10 Some Common Herbicides Used for Weed Control in Barley[a]

Herbicide	Crop stage	Weeds controlled	Remarks
Residual preemergence			
Triallate	Apply and incorporate in fall or spring	Wild oats	Granular formulations may be surface applied to fields with heavy residue
Trifluralin	Apply and incorporate in fall or postplant incorporate in spring	Green and yellow foxtail	Granular formulation should be used on fields with heavy residue
Residual preemergence and postemergence			
Chlorsulfuron	Apply from the 2-leaf stage until just before boot	Residual preemergence and early postemergence control of annual broadleaf weeds, and suppression of green foxtail, Canada thistle and hoarycress	Apply before annual weeds are over 2 inches tall or across, before foxtail is more than 1 inch tall or beyond the 1-to 2-leaf stage, and when Canada Thistle is in rosette stage through 6 inches tall
Metsulfuron	Apply from the 2-leaf stage until just before boot	Residual preemergence and postemergence control of annual broadleaf weeds and postemergence suppression of Canada thistle	Apply before annual weeds are over 2 inches tall or across and when Canada thistle is in the rosette stage through 6 inches tall
Postemergence			
Barban	Anytime	Wild oats	Apply when wild oats are in the 1 1/2-to 2-leaf stage
Bromoxynil	Apply after emergence until just before boot	Annual broadleaf weeds	Apply when weeds are in early seedling stage
Clopyralid+ 2,4–D	Apply from the 4-leaf stage until just before boot	Annual weeds and Canada thistle	
Dicamba	Apply before barley exceeds the 3-leaf stage	Certain annual broadleaf weeds and suppression of certain perennial broadleaf weeds	Applications later than recommended may result in crop injury

TABLE 10 (Continued)

Herbicide	Crop stage	Weeds controlled	Remarks
Diclofop	Apply before jointing	Wild oats, green foxtail, Persian darnel, Italian ryegrass and barnyardgrass	Apply when the majority of grassy weeds are 1-to 4-leaf stage or when Italian ryegrass is in the 3-to 5-leaf stage
Difenzoquat	Apply before flagleaf emergence	Wild oats	Apply when the majority of wild oats are in the 3-to 5-leaf stage
MCPA	Apply from the 3-leaf stage	Annual broadleaf weeds and suppression of certain perennial broadleaf weeds	
Metribuzin	Apply after fully tillered and development of secondary roots but before jointing	Certain annual grasses and broadleaf weeds	Apply before weeds are over 2 inches tall or across
Picloram	Apply at the 3-to 5-leaf stage	Certain annual broadleaf weeds and suppression of certain broadleaf weeds	
2,4–D	Apply amine formulations from the 4-leaf stage until just before boot and ester formulations from the 5-leaf stage until just before boot	Annual broadleaf weeds and suppression of certain perennial broadleaf weeds	

[a]Rates must be accordance with local and state regulations.
Source: Information courtesy of Dr. J. E. Nelson, Extension Weeds, Plant & Soil Science Department, Montant State University, Bozeman, MT, 59717.

TABLE 11 Effectiveness of Some Herbicides for Weed Control in Barley and Barley Tolerance[a]

Herbicide	Mustard	Wild buckwheat, annual smartweeds	Canada and perennial sowthistle	Pigweed, lambsquarters, common ragweed	Kochia, Russian thistle	Wild oats	Green foxtail	Barley tolerance
	General performance rating of herbicide on weeds							
Residual preemergence								
Triallate (Far–go, Avadex–BW)	N	N	N	N	N	G	N	G
Trifluralin (Treflan)	N	P	N	P–F	P	F	G	G
Residual preemergence and postemergence								
Chlorsulfuron (Glean)	G	G	F–G	G	G	P	F	G
Metsulfuron (Ally)	G	G	F–G	G	G	N	N	G
Postemergence								
Barban (Carbyne)	N	N	N	N	N	G	N	F–G
Bromoxynil (Buctril)	F	G	P–F	G	G	N	N	G
Clopyralid+2,4–D (Curtail)	G	G	G/F–G[b]	G	F	N	N	G
Dicamba (Banvel)	P	G	F–G	G	P–F	N	N	F–G
Diclofop (Hoelon)	N	N	N	N	N	G	G	F–G
Difenzoquat (Avenge)	N	N	N	N	N	G	N	G
MCPA amine	G	F	F	G	P	N	N	G
MCPA ester	G	F	F	G	F	N	N	G
Metribuzin (Sencor, Lexone)	G	P	P	G	F–G	P–F	F–G	G
Picloram (Tordon)	N	G	F–G	G	P/G[c]	N	N	G
2,4–D Amine	G	F	F	G	F	N	N	G
2,4–D Ester	G	F	F	G	F–G	N	N	G

[a]P = poor, F = fair, G = good, N = no control.
[b]G = Canada thistle, F–G = Perennial sowthistle.
[c]P = Kochia, G = Russian thistle.
Source: Ref. 7 and J. E. Nelson.

TABLE 12 Insects and Related Pests on Barley in North America and Their Control

Types	Identification	Control	Remarks
APHIDS			
Corn leaf aphid (*Rhopalosiphum maidis*)	Pear-shaped, greenish blue with darker spots surrounding the cornicles, cornicles short and broad, all black	Insecticides, tolerant cultivars, parasites, adverse weahter	Control with contact insecticides difficult
English grain aphid (*Sitobion avenae*)	Pear-shaped, green or yellow-green, cornicles long and narrow, all black	Insecticides, resistant cultivars, parasites, adverse weather	Vectors BYDV; causes damage in northern United States and Canada
Greenbug (*Schizaphis graminum*)	Pear-shaped, pale yellowish- to bluish-green with dark green stripe down the back, tips of cornicles black	Insecticides, resistant cultivars, parasites, dadverse weather	Causes more damage than most aphids, injects toxins during feeding, causing yellow-red area
Oat birdcherry aphid (*Rhopalosiphum padi*)	Pear-shaped, reddish-orange band on rear of abdomen, tips of cornicles black	Insecticides, adverse weather, parasites	Vectors BYDV; control usually not required
Russian wheat aphid (*Diuraphis noxia*)	Spindle-shaped, uniform pale green, antennae short, supracaudal process (= double tail)	Insecticides, adverse weather	Infected plants have white, yellow, or purple streaks on leaves and leaves are tightly curled, severe yield losses possible
Yellow sugarcane aphid (*Sipha flava*)	Bright lemon yellow cornicles no longer than wide	Insecticides, tolerant cultivars, parasites, adverse weather	Injury similar to that of greenbug, not a vector of viruses

MITES			
Banks grass mite (*Oligonychus pratensis*)	Half as large as brown wheat mite, lighter color than other mites	Contact insecticides, rainfall	Webbing on upper part of plant, damage found along margin of field
Brown wheat mite (*Petrobia latens*)	Large, oval, brownish mite with yellowish legs, front legs usually longer than rest	Irrigation or rainfall, insecticides	Vector of "new viruslike disease of barley," scatters when disturbed
Wheat curl mite (*Aceria tulipae*)	Very small, white, spindle shaped, only 4 legs, located anteriorally	Systemic insecticides, remove volunteer small grains	Vector of wheat streak mosaic virus, direct damage seldom important
Winter grain mite (*Penthaleus major*)	Dark brown with reddish legs, orange spot on top	Irrigation or rainfall, insecticides	Active in cold temp., early appearing, nocturnal
CHEWING INSECTS			
Armyworms (Pseudaleta *unipuntcta*)	Larvae pale green when young, and yellow to brownish green striped when mature	Insecticides, parasites	Nocturnal feeders, young larvae feed on leaf surface, older larvae eat all the leaf tissue
Cereal leaf beetle (*Oulema melanopus*)	Larvae yellowish covered by black fecal coating, adults metallic bluish-black wings & head, legs & thorax orange	Insecticides, parasites	Larvae eat leaf epidermis, damage mainly to spring grains

TABLE 12 (Continued)

Types	Identification	Control	Remarks
Cutworms, army (*Euxoa auxiliaris*)	Larvae tan with dark stripes, 3 cm long at maturity	Insecticides, diseases, parasties, predators	Larvae feed above ground and spend nonfeeding time below ground
Cutworms, pale western (*Agrotis orthogonia*)	Larvae white with brown head capsules, 3 cm long at maturity	Parasites, predators, insecticides	Larvae are below ground feeders, thrives in drought years
Grasshoppers (*Melanoplus* and *Camnula* species)	Body elongate and narrow, forewings leathery, hindwings membranous, very diverse with respect to size and color	The pathogen *Nosema locustae*, insecticides	Feed on leaves and chew stems causing head drop
Thrips (various species)	Dusky colored, cigar-shaped < 2 mm long	Insecticides	Flecking and light-colored streaks on leaves, may cause flower sterility
OTHER INSECTS			
Barley jointworm (*Harmolita hordei*)	Larvae pale green to lemon, average 3 mm in length	Resistant cultivars	Galls form on stems disrupting nutrient transport causing shriveled spikes

Hessian fly (*Mayetiola destructor*)	Reddish orange eggs on upper leaves, white larvae behind leaf sheath, overwinter as brownish puparium, small dark colored flies emerge in spring	Delayed fall planting, insecticides, crop rotation, sanitation	Larvae kill tillers and stunt plant growth in fall by feeding on the sap, in spring culms are broken over
Sawfly (*Cephus cinctus*)	Adults black & yellow wasps, larvae cream colored, 2 cm long with hardened head capsule	None	Barley less susceptible than wheat to damage
Wheat stem maggot (*Meromyza species*)	Maggots pale green with mouth parts 2 black hooks	Sanitation, early maturing cultivars	Damages seedling tillers causes "white head"
White grubs (*Phyllophaga* species)	Larvae whitish, G-shaped, dark posterior abdomen	Parasites, cultural practices, soil insecticides	Damage is severe pruning of roots and underground stems
Wireworms & false wireworms (*Agriotes mancus*, *Ctenicrea* sp. & *Eleodes* sp.)	Cylindrical, hard-shelled, segemented larvae, reddish-yellow to brown	Seed treatment, clean tillage	Damage seeds, seedlings, roots, and underground stems

Source: From Refs. 65, 74, and Dr. G. D. Johnson, Entomologist, Department of Entomology, Montana State University, Bozeman, MT 59717.

TABLE 13 Chemical Composition of Parts of Barley, Dry Matter Basis

Item (%)	Whole barley	Pearled barley	Hulls
Dry matter	91.21	90.67	92.64
Crude protein	11.73	12.45	13.31
Crude fat	1.52	1.20	4.51
Crude Fiber	6.56	1.85	17.98
Crude ash	2.52	1.53	5.58
Nitrogen-free extract[a]	77.67	82.97	58.62
Cellulose (included in crude fiber)	5.92	1.45	19.48

[a]Principally starch and free sugars, but includes part of the total or dietary fiber.
Source: Ref. 43.

TABLE 14 Grain Chemical Compositions of Some Barley Types with Different Starch Contents (%, w/w dry basis)

Constituent	Betzes	Shabet	Klages	Hiproly	Riso 13	Mt 1337–1
Carbohydrates						
Starch	64.4	58.1	57.9	47.1	25.2	21.2
Soluble sugars[a]	2.4	2.7	2.5	4.7	5.9	12.8
Nonstarch polysaccharides (NSP)	10.5	11.2	11.6	15.4	29.7	25.3
Arabinose residues	1.7	1.6	1.7	2.9	3.7	3.4
Xylose residues	2.3	2.6	2.9	3.2	5.7	4.7
Mannose residues	0.4	0.3	0.3	0.4	0.7	1.3
Galactose residues	0.2	0.2	0.2	0.3	0.4	0.4
Glucose residues	5.6	6.1	5.9	8.1	18.7	15.2
Uronic acid residues	0.2	0.4	0.5	0.5	0.6	0.4
Crude protein (N x 6.25)	13.6	15.4	14.8	21.0	16.1	20.5
Crude fat	3.9	3.7	3.7	4.6	5.9	7.3
Ash	2.5	2.3	2.4	2.4	3.0	2.5
Lignin	1.5	1.9	2.6	1.0	4.2	0.6
Total	98.8	95.3	95.5	96.2	90.0	90.2
Dietary fiber (NSP + lignin)	12.0	13.1	14.2	16.4	33.9	25.9

[a]Sum of glucose, fructose, sucrose, and fructans.
Source: Ref. 4.

TABLE 15 Compositions of Two-Rowed Barley Grain (%, w/w dry basis)

Cultivar	Glucose	Fructose	Sucrose	Fructans	Starch	Crude protein (N x 6.25)	Crude fat	Ash	Dietary fiber
Normal (covered)									
Betzes	0.1	0.1	1.9	0.7	58.1	15.4	2.2	2.3	19.3
Compana	0.2	0.1	2.1	0.9	59.5	13.1	2.4	3.0	18.8
Waxy Compana	0.2	0.1	2.2	0.5	57.1	12.5	1.9	2.9	22.6
Normal (hulless)									
Nubet[a]	0.1	0.1	1.6	0.6	64.4	13.6	3.9	2.5	13.3
Nupana[b]	0.1	0.1	1.4	0.5	65.2	12.1	2.7	2.3	15.6
Washonupana[c]	0.2	0.1	3.2	0.7	60.5	16.6	2.7	3.5	12.6
High-sugar (hulless)									
MT 1337–1[d]	0.5	0.4	8.3	3.6	21.2	20.5	7.3	2.5	35.8
MT 1337–2[d]	0.5	0.3	5.3	1.3	43.3	19.7	4.4	2.2	22.9
High-lysine									
Riso 7[e]	0.2	0.2	1.2	0.4	54.0	13.2	3.0	3.8	24.2
Riso 13[e]	0.2	0.4	4.2	1.6	25.0	15.0	5.8	3.4	44.4
Riso 29[f]	0.4	0.2	2.0	1.2	42.8	13.0	4.5	3.0	33.0
Riso 56[f]	0.4	0.4	2.0	0.2	53.9	15.4	2.9	3.2	21.7
Riso 86[f]	0.3	0.5	3.5	1.0	40.0	15.9	4.1	2.9	31.9
Riso 1508[e]	0.2	0.2	3.1	1.0	50.1	13.9	4.4	3.9	24.7

[a]Hulless line from backcross of Sermo on Betzes.
[b]Hulless line from backcross of Sermo on Compana.
[c]Hulless, short-awn, waxy line from backcross of Waxy Oderbrucker on Nupana.
[d]Chemical-induced mutants of Washonupana.
[e]High-lysine mutants of Bomi.
[f]High-lysine mutants of Carlsberg II.
Source: Ref. 4.

TABLE 16 Protein and Amino Acid Content of Two-Rowed and Six-Rowed Montana Barley Grain[a]

Item	Type 2–rowed (%)	6–rowed (%)
Protein (N x 6.25)	13.6	13.0
Alanine	3.65	3.69
Arginine	4.83	5.16
Aspartic acid	5.82	5.69
Cystine/2	2.17	2.53
Glutamic acid	27.02	26.23
Glycine	3.35	3.48
Histidine	2.37	2.32
Isoleucine	3.65	3.69
Leucine	6.90	7.06
Lysine	3.35	3.58
Methionine	1.68	1.79
Phenylalanine	5.23	5.27
Proline	12.03	11.59
Serine	4.54	4.64
Threonine	3.55	3.58
Tryptophan	1.97	1.90
Tyrosine	2.96	2.85
Valine	4.93	4.95

[a]Means represent 8 samples of each row type and amino acids are expressed as percent of protein. The two-rowed barley was Compana (CI 5438), and the six-rowed barley was Unitan (CI10421). All barleys were grown at Bozeman, MT in the same year.
Source: Ref. 50.

TABLE 17 Productivity and Seed Composition of Mutant No. 1508 and the Parent Variety Bomi

	Bomi	Mutant No. 1508	Relative value of mutant
Lysine content (g/16 g N)	3.75	5.20	139
Lysine yield (g/plot)	10.4	13.7	132
Protein content (6.25 x N%)	9.7	11.0	113
Protein yield (g/plot)	278	263	95
Seed size (mg)	40.8	37.5	92
Number of seeds/plot x 10^{-3}	70.7	63.4	90
Grain yield (kg/plot)	2.88	2.38	83

Source: Ref. 22.

TABLE 18 Minimum, Maximum, and Average Selected Nutrient Composition of Barley Grain Observed at the Montana Station

	Minimum observed	Maximum observed	Average
Protein (%)	9.9	24.1	12.8
Crude fiber (%)	1.9	9.9	5.3
Ether extract (%)	1.8	4.9	2.1
Lysine (%)	0.29	0.87	0.40
Methionine (%)	0.11	0.36	0.21
Cystine (%)	0.20	0.43	0.26
Threonine (%)	0.34	0.69	0.44
Tryptophan (%)	0.14	0.49	0.23
Phosphorus (%)	0.29	0.68	0.37
Kernel weight (mg)	25	51	46

Source: Ref. 50.

TABLE 19 Quantities (μg/g tissue) of Some Minerals in Naked Himalaya Barley and Some Fractions Prepared by Hand-Dissection

Element	Whole kernels	Germ	Hand-dissected endosperm
P	5,630	12,930	1,120
K	5,070	10,900	1,440
Mg	1,410	2,940	78
Ca	406	740	132
Na	254	—	58
Fe	36.7	56.5	10.0
Zn	23.6	71.3	4.7
Mn	18.9	69.4	10.0
Cu	15.1	—	3.5
Al	4.9	26.5	tr
Mo	1.35	2.32	0.41

tr = trace quantity detected.
Source: Ref. 15.

TABLE 20 Proximate Analyses of Some North American Barleys and Other Materials Used in Animal Feeds

Product	Moisture (%)	Protein (%)	Fat (%)	Fiber (%)	N-free extract (%)[a]	Ash (%)
Barley, lightweight	10.2	12.3	2.3	8.5	63.7	3.0
Barley, hulless or bald	9.8	11.6	2.0	2.4	72.1	2.1
Barley, bran, nearly all hulls	8.1	5.9	1.3	26.4	51.8	6.4
Barley feed, low grade	9.9	12.3	3.5	14.7	56.2	5.3.
Barley feed, high grade	8.0	13.2	3.5	8.4	61.0	4.0
Barley malt	6.6	12.7	2.1	5.4	70.9	2.3
Barley screenings	11.4	11.5	2.8	9.5	60.6	4.2
Brewers' grains, dried 25% protein or over	7.4	26.6	6.8	14.6	41.0	3.6
Brewers' grains, dried 23–25% protein	6.1	23.8	6.5	14.9	44.9	3.8
Brewers' grains, dried, below 23% protein	7.0	21.6	6.2	16.1	45.2	3.9
Brewers' grains, dried, from Californian barley	8.9	20.0	5.7	18.1	43.6	3.7

[a]Principally starch and free sugars, but includes part of the total or dietary fiber.
Source: Ref. 15.

TABLE 21 Analyses of Horsford Barley (covered, hooded, six-rowed, *H. vulgare*, L., var. horsfordianum) Cut at Different States of Maturity to Leave Approximately 11 cm (4 in.) of Stubble

Date harvested 1933	Height (in.)	Stage	Age (days)	Water (%)	Ash (%)	Crude protein (N x 6.25, %)	Crude fiber (%)	Nitrogen-free extract[a] (%)	Fat (%)	Calcium (%)	Phosphorus (%)	Fraction of whole plant (%)
				Composition on a water-free basis								
14 June	16		55	84.3	9.7	17.0	20.2	48.3	4.8	0.33	0.46	
21 June	26	Bloom on some heads	62	82.4	8.0	17.0	22.8	48.6	3.7	0.37	0.35	
28 June	35	Full bloom	69	79.5	7.2	9.9	28.4	52.2	2.3	0.23	0.33	
5 July	42–43	Kernels water stage	76	74.7	6.9	9.2	27.8	53.9	2.3	0.24	0.25	
8 July		Milk	79	15.7[b]	7.8	8.4	32.3	49.6	1.9	0.26	0.31	
14 July		Medium dough	85	14.2[b]	7.2	8.0	24.2	58.4	2.3	0.19	0.29	
26 July		Dead ripe	97	15.0[b]	7.5	7.3	19.3	64.4	1.5	0.14	0.30	
Whole plants:												
		Milk			7.0	10.2	29.8	50.9	2.1	0.26	0.30	100
		Dough			7.1	8.3	22.6	60.0	2.0	0.18	0.29	100
		Ripe			7.5	7.0	18.6	65.3	1.7	0.14	0.29	100
Leaves:												
		Milk			14.0	8.1	28.8	45.1	3.9	0.47	0.31	22.8
		Dough			16.8	10.1	26.6	42.7	3.9	0.44	0.29	20.7
		Ripe			15.0	8.2	29.8	43.0	4.0	0.51	0.27	20.9
Stems:												
		Milk			3.7	3.8	38.9	52.4	1.2	0.12	0.19	36.8
		Dough			4.2	3.1	36.5	55.1	1.2	0.11	0.15	30.2
		Ripe			6.1	2.2	47.9	43.1	0.84	0.06	0.07	15.2
Heads:												
		Milk			6.3	13.3	17.1	61.7	1.7	0.11	0.42	40.4
		Dough			5.6	10.2	11.8	70.9	1.5	0.10	0.35	49.3
		Ripe			4.8	9.0	8.8	76.1	1.4	0.09	0.39	63.9

[a]Principally starch and free sugars, but includes part of the total or dietary fiber.
[b]Samples dried.
Sources: Refs. 15, 72.

TABLE 22 Composition of Whole Barley and Silage (% of dry matter)[a]

	Barley	Silage
Dry matter	32.4	32.4
Ash	12.8	15.3
Inorganic matter	87.2	84.7
Crude protein	6.3	6.4
Ether extract	1.3	1.5
Crude fibre	24.3	24.8
Total N	1.0	1.0
Protein N	0.77	0.39
Nonprotein N	0.24	0.63
Volatile N	—	0.07
Water-soluble carbohydrates	25.3	11.7
Cellulose	25.6	24.6
Lignin	6.0	6.2
Acetic acid	—	1.6
Propionic acid	—	0.16
Butyric acid	—	Abs.
Succinic acid	—	0.09
Lactic acid	—	3.4
Ethanol	—	0.53
pH	5.8	3.9

[a]Plants of Ymer barley (1049 kg) cut at early milk-ripe stage and ensiled in small towers. No effluent was produced.
Source: Refs. 15 and 23.

TABLE 23 Chemical Composition (A) and Digestibility (B) of Barley Straw Treated in Various Ways and Used in Feeding Trials[a]

A.	Treatment	Crude Ash (% DM)	Neutral-detergent protein (% DM)	Acid-detergent fiber (% OM)	Acid-detergent fiber (% OM)	lignin (% OM)	Cellulose (% OM)	Gross energy (Mcal/kg DM)
(C)	Chopped	6.1	5.0	85.6	57.5	7.4	46.0	4.23
(M)	Milled	6.1	5.0	85.6	57.5	7.4	46.0	4.23
(CIL)	Chopped,1.5% NaOH	4.4	3.2	89.7	66.9	8.3	58.5	4.36
(CIH)	Chopped, 3.0% NaOH	3.0	2.4	93.4	74.7	8.2	63.7	4.40
(CS)	Chopped, sprayed	14.8	5.3	70.0	53.4	6.6	41.1	4.10
(MS)	Milled,sprayed	14.8	5.3	70.0	53.4	6.6	41.1	4.10

B.	Treatments C	M	CIL	CIH	CS	MS	S.E. of differences
Digestibility (%)							
Organic matter	45.4	44.9	71.1	69.6	60.8	63.5	± 2.89
Energy	40.4	39.1	65.0	64.4	57.3	60.2	± 3.23
Energy as straw	40.4	39.1	65.0	64.4	53.2	56.5	—
Acid-detergent fiber	49.4	47.9	71.9	73.6	59.0	62.9	—
Cellulose	58.4	57.1	81.5	81.6	73.1	73.8	—
Intake (per day)							
Dry matter (g)	501	696	742	869	950	1048	± 90
Organic matter (g)	471	654	704	841	809	893	± 78
Digestible organic matter (g)	218	293	494	587	492	568	± 54
Gross energy (Mcal)	2.13	2.94	3.21	3.82	3.90	4.30	± 0.37
Energy as straw (Mcal)	2.13	2.94	3.21	3.82	3.56	3.94	—
Intake (per kg W per day)							
Dry matter (g)	26.7	36.2	37.1	44.2	48.4	53.6	± 4.2
Digestible energy (Kcal)	46	60	105	125	114	132	—

[a]The treatments were: C—straw was chopped. M—straw was milled. CIL—chopped straw was soaked in 1.5% NaOH, then washed; dry weight loss 28%. CIH—chopped straw was soaked in 3% NaOH, then washed; dry weight loss 32%. CS—chopped straw sprayed with 16% NaOH; neutralized with propionic acid before feeding. MS—milled strawed treated as CS. DM—dry matter. OM—organic matter.

Sources: Refs. 15 and 18.

TABLE 24 Cultivar means for β-glucan and Brookfield Viscosity in 1983[a]

Identification	β-glucan (%)	Brookfield viscosity centipoise units
Opaque Glenn	3.7 a	153 a
Glenn	5.2 b	262 b
Karla	5.2 b	274 bc
Robust	5.2 b	285 bcd
Triumph	5.2 b	288 bcd
Gallatin	5.2 b	301 bcde
Morex	5.3 b	291 bcd
Hector	5.7 b	328 bcdef
Clark	5.8 b	326 bcdef
Steptoe	6.0 bc	399 f
Lewis	6.1 bc	341 bcdef
Harrington	6.2 bc	337 bcdef
Klages	6.2 bc	360 cdef
Short Betzes	7.0 c	490 g

[a]Means from 8 locations in Montana followed by the same letter are not significantly different at $p = 0.05$ by Newman-Keuls sequential studentized range test.
Source: Ref. 33.

TABLE 25 Beer Sales by United States Breweries in 1987

Company	Million barrels[a]	Percentage of total
Anheuser-Busch, Inc.	76.1	42.0
Miller Brewing Co.	39.3	21.7
The Stroh Brewery Co.	21.6	11.9
G. Heileman Brewing Co.	16.3	9.0
Adolph Coors Co.	15.7	8.7
Pabst Brewing Co.	6.1	3.4
Others	6.0	3.3
Total	181.1	100.0

[a]119 million liters
Source: Ref. 26.

TABLE 26 Malt Utilization and Barrels of Beer in U.S. Brewing Industry, 1959–1987

Year	Total utilization (1000 t)	Barrels of beer (million)	Malt/barrel (kg/hl)
1959	1186	91.0	11.2
1963	1246	98.0	10.9
1967	1485	116.6	10.9
1971	1670	134.1	10.6
1975	1918	157.9	10.4
1979	2220	183.5	10.3
1983	2190	195.7	9.7
1986	2156	194.0	9.5

Source: Data from Brewers Almanac 1987, U.S. Brewers Association.

TABLE 27 Levels of the Main Carbohydrates Found in Commercial Syrups, Compared with Wort from an All-Malt Mash[a]

Carbohydrates	Barley syrup	Green malt syrup	Malt extract	All malt brewing wort
Glucose (G) & fructose (F)	7–27	5.6–7.5	12.5–33.5	8–12 (G, 8–11; F, 1–4)
Sucrose	0–5.5	4.6–7.7	0–3.8	1–5
Maltose	46–60	48.4–55.1	18.1–54.0	42–50
Maltotriose	7.5–17.3	8.9–18.5	5.8–9.3	13–16
Dextrins	5.1–27	22.6–23.1	20.7–39.4	21–30
Total N (%)	0.62–0.80	0.64–0.70	0.7–1	(80–95 mg/100 ml)

[a]Carbohydrates expressed as % total carbohydrates. Nitrogen expressed as % syrup solids, except for wort .
Source: Refs. 15, 45 and 63.

TABLE 28 Spring Barley Acreage in the United States

State	Million hectares 1986	Million hectares 1987	Malting barley, 1987 (%)
North Dakota	1.50	1.22	79.7
Montana	0.97	0.93	37.8
Minnesota	0.49	0.49	96.0
Idaho	0.46	0.34	31.9
Washington	0.37	0.27	6.4
South Dakota	0.31	0.34	68.2
Colorado	0.15	0.09	37.1
Oregon	0.14	0.10	4.6
Wyoming	0.07	0.06	66.8

Source: American Malting Barley Association Inc., Gleanings, 1988.

TABLE 29 Recommended Values for Acceptable Malting Barley in United States

Factor	Values[a] Six-rowed	Two-rowed
Barley factors:		
Kernel shape	Ellipsoidal, axial ratio 2.0–2.5 to 1.	
Kernel size:		
On 2.78 mm (7/64") screen[b]	No maximum	60% mimimum–no maximum
On 2.38 mm (6/64")screen	70% minimum	85% minimum
Through 1.98 mm (5/64") screen	Less than 3%	Less than 3%
Germination	Uniform, rapid germination of at least 96% is important. Barley should mature rapidly and break dormancy quickly.	
Hull characteristics	Should be thin, bright, and adhere tightly to kernel during harvesting, cleaning, and malting.	
Protein	12.0–14.0%	11.5–13.5%
Water sensitivity	Equal to or less than standard cultivar.	
Skinned and broken	Less than 10%	Less than 10%
Malt factors:		
Total protein	12.0–13.5%	11.5–13.0%
Soluble protein	5.2% minimum	5.0% minimum
S/T protein ratio	40–44%	40–44%
Extract (fine grind)	Equal to standard cultivar with no upper limit.	
F-C difference[c]	Less than 2.0%	Less than 2.0%
Wort viscosity	Equal to or less than standard cultivar.	
Wort turbidity	Equal to or less than standard cultivar.	
Diastitic power	Equal to or greater than standard cultivar.	

[a]Cultivar standards: 'Morex' (six-rowed); 'Klages' (two-rowed).
[b]2.78 x 19 mm (7/64 x 3.4 in.) openings.
[c]Extract difference, fine grind minus coarse grind.
Source: Ref. 17.

TABLE 30 Composition of American Malts[a]

	Barley type[c]	
	Morex	Klages
Barley		
Protein, %	14.1	13.9
Kernel weight, mg	37.1	39.6
Plump, % on 0.24 mm screen	89.9	80.9
Color, agtron	69	66
Malt		
Extract, %	79.6	79.6
Extract difference (F-C), %	3.5	3.7
Wort color score	1.3	1.2
Wort protein, %	5.17	5.15
Protein ratio: soluble as % of total	37.5	37.8
Diastatic power, deg.	172	145
Alpha amylase, 20° dextrinizing units	43.8	41.7

[a]Analyzed by American Society of Brewing Chemists methods. Data from 1986 Western Regional Spring Barley Nursery, average of 8 locations, courtesy USDA-ARS, Cereal Crops Research Unit, Madison, WI.
[b]Morex is a six-rowed barley, and Klages is a two-rowed barley.

TABLE 31 Classes and Subclasses of Barley

Barley[a] class/subclass	Characteristics
Six-rowed barley	<10% two-row barley
six-rowed malting barley	>90% kernels with white aleurone layers
Six-rowed blue malt barley	>90% kernels with blue aleurone layers
Six-rowed barley	Does not meet previous classes
Two-rowed barley	<10% six-row barley
Two-rowed malting barley	>95% suitable matling types
Two-rowed barley	>90% two-row barley not meeting class above
Barley	Barley not meeting requirements for six-row or two-row barley, or with >10% black barley

[a]50% or more whole kernels of cultivated barley and less than 25% other grains.
Source: Ref. 65.

TABLE 32 Appearance, Temperature, and Humidity Requirements of a Few Common Pests of Stored Grain

Pest	Length and appearance	Optimum temperature	Minimum temperature	Minimum r.h.(%)[a]	Rate of increase in 4 weeks[b]	Comment
Granary weevil (*Sitophilus granarius*)	1.5–3.5 mm, dark brown, black	26–30°C (79–86°F)	15°C (59°F)	50	15	Over 35°C (95°F) life cycle not completed, adults cannot fly
Rice weevil (*Sitophilus oryzae*)	3.5 mm, dark; reddish spots on wing-cases	27–31°C (81–88°F)	17°C (63°F)	60	25	Adults can fly
Lesser grain borer (*Rhizopertha dominica*)	2.5–3 mm, dark brown, black, cylindrical, head turned down under thorax	32–35°C (90–95°F)	23°C (73°F)	30	20	At 27°C (80°F) the period from egg to adult is about 1 month
Rusty grain beetle (*Cryptolestes ferrugineus*)	1.6 mm, flattened, reddish-brown	32–35°C (90–95°F)	23°C (73°F)	10	60	Adult can fly
Red flour beetle (*Tribolium castaneum*)	4 mm, reddish-brown, similar in appearance to *T. confusum*	32–35°C (90–95°F)	22°C (72°F)	1	70	Adults fly
Confused flour beetle (*Tribolium confusum*)	4 mm, reddish-brown, shiny somewhat flattened	30–33°C (86–91°F)	21°C (70°F)	1	60	Adult does not fly, most abundant pest in flour mills in U.S.
Saw-toothed grain beetle (*Oryzaephilus surinamensis*)	2.5–3 mm, dark red-brown narrow, flattened; edges of thorax saw-toothed	31–34°C (88–93°F)	21°C (70°F)	10	50	Common pest in grain bins, elevators, mills, etc.; feeds on broken kernels and grain residue
Khapra beetle (*Trogoderma grananium*)	Adult 2–2.5 mm, light brown; larva 3.5 mm, hairy	33–37°C (91–99°F)	24°C (75°F)	1	12.5	Adults hard to detect, cast skins of hairy larvae more easily seen, adults do not fly
Angoumois grain moth (*Sitotroga cerealella*)	10–14 mm, buff-colored, stored grain moths	26–30°C (79–86°F)	16°C (61°F)	30	50	Injurious in southern U.S. where it attacks grain in fields as well as in storage
Australian spider beetle (*Ptinus tectus*)	About 2.5 mm, dark brown, globular	23–25°C (73–77°F)	10°C (50°F)	50	4	Nocturnal, rarely shows up in economic numbers in U.S.
Flour mite (*Acarus siro*)	0.4–0.5 mm, whitish, globular, eight brown legs; no division between head and body	21–27°C (70–81°F)	7°C (45°F)	65	2500	Taints infested grain, highly allergenic

[a]r.h. = required humidity.
[b]The rate of increase/4 weeks is for favorable conditions. Life cycles vary greatly in length according to the ambient conditions.
Sources: Refs. 15 and G. D. Johnson.

TABLE 33 United States Barley Exports by Country of Destination, 1983 and 1984

Region of destination	Barley, 1000 metric tons	
	1983	1984
North America[a]	0.4	32.2
EEC[b]	8.3	135.1
Belgium-Luxembourg	0	131.9
Italy	5.2	2.9
Western Europe (non-EEC)	18.1	0.8
Spain	18.1	—
Eastern Europe	129.8	147.7
East Germany	53.6	53.7
West Asia	598.0	1,383.5
Saudi Arabia	206.3	317.9
Central America	0.5	5.3
West Africa	0	31.5
North Africa	0	42.8
East Asia	561.3	631.8
Japan	287.4	413.1
Southeast Asia	189.0	107.1
Singapore	—	107.0
Caribbean	0.8	0.6
South America	9.0	2.6
World Total	1,515.2	2,521.0

[a]Exports to Canada and Mexico
[b]European Economic Community
Source: World Feed Grain Trade, 1962–1985, USDA Statistical Bulletin 755.

TABLE 34 United States Malt Exports by Country of Destination, 1984 and 1985

Country of destination	Malt (metric tons) 1984	1985
Canada	531	383
Costa Rica	12	76
Dominican Republic	11,302	8,560
Jamaica	6,804	5,158
Guyana	72	—
Peru	127	—
Venezuela	175	128
Other Americas	1,023	682
Japan	17,084	3,635
Other Countries	59,206	38,238
World total	96,336	56,860

Source: Data from Agricultural Statistics 1987, USDA.

REFERENCES

1. Åberg, E., and Wiebe, G. A., Taxonomic value of characters in cultivated barley, *USDA Tech. Bull*, 942 (1948).
2. Åhokas, H., Cytoplasmic male sterility in barley. I. Anther and pollen characteristics of *msm1*, restored and partially restored *msm1* genotypes, *Z. Pflanzenzucht, 81*:327–332 (1978).
3. Åhokas, H., Variation of kernel protein and lysine in the wild progenitor of barley, *Hereditas, 96*:29–37 (1982).
4. Åman, P., and Newman, C. W., Chemical composition of some different types of barley grown in Montana, U.S.A., *J. Cereal Sci., 4*:133–141 (1986).
5. Anderson, J. W., Fiber and health: An overview, *Nutrition Today* (Nov/Dec):22–26 (1986).
6. Anderson, M. K., and Reinbergs, E., Barley breeding. In *Barley*, ASA Mono. 26, Am. Soc. Agron., Madison, WI, 1985, pp. 231– 268.
7. Baldridge, D. E., Brann, D. E., Ferguson, A. H., Henery, J. L., and Thompson, R. K., Cultural practices. In *Barley*, ASA Mono. 26, Am. Soc. Agron., Madison, WI, 1985, pp. 457–482.
8. Benito, M. C., Sanchez, M., Shin, J., and Blake, T., A map of barley chromosome 2 using isozymic and morphological markers, *Biochem. Genet., 26*:387–394 (1988).
9. Berk, Z., *The Biochemistry of Foods*, Elsevier Scientific Publishing Co., Amsterdam, 1976.
10. Blake, T., Dept, Plant and Soil Science, Montana State University, Bozeman, MT.
11. Blum, A., *Plant Breeding for Stress Environments*, CRC Press, Inc., Boca Raton, FL, 1988.
12. Bose, R. D., and Dixit, P. D., Studies in Indian barley. II. The root system, *Indian J. Agric. Sci., 1*:90–108 (1931).
13. von Bothmer, R., Natural variation, phylogeny and genetic resources in *Hordeum*. In S. Yasuda and T. Konishi (Eds.), Barley Genet. V Proc. 5th Int. Barley Genet. Symp., Okayama, 6–11 October 1986. Sanyo Press Co., Ltd., Okayama, 1987, pp. 23–33.
14. von Bothmer, R., and Jacobsen, N., Origin, taxonomy, and related species. In *Barley*, ASA Monogr. 26, Am. Soc. Agron., Madison, WI, 1985, pp. 19–56.
15. Briggs, D. E., *Barley*, Chapman and Hall Ltd., London, 1978.

16. Brown, P. L., Black, A. L., Smith, C. M., Enz, J. W., and Caprio, J. M., Soil water guidelines and precipitation probabilities for barley and spring wheat in flexible cropping systems—Montana and North Dakota, *Bull. 356 Coop. Ext. Serv.*, Montana State Univ., Bozeman, MT, 1981.
17. Burger, W. C., and LaBerge, D. E., Malting and brewing quality. In *Barley*, ASA Mono. 26, Am. Soc. Agron., Madison, WI, 1985, pp. 367–401.
18. Carmona, J. F., and Greenhalgh, J. F. D., The digestibility and acceptability to sheep of chopped or milled barley straw soaked or sprayed with alkali, *J. Agric. Sci.*, (Camb.), *78*:477–485 (1972).
19. Ceccarelli, S., Tolerance to climatic stress. In S. Yasuda and T. Konishi (Eds.), Barley Genet. V Proc. 5th Int. Barley Genet. Symp., Okayama 6–11, October 1986. Sanyo Press Co., Ltd., Okayama, 1987, pp. 689–702.
20. Chapman, C. G. D., Barley genetic resources, the status of collecting and conservation. In S. Yasuda and T. Konishi (Eds.), Barley Genet. V Proc. 5th Int. Barley Genet. Symp., Okayama 6–11, October 1986. Sanyo Press Co., Ltd., Okayama, 1987, pp. 43–49.
21. Chojecki, J., ICI, Inc., Bracknell Berks, England.
22. Doll, H., and Koie, B., Evaluation of high lysine barleys. In *Breeding for Seed Protein Improvement Using Nuclear Techniques*, Vienna, IAEA, 1975, pp. 55–59.
23. Edwards, R. A., Donaldson, E. D., and MacGregor, A. W., Ensilage of whole-crop barley, I.—Effects of variety and stage of growth, *J. Sci. Fd. Agric.*, *19*:656–660 (1968).
24. Fedak, G., Wide Crosses in *Hordeum.* In *Barley*, ASA Mono. 26, Am. Soc. Agron., Madison, WI, 1985, pp. 155–186.
25. Fedak, G., Intergeneric hybrids in *Hordeum.* In S. Yasuda and T. Konishi (Eds.), Barley Genet. V Proc. 5th Int. Barley Genet. Symp., Okayama 6–11, October 1986. Sanyo Press Co., Ltd., Okayama, 1987, pp. 301–310.
26. Finnegan, T., Annual statistical studies, *Modern Brewery Age.*, *39*(11):4–20 (1988).
27. Harlan, H. V., and Martini, M. L., Problems and results in barley breeding, *USDA Agric. Yearb.*:313–345 (1936).
28. Harlan, J. R., On the origin of barley. In *Barley*, USDA Agric. Handbook 338, 1979, pp. 10–36.
29. Hayes, J. D., and Foster, C. A., Heterosis in self-pollinating crops, with particular reference to barley, *Eucarpia, 7*:239–256 (1976).
30. Heun, M., Lehrstuhl für Pflanzenbau und Pflanzenzüchtung, Technische Universität Müchen, Federal Republic of Germany.
31. Hockett, E. A., Hybrid barley research: history and potential, *Proc. Mont. Acad. Sci.*, *47*:27–35 (1987).
32. Hockett, E. A., Carroll, T. W., and Zaske, S. K., Registration of barley composite cross XXXI-A and B, *Crop Sci.*, *28*:722 (1988).
33. Hockett, E. A., McGuire, C. F., Newman, C. W., and Prentice, N., The relationship of barley beta glucan content to agronomic and quality characteristics. In S. Yasuda and T. Konishi (Eds.), Barley Genet. V Proc. 5th Int. Barley Genet. Symp., Okayama 6–11, October 1986. Sanyo Press Co., Ltd., Okayama, 1987, pp. 851– 860.
34. Hockett, E. A., and Nilan, R. A., Genetics. In *Barley*, ASA Mono. 26, Am. Soc. Agron., Madison, WI, 1985, pp. 187–230.
35. Hockett, E. A., and White, L. M., Simultaneous breeding for feed and malting quality. In M. Asher (Ed.), Barley Genet. IV Proc. 4th Inc. Barley Genet. Symp., Edinburgh, Scotland, 22–29 July, 1981. Edinburgh Univ. Press, Edinburgh, 1981, pp. 234–241.
36. Holwerda, B. C., Jana, S., Crosby, W. L., Chloroplast and mitochondrial DNA variation in Hordeum vulgare and Hordeum spontaneum, *Genetics, 114*:1271–1291 (1986).
37. Hough, M. N., Meteorological models for the effects of weather on barley development and yield, *Prog. Biometerology, 1*:240–251, 439–440 (1975).
38. Jackson, M., Weeds and weed control. In *The Montana Small Grains Guide*, MAES Bull. 364, 1985, pp. 63–65.

39. Jensen, C. J., Biotechnology: barley cell and tissue culture. In S. Yasuda and T. Konishi (Eds.), Barley Genet. V Proc. 5th Int. Barley Genet. Symp., Okayama 6–11, October 1986. Sanyo Press Co., Ltd., Okayama, 1987, pp. 493–501.
40. Kiesling, R. L., The diseases of barley. In *Barley*, ASA Mono. 26, Am. Soc. Agron., Madison, WI, 1985, pp. 269–312.
41. Kirby, E. J. M., and Appleyard, M., *Cereal Development Guide*, 2nd ed., Arable Unit, National Agric. Centre, Warwickshire, England, 1984.
42. Kleinhofs, A., Chao, S., and Sharp, P. J., Mapping of the nitrate reductase genes in barley and wheat. In P. E. Miller and P. M. D. Koebner (Eds.), Proc. 7th Int. Wheat Gen. Symp., Vol. I Cambridge, Inst. Plant Sci. Res., Cambridge, 1988, pp. 541–546.
43. Larson, L. M., and Oldfield, J. E., Improvement of barley rations for swine. III. Effect of fiber from barley hulls and purified cellulose in barley and corn rations, *J. Anim. Sci.*, *20*:440–444 (1961).
44. Mathre, D. E., *Compendium of Barley Diseases*, Phytopathol. Soc., St. Paul, MN, 1982.
45. MacWilliam, I. C., Composition of Brewing Syrups—A Review, *J. Inst. Brewing*, *77*:295–299 (1971).
46. Morrill, W., Small grain insects. In *The Montana Small Grains Guide*, MAES Bull. 364, 1985, pp. 81–86.
47. Moseman, J. G., Barley disease and their control. In *Barley*, USDA Agric. Handbook 338, 1979, pp. 55–77.
48. Moseman, J. G., and Smith, Jr., D. H., Germplasm resources. In *Barley*, ASA Mono. 26, Am. Soc. Agron., Madison, WI, 1985, pp. 57–72.
49. Munck, L., Barley for food, feed and industry. In Y. Pomeranz and L. Munck (Eds.), *Cereals: A Renewable Resource, Theory and Practice*, Am. Assn. Cereal Chem., St. Paul, MN, 1981, pp. 427– 459.
50. Newman, C. W., and McGuire, C. F., Nutritional quality of barley. In *Barley*, ASA Mono. 26, 1985, pp. 403–456.
51. Newman, R. K., Newman, C. W., Fadel, J., and Graham, H., Nutritional implications of beta-glucans in barley. In S. Yasuda and T. Konishi (Eds.), Barley Genet. V Proc. 5th Int. Barley Genet. Symp., Okayama 6–11, October 1986, Sanyo Press Co., Ltd., Okayama, 1987, pp. 773–780.
52. Nilan, R. A., The cytology and genetics of barley, 1951– 1962. Monog. Suppl. 3, Res. Stud. Vol. 32, No. 1, Washington State University press, Pullman, WA, 1964.
53. Percival, J., *Agricultural Botany*, 2nd ed., Duckworth, London, 1902.
54. Poehlman, J. M., Adaptation and distribution. In *Barley*, ASA Monogr. 26, Am. Soc. Agron., Madison, WI, 1985, pp. 1–17.
55. Pope, M. N., The course of the pollen tube in cultivated barley, *J. Am. Soc. Agron.*, *38*:432–440 (1946).
56. Poulsen, C., The barley chloroplast genome: Physical structure and transcriptional activity in vivo, *Carlsberg Res. Commun.*, *48*:57–80 (1983).
57. Ramage, R. T., Heterosis and hybrid seed production in barley, *Monogr. Theor. Appl. Genet.*, *6*:71–93 (1983).
58. Ramage, R. T., Cytogenetics. In *Barley*, ASA Mono. 26, Am. Soc. Agron., Madison, WI, 1985, pp. 127–154.
59. Ramage, R. T., A history of barley breeding methods, *Plant Breed. Rev.*, *5*:95–138 (1987).
60. Reid, D. A., Morphology and anatomy of the barley plant. In *Barley*, ASA Mono. 26, Am. Soc. Agron., Madison, WI, 1985, pp. 73–101.
61. Reid, D. A., Shands, R. G., and Suneson, C. A., Culture of barley in the United States. In *Barley*, USDA Agric. handbook 38, 1979, pp. 37–44.
62. Reid, D. A., and Wiebe, G. A., Taxonomy, botany, classification, and world collection. In *Barley*, USDA Agric. Handbook 38, 1979, pp. 78–104.
63. Roberts, R. H., and Rainbow, C., *Changes in Traditional Brewing Materials and Methods in Great Britain, Master Brewers Assoc. Am. Tech. Q.*, *8*(1):1–6 (1971).
64. Robertson, N. L., and Carroll, T. W., Virus like particles and a spider mite intimately associated with a new disease of barley, *Science*, *240*:1188–1190 (1988).

65. Schafer, W., Bauder, J., and Jones A. (Eds.), The Montana Small Grains Guide, *MAES Bull.*, 364 (1985).
66. Scholz, F., and Künzel, G., Hybrid barley—problems and advance, especially in developing genetic systems. In S. Yasuda and T. Konishi (Eds.), Barley Genet. V Proc. 5th Int. Barley Genet. Symp., Okayama, 6–11, October 1986, Sanyo Press Co., Ltd., Okayama, 1987, pp. 937–943.
67. Sharp, E. L., Breeding for pest resistance. In *Barley*, ASA Mono. 26, Am. Soc. Agron., Madison, WI, 1985, pp. 313–333.
68. Shin, J. S., Chao, S., Corpuz, L., and Blake, T. K., A partial map of the barley genome incorporating RFLP, PCR, isozyme and morphological marker loci. *Genome*, in press.
69. Singh, R. J., and Tsuchiya, T., An improved N-banding technique for the identification of barley chromosomes, *J. Heredity, 73*:227–229 (1982).
70. Smith, L., Cytology and genetics of barley, *Bot. Rev., 17*:1–51, 133–202, 285–355 (1951).
71. Sogaard, R., and von Wettstein-Knowles, P., Barley:genes and chromosomes, *Carlsberg Res. Commun., 52*:123–196 (1987).
72. Sotola, J., The chemical composition and nutritive value of certain cereal hays as affected by plant maturity, *J. Agric. Res., 54*:399–415 (1937).
73. Spiller, G. A. (Ed.), *CRC Handbook of Dietary Fiber in Human Nutrition*, CRC Press, Inc., Boca Raton, FL, 1986.
74. Starks, K. J., and Webster, J. A., Insects and related pests. In *Barley*, ASA Mono. 26, Am. Soc. Agron., Madison, WI, 1985, pp. 335–365.
75. Stolen, O., Kolster, P., and Monk, L., Multilines and mixtures of cultivars in barley. In S. Yasuda and T. Konishi (Eds)., Barley Genet. V Proc. 5th Int. Barley Genet. Symp., Okayama 6–11, October 1986, Sanyo Press Co., Ltd., Okayama, 1987, pp. 601–606.
76. Sundstol, F., and Owen E. (Eds.), *Straw and Other Fibrous By-Products as Feed*, Elsevier, Amsterdam, 1984.
77. Tsuchiya, T., Aneuploids and chromosome mapping in barley. In *Cytogenetics of Crop Plants*, MacMillan, India, 1983, pp. 251– 281.
78. Tsuchiya, T., Linkage maps of barley, *Barley Genet. Newsl., 16*:40–43 (1986).
79. Tsuchiya, T., List of barley genetic stocks, *Barley Genet. Newsl., 16*:81–121 (1986).
80. Tsuchiya, T., Gene analysis and linkage studies in barley. In S. Yasuda and T. Konishi (Eds.), Barley Genet. V Proc. 5th Int. Barley Genet. Symp., Okayama 6–11, October 1986, Sanyo Press Co., Ltd., Okayama, 1987, pp. 175–187.
81. Ward, D. J., Some evolutionary aspects of certain morphologic characters in a world collection of barleys, *USDA Tech. Bull.*, 1276 (1962).
82. Weaver, J. C., *American Barley Production*, Burgess Pub. Co., Minneapolis, 1950.
83. von Wettstein, D., Jende-Strid, B., Ahrenst-Larsen, B., and Sorensen, J. A., Biochemical mutant in barley renders chemical stabilization of beer superfluous, *Carlsberg Res. Commun., 42*:341–351 (1977).
84. Wiebe, G. A., Breeding. In *Barley*, USDA Agric. Handbook 338, 1979, pp. 117–127.
85. Wiebe, G. A., Genetics. In *Barley*, USDA Agric. Handbook 338, 1979, pp. 128–135.
86. Wiebe, G. A., Introduction of barley into the new world. In *Barley*, USDA Agric. Handbook 338, 1979, pp. 2–9.
87. Wiebe, G. A., and Reid, D. A., Classification of barley varieties grown in the United States and Canada in 1958, *USDA Tech. Bull.*, 1224 (1961).
88. Wilson, W. W., Production and marketing in the United States and Canada. In *Barley*, ASA Mono. 26, Am. Soc. Agron., Madison, WI, 1985, pp. 483–510.
89. Zohary, D., The origin of cultivated cereals and pulses in the near east, *Chromosomes Today, 4*:307–320 (1973).
90. Zohary, D., and Hopf, M., *Domestication of Plants in the Old World*, Clarendon Press, Oxford, 1988.

4

OATS

Michael S. McMullen
North Dakota State University, Fargo, North Dakota

I. HISTORY

A. Origin of Cultivated Oats

A probable chain of events leading to the development of modern cultivated oats was developed by Coffman (1). The exact sequence of events which resulted in development of cultivated oats will likely remain a mystery. The Asia Minor region is generally accepted as the place of origin of the wild hexaploid oat species, *Avena sterilis* and *A. fatua*, which are believed to be the progenitors of cultivated oats (1–3). Early investigators (4) assumed *A. fatua* was the wild progenitor of cultivated hexaploid oats. Coffman (2) presented evidence which indicated that *A. sterilis* is more likely to be the true progenitor. The issue still remains the subject of conjecture, and Ladizinsky (5) suggested that cultivated hexaploid oats originated as a secondary crop from the weedy form of *A. fatua* during the late Bronze and Iron Ages in central and western Europe.

Vavilov (6) suggested that oat culture expanded to the north from its Asia Minor center of origin as a weed admixture in cultivated emmer (*Triticum dicoccum* Schlub.) and einkorn (*T. monococcum* L.), which are less adapted to northern latitudes than oats and died out while oats thrived in the new environment. According to Coffman (7), Slavs and Scythians, warlike horsemen from the region of the Black Sea in central Asia, migrated over vast regions prior to A.D. 450, probably carrying with them cereal grain consisting mainly of barley, emmer, and spelt with oats present as weedlike mixtures. The oats in this mixture were assumed to be of *A. byzantina* C. Koch derivation similar to *A. sativa* L. *Avena sativa* types were probably better adapted to cooler northern climates and favored by natural selection resulting in *A. sativa* becoming the most widely cultivated type in northern Europe.

B. Expansion of Oat Cultivation

By the beginning of the seventh century oats were extensively established in western Europe as a grain and forage crop, and by A.D. 1024 oats had become an important crop in the British Isles (1). During the period of A.D. 1000–1500 a new system of agriculture evolved in northern Europe which included oats as an important component of crop rotations and contributed to utilization of the horse as a source of power allowing more efficient farming and transportation (8). The relationship of oats for feed to provide horsepower was established and greatly influenced increased culture of oats during the period when horsepower was important and contributed to decreased oat culture during the period of decreased dependence on horses for power.

Oat cultivation probably developed simultaneously in other regions of the world. Ancient Chinese historical records indicate that naked-seeded oats were grown in China prior to A.D. 1000 (1). Harlan (9) observed wild oats tolerated as mixtures in wheat and barley fields in Ethiopia. The wild oats were subsequently used by the farmers as a source of forage for livestock and food for humans.

Oats were first planted in the United States on Cuttyhunk, an island off the Massachusetts coast, in 1602 (1). Culture of oats became of some importance within 30 years after their initial introduction. Oat culture expanded slowly westward and little to the south since oats were not economically competitive with tobacco in the tobacco-growing regions.

The early farmers of Pennsylvania were general and livestock farmers, a system to which the culture of oats was well suited, resulting in expansion of oat production. Census reports of 1840 indicate production of 1.79 million tons of oats in the United States with New York, Pennsylvania, and Ohio ranking first, second, and third, respectively, in production. Nearly all oat production was east of the Mississippi River at this time. Immigrant farmers from northern Europe who settled in the upper Mississippi Valley were familiar with oat production and moved oat culture westward as new land was settled. By 1879, oat production had moved westward until the upper Mississippi Valley became the center of oat production in the United States. U.S. oat production continued to expand until the mid-1950s when acreage and production began to decline.

Culture of fall-sown oats developed simultaneously in the southern states. Many of the fall-sown cultivars were selected from a landrace variety, "Red Rustproof". According to Coffman (3), Red Rustproof was first grown by a farmer in South Carolina and reportedly would not rust. The landrace was originally called "Red Mexican Rust Proof Oats" and was brought back from Mexico by a soldier of the Mexican War in 1848.

C. Development of Oat Milling

Prior to 1850, oats were not used extensively as human food in the United States but were consumed in other areas of the world (10). Until this time oatmeal was imported into the Unites States from Scotland and Germany, where oat milling was well developed, and the products were dispensed by apothecaries in 1/2–lb packages to make an easily digestible food for invalids and convalescents. Ferdinand Schumacher, a pioneer miller of Akron, Ohio, began hand-grinding oats in a rear room of his grocery store in 1854. Demand for his product was high, and he organized the German Mills American Oatmeal Factory in 1856, which produced 20 180lb barrels of oatmeal daily with a water-powered stone mill. Raw oats were hulled by grinding them between two circular stones rotating in opposite directions. The groats were separated from the residue and crushed through another set of mill stones, resulting in a meal that was considered edible after 3–4 hours of cooking. A Schumacher employee, Asmus J. Errichsen, developed a method to produce steel-cut oats in 1877, which eliminated the mill stones and was the first significant modernization in oat milling. Steam and water power cut hulled oats into a fine meal by fine rotating knife blades. This process reduced the amount of flour produced thus increasing the amount of recovered product.

An influx of German and Irish immigrants in the 1870s and 1880s who were accustomed to oats as a staple food created increased market demand for oatmeal. In 1878 Schumacher imported porcelain rollers from England to manufacture rolled oats as a means of reducing the amount of unsalable powdery oat flour that previously had wasted one fourth of the oats in earlier processing methods. Rolled oats required considerably less cooking time than previous products, subsequently increasing the popularity of oatmeal as a breakfast food. As the demand for oatmeal increased, many other milling companies began manufacturing oatmeal and the modern oat milling industry of the United States began. Oat-processing machinery has since become much more efficient.

II. GENETICS AND BREEDING

A. Cytogenetic Relationship of Species Within Avena

Cultivated hexaploid oats (*A. sativa*) belong to the genus *Avena* within the family *gramineae* (3). The genus *Avena* consists of cultivated, wild, and weedy species with a basic chromosome number of $n = 7$. Species within *Avena* are classed cytologically according to the number of chromosomes in their cells as diploids ($2n$ = 14), tetraploids ($2n$ = 28), and hexaploids ($2n$ = 42) (11). Thomas and Rajhathy (12) classified 19 species of *Avena* on the basis of chromosome number, genome, and diaspore (unit of dispersal) (Table 1)

Leggett (13) used the same criteria to classify the diploid *A. atlantica* . Baum (14) used micromorphological characteristics and various clustering procedures to identify and describe 27 species within *Avena*. The genome of the perennial tetraploid appears somewhat related to that of *A. sativa*, but further work is needed to assign a genome designation (13).

Avena sativa and the wild hexaploid species are allopolyploids combining three related genomes designated A, C, and D, which have the same basic chromosome structure as *A. sativa* (15). Even though the hexaploids contain three sets of partially homologous chromosomes, meiotic behavior is diploid in nature, leading to disomic inheritance. The cytogenetic structure of oats is similar to hexaploid wheat with diploid like pairing under genetic control (16). The homologous relationship of chromosomes can be established through evaluation of nullisomatic-tetrasomic compensation (17). Due to the polyploid cytogenetic structure of *A .sativa,* chromosome deficiency and duplication are tolerated, which permits development of aneuploid lines (18,19). that are useful in studying the cytogenetic structure of the cultivated oat. Leggett (20) demonstrated cytoplasmic variability among six *Avena* species which is observed when the *A. sativa* nuclear genome is substituted into an alien source of cytoplasm. Robertson and Frey (21) provided evidence to suggest that cytoplasmic diversity provided by *A. sterilis* may be useful for improving productivity of cultivated oats.

B. Genetic Markers

Breeding of cultivated oats has been facilitated through increased understanding of the genetic control of desired characteristics of oat cultivars. The amount of genetic information available relative to a species is usually closely related to its economic importance and the ease with which genetic research can be performed. The allopolyploid nature of oats and difficulty in producing large numbers of hybrids relative to other crop species such as maize have limited the amount of oat genetic research possible. Jensen (22) extensively reviewed the genetics and inheritance of oats, and Murphy and Coffman (23) reviewed the genetics of disease resistance. A catalog of identified genes controlling characters of oats and a system of standardized gene and chromosome nomenclature was developed by Simons et al. (24). Although many genes have been characterized, an extensive linkage map of the oat genome is not available, partially due to the dearth of genetic markers, few multiple marker stocks available, and difficulty in working with oat aneuploids. Recent developments in the use of molecular markers for mapping plant chromosomes (25) should allow more efficient mapping of the oat genome.

C. Utilization of Germplasm Resources

Although *A. sativa* is the most widely cultivated species, the related species within the genus comprise a valuable genetic resource base for use in oat breeding. Nearly 20,000 accessions are included in the USDA World Oat Collection, of which 45% belong to wild and weedy species (26). The hexaploid accessions, including wild weedy species, comprise 95% of the collection and have been demonstrated to provide useful genetic variability for cultivar improvement. Minor structural differences are indicated by the meiotic behavior of hybrids between the hexaploids (27,28) but relatively free flow of genes between the hexaploid species occurs since there are no distinct isolation and sterility barriers. *Avena sterilis* germplasm has been used to improve crown rust resistance, stem rust resistance, barley yellow dwarf virus (BYDV) tolerance, grain protein content, and grain yield (26). Lawrence and Frey (29) described methods based on backcrossing for introgression of genes from *A. sterilis* germplasm into the *A. sativa* gene pool.

Utilization of germplasm from diploid and tetraploid species has been limited due to difficulties in achieving gene transfer to cultivated oats related to sterility and isolation barriers between the species. Understanding the cytogenetic relationship between *A. sativa* and the diploid and tetraploid species and application of appropriate chromosome manipulation techniques has allowed introgression of useful characteristics, especially disease resistance from the lower ploidy species into *A. sativa* (30–32).

D. Breeding

1. Breeding Objectives

Most breeding programs are directed toward producing cultivars with genetic potential to produce increased yields of grain with improved quality. The quality characteristics desired in a cultivar may vary relative to the end use of the crop. Special quality characteristics such as a white hull color and high test weight may vary relative to the end use of the crop. Most quality characteristics considered by breeders are related to nutritional value and milling yield. These characteristics include grain protein and oil content, hull content, and test weight. Recognition of nutritional attributes related to diabetes and control of blood cholesterol levels (33) has stimulated research activity to develop cultivars with increased levels of β-glucan or gum content. Increased β-glucan content may be undesirable for feeding certain classes of animals. Burrows (34) suggested methods for improving the yield and quality while providing protection from lodging, diseases and pests that may reduce yield, and quality short of the genetic potential.

Many diseases and pests can reduce productivity of oat cultivars, but BYDV, crown rust, and stem rust are generally considered the most important diseases of oats in North America. Brown and Jedlinsky (35) developed germplasm which provides good levels of protection from BYDV. Baltenberger et al. (36) demonstrated the utility of recurrent selection to improve BYDV tolerance. Specific strategies for providing stable protection from crown rust (37) and stem rust (38) have been suggested and incorporated into breeding programs.

2. Breeding Procedures

Genetic improvement of cultivated oats has been in progress since domestication of the crop nearly 2000 years ago (1). Oat breeding involves identification or creation of a population with useful genetic variation and application of appropriate breeding procedures utilize genetic variability to develop improved germplasm. Early improvement probably involved mass selection, which developed heterogeneous land races which were adapted to local areas. Oat breeding prior to 1930 mostly involved single plant selection from the land races to produce improved pure line cultivars. Shirreff described a technique for crossing oats in 1873. By 1885 hybridization was used in oat breeding by personnel of Gartons, Ltd. in England, but the success of selection from landraces in the United States precluded extensive use of hybridization of oats until the 1930s. Brown and Shands (39) described methods for crossing oats, and McDaniel (40) developed an approach crossing method which reduced labor requirements.

Many breeding procedures have been applied to oat improvement including pedigree selection, backcross breeding, bulk breeding, mass selection, and more recently recurrent selection. Stuthman (41) discussed oat breeding strategy issues and suggested design of a breeding program which involves (1) the issues of selection of parents to create desired genetic variability in a population and (2) the choice of procedures for selection among the progeny. Various methods for handling of progeny which advance populations more than one generation per year have been described (42–44). Methods that allow rapid generation advancement often do not allow effective selection for traits of low heritability but may allow more genetic gain per unit of time. Rodgers et al. (44) suggest that greater emphasis on rapid generation advance in oat breeding has contributed to recent yield improvements in oat cultivars.

To maintain progress in cultivar improvement, genetic variability must be available in the gene pool. Wych and Stuthman (45) evaluated oat cultivars adapted to Minnesota released in the period 1923–1979 and found a 0.9% increase in groat yield per year. Langer et al. (46) analyzed production and stability characteristics of oat cultivars released over a 40-year period in the United States and concluded that the oat germplasm introduced into the Upper Mississippi Valley of North America from northern Europe contained little genetic variation for yield. They suggested broadening the cultivated oat germplasm base through the use of germplasm from *A. sterilis* and *A. sativa.* Rodgers et al. (44) suggested that recent oat yield increases were due to use of a wider array of germplasm in breeding programs. Souza and Sorrells (47) suggested use of cluster analysis of coefficient-of-parentage calculated from pedigree analysis of North American oat cultivars, released from 1951 to 1985, to identify parents of divergent clusters that are most likely to produce transgressive segregation for desired characteristics. They also found a trend towards broadening of the oat germplasm base over time. From their analysis they concluded that the oat cultivars studied appear to be more diverse than either soft or hard red winter wheat.

The genetic base for improving oats appears quite diverse, suggesting that adequate variability is present in the gene pool for much further improvement of desired characteristics.

III. STRUCTURE

The morphology and development of the oat plant was well described by Bonnett (48). Fulcher (49) has thoroughly described the morphological and chemical organization of the oat kernel.

A. The Oat Plant

The oat plant is an annual grass with structure similar to the other cereals. The leaf consists of the blade, sheath and ligule. The absence of auricles (Fig. 1) may be used during vegetative stages of growth to differentiate oats from other small grains. The elongated internodes of mature stems have hollow centers, and the nodes are solid. The leaves are solitary, alternate, distichous, and sessile. (48). Tillers arise from axillary buds in the axils of

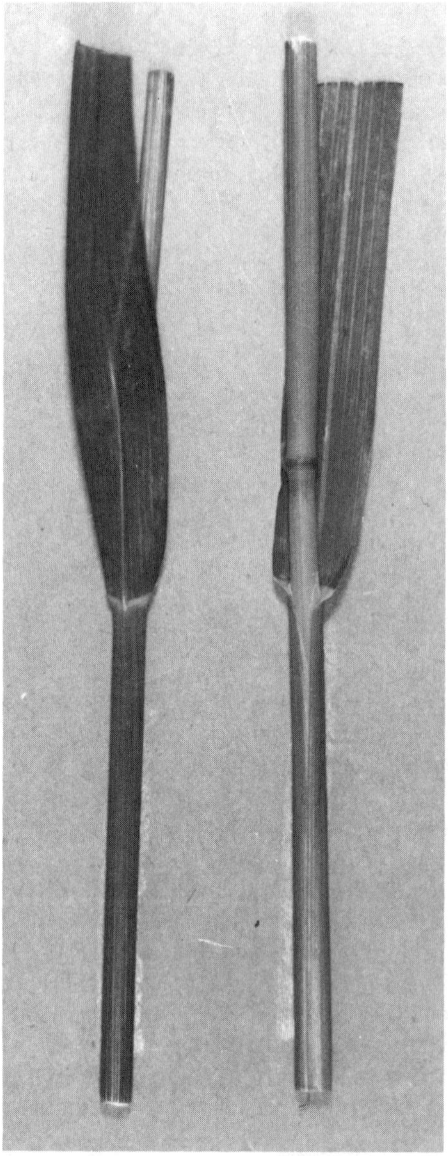

FIGURE 1 Culm and leaf sheath of *Avena sativa.*

the foliage leaves. Under normal growing conditions, the oat plant commonly matures 2–3 tillers although the number of tillers may be influenced by genotype, day-length, and many other environmental conditions. The culm usually has 4–5 nodes and internodes with the upper internode called the peduncle.

The oat plant has two root systems: the seminal roots, which originate during development of the embryo, and the adventitious roots, which arise in the nodes of the main stem and tillers just beneath the soil surface.

The inflorescence is a panicle composed of the rachis, rachis branches, and spikelets (Fig. 2). Panicle size and shape varies greatly with genotype and environment. The main

FIGURE 2 Panicle of *A. sativa.*

axis or rachis of the panicle is a continuation of the stem and ends in a single pedicellate spikelet. Each rachis branch terminates in a pedicellate spikelet. Spikelets are subtended by two empty glumes or bracts on a rachilla, which usually bears 1–3 fertile florets (Fig. 3). The glumes are membranous and generally longer than the spikelet. A floret includes the rachilla segments, lemma, palea, and the sexual organs or later the mature caryopsis. The rachilla segments of naked or hull-less varieties are extremely long (see Figs. 3,4) and three or more fertile florets per spikelet are normally produced. Mature spikelets separate from their pedicels and florets from rachilla segments. Awns, if present, arise dorsally on the lemma.

FIGURE 3 Spikelet and florets of mature *A. sativa*: normal oats (left), hull-less oats (right).

B. The Mature Grain

At maturity the caryopsis, usually called the groat, is tightly enclosed within the fibrous lemma and palea, which persist after threshing during harvesting operations and form the hull except in the case of hull-less oats (see Fig. 3). The caryopsis usually accounts for 65–75% and the hull 25–35% of the whole oat. The ratio of groat to hull is influenced by environment and genotype. Youngs (50) found a range of 68.2–76.4% groat and a range in groat weight of 18.7–22.7 g per 1000 kernels among seven genotypes grown in the same environment (Table 2). The lemma and palea of hull-less oats are more membranous and are mostly removed from the caryopsis during threshing and cleaning. A crease in the groat on the side opposite the embryo extends the entire length of the caryopsis. The caryopsis is covered with monocellular hairs called trichomes (Fig. 5).

The wall of the caryopsis consists of several cell layers derived from three different sources: the pericarp, from the ovary wall; the testa, from the inner integument; and the epidermis, from the nucellus (48). The endosperm is located inside the wall layers of the caryopsis and is composed of the aleurone cells and the starchy endosperm (see Fig. 5). The layer or two of cells which comprise the aleurone layer are more rectangular in shape and thicker walled than the starchy endosperm cells and contain aleurone grains. The aleurone cells secrete a variety of hydrolytic enzymes during germination which digest and mobilize the starchy endosperm reserves. The wall and aleurone layers form the milling fraction referred to as the bran. Youngs' (50) hand-dissection of groats of seven different genotypes found a range of 28.7–41.4% bran content (Table 3). Variation for bran thickness was found to vary among genotypes (see Table 2). Lower bran yield was recovered using an experimental mill.

The starchy endosperm is composed of large thin-walled parenchyma cells filled with starch contributing approximately 55.8–68.3% of the weight of the mature kernel (see Table 3) and comprises the commercial milling fraction called the starchy endosperm. The starchy endosperm is the primary source of nutrients for use by the embryo during germination.

The oat embryo, comprised of scutellum and the embryonic axis, is located on the anterior side, near the base of the caryopsis (see Fig. 5). The epidermal layer of the scutellum or epithelium is in contact with the endosperm and secretes enzymes, which digest contents of the endosperm during germination (48). The embryonic axis consists of the plumule and the radicle. The plumule is made up of the coleoptile, which encloses the shoot apex and two leaf primordia. The root is comprised of the radicle and two or three adventitious roots,

FIGURE 4 Panicle of hull-less oat.

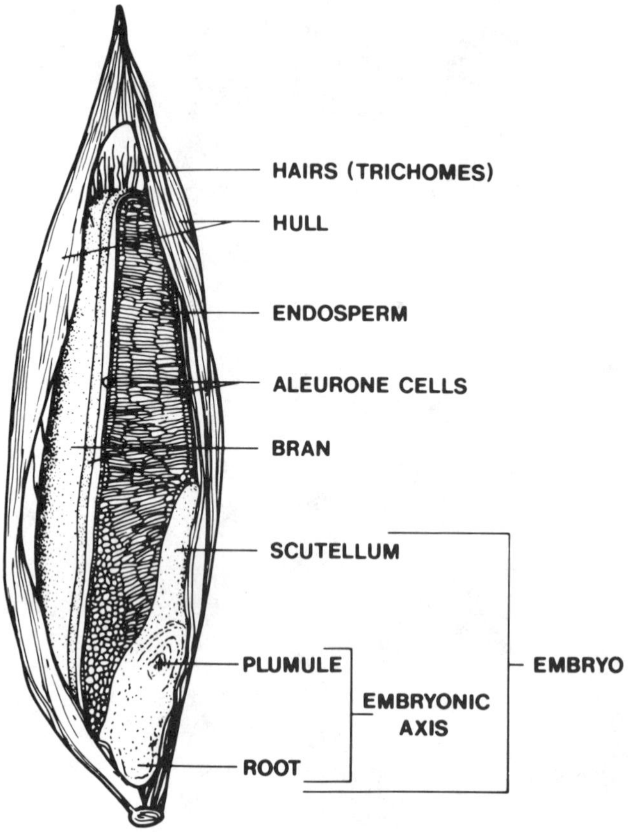

FIGURE 5 Mature oat floret indicating source of groat fractions.

which produce the seminal root system. Youngs (50) determined that the embryonic axis and the scutellum comprise 1.0–1.4% and 1.7–2.6%, respectively, of the groat.

IV. PRODUCTION

A. World Production

Oats are a cool season crop best adapted to areas between latitudes of 35–50° north and 20–40° south, which is reflected by the areas of greatest production (Table 4) (51, 52). Oats account for approximately 5–7% of world coarse grain production and rank sixth in world cereal production after wheat, maize, rice, barley, and sorghum (53). World production of oats has been maintained at a level near 50 million tons since the early 1960s. In recent years oat production has generally increased in the U.S.S.R., West Germany, and China, while declining in the United States, Canada, and Poland. The U.S.S.R. has the greatest number of hectares under oat production and the most oat production, accounting for more than 40% of total production (see Table 4). The 10 leading oat-producing countries account for more than 80% of world production. Recent average highest yields per hectare have been achieved in the Netherlands, Switzerland, West Germany, the United Kingdom, Ireland, Sweden, and France. Grain yield per unit of land has nearly doubled since World War II because of improved production practices and improved disease-resistant cultivars (53).

B. U.S. Oat Production

Oat production in the United States has declined since the 1950s when more than 17 million tons were produced annually (53) until reduced planting and stress induced yield reductions during 1988 resulted in record low production of 3.19 million tons. Grain yield has been increasing at a steady rate, but the area harvested for grain has decreased by about 387 thousand hectares per year since 1960. Although oats are produced in many regions of the United States (Table 5), production has been concentrated in the north central states of Iowa, Minnesota, North Dakota, South Dakota, and Wisconsin, which account for about 65% of oats harvested for grain each year (52).

Demand for oats for human consumption has steadily increased during a period when production has steadily declined, leading to a greatly increased proportion of U.S. oat production used for human food (Table 6). Prior to 1982, the United States was a net exporter of oats with a market share of world exports averaging 16% in the 1960s (53). During 1970–1974, the U.S. export share rose to 21.7%, but declined to 5% during 1980–1985, and the United States has been a net importer of oats since 1982. During this period imports have originated mainly in Canada, Sweden, and Finland, with the Netherlands, Australia, and Argentina contributing smaller amounts. Increased demand for oats for human consumption and for quality feed for horses has renewed interest in oat production, however, commodity target price levels designated by the Food Security Act of 1985 apparently are hindering expansion of oat production to meet domestic demand.

V. CHEMICAL COMPOSITION

A. Protein

The protein quality and content of oats is recognized as superior to other cereals for livestock feed and human nutrition. Peterson and Brinegar (55) extensively reviewed the characteristics of oat storage protein which make oats nutritionally valuable. Frey (56) has published a review of the genetics controlling oat protein content.

1. Protein Content and Distribution

Protein concentration of oat groats is considerably higher than that of other cereals. Groat protein concentration of cultivated genotypes is typically 15–20%. Most of the protein of the whole oat is located in the groat, and a low concentration of found in the hull (Table 8).

Although the bran comprises a smaller portion of the groat than the starchy endosperm (see Table 3), the higher protein concentration of the bran (Table 9) causes it to contribute an amount of protein to the kernel similar to the contribution of the starchy endosperm (Table 10). The scutellum and embryonic axis, even though relatively high in protein, contribute little to the total protein of the groat due to their small proportion relative to the mass of the groat.

Pomeranz et al. (57) observed a much wider range in protein content among *Avena* species other than *A. sativa* with groats of samples of some accessions reaching 37% protein. A large amount of genetic variability for improvement of grain protein content in cultivated oats is available in both *Avena sativa* and the related wild and weedy species.

2. Solubility Classification

Peterson (59) discussed extraction and classification of storage protein using the classic technique of Osborne (58). Oats and rice, unlike other cereals, have a low prolamin (alcohol-soluble fraction) and high globulin (salt-soluble fraction) content relative to other cereals. The globulin fraction is relatively rich in lysine. Using Osborne classification of the protein of five oat cultivars, Peterson (59) reported a prolamin content ranging from 7 to 13% of

total protein; the water-soluble albumins, 10–19%; the salt-soluble globulins, 52–56%; and glutelins, 21–27%. The actual globulin concentration may be as high as 75% since the Osborne method probably does not completely extract the globulins.

Peterson (59) found that increased protein in cultivars grown in diverse environments was related to increased levels of the globulin fraction. Since the amino acid profile of globulins is similar to that of the total protein, the amino acid profile of oats, unlike other cereals, remains relatively constant over a broad range of protein levels.

3. Amino Acid Composition and Distribution

Robbins et al. (60) surveyed 289 groat samples of cultivars grown in the United States and Canada from 1900 to 1971. Little variability was detected among the nutritionally limiting amino acids lysine, threonine, and methionine (Table 11). The average lysine and threonine concentrations of 4.2 and 3.3 g/100 g of amino acids recovered are higher than in other cereals, but below the FAO reference standard of 5.5 and 4.0 g/100g g, respectively, for lysine and threonine (61). A similar amino acid composition was found for commercially milled oats (see Table 8). Amino acid composition of oats is considered quite constant over a wide range of protein contents.

Separating groats into milling fractions, Pomeraz et al. (57) found a higher concentration of lysine in the embryonic axis and scutellum than in other portions of the kernel (Table 12). Lysine, threnonine, and methionine concentrations in the bran were similar to those of the endosperm. Fulcher (49) found approximately 25% higher lysine content in the bran of groats of the cultivar Scott. This suggests that significant differences exist in the amino acid composition between the bran and starchy endosperm of some cultivars.

B. Lipids

Oats are unique among the cereals for their high lipid content. Although oat oils are not used commercially, they are important nutritionally and must be considered in commercial milling operations because of problems associated with rancidity caused by lipid-related enzymes which are activated during milling operations. Genetic manipulation of lipid content may become more important in the future to improve milling quality or to increase energy of oats for livestock feeding. Youngs (62) has published a recent review concerning oat lipids and lipid-related enzymes which will be useful to the reader interested in oat lipids.

1. Lipid Content and Distribution

Oat groats have the highest lipid concentration among the cereals. Oat lipids can add a significant amount of energy to livestock feed and are important in human nutrition because they are highly unsaturated and contain considerable amounts of an essential fatty acid for humans, linoleic acid. An extensive survey of lipid content among oat genotypes (63) found a range in lipid concentration of 3.1–11.6%. Other studies have found similar ranges in lipid concentration (62). Youngs et al. (64) used hand-dissection of groats of two cultivars to determine lipid distribution within the oat. The greatest lipid concentration was in the scutellum (Table 13), but since the scutellum accounts for only a small portion of the mass of the kernel, it contains only a small amount of the total lipid in the groat (62). In both cultivars, the starchy endosperm and bran contained in excess of 90% of the total groat lipid even though lipid concentration is relatively low in these fractions. Most of the lipid is not bound and is easily extracted.

2. Lipid Composition

Youngs (65) used hand-dissection of groats to determine the distribution of lipid components in the groat. Thin layer chromatography was used to separate and quantify 12 lipid components (Table 14). The percentage of triglyceride was least in the bran and greatest in the embryo. Youngs (65) determined the relative fatty acid composition of 15 oat strains. The mean relative composition was: myristic, 0.6; palmitic, 18.9; stearic, 1.6; oleic, 36.4; and linoleic, 40.5. Other workers have obtained similar results. Wide variation in individual fatty acid concentrations occurs among cultivars, which should allow genetic manipulation of fatty acid composition.

C. Polysaccharides

1. Starch

Oat starch requires different extraction techniques than that of wheat. Wet milling techniques have been developed which produce starch yields in the range of 43–61% with protein and lipid content less than 0.4% and 0.3%, respectively (66).

2. β-Glucan

A nonstarch polysaccharide which dissolves in water to form a viscous gum is found concentrated mainly in the cell wall of the subaleurone layer of the groat (69). This polysaccharide is a mixed linkage β-D-glucan. Oat β-glucan was recently reviewed by Wood (70). The mixed-linkage β-glucan of oat is an unbranched polysaccharide composed of (1-4) and (1-3) linked β-D-Glucpyranosyl units in varying proportions. Since the β-glucan is concentrated mainly in the subaleurone layer, oat bran provides an enriched source of gum and has been demonstrated to produce beneficial effects in lowering serum blood cholesterol and moderating glucose metabolism in diabetics. These nutritional benefits attributed to β-glucan have recently greatly increased interest in oats as a human food.

Obtaining complete extraction of oat β-glucan free from contaminating starch, pentosan, and protein for quantitative purposes is difficult. The nonenzymatic Calcofluor method of Wood and Weisz (71) and hydrazinolysis method described by Martin and Bamforth (72) appear useful for quantifying β-glucans in oats. Welch and Lloyd (73) used modified Calcofluor and hydrazinolysis methods to evaluate β-glucan content in 100 oat genotypes. They observed a range in β-glucan content from 3.2 to 6.3% among cultivars.

D. Minerals

Morgan (74) surveyed the mineral content of 171 oat samples and found that oats can contribute significantly to mineral nutrition (Table 15). Peterson et al. (75) determined the mineral composition of the bran and endosperm of oats and found that the mineral components are concentrated in the bran fraction (Table 16). Since oats are normally consumed as a whole grain. oat products can contribute significantly to human mineral nutrition. Peterson's study also indicated genetic variability among cultivars for mineral content, suggesting that mineral nutrition could be improved by breeding.

E. Vitamins

Shukla (76) summarized vitamin composition information available for groats and rolled oats (Table 17). Oat products can contribute significant amounts of thiamine and pantothenic acid to human nutrition.

VI. PROCESSING AND UTILIZATION

A. Utilization

Utilization of oats for food and other industrial products was extensively reviewed by Shukla (76). Webster (77) published a more recent review of oat utilization.

Although recent interest has stimulated demand for oats as a human food, the largest portion of the oat crop is still used as a livestock feed (see Table 6). Many livestock producers recognize the unique nutritional value of oats especially for classes of livestock which require feed with a relatively high level of good quality protein but with lower energy content. Most horse owners prefer oats over other grains because their bulky hulls produce a loose mass which is easily digested (78).

Most oats used as food are consumed as whole grain products so that the nutritional quality characteristics of the groat are reflected in the food product. The major food uses of oats are in hot cereals, cold cereals, bread products, cookies and infant foods (77). Rolled oats is the most important form of oats in hot cereals, but recognition of the nutritional quality of oat bran has greatly increased its consumption as a hot cereal. Oat flour is often used as an ingredient in cold cereals. Oat products are often used in bakery items to produce desired texture or to increase moisture retention. Demand for food containing oat products has increased substantially since health benefits were demonstrated (33). The oat bran is rich in soluble fiber that is active in lowering blood serum cholesterol levels. D'Appolonia and Youngs (79) found that addition of oat bran to bread dough increased farinograph absorption and produced doughs with good stability, but decreased loaf volume.

Unique antioxidant properties have been attributed to oat flour (62). Oat flour was formerly used commercially as an antioxidant, but has been largely replaced by butylated hydroxyanisole (BHA) and butylated hydroxytoluene (BHT). Compounds which contribute to oat antioxidant properties include glyceryl esters of hydroxycinnamic acid, ferulic, and caffeic acids.

Wood et al. (80) found that oat gums produce viscosities in solution that equal or exceed other biological colloids. When dissolved in hot water the gums produce highly viscous solutions that slowly lose viscosity on standing and may have potential for use as an industrial hydrocolloid.

Oat hulls contain an average of 29.5% pentosans which are used in the manufacture of furfural through a series of reactions initiated by the hydrolysis of pentosans and completed by a complex cyclodehydration to fufural (81). Oat hulls have cariostatic properties which may lead to sweeteners for chewing gum that would provide protection from dental caries.

B. Processing

Deane and Commers (82) provided a detailed description of oat cleaning and processing. Much of the following information is taken from their description. Many different procedures can be used in the cleaning and processing of oats, and each processor, depending on the size of operation and market requirements, has preferred methods to achieve maximum efficiency. Even with the great diversity of approaches taken by processors, some operations must be addressed for efficient processing and production of a quality product. The general steps in oat processing are cleaning, hulling, steaming, and flaking. Oats must be cleaned and graded for size prior to hulling for efficient operation of the huller. Groats must be steamed to inactivate lipolytic enzymes which will result in rancidity and short shelf life of the product if the enzymes are activated during cutting and rolling of the groats. The products of oat milling and average percentages expected from the milling process are: rolled oat, 45–60%; oat hulls, 24–27%; feed oats, 10–20%; mixed grains and seeds, 2–3%;

and fines, 2–5% (83). These values will vary with the quality of the oats and the overall efficiency of the milling operation.

1. Cleaning

Preliminary cleaning of field oats received at the processing plant usually involves passing the oats over a perforated inclined screen of a receiving separator or horizontal, slowly rotating, coarse wire-mesh reels or cylinders of a scalper-reel type receiving separator. The perforations allow the oats to fall through while the coarse impurities such as straw and sticks are overtailed. Aspiration is usually used in the preliminary cleaning process to remove some light impurities. The preliminary cleaning is intended to be rapid enough for use when unloading trucks or railroad cars. After the preliminary cleaning, the oats are usually graded and binned by degree of contamination, milling yield, and moisture content so that the oats in each bin are relatively uniform. The processing equipment will require less frequent adjustment if the oats being processed have constant quality. The binned oats are stored until needed for further processing.

After preliminary cleaning, the first step in oat processing is a specialized cleaning system that may involve various types of cleaners to remove other grain and seed impurities. Other undesirable materials are separated from the oats before milling. Oats that are not suitable for milling are also removed in this step. These include bosom oats, pin oats, light oats, and other undesirable types.

The next step usually involves a milling separator combining coarse and fine screening with aspiration. In the milling separator the oats and fine material fall through the top sieve layer, which is clothed with perforated sheet metal or wire mesh to provide a close scalping operation onto the lower sieve layer with finer screens for removal of fines. After the milling separator cleaning, various specialized cleaning machines for further cleaning and grading are used. These machines include disk and indented cylinder separators that perform length separation of particles, a width sizer that separates grains when their length is the same but the width is different, a gravity separator that separates impurities that have a different specific gravity than oats, and a paddy separator which is used to separate materials of nearly equal size, shape, and specific gravity, but with a different surface texture. The cleaning machines can be used in various combinations, but usually involving the use of the disc or indented cylinder separators first to separate the grain into large and small fractions. From the disc separators, the large and small fraction pass to the width sizer, and then to the gravity table. The objective of the cleaning and grading is to produce well-cleaned oats graded according to size for hulling or removal of the hull.

2. Drying and Cooling

Oats are heated prior to dehulling to partially inactivate the lipase enzymes to prevent development of undesirable flavors during and after processing. The heating also develops a slightly roasted flavor considered desirable by most processors. Heating also makes the oat hulls more brittle for easier hulling. During the heating, temperature of the oats usually ranges from 88 to 93° C and moisture content is reduced to the 7–10% range. The oats are cooled before further processing.

3. Hulling

Impact hullers have replaced stone hullers. The dried, graded oats enter the center of a high speed rotor fitted with fins that centrifugally throw the oats against a carborundum or hard rubber ring on the machine housing where the hulls are separated from the groats by impact and abrasion. Rotor speed, which is usually between 1400 and 2000 rpm, is adjusted to obtain maximum efficiency depending on the condition of the oats. The hard rubber ring on the dehuller is preferred by some millers to reduce the breakage of groats and less formation of fines. The hulls and fines are separated from the stream by aspiration. The groats are polished with mild abrasion in a scourer, which is a horizontal cylinder lined with a moderately rough material. A horizontal shaft with inclined blades conveys the groats against the rough surface, removing any small pieces of husk still adhering to the groats. Aspiration removes the husk pieces. Disc or indented cylinder separators are used to separate the unhulled oats from the groats. Use of the paddy separator may be necessary since the unhulled oats are often clipped in the hulling operation so that they are nearly the same length as the groats. The recovered unhulled oats are passed through the huller a second time.

4. Cutting and Flaking

Whole groats produce very large flakes when rolled, so groats are usually cut into 2–4 uniform pieces before rolling. Cutting in a rotary granulator is achieved by passing the groats through a rotating drum with countersunk holes that align the oats so that the end passes through the hole against stationary knives on the outside and bottom of the drum. As the tip of the groat passes through the perforation, it is sheared off by the knife. The fines produced by the cutting operation are removed using a shaker screen. The fines consist of oat middlings and flour and are used to produce high quality animal feed. The cut groats are conditioned for flaking by steaming with live steam. The heat and moisture serve to bind the flattened groat together and prevents disintegration during the flaking process. Consistent and homogeneous heat and moisture distribution are essential to produce a good yield of quality flakes. The steaming also completes the inactivation of the lipase enzymes. The cut groats are retained in a steamer for about 12–15 minutes during which time the temperature of the groats is increased to 99–104° C.

The steamed groats pass directly to the rolls from the steamer. The rolls run at the same speed in the same direction and rotate at 250–450 rpm. Cut groats are rolled into thin flakes (0.10–0.015 in.) for instant or quick-cooking oatmeal. Flakes marketed as regular oatmeal about 50– 75% thicker. After rolling the flakes pass over a shaker screen where fines are removed and agglomerations of overcooked flakes are scalped off. The flakes are cooled to around 43° C and are ready for packing.

5. Oat Flour

The same procedures involved in producing oat flour are used to produce flakes. The oats are cleaned and graded, dried, and hulled and cleaned to produce groats. The groats are steamed before milling to improve the stability and complete the inactivation of lipolytic enzymes. A rotary granulator is used to cut the groats after they are steamed. The cut groats are ground with an impact-type hammer mill, and the ground product is shifted on a gyratory sifter equipped with wire meshes to separate the oat flour fraction corresponding to the granulation requirements.

TABLE 1 The Species of *Avena* and Their Genomic Compositions

Ploidy	Species	Genome
Diploid		
($2n = 2x = 14$)	*clauda*	CpCp
	pilosa	
	ventricosa	CvCv
	prostrata	ApAp
	longiglumis	AlAl
	damascena	AdAd
	canariensis	AcAc
	weistii	AsAs
	hirtula	
	strigosa	
	atlantica[a]	
Tetraploid		
($2n = 4x = 28$)	*barbata*	AABB
	vaviloviana	
	abvssinica	
	morocanna	AACC
	nurphyi	
	macrostachya	?
Hexaploid		
($2n = 6x = 42$)	*fatua*	AACCDD
	sterilis	
	byzantina	
	sativa	

[a]From Ref. 13.
Source: Ref. 12.

TABLE 2 Physical Characteristics of Oats (*A. sativa*)

Variety	Groats (%)	Groats (1000 Kwt.) (g)	Groats, bran thickness (mm)
Orbit	71.9	21.5	0.063
Lodi	68.2	20.7	0.058
Garland	72.0	14.4	0.065
Froker	74.9	22.7	0.079
Portal	73.4	16.7	0.075
X–1289	76.4	19.5	0.087
X–1656	73.4	20.3	0.101

Source: Ref. 50.

TABLE 3 Weight Distribution of Groats (*A. sativa*)

Variety	Embryonic axis (%)	Scutellum (%)	Bran (%)	Starchy endosperm (%)
Orbit	1.2	1.8	28.7	68.3
Lodi	1.1	2.0	33.6	63.3
Garland	1.4	2.1	33.9	62.7
Froker	1.0	1.8	30.2	67.0
Portal	1.2	1.6	39.8	57.4
X–1289	1.1	2.6	34.4	61.8
X–1656	1.1	1.7	41.4	55.8

Source: Ref. 50.

TABLE 4 Area, Yield and Production of Oats in the World, 1984–1986

Country	Area harvested (1000 ha)			Yield per ha (1000 t)			Production (1000 t)		
	1984	1985	1986	1984	1985	1986	1984	1985	1986
U.S.S.R.	12,806	12,604	13,173	1.50	1.63	1.67	19,200	20,500	21.929
United States	3,304	3,309	2,780	2.08	2.28	2.01	6,875	7,559	5,582
Canada	1,406	1,411	1,555	1.90	2.12	2.51	2,670	2,997	3,906
West Germany	555	692	605	4.52	4.74	4.44	2,507	3,278	2,687
Poland	934	995	924	2.79	2.70	2.69	2,604	2,682	2,486
Sweden	428	445	455	4.44	3.75	3.25	1,904	1,668	1,478
Australia	1,041	1,068	1,255	1.31	1.25	1.31	1,367	1,330	1,681
France	444	433	315	4.22	4.09	3.48	1,875	1,770	1,096
Finland	419	411	403	3.16	2.96	2.91	1,321	1,218	1,175
East Germany	161	178	163	4.36	4.20	4.01	700	746	666
Spain	479	459	384	1.64	1.48	1.10	788	680	422
Argentina	434	333	400	1.65	1.20	1.40	717	400	560
United Kingdom	106	134	97	4.88	4.58	4.85	517	614	470
Norway	123	127	127	4.90	4.07	3.15	601	518	400
People's Republic of China	300	300	300	1.67	1.63	1.67	500	490	500
Italy	190	182	184	2.26	2.00	2.16	433	363	397
Czechoslovakia	129	121	115	3.70	3.90	3.65	479	473	419
Turkey	172	167	162	1.84	1.88	1.85	316	314	300
Austria	77	75	73	3.77	3.77	3.68	292	284	270
Yugoslavia	153	151	152	1.67	1.67	1.71	256	252	260
Korea DPR	90	95	98	1.89	1.95	1.95	170	185	192
Chile	96	85	64	1.69	2.01	1.95	163	170	124

TABLE 4 (Continued)

	Area harvested (1000 ha)			Yield per ha (1000 t)			Production (1000 t)		
Country	1984	1985	1986	1984	1985	1986	1984	1985	1986
Denmark	34	42	25	4.67	3.97	4.44	158	168	111
Hungary	44	44	41	3.56	3.03	3.01	156	133	124
Portugal	194	190	194	0.78	0.63	0.71	152	119	134
Brazil	114	142	129	1.00	1.11	0.94	114	157	121
Mexico	117	90	100	1.55	0.94	1.00	180	85	100
Belgium–Luxembourg	27	29	21	4.37	4.24	4.00	118	123	84
Ireland	25	23	21	5.66	4.55	4.88	141	106	102
Romania	67	72	70	1.41	1.41	2.14	94	102	150
Algeria	125	150	150	0.51	0.72	0.53	64	108	80
Greece	44	43	42	1.53	1.51	1.87	67	65	79
New Zealand	20	13	13	3.79	3.79	3.72	75	51	48
South Africa	496	436	449	1.63	0.28	1.63	81	12	55
Morocco	55	47	53	0.81	0.77	1.45	44	36	77
Netherlands	12	11	7	4.79	5.16	6.11	58	58	40
Mongolia	56	44	49	0.69	1.20	1.04	38	53	50
Switzerland	10	10	7	5.09	5.23	4.22	53	52	31
Uruguay	50	33	43	0.98	1.15	0.79	48	38	34
Others	112	121	121				98	122	133
World	25,449	25,315	25,319	1.89	1.98	1.92	47,994	50,079	48,556

Source: FAO monthly bulletin of statistics, vol. 10, 1987.

TABLE 5 Area, Yield and Production of Oats in the United States, 1984–1986

State	Area harvested (1000 ha)			Yield per ha (1000 t)			Production (1000 t)		
	1984	1985	1986	1984	1985	1986	1984	1985	1986
S. Dak.	628	575	425	2.01	2.01	1.58	1,261	1,155	671
Minn.	486	445	344	2.33	2.51	1.83	1,133	1,119	630
Wis.	348	316	344	2.22	2.37	25.25	775	748	766
Iowa	300	308	255	2.30	2.73	2.15	688	839	549
N. Dak.	397	340	283	1.83	1.90	1.97	726	647	559
Mich.	142	158	109	2.23	2.41	2.26	315	380	247
Nebr.	130	170	146	1.76	2.18	2.12	228	372	309
Penn.	118	122	109	2.05	2.51	2.23	232	3.5	243
Ohio	89	126	65	2.26	3.05	2.72	201	382	177
N.Y.	73	93	77	2.12	2.76	2.33	154	257	179
Ill.	67	65	81	2.47	2.80	2.47	165	181	201
Tex.	101	122	81	1.26	1.79	1.51	127	218	122
Kans.	49	76	81	1.9	2.01	1.93	92	153	157
Oreg.	30	41	32	3.16	3.30	3.41	96	134	110
Ind.	32	45	37	2.23	2.47	2.55	72	110	93
Mo.	26	43	41	1.72	1.97	1.80	23	84	73
Okla.	32	26	40	1.65	1.54	1.58	54	41	64
Mont.	43	28	37	1.32	1.19	1.65	56	34	60
Calif.	20	18	18	2.47	1.41	2.51	50	44	46

TABLE 5 (Continued)

State	Area harvested (1000 ha)			Yield per ha (1000 t)			Production (1000 t)		
	1984	1985	1986	1984	1985	1986	1984	1985	1986
N.C.	28	25	22	2.08	1.51	1.44	57	38	32
Maine	16	19	16	2.01	2.62	2.33	33	49	38
Wyo.	28	18	22	2.08	2.29	2.01	47	29	39
Colo.	20	22	16	1.97	1.90	1.97	40	42	32
Idaho	18	16	12	2.44	1.90	2.47	44	31	30
Ga.	24	18	14	1.97	1.61	1.40	48	29	20
Wash.	12	13	13	44	2.33	2.33	30	31	31
Ark.	11	7	14	2.51	2.33	2.41	29	16	32
S. C.	16	17	11	2.08	1.36	1.29	34	253	15
Ala.	12	14	12	1.72	1.47	1.44	21	21	17
Utah	5	5	5	2.41	2.55	2.59	13	13	13
Md.	6	6	6	2.05	2.15	1.97	12	13	12
Va.	5	5	4	1.69	1.69	1.65	8	10	7
W. Va.	3	3	4	1.83	2.19	1.80	6	7	7
Ky.	2	4	2	1.51	1.62	1.51	4	6	4
N. J.	2	2	2	2.01	2.26	1.94	5	5	3
U. S.	3,306	3,312	2,782	2.08	2.29	2.01	6,881	7,566	5,586

Source: USDA Agricultural Statistics, U.S. Govt. Printing Office, Washington, D.C., 1987.

TABLE 6 Supply and Disappearance of Oats in the United States (million metric tons)

Crop year	81/82	82/83	83/84	84/85	85/86	86/87	87/88	88/89
Supply								
Production	7.41	8.62	6.93	6.89	7.57	5.61	5.43	3.19
Imports	0.03	0.06	0.44	0.49	0.39	0.52	0.67	0.87
Disappearance								
Exports	0.10	0.23	0.03	0.02	0.03	0.03	0.03	0.03
Food	0.81	0.83	0.86	0.87	0.88	0.93	1.09	1.31
Feed	6.38	6.07	6.60	6.01	6.45	5.27	4.78	2.91
Seed	0.51	0.74	0.45	0.47	0.53	0.65	0.51	0.43
Food use								
% of production	10.9	9.6	12.3	12.7	11.8	16.4	20.1	41.1

Source: P. Sisson, The Quaker Oats Company.

TABLE 7 Protein Content of Oats (*A. sativa*)

Variety	Groats (%)	Hulls (%)
Orbit	13.8	1.7
Lodi	14.6	1.6
Garland	14.8	1.4
Froker	15.5	1.4
Portal	16.5	1.9
X–1289	20.8	1.5
X–1656	22.5	1.9

Source: Ref. 50.

TABLE 8 Protein Content and Amino Acid Composition of Commercially Milled Oat

Protein or amino acid	Heavy oats	Light oats	Groats	Hulls	Flakes
Protein (%)	13.4	9.6	18.9	5.7	17.6
Lysine	4.2	5.2	3.9	4.9	4.1
Histidine	2.4	2.7	2.3	2.4	2.4
Ammonia	3.4	3.8	3.2	3.6	3.2
Arginine	6.4	6.3	6.2	6.8	6.0
Aspartic acid	9.2	11.1	9.0	10.5	9.0
Threonine	3.3	4.1	3.1	4.1	3.1
Serine	4.0	4.5	3.9	4.6	4.0
Glutamic acid	21.6	20.0	22.4	20.3	22.7
Proline	5.7	3.1	6.2	2.4	6.1
Cystine	1.7	0.4	2.0	0.5	1.7
Glycine	5.1	6.0	5.0	6.1	5.0
Alanine	5.1	5.5	5.0	5.4	5.0
Valine	5.8	6.2	5.7	6.4	5.8
Methionine	2.3	1.5	2.5	1.5	2.4
Isoleucine	4.2	4.5	4.3	4.5	4.3
Leucine	7.5	7.6	7.4	7.8	7.5
Tyrosine	2.6	2.4	2.5	2.9	2.1
Phenylalanine	5.4	5.3	5.5	5.3	5.6

Protein on dry matter, N x 6.25. Grams amino acid per 100 g amino acid recovered.
Source: Ref. 57. Protein on dry matter.

TABLE 9 Protein Content of Groat Fractions (*A. sativa*)

Variety	Embryonic axis (%)	Scutellum (%)	Bran (%)	Starchy endosperm (%)
Orbit	44.3	32.4	18.8	9.6
Lodi	36.5	26.2	19.6	10.7
Garland	40.5	28.9	18.5	10.9
Froker	26.3	28.0	20.7	9.7
Portal	35.3	29.1	23.0	10.3
X–1289	40.9	24.2	26.5	13.5
X–1656	40.7	32.4	32.5	17.0

Source: Ref. 50.

TABLE 10 Distribution of Total Protein in Groats (*A. sativa*)

Variety	Embryonic axis (%)	Scutellum (%)	Bran (%)	Starchy endosperm (%)
Orbit	4.0	4.5	41.3	50.2
Lodi	2.8	3.6	46.2	47.4
Garland	4.0	4.2	43.9	47.8
Froker	1.9	3.7	46.3	48.1
Portal	2.7	2.9	57.3	37.2
X–1289	2.4	3.4	49.2	45.0
X–1656	1.9	2.3	56.2	39.7

Source: Ref. 50.

TABLE 11 Mean, Maximum Value (Max.) Minimum Value (Min.) and Standard Deviation (s) for Crude Protein and Amino Acid Composition of 289 Oat Samples

Protein or amino acid	Max	Min	Mean	s
Percent protein	24.4	12.4	17.1	2.010
Lysine (Lys)	5.2	3.2	4.2	0.324
Histidine	3.1	1.2	2.2	0.316
Ammonia	3.0	2.5	2.7	0.079
Agrinine	7.8	6.2	6.9	0.231
Aspartic acid	9.9	8.3	8.9	0.278
Threonine (Thr)	3.5	3.0	3.3	0.098
Serine	4.8	3.8	4.2	0.187
Glutamic Acid	26.9	21.9	23.9	0.938
Proline	5.8	3.8	4.7	0.416
Half cystine	2.6	0.6	1.6	0.442
Glycine	5.5	4.4	4.9	0.210
Alanine	5.5	4.2	5.0	0.187
Valine	5.7	4.9	5.3	0.106
Methionine (Met)	3.3	1.0	2.5	0.347
Isoleucine	4.1	3.4	3.9	0.088
Leucine	7.8	4.8	7.4	0.205
Tyrosine	4.4	2.3	3.1	0.232
Phenylalanine	5.7	4.9	5.3	0.138
Sum of Lys + Thr + Met	11.1	8.2	10.0	0.487

Proteins given as N x 6.25(db), %; amino acids given as g/100 g of amino acids and ammonia recovered.
Source: Ref. 60.

TABLE 12 Protein Content and Amino Acid Composition of Hand-Separated Fraction of Groats of "Orbit"

Protein or amino acid	Whole groats	Embryonic axis	Scutellum	Bran	Endo-sperm
Protein (%)	13.8	44.3	362.4	18.8	9.6
Lysine	4.5	8.2	6.9	4.1	3.7
Histidine	2.4	3.9	3.6	2.2	2.2
Ammonia	2.7	1.9	1.8	2.5	2.9
Arginine	6.8	8.3	9.0	6.8	6.6
Aspartic acid	8.7	10.2	9.7	8.6	8.5
Threonine	3.4	5.0	4.7	3.4	3.3
Serine	4.6	4.8	5.0	4.8	4.6
Glutamic acid	21.7	14.2	14.9	21.1	23.6
Proline	5.5	3.3	3.6	6.2	4.6
Cystine	2.1	0.5	1.0	2.4	2.2
Glycine	5.2	6.3	6.2	5.4	4.7
Alanine	5.0	7.2	6.9	5.1	4.5
Valine	5.5	6.0	6.2	5.5	5.5
Methionine	2.2	2.2	2.1	2.1	2.4
Isoleucine	3.9	3.9	3.8	3.8	4.2
Leucine	7.6	7.1	7.1	7.4	7.8
Tyrosine	3.0	2.9	3.0	3.5	3.3
Phenylalanine	5.2	4.2	4.4	5.1	5.6

Protein content on dry matter basis, N x 6.25. Amino acid content, g amino acid per 100 g amino acid recovered.
Source: Ref. 57.

TABLE 13 Lipid Concentration in the Groats and Oat Fractions of Two Oat Cultivars

Source	Ether extraction[a] (free lipids) % + SD		WSB[b] Extraction[a] (bound lipids) % + SD	
Dal				
Groats	8.0	0.75	1.6	0.22
Hulls	2.3	0.71	0.6	0.06
Bran	9.5	0.21	1.2	0.14
Endosperm	6.8	0.25	1.0	0.11
Scutellum	20.6	2.04	2.8	0.80
Embryonic axis	12.6	1.63	3.3	1.00
Froker				
Groats	5.5	0.30	1.4	0.35
Hulls	2.0	0.11	0.6	0.06
Bran	6.4	1.41	1.3	0.14
Endosperm	5.2	0.35	1.0	0.21
Scutellum	20.4	0.91	4.2	0.37
Embryonic axis	10.6	0.61	4.1	0.63

[a]Dry basis.
[b]Water saturated n-butanol.
Source: Ref. 64.

TABLE 14 Distribution of Lipid Components in Oat Groat Fractions from Two Cultivars in Two Crop Years (% of total lipids)

Component	Groats	Bran	Endo-sperm	Scutellum	Embryonic axis
Triglycerides	41	39	41	50	58
1,3–Diglycerides	2	2	2	1	2
1,2–Diglycerides	1	1	1	2	2
Free fatty acids	5	3	3	2	2
Sterols	1	1	1	1	1
Sterol glucosides	1	1	1	1	1
Monogalactosyl–monoglycerides	4	4	5	—	—
Digoloatosyl–diglycerides	7	9	8	—	—
Phosphatidyl–ethanolamine	2	3	2	1	1
Lysophosphatidyl–ethanolamine	1	2	2	—	—
Phosphatidyl–choline	5	4	4	3	3
Lysophosphatidyl–choline	2	3	3	—	—
Components not measured	28	28	27	39	30

Source: Ref. 65.

TABLE 15 Mineral Content of 171 Welsh Oat Samples

Mineral	Mean	Range
Ca (%)	0.11	0.07–0.18
Mg (%)	0.13	0.10–0.18
K (%)	0.47	0.31–0.65
Na (%)	0.02	0.004–0.06
P (%)	0.38	0.29–0.59
Cl (%)	0.09	0.04–0.18
Cu (ppm)	4.7	3.0–8.2
Co (ppm)	0.05	0.02–0.17
Mn (ppm)	45	22–79
Zn (ppm)	37	21–70

Source: Ref. 74.

TABLE 16 Mineral Concentration of Bran and Endosperm Fractions of "Dal" Oats

Mineral	Bran	Endosperm
P (%)	1.02	0.26
K (%)	1.00	0.16
Mg (%)	0.38	0.07
Ca (%)	0.11	0.02
Fe (ppm)	89.8	18.2
Zn (ppm)	58.5	24.3
Mn (ppm)	87.8	30.8
Cu (ppm)	4.93	2.82
B (ppm)	2.82	0.00
Ba (ppm)	2.50	0.10

Source: Ref. 75.

TABLE 17 Vitamin Content of Groat and Rolled Oats

Vitamin	Groat (mg/g)	Rolled oats (mg/g)
Thiamine (B_1)	0.77	0.67
Riboflavin (B_2)	0.14	0.14
Niacin	0.97	0.98
Pantothenic acid	1.36	1.48
Pyridoxine (B_6)	0.12	0.13
Folic Acid	0.06	—
αTocopherol (E)	—	1.94

Source: Ref. 76.

REFERENCES

1. Coffman, F. A., Origin and history. In Franklin A. Coffman (Ed.) *Oats and Oat Improvement*, American Society of Agronomy Publ., Madison, WI, 1961, pp. 15–40.
2. Coffman, F. A., Origin of cultivated oats, *Am. Soc. Agron. J., 38:*983–1002 (1946).
3. Coffman, F. A., Oat history, identification, and classification, USDA-ARS Technical Bull. No. 156, 1977.
4. Hunter, H., Oats: *Their Varieties and Characteristics*. Ernest Benn, Ltd., London, 1924.
5. Ladizinsky, G., The domestication and history of oats. Third International Oat Conference, Lund, Sweden, July 4–8, 1988. Abstr.
6. Vavilov, N. I., Studies on the origin of cultivated plants. *Bul. Appl. Bot. and Plant Breeding 17:* 139–245 (1926).
7. Coffman, F. A., *Avena Sativa* L. ems Probably of Asiatic origin, *Agron. J.*, 47:281 (1955)
8. Tannahill, R. *Food in History*, Stein and Day, New York, 1973.
9. Harlan, J. R., *Crops and Man*, American Society of Agronomy, Madison, WI, 1975.
10. Marquette, A. F., *Brands, Trademarks and Goodwill, The Story of The Quaker Oats Company*, McGraw-Hill, New York, 1967.
11. O'Mara, J. G., Cytogenetics. In *Oats and Oat Improvement*, F. A. Coffman (Ed.), American Society of Agronomy, Madison, WI, 1961, pp. 112–124.
12. Rajathy, T., and Thomas, H., Cytogenetics of Oats *(Avena* L.), Miscellaneous Publications of The Genetics Society of Canada, No. 2, 1974.
13. Leggett, J. M., Interspecific hybrids involving the recently described diploid taxon *Avena atlantica, Genome, 29:* 361–364 (1987).
14. Baum, B. R., *Oats: Wild and Cultivated, A Monograph of the Genus Avena* L. (Poaceae), Minister of Supply and Services Canada, Ottawa, Canada, 1977.
15. Rajhathy, T., and Morrison, J. W., Genome homolgy in *Avena*, *Can. J. Genet. Cytol.,2*:278–285 (1960).
16. Thomas, H. and Bhatti, I. M., Notes on the cytogenetic structure of the cultivated oat *Avena sativa* (2n=6x=42), *Euphytica, 24*:149–157 (1975).
17. Ansari, N.,and Thomas, H., A study of homeologous relationships in the cultivated oat *Avena sativa* (2n=6x=42). *Theo. Appl. Genet., 66*:303–305 (1983).
18. McGinnis, R. C., Establishing a monosomic series in *Avena sativa* L., In *Chromosome Manipulations and Plant Genetics*, R. Riley and K.R. Lewis (Eds.) Oliver and Boyd, Edinburgh, 1966, pp. 296–301.
19. Hacker, J. B., and Riley, R. Morphological and Cytological effects of chromosome deficiency in *Avena sativa, Can. J. Genet. Cytol.* ,*7*:304–315 (1966).
20. Leggett, J. M., Cytoplasmic substituions involving six *Avena* species, *Can. J. Genet. Cytol., 26*:698–700 (1984).
21. Robertson, L. D., and Frey, K. J. Cytoplasmic effects on plant traits in interspecific matings of *Avena, Crop Sci.,24*:200–204 (1984).
22. Jensen, N. F., Genetics and inheritance in oats, In *Oats and Oat Improvement*, F. A. Coffman (Ed.), American Society of Agronomy, Madison, WI, 1961, 125–206.
23. Murphy, H. C., and Coffman, F. A., Genetics of disease resistance, In *Oats and Oat Improvement*, F.A. Coffman, (Ed.), American Society of Agronomy, Madison, WI, 1961, pp. 207–226.
24. Simons, M. D., Martens, J. W., McKenzie, R. I. H., Nishiyama, I., Sadanaga, K., Sebasta, J., and Thomas, H., Oats: A standardized system for nomenclature for genes and chromosomes and catalog of genes governing characters. USDA Handbook Number 509, 1978.
25. Tanksley, S. D., Miller, J., Paterson, A., and Bernatzky, R., Molecular mapping of plant chromosomes, In *Chromosome Structure and Function*, J. P. Gustafson and R. Appels (Eds.), Plenum Press, New York, 1987, pp. 157–173.
26. Frey, K. J., Genetic resources and their use in oat breeding, In *Proceedings of the Second International Oats Conference*, D. A. Lawes and H. Thomas (Eds.), Martinus Nijhoff, Dordrecht, 1986, pp. 7–15.

27. Ladizinsky, G., Chromosome rearrangements in the hexaploid oats, *Heredity, 25*:457–461, (1970).
28. McMullen, M. S., Phillips, R. L., and Stuthman, D. D., Meitotic irregularities in *Avena sativa* L./*A. sterilis* L. hybrids and breeding implications, *Crop Sci., 22*:890–897 (1982).
29. Lawerence, P. K., and Frey, K. J., Backcross variability for grain yield in oat species crosses (*Avena sativa* L. x *A. sterilis* L.) *Euphytica, 24*:77–85 (1975).
30. Aung, T., Thomas,H., and I. T. Jones, The transfer of the gene for mildew resistance from *Avena barbata* (4x) into the cultivated *A. sativa* by an induced translocation, *Euphytica, 26*:623–632 (1977).
31. Sharma, D.C., and Forsberg, R. A., Spontaneous and induced intraspecific gene transfer for crown rust resistance in *Avena*, *Crop Sci., 17*:885–860 (1977).
32. Rothman, P. G., Adequate resistance in oats, In *Proceedings of the Second International Oats Conference*, D. A. Lawes and H. Thomas (Eds.), Martinus Nijhoff, Dordrecht, 1986, pp. 72–76.
33. Anderson, J. W., and Chen, W. J., Cholesterol-lowering properties of oat products, In *Oats: Chemistry and Technology*, F. H. Webster (Ed.), American Association of Cereal Chemists, Inc., St. Paul, MN, 1986, pp. 309–333.
34. Burrows, V. D., Breeding oats for food and feed: conventional and new techniques and materials, In *Oats: Chemistry and Technology*, F. H. Webster (Ed.), American Association of Cereal Chemists, Inc., St. Paul, MN, 1986, pp. 13–46.
35. Brown, C. M., and Jedlinsky, H., Registration of 13 germplasm lines of oats, *Crop Sci. 18*:1098 (1978).
36. Baltenberger, D. E., Ohm, H. W., and Foster, J. E., Recurrent selection for tolerance to barley yellow dwarf virus in oat, *Crop Sci., 28*:477–480 (1988).
37. Frey, K. J., Management systems for host genes to control disease loss, *Ind. J. Genet. Pl. Breeding, 39*:10–21 (1979).
38. Martens, J. W., Oat stem rust, In *The Cereal Rusts, vol. II*, A. P. Roelfs and W. R. Bushnell (Eds.), Academic Press, Inc., New York, 1985, pp. 102–129.
39. Brown, C. M., and Shands, H. L. Factors influencing seed set of oat crosses, *Agron. J., 48*:173–176 (1956).
40. McDaniel, M. E., Kim, H. B., and Hathcock, Approach crossing of oats, *Crop Sci., 7*:538–540 (1967).
41. Stuthman, D. D., Development of oat breeding methods, In *Proceedings of the Second International Oats Conference*, D. A. Lawes and H. Thomas (Eds.), Martinus Nijhoff, Dordrecht, 1986, pp.105–111.
42. Cisar, G., Howey, A. E., and Brown, C. M., Optimal population density and random vs. selective elimination of genotypes under a modified single-seed descent method with spring oats, *Crop Sci., 22*:576–579 (1982).
43. Valentine, J., Accelerated pedigree selection: An alternative to individual plant selection in the normal pedigree breeding method in the self-pollinated cereals, *Euphytica, 33*:943–951 (1984).
44. Rodgers, D. M., Murphy, J. P. and Frey, K. J., Impact of plant breeding on the grain yield and genetic diversity of spring oats, *Crop Sci., 23*:737–740 (1983). 45. Wych, R. D., and Stuthman, D. D., Genetic improvement inMinnesota-adapted oat cultivars released since 1923, *Crop Sci., 23*:879–881. (1983).
46. Langer, I., Frey, K. J., and Bailey, T. B., Production response and stability characteristics of oat cultivars developed in different eras, *Crop Sci. 18*:938–942 (1978).
47. Souza, E., and Sorrells, M. E., Pedigree analysis of North American oat cultivars released from 1951 to 1985, *Crop Sci., 29*:595–601 (1989).
48. Bonne(??? illegible) O. T.,. Morphology and development, In *Oats and Oat Improvement*, F. A. Coffman (Ed.), American Society of Agronomy, Madison, WI, 1961, pp. 41–74.
49. Fulcher, R. G., Morphological and chemical organization of the oat kernel, In *Oats: Chemistry and Technology*, F. H. Webster (Ed.), American Association of Cereal Chemists, Inc., St. Paul, MN, 1986, pp. 47–74.
50. Youngs, V. L., Protein distribution in the oat kernel, *Cereal Chem., 49*:407–411 (1972).

51. Schrickel, D. J., World oats use and marketing, In *Proceeding of the Second International Oats Conference*, D. A. Lawes and H. Thomas (Ed.), Martinus Nijhoff, Dordrecht, 1986, pp. 197–199.
52. Schrickel, D. J., Oats production, value, and use, In *Oats: Chemistry and Technology*, F. H. Webster (Ed.), American Association of Cereal Chemists, Inc., St. Paul, MN, 1986, pp. 1–11.
53. Hoffman, L. A., and Livezey, J., The U.S. oats industry, USDA-ERS Agricultural Economic Report Number 573, 1987.
54. Hoffman, L. A., and Scronce, P. W., Oats imports, the U.S. market, and government programs, USDA-ERS Commodity Economics Division Staff Report No. AGE870601, 1987.
55. Peterson, D. M., and Brinegar, A. C. Oat storage proteins, In *Oats: Chemistry and Technology*, F. H. Webster (Ed.), American Association of Cereal Chemists, Inc., St.Paul, MN, 1986, pp. 153–203.
56. Frey, K. J., Protein of oats, *Z. Pflanzenzuecht.*, *78*:185– 215 (1977).
57. Pomeraz, Y., Youngs, V. L., and Robbins, G. S., Protein content and amino acid composition of oat species and tissues, *Cereal Chem.*, *50*:702–707 (1973).
58. Osborne, T. B., Die Pflanzenproteins, *Ergeb. Physiol.*, *10*:47–215 (1910).
59. Peterson, D. M., Protein concentration, concentration of protein fractions, and amino acid balance in oats, *Crop Sci.*, *16*:663–666 (1976).
60. Robbins, G. S., Pomeranz, Y., and Briggle, L. W., Amino acid composition of oat groats, *J. Agric. Food Chem.*, *19*:536– 539 (1971).
61. FAO, Energy and protein requirements, FAO Nutr. Rep. Ser. 52, Food and Agric. Org. of the U.N., Rome, 1973.
62. Youngs, V. L., Oat lipids and lipid-related enzymes, In *Oats: Chemistry and Technology*, F. H. Webster (Ed.), American Association of Cereal Chemists, Inc., St. Paul, MN, pp. 205–226.
63. Brown, C. M., and Craddock, J. D., Oil content and groat weight of entries in the world oat collection, *Crop Sci.*, *12*:514–515 (1972).
64. Youngs, V. L., Puskulcu, M., and Smith, R. R., Oat lipids. I. Composition and distribution of lipid components in two oat cultivars, *Cereal chem.*, *54*:803– 812 (1977).
65. Youngs, V. L., Oat lipids, *Cereal Chem. 55*:591–597 (1978).
66. Paton, D., Oat starch: Physical, chemical, and structural properties, In *Oats: Chemistry and Technology*, F. H. Webster (Ed.), American Association of Cereal Chemists, Inc., St. Paul, MN, 1986, pp. 93–120.
67. Morrison, W. R., Milligan, T. P., and Azudin, M. N., A relationship between amylose and lipid contents of starches from diploid cereals, *J. Cereal Sci.*, 2:257–271 (1984).
68. MacArthur, L. A., and D'Appolonia, B. L., Composition of oat and wheat carbohydrates. II, *Starch Cereal Chem.* , *56*:458–461 (1979).
69. Wood, P. J., Dye interactions: A basis for specific detection and histochemistry of polysaccharides, *J. Histochem. Cytochem.*, *31*:823–826 (1983).
70. Wood, P. J., Oat β-glucan: Structure, location, and properties, In *Oats: Chemistry and Technology*, F. H. Webster (Ed.), American Association of Cereal Chemists, Inc., St. Paul, MN, 1986, pp. 121–152.
71. Wood, P. J., and Weisz, J., Use of Calcofluor in analysis of oat beta-d-glucan, *Cereal Chem.*, *61*:73–75 (1984).
72. Martin, H. L., and Bamforth, C. W., An enzymatic method for the measurement of total and water-soluble β-glucan in barley, *J. Inst. Brew.* London, 87:88–91 (1981).
73. Welch, R. W., and Lloyd, J. D., Kernel (1→3) (1→4)-β-D glucan content of oat genotypes, *J. Cereal Sci.*, *9*:35–40 (1989).
74. Morgan, D. E., Note on variations in the mineral composition of oat and barley grown in Wales, *J. Sci. Food Agric.*, *19*:393–395 (1968).
75. Peterson, D. M., Senturia, J., Youngs, V. L., and Schrader, L. E., Elemental composition of oat groats, *J. Agr. Food Chem.*, *23*:9–13 (1975).
76. Shukla, T. P., Chemistry of oats: Protein foods and other industrial products, *Crit. Rev. Food Sci. Nutr.*, *6*:383–431 (1975).

77. Webster, F. H., Oat utilization: past present, and future, In *Oats: Chemistry and Technology*, F. H. Webster (Ed.), American Association of Cereal Chemists, Inc., St. Paul, MN, 1986, pp.413–424.
78. Western, D. E., and Graham, W. R., Marketing, processing, uses and composition of oats and oat products, In *Oats and Oat Improvement*, F. A. Coffman (Ed.), American Society of Agronomy, Madison, WI, 1961, pp.552–578.
79. D'Applonia, B. L., and Youngs, V. L., Effect of bran and high-protein concentrate from oats on dough properties and bread quality, *Cereal Chem.*, *55*:1038–1049 (1978).
80. Wood, P. J., Siddiqui, I. R., and Paton, D., Extraction of high-viscosity gums from oats, *Cereal Chem.*, *55*:1038– 1049 (1978).
81. MacArthur-Grant, L. A., Sugars and nonstarchy polysaccharides in oats, In *Oats: Chemistry and Technology*, F. A. Webster (Ed.), American Association of Cereal Chemists, St. Paul, MN, 1986, pp. 75–91.
82. Deane, D., and Commers, E., Oat cleaning and processing, In *Oats: Chemistry and Technology*, F. H. Webster (Ed.) American Association of Cereal Chemists, Inc., St. Paul, MN, 1986, pp. 371–412.

5

Sorghum

Lloyd W. Rooney and Sergio O. Serna-Saldivar
Texas A&M University, College Station, Texas

I. INTRODUCTION

Sorghum (*Sorghum bicolor L. Moench*) is a unique cereal due to its drought tolerance and adaptation to tropical conditions. Because of these characteristics, sorghum is mainly utilized as a subsistence and cash crop in the arid regions of the world. To assure grain supply in semi-arid tropical agricultural systems, sorghum is sometimes intercropped with maize, legumes, and millets. Among cereals, sorghum ranks fifth with 59 million metric tons of total production in 1987 (25). The major producing centers are in Asia and Africa (47% of total production), although about 32% of the world sorghum production is grown in the United States (Table 1). Production in the United States is concentrated in Kansas, Texas, Nebraska, and Missouri. In 1985, Kansas and Texas produced 48% of the total U.S. production (83).

Sorghum is the major food staple in many African countries and India where agricultural and environmental conditions are unfavorable for the production of other crops. In the Western Hemisphere, where 49.8% of sorghum is produced, sorghum is a major ingredient in livestock feeds. In Latin America, sorghum grits are often used as an adjunct for brewing European barley beers. Sorghum utilization directly for human foods is limited in the Western Hemisphere, but it is increasing. Sorghum tortillas are consumed in some areas of Mexico and Central America. Sorghum flour and grits have been used in breads and as a ricelike product. Pitimi, a thick porridge from boiled sorghum grits, is used extensively in Haiti.

Sorghums are referred to as grain sorghums, forage sorghums, sweet sorghums, and grassy or grazing types. Grain sorghum refers to short mechanically harvested sorghum varieties, whereas forage sorghums are tall and yield fodder and grain for livestock and foods. Waxy sorghums containing starch with 100% amylopectin have unique industrial uses and improved nutritional value. Sweet sorghums produce a juice high in soluble carbohydrates that can be further processed into syrups, sugar or fermented into alcohol. Grassy types of sorghums are used for grazing, silage, or hay crops. The stalks and roots of tall

Technical contributions number TA 24102 from the Texas Agricultural Experimental Station, Texas A&M University, College Station, Tx.

sorghums grown in Africa and India are used as fuel and building material. Sorghum ashes are leached to produce alkali that is utilized in some traditional food systems.

Sorghum has been classified as high and low tannin types. Brown or high tannin sorghums have a reduced nutritional value and are grown because of their agronomic advantages including bird resistance and decreased weathering, mold infestation, and grain sprouting. Because the low tannin types do not contain any condensed tannins, they should be referred to as sorghums without tannins. These sorghums have a nutritional value similar to that of maize (*Zea mays*). White sorghums without a pigmented testa are the most suitable for use in most traditional food systems.

II. ORIGIN

Sorghum originated in Africa and Asia where it has been grown for more than 2,000 years. More than 22,000 selections are present in the world Sorghum Collection in India. This gene bank is used by plant breeders to improve the crop. Sorghum was introduced to the United States in the middle of the last century and was first cultivated on the Atlantic coast. By 1900, sorghum producing areas spread to the southwest and California (57). The successful development of sorghum hybrids in the 1950's increased production in the United States, Argentina, and Mexico where sorghum is the second-most-important cereal. Hybrid sorghum is able to outyield its main competitor, corn, in regions where irrigation water is scarce. In Mexico, sorghum is grown under irrigation because the hybrids are more productive than corn hybrids.

III. STRUCTURE AND PHYSICAL PROPERTIES

The sorghum kernel is a naked caryopsis with a thousand kernel weight which varies from 3 to 80 g. The size and shape of the grain varies widely among sorghums in the world collection. Grain of commercial sorghum hybrids has a flattened-spherical shape 4 mm long, 2 mm wide, and 2.5 mm thick with a thousand kernel weight of 25–35 g. Test weight and grain density generally range from 55 to 61 lb/bu and 1.26–1.38 g/cc, respectively.

The sorghum kernel is composed of three main anatomical parts: pericarp (outer layer), endosperm (storage tissue), and germ (embryo) (Fig.1). The relative proportion of these components varies, but in most cases it is 6, 84, and 10% of the kernel, respectively. The pericarp originates from the ovary wall and is subdivided into three parts: epicarp, mesocarp, and endocarp. The epicarp is the outermost layer and is generally covered with a thin waxy film. The middle structure or mesocarp varies in thickness from a few remnant cells with few starch granules to three to four cell layers containing many starch granules. Sorghum is the only cereal grain known to have starch in this anatomical part. The endocarp is composed of cross and tube cells. A stylar area is located on the tip of the caropsis, opposite the germ.

The hilum or black layer is the scar tissue that develops when the caryopsis reaches physiological maturity. The seed coat or testa is a thick pigmented layer present only in sorghums with dominant B_1–B_2-genes. Sorghums with B_1–B_2-genes are higher in tannins when a dominant S_gene is present. The "low tannin" sorghums do not contain a pigmented testa, which is the major way of identifying brown sorghums.

The endosperm, composed of the aleurone layer, peripheral, corneous, and floury areas, is the major storage tissue. The aleurone layer is composed of a single layer of rectangular cells adjacent to the tube cells or pigmented testa (if present). These cells contain large amounts of protein bodies with inclusions containing phytin, oil bodies, minerals, and enzymes.

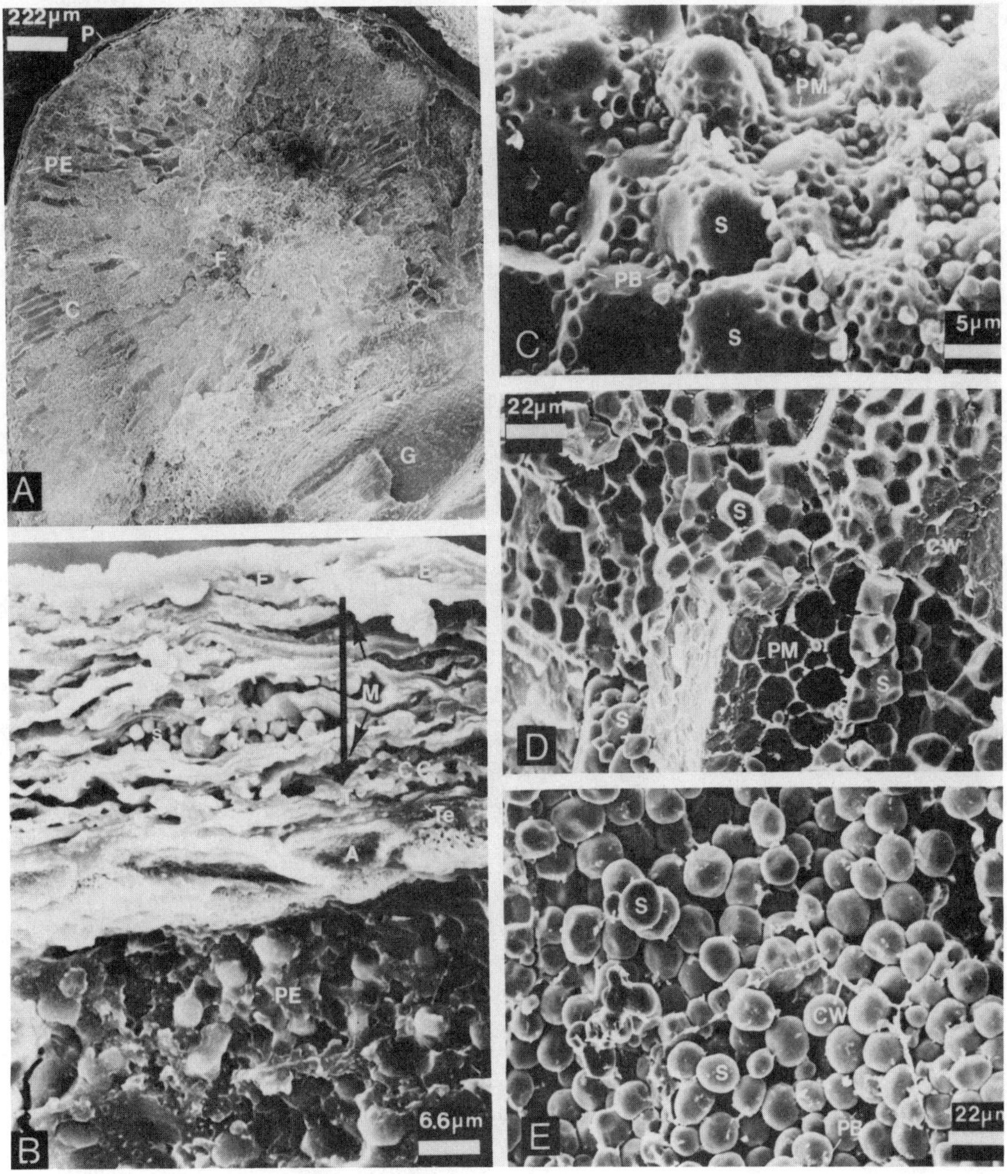

FIGURE 1 Structure of the mature sorghum kernel. (A) Cross section, (B) pericarp, (C) peripheral endosperm, (D) corneous endosperm, (E) floury endosperm. P = pericarp; PE = peripheral endosperm; C = corneous endosperm; F = floury endosperm; G = germ; E = epicarp; M = mesocarp; CC = cross cells; T = tube cells; Te = testa; A = aleurone; S = starch granule; PB = protein body; PM = protein matrix; Cw = cell wall.

The relative proportion of protein and starch in the endosperm is the most important factor affecting grain hardness and density. Each endosperm cell is composed of a thin cell wall, protein matrix, protein bodies, and starch granules. The size of the starch granules varies from 4 to 25 μm (average of 15 μm). In the corneous endosperm, the protein has a continuous interface between the starch granules with protein bodies embedded in the matrix. The outer portion of the corneous endosperm, called peripheral endosperm, consists of several layers of densely packed cells containing large quantities of protein and small starch granules. Starch granules in the corneous endosperm are polygonal and often contain dents where protein bodies were embedded in the starch. The corneous endosperm is translucent (vitreous). On the other hand, the floury endosperm has a discontinous protein network with loosely packed starch granules in the endosperm cells. Thus, small voids occur between the spherical starch granules and particles of protein matrix which diffract light and give an opaque or chalky appearance.

The germ is composed of two major parts: the embryonic axis and the scutellum. The embryonic axis, plumulae, and primary root develop into the new plant, whereas the scutellum is a reserve tissue with large amounts of oil (spherosomes), protein, enzymes, and minerals.

IV. APPEARANCE OF SORGHUM GRAIN AND ITS GENETICS

Several interacting factors influence the color and appearance of the sorghum kernel. The genetics of pericarp color, pericarp thickness, presence of the pigmented testa and endosperm, testa and glume color are the main genetic factors affecting grain appearance (69). Grain appearance is also affected by environmental factors such as insect damage and mold infestation.

Pericarp color is controlled by R and Y genes. The combination of these genes can produce white (R_yy or rryy), lemon-yellow (rrY_), or red (R_Y_) color. Grain color and appearance are influenced by the presence of the testa and the condition of the spreader gene. The testa is controlled by B_1-B_2-genes. The pigmented testa is absent when the genes are homozygous recessive ($b_1b_1b_2b_2$, B_1- b_2b_2, $b_1b_1B_2$-).

Another gene, tp tp, controls color of the testa, which is brown or purple. Sorghums with pigmented testa (B_1-B_2-; S_are classified as type III and are the brown, bird-resistant sorghums. The Z gene controls pericarp thickness, which affects kernel appearance. Sorghums with a homozygous recessive (zz) have a thick pericarp and chalky kernel appearance, which mask testa and endosperm color (69).

Endosperm color (i.e., yellow endosperm) can affect grain appearance, especially in those sorghums with thin pericarp without a pigmented testa. In the United States, yellow endosperm sorghums are referred to as cream, white, or bronze hybrids depending on pericarp color or thickness and intensity of yellow endosperm. Heteroyellow endosperm sorghums are hybrids with one yellow and one nonyellow parent.

Several environmental factors affect grain appearance. An environment that is hot and humid during grain maturation negatively affects grain quality due to grain weathering and deterioration, which results from insect and mold attack. Insect attacked grain secretes phenolic compounds, which produce a pigmented spot in the affected area. Molds discolor grain, break down the kernel, reduce hardness, and significantly affect processing qualities.

V. SORGHUM MARKET CLASSES

The U.S. Federal Grain Inspection Service (F.G.I.S.) currently recognizes four market classes of sorghum: brown sorghums have kernels with a thick colored testa; white sorghums have kernels with white pericarps without pigmented testa, which contain no more than 2% of kernels with colored pericarp; yellow sorghums, which contain kernels of any pericarp color but cannot contain more than 10% brown sorghum kernels; and the mixed class is sorghum that does not meet requirements for any of the three previous classes. Most sorghum marketed in the United States is classified as yellow sorghum. Table 2 shows the current grade requirements for all classes of sorghum (86). Grade is determined by bushel weight, grain moisture, damaged and broken kernels, foreign material, and other factors. Brown sorghums contain condensed tannins that negatively affect nutritional value. The white and yellow classes of sorghum do not contain condensed tannins.

The F.G.I.S. is currently revising the classes and standards for sorghum. The new standards, likely to be effective in 1991 or 1992, propose four classes; 1) Sorghum, 2) Tannin Sorghum, 3)White Sorghum and 4) mixed sorghum. The sorghum class would contain grain of any pericarp color without a pigmented testa and less than 3% of sorghum with a pigmented testa. The tannin sorghum class would contain 90% or more of kernels with a pigmented testa. White and mixed classes would remain the same. The level of foreign material allowed in number 1, 2, and 3 grades would be decreased.

VI. COMPOSITION

Sorghum composition (Table 3) varies significantly due to genetics and environment. For example, high nitrogen fertilizer levels increase grain protein content and decrease the amount of starch (NFE) in the grain. The prolamine protein fraction, which is very low in lysine, increases predominantly with increased grain protein content. Therefore, the lysine proportion of the higher protein grain does not increase. High lysine types of sorghum exist, but they are not grown because of serious yield impairment and soft-floury endosperm characteristics.

Sorghum is similar in composition to corn. Starch is the major grain component followed by protein. Most sorghums contain 70–80% branched amylopectin and 20–30% amylose. However, waxy or glutinous sorghum contains starch with 100% amylopectin and has unique properties similar to waxy corn for industrial and feed use.

Generally, sorghum contain 1% less fat and more waxes than corn. Protein content of sorghum is more variable and usually 1–2% higher than corn.

Approximately 80%, 16%, and 3% of the protein is in the endosperm, germ, and pericarp, respectively (82). The alcohol-soluble proteins, kafirins, comprise 50% of the protein. The kafirins are hydrophobic, rich in proline, aspartic acid and glutamic acid, and contain little lysine. Kafirins are found mainly in the protein bodies which increase with increased levels of protein in the grain. Glutelins are the second major protein fraction and the most difficult to extract. Glutelins are isolated after being solubilized in dilute alkali or acid solutions. Better fraction yields are obtained when mercaptoethanol is utilized to break disulfide bonds. Glutelins are high molecular weight proteins which form the structure of the endosperm (protein matrix). Albumins and globulins are soluble in water and dilute salt solutions, respectively. These fractions predominate in the germ and have the highest quantities of lysine.

Nutritionally, sorghum protein is first limiting in lysine followed by threonine. Lysine of sorghum protein provides about 45% of the recommended FAO/WHO requirement

(5.44g /100g protein) (24). Although not commercially grown, high lysine sorghum seeds contain approximately 52% more lysine than its normal counterpart (51).

Table 4 compares soluble sugars of sorghums with different endosperm types including high lysine sorghum. The total sugar content varies according to the stage of grain development. The sugar content decreases as the grain approaches physiological maturity. Sucrose, glucose and fructose are the major soluble sugars found in physiologically matured kernels (81). Maltose is a minor component of the soluble sugar fraction. Sugary sorghums contain approximately twice as many sugars as normal sorghum. Raffinse and glucose/fructose are present in high amounts in sugary types compared to normal sorghum (51).

The major pentosan fraction of sorghum is water soluble. The whole grain contained 0.9% and 0.42% water soluble and alkali soluble pentosans, respectively. Most of the sorghum pentosans are located in the pericarp. Pentosans in the pericarp consist almost entirely of the alkali soluble fraction. The carbohydrate content of the pentosans ranged from 68 to 85%; glucose and arabinose were the carbohydrates found in larger quantities (34).

Most of the crude fiber is present in the pericarp and endosperm cell walls. The fiber is composed mainly of cellulose, hemicellulose, and small quantities of lignin. This fraction is generally associated with phenolic compounds such as ferulic and caffeic acid.

Germ and aleurone layers are the main contributors to the lipid fraction. The germ itself provides about 80% of the total oil. The fatty acid composition consists primarily of linoleic, oleic, and palmitic acids (65) (Table 5). Refined commercial sorghum oil and corn oil can be used interchangable.

Table 6 shows the average mineral and vitamin composition of sorghum grain. The calcium content is low; most of the phosphorus is in the phytic acid molecules.

VII. TANNINS AND POLYPHENOLS: EFFECTS ON SORGHUM QUALITY AND NUTRITIONAL VALUE

All sorghums contain phenols, which affect the color, appearance, and nutritional quality of the grain. The phenolic compounds can be subdivided into three groups: phenolic acids, flavonoids, and tannins. All sorghums contain phenolic acids and most contain flavonoids, but only brown sorghums (Type II and III) contain condensed tannins. The phenolic acids are derivatives of benzoic or cinnamic acid (Fig. 2). Flavonoids consist of two units: a C6–C3 fragment from cinnamic and a C6 fragment from malonyl-Co A (see Fig. 2). The major groups of flavonoids in sorghum are the flavans. Flavan-3-en-3-ols (double bond between C3 and C4; hydroxylated at C3) are anthocyanidins (28). Tannins are oligomers of five to seven flavan-3-ol units (Fig. 2). In strong acids such oligomers depolymerize to yield anthocyanidin.

Tannins protect the kernel against attack by insects, birds, and microorganisms. In addition, the rate of preharvest germination is significantly lower for most high tannin sorghums. Unfortunately, these agromomic advantages are accompanied by nutritional disadvantages and lower food qualities (6). For example, alkaline food products (i.e., tortillas and tô) produced from high tannin grains or flour usually have undesirable off-colors.

Several techniques are currently used to characterize phenolic compounds in sorghum grain (6,28,45). Most tannin assays measure the level of phenols in sorghum, which may or may not be condensed tannins (Table 7). These chemical assays usually correlate well with the time-consuming measurements of antinutritional effects as determined with feeding tests (6), but they are not accurate enough to use to discount price of high tannin grains. In Table 8, the catechin equivalents for different types of sorghums (I, II, III) measured by different techniques (20) show that significant levels of "tannins" are reported in type I sorghums even though it is known that condensed tannins are not present in type I sorghums.

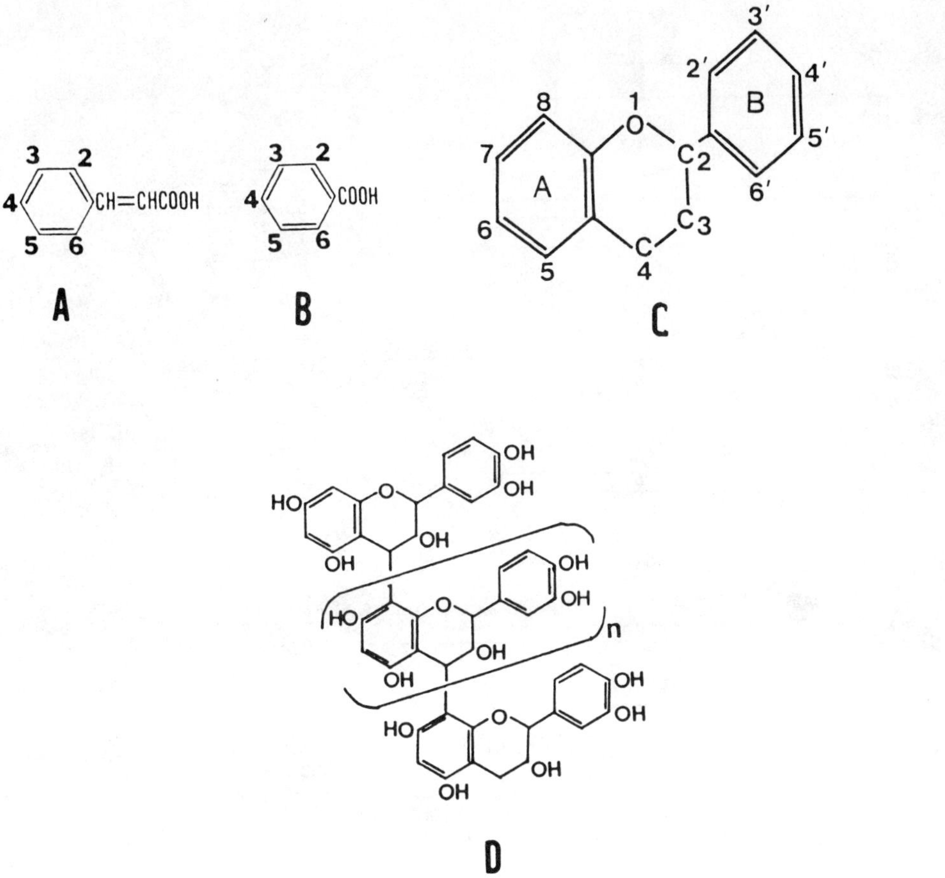

FIGURE 2 Basic structure of phenolic acids (A and B), flavonoids (C), and tannins (D). A = cinnamic acid; B = benzoic acid. The major flavonoid groups are: flavanones (carbonyl at 4), flavonols (carbonyl at 4, hydroxyl at 3), flavones (carbonyl at 4 double bond between 2 and 3), and flavans (no carbonyl at 4). The major flavans are leucoanthocyanidins (hydroxyls at 3 and 4) and catechin (hydroxyl at 3, double bond between 3 and 4). D = proanthocyanidin (tannin) polymer (n = 2–4). (Adapted from Ref. 28.)

VII. INDUSTRIAL UTILIZATION

A. Wet Milling

Sorghum was commercially processed into traditional wet milling products at Corpus Christi, Texas for 22 years. In 1975, sorghum wet milling was discontinued because sorghum prices increased to a level similar to maize. Recently, Caransa and Bakker (7) reported the establishment of a modern sorghum starch plant in the Sudan with a production capacity of 150 tons/day.

The procedure for sorghum wet milling is very similar to the one used for maize (88). The diagram of the process is shown in Figure 3. According to Watson (88), the major difference between maize and sorghum wet milling is the difficulty of separating sorghum

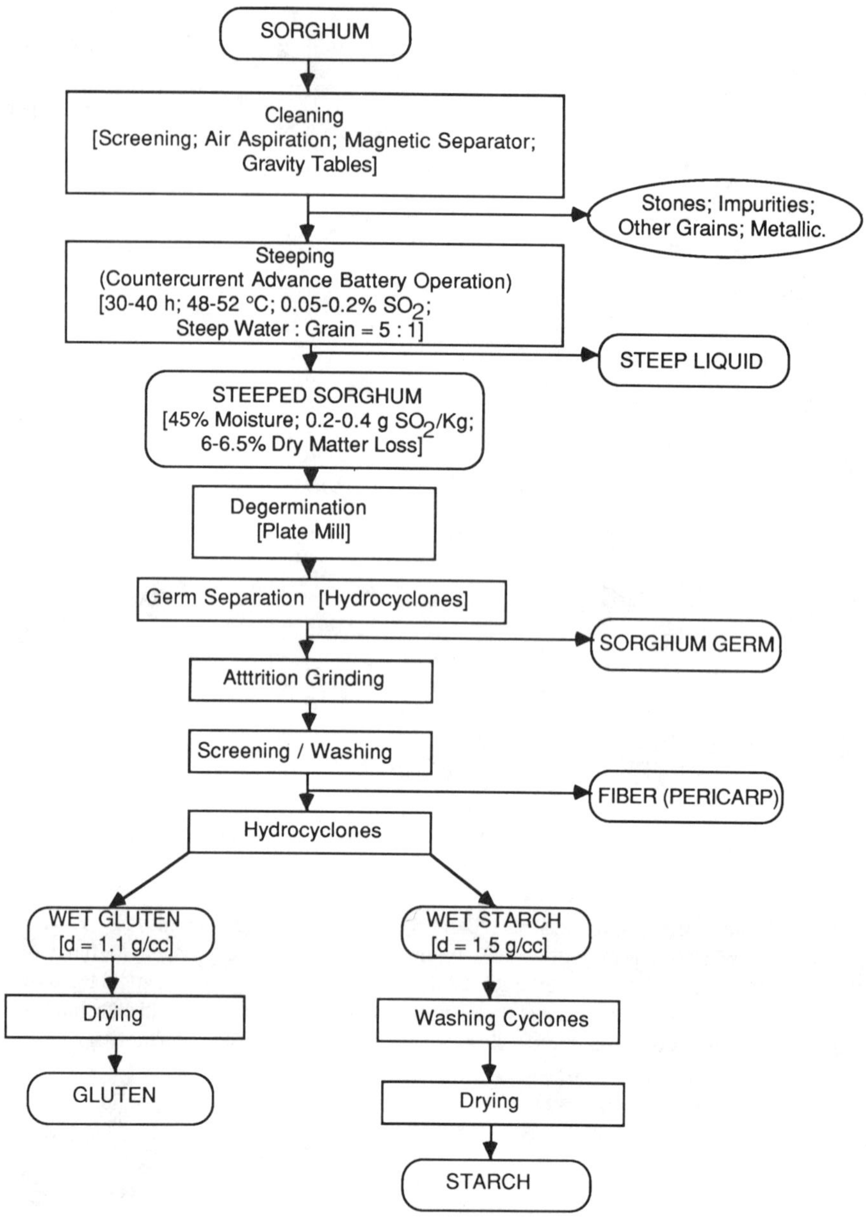

FIGURE 3 Flow chart of the commercial wet milling process for sorghum grain. (Adapted from Ref. 88.)

starch and gluten. The sorghum pericarp is more fragile than the pericarp or maize, thus small pericarp particles impede the seperation of the starch and protein and cause off-colored starch. The properties of sorghum starch and oil are similar to those of corn. However, sorghum starch must be bleached to remove polyphenols, takes slightly more energy to cook, and is more difficult to recover in high yields. Other differences include reduced yields of sorghum oil compared to corn oil, more refining of sorghum oils, and lack of carotenoid pigments in the gluten. The intense yellow color of corn gluten is highly desired by the broiler industry for pigmentation prior to slaughtering. Even yellow endosperm sorghums lack sufficient pigmentation.

Due to its smaller kernel size, sorghum requires more grinding teeth in the plates of the degerminating mills and changes in the steeping time to increase the disruption of the sorghum during grinding (7).

The economics of using sorghum for wet milling could be improved if new cultivars with improved properties are developed. Yellow endosperm sorghum hybrids with significantly improved wet-milling properties have been identified (54). A plentiful supply of sorghum at low cost compared to corn is necessary for economical wet milling of sorghum.

B. Sorghum Starches

Nonwaxy and waxy sorghum starches were produced commercially in the United States in the 1950's and 1960's. The sorghum starch had properties and uses similar to corn starch (38). Table 9 summarizes starch properties of waxy and normal sorghum starches. Starch from waxy sorghum is notable for past clarity, high water binding capacity, and resistance to gel formation and retrogradation. The pastes are rather stringy and cohesive (31,40,56,74).

Enzymatic conversion of sorghum starch to liquid glucose syrup can be achieved by heating a starch slurry (30% w/v) at 105°C/5 min, liquifying the gelatinized slurry with termamyl (heat-stable α amylase) at 95% C/2 hr and ph 6.5, cooling the slurry to 60°C, saccharifying the liquified media with amyloglucosidase for 72 hr, and treating with activated charcoal to remove pigments (38).

C. Dry Milling

The most highly refined sorghum products are produced by abrasive removal of the pericarp, followed by degermination and physical separation and/or classification of the dry milled fractions (Table 10). Most of the traditional African and Indian foods are produced from decorticated sorghum which is milled into flour. Industrial sorghum decortication involves the use of mills with abrasive disks or carborundum stones (62,64). Whole sorghum flour can be produced by the use of stone, hammer, pin or roller mills. Indian villages still use stone mills to produce a coarse flour that is used for Roti and Sankati. Hammer, pin and roller mills are used to produce flours. Many sorghum flours impart a sandy texture to products unless care is taken to control particle size. Most of the sorghum milled for production of brewers adjuncts is first decorticated, degerminated by impactation, and seperated into grits by gravity tables. Mechanical dehulling improves both the quantity and quality of sorghum grits or flour (62).

D. Alcohol Production

Utilization of sorghum grain or sweet sorghum biomass (Fig. 4) for ethanol production has been reviewed (11,15,41). Data provided by Coble et al. (11) indicated that one ton of maize grain produces 387 L of 182-proof alcohol, whereas the same amount of sorghum

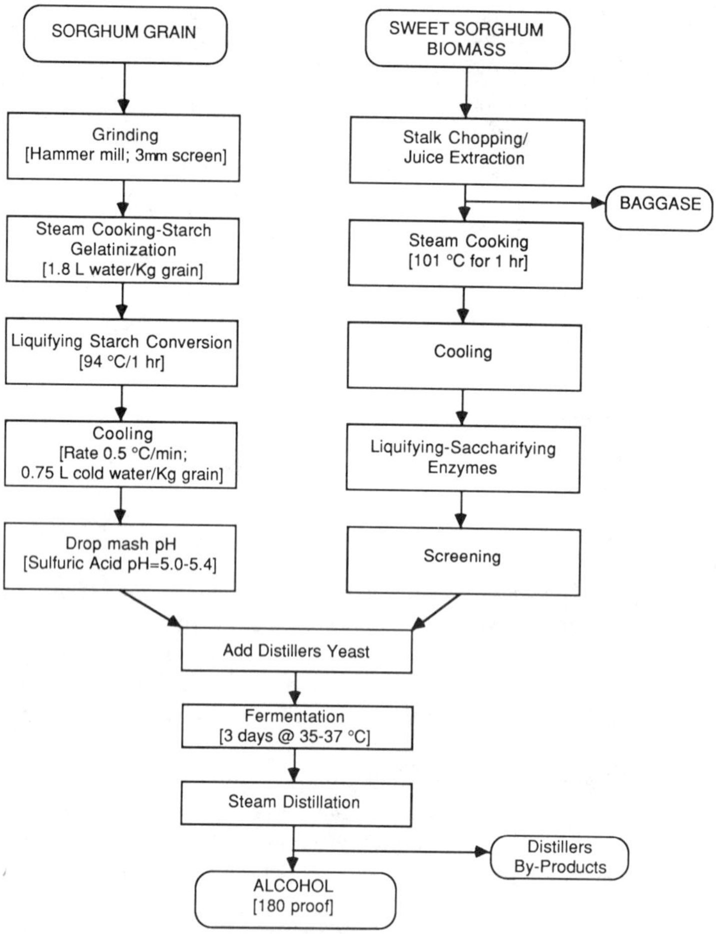

FIGURE 4 Flow chart of the process to produce ethanol from sorghum grain or sweet sorghum biomass. (Adapted from Refs. 11, 15, 41.)

grain produces 372 L. A bushel of sorghum (56 lb) produces about 2.5 gallons of alcohol. The fermentation of the grain produces distillers grain, which has a protein content of about 30% (Nx6.25).

The commercial technology required to ferment sweet sorghum biomass into alcohol has been perfected in Brazil (73). Sweet sorghum crops have the potential to yield from 22 to 45 tons per hectare in 110 days. The percentage of fermentable solids is 11.2% (2.5–5 tons/ha, which are composed of 80% soluble sugars and 20% starch. Therefore, to optimize ethanol production, both liquifying and saccharifying enzymes are required (11,41). One ton of sweet sorghum stalks has the potential to yield 74 L of 200 Proof alcohol.

E. Use of Sorghum for Beer and Malt

1. Lager Beer

Major breweries in Mexico, Africa, and Asia use sorghum grits as an inexpensive source of fermentable carbohydrates in brewing barley beer. The most desirable grit has light color, bland flavor, low oil content, and high extract levels. In most cases, red sorghums are used, but white sorghums with tan plant color and straw-colored glumes are preferred due to their improved milling properties, which produce high yields of excellent quality grits. Major problems with sorghum grits are variation in run-off time, level of phenols and grit yields. These problems can be overcome by selecting sorghum samples based on hardness and milling performance. Weathering of sorghum reduces grit yields significantly and affects brewing properties.

2. Sorghum Malt

Sorghum malt is produced by traditional village processes and on an industrial basis in Southern Africa (55) and India (39,44). A good sorghum variety for malting should produce high diastatic activity with proper modification of the endosperm after germination. The sorghum diastatic activity developed during malting depends on temperature, moisture content, duration of malting, germination level, and type of grain used. Between 18 and 50% of the amylolytic activity of sorghum malts is attributed to the action of B-amylase. In contrast, most of the diastatic activity of barley malt is attributed to α–amylase.

Tribal sorghum malting involves the following steps: (1) steeping (1–3) days, (2) germination (2–6 days), (3) sun drying and grinding with a mortar and pestle. Essentially the same process is used to produce sorghum malt industrially (Fig. 5). In contrast, the industrial process steeps the sorghum in special tanks which are aerated. Germination is done in a battery operation on concrete floors where the grain is turned regularly and transferred daily to another bed and malt is dried with hot air (55).

Roasted sorghum malt with a strong flavor and aroma is used for breakfast cereals in southern African countries. In India, one way of using surplus sorghum grain is to produce malt, which is used in weaning foods and other products (39,44).

3. Clear Sorghum Beer

Some African governments (i.e., Nigeria) have restricted the importation of barley and barley malt so new technologies have been developed to produce European type beers (Clear Beer) from sorghum malt or a combination of barley and sorghum malt. Clear beer (Fig. 6) is brewed using sorghum malt, sorghum grits, and additional saccharifying enzymes. The gelatinized adjuncts are solubilized by the sorghum malt and enzymes during mashing. After filtration, the wort is fermented with yeast (*Saccharomyces cerevisiae*) to yield green beer which is aged, carbonated, filtered, and bottled. The alcohol content (3.9%), ph (4.6), color, and foam stability of sorghum clear beer is reported similar to barley lager beer (37). The beverage has an unique acceptable taste but it is certainly not similar to European beer. Its future is directly related to Governmental importation policies.

4. Sour, Opaque Beer

The opaque sorghum beer industry in southern Africa is of major significance as an industrial user of sorghum. Tribal brewing evolved into commercial production of opaque beer. Different types of opaque beer are produced (i.e., Reef, iJuba, Kimberley). Some commercial large-scale operations produce dry sorghum beer powder, which can be used to produce beer by dissolving the powder in water to initiate fermentation.

The general procedure for the production of opaque sorghum beer is presented in Figure 7. In contrast with European brewing, in sour, opaque African beer there are two mashing

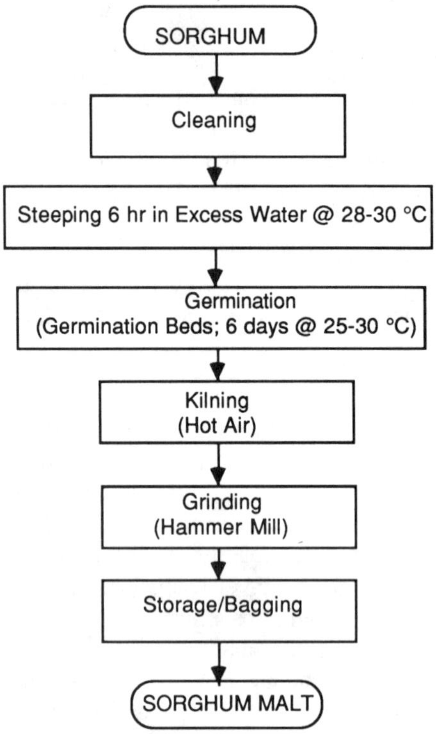

FIGURE 5 Typical industrial process for production of sorghum malt. (Adapted from Ref. 55.)

and two fermentation steps, which are not completely seperated. A mixture of sorghum malt and water is soured by lactic acid fermentation, which decreases the pH. The second mashing step starts when sorghum malt is added to convert the cooked adjunct (usually maize grits) into fermentable carbohydrates. Control of the mash pH is critically important because it affects viscosity, sugar concentration, and yield of alcohol. The wort is obtained after a differential centrifugation in which the very coarse cereal particles (i.e., pericarp, plumules) are seperated. In contrast to European brewing, the wort is not pasteurized and hops are not utilized. The wort is inoculated with bakers yeast to undergo alcoholic fermentation (55).

Opaque sour beer contains 2–4% alcohol, 0.3–0.6% lactic acid, and 4–10% solids with an acidic pH (3.3–3.5). It is consumed while undergoing active fermentation. The beer is opaque with a pale buff to pinkish-brown color. Opaque beer is more viscous than barley beer and is generally served warm. Throughout storage, beer after-souring can occur due to residual enzyme activity, which promotes growth of mesophilic heterofermentative organisms that produce acetic acid. Opaque sorghum beer is a good source of vitamins, minerals, proteins, and carbohydrates that are solubilized during malting and brewing (55). Generally, sorghums with bright red pericarp without a pigmented testa with an intermediate endosperm are preferred for malting. Sometimes, brown sorghums are malted after treatment with formaldehyde to eliminate condensed tannins.

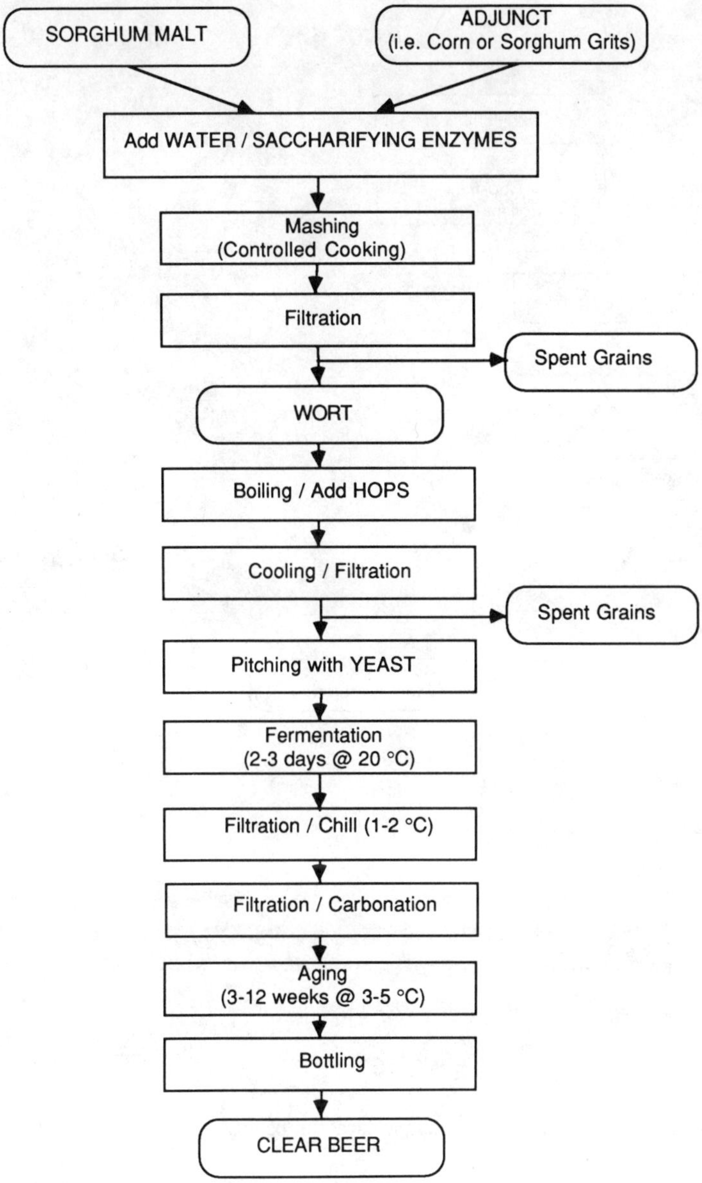

FIGURE 6 Brewing process for the production of clear lager sorghum beer. (Adapted from Ref. 37.)

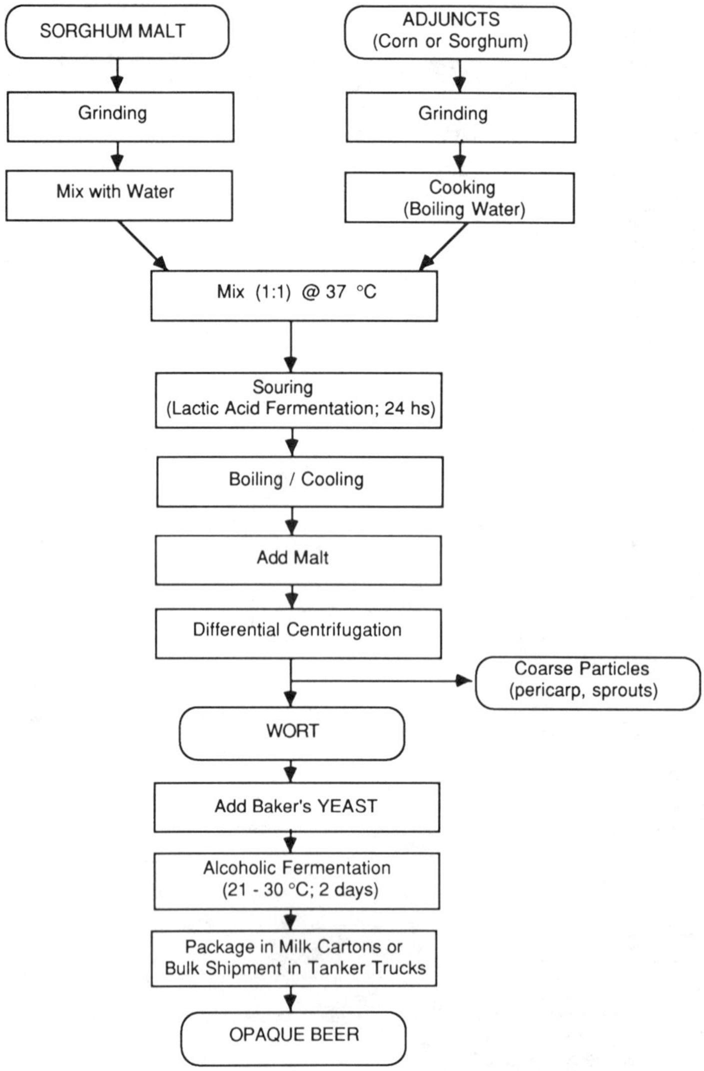

FIGURE 7 Brewing process for the production of opaque sorghum beer (Adapted from Ref.55.)

IX. PROCESSING FOR USE IN FEEDS

Sorghum and maize are major ingredients in swine, poultry, and cattle feeds in the Western Hemisphere. Generally, U.S. No. 2 yellow sorghum is utilized. Most of the sorghum fed to monogastrics is ground by hammer-milling. In the United States, large feedlots containing 200,000 cattle use sophisticated processes to improve palatability, enhance consumption, and improve nutritional efficiency of high concentrate rations (Table 11). Steam flaking is the most widely used method of processing sorghum and corn in feedlots. For sorghum, optimum feed efficiency is achieved with a very thin flake in which the endosperm is thoroughly disrupted. Of all cereals, sorghum requires more thorough processing to achieve optimum results. Some of the yellow endosperm and new food type sorghums have improved feed-processing properties.

X. PROCESSING FOR FOOD

A. Traditional Food Systems

Sorghum is processed into many different traditional foods around the world (Table 12). It has been estimated that about 30% of world sorghum production is consumed directly by humans.

For the production of most African traditional foods, sorghum is decorticated using a wooden mortar and pestle. Hand decortication is a laborious chore generally done by housewives. Sorghums with thick pericarp and hard endosperm are preferred because they are easier to decorticate. In some instances, mechanical dehullers are used to service small villages and urban areas. When using mechanical dehullers, milling yields are related to kernel hardness, size, and shape. Most of the sorghums are milled to remove 10–30% of the original weight. The use of abrasive mills for dehulling has been increasing slowly.

B. Sorghum in Baked and Pasta Products

Sorghum flour does not contain proteins that produce the viscoelastic gluten of wheat; therefore, acceptable yeast leavened products from sorghum flour are difficult if not impossible to obtain. Sorghum and wheat flour blends have been used to produce an array of baked products including yeast leavened breads, cakes, muffins, cookies, biscuits, flour tortillas, and others (18,62,67). The level of sorghum substituted for wheat flour depends upon the strength and quality of gluten in the wheat flour, the baking procedure, the definition of acceptable bread, the color and particle size of the sorghum flour, and the use of gums, emulsifiers, and other additives. Usually only 5–20% sorghum flour is substituted for wheat flour. Particle size of sorghum affects grittiness of the products. Tempering and other modifications during processing can partially modify the sandiness of sorghum products.

Faure (23) summarized research designed to produce pasta products from sorghum-wheat composite flours. Sorghums with soft texture, yellow endosperm, with white pericarp and without pigmented testa produced the best pasta products. Obtaining optimum gelatinization of sorghum flour and avoiding the oxidation of phenolic pigments that produce brown color in the pasta are major problems.

Composite blends of maize and sorghum masa have been shown to produce acceptable tortillas when white pericarp grains from tan plant sorghums with a bland flavor are used. These are used commercially by villagers in Central America and southern Mexico (9). Serna-Saldivar et al. (75) produced acceptable tortilla chips from blends containing 50% sorghum/50% maize. Flour from white sorghum can replace up to 20% of wheat flour for production of flour tortillas (85). The use of sorghum flour in baked products is directly related to its cost versus that of wheat flour, and governmental policies.

C. Sorghum Syrup, Molasses, and Sugar

Sweet sorghum varieties which produce extra quantities of sugar in the juice are used to produce sorghum syrup or molasses (Fig. 8) in the southern United States. Sweet sorghum plants are generally harvested when the grain is in the dough stage (12). Earlier harvesting usually causes problems during syrup clarification due to excess chlorophyl pigments. Therefore, the color, clarity, and viscosity of the syrup improves with plant maturity. Sorghum stalks should be processed immediately after harvesting to avoid respiration losses

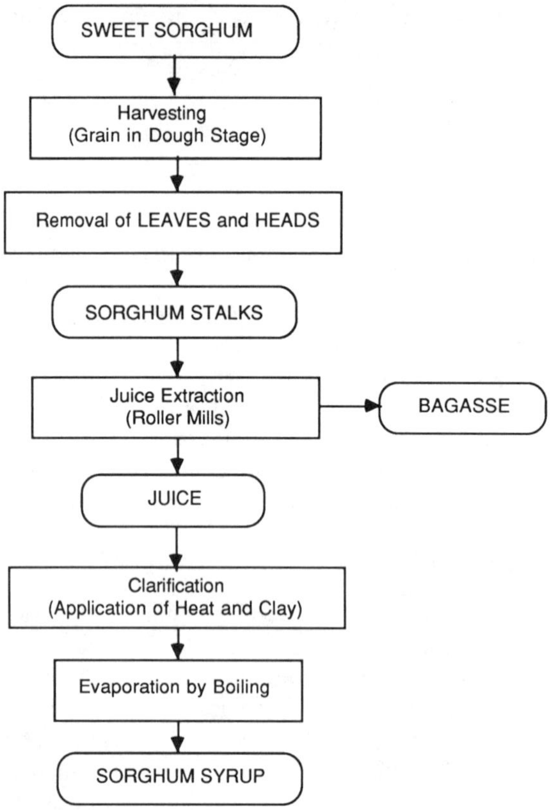

FIGURE 8 Flow chart of sorghum syrup production. (Adapted from Ref. 12.)

and sucrose inversion. The juice is extracted from the stalks that have been previously stripped of leaves. Generally, roller mills are used to extract the juice. Clarification is usually done with heat, although some processors use clay to improve syrup clarity. Solids are concentrated by evaporating part of the juice until a syrup is obtained. The finished product usually contains 74–78% solids, 68% carbohydrates, and 2.4% ash (80). A good quality sorghum syrup should be mild and sweet with a light color.

Special sweet sorghums have been used to produce crystalline sugar experimentally in the United States. Smith (80) summarized the technology, procedures, problems, and potential of using sweet sorghum for sugar. The characteristics of sweet sorghum varieties for raw sugar production are different from those desired for production of syrup or alcohol. For example, the sweet sorghum must have a high content of sucrose in the stalks and low levels of starch and aconitic acid. Sweet sorghum fields are not burned because the leaves are too green. The main difference between sweet sorghum and beet and sugar cane refining is the need to significantly reduce levels of starch and aconitic acids, which impede sorghum sugar crystallization. Juice extraction is identical to that used for sugar cane. However, the raw juice from sweet sorghum is clarified at higher pH (7.8) and at lower temperature (50–58°C.) Starch is flocculated and then removed by continuous horizontal desludging centrifuges. About 90% of the starch is removed during the first flocculation. More starch is removed by adding more flocculant before syrup concentration. Sorghum syrup is concentrated by the use of multiple effect evaporators. At 60–65% solids, the concentrated syrup is adjusted to 8.3 pH, treated with calcium chloride, and heated to 80–85°C to remove aconitic acid. The insoluble calcium aconitate is removed and the clear syrup is further evaporated and crystallized into sugar. One ton of stalks produces 180–200 lb of sugar. Use of sweet sorghums could extend the processing season for sugar cane refineries, which could enhance the economics of sugar production in some areas.

XI. NUTRITIONAL VALUE

Sorghum has proximate composition, amino acid contents, and nutritional value similar to corn. However, due to its lower fat content, sorghum usually has lower gross, digestible, and metabolizable energy levels than corn. In general, yellow sorghum has 95% the value of good yellow dent corn for all livestock species. Feeding value of sorghum hybrids have been improved relative to that of corn hybrids during the past 20 years when brown, bird-resistant high tannin hybrids are excluded (70).

A. Nutritional Value of Sorghum as Livestock Feed

Animal feeding trials with swine, poultry, and feedlot steers have demonstrated that brown sorghums (type III) have a lower protein digestibility and efficiency of feed conversion than type I sorghums or corn (Table 13). For monogastrics, U.S. yellow (type I) and brown (high tannin, Type III) sorghums have about 95% and 90% of the nutritional value of corn. Brown sorghums generally had about 5% and 8% lower protein digestibility than yellow sorghum and corn, respectively. Condensed tannins bind dietary proteins and inhibit digestive enzymes.

In rumimant diets, sorghum is generally processed more rigorously than corn to disrupt the endosperm and improve rates of nutrient digestibilities. In ruminants, processed yellow sorghum had about 93–96% of the value of corn. Ruminants fed brown sorghums have lower efficiency of feed conversion than counterparts fed yellow sorghum or corn (70,79).

Both monogastrics and ruminants fed waxy (100% amylopectin) sorghum perform better than animals fed nonwaxy (75% amylopectin; 25% amylose) sorghums (Table 14). Animals gain weight faster and more efficiently due to better rates of nutrient digestibilities. Sorghum nutritional value has been improved significantly over the last 20 years.

Hoseney et al. (32) have presented an excellent review of the nutritional value of sorghum grain. Following is a summary of nutrition studies and the effects of processing on the nutritional value of sorghum.

1. Human Digestibility Studies

Both livestock and human nutritionists recognize that protein digestibility of sorghum grain is slightly lower than that of other major cereals (33,70). The proper processing of sorghum is required to improve its digestibility. MacLean et al. (42,43), working with preschool children, demonstrated that protein digestibility of sorghum is lower than other cereals and that sorghum protein digestibility was considerably improved when the grain was decorticated and extruded.

Sorghum proteins have reduced digestibility due to (a) protein cross linking (35), which lowers protein solubility; (b) a stronger association of proteins with undigestible fiber components (5); and (c) the presence of a high proportion of peripheral endosperm with high levels of proteins (70).

2. Effect of Processing

a. Effect of sorghum decortication Most African and some Indian foods are processed from decorticated sorghum, which reduces the amount of fiber, minerals, proteins, and lysine significantly (21,76–77). Eggum et al. (21) and Serna-Saldivar et al. (76–77) reported 40% and 14% losses of lysine in manual sorghum decortication (yield = 75–80%) and mechanical sorghum decortication (yield = 90%), respectively. Nutrient digestibilities of decorticated sorghums are slightly higher than those of the parent whole grains, but nitrogen retention and protein efficiency ratios are considerably lower in decorticated grain because germ, the grain fraction with the highest amount of lysine (82), is partially lost.

Decortication of brown sorghums significantly reduces the amount of condensed tannins and presumably can overcome the deleterious effect of tannins on the nutritional value of sorghum grain (53). Stepwise removal of tannins via decortication gradually increased the percent of nitrogen solubilized by pepsin and by trypsin-chymotrypsin. The decortication process reduced the tannin content from 4.5 catechin equivalents to 0.2 in a kernel in which 37% of its original weight was removed (8).

b. Effect of cooking When sorghum is cooked, the solubility of the proteins, especially that of prolamins, is altered. The amount of prolamins extracted from sorghum is considerably reduced by cooking (from 42 to 6%). Other cooked cereals have a reduced protein solubility, but the effect is considerably lower. Using SDS-PAGE and gel filtration chromatography, sorghum proteins were shown to form intermolecular disulfide bonds during cooking. The cross-linked protein polymers occurred to a greater extent among glutelins followed by prolamins. Treatment of glutelin and prolamin extracts with mercaptoethanol improved protein digestibility of cooked sorghum to a level similar to that of the raw grain (35).

c. Effect of alkaline or acid treatment Sorghum is cooked with alkali (i.e., ashes and lime) for the production of popular traditional food systems such as tô, nixtamalized porridges, and tortillas. Such cooking conditions have a slightly detrimental effect on protein quality or availability (76–77). During alkali cooking unavailable peptides (i.e., lysinoalanine, orithoalanine, and lanthionine) that lower protein quality are produced (10,17). Sorghum tortillas and Ugali have lower protein digestibilities than their corresponding raw grain. On the other hand, it has been demonstrated that lime cooking increases the bioavailability of niacin in corn (36), but there are no studies conducted on sorghum. Lime cooking also notably increases the amount of calcium, on of the most important and limiting minerals in childhood nutrition.

Porridges cooked in an acidic medium (Acid Tô, Acid Ugali, etc.) show no decrease in protein digestibility compared to that of the uncooked sorghum flour (22).

d. Effects of fermentation According to Graham et al. (27), a traditional Sudanese fermented food called Nasha had a better nutritional value than the original grain source. The increased nitrogen utilization was due to an increased rate of protein digestibility. Axtell et al. (4) found that Kisra and Abrey, two fermented sorghum products, were more digestible than the unfermented grains. Aliya and Geervani (2) concluded that rats fed fermented sorghum had better protein digestibilities, nitrogen retention (biological value and net protein utilization), and PERs than counterparts fed the unfermented grain.

e. High tannin sorghum Several studies have demonstrated that high tannin sorghums have lower nutritional value than sorghums without tannins (33,46,49). Mitaru et al. (49), working with ileum-cannulated swine, found that reconstitution of sorghum grain improved nutrient digestibilities in high tannin grain. Hydrophobic bonding between proteins and tannins is important in formation of undigestible complexes in the digestive tract of pigs. The treatment of high tannin sorghum grain with CaO, $K2CO_3$, NH_4OH, or $NaHCO_3$ lowers the amount of assayable tannins and improves the nutritional value of the grain to a point that is equivalent to that of sorghum without tannins (6,58–59). Grain germination also lowers tannins with the corresponding improvement in its nutritional value (6).

f. Effects of protein fortification The protein quality of sorghum is limited by the amount and bioavailability of lysine. Addition of synthetic lysine to increase dietary lysine content to 0.75% considerably improved protein efficiency ratio of a sorghum-based diet (1.36 vs. 2.11). The addition of roasted or boiled Indian legumes (Bengal gram, *Cicer-arientinum*; Black gram, *Phaseolus mungus*; Red gram, *Cajanus cajun*; Cow pea, *Vignia unguiclata*; soybeans, *Glycine max*; and Green gram, *Phaseolus radentus*) to a sorghum-based diet in which sorghum provided two thirds of the protein and the legume the other third considerably improved dietary protein quality (61). In most cases, addition of raw legumes to the basal diet did not improve the overall protein quality. Further studies with Indian children demonstrated that addition of Black gram or Red gram to a sorghum-based diet considerably improved both the nutritional status and anthropometric measurements of the children on the supplemented diets (63). Addition of 8% defatted soybean meal to maize/sorghum tortillas doubled the amount of lysine, improved nitrogen retention values, and considerably improved the protein efficiency ratio (1.02 vs. 2.25) of tortillas fed to weanling rats (78).

Sorghum is a little lower in nutritional value than most other cereals, and it must be properly processed to improve its digestibility. Large numbers of sorghum consumers in Africa and Asia demonstrate that sorghum is a useful staple cereal.

ACKNOWLEDGMENT

The authors are grateful for the long-term support of scientists in the Texas Sorghum Improvement Program at Texas A&M University. This work was part of the INTSORMIL Title XII Sorghum and Millet Collaborative Research Support Program supported in part by Grant AID/DSAN/1254 G-TS-50-65 from the Agency for International Development, Washington, DC.

TABLE 1 Statistics of Sorghum Production, 1987

	Area Harvested (1000 ha)	Yield (kg/ha)	Total Production (1000 MT)
Africa	14,658	758	11,106
Nigeria	4,400	1,023	4,500
Sudan	3,360	387	1,300
Ethiopia	910	1,099	1,000
North-Central America	6,844	3,516	24,066
United States	4,291	4,386	18,819
Mexico	2,108	2,277	4,800
South America	1,958	2,701	5,288
Argentina	1,005	3,025	3,040
Asia	19,380	866	16,777
India	16,000	587	9,400
China	1,898	3,213	6,100
Europe	120	3,799	455
Oceania	768	1,544	1,186
Australia	767	1,544	1,186
U S S R	100	1,000	100
World	43,828	1,346	58,978
Developed countries	5,592	3,760	21,026
Developing countries	38,236	993	37,952

Source: Ref. 25.

TABLE 2 Grade Requirements for Sorghum[a]

Grade	Minimum test weigh (lb/bu)	Moisture (%)	Damaged kernel		Broken kernels foreign material, and other grains (%)
			Total (%)	Heat-damaged (%)	
US No 1	57.0	13.0	2.0	0.2	4.0
US No 2	55.0	14.0	6.0	0.5	8.0
US No 3[b]	53.0	15.0	10.0	1.0	12.0
US No 4	51.0	18.0	15.0	3.0	15.0

[a]US sample grade is sorghum that: a) does not meet the requirement for US grades 1, 2, 3, or 4 or b) contains 8 or more stones which have an aggregate weight in excess of 0.2% of the sample weight, 2 or more pieces of glass, 3 or more crotalaria seed (*Crotalaria spp*), 2 or more castor beans (*Riccinus comunis*), 4 or more particles of an unknown foreign substance(s) or a commonly recognized harmful or toxic substance(s), 8 or more cocklebur (*Xanthium spp*), or similar seeds singly or in combination, 10 or more rodent pellets, bird droppings, or equivalent quantity of animal filth per 1,000 grams of sorghum; or c) has a musty, sour, or commercially objectionable foreign odor (except smut odor); or d) is badly weathered, heating or distinctly low quality.

[b]Sorghum that is distinctly discolored shall be graded not higher than US No. 3.

Source: Ref. 86.

TABLE 3 Typical Composition of Sorghum Grain

	Mean	Range	Number of samples
Proximate analysis			
Protein (Nx6.25), %	11.6	8.1–16.8	1463
Ether extract, %	3.4	1.4–6.2	1462
Crude fiber, %	2.7	0.4–7.3	1383
Ash, %	2.2	1.2–7.1	1436
Nitrogen free extract[a], %	79.5	65.3–81.0	1372
Fiber, %			
Dietary insoluble	7.2	6.5–7.9	2
Dietary soluble	1.1	1.0–1.2	2
Acid detergent	3.3	2.9–3.6	2
Protein fractionation			
Prolamine, %	52.7	39.3–72.9	5
Glutelins, %	34.4	23.5–45.0	5
Albumins, %	5.7	1.6–9.2	5
Globulins, %	7.1	1.9–10.3	5
Essential amino acids[b] *(g AA/100 g protein)*			
Lysine	2.1	1.6–2.6	640
Leucine	14.2	10.2–15.4	582
Phenylalanine[c]	5.1	3.8–5.5	582
Valine	5.4	.0–5.8	582
Tryptophan	1.0	0.7–1.3	92
Methionine[d]	1.0	0.8–2.0	582
Threonine	3.3	2.4–3.7	582
Histidine[e]	2.1	1.7–2.3	582
Isoleucine	4.1	2.9–4.8	582

All values are expressed on dry matter basis.

[a]Calculated by difference

[b]FAO/WHO (24) suggested pattern (g AA/100 g protein):
lysine, 5.44; leucine, 7.04; phenylalanine+tyrosine, 6.08; valine, 4.96; tryptophan, 0.96; methionine +cysteine, 3.52; threonine, 4.0; isoleucine, 4.0.

[c]Phenylalanine requirement can be partially spared by tyrosine.

[d]Methionine requirement can be partially spared by cysteine.

[e]Histidine is considered essential amino acid only for children.

Source: Adapted from Refs. 5 and 66.

TABLE 4 Soluble Sugar Composition of Sorghum Grains[a]

	Total	Stachyose	Raffinose	Sucrose	Glucose and Fructose
Sorghum Type					
Normal[b]	2.25	0.10	0.23	1.68	0.25
Normal[c]	1.34	0.06	0.15	0.61	0.52
Sugary[c]	2.21	0.06	0.39	0.81	0.95
High Lysine[c]	2.57	0.11	0.39	0.94	1.13

[a]Values are expressed on percentage (g/100 g). Sugars were extracted with 80% ethanol and estimated colormetrically by the phenol-sulfuric acid method.
[b]Information obtained from Subramaniam et al. (81). Average of 10 sorghum cultivars (range of hard to soft endosperm) grown at ICRISAT, Patancheru, India.
[c]Information obtained from Murty et al. (51). High lysine data is the average of two lines.

TABLE 5 Typical Fatty Acid Composition of Sorghum Lipids

Component	Sorghum (%)
Ether extract	3.4
Palmitic (16:0)	14.3
Palmitoleic (16:1)	1.0
Stearic (18:0)	2.1
Oleic (18:1)	31.0
Linoleic (18:2)	49.0
Linolenic (18:3)	2.7
Arachidic (20: 0)	0.2

Fatty acid composition is expressed as percent of total.
Source: Ref. 65.

TABLE 6 Mineral and Vitamin Composition of Sorghum Grain

	Sorghum	
	Whole	Decorticated[a]
Minerals		
Calcium, %	0.05	0.03
Phosphorus, %	0.35	0.35
Potassium, %	0.38	0.37
Sodium, %	0.05	—
Magnesium, %	0.19	0.02
Iron, mg/kg	50.00	50.00
Cobalt, mg/kg	3.10	—
Copper, mg/kg	10.80	7.92
Manganese, mg/kg	16.30	18.70
Zinc, mg/kg	15.40	—
Vitamins		
Thiamine, mg/kg	4.62	4.84
Riboflavin, mg/kg	1.54	1.76
Niacin, mg/kg	48.40	53.24
Pyridoxine, mg/kg	5.94	10.34
Pantothenic Ac, mg/kg	12.54	12.76
Choline, mg/kg	761.20	—
Biotin, mg/kg	2.90	—
Folic acid, mg/kg	0.20	—
Carotene, mg/kg[b]	1.32	—

All values are expressed on dry matter basis.
[a]The extent of decortication is not reported.
[b]Sorghums with yellow and heteroyellow endosperms have a higher amount of carotenes. Carotenes are precursors of vitamin A.
Source: Adapted from Refs. 14 and 48.

TABLE 7 Assays Used to Determine Phenols and Tannins of Sorghum

Method	Standard	Reagents	Extraction	Time	Assayed compounds
Vanillin	Catechin	4% HCl, 1% vanillin in methanol	Methanol 1% HCl in methanol	24 hr 20 min	Leucoantho- cyanidins proanthocyanidins (Tannins)
Prussian Blue	Catechin	$FeCl_3$ in HCl	Methanol	1 min	Total phenols
Folin-Denis	Tannic acid	Folin-Denis Reagent	Water	5 hr	All reducing substances
Protein precipitation	Tannic acid	$FeCl_3$, alkaline detergent	Methanol	20 min	Proanthocyanidins (tannins)
Folin-Ciocalteu	Catechin or Gallic acid	Folin-Ciocalteu Reagent	Methanol 1% HCl in methanol, dimethyl- formamide	1 hr 2 hr	Total phenols
Relative degree of polymerization		4% HCl, 0.5% vanillin in acetic acid, 4% HCl in acetic acid	Methanol	20 min	Anthocyanidins, proanthocyanidins leucoantho- cyanidins

Source: Ref. 28.

TABLE 8 Polyphenol Levels of Three Groups of Sorghum

Sorghum Type[a]	Prussian Blue	Vanillin-HC1		Modified vanillin-HC1			α-Amylase inhibition
		With blank	Without blank	With blank	Without blank	24 hr	
Group I							
Mean	0.07	0.03	0.12	0.02	0.37	0.55	3.7
Range	0.04–0.09	0.01–0.04	0.05–0.24	0.01–0.03	0.20–0.58	0.33–0.96	1.8–15.4
Group II							
Mean	0.07	0.03	0.19	0.07	0.52	0.98	6.0
Range	0.05–0.09	0.02–0.04	0.09–0.52	0.02–0.19	0.34–0.88	0.68–0.98	2.3–15.8
Group III							
Mean	0.39	2.39	2.68	1.10	1.75	2.14	31.7
Range	0.24–0.99	0.88–6.67	1.12–7.45	0.36–3.48	1.00–4.85	1.49–4.85	13.9–77.0

All values except for % α amylase inbibition are expressed in mg catechin per 100 mg sample.
[a]Group I are sorghums without testa; Group II are sorghums with pigmented testa without a dominant spreader gene; Group III are sorghums with pigmented testa and dominant spreader gene. The group I sorghums do not contain condensed tannins.
Source: Ref. 20.

TABLE 9 Properties of Starches from Normal and Waxy Sorghum

	Sorghum		Corn	
	Normal	Waxy	Normal	Waxy
Gelatinization[a]				
Temp. Range, °C	68.5–75.0	67.5–74.0	62.0–72.0	63.0–72.0
Swelling Power[b]				
@95°C	22	49	24	64
Solubility (b)				
@95°C	22	19	25	23
Amylograph Behavior, Brabender units				
Peak Viscosity[c]	386	896	295	680
Peak Viscosity[c]	750	1200	—	—
Viscosity @ 95°C[d]	600	380	—	—
Viscosity after 1 hr				
@ 95°C	400	290	—	—
Viscosity @ 50°C	580	390	—	—
Viscosity after 1 hr				
@ 50°C	520	350	—	—

[a]Kofler gelatinization temperature range. Information from Ref. 40.
[b]Estimated after 30 min heating at 95°C. Information from Ref. 40.
[c]Information from Ref. 31. Amylograph of a paste containing 6.6% solids. Nonwaxy and waxy sorghum values are average of 5 and 3 cultivars, respectively.
[d]Information from Ref. 56. The nonwaxy and waxy slurries contained 34 and 35 g solids/500 ml. respectively.

TABLE 10 Common Dry Milling Procedures for Sorghum

Milling operation	Process	Products	Composition
Decortication degermination	The grain is tempered and decorticated via an abrasive process. Decorticated kernels are tempered and degerminated by impaction or pin milling. Fractions are separated by sieving and gravity separators.	16, 7, 4, 23, and 50% yields of feed, germ, flour, +20 grits, and +14 grits are obtained respectively from yellow sorghum.	Fat content was 9,15 , 3.0, 0.5 and 0.8 dry wt for feed, germ, flour +20 grits, and +14 grits, respectively.
Decortication grinding	Sorghum is tempered and decorticated by an abrasive mill. Mortar and pestle or mechanical dehullers (rice mills; abrasive disk mills). Tempering permits greater retention of the germ tissue with the kernel	Decorticated whole grain, brokens, flour and/or meal depending upon the method of grinding. Normally, around 10–15% of kernel weight mainly pericarp tissue is removed.	Composition depends upon the degree of decortication. Usually a significant amount of germ remains with the endosperm so the products contain 2% fat or more. Keeping properties of the flour are poor.
Roller milling	Sorghum is tempered and roller milled with wheat milling equipment.	High extraction flours (90% yield) and lower extraction flours (70%) produced.	The 70% and 90% extraction flours have 2 and 2.8% fat, respectively.
Semi–moist roller milling	Sorghum is tempered to 30–35% moisture, and milled using wheat flour rollers.	A whiter improved flour is produced even from brown sorghums	The method is experimental currently but may be useful and practical since wheat flour mills exist.

Source: Refs. 3, 50, 52.

TABLE 11 Common Methods of Processing Sorghum for Use in Livestock Feed[a]

Category	Type of process	Procedure	Characteristics
Mechanical action	Grinding/Rolling	Particle size reduction using hammer, plate, pin, or roller mills.	Most commonly used, least expensive. Increase feed efficiency and digestibility by 10–20% of whole grain.
Wet process	Reconstitution	Increase grain moisture to 25–30%. Wet grain is anaerobically stored for 2–3 weeks prior to grinding and feeding.	Improves feed efficiently about 10–15% over dry ground grain due to higher protein and energy digestibility.
	Early harvest	Grain is harvested at 20–30% moisture and stored anaerobically or with organic acids (i.e., propionic). Grain is ground prior to or after storage.	Similar to reconstitution.
	Soaking	Soak grain in water for 12–24 hr. Feed whole or crush.	Tendency for grain to ferment or sour. Only limited use.
Heat and moisture	Steam rolling	Grain subjected to live steam (180°F) 3–5 min then rolled.	Slight increase over dry rolling. Reduces fines and dust.
	Steam flaking	Grain exposed to high moisture steam for 5–15 min to reach 18–20% moisture. Then grain is rolled to desired flake thickness.	Most common method in feedlots. Thin flaking of sorghum increases digestibility and feed efficiency equal to that of reconstitution.
	Pelleting	Ground grain is conditioned with steam, forced through a die and pellets are cooled.	Reduces dust, improves palatability, uniformity, and handling of feeds. Prevents segregation of micronutrients.
	Exploding	Grain exposed to high pressure steam, the starch is gelatinized, the pressure is decreased and rapid expansion of the kernel occurs.	Similar to puffing of cereals for breakfast foods. Feed efficiency is similar to steam flaked or reconstituted grain.
Hot dry heat	Popping	Hot, dry air expansion of grain. Bulk density is low. Density is increased by spraying with water and rolling sometimes.	Ruptures endosperm increasing starch availability. Feed efficiency is similar to steam flaking or reconstitution.
	Micronizing	Heat grain with gas-fired infrared burners to the point of eversion followed by rolling through a roller mill.	Feed efficiency similar to steam flaking, exploding or popping. Bulk desnity similar to steam-flaked grain.

Source: From Refs. 29, 30, 70, 71.

TABLE 12 Traditional Foods Made with Sorghum

Type of food	Common names	Countries	Description
Unfermented Bread	Chapati, roti, rotte	India	Flour (95–100% extraction) is mixed with water and kneaded into a dough, dough is formed into a thin (1.3–3 mm) circular piece about 20–25 cm in diameter. The dough piece is baked on a hot grill at 210°C for 25 sec. Then the outside is sprinkled with water; the dough piece is turned over and cooked for 25 sec or until it puffs. Sorghums differ in extent of puffing, texture, color, taste, and keeping quality. White sorghum makes the best chapati.
	Tortilla	Central America, Mexico	A 3:1 ratio of lime solution to sorghum grain is cooked for 3–10 min at the boiling point and steeped for at least 4 hr. The lime solution contains 0.5% lime based on grain weight. The cooked soaked nixtamal is cooked, washed with water, ground into masa, and pressed into tortillas. The tortilla is cooked on each side for 30 sec at about 210°C on a grill. Sorghum tortillas are off-colored compared to those made with white maize. Usually mixtures of maize and sorghum are used.
Fermented bread	Kisra, dosa, dosai	Africa, Sudan, India	Whole sorghum is ground into a fine flour. A starter, which contains a combination of yeast and *Lactobacillus*, usually obtained by saving part of the last batter, is mixed with water and flour (1:2:9 ratio, respectively) to form a paste, which is fermented overnight. The fermented batter is fluid enough to be spread in a thin layer on the surface of a hot griddle. The kisra is cooked for about 30 sec and removed from the grill. Good kisra is paper thin with a sour taste and fermented flavor.

TABLE 12 (Continued)

Type of food	Common names	Countries	Description
	Injera	Ethiopia	Whole sorghum flour is mixed with about 40% of the total water and inoculum.Usually starter from a previous batch is used. The mixture is kneaded into a dough, which is fermented 12–48 hr. Then about 10% of the fermented dough is mixed with water and cooked. The cooked material is added to the remaining dough with additional water. The batter is allowed to undergo vigorous fermentation for 2 hr; then it is poured in a thin layer onto a hot griddle or metad and baked. The injera is large (60 cm in diameter), and thin layered with a very soft spongy texture and numerous "fish eyes" (4 eyes/sq. cm) on the surface. Injera batter is thicker than that of kisra.
Stiff porridge	Ugali, tuwo, saino, dalaki, aceda, atap, bogobe, ting, tutu, kalo, kwon, karo, nshimba, nuchu, tô, tuo, zaafi, mato, asidah, sadza	Africa, India	Decorticated sorghum flour is stirred into boiling water until a very stiff paste is formed. The paste is stirred vigorously while it is cooked. The porridge is placed in a calabash (gourd), cooled for 1 hr, and eaten by hand with a sauce. Sometimes acid or base is mixed into the cooking water. In Burkina Faso, tamarind pods are steeped overnight in the cooking water. In Mali, alkali is added to the cooking water. Soured, fermented porridges are common in most countries. Granulation of the flour, composition of the sauce and consistency of the porridge vary among countries and tribes within a country.
Thin porridge	Uji, ambali, edi, eko, kamu, nasha, obungi, bwa, kal, obushera, atole	Africa, India, Mexico, Central America	Decorticated sorghum flour in cool water is mixed into boiling water with stirring until a thin paste is formed. The consistency is similar to that of cream. It is made with acid or alkali in the cooking water. Sometimes a part of the flour is fermented 24 hr before cooking. Particle size varies. Flour is made from sprouted grain, pearled grain, or whole grain. The porridge is served with milk, sour milk, sugar, honey, fruit, and other variations, depending on local preference.

Thin porridge	Ogi, oko, akamu, kafa, koko, akasa	Nigeria, Ghana	Sorghum grains are steeped in water for 2–4 days at room temperature. The fermented grain is milled, and the bran is removed by screening. The sediment in the pot is cooked in water to produce a porridge that is consumed warm or cooled to form a gel or pudding. It is preferred as a weaning food and for elderly people.
Steamed cooked products	Couscous	West Africa	Finely ground sorghum or millet flour is kneaded with water until the flour particles agglomerate. Then the particles are forced through a coarse screen. The particles are steamed in a container with a perforated bottom, which is placed on top of a pot filled with boiling water.Several times during steaming, the couscous is removed from the pot, stirred, sieved, and returned to the cooker. Usually, ground baobab leaves, peanut butter, okra, or some other additive is mixed with the couscous during the final steaming. The cooked product is consumed with a sauce or milk. Sometimes it is dried, stored, and used as a convenience food.
Boiled whole or pearled	Acha, sankati, mudde, kali, piti	Africa, India, Haiti	Sorghums are pearled to remove the pericarp and cooked like rice. Special varieties of sorghum with a high proportion of corneous endosperm are preferred. Whole grains are sometimes boiled, parched or steamed and consumed with legumes or a sauce. Sorghum is mixed with rice as an extender.
Snack foods		Worldwide	Sorghums are popped, puffed, and parched. They may be consumed directly or ground and mixed with other ingredients. Fried snacks are popular. There are numerous variations in snacks.

TABLE 12 (Continued)

Type of food	Common names	Countries	Description
Alcoholic beverages	Burukutu, dolo, pito, talla	West Africa	Sorghum is steeped, germinated, and dried. Mash is prepared by mixing the ground malt with water. The mixture is filtered to remove the bran. The wort is boiled and inoculated with yeast from a previous batch of beer. Fermentation occurs overnight. The actively fermenting beer is drunk the next day. Beer is a relatively clear red liquid with a sweet pleasant taste and low solids content. Continued fermentation produces a sour taste. The beer is about 1–5% alcohol.
Sour/opaque beers	Marisa, busaa, merrisa, urwaga mwenge, munkoyo, banutu beer, kaffir beer, sorghum beer, utshwala, utywala, ikigage	Africa	Ground sorghum malt is mixed with water, allowed to sour, boiled with adjunct (corn grits), cooled to 60°C, and saccharified with more sorghum malt. Maize grits are often unavailable, so sorghum is used in many areas. The mixture is filtered to remove the larger particles and fermented with top-fermenting yeast. Beer has a high solids content with a sour taste and bright pink color. It has the consistency of a milk shake. The alcohol depends on fermentation time and ranges from 1 to 8% by volume.
Nonalcoholic beverages	Mahewu, amaheu, marewa, magou, leting	Africa	Ground sorghum, sometimes mixed with sprouted sorghum, is held at elevated temperatures for up to 20 hr. *Lactobacilli* and other microorganisms cause souring of the mixture. Very little alcohol is produced. Sour nonalcoholic beverages have lower dry matter contents than sour thin porridges.

Source: Refs. 1, 9, 16, 26, 68, 84, 87

TABLE 13 Comparison of the Nutritional Value of Yellow and High Tannin, Sorghums with Maize Diets Fed to Swine and Poultry[a]

	Sorghum		
	Type I	Type III	Maize
Swine			
Tannin content[b]	0.85	3.28	0.01
Digestibility			
Energy, %	88.8	84.9	88.8
Protein. %	83.9	79.0	86.9
Digestibile energy, kcal/g	3.97	3.82	4.02
Metabolizable energy, kcal/g	3.89	3.75	3.95
Feed to gain ratio	3.11	3.44	2.99
Poultry			
Tannin Content[b]	0.03	4.80	0.00
Body Weight Gain, g/bird	512	473	506
Feed Intake, g/bird	779	781	831
Feed to gain ratio, 1 day - 21days	1.52	1.65	1.64

[a]Swine and poultry data adapted from Refs. 13 and 19, respectively.
[b]Expressed as catechin equivalents without blanks subtracted. The type I sorghums did not have any condensed tannins.

TABLE 14 Comparison of the Nutritional Value of Waxy and Nonwaxy Sorghum Grain Fed to Swine and Feedlot Steers (72)

	Gain (g/day)	Feed/Intake (g/day)	Feed/Gain	Reference
Swine				
Nonwaxy endosperm	760	2550	3.35	60
Waxy endosperm	840	2620	3.11	
Feedlot steers				
Nonwaxy endosperm	878	7052	8.08	
Waxy endosperm	1012	7073	6.95	47
Nonwaxy endosperm	1398	12485	8.93	
Waxy endosperm	1371	11350	8.28	79
Nonwaxy endosperm	1016	8476	8.35	
Waxy endosperm	1144	8303	7.27	47

REFERENCES

1. Akingbala, J. O., Faubion, J. M., and Rooney, L. W., A laboratory procedure for the preparation of Ogi, a Nigerian fermented food, *J. Food. Sci., 46*: 1523 (1981).
2. Aliya, S., and Geervani, P., An assessment of the protein quality and vitamin B content of commonly used fermented products of legumes and millets, *J. Sci. Food. Agric., 32*:837 (1981).
3. Anderson, R. A., and Burbridge, L. H., Integrated process for dry milling grain sorghum, *Northwestern Miller, 6*:278 (1971).
4. Axtell, J. D., Kirelis, A. W., Hassen, H. M., Manson, N. O., Mertz, E. T., and Munck, L., Digestibility of sorghum proteins, *Proc. Nat. Academy of Sci. USA, 78*:1333 (1981).
5. Bach Knudsen, K. F., and Munck, L., Dietary fiber content and composition of sorghum and sorghum based foods *J. Cereal Sci., 3*:153 (1985).
6. Butler, L. G., Polyphenols and their effects on sorghum quality. In *International Symposium on Sorghum Grain Quality*, L. W. Rooney and D. S. Murty (Eds.), ICRISAT. Patancheru, A. P., India, 1982, p. 294.
7. Caransa, A., and Bakker, W. G. M., Modern process for the production of sorghum starch, *Starch/Starke, 39* (11):381 (1987).
8. Chibber, B. A. K., Mertz, E. T., and Axtell, J. D., In vitro digestibility of high tannin sorghum at different stages of dehulling, *J. Agric. Food Chem., 28*:160 (1980).
9. Choto, C. E., Morad, M. M., Rooney, L. W. The quality of tortillas containing whole sorghum and pearled sorghum alone and blended with yellow maize, *Cereal Chem., 62*:51 (1985).
10. Chu, N. T., Pellet, P. L., and Nawar, W. W., Effect of alkali treatment on the formation of lysinoalanine in corn, *J. Agric. Food Chem., 24*: 1084 (1976).
11. Coble, C. G., HILER, E. A., Sweeten, J. M., O'Neal, H. P., Reidenback, V. G., LePori, W. H., Schelling, G. T., and Kay, R. D., Small scale ethanol production from cereal feedstocks. In *Cereals: A Renewable resource,* Y. Pomeranz and L. Munck (Eds.), The American Association of Cereal Chemists, St. Paul, MN, 1981, p. 611.
12. Coleman, O. H., Syrup and sugar from sweet sorghum. In *Sorghum Production and Utilization*, J. S. Wall and W. M. Ross (Eds.), The AVI Publishing Co., Westport, CT, 1970.
13. Cousins, B. W. Tanksley, T. O., Knabe, D. A., and Zebrowska, T., Nutrient digestibility and performance of pigs fed sorghums varying in tannin concentration, *J. Anim. Sci. 53*: 1524 (1981).
14. Crampton, E. W., and Harris, L. E., *Applied Animal Nutrition. The Use of Feedstuffs in the Formulation of Livestock Rations*, W. H. Freemann and Co., San Francisco, CA, 1974.
15. Creelman, R. A., Rooney, L. W., and Miller, F. R., Sorghum. In *Cereals; A Renewable Resource*, Y. Pomeranz and L. Munck (Eds), The American Association of Cereal Chemists, St. Paul, MN. 1981.
16. Da, S., Akingbala, J. O., Rooney, L. W., and Miller, F. R. Evaluation of tô quality in a sorghum breeding program. In *International Symposium on Sorghum Grain Quality*, L. W. Rooney and D. S. Murty (Eds.), ICRISAT, Patancheru, A. P., India, 1982, pp. 11–23.
17. DeGroot, A. P., and Slump, Pl, Effect of severe alkali treatment of proteins on amino acid compostion and nutritive value, *J. Nutr., 98*:45 (1969).
18. Dendy, D. A. V., Composite flour: Past, present and future. A review with special emphasis on the place of composite flour in the semi-arid zones. Presented at the International Workshop on Policy & Potential Relating to Uses of Sorghum and Millets, SADCC/ICRISAT. Bulawayo, Zimbabwe, June 8–12, 1988.
19. Douglas, J. H., and Sullivan, T. W., Evaluation of different varieties of grain sorghum and yellow corn in starting broiler diets, *Poultry Sci., 65* (Supplement 1):36 (1986).
20. Earp, C. F., Akingbala, J. O., Ring, S. H., and Rooney, L. W., Evaluation of several methods to determine tannins in sorghums with varying kernel characteristics *Cereal Chem. 58*: 234 (1981).

21. Eggum, B. O., Bach Knudsen, K. E., Munck, L., Axtell, J. D., and Mukuru, S. Z., In *International Symposium on Sorghum Grain Quality*, L. W. Rooney and D. S. Murty (Eds.), ICRISAT, Patancheru, A. P., India, 1982, p. 211.
22. Eggum, B. O., Monowar, L., Bach Knudsen, K. E., Munck, L., and Axtell, J., Nutritional quality of sorghum and sorghum foods from Sudan, *J. Cereal Sci., 1*:127 (1983).
23. Faure, J., Sorghum utilization in pasta production. Presented at the International Workshop on Policy & Potential Relating to Uses of Sorghum and Millets, SADCC/ICRISAT. Bulawayo, Zimbabwe, June 8–12, 1988.
24. Food Agriculture Organization/World Health Organization, Energy and protein requirements, FAO Nutrition Reports Series, 52, 1973.
25. Food Agriculture Organization, *Production Yearbook.*, FAO Statistics Vol. 41, Rome, Italy, 1988.
26. Galiba, M., Rooney, L. W., Waniska, R. D., and Miller F. R. The preparation of sorghum and millet couscous in West Africa, *Cereal Foods World, 32*:878 (1987).
27. Graham, G. G., Maclean, W. C., Morales, E., Hamaker, B. R., Kirelis, A. W., Mertz, E. T., and Axtell, J. D., Digestibility and utilization of protein and energy from nasha, a traditional Sudanese fermented weaning food, *J. Nutr., 116*:978 (1986).
28. Hahn, D. H., Rooney, L. W., and Earp, C. F. Tannins and phenols of sorghum, *Cereal Foods World*, 29:776 (1984).
29. Hale, W. H., Influence of processing on the utilization of grains (starch) by ruminants, *J. Anim. Sci. 37*:1075 (1973).
30. Hale, W. H., and Theurer, C. B., Feed preparation and processing. In *Digestive Physiology and Nutrition of Ruminants*, Vol 3, Ch. 4, Church, D. C., (Ed.), Oregon State University, Corvallis, OR, 1972.
31. Horan, F. E., and Heider, M. F., A study of sorghum and sorghum starches, *Cereal Chem., 23*: 492 (1946).
32. Hoseney, R. C., Andrews, D. U., and Clark, H., Sorghum and pearl millet. In *Nutritional Quality of Cereal Grains: Genetic and Agromomic Improvement*, Ch. 9, R. A., Olson and K. Frey (Eds.), American Society of Agronomy, Inc., Crop and Soil Science Societies of America, Inc., Madison, WI, 1987.
33. Jambunathan, R., and Mertz, E. T., Relationship between tannin levels, rat growth and distribution of proteins in sorghum, *J. Agric. Food Chem., 21*:692 (1973).
34. Karim A., and Rooney, L. W., Characterization of pentosans in sorghum grain, *J. Food Sci., 37*:369 (1972).
35. Kirelis, A. W., Effects of processing on sorghum protein digestibility, Presented at the INTSORMIL/ICRISAT Sorghum Workshop. Tegucigalpa Honduras, December, 1987.
36. Koetz, R., and Neukom, H., Nature of bound nicotinic acid in cereals and its release by thermal and chemical treatment. In *Physical, Chemical and Biological Changes in Food Caused by Thermal Processing*, T. Hoyden and O. Kuale (Eds.), Applied Science Publishers, London, 1977, p. 305.
37. Koleoso, O, A., Sorghum malt/adjunct replacement in clear lager beer: Policy and practice in Nigeria, Presented at the International Workshop on Policy & Potential Relating to Uses of Sorghum and Millets, SADCC/ICRISAT, Bulawayo, Zimbabwe, June 8–12, 1988.
38. Kulkarni, D. N., Kshirsagar, S. S., and Ingle, U. M., Processing of sorghum grain for starch and glucose syrup. In *Technology and Applications for Alternative Uses of Sorghum*, Proceedings of the National Seminar, Marathwada Agricultural University, Parbhani, India, 1987, p. 115.
39. Kulkarni, D. N., Kshirsagar, S. S., and Ingle U. M., Product application studies on sorghum malt: cookies, weaning food and beer wort. In *Technology and Applications for Alternative Uses of Sorghum*, Proceedings of the National Seminar, Marathwada Agricultural University, Parbhani, India, 1987, p. 130.
40. Leach, H. W., Gelatinization of starch. In *Starch: Chemistry and Technology*, Ch. XII, Vol. I *Fundamental Aspects*, R. L. Whistler and E. F. Paschali, (Eds.), Academic Press, New York, 1965.

41. Lipinski, E. S., Systems study of fuels from sugarcane, sweet sorghum and sugar beets. In *Agricultural Consideration. Vol. II,* Battelle Columbus Laboratories, U. S. Energy Research and Development Administration, 1976.
42. MacLean, W. C., Lopez de Romana, G., Placko, R. P., and Graham, G. G., Protein quality and digestibility of sorghum in pre-school children: Balance studies and plasma free amino acids, *J. Nutr., 111*:1928 (1981).
43. MacLean, W. C., Lopez de Romana, G., and Graham, G. G., The effect of decortication and extrusion on the digestibility of sorghum by preschool children, *J. Nutr., 113*:2171 (1983).
44. Malleshi, N. G., Malting of sorghum for food uses. *In Technology and Application for Alternative Uses of Sorghum,* Proceedings of the National Seminar, Marathwada Agricultural University, Parbhani, India, 1987, p. 144.
45. Maxson, E. D., and Rooney, L. W., Evaluation of methods for tannin analysis in sorghum, grain, *Cereal Chem., 49*:719 (1972).
46. Maxson, E. D., Rooney, L. W., Lewis, R. W., Clark, L. E., and Johnson, J. W., The relationship between tannin content, enzyme inhibition, rat performance and characteristics of sorghum grain., *Nutr. Reports Int., 8*:145 (1973).
47. McCullough, R. L., Drae, C. L., Roth, G. M., Brent, B. E., Riley, J. G., and Schalles, R. R., Several reports in nutritive value of hybrid sorghum grains at 59th Cattlemen's day, Kansas Agric. Exp. Station, Manhattan, KS, Bulletin No. 557, 1972.
48. Miller, D. F., *Composition of Cereal Grains and Forages,* Publication 585, National Academy of Sciences-National Research Council, Washington, DC, 1958.
49. Mitaru, B. N., Reichert, R. D., and Blair, R., The binding of dietary protein by sorghum tannins in the digestive tract of pigs, *J. Nutr., 114*:1787 (1984).
50. Munck, L., Back Knudsen, K. E., and Axtell, J. D., Industrial milling of sorghum for the 1980's In *Sorghum in the Eighties II,* Proc. Int. Symposium on Sorghum, L. R. Hause, L. K. Mughogho, J. M. Peacock, and J. V. Mertin (Eds.), ICRISAT, Patancheru, A. P., India, 1982, pp. 564–570.
51. Murty, D. S., Singh, U., Suryaprakash, S., and Nicodemus, K. D., Soluble sugars in five endosperm types of sorghum, *Cereal Chem., 62*:150 (1985).
52. Murty, D. S., and Subramanian, V., Sorghum roti: Traditional methods of consumption and standard procedures for evaluation. In *International Symposium on Sorghum Grain Quality,* L. W. Rooney and D. S. Murty (Eds.), ICRISAT, Patancheru, A. P., India, 982, pp. 73–78.
53. Mwasaru, M. A., Reichert, R. D., and Mukuru, S. Z. Factors affecting the abrasive dehulling efficiency of high tannin sorghum, *Cereal Chem., 63*:171 (1988).
54. Norris, J. W., and Rooney, L. W., Wet milling properties of four sorghum parents and their hybrids, *Cereal Chem., 47*:64 (1970).
55. Novellie, L., and de Schaepdrijver, P., Modern development in traditional African beers. In *Progress in Industrial Microbiology,* Vol. 23 M. R. Adams (Ed.) Elsevier Sci. Publishers, The Netherlands, 1986, pp. 73–157.
56. Otterbacher, T. J., and Kite, F. E., The milo starches, *The Bakers Digest,* October, 1958.
57. Poehlman, J. M., *Breeding Field Crops,* Henry and Holt Co, Inc., New York, 1959.
58. Price, M. L., Butler, L. G., Featherston, W. R., and Rogler, J. C., Detoxification of high tannin sorghum grain, *Nutr. Reports Int., 17*:229 (1975).
59. Rice, M. L., Butler, L. G., Rogler, J. C., and Featherston, W. R. Overcoming the nutritionally harmful effects of tannin in sorghum grain by treatment with inexpensive chemicals, *J. Agric. Food Chem., 27*:441 (1979).
60. Purser, K. W., The effect of sorghum endosperm type on performance of growing-finishing swine and digestibility of selected nutrients measured at the terminal ileum and in the feces, Ph. D. dissertation, Texas A&M University, College Station, TX. 1979.
61. Pushpamma, P., and Anjali Devi, C., Nutritional quality of sorghum and legume based food mixture for infants and pre-school children, I. *Nutr. Reports Int., 19*: 635 (1979).
62. Pushpamma, P., and Mrudula Kalpalathika, P. V., Alternative uses of sorghum in small scale food industry. In *Technology and Applications for Alternative Uses of Sorghum,* Proceedings of the National Seminar, Marathwada Agricultural University, Parbhani, India, 1987, p. 81.

63. Pushpamma, P., Ratnakumarti, A., and Geervani, P., Nutritional quality of sorghum and legume based food mixture for infants and pre-school children. II. *Nutr. Reports Int., 19*:643 (1979).
64. Reichert, P. D., Sorghum dry milling. In *Sorghum in the Eighties*, Vol II, ICRISAT, Patancheru, A. P., India, 1982, p. 547.
65. Rooney, L. W., Sorghum and pearl millet lipids, *Cereal Chem., 55*:584 (1978).
66. Rooney, L. W., Earp, C. F., and Khan, M. N., Sorghums and millets In *CRC Handbook of Processing and Utilization in Agriculture*, Vol. II, I. A. Wolf (Ed.), CRC Press, Boca Raton, F. 1982, p. 123.
67. Rooney, L. W., Khan, M. N., and Earp C. F. The technology of sorghum products. In *Cereals for Food and Beverages: Recent Progress in Cereal Chemistry*, G. E. Inglett and L. Munck (Eds.), Academic Press Inc., New York, 1980.
68. Rooney, L. W., Kirelis, A. W., and Murty, D. S., Traditional foods from sorghum: their production, evaluation and nutritional value. In *Advances in Cereal Science and Technology*, Vol VIII, Pomeranz (Ed.), American Association of Cereal Chem., St. Paul, MN, 1986.
69. Rooney, L. W., and Miller, F., Variation in the structure and kernel characteristics of sorghum. In *Interational Symposium on Sorghum Grain Quality*, L. W. Rooney and D. S. Murty (Eds.), ICRISAT, Patancheru, A. P., India, 1982, p. 143.
70. Rooney, L. W., and Pflugfelder, R. L., Factors affecting starch digestibility with special emphasis on sorghum and corn, *J. Anim. Sci., 63*:1607 (1986).
71. Rooney, L. W., and Riggs, J. K., Utilization of sorghum, Progress Report No 2945 and 2946 in Grain Sorghum Research in Texas, Texas Agric. Exp. Station, College Station, TX. 1971.
72. Rooney L. W., and Sullins, R. D., Differences in feed lot performance of ruminants fed sorghum grains with different endosperm types. A review, Progress Report 3242, Texas Agric. Exp. Station, College Station, TX. 1974.
73. Schaffert, R. E., Sweet sorghum substrate for industrial alcohol, Presented at the International Workshop on Policy & Potential Relating to Uses of Sorghum and Millets, SADCC/ ICRISAT, Bulawayo, Zimbabwe, June 8–12, 1988.
74. Schoch, T. J., Starch in bakery products, *Bakers Digest, 39*(2):48(1965).
75. Serna-Saldivar, S. O., Tellez-Giron, A., and Rooney, L. W., Production of tortilla chips from sorghum and maize, *J. Cereal Sci., 8*:275 (1987.)
76. Serna-Saldivar, S. O., Knabe, D. A., Rooney, L. W., and Tanksley, T. D., Effects of lime cooking on energy and protein digestibilities of maize and sorghum, *Cereal Chem., 64*:247 (1987).
77. Serna-Saldivar, S. O., Knabe, D. A., Rooney, L. W., Tanksley, T. D., and Sproule, A. M., Nutritional value of sorghum and maize tortillas, *J. Cereal Sci. 7*:83 (1987).
78. Serna-Saldivar, S. O., Cannett, R., Vargas, J., Gonzalez, M., Bedolla, S., and Medina, C., Effect of soybean and sesame addition on the nutritional value of maize and decorticated sorghum tortillas produced by extrusion cooking, *Cereal Chem., 65*:44 (1988).
79. Sherrod, L. B., Albin, R. C., and Furr, R. D. Net energy of regular and waxy sorghum grains for finishing steers, *J. Anim. Sci., 29*:997 (1969).
80. Smith, B. A. Sweet sorghum In *CRC Handbook of Processing and Utilization in Agriculture*, Vol. II, I. A. Wolf (Ed.), CRC Press, Boca Raton, FL. 1982, p. 611.
81. Subramanian, B., Jambunathan, R., and Suuryaprakash, S., Note on the soluble sugars of sorghum, *Cereal Chem., 57*:440 (1980).
82. Taylor, J. R. N., and Schussler, L., The protein composition of the different anatomical parts of sorghum grain, *J. Cereal Sci., 4*:361 (1986).
83. Texas Department of Agriculture–United States Department of Agriculture, United States Department of Agriculture, Field Crop Statistics, D. S. Findley, State Statistician, 1986.
84. Thorat, S. S., Satwadeher, P. N., Kulkarni, D. N., Choudhari, S. D., and Ingle, U. M. Varietal differences in popping quality of sorghum grains In *Technology and Applications for Alternative Uses of Sorghum*, Proceedings of the National Seminar, Marathwada Agricultural University, Parbhani, India, 1987, p. 153.

85. Torres, P. I., Ramirez, B., Serna-Saldivar, S. O., Rooney, L. W., and Sweat, V., Effect of sorghum addition on wheat flour dough and tortilla textural shelf-life, Presented at the 49th Institute of Food Technologist Annual Meeting, New Orleans, LA, 1988.
86. United States Department of Agriculture, *The Official United States Standards for Grain*, Federal Grain Inspection Service, Inspection Division, Washington, DC, 1987.
87. Vivas, N. E., Waniska, R. D., and Rooney, L. W., Thin porridges (atole) prepared from maize and sorghum, *Cereal Chem., 64*:390 (1987).
88. Watson, S. A., Corn and sorghum starches: Production. 2nd ed., In *Starch Chemistry and Technology (Eds.),* Academic Press, Orlando, FL, 1984.

6

THE MILLETS

Sergio O. Serna-Saldivar, Cassandra M. McDonough, and Lloyd. W. Rooney
Texas A&M University, College Station, Texas

I. Introduction

The grasses known collectively as millets are a set of highly variable small seeded plant species indigenous to many areas of the world (Table 1). They are well adapted to grow under low soil fertility, low moisture, and hot environmental conditions. Millets are of value especially in semiarid regions because of their short growing season and higher productivity under heat and drought conditions. Often millets are produced in areas where maize and sorghum crops usually fail. The grain and forage are valuable as food and feed resources in Africa, Russia, India, and China.

Pearl millet is the most widely grown millet and is a very important crop in India (38) and parts of Africa (26). There are more than 15,000 pearl millet lines in the World Germplasm Collection located in India. Finger millet is popular in East Africa and India (26,41). Foxtail and proso millets are cultivated primarily in the Near East and China (59).

Millets are important cereals in the diet of many people in Africa and India (17), where millets and sorghum are used interchangeably in the same traditional food systems. The FAO (31) reported that people in Mali and Niger, West Africa, consumed approximately 300 g millet per capita per day in 1977. In the Western Hemisphere, millets are grown as emergency forage and catch crops. Proso millet is the only grain millet of economic importance grown in the United States, where it is grown for birdseed (106). Some yellow endosperm proso millets are decorticated and sold in health food stores in the United States or puffed and used in special breakfast cereals.

World production of millets has been stable during the last decade. According to the FAO (32), 39.4 million hectares of millet were planted in 1987 with an average production of only 704 kg/ha; world millet production accounted for 27.7 million metric tons (Table 2). Africa, India, and China harvested 36, 25, and 19%, respectively, of the total world production. Nigeria produced 30% of all African production (Table 2). Before the advent of pearl millet hybrids, grain yields in India averaged 350–400 kg/ha. With the use of fertilizer, hy-

Technical contribution number (TA 25533) from the Texas Agricultural Experimental Station, Texas A&M University, College Station, TX.

brids, and limited irrigation, Indian farm demonstrators were able to harvest 3500–8000 kg/ha (17).

Stoskopf (95), Hulse et al. (44), Rachie and Majmudar (81), Rooney et al. (89), Hoseney et al. (42,43) have reviewed the agronomy, chemical composition, food uses and nutritional value of millets.

II. Origin

The millets originated primarily in east and west Africa, Eurasia, India, and China from wild seed stock (101). Pearl millet is believed to be one of the earliest domesticated millets; carbonized grains have been found in east and west African sites inhabited 4,000–5,000 years ago (16, 37). Finger millet appeared in east Africa and India around the same time (40). Both grains were introduced into other areas via trade routes as early as 4,000 years ago (41,103). Foxtail and proso millets originated in Eurasia and China and spread via trade routes to India and Africa around 2,000 years ago; proso millet was used as a bread grain in Europe until medieval times (81).

III. Structure and Physical Properties

Kernel characteristics of the various millets are extremely diverse. Pearl millet varieties from the World Collection probably have more variation in physical characteristics than any other millet (70,71). Kernel colors range from white, yellow and tan to gray, green, purple, and black. Kernel shape has five different classifications: obovate, hexagonal, lanceolate, globular, and elliptical. In west Africa, pearl millet is classified as either Sanio (globular), or Souna (obovate, hexagonal, and lanceolate) varieties. Kernel density ranges from 1.28 to 1.42 g/cc. By comparison, 1000 kernel weight values for finger millet, proso millet, teff, fonio, and foxtail millet are 2.6, 4.7–7.2, 0.35–0.42, 0.63, and 1.86 g, respectively (70).

Most of the millets are very similar to each other in basic structure, although there are some very specific differences (Table 3; Fig. 1). The millets can be separated into two types of seeds: utricles and caryopses (Table 3, Fig. 1). In an utricle, the pericarp surrounds the seed like a sac, but is attached to the seed at only one point. Therefore, the protective layer for utricles is the testa or seed coat. Finger, proso, and foxtail millets are utricles. In these millets, the pericarp usually breaks away from the seed coat or testa, which is the effective protection barrier. Thus, the seed coat or testa of utricles is well developed, thick, and forms a strong barrier over the endosperm. In a caryopsis, the pericarp is completely fused to the seed. Pearl millet, fonio, and teff are caryopses. For pearl millet, the kernels are composed of the pericarp, endosperm, and germ, which comprise 8.4, 75.1, and 16.5% of the total kernel weight, respectively (2). The structure of the finger millet has been characterized by McDonough et al. (72). Angold (4) also reviewed the structure of the finger millet, as well as proso, foxtail, and pearl millets.

Of all the millets, pearl millet is the most widely used throughout the world. Its structure has been reviewed extensively by McDonough and Rooney (72), Rooney and McDonough (90), Zeleznak and Varriano–Marston (107), and Sullins and Rooney (98). The structure of the pearl millet kernel is similar to that of most of the millets with caryopses (Table 3). Pearl millet has a thick 1–2 layer epicarp, a mesocarp that varies in thickness due to genetic factors, and an endocarp layer that contains cross and tube cells. Teff, fonio, pearl, proso, and foxtail millets have some pigments in the pericarp (70).

Teff, fonio, and finger millet have a pigmented testa, whereas pearl has a thin pigmented or unpigmented seed coat beneath the endocarp. As with sorghum, pearl millet varieties with a thick mesocarp are preferred for traditional milling. During decortication, the pericarp usually separates from the rest of the kernel at the endocarp or aleurone layer (71,72). Many of the other millets, especially finger millet, are ground into flour without decortication, which can produce some undesirable nutritional effects.

The endosperm comprises the majority of the kernel weight for all millets. There are four structural parts of the endosperm: the aleurone layer, and the peripheral, corneous, and floury endosperm areas. All millets have a single layer aleurone that completely encircles the endosperm. The aleurone cells are rectangular with thick cell walls, and they contain protein, oil, minerals, and enzymes. The peripheral, corneous, and floury endosperm areas are beneath the aleurone in that order. The endosperm cells contain starch granules that are embedded in a protein matrix; protein bodies are distributed throughout the matrix. The starch granules are spherical in the floury area and become progressively more polygonal in the corneous and peripheral endosperm areas. Teff, fonio, and foxtail millet tend to have very little floury endosperm. Finger and proso millets are intermediate in texture. Pearl millet varieties range from those with all the floury endosperm to those with all corneous endosperm. Regardless of endosperm type, all millet varieties have at least one layer of peripheral endosperm just beneath the aleurone layer.

The germ can be large relative to the endosperm, as in pearl millet, or very small, as in proso and finger millets. The pearl millet germ is similar to that found in sorghum and contains 24.5, 32.8, and 7.2% of the total protein, fat, and ash found in the kernel, respectively (2).

IV. Composition

The mean values and variation in proximate composition of six types of millet are presented in Table 4. The proximate composition is affected by both environment and genetics. The amount of protein depends upon the nitrogen level in the soil and total grain yield, which is usually negatively correlated with protein content. Pearl millet usually has higher protein and fat contents than sorghum or the other millets because the kernel is a naked caryopsis.

For all millets, the essential limiting amino acid is lysine (Table 5). The endosperm protein fraction of pearl millet contains low levels of lysine (1.4 g/100 g protein), whereas the germ contains 5.16 g/100 g protein (2). Among millets, teff and pearl millet usually have the most lysine; proso millet has the lowest amount or poorest essential amino acid composition.

The prolamine fraction (soluble in alcohol) is the major protein fraction in pearl, foxtail, and finger millets (Table 6). Pearl millet resembles maize in its distribution of proteins especially with regard to prolamines and cross–linked prolamines (fractions II and III) (30). Pearl millet proteins contain a lower proportion of cross–linked prolamines than sorghum (30,47). The reduced level of cross–linked prolamines may be the explanation for pearl millet proteins having higher in vitro digestibility values than sorghum proteins (30). Because of its higher lysine content and its generally improved protein digestibility, pearl millet has a better protein quality than sorghum grain. Vincent–Monteiro et al. (102) determined the amino acid composition of the protein fractions of foxtail millet (Table 6). As in other cereals, lysine was present in the highest amounts in the albumin/globulin fraction. Its prolamine fraction contained only small quantities of lysine, the most limiting amino acid.

The starch content of pearl millet varies from 56 to 65% and amylose ranges from 17 to 29% (12,70). Table 7 summarizes starch properties of several millets. Lorenz and Hinze (61) reported that the starch of proso and foxtail millets had higher water–binding capacity than starch isolated from wheat. Amylograph viscosities of millet starches were higher than those of the wheat starch at all reference points in the amylograph curve.

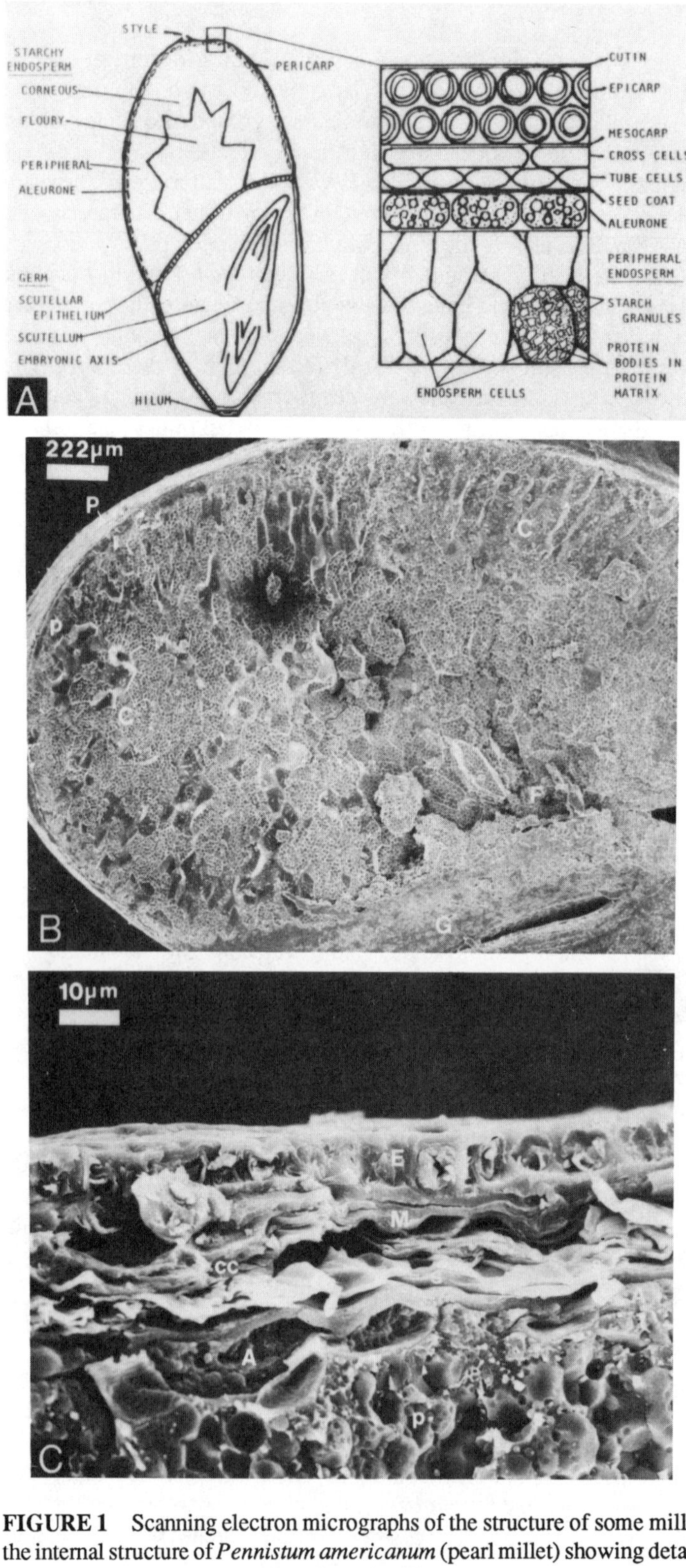

FIGURE 1 Scanning electron micrographs of the structure of some millets. (A) Line drawing of the internal structure of *Pennistum americanum* (pearl millet) showing details of the pericarp. (B-C) Scanning electron micrographs of pearl millet.

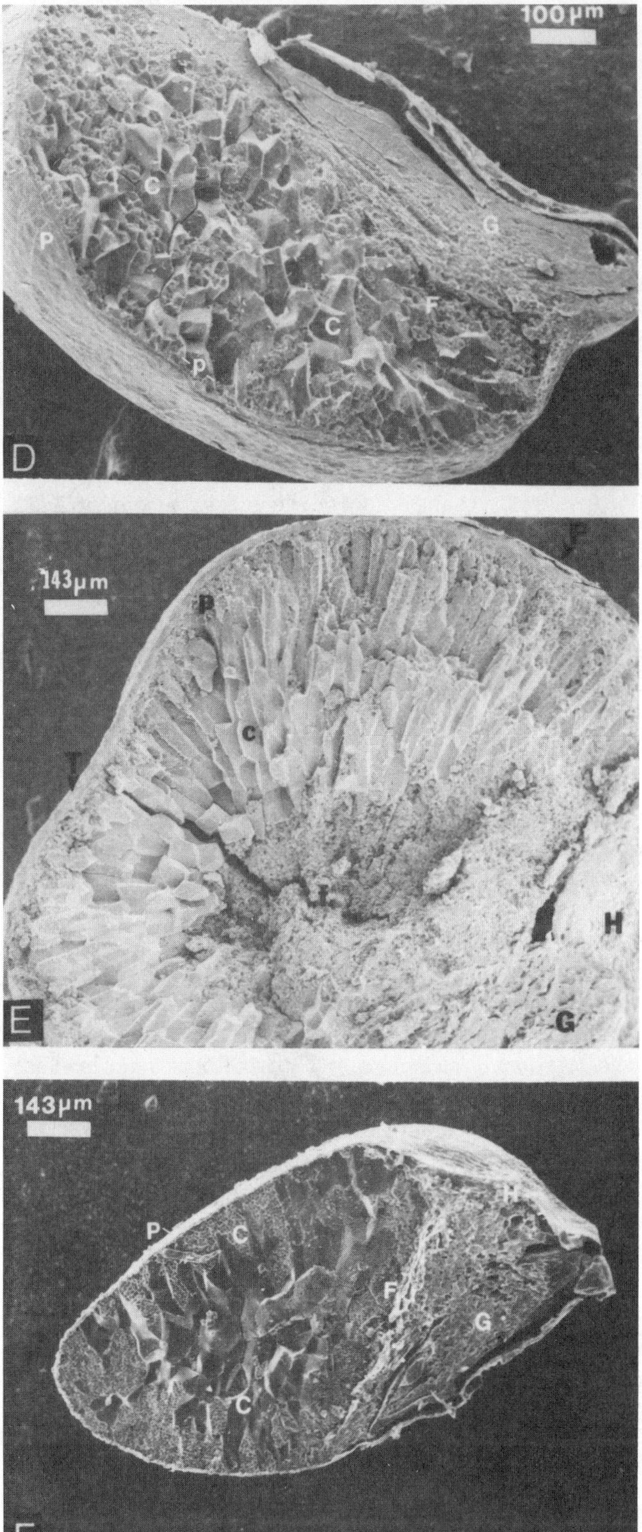

FIGURE 1 (Continued) (D) *Ergrastostis tef* (red teff). (E) *Eleusine coracana* (finger millet or ragi). (F) *Digitaria exilis* (fonio). P--pericarp; p--peripheral endosperm; C--corneous endosperm; F--floury endosperm; G--germ; A--aleurone; M--mesocarp; cc--cross cells; s--seed coat; H--hilum.

Soluble carbohydrates in pearl millet range from 2.3 to 2.7%. About 63% and 29% of the soluble carbohydrates are sucrose and raffinose, respectively. Other carbohydrates include stachyose, sucrose, glucose, and fructose (94) (Table 8). Sucrose and glucose constitute 33 and 12.5% of the soluble carbohydrates of finger millet (105).

Pentosans constitute 2–3% of the whole pearl millet grain. As in other cereals, the occurrence of these polysaccharides is mainly in cell walls. Bailey et al. (9) isolated pearl millet pentosans into four fractions. Ribose was only found in the fraction extracted with 80% ethanol. The ethanol–extracted fraction contained mainly glucose, galactose, rhamnose, and fucose, whereas the alkali–extracted fraction contained various amounts of rhamnose, fucose, arabinose, xylose, mannose, and glucose. Two water–soluble fractions (one extracted at 25 and other at 50°C) contained rhamnose, fucose, arabinose, xylose, mannose, galactose, and glucose. Sorghum grain has a higher degree of branching. This study demonstrated that pearl millet pentosans are associated with proteins. The alkali–extracted fraction contained the highest amount of protein.

Rooney (88) reported that the free lipids extracted from pearl millet vary from 3 to 7.4%. The fatty acid composition of pearl millet reported by Rooney (88) is very similar to the one given by Jellum and Powell (49) (Table 9). Lai and Varriano Marston (56) reported free and bound lipid contents of 5.5–7% and 0.57–0.9% for pearl millet. 70.3% and 51.7% of the free and bound fatty acids in pearl millet are unsaturated. According to Mahadevappa and Raina (66), the range of lipid content of seven breeding varieties of finger millet was 1.85–2.1%. The lipid fraction consists of 70–72% neutral lipids, 10–12% glycolipids, and 5–6% phospholipids. The average fatty acid composition of finger and kodo millets is also reported in Table 8.

Pearl millet is probably the cereal grain that most rapidly develops off-odors and flavors after milling. The possible reasons for the fast deterioration are:(1) the fat content is high, (2) the unsaturated fatty acids are higher than in other cereals, (3) millet does not have naturally occurring antioxidants, and (4) enzymatic–hydrolytic activity is greater (42,50). Kaced et al. (50) suggested that the fat content is the major contributing factor for the rapid increase in fat acidity in ground millet. Hexanal is a major product of oxidative degradation of lipids. Chaudhany and Kapoor (25) reported that the whole grain flour of three varieties of pearl millet became rancid after 6–10 days storage and inedible after 11–14 days storage. Moisture, sugar, free fatty acids, fat acidity, and peroxide levels increased throughout storage. Flours stored in polyethylene bags deteriorated more slowly than flours stored in gunny sacks. Pearl millet flour stored in polyethylene bags for 15 days did not show any hexanal production (50). Reddy et al. (84) indicated that the characteristic mousy, acidic odor generated in ground pearl millet during storage was not associated with oxidative rancidity. The generation of off–odors required a high moisture content in the grits, suggesting that the deterioration might be enzymatic. The odor precursor was extracted with methanol and had similar characteristics to apigenin, the aglycone of the major c-glycosylflavone of pearl millet.

Millets generally have a higher ash content than other cereals. The mineral composition of some types of millets is shown in Table 10. The major minerals found in foxtail and proso millets are: potassium, phosphorus, magnesium, and calcium. Finger millet ash is high in phosphorus, calcium, zinc, and manganese (62,99).

A. Polyphenols and Antinutritional Factors

Phenols and condensed tannins bind with and precipitate proteins in food systems, thus decreasing digestibility. Polyphenol and tannin levels of several millets are presented in Table 11. Many of the millets test positive in the vanillin assay for trace amounts of catechin equivalents, but only finger millet contains condensed tannins. Lorenz (58) stated that whole grain proso millet contained 0.055–0.178% tannins (catechin equivalents), but after decortication, the grains contained only trace amounts of 0.23–0.034%. Several compounds in proso millet react with vanillin to produce catechin equivalents, but the compounds are not condensed tannins.

Ramachandra et al. (82) reported that finger millet contained condensed tannins at levels ranging from trace amounts (0.04% catechin equivalents) to levels comparable to some high tannin sorghum varieties (3.47%). The same study reported a range of 55.4– 85.1% in vitro protein digestibility (IPVD) for high and low tannin varieties, respectively. They concluded that IPVD could be increased by 69.1% after dehulling the grain.

Phenolic acid content, determined by HPLC of teff, pearl, finger, and foxtail millets, is presented in Table 10. Ferulic, coumaric, cinnamic, and gentisic acids were present in higher levels than other phenolics. Foxtail millet contained the highest level of total phenolic acids, but the sample was not decorticated. The decorticated grain contains fewer phenolic acids than whole kernels and similar levels to whole wheat (42).

According to Reichert (85), pH-sensitive pigments that are extracted with methanol are responsible for color changes in millet flour-water pastes. Pigments highly sensitive to changes in pH were glycosylvitexin, glycosylorientin, and alkali-labile ferulic acid. These compounds are responsible for the intense yellow–green discoloration of the flour in the presence of alkali and may be responsible for the natural gray color of the peripheral endosperm of the kernel. The concentration of C-glycosylflavones and alkali-labile ferulic acid decreased markedly after dehulling the grain.

Soaking pearl millet kernels in solutions containing acid (i.e., sour milk, tamarind pods) markedly reduces the color of the grain. Reichert and Youngs (87) mentioned that the decolorizing effect is pH dependent. The acidic solution penetrates the whole grain through areas around the embryo. Dehulled grains decolorized faster than whole grains because the acidic solution penetrated the grain at a faster rate.

Phytate phosphorus is undesirable in foods because it decreases the bioavailability of divalent metal ions. Lorenz (58) reported that the phytate was present at levels of 0.17–0.47% in whole grain proso millet, but was reduced to 0.17–0.33% after decortication. Rao and Deosthale (83) reported that malting reduced the phytate content in both pearl millet and finger millet from 172 to 69 and 132 to 88 mg/100 g grain, respectively. Carr (21) also reported high levels of phytate in finger millet.

Chandrasekhar and Pattabiraman (24) isolated two trypsin inhibitors in pearl millet with molecular weights around 11,000. The inhibitors were extracted with 0.1 N HCl, fractionated with ammonium sulfate and identified on CM cellulose and gel chromatography.

A higher incidence of goiter among millet eaters in West Africa has been attributed to a goitrogen found in pearl millet. Epidemiological studies have suggested that pearl millet might be at least partially responsible for the higher goiter incidence in Sudan. Osman et al. (76) isolated and attributed the goitrogenic effect of pearl millet to a thionamide. Klopfenstein et al. (52,53) reported that the millet goitrogen is found in the bran and endosperm and that it inhibits the normal conversion of thyroxine (T4) to tri-iodothyroxine (T3). The T4 level was higher and T3 level lower in rats fed millet than for those fed sorghum diets. Millet feeding resulted in enlarged thyroid colloid follicles and flattened cells. Moreover, histological changes in the thyroid glands and a distortion of the thyroid hormone patterns in rats consuming millet have been reported (75,76). Addition of supplemental iodine to the diet did not improve the physiological status of rats. Grain fermentation did not reduce the levels or activity of the antinutritional factor. However, animals fed autoclaved or intensely heated millet had a normal thyroid hormone pattern (53). In most pearl millet-consuming areas, the incidence of goiters has not been documented to be higher than for similar non–

millet-consuming populations. Thus, the importance of this goitrogenic effect awaits determination.

B. Enzymes

Chandrasekhara and Swaminathan (23) found that the amylase enzymatic activity of finger millet malt was approximately twice as much as that of ungerminated grain. The optimum pH for the amylases was 4.6–5.0. The same authors reported no protease activity in ungerminated finger millet. Activity of malt proteases was optimum at pH 4.4.

Skovron and Lorenz (94) studied the enzymatic activity of eight cultivars of proso millet. All cultivars showed β-amylase, protease, cellulase, and hemicellulase activity. The optimum pH for β-amylase was 5.0. Protease activity was best at pH 3 and 5. Cellulase and hemicellulase activities were also present in proso millet.

V. Postharvest Technology

After harvesting, millet grains are almost always processed and consumed within regions of production. Pearl millet heads (Fig. 2A) are harvested, dried and stored intact in storage bins (Fig. 2B). Sometimes a group of villagers have a community threshing (Fig. 2C) when large amounts of grain are needed. However, usually the heads are pounded in a mortar and pestle, winnowed, and the grain required for daily consumption is further processed (dehulled and ground) in the mortar and pestle (Fig. 3A). The early millets are consumed immediately after harvest since they represent the first grain of the new crop year. The late millets mature in the dry season and are air–dried outside.

FIGURE 2 (A) Preparing pearl millet spikes for transport to village grainaries.

FIGURE 2 (B) Typical village grainaries in Northern Mali.

FIGURE 2 (C) A community threshing operation in Central Mali.

FIGURE 3 Common millet-milling procedures. (A) A typical mortar and pestle milling operation in West Africa. The mortar is used for threshing, decortication, and pounding to produce flour.

A. Milling

In developing countries, millets are still decorticated and ground with a mortar and pestle (Fig.3A) or with grinding stones followed by winnowing or washing at various stages of grinding to remove bran, coarse particles, and fine particles. These milling techniques are labor intensive. For example, in Senegal one person spends up to six hours per day milling whole millet kernels into flour required to prepare one day's food for a family (100). In India, the implement most widely used for grinding is a stone hand-grinder, consisting of two round (60–cm diameter) stones rotating horizontally against each other. In some villages and urban areas, millets are decorticated with abrasive disks in mechanical dehullers (Fig. 3B,C) and ground into flour with diesel-powered attrition or hammer mills (22). Milling procedures and equipment are similar to the ones used for sorghum (see Chapter 5, Table 10). Reichert and Youngs (86) monitored changes in protein, ash, and oil in pearl millet decorticated by traditional village procedures, mechanical dehullers, and attrition mills. Grains were decorticated to remove from 0 to 45% of the total kernel weight. For protein, ash, and oil, losses increased as the level of decortication increased. There were small differ-

FIGURE 3 (B) A batch-type mechanical decorticator. The machine is gaining increased acceptance in Africa. It is most often diesel-powered.

FIGURE 3 (C) Decortication is caused by abrasive action of the discs and abrasive kernel-to-kernel friction.

FIGURE 4 Scanning electron micrograph of starch granules isolated from pearl millet. (Courtesy of R. D. Sullins.)

ences in nutrient losses between the different decortication processes when samples were extracted to the same level.

Carr (21) reported milling extraction rates of 75% and 80% for pearl and finger millet, respectively, in Rhodesia (Zimbabwe). Milling resulted in losses of some nutrients in both types of millets. Millet flour has the tendency to become rancid after being ground into flour for the reasons stated before. An effective milling operation should be able to remove most of the germ to increase flour shelf life. The abrasive decortication equipment used for milling sorghum has proven effective for pearl millet, although the germ is not fully removed.

The best pearl millet for dry milling has a spherical kernel, intermediate to thick pericarp, with a high proportion of corneous endosperm. The spherical kernels consistently have the highest milling yields compared to long hexagonally shaped grains (90).

Experimentally, modern wheat flour milling equipment has been found to be suitable for milling proso millet into flour (75% extraction) (7). Lorenz and Dilsaver (59) produced proso millet flour by decorticating the grain in a barley pearler and then reducing the decorticated kernel into flour with a roller mill. Compared to wheat flour, proso flours were higher in ash and fat with lower lysine content and in vitro protein digestibility.

B. Wet Milling

Starch yields from pearl millets were significantly lower than those obtained from corn or sorghum (33). A lower protein content and higher starch content in the pearl millet by-products indicated poorer starch-protein separation. However, the cooking properties of pearl millet starch were in general similar to those of starch from corn or sorghum. Figure 4 shows native starch granules isolated from pearl millet.

VI. Food Uses

Millets have been utilized for human food from prehistoric times. Currently, millets are consumed in northern China, India, Africa, and southern U.S.S.R., with about 80% of the crop consumed directly as human food (60). In India, virtually all of the pearl millet and most of the finger millet is directly consumed as human food. Recent pearl millet surpluses in India have promoted a search for alternative uses for the grain. The most important use of pearl millet flour in India is in baking chapatis or rotis (96). Millet grain is also used for production of ricelike products and porridges (Fig.5A). In Africa, millet is consumed primarily in the form of thick and thin porridges (fermented or unfermented), flat breads (fermented or unfermented), steamed or boiled cooked products, snacks, alcoholic beverages, and in composite flours for bread, cookies, noodles, etc. In Nigeria, millets are dry milled into flour for production of Tuwo and Barabusco or mixed with wheat flour for production of biscuits and bread (73), (see Chapter 5, Table 11).

Pearl millet has been successfully used to produce malt and alcoholic beverages (Fig. 5B). Opaque sour beers are commonly prepared in Africa from sorghum and/or millets (see Chapter 5). For production of malt the grain is generally steeped in water for 24 hr, germinated for 72 hr at 23–27°C and dried in the sun or with hot air (45°C). Sprouts are removed from the dry malt before grinding (23). Pearl millet malt can replace 25% of barley malt and produce beer with analytical and organoleptic characteristics comparable to the control beer. Higher concentrations of pearl millet malt slightly decreases wort extract (29). The diastatic activity of pearl millet reaches a maximum value after 32–hr germination. The activity then rapidly declines to practically nil after 72 hr (46).

Millet malt is used for the production of alcoholic beverages and for the preparation of malt foods. Malted millet flour, especially that of finger millet, is traditionally preferred and used for production of weaning foods in India. The grain is generally malted, mixed with legume flours, and offered as a porridge to infants (11,68). The malt enzyme decreases the viscosity of foods, improving the palatability of the products for children.

Yeast-leavened pan bread cannot be made from 100% millet flour because millets do not have gluten (22). However, proso millet flour has been successfully used to replace 20–35% of wheat flour to produce breads that were considered to be quite acceptable (60). The quality of breads containing proso millet flour has been considerably improved with the use of dough conditioners (5,6). Pearl millet produced cookies with spread characteristics equal to soft wheat flour cookies when the flour was previously hydrated with water, dried, and supplemented with 0.6% unrefined soy lecithin (8). Finger millet or ragi is puffed for production of snack foods. The optimum conditions for puffing are: conditioning the grain to 19% moisture; equilibrating for 4 hr; and puffing in hot sand at 270°C. Large differences in puffing ability of different cultivars were reported (67).

VII. Nutritional Value

A. Feed Use

Feeding trials with different domestic animals have shown that millets have comparable or even better nutritional value than other major cereals (91–93). Animals fed pearl millets usually perform better than animals fed sorghum. Millets produce better growth because

A

B

FIGURE 5 (A) Preparation of tô, a thick porridge, in a village near Bamako, Mali. Tô is consumed with a spicy sauce. (B) A traditional brewery in Burkina Faso. Green malt is drying in the foreground, and wort is being sterilized in the background prior to fermentation.

they have more calories and better protein quality. Nutrient digestibility of millets are comparable to other cereals.

Sharma et al. (92) reported better weight gains in poultry fed pearl millet than birds fed sorghum or wheat. The efficiency of feed conversion was better for chicks fed pearl millet than counterparts fed wheat, maize, or sorghum. Another study by Simhaee et al. (93) compared maize, wheat, pearl millet, and their combinations as the main source of energy in broiler diets. No differences among grains was found in body weight gain. However, millet supported slightly poorer feed efficiency than either maize or wheat. Sanford et al. (91) found that the performance of chicks fed millet or sorghum was similar when isonitrogenous diets were formulated. However, pearl millet usually has a higher protein content than sorghum grain. Abate and Gomez (1) found that chicks fed pearl millet outperformed chicks on maize or finger millet. Chicks consuming finger millet had similar weight gains and feed conversion to chicks fed maize. Luis et al. (63–65) have reviewed the feeding value of proso millet in broiler and turkey diets. Proso millets supported equivalent egg production, egg weight, and feed consumption compared to sorghum or maize. Pullets fed proso millet had body weight gains similar to pullets fed maize but greater than birds fed sorghum. Turkeys fed proso millet had better or comparable performance than those fed maize or sorghum (64).

Early swine feeding studies conducted by Calder (18–20) in Zimbabwe demonstrated that the replacement of corn by finger millet in growing and finishing pig diets had minimum effects on pig performance. Swine fed finger millet performed better or equally to animals fed corn. Carcass weight and length and fat color were not affected by addition of finger millet in the diet. Danielson and Grabouski (28) compared the performance of swine fed corn or proso millet. Feed efficiency was slightly lower for animals fed proso millet (2.88 vs. 3.14 lb feed/lb gain). Addition of lysine (0.0005–0.0015%) to the proso millet based diet significantly improved feed conversion to equal that of swine fed corn.

Brethour (13) conducted a trial in which rolled pearl millet completely replaced sorghum grain in a steer finishing ration. Steers gained equally when fed pearl millet or sorghum grain. The estimated net energy of pearl millet was 3–4% higher than finely rolled sorghum grain (12,13). In another experiment, a ration containing sorghum and 6 lb of pearl millet was compared with a control ration comprised of sorghum, soybean meal, and urea. Steer performance was slightly better when the ration containing pearl millet was fed. The author concluded that millet provided excellent protein for beef cattle rations (14).

B. Nutritional Value

The most limiting amino acid in all millets is lysine (Table 4). Protein digestibility and lysine content of pearl millet are higher than those of sorghum and comparable to those of maize. In contrast to sorghum, pepsin digestibilities of pearl millet and maize did not decrease as much upon cooking. The amino acid profile of pearl millet is better than that of sorghum and maize and is comparable to wheat, barley and rice (30).

The protein efficiency ratios (PERs) for foxtail, proso, pearl, and finger millets were 0.8, 1.1, 1.6, and 2.0, respectively. Addition of 40% chickpea flour (*Cicer arietinum*) to millet-based diets considerably improved rat performance. Chickpea protein is rich in lysine and threonine and, therefore, complemented millet proteins exceptionally well (34).

Rats fed pearl and finger millets had PERs of 1.84 and 1.74, respectively, whereas rats on sorghum and maize diets had PERs of 1.36 and 1.46, respectively (79). Other authors (54) also reported that pearl millet was a more efficient and digestible food and produced higher weight gains in rats than sorghum grain.

In a rat bioassay, the PER, net protein ratio (NPR), net protein utilization (NPU), and biological value (BV) obtained for a traditional Nigerian weaning food produced from millet were 1.22, 2.51, 49.2, and 53.7, respectively. Fortification of the traditional food with 22.5% heat–treated soybeans considerably improved the nutritive value. The PER, NPR, NPU, and BV of the fortified porridge increased to 2.2, 2.76, 67.6, and 79, respectively (45).

Goswami et al. (35) added lysine to a pearl millet–based diet, which significantly improved both protein score and PER. Rats consuming diets containing a blend of 75.5% pearl millet and 5% chickpea (*Cicer arietinum*) had a PER of 2.4. Addition of lysine to the control diet raised the PER to 3.38. A similar study was conducted by Jansen et al. (48) in which a teff-based diet was supplemented with 0.25 or 0.50% lysine. Addition of 0.25 and 0.50% lysine improved PERs from 1.95 to 2.78 and 3.27, respectively. Rats fed the teff diet supplemented with 0.5% lysine had comparable PERs to rats fed the control-casein diet (PER = 3.47). Klopfenstein et al. (54) reported that a nutritious pearl millet-based weanling food is used in Senegal. The dehulled grain is made into couscous and supplemented with skim milk powder, vegetable oil and sugar. The supplemented weanling food was found to be equivalent to casein in feed efficiency, digestibility, weight gains, and PER estimated with growing rats.

C. Human Studies

Pearl millet is a nutritious and well-digested source of calories and proteins for humans. Kurien et al. (55) studied the effect of partial or complete replacement of rice by pearl millet in the diet of malnourished children. Pearl millet contained more protein, calcium, fiber, and phytate than milled rice. Rice had a higher protein digestibility than pearl millet but children fed pearl millet performed as well as children fed the control rice diet.

A nitrogen balance study found that children fed refined finger millet flour digested the protein better and retained more nitrogen than children fed a whole finger millet flour. Addition of lysine and threonine to a finger millet-based diet offered to young girls significantly improved nitrogen retention and biological and net protein utilization values (27). Another study with eight young females in which the typical rice-based diet was replaced with different levels of finger millet indicated that both protein digestibilities and nitrogen balance slightly decreased as the level of finger millet increased in the diet. Nevertheless, even when finger millet was the only source of nitrogen, the young girls maintained a positive nitrogen balance.

D. Effect of Decortication on Nutritional Value

Decortication reduces the amount of important nutrients but also increases nutrient digestibilities. For instance, Awadalla and Slump (5) reported losses of fat (4.6–2.6%), fiber (5.1–0.65%), ash (2.5–0.9%), calcium (0.04–0.01%), iron (28–13 mg), lysine (182–143 mg/g N), and methionine (129–110 mg/g N) when proso millet was decorticated. When whole and decorticated proso millet flours replaced 20% of refined wheat flour, the protein quality of breads containing the proso flour was slightly lower than that of the whole wheat flour control bread. The protein digestibility of breads containing decorticated proso millet flour was 3% higher than that of the control bread or the bread containing proso millet whole flour. However, the chemical score, BV and NPU values of the bread with 20% decorticated proso millet flour was slightly lower due to the loss of lysine during decortication.

Decortication of finger millet for production of Vadragi (an Indian weaning food produced from dried-roasted millet flour) gave a product with lower fiber, calcium, and protein than the original grain source. Rats fed Vadragi had reduced weight gains when compared with counterparts fed whole or sprouted grain. However, rats fed Vadragi absorbed and retained more calcium than rats fed original grains, even though the Vadragi had lower calcium content. The higher calcium retention value was attributed to the lower fiber and phytic acid content (30).

E. Effect of Germination and Fermentation on Nutritional Value

The processes of germination and fermentation are commonly used for production of traditional foods. Germination improves the vitamin content of the grain and lowers levels of lipids, phytate, and oxalate (74). The concentration of free sugars, amino nitrogen, B vitamins, and ascorbic acid are increased as well (3,36) due to the partial loss of soluble carbohydrates. The partial protein digestion via intrinsic grain enzymes or extrinsic (microorganisms) proteolytic enzymes improves protein quality and digestibility. The availability of lysine, relative nutritive value, and T'PER (PER estimated with a growing protozoa) of fermented millet was higher than the control grain. The average weight gains of rats fed an unsupplemented diet of sprouted finger millet flour was considerably higher than weight gains of rats fed whole flour. The same study carried out with vitamin and mineral supplements showed that the sprouted and whole grain supported equal rat growth (39). Fermented pearl or finger millet had slightly higher protein digestibilities, BV, NPU, and PER values than raw grains. The higher protein digestibility (+3.0%) was reflected by a better utilization of the protein and increased rat weight gains.

F. Relationship Between Millet and Incidence of Pellagra

Millet consumption has been associated with a lower incidence of pellagra, a niacin-deficiency disease. An early report (69) indicated that the incidence of pellagra was lower in groups consuming millets than in similar groups consuming maize. Possible explanations for the observed differences are that (1) millet has a higher tryptophan content, which can be transformed into niacin in the human system, and (2) niacin in raw corn is unavailable because it is bound to a polypeptide called niacinogen (51).

VIII. Current Status and Future

The millets are a diverse group of cereals that are generally grown in harsh environments or as early maturing crops. They are critically important food cereals for many people in Asia and Africa. Millets are used in numerous thick and thin porridges, fermented and unfermented breads, alcoholic and nonalcoholic beverages, steamed products, and snacks. Millet food products generally have strong flavor and aroma. Millets vary in composition and nutritional value among species.

Pearl millet products have a nice flavor with good nutritional quality. The grain is a major source of food in India and Sahelian Africa. Pearl millet can be used efficiently in livestock feeds but very little is processed industrially. Recently, production in India has produced surplus grain, which has stimulated research on industrial and feed utilization.

The yield potential of millets is lower than other cereals, so they will continue to be used mainly in harsh environments or as a short season catch crop. Pearl millet has excellent feeding value for swine and poultry because of its higher oil, protein, and lysine contents. Pearl millet improvement programs at the Kansas Fort Hays Experimental Station have already made significant progress in developing hybrids with adaptation to western Kansas. Pearl millet hybrids may be grown in selected, dry, sandy soil areas of Kansas as an alternative crop, which could be a useful ingredient in millet grain breads and other products.

More work on millets for livestock feed and for industrial utilization will secure markets for millets, especially pearl millet. Pearl millets with improved dry-milling properties are needed to provide shelf-stable food products. Pearl millets especially have potential for new food products with unique flavor.

ACKNOWLEDGMENTS

The authors wish to express their appreciation to Dr. Cheryl Earp for critically reviewing the manuscript. Authors are grateful for the long-term support of scientists in the Texas Sorghum Improvement Program at Texas A&M University. This work was part of the INTSORMIL Title XII Sorghum and Millet Collaborative Research Support Program supported in part by Grant AID/DSAN/1254 G–TS–50–65 from the Agency for International Development, Washington, DC.

TABLE 1 Common and Scientific Names of the Major Types of Millets

Scientific names	Common names	Locations
Pennisetum americanum *P. typhoides*	pearl millet, bajra, cattail millet	Africa, India
Panicum milaceum	proso millet, broomcorn millet, hog millet, samai, panivarigu, common millet	China, U.S.S.R. United States
Eleusine coracana	finger millet, ragi, bird's foot millet, African millet	Africa, India China
Seteria italica	foxtail millet, navane, Italian millet, German millet	China, Near East
Digitaria exilis	fonio	West and North Africa
Eragrostis tef	teff	East Africa, Ethopia
Paspalum commersoni *P. scrobiculatum*	kodo millet, varagu	Southern Asia
Echinochloa crusgalli *E. utilis* *E. frumentacea*	Japanese barnyard, Japanese millet, barnyard millet	Orient

TABLE 2 Statistics of Millet Production, 1987

	Area harvested (1,000 ha)	Yield (kg/ha)	Total production (1,000 MT)
Africa	15,504	652	10,109
Nigeria	3,700	811	3,000
Mali	1,540	736	1,134
Niger	3,000	340	1,020
Asia	20,694	654	13,528
India	16,702	413	6,900
China	2,692	1,969	5,301
U.S.S.R.	3,000	1,300	3,900
World	39,360	704	27,711
Developed countries	75	906	68
Developing countries	33,139	539	17,870

Source: Ref. 32.

TABLE 3 Structural Comparisons of Some Millets[a]

	Finger	Pearl	Proso	Fonio	Teff
Seed type	utricle	caryopsis	utricle	caryopsis	caryopsis
Pericarp	unattached	attached	unattached	attached	attached
Seed coat	5 layers	1 layer	1 layer	1 layer	1 layer
pigmented	yes	sometimes	no	no	yes
thickness	10.8x24.2	0.4	0.2–0.4	0.5–1.5	1.5–2.0
Aleurone	1 layer	1 layer	1 layer	1 layer	1 layer
cell size (L x W)	18.0x7.6	16.0–30.0 x5.0–15.0	12.0x6.0	12.0x4.0	12.0x4.0
Starch granules					
peripheral	8.0–16.5	6.4	3.9	7.2	5.6–7.2
corneous	3.0–19.0	6.4	4.1	7.8	15.1–22.4
floury	11.0–21.0	7.6	4.1	6.5	10.2–15.8
Type	simple/ compound	simple	simple	simple	simple/ compound
Protein bodies					
size	2.0	0.6–0.7	0.5–1.7	1.2–2.4	1.2–2.4
location	peripheral/ corneous	all areas	peripheral	peripheral	all areas
Germ[b]					
size (L × W)	980 × 270	1420 × 620	1100 × 310	517 × 977	160 × 725
endosperm germ ratio	11:1	2.5:1	12:1	6:1	3:1

[a]All values expressed as μm.

[b]Measurements taken of surface area of longitudinally cut seeds: L refers to length of longitudinal axis of germ and W to the width.

Source: All data determined at Cereal Quality Lab, Texas A&M University. Yanez and Walker (106) reported proso millet starch granules to average 7 μm in diameter.

TABLE 4 Proximate Analysis of Millets

Millet Type	Protein[a]	Ether Extract	Crude Fiber	Ash	NFE[b]
Pearl millet	14.5	5.3	2.2	2.1	72.5
	6.9–20.9	3.3–6.9	0.9–3.6	0.3–5.1	59.8–80.6
Proso millet	11.6	4.2	12.0	3.6	70.4
	11.3–12.7	3.8–4.9	7.9–19.2	3.1–4.2	66.3–73.4
Japanese millet	11.6	5.8	14.7	4.7	63.2
	11.2–11.8	5.5–6.3	11.6–16.3	3.0–5.6	60.9–67.8
Foxtail millet	13.0	4.8	10.0	4.0	68.2
	12.6–14.0	4.6–5.0	9.3–10.6	3.7–4.3	66.1–69.7
Finger millet	8.0	2.0	3.6	3.0	72.4
	4.9–11.3	0.9–7.7	0.7–8.0	2.0–5.0	69.3–73.4
Teff[c]	7.7				
	6.5–9.3				
Kodo[d]	8.4			3.0	

Mean values and ranges. All values are expressed on dry matter basis.

[a]N × 6.25.

[b]Nitrogen free extract

[c]Lester and Bekele (57).

[d]Paramahans et al. (77).

Source: Adapted from Ref. 89.

TABLE 5 Essential Amino Acid Composition of Some Millets

	FAO/WHO pattern	Millets				
		Pearl	Proso	Finger	Foxtail	Teff
Lysine	5.44	2.6	1.5	2.0	2.6	3.7
Leucine	7.04	17.4	12.2	6.4	16.5	8.5
Phenylalanine[a]	6.08	4.9	5.5	3.2	5.1	5.7
Valine	4.96	5.7	5.4	5.8	5.4	5.5
Tryptophan	0.96	2.3	0.8	1.5	0.6	—
Methionine[b]	3.52	2.5	2.2	2.3	2.0	4.1
Threonine	4.00	4.9	3.0	2.7	3.8	4.3
Isoleucine	4.00	4.3	4.1	3.7	4.2	4.0
Histidine[c]	—	2.1	2.1	0.3	3.1	3.2
Chemical Score[d]	100	47.8	27.6	36.7	47.8	68.0

[a]Phenylaline requirements can be partially spared by tyrosine.
[b]Methionine requirement can be partially spared by cysteine.
[c]Histidine is considered essential amino acid only for infants.
[d]For all millets the limiting essential amino acid is lysine.
Source: Adapted from Ref. 57, 89 and 102.

TABLE 6 Protein Fractions in Sorghum and Some Millets

Fraction	Sorghum[a]	Millet		
		Pearl[b]	Foxtail[c]	Finger[d]
I Albumin/Globulins	17.4	25.0	17.4	I 9.6–12.2
II Prolamins	6.4	28.4	57.2	II 22.1–29.7
III Cross-linked Prolamins	18.8	2.7	9.1	
IV Glutenlike	4.0	5.5	9.4	
V Glutelin	35.7	18.4	6.8	III 14.5–16.2
VI Residue	10.6	3.9	—	
Total	92.9	83.9	100.0	48.8–58.1

[a]Based on three genotypes.
[b]Based on four genotypes.
[c]Mean value of 14 varieties expressed as percent of the total protein extracted (102).
[d]Mean value of five genotypes (82).
Source: Adapted from Ref. 47.

TABLE 7 Starch Properties of Some Millets

Type of Millet	Amylose (%)	Gelatinization Initial (°C)	Gelatinization Final (°C)	Water-binding Capacity (%)	Swelling at 90°C	Solubility at 90°C	Amylograph[a] (Brabender Units) A	B	C	D
Pearl[b]	21.1	61.1	68.7	87.5	13.1	9.14	460	396	568	536
Proso[c]	28.2	56.1	61.2	108.0	12.0	6.89	688	520	826	1203
Foxtail[d]	—	53.5	59.5	128.5	11.2	4.65	840	620	1100	1220
Foxtail[e]	17.5	55.0	62.0	—	9.8	4.80	1780	1540	2000	—
Finger[e]	16.0	64.3	68.3	—	11.4	6.50	1633	1286	1796	—
Kodo[f]	24.0	57.0	68.0	—	12.0	5.50	300	270	390	—

[a]A = Viscosity at 93–95°C; B = viscosity after holding sample at 95°C; C = viscosity cooling to 35 or 50°C; D = viscosity after holding sample at 35 or 50°C.

[b]Data obtained from Beleia et al. (15). The amylograph slurry contained 9% starch (14% mb) and was suspended in disodium phosphate/citric acid buffer. First holding period (B) = 1 hr; second holding period (D) = 1 hr. Temperature of the slurry was cooled to 50°C.

[c]Amylose content obtained from Yanez and Walker (106). Other information was obtained from Lorenz and Hinze (61). The amylograph slurry contained 9 parts of starch solids per 100 parts (420 ml) water. First holding period (B) = 30 min; second holding period (D) = 30 min. Temperature of the slurry was cooled to 35°C.

[d]Data obtained from Lorenz and Hinze (61). The amylograph slurry contained 9 parts of starch solids per 100 parts (420 ml) water. First holding period (B) = 30 min; second holding period (D) = 30 min. Temperature of the slurry was cooled to 35°C.

[e]Data obtained from Wankhede et al. (105). The amylograph slurry contained 9.5% parts of starch solids per 100 parts water. First holding period (B) = 30 min. Temperature of the slurry was cooled to 50°C.

[f]Data obtained from Paramahans et al. (77). Values were extrapolated from amylograph curve. Peak viscosity was achieved at 83.5°C.

TABLE 8 Saccharide Composition of Some Millets

	Total sugar	Stachyose	Raffinose	Sucrose	Glucose + fructose
Pearl[a]	2.56	0.09	0.71	1.64	0.11
Proso[b]	—	—	0.08	0.66	—
Foxtail[c]	—	—	0.10	1.01	—
Foxtail[d]	0.46	—	0.04	0.15	0.10
Finger[d]	0.65	—	0.07	0.21	0.18

All values are expressed on a percent, dry matter basis.
[a]Data obtained from Subramanian et al. (97). Average of nine cultivars.
[b]Data obtained from Becker and Lorenz (10). Average of six cultivars.
[c]Data obtained from Becker and Lorenz (10). Average of two cultivars.
[d]Data obtained from Wankhede et al. (104). Foxtail data is the average of three cultivars. Finger millet data is the average of one cultivar. The authors reported a maltose content of 0.04% and 0.07% for foxtail and finger millets, respectively.

TABLE 9 Fatty Acid Composition of Some Millets

Fatty Acid		Pearl (%)		Finger[c] (%)	Kodo[d] (%)	
Palmitic	C 16:0	20.5[a]	19.0[b]	24.7	18.4	19.9
Palmitoleic	C 16:1	0.6	0.6	—	—	—
Stearic	C 18:0	3.9	5.0	Trace	1.0	—
Oleic	C 18:1	25.6	25.0	49.8	37.7	38.7
Linoleic	C 18:2	45.1	46.0	24.2	43.0	41.5
Linolenic	C 18:3	3.7	3.2	1.3	—	—
Arachidic	C 20:0	0.6	0.5	—	—	—

[a]Data in this column was from Jellum and Powell (49). Average of 65 lines grown at Tifton Georgia in 1968.
[b]Data in this column from Rooney (88).
[c]Data from Mahadevappa and Raina (66).
[d]Data from Paramahans and Tharanathan (78). First and second columns correspond to unpolished (endosperm + bran) and polished kernels, respectively.

TABLE 10 Mineral Composition of Some Millets

	K (%)	Mg (%)	Ca (%)	Na (ppm)	Fe (ppm)	Mn (ppm)	Zn (ppm)	Cu (ppm)
Pearl[a]	0.44	0.13	0.01	40.9	74.9	—	29.5	—
Proso[b]	0.21	0.12	0.01	46.9	33.1	18.1	17.2	8.3
Foxtail[c]	0.27	0.13	0.01	54.6	32.6	21.9	21.4	9.2
Finger[d]	0.43	0.11	0.28	50.0	46.0	7.5	—	0.3

All values are expressed on a dry matter basis.
[a]Data from Varriano-Marston and Hoseney (99). Average of 15 lines and 4 bulk populations.
[b]Data from Lorenz et al. (62). Average of 6 cultivars.
[c]Data from Lorenz et al. (62). Average of 2 cultivars.
[d]Data from Pore and Magar (80). Average of 2 cultivars.

TABLE 11 Phenolic Acid Composition of Some Millets

	Finger	Pearl	Proso	Teff	Fonio	Foxtail
Phenolic assay[b]						
Folin/Ciocalteu	0.55–0.59	0.19–0.33	0.05–0.10	0.09–0.15	0.14	0.12
Vanillin-HCl	0.17–0.32	0.05	0.0	0.0	0.0	0.0
Phenolic acids[c]						
Protocatechuic	23.1	11.8	—	25.5	—	—
Gentisic	61.5	96.3	—	15.0	—	21.5
p-OH Benzoic	8.9	22.0	—	—	—	14.6
Vanillic	15.2	16.3	—	54.8	—	87.1
Caffeic	16.6	21.3	—	3.9	—	10.6
Syringic	7.7	17.3	—	14.9	—	93.6
Coumaric	56.9	268.2	—	36.9	—	2133.7
Ferulic	387.0	679.7	—	285.9	—	765.8
Cinnamic	35.1	345.3	—	46.0	—	781.7

[a]Lemma and palea attached.
[b]Values are expressed on mg/100 mg catechin equivalents, dry weight basis. All millets have phenols but only finger millet has condensed tannins. Pearl millet gives a reading with the vanillin assay, but does not contain condensed tannins.
[c]μg phenolic acid/mg samples, as is moisture basis.
Source: Ref. 70.

REFERENCES

1. Abate, A. N., and Gomez, M. Substitution of finger millet (*Eleusine coracana*) and bulrush millet (*Pennisetum typhoides*) for maize in broiler feeds. *Anim. Feed Sci. Technol.*, *10*:291 (1984).
2. Abdelrahman, A., Hoseney, R. C., and Varriano–Marston, E., The proportions and chemical compositions of hand–dissected anatomical parts of pearl millet. *J. Cereal Sci.*, *2*:127 (1984).
3. Aliya, S., and Geervani, P., An assessment of the protein quality and vitamin B content of commonly used fermented products of legumes and millets. *J. Sci. Food. Agric.*, *32*:837 (1981).
4. Angold, R. E., Cereal and bakery products. In *Food Microscopy*, R. G. Vaughan (Ed.). Academic Press, London, 1979, p. 75.
5. Awadalla, M. Z., and Slump, P., Native Egyptian millet as supplement of wheat flour bread. I. Nutritional studies. *Nutr. Reports Int.*, *9*:59 (1974).
6. Awadalla, M. Z., and Slump, P., Native Egyptian millet as supplement of wheat flour bread. II. Technological studies. *Nutr. Reports Int.*, *9*:69 (1974).
7. Badi, S. M., Hoseney, R. C., and Finney, P. L., Pearl millet. II. Partial characterization of starch and use of millet flour in bread making. *Cereal Chem.*, *53*:733 (1976).
8. Badi, S. M., and Hoseney, R. C., Use of sorghum and pearl millet flours in cookies. *Cereal Chem.*, *53*:733 (1976).
9. Bailey, A. V., Sumrell, G., and Burton, G. W., Pentosans in pearl millet. *Cereal Chem.*, *56*:295 (1979).
10. Becker, R., and Lorenz, K., Saccharides in proso and foxtail millets. *J. Food Sci.*, *43*:1412 (1978).
11. Brandtzaeg, B., Malleshi, N. G., Svanberg, U., Desikachar, H. S. B., and Mellander, O., Dietary bulk as a limiting factor for nutrient intake with special reference to the feeding of preschool children III. Studies of malted flour from ragi, sorghum and green gram. *J. of Tropical Pediatrics*, *27*:184 (1981).
12. Brethour, J. R., Rolled pearl millet for beef cattle. Report of Progress 417, Roundup 69. Hays Branch, Agricultural Experimental Station, Kansas State University, Manhattan, KS, 1983.
13. Brethour, J. R., Pearl millet for beef cattle. Report of Progress 432, Roundup 70. Hays Branch, Agricultural Experimental Station, Kansas State University, Manhattan, KS, 1983.
14. Brethour, J. R., Pearl millet for beef cattle. Report of Progress 452, Roundup 71. Hays Branch, Agricultural Experimental Station, Kansas State University, Manhattan, KS, 1984.
15. Beleia, A., Varriano–Marston, E., and Hoseney, R. C., Characterization of starch from pearl millets. *Cereal Chem.*, *57*:300 (1980).
16. Brunken, J., de Wet, J. M. J., and Harlan, J. R., The morphology and domestication of pearl millet. *Econ. Bot.*, *31*:163 (1977).
17. Burton, G. W., Wallace, A. T., and Rachie, K. O., Chemical composition and nutritive value of pearl millet (*Pennisetum typhoides*). *Crop Sci.*, *12*:187 (1972).
18. Calder, A., Value of munga (millet) for pig feeding. *Rhodesia Agr. J.*, *52*:161 (1955).
19. Calder, A., The value of rupoko (millet) for pig feeding. *Rhodesia Agr. J.*, *57*:116 (1960).
20. Calder, A., The production of pork pigs comparing maize, munga (millet) and pollards, *Rhodesia Agr.*, *58*:362 (1961).
21. Carr, W. R., Observations on the nutritive value of traditionally ground cereal in southern Rhodesia. *Br. J. Nutr.*, *33* ?? (1961).
22. Casey, P., and Lorenz, K., Millet functional and nutritional properties. *Bakers Digest*, *51*:45 (1977).
23. Chandrasekhara, M. R., and Swaminathan, M., The enzymes of pearl millet (Pennisetum typhoideum) malt. I. Amylases. *J. Sci. Ind. Res.*, *160*:35 (1957).
24. Chandrasekher, G., and Pattabiraman, T., Natural plant enzyme inhibitors: isolation, and characterization of two trypsin inhibitors from bajra (*Pennisetum typhoideum*). *Ind. J. Biochem. and Biophysics*, *19*:1 (1981).

25. Chaudry, P., and Kapoor, A. C., Changes in the nutritional value of pearl millet flour during storage. *J. Sci. Food Agric.*, *35*:1219 (1984).
26. Dako, D. Y., Cereal utilization in west Africa. In *Amino Acid Composition and Biological Value of Cereal Proteins*, R. Lasztity and M. Hidvegi (Eds.). D. Reidel Publishing Co., Budapest, Hungary, 1983, p. 27.
27. Daniel, V. A., Kurien, S., Narayanaswamy, D., and Swaminathan, M., Supplementary relations between the proteins of bengal gram, rice and ragi. *Indian J. Nutr. Dietet.*, *11*:137 (1974).
28. Danielson, D. M., and Grabouski, P. H., Diets compared. Wheat, sorghum, corn, millet. In Nebraska Swine Report No 219. North Platte Station. University of Nebraska, 1970, p. 8.
29. Dhamijia, S. S., and Singh, D. P., Adjuncts in brewing. I. Bajra and sorghum. *J. Food Sci. Tech.*, India, *15*:197 (1978).
30. Ejeta, G., Hassan, M. M., and Mertz, E. T., In vitro digestibilities and amino acid composition of pearl millet (Pennisetum typhoides) and other cereals. *Applied Biology*, *84*:6016 (1987).
31. FAO, *Production Year Book*, Book No 15, Rome Italy, 1981.
32. FAO, *Production Year Book*, FAO Statistics Series, Vol. 41. Rome, Italy, 1988.
33. Freeman, J. E., and Bocan, B. J., Pearl millet: a potential crop for wet milling. *Cereal Sci. Today*, *18*:69 (1973).
34. Geervani, P., The influence of home processing on the quality of cereal and millet proteins. In *Amino Acid Composition and Biological Value of Cereal Proteins*, R. Lasztity and M. Hidvegi (Eds.). D. Reidel Publishing Co., Budapest, Hungary, 1983, p. 495.
35. Goswami, A. K., Sherma, K. P., and Sehgal, K. L., Nutritive value of proteins of pearl millet of high yielding varieties and hybrids. *Brit. J. Nutr.*, *23*:913 (1969).
36. Hamad, A. M., and Fields, M. L., Evaluation of the protein quality and available lysine of germinated and fermented cereals. *J. Food Sci.*, *44*:456 (1979).
37. Hanna, W. W., Utilization of wild relatives of pearl millet. In *Proceedings International Pearl Millet Workshop*, J. R. Witcombe and S. R. Beckerman (Eds.) ICRISAT, Patancheru, A. P. India, 1987, p. 33.
38. Harinarayana, G. Pearl millet in Indian agriculture. In *Proceedings International Pearl Millet Workshop*, J. R. Witcombe and S. R. Beckerman (Eds.). ICRISAT, Patancheru, A. P. India, 1987, p. 5.
39. Hemanalini, G., Padma Umapathy, K., Rao, J. R., and Sarawathi, G., Nutritional evaluation of sprouted ragi. *Nut. Rep Int.*, 22:271 (1980).
40. Hilu, K. W., and de Wet, J. M. J., Domestication of *Eleusine coracana*. *Econ. Bot.*, *30*:199 (1976).
41. Hilu, H. W., de Wet, J. M. J., and Harlan, J. R., Archaeobotanical studies of *Eleusine coracana* ssp. *coracana* (finger millet). *Amer. J. Bot.*, *66*(3):330 (1979).
42. Hoseney, R. C., Varriano–Marston, E., and Dendy, D. A. V., Sorghum and millets. In *Advances in Cereal Sciences and Technology*, Vol. IV, Y. Pomeranz (Ed.). American Association of Cereal Chemists, ST. Paul, MN, 1981, Chap. 3.
43. Hoseney, R. C., Andrews, D. J., and Clark, H., Sorghum and pearl millet. In *Nutritional Quality of Cereal Grains: Genetics and Agronomic Improvement*, R. A. Olson and K. J. Frey (Eds.). American Society of Agronomy, Inc., Crop and Soil Science Societies of America, Inc., Madison, WI, 1987, Chap. 9.
44. Hulse, J. H. Laing, E. M., and Pearson, O. E, *Sorghum and the Millets: Their Composition and Nutritive Value*. Academic Press, New York, 1980.
45. Ifon, E. T., Biological evaluation of the nutritive value of the millet porridge – A traditional Nigerian weanling food before and after fortification with soya proteins. *Nut. Rep. Int.*, 22:109 (1980).
46. Jain, A. K., and Date, W. B., Relative amylase activity of some malted cereal grains. *J. Food Sci. Tech*, India *12*:131 (1975).
47. Jambunathan, R. Singh, U. and Subramanian, V. Grain quality of sorghum, pearl millet, pigeon pea and chick pea. In *Interfaces Between Agriculture, Nutrition and Food Science*.

Proceedings of a Workshop, Nov. 10–12, 1981, Hyberabad, India. K. T. Achaya (Ed.). Food and Nutrition Bulletin, Supplement No. 9. Tokyo, Japan, 1984, pp. 47–60.

48. Jansen, G. R., Dimuo, L. R., and Hause, N. L., Amino acid composition and lysine supplementation of teff. *J. Agric. Food Chem.*, *10*:62 (1962).
49. Jellum, M. D., and Powell, J. B., Fatty acid composition from pearl millet seed. *Agron. J.*, *63*:29 (1971).
50. Kaced, I., Hoseney, R. C., and Varriano–Marston, E., Factors affecting rancidity in ground pearl millet (*Pennisetum americanum L. Leeke*). *Cereal Chem.*, *61*:187 (1984).
51. Koetz, R., and Neukom, H., Nature of bound nicotinic acid in cereals and its release by thermal and chemical treatment. In *Physical, Chemical and Biological Changes in Food Caused by Thermal Processing*, T. Hoyden and O. Kaule, (Eds.). Applied Sci. Publishers, London, 1977, p. 305.
52. Klopfenstein, C. F., Hoseney, R. C., and Leipold, H. W., Goitrogenic effects of pearl millet diets. *Nutr. Rep. Int.*, *27*:1039 (1983).
53. Klopfenstein, C. F., Hoseney, R. C., and Leipold, H. W., Further studies on the goitrogenic effects of pearl millet diets. *Nutr. Rep. Int.*, *28*:1137 (1983).
54. Klopfenstein, C. F., Hoseney, R. C., and Leipold, H. W., Nutritional quality of pearl millet and sorghum grain diets and a pearl millet weanling food. *Nut. Rep. Int.*, *31*:287 (1985).
55. Kurien, S., Naratanaswamy, D., Daniel, V. A., Swaminathan, M., Rajalakshmi, D., and Parpia, H. A. B., Improvement of protein value of poor pearl millet (*Pennisetum typhoideum*) diet by supplementation with limiting amino acids. *Nutr. Reports Int.*, *3*:357 (1971).
56. Lai, C. C., and Varriano–Marston, E., Lipid content and fatty acid composition of free and bound lipids in pearl millets. *Cereal Chem.*, *57*:271 (1980).
57. Lester, R. N., and Bekele, E., Amino acid composition of the cereal teff and related species of Eragrostis. *Cereal Chem.*, *58*:113 (1981).
58. Lorenz, K., Tannins and phytate content in proso millet (*Panicum milaceum*). *Cereal Chem.*, *60*:424 (1983).
59. Lorenz, K., and Dilsaver, W., Proso millets. Milling characteristic, proximate compositions, nutritive value of flours. *Cereal Chem.*, *57*:16 (1980).
60. Lorenz, K., and Dilsaver, W., Rheological properties and food applications of proso millet flours. *Cereal Chem.*, *57*:21 (1980).
61. Lorenz, K., and Hinze, G., Functional characteristics of starches from proso and foxtail millets. *J. Agr. Food Chem.*, *24*:911 (1976).
62. Lorenz, K., McFarland, G. and Hinze, G. The mineral composition of proso and foxtail millets. *Lebensm. Wiss. u. Technol.*, *9*:357 (1976).
63. Luis, E. S., Sullivan, T. W., and Nelson, L. A., Nutrient composition and feeding value of proso millet, sorghum grain and corn in broiler diets. *Poultry Sci.*, *61*:311 (1982).
64. Luis, E. S., Sullivan, T. W., and Nelson, L. A., Nutritional value of proso millet, sorghum grain and corn in turkey starter diets. *Poultry Sci.*, *61*:321 (1982).
65. Luis, E. S., Sullivan, T. W., and Nelson, L. A., Nutritional value of proso millet in layer diets. *Poultry Sci.*, *61*:1176 (1982).
66. Kahadevappa, V. G., and Raina, P. L., Lipid profile and fatty acid composition of finger millet (*Eleusine coracana*). *J. Food Sci. & Tech.*, *15*:100 (1978).
67. Malleshi, N. G., and Desikacher, H. S. R., Varietal differences in puffing quality of ragi (*Eleusine coracana*). *J. Food Sci. and Tech.*, *18*:30 (1981).
68. Malleshi, N. G., and Desikacher, H. S. R., Formulation of a weaning food with low hot paste viscosity based on malted ragi (*Eleusine coracana*) and green gram (*Phaseolus radiatus*). *J. Food Sci. and Tech.*, *19*:193 (1982).
69. Mangay, A. A., Pearson, W., and Darby, W. J., Millet (*Seteria italica*): its amino acid and niacin content and supplementary nutritive value for corn (maize). *J. Nutr.*, *62*:377 (1957).
70. McDonough, C. M., and Rooney, L. W., Structure and phenol content of six species of millets using fluorescence microscopy and HPLC (Abstract). *Cereal Foods World.*, *30*:550 (1985).

71. McDonough, C. M., and Rooney, L. W., Structural characteristics of *Pennisetum americanum* using scanning electron and fluorescence microscopies. *Food Microstr.*, *8*:137 (1989).
72. McDonough, C. M., Rooney, L. W., and Earp, C. F., Structural characteristics of *Eleusine coracana* (finger millet) using scanning electron and fluorescence microscopy. *Food Microstr.*, *5*:247 (1986).
73. Olatunji, O., Edwards, E., Bamiro, A., and Koleoso, O. E., Dry milled millet and uses as "tuwo," "barabusco," biscuits and bread (Abstract). *Cereal Foods World.*, *24*:456 (1970).
74. Opoku, A. R., Ohenehen, S. O., and Ejiofor, N., Nutrient composition of millet (*Pennisetum typhoides*) grains and malts. *J. Agric. Food Chem.*, *29*:1247 (1981).
75. Osman, A. K., Bulrush millet (*Pennisetum typhoides*) a contributory factor to the endemicity of goitre in western Sudan. *Ecol. Food Nutr.*, *11*:121 (1981).
76. Osman, A. K., Basu, T. K., and Dickerson, J. W. T., A goitrogenic agent from millet (*Pennisetum typhoides*) in Durfur province western Sudan. *Ann. Nutr. Metab.*, *27*:14 (1983).
77. Paramahans, S. V., Wankhede, D. B., and Tharanathan, R. N., Studies on varagu starch. *Starch/Starke.*, *32*:109 (1980).
78. Paramahans, S. V., and Tharanathan M., Carbohydrate composition of the millet varagu. *Starch/Starke.*, *32*:73 (1980).
79. Pushpamma, S., Parrish, D. B., and Deyoe, C. W., Improving protein quality of millet, sorghum and maize diets by supplementation. *Nutr. Reports Int.*, *5*:93 (1972).
80. Pore, M. S., and Magar, N. G., Nutrient composition of hybrid varieties of finger millet. *Indian J. Agric. Sci.*, *49*:526 (1979).
81. Rachie, K, O., and Majmudar, J. V., *Pearl millet.* Pennsylvania University Press, University Park, PA, 1980.
82. Ramachandra, G., Virupuska, T. K., and Shadaksharaswamy, M., Comparison of the protein fractions of finger millet. *Phytochem.*, *17*:1487 (1978).
83. Rao, D. S., and Deosthale, Y. G., Mineral composition, ionisable iron and soluble zinc in malted grains of pearl millet and ragi. *Food Chem.*, *11*:217 (1983).
84. Reddy, V. P., Faubion, J. M., and Rooney, R. C., Odor generation in ground, stored pearl millet. *Cereal Chem.*, *63*:403 (1986).
85. Reichert, R. D., The pH sensitive pigments in pearl millet. *Cereal Chem.*, *56*:291 (1979).
86. Reichert, R. D., and Youngs, C. G., Dehulling cereal grains and grain legumes for developing countries. II. Chemical composition of mechanically dehulled and traditionally dehulled sorghum and millet. *Cereal Chem.*, *54*:174 (1977).
87. Reichert, R. D., and Youngs, C. G., Bleaching effect of acid in pearl millet. *Cereal Chem.*, *56*:287 (1979).
88. Rooney, L. W., Sorghum and pearl millet lipids. *Cereal Chem.*, *55*:584 (1978).
89. Rooney, L. W., Earp, C. F., and Khan, M. N., Sorghums and millets. In *CRC Handbook of Processing and Utilization in Agriculture*, Vol. II, I. A. Wolf (Ed.). CRC Press, Boca Raton, FL, 1982, p. 123.
90. Rooney, L. W., and McDonough, C. M., Food quality and consumer acceptance of pearl millet. In *Proceedings, International Pearl Millet Workshop.*, J. R. Witcombe and S. R. Beckerman (Eds.). ICRISAT, Patancheru, A.P. India, 1987, p. 43.
91. Sanford, P. E., Camacho, F., Nake, R. P., Deyoe, C. W., and Cassidy, A. J., Performance of meat strain chicks fed pearl millet as a source of energy and protein. *Poultry Sci.*, *52*:2081 (1973).
92. Sharma, B. D., Sadagopan, V. R., and Reddy, V. R., Utilization of different cereals in broiler diets. *Brit. Poultry Sci.*, *20*:371 (1979).
93. Simhaee, E., Ghorban, K., and Makarechian, M., Comparison of nutritive value of corn, wheat, and millet in broiler diets, *Iran J. Agric. Res.*, *1*:51 (1971).
94. Skovron, J., and Lorenz, K., Enzymatic activities in proso millets. *Cereal Chem.*, *56*:559 (1979).
95. Stoskopf, N. C., *Cereal Grain Crops.* Reston Publishing Co., Inc., Reston, VA, 1980.

96. Subramanian, V., Jamunathan, R., and Ramaiah, C. D., Physical and chemical characteristics of pearl millet grains and their relationship to roti quality. *J. Food Sci.*, *51*:1005 (1986).
97. Subramanian, V., and Suryaprakash, S., Sugars of pearl millet (*Pennisetum americanum* [L.] *Leeke*). *J. Food Sci.*, *46*:1614 (1981).
98. Sullins, D., and Rooney, L. W., Pericarp and endosperm structure of pearl millet (*Pennisetum typhoides*). In *Proceedings of a Symposium on Sorghum and Millets for Human Food*, Dendy, D. A. V., (Ed.). Tropical Products Institute, London, 1977, p. 79.
99. Varriano–Marston, E., and Hoseney, R. C., Note on mineral content and location in pearl millet. *Cereal Chem.*, *57*:150 (1980).
100. Varriano–Marston, E., and Hoseney, R. C., Barriers to increased utilization of pearl millet in developing countries. *Cereal Food World.*, *28*:392 (1983).
101. Vavilov, N. I., *The Origin, Variation, Immunity and Breeding of Cultivated Plants*, Vol. 13, Chronica Botanica, F. Verdoorn (Ed.). Chronica Botanica Co., Waltham, MA, 1951.
102. Vincent–Monterio, P., Virupaksha, T. K., Rajagopol Rao, D., Proteins of Italian millet: Amino acid composition, solubility fractionation and electrophoresis of protein fractions. *J. Sci Food Agric.*, *33*:1072 (1982).
103. Vishnu–Mittre, Protohistoric records of agriculture in India. *Tans. Bose Res. Inst.*, *31*:87 (1968).
104. Wankhede, D. B., Shehnaz, A., and Raghavendra Rao, M. R., Carbohydrate composition of finger millet (*Eleusine coracana*) and foxtail millet (*Seteria italica*). *Qual. Plant. Pl. Fds. Hum. Nutr.*, *28*:293 (1979).
105. Wankhede, D. B., Shehnaz, A., and Raghavendra Rao, M. R., Preparation and physiochemical properties of starches and their fractions from finger millet (*Eleusine coracana*) and foxtail millet (*Seteria italica*). *Starke*, *31*:153 (1979).
106. Yanez, G. A., and Walker, C. E., Effect of tempering parameters on extraction and ash of proso millet flours, and partial characterization of proso starch. *Cereal Chem.*, *63*:164 (1986).
107. Zeleznak, K., and Varriano–Marston, E., Pearl Millet (*Pennisetum americanum* [L.] *Leeke*) and grain sorghum (*Sorghum bicolor* [L.] *Moench*) ultrastructure. *Am. J. Bot.*, *69*:1306 (1982).

7

RICE: PRODUCTION, PROCESSING, AND UTILIZATION

Roy N. Sharp
University of Arkansas, Fayetteville, Arkansas

I. HISTORY

The geographical site of original rice (*Oryza sativa*) domestication is yet obscure. The general consensus is that rice domestication occurred independently in three areas—India, Indonesia, and China—thereby giving rise to three races of rice—*indica*, *javonica*, and *sinica* (also known as japonica), respectively. There are indications that rice was cultivated in India between 1500 and 2000 B.C., and in Indonesia around 1084 B.C. (1). In various areas of the world, rice can be found growing under extremely diverse conditions, from dryland to as deep as 6 meters of water (2). The most productive areas are those where rice is grown standing in water in which the depth is controlled.

The growth of the shipping industry no doubt played a part in the dissemination of rice grain as we know it today. Rice reportedly was first introduced into the United States in 1686, when a ship from Madagascar was forced to shore in Charleston, S.C. Upon departure the captain presented a bag of rice to the townspeople in appreciation for the hospitality shown him. (3). The slave trade was responsible for greater movement of rice into the United States via the Caribbean Sea area and the Louisiana gulf area. Although rice was grown in the United States as early as 1609 in Virginia and was an established crop in South Carolina in 1690, it did not become an established crop in Louisiana and Texas until approximately 1888, in Arkansas in 1904, and in California in 1912 (4). Since 1942, rice production has been of considerable importance in the Mississippi delta area.

II. RICE PRODUCTION

Rice production is widely scattered throughout the world, with the FAO (5) reporting significant rice production in countries on six continents in 1985. Total rice production for 1985 was placed at 465,970,000 metric tons (MT) (Table 1) after having increased steadily during the last 30 years. The total production increase was partly due to an average annual increase of 1% in production area, but more significantly to an average annual yield increase of 2.6% (Table 1). The production by rice-producing countries showed little variations during the period 1975–1985 (Table 2). The countries producing the most rice are China and India.

The distribution of rice production has no relation to the country of origin of the rice entering the international market. The share of the total export market for Thailand and the United States is approximately 35 and 20%, respectively (Tables 3 and 4), in contrast to their share of total world production, which is only about 4.2 and 1.5% (Table 2), respectively. In general terms, Thailand and the United States export 30 and 50% of their rice, respectively.

Within the United States, measurable rice production occurs in seven states: Arkansas, California, Florida, Louisiana, Mississippi, Missouri, and Texas (Table 5), with the majority of the rice area seeded with varieties of long grain rice. In 1988, 11 long grain varieties were grown in significant amounts. Two of those varieties (Rexmont and Toro 2) are specialty rices that receive special handling to preclude them from becoming mixed with other varieties (Table 6). Much less land area was devoted to the production of medium grain types although eight varieties were reportedly grown commercially. Only three varieties of short grain rice were planted in sufficient amounts to be reported, with the total hectares seeded to short grain quite small (Table 6) (10). Table 7 shows the main variety in each of the rice-producing states. Research scientists at each of the state agricultural experiment stations inform producers of new developments in rice production. Seeds of newly released varieties are always in great demand by the producers. Therefore, popularity of a variety may be short lived due to the release of new varieties more suitable to the needs of the producer as well as the end user (Table 8). Interdisciplinary cooperation between scientists involved in rice production resulted in the development of an interactive computerized production system for Arkansas based on variety, weather conditions, and emergence date. As the season progresses, other parameters provide interactive data as predictive factors (12). One half of the rice produced in the state of Arkansas is monitored by this system.

III. ANATOMICAL STRUCTURE

Rice is harvested as a covered grain in which the kernel (caryopsis) is encased in a protective hull or husk. The hull is a two-piece structure (lemma and palea) which tightly encompasses the kernel, providing security against many insects and a barrier to reduce the rapid change in moisture content of the kernel as changes in humidity occur. Rice grain, in the harvest state commonly known as rough or paddy rice, is 18–20% hull (Fig. 1). After the hull is removed, the caryopsis is known as brown rice regardless of its visual hue (color).

The mature brown rice kernel is approximately 93% endosperm (Fig. 2). The endosperm, the starchy inner portion of the kernel, is composed of tightly packed compound starch granules with protein interspersed as spherical bodies and crystalline structures. The major constituent of the endosperm is starch (approx. 90%). The outermost layer of the endosperm is the aleurone. This structure is botanically part of the endosperm; however, rice millers consider it part of the bran because it is mostly removed in the milling process (13).

The area between the hull and endosperm is composed of three distinct layers: pericarp, tegmen (seed coat), and nucellus (Fig. 3). These three layers plus the aleurone account for approximately 3% of the brown rice weight (Fig. 2) and together are known as the bran. The remaining 4% of the brown rice kernel is comprised of the germ (embryo).

IV. RICE TYPES AND THEIR DETERMINATION

Throughout the world, rice varieties are divided into different types according to custom and usage. Rice varieties commonly used in commercial rice production in the United

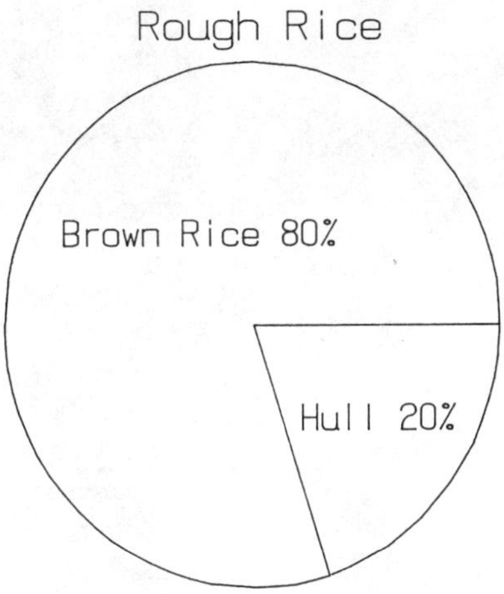

FIGURE 1 Anatomical composition of rough rice.

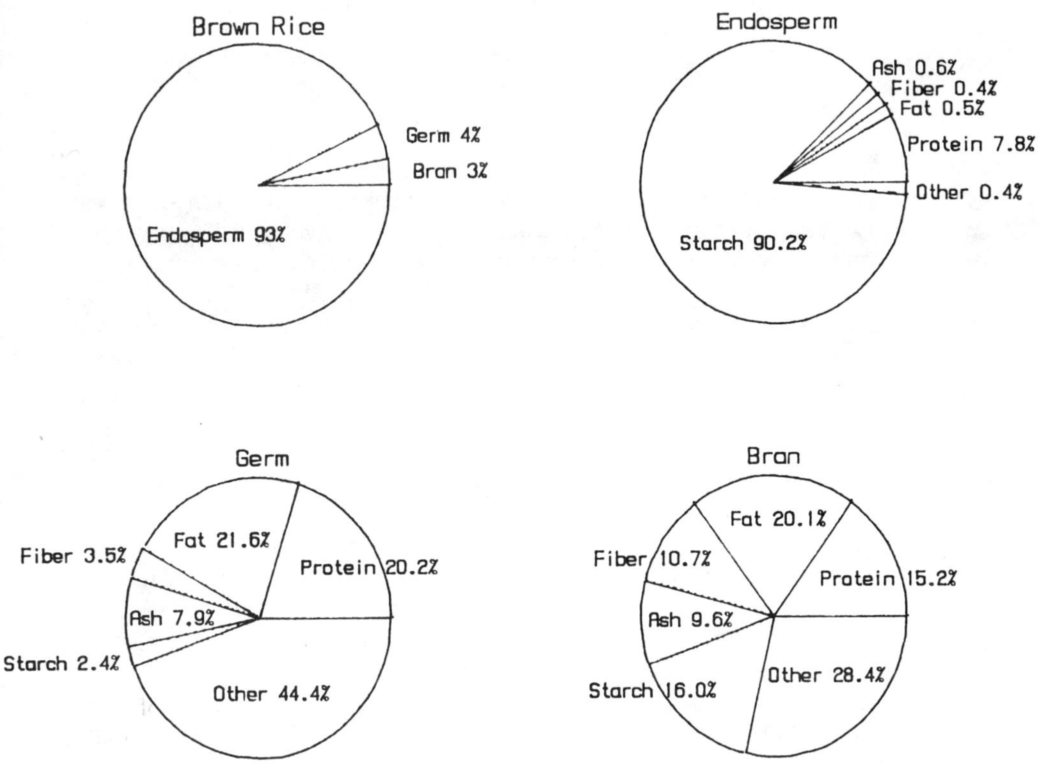

FIGURE 2 Anatomical composition of brown rice.

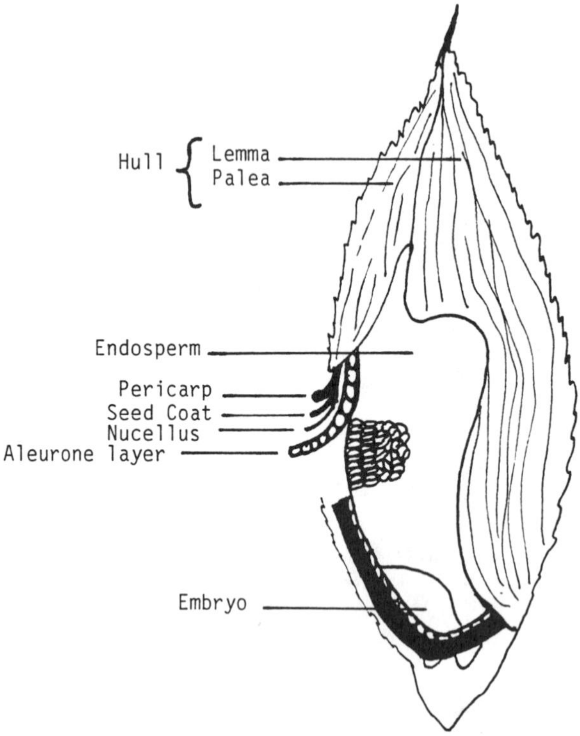

FIGURE 3 Structure of the rice kernel. (From Ref. 13.)

States are of three grain types according to the physical measurements of length and width of the grain (Fig. 4; Table 9) (14). By agreement between plant breeders of the USDA and state experiment stations, a new variety will not be released for commercial production unless it conforms to specific quality attributes typical of its physical measurements (Table 10) (15). Kernel measurements are normally taken as the average length and width of 15 kernels (Fig. 5) (14). Occasionally, varieties for special uses, such as Della and Toro 2, are released, recognizing that these varieties are not typical of long, medium, or short grain types. Care must be exercised to prevent the comingling of seeds, or else the reliability of the system will not be maintained.

V. RICE QUALITY AND GRADING STANDARDS

There are four broad categories of rice quality: (1) milling quality, (2) cooking, eating, and processing quality, (3) nutritional quality, and (4) specific quality designations regarding cleanliness, soundness, and purity. Market criteria are combinations of milling quality and standards for cleanliness, soundness, and purity. These have little relation to cooking, eating and processing qualities and no relation to nutritional quality.

Milling quality is the degree to which the endosperm remains intact at the end of bran removal and how much of the bran is removed. Milling quality, indicated by total milling yield or head rice yield, is expressed as a percentage of rough rice. Head rice is the milled kernels that are 3/4 or more of the endosperm length. Therefore, the difference between total milling yield and head rice yield is the amount of broken kernels. It is important to assess the

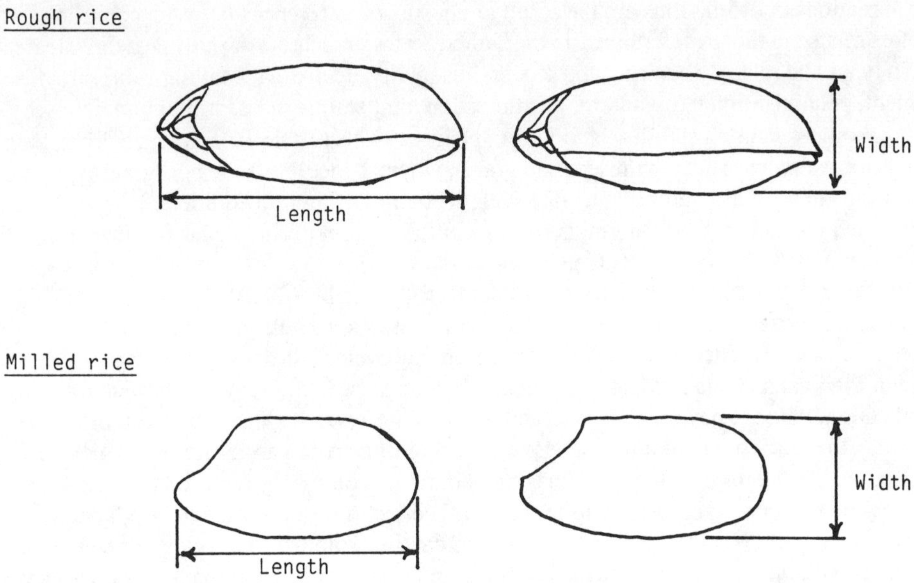

FIGURE 4 Physical measurement of the rice kernel. (From Ref. 14.)

degree of milling, because the stress placed on the kernels to remove the bran causes a large amount of breakage. Oil in bran is more susceptible to oxidation, therefore, excessive bran remaining on the kernel decreases the shelf life of the rice.

Cooking, eating, and processing quality relates to the suitability of the rice for the intended use. In the United States, rice types are generally selected for their uses, e.g., cooking, eating, and processing. Objectively, the quality is assessed in terms of amylose percentage, the most important index of suitability of end use (16). The United States was the first country to establish a quality evaluation program for cooking and processing quality (Table 11). Since 1955 many countries have initiated programs to evaluate cooking, eating, and/or processing qualities of rice.

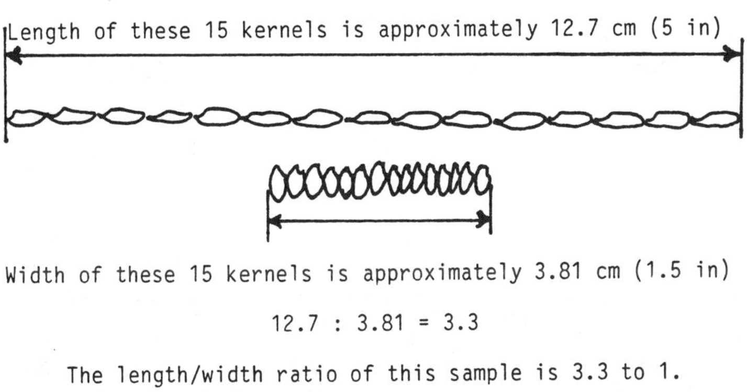

FIGURE 5 Measurement method to determine rice grain type. (From Ref. 14.)

Variations in evaluations are the result of consumer preferences between countries and factors related to those preferences. In the United States, indicators of rice that delineate rice grain types and relate to cooking and processing quality are alkali spreading value, amylose content, gelatinization temperature, gelatinization temperature class, amylographic viscosity, parboiling-canning stability, and water uptake at 77° C (Table 10). Upon cooking, long grain rice is dry and fluffy with seperate grains, whereas medium and short grain types are moist and chewy with grains that tend to stick or clump together. In addition to those above-mentioned characteristics, the grains must meet the criteria of the physical dimensions of long, medium, and short grains (see Table 9).

Brown rice contains more nutrients (minerals and vitamins) than does milled rice (Table 12), but the nutrient content alone is insufficient to assess the nutritional value of food. Although brown is higher than milled rice in nutrient content, it does not necessarily have greater food value. Data relating to the nutrient value of rice are extensive, have conflicting implications, and are beyond the intent and scope of this review. One of the most impressive points in the nutrient content of rice is the vast pool of genetic variability in the world rice collection. One example is that protein content of rice is commonly reported to be 6.9–7.3% (13), while low and high values within the world collection are reported to be 4.3 and 18.2, respectively (26). Whether these characteristics can be exploited to provide more essential nutrients without a loss of the various cooking and eating qualities remains a monumental challange.

As with other grains in the United States, a system of grades and standards to facilitate marketing has been developed for rice. This system is governed by regulations (Agricultural Marketing Act of 1946, as amended; the Code of Federal Regulations, Title 7, Part 68, Subpart A) and has specified standards for rice (Code of Federal Regulations, Title 7, Part 68, Subparts C, D, and E). The Federal Grain Inspection Service, an agency of the United States Department of Agriculture, administers the system of U.S. Standards for Rice. Guidelines for sampling, inspection, grading, and certification of rough rice, brown rice for processing, and milled rice are given in the Rice Inspection Handbook (14).

The U.S. Standards recognize three types of rice (long, medium and short), based on the length/width ratio of unbroken rice (see Table 9). Requirements for numerical grades of rough, brown, and milled rices are shown in Tables 13, 14, and 15, respectively.

Special grades are provided for specific qualities or conditions of rice that affect marketability. These special grades are:

Rough rice: Parboiled rough rice
Smutty rough rice
Weevily rough rice
Brown rice: Parboiled brown rice for processing
Smutty brown rice for processing
Milled rice: Coated milled rice
Granulated brewers rice
Parboiled milled rice
Undermilled milled rice

Grade designation follows this order: (1) The letters "U.S."; (2) the number of the grade; (3) the class; (4) each applicable special grade; and (5) a statement of the milling yield. For example, grade designation for a specific lot of rice may read " U.S. No. 3, long grain rough rice, parboiled, milling yield 50%– 70%." The 50% and 70% indicate head rice and total milling yields respectively. Item number 5 in the grade designation is required for rough rice and is permissable for brown rice (14).

Detailed guidelines for grading are found in the Rice Inspection Handbook (14).

VI. DIRECT UTILIZATION OF RICE AS FOOD

In 1985, world rice production was 465,970,000 MT (see Table 1) and the total world exports in 1984 were 12,518,000 MT (see Table 3) or approximately 2.7%. Thus, most of the world rice production is consumed in the country of origin. In 1975, the world per capita annual consumption of rice was 58.4 kg (128 lb), with South Vietnam and Thailand having the greatest per capita consumption of 247 and 212 kg (543 and 466 lb), respectively, compared to 6.4 kg (14.1 lb) rough rice basis for the United States (6).

Due to the low per capita consumption rate, the United States exports a relatively large percentage of its annual production. One estimate (Arkansas Rice Council Annual Meeting, January 22, 1987, Forest City, AR) placed the U.S. domestic rice consumption at 50% of production.

Domestic rice consumption in the U.S. has steadily increased in recent years (Table 16). The per capita consumption of rice (Table 17) and the rice grain types consumed vary from region to region (Table 18). Because the Pacific region uses a greater proportion of rice from California, medium grain rice is preferred in that area, whereas all other regions use a greater proportion of rice from Arkansas and Texas, indicating preference for long grain.

Utilization of rice as food can be categorized three ways: direct food use, processed foods, and brewing. Rice for direct consumption includes regular milled rice, parboiled rice, precooked rice, brown rice, and specialty products such as flour and flavored rice. Rice used for processed foods includes that for cereals, soup, baby food, packaged mixes, and minor unclassified uses. Rice used for brewing is primarily for the production of beer (27). Due to the unavailability of information for areas other than the United States, food distribution patterns will, by necessity, address only those of the United States.

The use of rice for direct food is by far the greatest of the three uses. All three uses have increased steadily for several years (Table 19). Although direct food use accounts for the largest domestic consumption, a significant quantity of rice is used in processed food. The functions performed by rice in processed foods are very important. The strength of the paste and the resulting gel (Fig. 6) play importnt roles in processed foods.

The use of rice by-products as human food should not be overlooked. Utilization of by-products requires a specialized technology. Heat generated during milling triggers enzymatic activity, resulting in hydrolysis of lipids, and oxidative changes lead to rancidity. Immediate stabilization of the bran prevents oil breakdown and helps control growth of insects and microorganisims (29).

Attempts to stablilize rice bran include dielectric heating (30), treatments with hydrochloric acid (31), and treatment with sodium metabisulfite (32). The process that provides the most promising solution is extrusion cooking. Commercial use of this method has been initiated. The process is a new venture and has not been widely accepted in the United States rice-milling industry.

The potential for incorporating rice bran as a fiber ingredient into baked goods has been demonstrated (33–35).

VII. RICE AS LIVESTOCK FEED

Rice is utilized throughout the world as a food commodity and, with the exception of rice bran and germ, is rarely used in animal feeding. In countries where a greater proportion of the world rice is grown and where human food uses essentially require the entire cereal grain production, ruminants are feed rice straw. However, due to the high feeding perform-

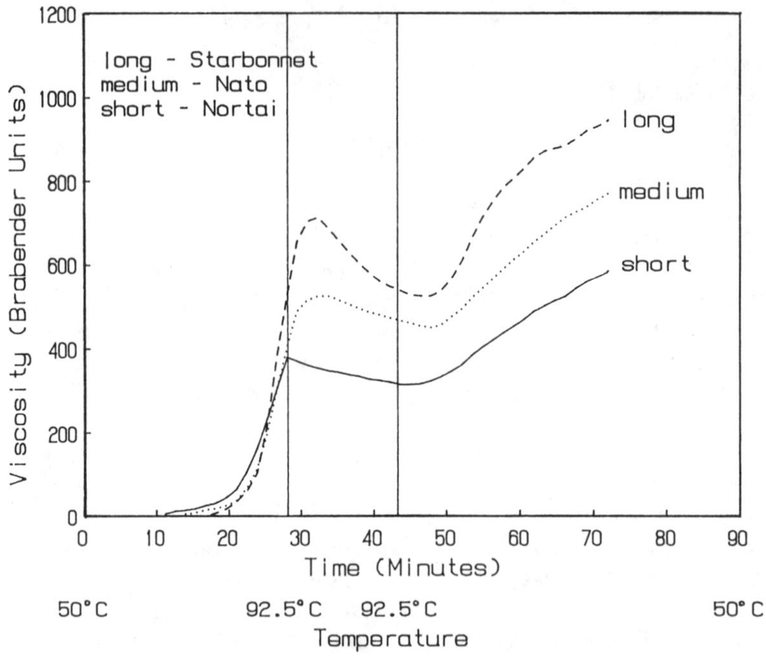

FIGURE 6 Visco-amylograph curves for one variety of each rice grain type.

ance required in Canada and the United States, only a small portion of the animal diet can be replaced with forages of such poor quality. For overwintering steer calves, the replacement of more than 25% of the prairie hay in the diet resulted in a reduced growth rate (36). Overwintering beef cattle fed a rice hull and soybean meal feed formulated to 12% protein caused erratic eating patterns (37).

Rice can be used to replace up to 50% of the corn in growing and finishing diets for swine if the diet is pelleted, but if fed in meal form, more than 35% replacement results in depressed weight gain. When pigs are young, only 20% rice bran can be used and then only when pelleted (38). Processing of rice bran by cooking, ensiling, extrusion, pelleting, cellulase, and amylase treatments did not improve utilization beyond the 40% corn replacement level (39, 40). Feeding studies with poultry have yielded similar findings. It is not economical to feed rice to livestock unless the rice has been severely damaged such as when stack burned.

VIII. OTHER USES FOR RICE HULLS

Rough rice contains approximately 20% hull, therefore, the total mass of hulls from 500 billion MT of rice presents a sizable waste-disposal problem. Rice hulls have been used for a number of purposes. Particularly in the southern United States, rice hulls have been used extensively as littering materials for poultry houses. Rice hulls are used as a soil amendment for potting plants and as bedding material for plants. Because rice hulls resist enzymatic decomposition, it is not a good material for composting or biomass conversion.

Although the burning of rice hulls to heat water for parboiling purposes has been practical for many years, controlled burning of masive amounts of the hull is a relatively new technology with the greatest sophistication being extensive pyrolysis that yields ultrapure silica suitable for manufacture of computer microchips and solar cells. The resultant heat from the process is currently used for steam generation and rice drying in Arkansas and for cogeneration of electricity in Louisiana.

IX. SPECIALTY RICES

Specialty rices are those rice types that fall outside the realm of traditional rice. In some geographical locales, some of these rices may be very common with production utilizing a sizable land mass. These rice types meet specific needs that traditional rices canot fulfill.

One specialty rice that is an exception to the typical type of rice is aromatic (scented) rice. There are many cultivars of aromatic rice in each geographical area to which aromatic is native, yet differences in agronomic characteristics have only recently begun to be identified. Characteristics relating to cooking and eating variability are yet to be ascertained, but some progress has been made in evaluation methodology (Table 20). Plant breeders at Louisiana Agricultural Experiment Station registered an aromatic cultivar of rice, Della, in 1973 (41). Dela is a cross between a typical U.S. long grain type and a Basmati (India) cultivar. Comparisons of cooking quality between Della, Basmati (India), and Jasmine (aromatic type from Thailand) showed Della elongated to a lesser extent than either Basmati or Jasmine but more nearly resembled the cooking quality of typical U.S. long grain rice with flavor like the other aromatics (42).

When cooked, aromatic rice emits an aroma similar to roasted popcorn or nuts. This unique flavor compound has been identified as 2-acetyl-1-pyrroline and is found in small amounts during cooking of the more traditional rices (43). This flavor compound is stable during normal cooking temperatures but is destroyed when subjected to high heat experienced in parboiling.

Waxy rice is characterized by its low anylose content (less than 2.0%) (16) and is known as glutinous or sweet rice. Very little waxy rice is produced in the United States, but in some parts of the world it is common enough to be consumed daily. In the U.S. this rice performs specific functions in several commercial products such as desserts, gravies, certain bakery products, and salad dressing.

When marketing designations are agreed upon that combine physical measurement with specific cooking and processing criteria, little time elapses before there is a demand for exception. Toro rice is an example of that exception. It is a long grain rice (physical measurements) that has a low amylose content, low gelatinization temperature, and the cooking characteristics of a medium grain (44). This type of rice is popular in southern Louisiana.

The need for a rice with superior processing quality that will withstand the rigors of thermal processing required for canned products prompted the development of the variety Newrex. This variety was developed specifically to fill an industry need and displays a very low hot paste breakdown, a very high set back viscosity, and a high (28%) amylose value (45). As is the case with many traditional rice varieties, this variety has already been replaced by an improved variety, Rexmont.

X. RICE MILLING

The primary objective in rice milling (Fig. 7) is to remove the hull and bran with minimum breakage of the endosperm. The process used to accomplish this task differs throughout the world. The methods have many variations, ranging from small batch milling to the continuous milling of large lots. Regardless of the methods used, the process follows the same path of hull removal and subsequent bran removal. The bulk of marketable rice is commercially milled in a continuous process (Fig. 8), starting with rice that has been harvested and dried

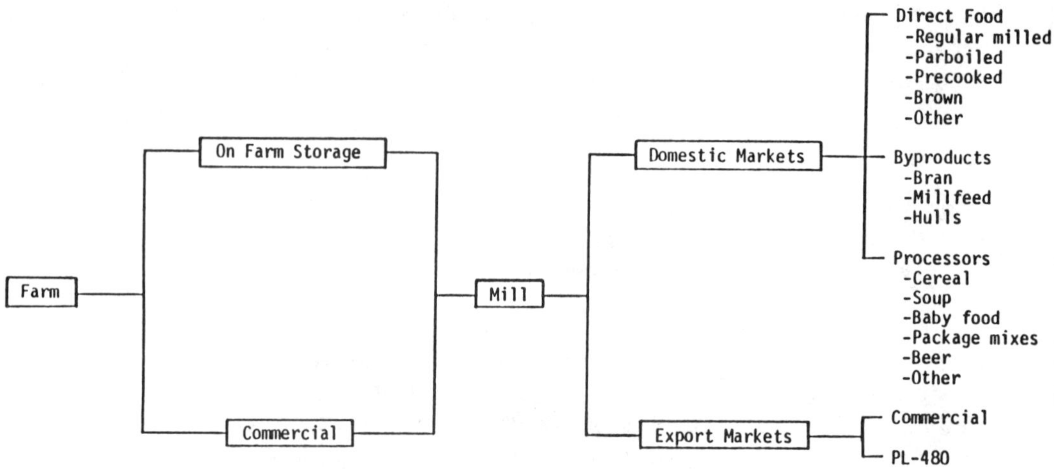

FIGURE 7 Flow diagram of movement of rice from production to consumption. (From Ref. 27.)

to approximately 11–12% moisture. Variations of the process will occur due to different equipment uses and capabilities of the mill.

The miller must endeavor to maximize ecomonic return. A recent study (46) reported that rice millers in the United States are responsive to changes in production, consumption, and technology. Such changes are keeping the U.S. rice-milling industry competitive.

Rice milling does remove a significant amount of nutrients (see Table 12), therefore, the restoration of some of the lost nutrients has become a practice in some countries. The United States does not require that milled rice be enriched, but if the label declares enrichment, there are specified requirements for niacin, thiamine, riboflavin, and iron (47). Prevention of appreciable nutrient loss can be accomplished by parboiling the rice prior to milling; however, in the United States there are other reasons considered more important for parboiling rice, and parboiling does not accomplish the objective of enrichment. Parboiling can be accomplished in a variety of ways, the details of which are normally not divulged by the processors. The general scheme is to hydrate rough rice to 32–38% moisture and gelatinize or partially gelatinize the starch by heating. The gelatinization in some instances is accomplished at atmospheric pressure, but more commonly is carried out in a steam pressurized vessel (Fig. 9). Parboiling causes physical and chemical alterations in the grain leading to changes such as improved milling yields, increased resistance to insects, and firmer cooked rice texture accompanied by a darker and more yellow endosperm.

Sixty-six active rice mills were identified in the United States in 1985;(46), however, 25 mills (Table 21) process the majority of the rice.

Since the establishment of the International Rice Research Institute at Los Banos, Laguna, Philippines in 1960, knowledge of rice has rapidly expanded. With additional information, worldwide advances have been attained in production, processing, and utilization. These have not occurred solely as the result of new findings but because findings were shared internationally for the benefit of all people.

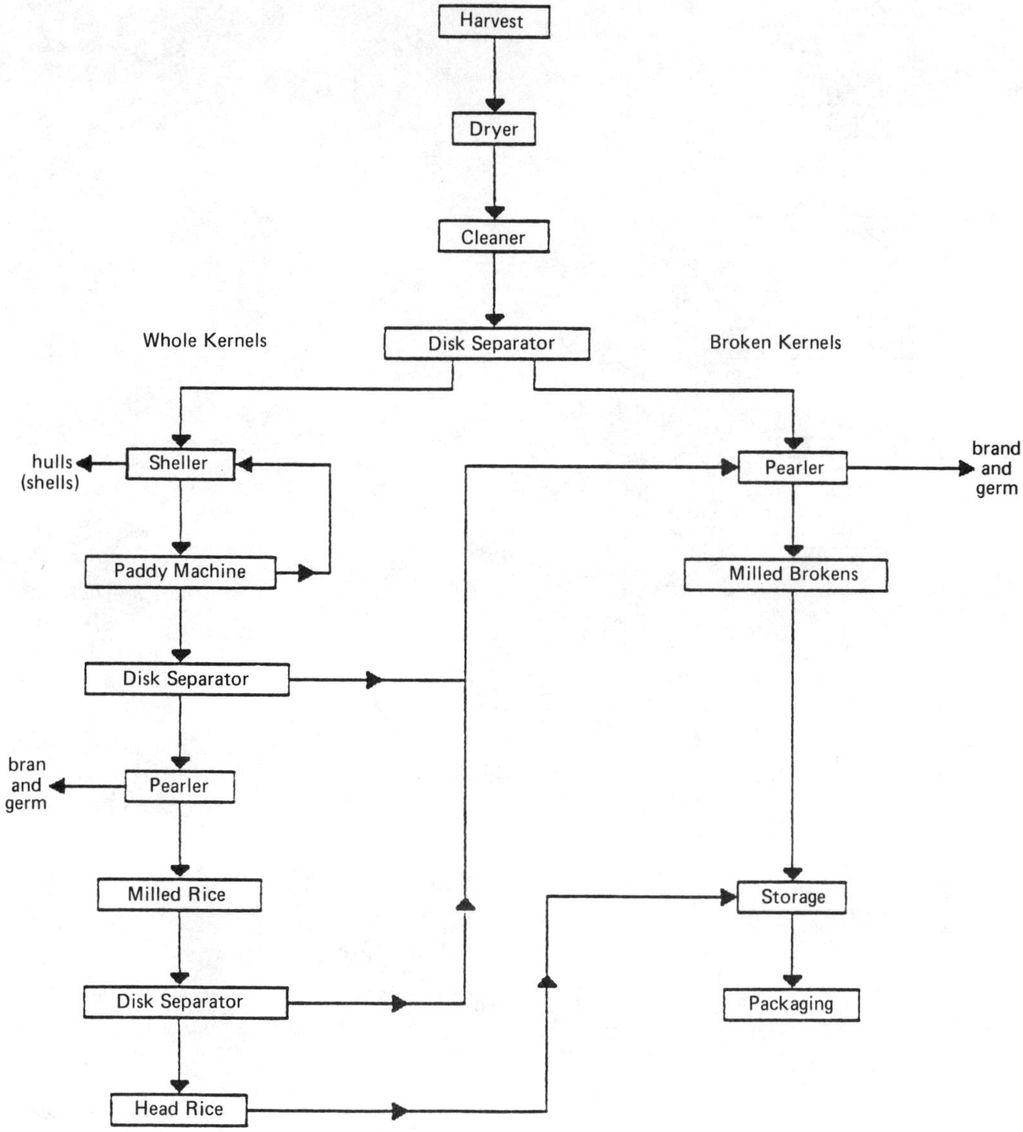

FIGURE 8 Process flow for typical rice milling.

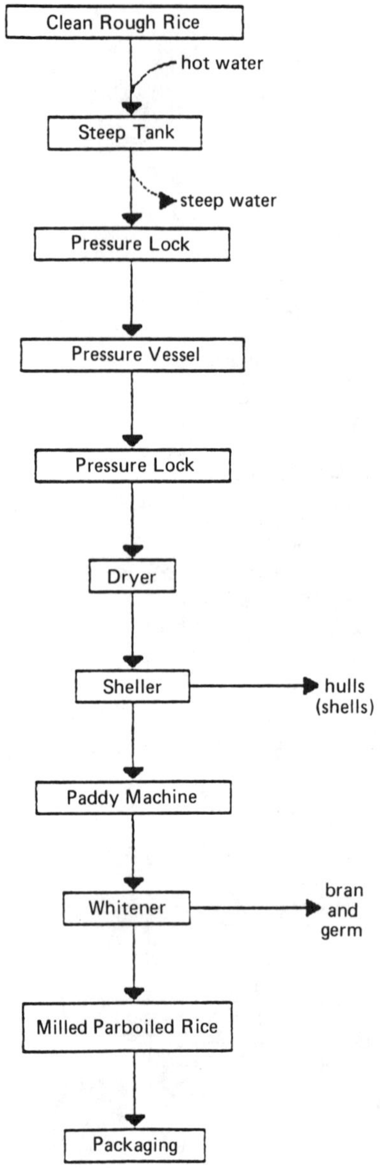

FIGURE 9 Process flow for commercial parboiling.

TABLE 1 World Rice Area, Yield and Production, 1955–1985

Year	Area (1000 ha)	Yield (kg/ha)	Production (1000 MT)
1955	108,783	1,798	195,640
1960	120,172	1,950	234,749
1965	123,902	2,070	256,480
1970	131,174	2,380	311,830
1975	143,056	2,520	360,575
1980	143,746	2,755	396,062
1985	144,674	3,221	465,970

Source: Adapted from Refs. 5,6.

TABLE 2 World Rice Production: Percentage of World Total by Continent and Selected Countries, 1975–1985

Continent/country	1975	1977	1979	1981	1983	1985
Africa	2.3	2.1	2.2	2.1	2.0	2.0
North America	2.2	1.7	2.1	2.6	1.2	1.9
South America	3.3	3.5	3.3	3.2	2.8	3.1
Asia	91.0	91.4	91.0	90.9	92.7	91.8
Europe	0.5	0.4	0.5	0.4	0.4	0.4
Oceania	0.1	0.1	0.2	0.2	0.1	0.2
U.S.S.R.	0.5	0.6	0.6	0.6	0.6	0.6
Australia	0.1	0.1	0.2	0.2	0.1	0.2
Bangladesh	5.3	5.0	5.1	5.0	4.8	4.7
Brazil	2.2	2.4	2.0	2.0	1.7	1.9
Burma	2.6	2.5	2.8	3.4	3.2	3.3
China	35.9	35.5	39.0	35.7	38.2	36.8
India	20.4	21.2	16.8	19.4	19.9	19.6
Indonesia	6.2	6.3	7.0	8.0	7.8	8.3
Italy	4.8	0.2	0.3	0.2	0.2	0.2
Japan	4.8	4.6	4.0	3.1	2.9	3.1
Thailand	4.3	3.7	4.2	4.3	4.1	4.2
United States	1.6	1.2	1.6	2.0	1.0	1.3
Vietnam	2.9	2.9	2.9	3.0	3.3	3.3

Source: Adapted from Refs. 5, 7.

TABLE 3 Share (percentage) of Total World Exports for Major Rice-Exporting Countries, 1982–1984

Country	1982	1983	1984
Burma	5.8	7.4	5.8
China	6.4	9.6	9.7
Pakistan	7.8	7.8	10.1
Thailand	31.1	30.1	36.9
United States	20.9	20.6	17.1
Total world export (1000 MT)	12,147	11,591	12,518

Source: Adapted from Ref. 8.

TABLE 4 United States Share of Total World Production and Exports, 1970–1987

	U.S. share of world total (%)		
Year	Production	Exports	Ending Stocks
1970	1.3	16.5	3.4
1971	1.3	22.4	2.4
1972	1.4	18.9	1.6
1973	1.3	22.2	2.0
1974	1.6	28.1	2.2
1975	1.7	24.2	6.3
1976	1.6	21.4	7.2
1977	1.2	23.6	3.9
1978	1.6	19.0	3.6
1979	1.7	23.4	3.6
1980	1.8	23.2	2.5
1981	2.1	21.1	7.6
1982	1.7	19.3	13.8
1983	1.0		
1984	1.3	17.0	3.7
1985	1.3	16.6	4.7
1986	1.3	18.8	3.3
1987	1.3	19.4	2.5

Source: From Ref. 9 with permission.

TABLE 5 Distribution of Rice Production in the United States for 1988

State	Long	Medium	Short	Specialty rice[a]	Share of U.S. total (%)	Total hect-arage
Arkansas	90.4	9.3	0.3	—	42.2	477,546
California	14.3	67.1	13.1	5.5	15.0	169,974
Florida	100.0	—	—	—	0.5	5,840
Louisiana	77.5	22.5	—	—	18.1	204,664
Missouri	99.7	0.3	—	—	2.7	30,494
Mississippi	100.0	—	—	—	7.8	88,361
Texas	99.3	0.6	0.1	—	13.6	153,650

[a]Includes proprietary varieties and waxy rice.
Source: Adapted from Ref. 10.

TABLE 6 Most Popular Varieties of Rice Grown in the United States Based on Hectarage Planted, 1988

Variety	1987	1988
Long		
CB801	1,610	6,671
Gulfmont	13,303	59,005
Labelle	11,156	9,551
Lebonnet	9,721	9,773
Lemont	255,623	411,077
L202	a	24,718
Newbonnet	282,545	180,218
Rexmont	413	2,025
Skybonnet	15,837	17,736
Tebonnet	91,663	157,609
Toro 2	343	1,760
Medium		
Mars	99,549	82,977
Mercury	70	4,731
Rico	a	472
Saturn	2,775	2,906
M102	a	4,047
M201	a	26,467
M202	a	72,846
M401	a	10,765
Short		
Nortai	a	<2,000[b]
S101	a	8,094
S201	a	14,165

[a]data incomplete.
[b]estimated.
Source: Adapted from Ref. 10.

TABLE 7 Predominant Rice Varieties for the Major Rice-Producing States in the United States, 1988.

State	Variety	Grain type	% of State hectarage
Arkansas	Newbonnet	long	32.6
California	M202	medium	42.9
Florida	Lebonnet	long	44.8
Louisiana	Lemont	long	46.8
Missouri	Lemont	long	70.8
Mississippi	Lemont	long	70.8
Texas	Lemont	long	63.6

Source: Adapted from Ref. 10.

TABLE 8 Characteristics of Popular U.S. Rice Varieties

Variety	Grain type	Grain yield index[a]	Milled kernel[b] weight (g)	% Head rice yield	% Total milling yield	Parboiling suita–bility[d]
Gulfmont	long	105	19.3	57	71	0
Lebonnet	long	102	19.3	56	70	–
Lemont	long	106	20.0	58	71	–
L202	long	115	21.1	55	70	0
Mars	med.	117	18.9	63	72	–
Mercury	med.	94	18.5	c	c	–
Newbonnet	long	100	17.5	59	70	0
Nortai	short	110	18.3	59	71	–
Rexmont	long	97	17.2	51	68	+
Skybonnet	long	102	18.8	57	70	0
Tebonnet	long	103	18.7	58	71	0

[a]Based on Arkansas production only.
[b]Weight of 1000 kernels.
[c]Data unavailable.
[d]0=average; +=above average; –=below average.
Source: Adapted from Ref. 11.

TABLE 9 Physical Measurement Criteria for Rice Types

Grain type	Rough	Brown	Milled
Long	3.4 or more	3.1 or more	3.0 or more
Medium	2.3–3.3	2.1–3.0	2.0–2.9
Short	2.2 or less	2.0 or less	1.9 or less

Source: Ref. 14.

TABLE 10 Cooking and Processing Characteristics of Typical U.S. Long, Medium, and Short Grain Types of Milled Rice

Characteristics	Long	Medium	Short
Alkali spreading value	3–5	6–7	6–7
Amylographic viscosity (BU)			
peak viscosity	765–840	890–980	820–870
hot paste viscosity	400–500	370–420	370–400
cool paste viscosity	770–880	680–760	680–690
Amylose content (%)	23–26	15–20	18–20
Gelatinization temperature (°C)	71–74	65–68	65–67
Gelatinization temperature class	Intermediate	low	low
Parboiling-canning stability (% solids loss)	18–21	31–36	30–33
Water uptake @ 77°C (ml/100 g)	121–136	300–340	310–360

Source: Adapted from Ref. 15.

TABLE 11 Initiation of Rice Cooking and Processing Quality Program in Selected Countries

Country	Year	Ref.
Australia	1971	17
Bangladesh	1960	18
France	mid-1970s	19
India	1968	20
Japan	mid-1960s	21
Philippines	1961	22
Spain	1968	23
Sri Lanka	1978	24
Thailand	1968	25
United States	1955	15

TABLE 12 Content of Selected Nutrients of Brown and Milled Rice at 12% Moisture

Nutrient	Brown	Milled
Crude ash (%)	1.0	0.5
Crude fat (%)	1.9	0.3
Crude fiber (%)	0.7	0.3
Crude protein (N x 5.95) (%)	7.2	5.8
Neutral detergent fiber(%)	2.7	0.8
Starch (%)	57.0	67.0
Calcium (mg/g)	0.3	0.2
Iron (μg/g)	24.0	13.0
Total phosphorus (mg/g)	2.5	1.2
Phytin phosphorus (mg/g)	2.0	0.5
Potassium (mg/g)	1.7	1.0
Sodium (μg/g)	315.0	45.0
Niacin (μg/g)	43.0	18.0
Riboflavin (μg/g)	0.9	0.4
Thiamine (μg/g)	4.4	0.7
Tocopherol (μg/g)	17.0	tr

tr = trace
Source: Adapted from Ref. 13.

TABLE 13 Grades and Grade Requirements for the Classes of Rough Rice.

	Seeds and heat-damaged kernels			Chalky kernels (%)			
Grade	Total (singly or combined) in 500 g	Heat damaged kernels and objectionable seeds (singly or combined) in 500 g	Red rice and damaged kernels (singly or combined %)	In long grain rice	In medium or short grain rice	Other types (%)	Color requirements
U.S. No. 1	4	3	0.5	1.0	2.0	1.0	Shall be white or creamy
U.S. No. 2	7	5	1.5	2.0	4.0	2.0	May be slightly gray
U.S. No. 3	10	8	2.5	4.0	6.0	3.0	May be light gray
U.S. No. 4	27	22	4.0	6.0	8.0	5.0	May be gray or slightly rosy
U.S. No. 5	37	32	6.0	10.0	10.0	10.0	May be dark gray or rosy
U.S. No. 6	75	75	15.0	15.0	15.0	10.0	May be dark gray or rosy

U.S. sample grade does not meet the requirements for any of the grades from U.S. No. 1 to U.S. No. 6.

All values are maximum limits.
Source: Adapted from Ref. 14.

TABLE 14 Grades and Grade Requirements for the Classes of Brown Rice for Processing

Grade	Paddy kernels (%)	Paddy kernels (in 500 g)	Seeds and heat-damaged kernels: Totally (singly or combined) (in 500 g)	Seeds and heat-damaged kernels: Heat-damaged kernels (in 500 g)	Seeds and heat-damaged kernels: Objectionable seeds (in 500 g)	Red rice and damaged kernels (singly or combined) (%)	Chalky kernels (%)	Broken kernels removed by a 6 plate or a 6 1/2 sieve (%)	Other types (%)	Well milled kernels (%)
U.S. No.1	—	20	10	1	2	1.0	2.0	1.0	1.0	1.0
U.S. No.2	2.0	—	40	2	10	2.0	4.0	2.0	2.0	3.0
U.S. No.3	2.0	—	70	4	20	4.0	6.0	3.0	5.0	10.0
U.S. No.4	2.0	—	100	8	35	8.0	8.0	4.0	10.0	10.0
U.S. No.5.	2.0	—	150	15	50	15.0	15.0	6.0	10.0	10.0

U.S. sample grade does not meet the requirements for any of the grades from U.S. No. 1 to U.S. No. 5.

All values are maximum limits.
Source: Adapted from Ref. 14.

TABLE 15 Grades and Grade Requirements for the Classes; Long Grain Milled Rice, Medium Grain Milled Rice, Short Grain Milled Rice, and Mixed Milled Rice

	Seeds, heat-damaged, and paddy kernels (Singly or combined)			Chalky kernels		Broken kernels				Other types			
Grade	Total (in 500 g)	Heat-damaged kernels and objectionable seeds (in 500 g)	Red rice and damaged kernels (singly or combined) (%)	In long-grain rice (%)	In medium or short grain rice (%)	Total (%)	Removed by a 5 plate (%)	Removed by a 6 plate (%)	Through a 6 sieve (%)	Whole kernels (%)	Whole and broken kernels (%)	Color requirements	Milling requirements (minimum)
U.S. No. 1	2	1	0.5	1.0	2.0	4.0	0.04	0.1	0.1	—	1.0	Shall be white or creamy	Well milled
U.S. No. 2	4	2	1.5	2.0	4.0	7.0	0.06	0.2	0.2	—	2.0	May be slightly gray	Well milled
U.S. No. 3	7	5	2.5	4.0	6.0	15.0	0.1	0.8	0.5	—	3.0	May be light gray	Reasonably well milled
U.S. No. 4	20	15	4.0	6.0	8.0	25.0	0.4	2.0	0.7	—	5.0	May be gray or slightly rosy	Reasonably well milled

TABLE 15 (Continued)

Grade	Seeds, heat-damaged, and paddy kernels (Singly or combined)		Red rice and damaged kernels (singly or combined) (%)	Chalky kernels		Broken kernels				Other types		Color require ments	Milling require-ments (minimum)
	Total (in 500 g)	Heat-damaged kernels and objec tionable seeds (in 500 g)		In long-grain rice (%)	In medium or short grain rice (%)	Total (%)	Removed by a 5 plate (%)	Removed by a 6 plate (%)	Through a 6 sieve (%)	Whole kernels (%)	Whole and broken kernels (%)		
U.S. No. 5	30	25	6.0	10.0	10.0	35.0	0.7	3.0	1.0	10.0	—	May be dark gray or rosy	Lightly milled
U.S. No. 6	75	75	15.0	15.0	15.0	50.0	1.0	4.0	2.0	10.0	—	May be dark gray or rosy	Lightly milled

U.S. sample grade does not meet the requirements for any of the grades from U.S. No. 1 to U.S. No. 6.

All values are maximum limits.
Source: Adapted from Ref. 14.

TABLE 16 U.S. per Capita Consumption of Rice for Selected Years

Year	kg	lb
1974–1975	4.14	9.10
1979–1980	6.20	13.65
1982–1983	6.95	15.30
1984–1985	7.90	17.40

Source: Ref. 9

TABLE 17 Proportion of Rice for Direct Food Use Distributed to Major U.S. Regions by Specified Year

	Amount per Capita		
Region	75/76	80/81	84/85
New England	1.77	1.77	2.41
Middle Atlantic	3.64	5.05	6.41
E. N. Central	1.23	1.68	1.77
W. N. Central	1.05	2.27	2.32
South Atlantic	2.95	3.77	3.59
E. S. Central	1.05	1.41	1.45
W. S. Central	5.52	5.01	6.36
Mountain	1.27	2.64	1.55
Pacific	5.09	6.23	5.64
U. S. Total	2.82	3.64	3.91

Source: Ref. 9.

TABLE 18 Regional Summary of Proportional Share of Rice Distributed for Direct Food Use, by Mill Area and Region for 1974, 1980, and 1984

Mill area	New England	Middle Atlantic	East-North Central	West-North Central	South Atlantic	East-South Central	West-South Central	Mountain	Pacific	Percentage of total U.S.
1974/1975										
Arkansas-Mississippi	11.5	36.8	60.8	47.8	31.8	39.2	8.3	33.2	21.9	30.5
Louisiana	10.5	17.9	.1	10.4	.3	6.0	21.7	1.2	5.0	10.1
Texas	76.8	43.4	35.1	41.1	66.8	54.3	69.1	61.5	24.5	47.2
California	1.2	1.9	4.0	.7	1.1	.5	.9	4.1	48.6	12.2
1980/1981										
Arkansas-Mississippi	22.4	52.2	62.2	33.9	45.2	46.5	24.1	41.7	15.3	36.4
Louisiana	0	4.4	0	.5	5.3	16.1	10.1	.1	.4	3.9
Texas	77.5	41.6	34.4	65.4	47.9	37.0	62.8	19.7	28.1	43.0
California	.1	1.8	3.0	.1	1.6	.4	3.1	38.5	56.2	16.7
1984/1985										
Arkansas-Mississippi	46.7	53.9	56.3	61.7	42.1	28.1	43.9	25.2	11.6	46.7
Louisiana	16.3	22.1	8.7	4.7	23.1	19.9	15.4	11.9	9.2	16.3
Texas	35.1	18.5	32.4	25.0	33.3	52.1	38.8	43.4	30.8	35.1
California	1.9	5.6	2.7	8.6	1.5	0	1.8	19.5	48.5	1.9

Source: From Ref. 9 with permission.

TABLE 19 Distribution of U.S. Milled Rice by Principal Uses

Year	Directly for food	Processed foods	Brewing	Total
1974	572	114	277	962
1975	588	129	211	928
1978	690	169	370	1229
1980	852	204	348	1404
1982	870	152	413	1434
1984	983	225	319	1527

All values are 1000 MT.
Source: Adapted from Ref. 9.

TABLE 20 A Simplified Method to Evaluate Elongation in Rice

1. Place 20–100 kernels of whole grain milled rice in a wire basket.
2. Presoak rice in water for 30 min.
3. Place in boiling water for 10 min, remove, and cool by immersing in a water bath at ambient temperature.
4. Express elongation as a ratio of length of cooked kernels to raw kernels. Measure the length of 10 cooked kernels and 10 raw kernels (before soaking), and divide mean length of cooked kernels by mean length of raw kernels.
5. Because kernels tend to curl and disintegrate when overcooked, undercooked rice is preferred to overcooking in elongation tests.

Source: Adapted from Ref. 28.

TABLE 21 Principal U.S. Rice Millers

Mill name and location	Products[a]
ADM Milling Company Rice Milling Division P.O. Box 368 Weiner, AR 72479 *Mill locations*: Otwell, AR; Weiner, AR; Crowley, LA.	BR, MR, BK, MF, B, H, LG, MG
Cormier Rice Milling Co., Inc. P.O. Box 152 Dewitt, AR 72042	BR, MR, BK, OB, B, H, LG, MG, SG
Producers Rice Mill, Inc. P.O. Box 461 Stuttgart, AR 72160 *Mill locations*: Dewitt, AR; Fair Oaks, AR; Stuttgart, AR.	BR, MR, PR, BK, MF, B, H, LG, MG, RM
Riceland Foods, Inc. P.O. Box 927 Stuttgart, AR 72160 *Mill locations*: Jonesboro, AR; Stuttgart, AR.	BR, MR, PR, BK, MF, B, H, RF, RC, LG, MG, SG
California Pacific Rice Milling, LTD P.O. Box 725 Arbuckle, CA 95912	BR, MR, BK, B, H, LG, MG, SG
Farmers' Rice Cooperative P.O. Box 15223 Sacramento, CA 95851 *Mill locations*: Dos Palos, CA; West Sacramento, CA.	BR, MR, BK, B, H, LG, MG, SG
Koda Farms P.O. Box 88 South Dos Palos, CA 93665	MR, BK, MF, B, RF, MG
The Rice Growers Assn. of California P.O. Box 958 Sacramento, CA 95812 *Mill locations*: Biggs, CA; Woodland, CA.	BR, MR, BK, MF, B, H, LG, MG,SG
Sem-Chi Rice Products Corp. P.O. Box 1097 Loxahatchee, FL 33470	BR, MR, SR, BK, OR, MF, B, H, MG
Broussard Rice Mill, Inc. P.O. Box Drawer 160 Mermentan, LA 70556	BR, MR, BK, MF, B, H
Conrad Rice Mill, Inc. P.O. Box 10640 New Iberia, LA 70562	BR, MR, BK, B, H, RC, RM, LG, MG
Falcon Rice Mill, Inc. P.O. Box Drawer 771 Crowley, LA 70527	BR, MR, BK, SR, B, H, LG, MG
Liberty Rice Mill, Inc. P.O. Box 218 Kaplan, LA 70548	BR, MR, BK, MF, B, H, LG, MG

TABLE 21 (Continued)

Mill name and location	Products[a]
The Powell Group P.O. Box Drawer P Lake Charles, LA 70602	BR, MR, BK, B, LG, MG
Supreme Rice Mill, Inc. P.O. Box 490 Crowley, LA 70527	BR, MR, BK, MF, B, H, LG, MG, QW, QB
Busch Agricultural Sources, Inc. 12855 Flushing Meadow Drive Suite 200-B St. Louis, MO 63131 *Mill locations*: Jonesboro, AR; Woodland, CA.	BR, MR, BK, MF, B, H, LG, MG, SG
Affiliated Rice Milling, Inc. P.O. Box 1446 Alvin, TX 77512	BR, MR, BK, MF, B, H, SG
American Rice, Inc. P.O. Box 2587 Houston, TX 77252	BR, MR, BK, MF, B, H, PR, QW, RM, LG, MG
Beaumont Rice Mills, Inc. P.O. Box 3111 Beaumont, TX 77704	BR, MR, BK, MF, B, H, LG, MG
Comet Rice, Inc. P.O. Box 1953 Houston, Tx 77251 *Mill locations*: Maxwell, CA; Greenville, MS; Arecibo, Puerto Rico.	BR, MR, BK, MF, B, H, PR, LG, MG, SG
El Campo Rice Milling Co. P.O. Box 110 El Campo, TX 77437 *Mill location*: Louise, TX.	BR, MR, BK, MF, B, H, PR, LG
P & S Rice Mills, Inc. P.O. Box 55040 Houston, TX 77255	BR, MR, BK MF, B, H, PR, RF, LG, MG
Riviana Foods, Inc. P.O. Box 2636 Houston, TX 77252 *Mill locations*: Carlisle, AR; Abbeville, LA; Memphis, TN; Houston, TX.	BR, MR, BK, MF, B, H, PR, QW, RF, RM, SB, LG, MG, SG
Sunrice, Inc. P.O. Box 19567 Houston, TX 77224 *Mill locations*: McGehee, AR; Houston, TX.	BR, MR, BK, MF, B, H, LG, MG
Uncle Ben's Inc. P.O. Box 1752 Houston, TX 77251 *Mill locations*: Greenville, MS; Houston, TX.	BR, PR, B, H, RM, LG

[a]BR = Brown Rice; MR = milled rice; BK = broken rice; MF = mill feed; B = bran; H = hulls; LG = long grain type; MG = medium grain type; SG = short grain type; OB = organic brown rice; PR = parboiled rice; RM = rice mixes; RF = rice flour; SR = scented rice; QW = quick cook white rice; QB = quick cook brown rice; SB = stabilized bran; RC = rice cakes.
Source: Adapted from Ref. 48.

REFERENCES

1. Chang, T. T., The origin, evolution, cultivation, dissemination, and diversification of Asian and African rices. *Euphytica*, 25:425–441 (1976).
2. Greenland, D. J., Exploited plants. *Rice Biologist*, 31:219– 225 (1984).
3. Dethloff, H. C., The colonial rice trade. *Agri. Hist.*, *56*:231–243 (1982).
4. Adair, C. R., Introduction. In *Rice in the United States: Varieties and Production*. ARS, USDA, Agriculture Handbook 289. Washington, D.C., 1973, pp. 1–5.
5. FAO, *FAO Production Yearbook 1985*. Food and Agriculture Organization, Rome, Italy, 1986, p. 112.
6. Lu, J. J., and Chang, T. T., Rice in its temporal and spatial perspectives. In *Rice: Production and Utilization*, B. S. Luh (ed.). AVI Publishing Co., Westport, CT, 1980, pp. 1–74.
7. United Nations, *Statistical yearbook 1983/1984*, 24th issue. United Nations, New York, 1986.
8. FAO, *FAO Production Yearbook 1984*. Food and Agriculture Organization, Rome, Italy, 1985, p. 118.
9. Smith, R. K., The market structure of the U.S. rice industry. MS Thesis, University of Arkansas, Fayetteville, AR, 1989.
10. Matlick, D. H. 1988 Rice variety acreage survey. *Rice J.*, *91* (6):17–23 (1988).
11. Huey, B. A., Moldenhauer, K. K., and Helms, R. S., Rice varieties in Arkansas. University of Arkansas Extension Service, Leaflet No. 518, Little Rock, AR, 1988.
12. Keisling, T. C., Wells, B. R, and Davis, G. L., Rice management decision aids based upon thermal time base 50 degrees Fahrenheit. University of Arkansas Cooperative Extension Service, Computer Bulletin No. 1, Little Rock, AR, 1984.
13. Juliano, B. O., and Bechtel, D. B., The rice grain and its gross composition. In *Rice: Chemistry and Technology*, B. O. Juliano (ed.). Amer. Assoc. Cereal Chem., Inc., St. Paul, MN, 1985, pp. 17– 57.
14. United States Department of Agriculture, *Rice Inspection Handbook*, FGIS, U.S. Department of Agriculture, Washington, D.C., 1982.
15. Webb, B. D., Bollich, C. N., Johnston, T. H., and McIlrath, W. O., Components of rice quality: their identification, methodology, and stage of application in United States breeding Programs. In *Chemical Aspects of Rice Grain Quality*. IRRI, Los Banos, Philippines, 1979, pp. 191–205.
16. Juliano, B. O. Amylose analysis in rice—a review. In *Chemical Aspects of Rice Grain Quality*. IRRI, Los Banos, Philippines, 1979, pp. 251–260.
17. Blakney, A. B., Rice grain quality evaluation in Australia. In *Chemical Aspects of Rice Grain Quality*. IRRI, Los Banos, Philippines, 1979, pp. 115–121.
18. Choudhury, N. H., Studies on quality of rice in Bangladesh. In *Chemical Aspects of Rice Grain Quality*. IRRI, Los Banos, Philippines, 1979, pp. 123–127.
19. Feillet, P., and Marie, R., Rice breeding for grain quality in France. In *Chemical Aspects of Rice Grain Quality*. IRRI, Los Banos, Philippines, 1979, pp. 129–133.
20. Bhattacharya, K. R., Status of rice breeding for quality in India. In *Chemical Aspects of Rice Grain Quality*. IRRI, Los Banos, Philippines, 1979, pp. 135–148.
21. Suzuki, H., Ikehashi, H., and Kushibuchi, K., Rice grain quality evaluation in Japan. In *Chemical Aspects of Rice Grain Quality*. IRRI, Los Banos, Philippines, 1979, pp. 149–159.
22. Merca, F. E., Masajo, T. M, and Bustrillos, A. D., Rice grain quality evaluation in the Philippines. In *Chemical Aspects of Rice Grain Quality*. IRRI, Los Banos, Philippines, 1979, pp. 161– 165.
23. Barber, S., and Tortosa, E., Rice grain quality evaluation in Spain. In *Chemical Aspects of Rice Grain Quality*. IRRI, Los Banos, Philippines, 1979, pp. 167–173.
24. Beckenridge, C., Rice grain quality in Sri Lanka. In *Chemical Aspects of Rice Grain Quality*. IRRI, Los Banos, Philippines, 1979, pp. 175–181.
25. Kongseree, N., Physicochemical properties of Thai rice varieties and methodology used in quality improvement. In *Chemical Aspects of Rice Grain Quality*. IRRI, Los Banos, Philippines, 1979, pp. 183–189.

26. Nanda, J. S., and Coffman, W. R., IRRI's efforts to improve the protein content in rice. In *Chemical Aspects of Rice Grain Quality*. Los Banos, Philippines, 1979, pp. 33–47.
27. Holder, S. H., and Grant, W. R., U.S. rice industry. AER No. 433, ESCS, U.S. Department of Agriculture, Washington, D.C., 1979.
28. Juliano, B. O., and Perez, C. M., Results of a collaborative test on the measurement of grain elongation of milled rice during cooking. *J. Cereal Sci.*, *2*:281–292 (1984).
29. Sayre, R. N., Saunders, R. M., Enochian, R. V., Schultz, W. G., and Beagle, E. C., Review of rice bran stabilization systems with emphasis on extrusion cooking. *Cereal Foods World*, *27*:317–322 (1982).
30. Sreenarayanan, V. V., and Chattapadhyay, P. K., Rice bran stabilization by dielectric heating. *J. Food Process. Pres.*, *10*:89– 97 (1986).
31. Prabhakar, J. V., and Venkatesh, K. V. L., A simple chemical method for stabilization of rice bran. *JAOCS*, *63*:644–646 (1986).
32. Azeemoddin, G., Mallikharjuna Rao, D. C., Krishna Reddy, N., Rama Rao, G., and Thirumala Rao, S. D., Stabilization of rice bran with sodium metabisulfite. *JAOCS*, *56*:589 (1978).
33. James, C., Sloan, S. and Gadbury, D., Utilization of rice bran in baked goods. *Ark. Farm Res.*, *32*(2):11 (1983).
34. James, C., and Sloan, S., Functional properties of edible rice bran in model systems. *J. Fd. Sci.*, *49*:310–311 (1984).
35. Skurray, G. R., Wooldridge, D. A., and Nguyen. M., Rice bran as a source of dietary fiber in bread. *J. Food Technol.*, *21*:727–730 (1986).
36. Noland, P. R., and Ford, B. F., For wintering steers—Rice hulls and rice mill feed. *Ark. Farm Res.*, *3*(3):8 (1954).
37. Ray, M. L., and Child, R. D., Rice hulls in wintering rations. *Ark. Farm Res.*, 12(5):2 (1963).
38. Serra, A., Johnson, Z. B., Smith, J., and Noland, P. R., Processing rice bran for pigs. *Ark. Farm Res.*, *32*(4):10 (1984).
39. Tangendjaja, B., Johnson, Z. B., and Noland, P. R., Effect of cooking and addition of enzymes on feeding value of rice bran for swine. *Nutr. Rep. Int.*, *37*:449–458 (1988).
40. Tangendjaja, B., Johnson, Z. B., and Noland, P. R., Effect of cooking, ensiling, water treatments, extrusion, and pelleting on rice bran as feed ingredient for pigs. *Nutr. Rep. Int.*, *37*:939–949 (1986).
41. Jodan, N. E., and Sonnier, E. A., Registration of Della rice. *Crop Sci.*, *13*:773 (1973).
42. Sharp, R. N., Quality evaluation of milled aromatic rice from India, Thailand and the United States. *J. Food Sci.*, *51*:634– 636 (1986).
43. Buttery, R. G., Ling, L. C., and Juliano, B. O., 2-Acetyl-1-pyrroline: an important aroma component of cooked rice. *Chem. Ind. London.* 958–959 (1982).
44. McKenzie, K. S., Jodon, N. E., Brandon, D. M, Rush, M. C., Robinson, J. E., and Miller, M. F., Toro 2: a new special purpose rice variety. *La. Agr.*, *28*(1):16 (1984).
45. Bollich, C. N., Webb, B. D., Marchetti, M. A., and Scott, J. E., Newrex—a new rice variety with superior cooking and processing qualities. Texas Agricultural Experiment Station, MP–1446, College Station, TX, 1980.
46. Wailes, E. J., and Holder, S. H., Cost models and an analysis of modern rice mills. University of Arkansas Agricultural Experiment Station, Bulletin 907, Fayetteville, AR, 1987.
47. FDA, Enriched rice. In *Code of Federal Regulations*, Title 21, Part 137, Sec. 137:350. U.S. Food and Drug Administration, Washington, D.C., 1987.
48 Directory of the U.S. rice mills. *Rice J.* *91*(6):25–30 (1978).

8

Rye

Klaus J. Lorenz
Colorado State University, Fort Collins, Colorado

I. HISTORY

Rye (*Secale cereale*) seems to be of a more recent origin than most other cereal grains. It was not found in any of the Egyptian ruins or Swiss lake dwellings. The Greek and Romans knew rye, however (35). The grain is mentioned in early northern European writings, suggesting that it was cultivated in this area. Rye samples found in the Neolithic sites of Austria and Poland are considered to be of "wild" origin (6).

Cultivated rye (*Secale cereale*) was probably domesticated from the wild rye *Secale montanum* of the Mediterranean region, although some agronomists and plant breeders regard *Secale anatolicum* of southwestern Asia as the more likely ancestor. A possible migration route of *S. montanum* during the first millennium B.C. would be from Turkey across the Balkan Peninsula into Europe and for *S. anatolicum* northward into Russia and then westward into Poland and Germany. Rye gradually spread throughout Europe and was brought to North and South America by European settlers in the sixteenth and seventeenth centuries. During the nineteenth century it was introduced into South Africa and Australia (6). Today, it has the widest distribution of all cereal crops.

II. CLASSIFICATION OF RYE AND GENETICS

Rye belongs to the grass family of the genus Secale. Today only *Secale cereale* L. is extensively cultivated. The cereal is a diploid ($2n = 14$) with seven pairs of somatic chromosomes. A tetraploid type with 14 pairs of chromosomes has been developed. It is grown only in Europe and to a very limited extent. Evans and Scoles (12) provide a detailed discussion of the cytogenetics and breeding of rye. Because rye is a cross-pollinated species and inbred lines usually lack vigor, genetic analyses have been restricted. However, the recent production of the rye chromosome additions to or substitutions into wheat has permitted specific characteristics to be attributed to specific rye chromosomes.

The inheritance of perennial versus annual growth habit has been studied using interspecific hybrids due to the lack of perennial forms of *S. cereale*. In such crosses, perennial habit is at least partially dominant. In both diploid and tetraploid cultivars, spring habit was found to be dominant and controlled by three dominant genes in the diploids; two dominant genes were indicated in the tetraploids. Rachis fragility is a characteristic of wild rye which is detrimental to a domesticated crop and has been eliminated from *S. cereale*. It is a dominant character controlled by a single gene. Earlier work indicated that brittle rachis and perennial habit were closely linked.

Ear branching is controlled by a single recessive gene. It is very sensitive to environmental conditions with maximum expression requiring ideal conditions.

Anthocyanin pigments may occur in the aleurone layer of the grain, the coleoptile, first leaf, stem base, and nodes, the upper internode, and in the anthers. Grain color is controlled by two complementary genes, A and B, and anthocyanin pigmentation in other organs is controlled by the action of gene A and an additional gene R. Numerous chlorophyll deficiencies following inbreeding have been reported. They are caused by recessive factors (12).

As in other cereal crops, dwarfing genes have been sought as a tool in producing improved varieties. Inbreeding of rye usually results in reduced height as a result of inbreeding depression. The addition to or substitution of rye chromosomes into wheat has permitted analyses of the effects of different rye chromosomes and the location of genes for various characters on particular chromosomes. Chromosome 1R was found to be involved in the genetic control of endosperm protein. Chromosomes 3R, 4R, 6R, and 7R increased total protein 3–4.5%. Chromosome 5R increased lysine by 8.7% and cystine by 10.7%. Chromosome 6R increased proline by 9.1% and decreased aspartic acid content by 8.6%. The other chromosomes had some influence on other amino acids. The gene for hairy neck of rye is located on chromosome 5R. All chromosomes with the possible exception of 2R are involved in shrivelling (12).

III. RYE BREEDING

Because of the relatively small acreage, little effort has gone into rye breeding. Improvements have been made in seed size, yield, winter hardiness, and plant height by selection within heterogeneous populations.

Disease is not a major problem with rye, with the possible exception of ergot caused by the organism *Claviceps purpurea*. No source of resistance has yet been identified within the genus, so breeding for ergot resistance is not possible.

The breeding procedures for rye are as follows: intraspecific breeding, interspecific, intergeneric, and polyploidy and mutation. Intraspecific breeding refers to hybridization of lines or cultivars within the species. This is the most commonly used breeding procedure for rye. A germplasm source population has been established by Qualset and Hoskinson (32) utilizing 43 different North American sources of the Balboa cultivar and 10 other North American cultivars. The 10 cultivars were used as a pollen source only, and the open-pollinated seed of the Balboa strains was composited. This source population hopefully will permit selection for such characteristics as winter hardiness, disease resistance, and agronomic suitability.

Several investigators have crossed *S. cereale* with other rye species, particularly *S. montanum,* in an attempt to improve winter hardiness, drought resistance, and disease resistance. Intergeneric hybrids have been of no significance to rye breeding but may be of value in producing cytoplasmic male sterile lines for use in hybrid rye breeding. Interspecific breeding has been used to produce rye with high protein content. Several other tetraploid cultivars have been produced. Some have been released in Europe for commercial production, but they have not been of economic significance (12).

IV. MORPHOLOGY AND KERNEL CHARACTERISTICS

The rye plant is an annual, although it can maintain itself by sprouting anew from the stubble (35). Its abundant adventitious root system is one of the most profuse among cereal grains. The stems are slender and tough. Both stems and leaves have a waxy epidermal coating. The inflorescence is an elongated spike with approximately 30 three-flowered spikelets. The uppermost of the three florets is usually abortive. Lemmas are awned. The grain is "bearded." Generally the flowers are self-sterile. Simmonds and Campbell (38) provide more detail on the morphology of rye.

The mature rye grain is more slender than that of wheat. It is grayish-yellow in color. Kernels of rye vary in length from 4.5 to 10 mm and in width from 1.5 to 3.5 mm. Average grain weight is 21 g per 1000 kernels. A diagrammatic view of the grain in longitudinal and in transverse midsection is shown in Figure 1. The proportions of parts in the rye kernel are given in Table 1.

V. GROWING CONDITIONS

Rye is typically a plant of cool, nonhumid climates. It can stand cold winter temperatures better then most other cereals. No other winter cereal can be grown as far north. It is a reli-

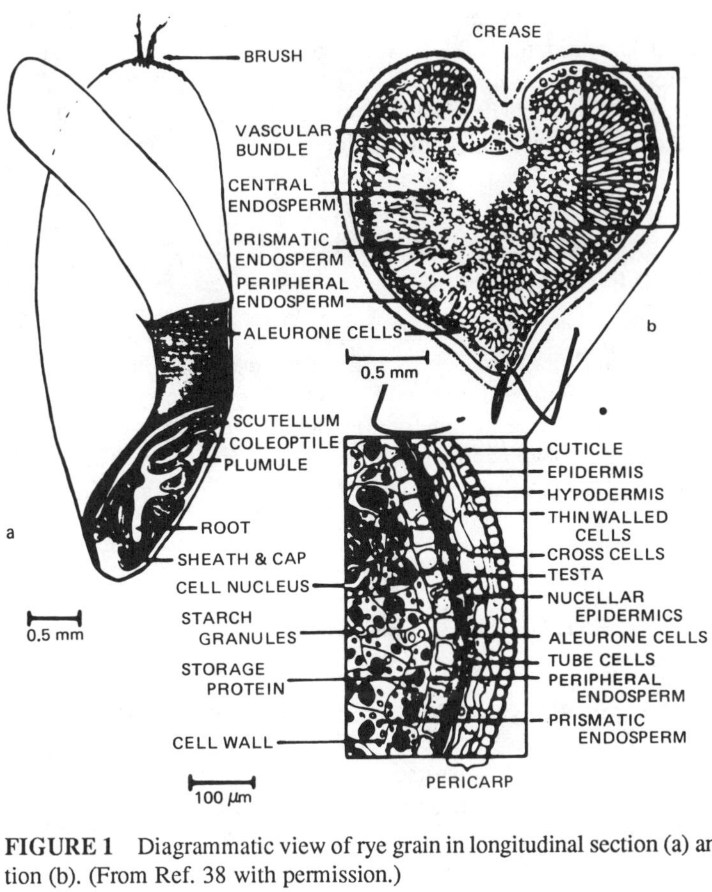

FIGURE 1 Diagrammatic view of rye grain in longitudinal section (a) and in transverse midsection (b). (From Ref. 38 with permission.)

able crop on almost any soil. Rye has been termed "poverty grain" because of its ability to grow on soils too poor for other crops (24).

Rye is primarily a winter-type grain, it is sown in the fall, becomes dormant in the winter, and is one of the first crops harvested during the next harvest season. Occasionally, rye is sown in the spring in areas of the world where winters are too severe for winter rye production. Spring varieties are considered inferior to winter varieties (6).

VI. PRODUCTION OF RYE

Areas of rye harvested in different parts of the world are given in Tables 2 and 3. Ninety percent of all rye is harvested in Europe. Russia leads the world in area under cultivation, followed by adjacent northern countries. Amounts of grain produced in various areas of the world are given in Tables 4 and 5. Europe is the largest producer of rye in the world. Russia is the largest producer of rye in Europe and in the world, followed by Poland, Germany, and Czechoslovakia. There are no obvious rye production trends in Europe. Production in most European countries fluctuates as the result of climate and demand. Argentina is the largest producer in South America. In Asia, China and Turkey are major producers. Rye grain production in the United States and Canada show no consistent trends.

Rye yields (kg/ha) (Tables 6 and 7) vary considerably from one country to another. They are highest in Europe, followed by North America and Asia. Rye yields are lowest in Africa. Yields between European countries differ significantly depending on the importance of rye in a country. The highest yields in Europe are reported from Denmark, Switzerland, West Germany, and the United Kingdom due to very intensive agricultural practices. Yields of the largest producer of rye in the world, the U.S.S.R., are surprisingly low, considering that rye is a very important cereal in Russia.

The area of cultivated land of the world devoted to growing of rye and grain production decreased substantially between 1960 and 1975, but has shown an increase since that time, as shown in Figure 2. There has been a steady increase in worldwide yields of rye over the same period of years.

VII. RYE EXPORTS AND IMPORTS

Rye exporters are shown in Table 8. Canada is the largest exporter of the grain in the world. The largest exporting countries in Europe are Denmark, Sweden, and West Germany. Rye imports in metric tons are given in Table 9. Japan and Korea the major importers of rye in the world today. Many countries of Europe are importers as well as exporters of rye. Poland and West Germany, which are European exporters of rye, are also importers. The major rye-importing countries of Europe in 1984 also include Norway, the Netherlands, England, and Switzerland.

Unit weights for rye in different countries are given in Table 10.

VIII. RYE STORAGE AND RYE GRAIN RESERVES AND DISAPPEARANCE

Rye grain storage, as the storage of other cereal grains, requires that the grain be dried to a low moisture level (approx. 12%) and stored under cool conditions to prevent spoilage.

Rye flours deteriorate faster than wheat flours under normal storage conditions. Whole rye flour cannot be stored safely as long as patent or straight-grade flours. High temperatures and/or high humidities accelerate deterioration during storage of rye (11).

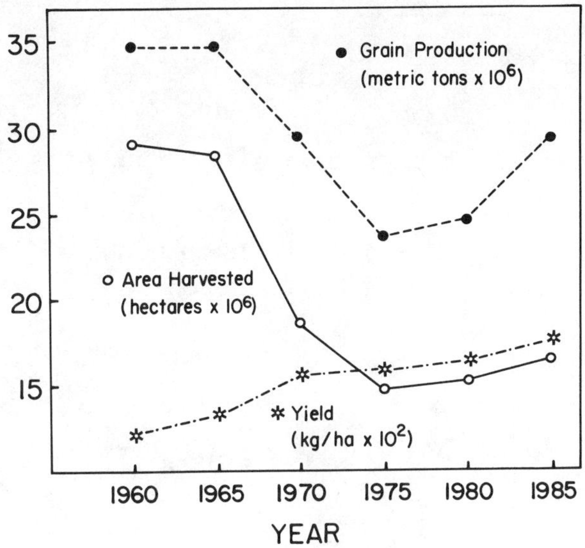

FIGURE 2 Area of rye harvested, grain production, and yields of rye in the world between 1960 and 1985. (*Source*: FAO Production Yearbooks 1978 and 1985.)

Yearly stocks of rye in the United States on-and off-the-farm have decreased significantly since 1971, as shown in Table 11. The disappearance of rye in the United States and in other countries is shown in Tables 12 and 13. About two thirds of the rye in the United States goes into stock feed (36).

IX. RYE STANDARDS AND GRADES

Grading systems or special classifications for rye have been established in only a few countries. Rye grades in North America are given in Table 14. The grades are based on external characteristics of the grain, but α-amylase activity is sometimes included (11).

Standard grades of rye for the United States are defined in the Official Grain Standards established by the U.S. Department of Agriculture (42). Only grades No. 1 and No. 2 are milled into flour for baking. Lower grades are used for animal feed. Standard grades of rye in Canada include four grades and a special grade for ergot rye. generally, only grades No. 1 and No. 2 are milled into flour for baking.

Rye classifications and grading factors for rye in the U.S.S.R. and the European Economic Community (EEC) are shown in Table 15. Only Sweden, East Germany, West Germany, and the U.S.S.R. have started to include some baking quality tests into their rye-grading systems (11,33).

X. RYE MILLING

In North America, rye is cleaned prior to milling using magnets, milling separators, disk machines, entoleters, scourers, aspirators, and stoners. A diagram of a rye-cleaning operation is shown in Figure 3. The cleaned rye is steeped for 6–15 hrs. to 14.5%–15.5% moisture depending on grain softness and vitreousness and milled into a flour without strict quality requirements (33). A typical mill will have 5 breaks, 1 bran buster, 1 sizing roll, 7 reduction rolls, and 1 tailing roll, as shown in the flow diagram in Figure 4. All rolls are corrugated.

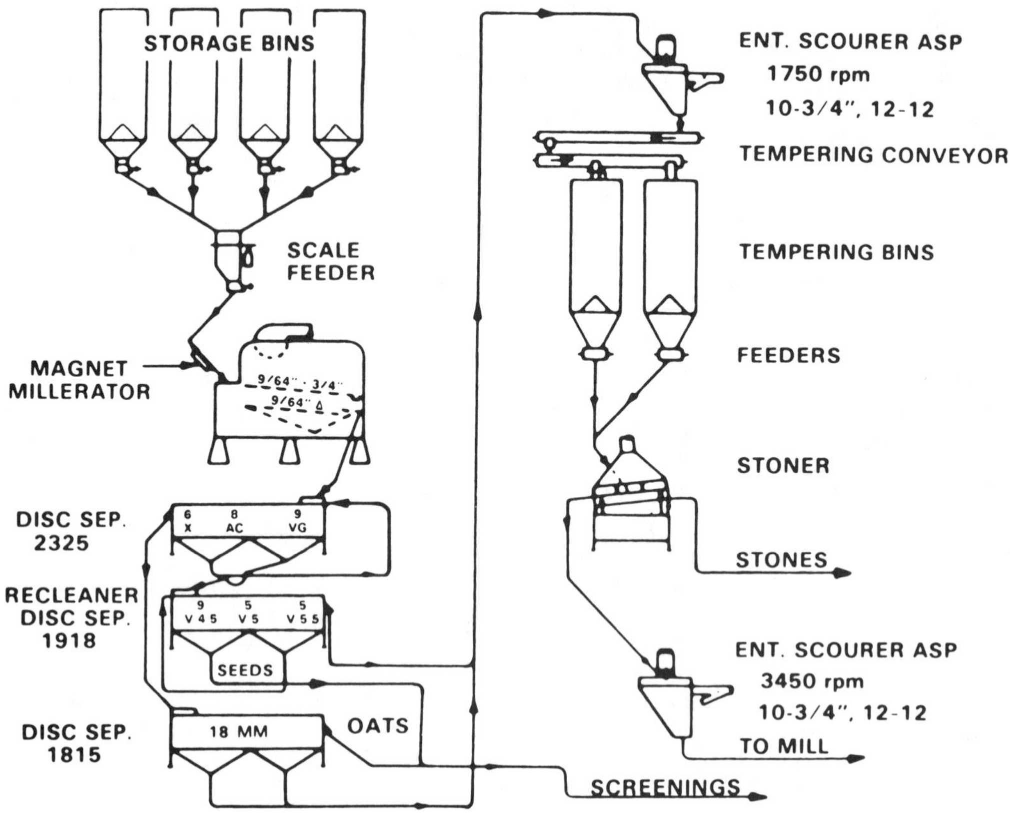

FIGURE 3 Rye-cleaning process diagram for a typical North American rye mill (1200 cwt/24hr). (From Ref. 33 with permission.)

Rye cleaning in western Europe is similar to that described briefly for a North American rye-milling operation. Typical western European rye milling starts with smooth crushing rolls, followed by 5–7 breaks, 2 or more bran finishers, 4–6 reduction rolls, and 1 or 2 tailing rolls. All rolls, except for crushing rolls, are corrugated (33). There is a gradual introduction of bran and shorts finishers in rye milling in western Europe. European rye milling has been improved by the successful introduction of mills built to mill both wheat and rye.

Rye-milling technology in eastern Europe and in the U.S.S.R. is highly developed. Rye cleaning is similar to that used in North America. Standards for the construction and operation of rye mills and product yields for those mills have been described by Rozsa (33).

XI. RYE FLOURS

The basic grades of rye flours produced in various countries are shown in Table 16. There are no standards of identity for rye flours in the United States or Canada. An average flour extraction is about 83% in the United States and 67% in Canada. In the United States, only white rye flour receives a slight flour treatment—addition of 0.19–0.31 g of chorine per kg of flour—for color improvement (33). Basic specifications include ash contents and in some countries protein content and flour color.

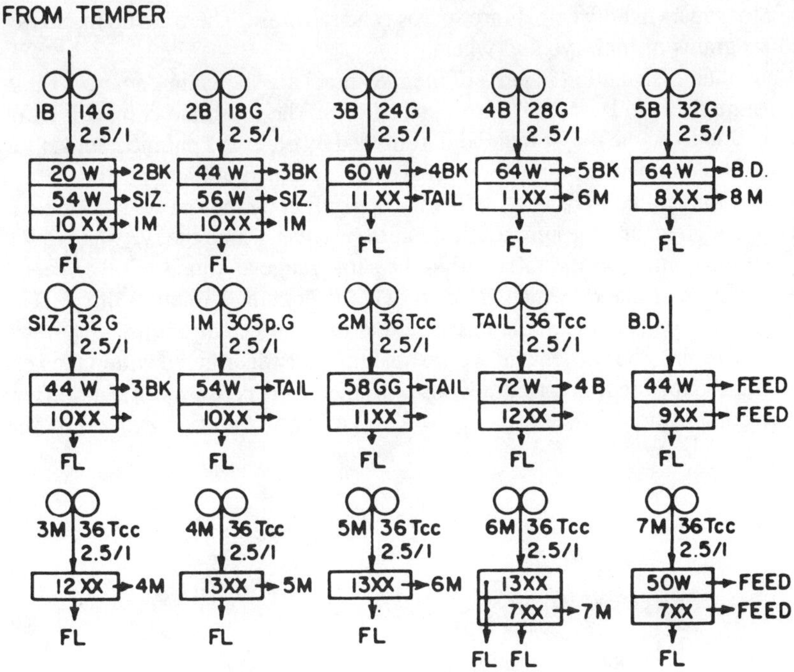

FIGURE 4 Rye-milling process diagram of a typical North American rye mill (1200 cwt/24 hr). (From Ref. 33 with permission.)

XII. NUTRIENT COMPOSITION OF RYE

The composition of rye grain, as reported from different countries, is given in Table 17. Grain protein, pentosan content, and 1000 kernel weight can vary considerably depending on the cultivar and the agronomic and climatic conditions under which it was grown.

The various flour-milling streams as produced by a commercial rye mill differ significantly in composition, as shown in Table 18. The miller will blend various streams into flour with compositions as required by the baker.

Component fractions of rye proteins and their amino acid compositions are given in Tables 19 and 20.

The amino acid compositions of rye cultivars grown in different countries are given in Table 21. Since protein is not uniformly distributed throughout the rye kernel, the milling fractions of rye will vary in amino acid composition, as shown in Table 22. Lysine is the first limiting amino acid in rye. There are, however, several varieties and/or species with relatively high lysine and protein contents, as seen in Table 23.

Comparing the average essential amino acid composition of rye with that of wheat and the 1973 WHO reference standard (Table 24) indicates that rye is a better source of lysine that wheat, but does not meet the reference standard. Among the essential amino acids in rye, only methionine and cystine, valine, and phenylalanine and tyrosine meet the WHO standard. The PER of whole rye is 1.6.

Rye starch characteristics compared with those of hard red spring wheat are given in Table 25. The differences in characteristics between the starches from these two cereals are minor. A comparison of the saccharide contents of rye and hard red spring wheat is shown in

Table 26. The major sugars in both cereals are sucrose and raffinose. There are no monosaccharides in mature grains of both rye and wheat.

Total lipids and their distribution in parts of the rye kernel are shown in Table 27. Lipid composition in comparison with wheat is given in Table 28. The fatty acid composition of rye shows linoleic acid to be the major fatty acid, followed by oleic and palmitic acids (Table 29). The vitamin and mineral compositions of rye are presented in Figure 5 and Tables 30–32. Vitamin and mineral compositions will vary depending on the rye variety, the degree of maturity of the grain, and the agronomic conditions under which the rye was grown. Generally, the vitamin content of rye falls within the same range as found in other cereal grains with the exception of niacin, which is found only in very small amounts of rye. The mineral composition of rye is comparable to that found in other cereal grains.

As in other cereal grains, the nutrients in rye are not uniformly distributed within the kernel. Milling the grain to different flour extractions will, therefore, cause significant differences in nutrient composition, as seen in Table 33. As flour extraction decreases, the

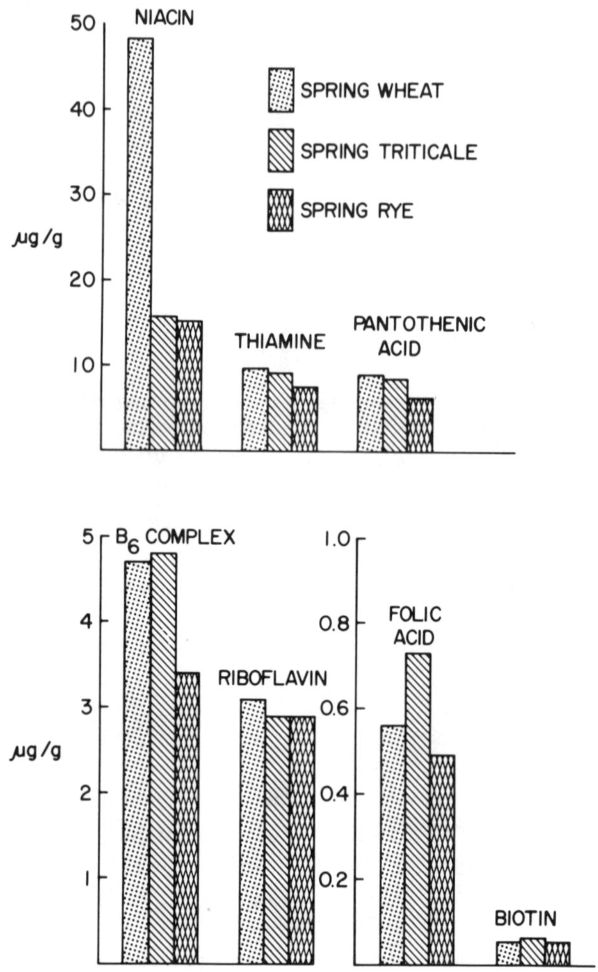

FIGURE 5 Vitamin contents of whole grains of spring cultivars of wheat, rye, and triticale. (From Ref. 27 with permission.)

amounts of nutrients in the flour decrease with the exception of total carbohydrates, which increase.

XIII. ANTINUTRITIONAL FACTORS IN RYE

The poisonous substances and the nutritional inhibitors in rye, which may interfere with the digestibility and availability of certain nutrients to humans and/or animals, include ergot, resorcinols, phytates, and enzyme inhibitors.

Ergot is due to an infection of cereal grains by the fungus *Claviceps purpurea,* which produces alkaloid-containing sclerotia (Fig. 6). Rye is notorious for harboring ergot. Epidemics supposedly caused by the consumption of ergot containing breads ravaged Europe from the tenth century to modern times and resulted in many deaths. Lorenz (23) discussed in detail the problem of ergot on cereal grains. Tolerance levels for ergot have been established in most countries.

Resorcinols seem to be responsible for reduced food intake and rate of weight increase in cattle, sheep, pigs, poultry, and horses. Rye contains higher amounts of resorcinols than other cereal grains. Milling of rye will channel most of these resorcinols into the feed frac-

FIGURE 6 Ergot sclerotium on rye.

tions as shown in Table 34. Approximately 20% of the alkylresorcinols are destroyed during the baking process (45).

Phytic acid represents 35–97% total phosphorous and is regarded as a poor source of phosphorous. Phytic acid also forms insoluble compounds with mineral elements, making these elements unavailable (17). The phosphorous and phytic phosphorous content of rye and rye flours are shown in Table 33. During fermentation of the bread-baking process, some of the phytic acid content in the rye flour is destroyed due to the enzyme phytase. Trypsin and chymotrypsin inhibitors have been extracted from rye. Heating of the rye flour will destroy part of the activity of the inhibitors.

XIV. FOOD USES OF RYES

Many types of rye breads and rolls are produced in different areas of the world. West Germany alone produces more than 200 different varieties of rye breads (37). Formulations and processing conditions and steps for many of these varieties have been compiled by Drews and Seibel (11), Seibel et al. (37), Pomeranz and Shellenberger (30), Lorenz (25).

In the United States rye breads are produced in a large number of varieties to meet the demands of local consumer preferences. Rye breads vary in crumb color from practically white to a very dark color, in shape from round to elongated loaves, and in taste from a mildly sour flavor to strong, distinctive acid taste. Rye breads may contain only the basic ingredients— wheat and rye flours, water, yeast, and salt— or may incorporate a number of additional ingredients, such as molasses, potato flour, sugar, shortening, buttermilk, dry mild solids, all designed to enhance either the flavor, color, or keeping quality of the product. It is possible for the baker to modify his rye bread formula to produce the type of product that will find greatest appeal among his customers. Uses of the basic rye flour types available in the United States are given in Table 35.

The popularity of rye breads and rye rolls is due largely to their distinctive flavor, taste, and eating quality. They satisfy the consumer's demand for variety in bakery products.

Rye flours are used occasionally in snack-type crackers. A very light rye has been recommended for use in soda crackers and some cookies as partial replacement for wheat flour. Rye flours are also used in some pancake and waffle mixes to produce a special flavor.

Rye flours are used as fillers for sauces and soups. Rye is also flaked in the United States to make a hot breakfast cereal. Rye flour can be fractionated by air-classification to produce high and low protein fractions (19). The low protein fractions might be useful in chocolate and darker prepared flour mixes. Meat packers and processors use some rye flours as fillers and binders in sausages (33).

Rye grain may be fermented to produce alcoholic beverages or industrial alcohol. U.S. Whiskey is generally made from maize or rye. Straight rye whiskey is made from a fermented mash containing a minimum of 51% rye. Irish whiskey is made from malted barley alone or with admixtures of unmalted barley, wheat, rye, or oats. Canadian whiskey is produced from maize, wheat, rye, or barley (19).

Annual consumption of rye (kg/person) has decreased steadily. This trend is illustrated for a few European countries in Figure 7.

XV. INDUSTRIAL USES OF RYE

Rye is used in the manufacture of paper. The soluble and insoluble gums of rye are good substitutes for other gums as wet-end additives (33). Rye starch is the main ingredient of adhesives because of its high waterbinding capacity (11,19). Small amounts of rye are used in the glue, match, and plastic industries or as foundry core binders and ore pellet binders (33).

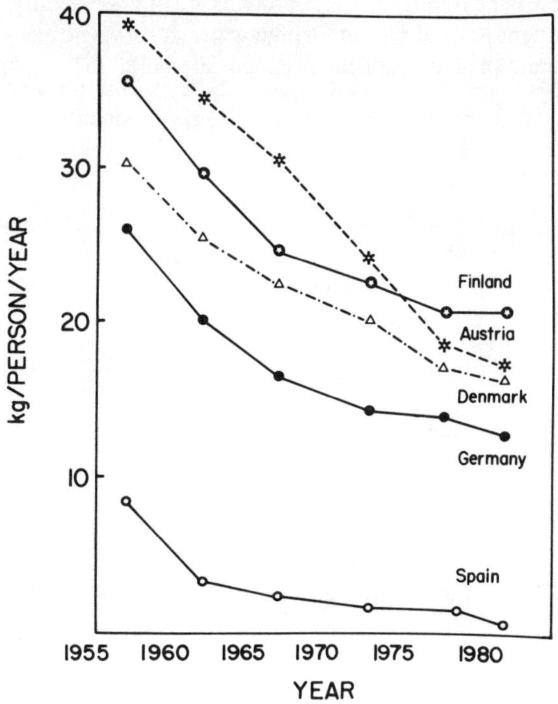

FIGURE 7 Annual consumption of rye (kg/person) in European countries. (*Source*: Organization of Economic Cooperation and Development. Food Consumption Statistics 1955–1973. Paris, France.)

XVI. RYE AS ANIMAL FEED

In some areas of the world, the major portion of the rye crop goes into stock feed. The rye plant is also often used for hay and pasturage, as cover crop and green manure, being plowed under as fertilizer for a crop of higher economic value. The long, fine, highly resistant straw is being used for animal bedding (35).

Rye straw is not used extensively in livestock feed because it is tough and fibrous. When the production of rye exceeds the demand and the price of rye falls below the price of barley, rye becomes attractive as a feed grain although it is of relatively low quality as a feed grain. When fed in large amounts to cattle, sheep, horses, pigs, chicks, and poultry, rye leads to slower growth in comparison with feeding of other cereal grains. The diminished growth rate is caused primarily by a decrease in feed intake. These palatability problems and various growth-inhibiting factors of rye have restricted its sue. The compounds causing the feeding problems have not yet been positively identified, but ergot on rye, trypsin inhibitors, and alkylresorcinols in rye have been mentioned as possibilities (22,45).

Rye, therefore, is used as part of a feed mixture with other grains. The technology of feed manufacture has been described by Smith (39).

Rye flour has been suggested for use in pet foods (33). Rye proteins are of good quality, the starch is readily digestible, and the pentosans of rye have a high water-absorbing capacity. The color of the flour is not a factor, since most of the pet foods are dark (33).

Table 1 Proportions of Morphological Parts of Rye Grain

Morphological part	% of Grain
Pericarp	11.14 – 13.00
Aleurone	10.86 – 12.20
Embryo	1.45 – 1.80
Scutellum	1.73 – 2.11
Endosperm	71.16 – 74.56

Source: Ref. 21.

TABLE 2 Area of Rye Harvested—Regional Totals (1000 ha)

	1969–71	1979–81	1983	1984	1985
Africa	24	35	40	40	40
North America	959	658	791	767	662
South America	479	227	166	168	142
Asia	697	1,210	1,014	992	994
Europe (incl. U.S.S.R.)	16,648	14,959	16,111	15,389	14,862
Oceania	39	28	52	33	26
Developed countries	17,666	13,678	16,992	16,227	15,588
Developing countries	1,180	1,440	1,183	1,162	1,138
World	18,846	15,117	18,174	17,389	16,726

Source: Ref. 14.

TABLE 3 Area of Rye Harvested in Different Countries (1000 ha)

	1948–52	1969–71	1979–81	1983	1984	1985
North America						
United States	686	603	295	363	397	290
Canada	571	356	364	428	370	372
South America						
Argentina	717	440	199	153	157	120
Brazil	23	23	16	4	4	13
Chile	7	9	8	5	3	5
Africa						
South Africa	29	20	33	38	38	38
Morocco	6	4	2	2	2	2
Asia						
China	n.a.	n.a.	733	700	700	700
Korea, DPR	n.a.	27	34	37	40	42
Turkey	493	663	441	275	250	250
Oceania						
Australia	28	39	27	52	33	26
Europe						
Albania	12	9	10	11	11	11
Austria	230	143	105	93	94	88
Belgium–Luxembourg	91	23	12	7	7	6
Bulgaria	226	22	21	25	26	25
Czechoslovakia	638	243	171	203	192	179
Denmark	155	42	59	77	122	126
Finland	133	67	44	47	44	31
France	496	139	121	101	102	90
East Germany	1293	679	671	713	718	687
West Germany	1388	868	532	445	439	426
Greece	57	7	4	7	9	9
Hungary	592	155	72	72	75	85
Italy	97	34	15	11	9	9
Netherlands	176	60	10	7	6	5
Poland	5063	3766	2970	3448	3545	3083
Portugal	270	227	205	133	183	163
Romania	184	45	33	35	32	35
Spain	622	320	219	217	233	222
Sweden	128	78	58	62	62	45
Switzerland	14	12	8	4	5	5
United Kingdom	25	5	6	7	6	6
U.S.S.R.	23592	9588	7557	10334	9420	9480
Yugoslavia	269	115	56	51	47	45

n.a. = not avaliable.
Source: Ref. 14.

TABLE 4 Rye Grain Production—Regional Totals (1000 metric tons)

	1969–71	1979–81	1983	1984	1985
Africa	10	8	8	11	12
North America	1,458	1,112	1,516	1,488	1,122
South America	305	195	140	150	133
Asia	832	1,786	1,759	1,843	1,448
Europe (incl. U.S.S.R.)	26,924	21,498	21,527	27,843	26,841
Oceania	19	12	24	15	11
Developed countries	28,409	22,628	33,074	29,356	27,983
Developing countries	1,139	1,983	1,901	1,995	1,584
World	29,548	24,611	34,975	31,351	29,567

Source: Ref. 14.

TABLE 5 Rye Grain Production in Different Countries (1000 metric tons)

	1948–52	1969–71	1979–81	1983	1984	1985
North America						
United States	524	984	474	689	825	524
Canada	469	474	638	828	664	598
South America						
Argentina	526	271	169	130	140	105
Brazil	17	19	15	3	3	13
Chile	5	11	10	4	4	11
Africa						
South Africa	10	7	6	6	9	9
Morocco	4	3	2	2	2	3
Asia						
China	n.a.	n.a.	1,167	1,300	1,400	1,000
Korea, DPR	n.a.	40	57	75	80	85
Turkey	500	781	558	380	360	360
Oceania						
Australia	12	18	11	24	15	10
Europe						
Albania	9	8	9	10	10	10
Austria	343	417	327	348	381	339
Belgium–Luxembourg	232	82	43	27	37	25
Bulgaria	240	27	29	31	37	43
Czechoslovakia	1,110	586	534	751	710	620
Denmark	365	137	2521	315	608	560
Finland	201	130	88	116	92	72
France	573	297	368	293	349	298
East Germany	2,516	1,594	1,848	2,092	2,510	2,500
West Germany	3,066	2,862	1,980	1,599	1,931	1,821
Greece	47	9	7	13	16	15
Hungary	732	193	117	138	193	164
Italy	123	65	35	27	24	23
Netherlands	455	196	39	26	25	19
Poland	6,374	7,142	6,166	8,781	9,540	7,600
Portugal	162	164	128	92	115	99
Romania	177	52	38	45	50	50
Spain	482	292	239	253	325	295
Sweden	258	239	198	237	248	157
Switzerland	34	48	35	19	29	24
United Kingdom	52	14	24	27	28	28
U.S.S.R.	17,961	12,235	8,942	16,200	10,500	12,000
Yugoslavia	135	132	78	83	81	77

n.a. = not avaliable.
Source: Ref. 14.

TABLE 6 Rye Yields – Regional Totals (kg/ha)

	1969–71	1979–81	1983	1984	1985
Africa	418	232	204	279	299
North America	1,521	1,689	1,917	1,941	1,695
South America	637	861	844	891	933
Asia	1,194	1,475	1,734	1,858	1,457
Europe (incl. U.S.S.R.)	2,028	2,325	2,653	2,906	2,757
Oceania	489	428	469	460	407
Developed countries	2,081	1,654	1,946	1,890	1,795
Developing countries	954	1,378	1,607	1,717	1,392
World	1,568	1,628	1,924	1,803	1,768

Source: Ref. 14.

TABLE 7 Rye Yields in Different Countries (kg/ha)

	1969–71	1979–81	1983	1984	1985
North America					
United States	1633	1610	1900	2077	1806
Canada	1332	1754	1932	1795	1608
South America					
Argentina	616	847	850	893	875
Bolivia	435	488	500	500	500
Brazil	836	945	795	756	1061
Chile	1300	1171	912	1292	2282
Peru	872	779	802	813	813
Africa					
South Africa	338	182	158	237	237
Morocco	800	956	1048	1048	1429
Asia					
China	n.a.	1591	1857	2000	1429
Japan	1729	2265	600	600	600
Korea, DPR	1481	1667	2027	2000	2024
Korea, RER	1681	1776	1650	1696	1737
Turkey	1177	1267	1382	1440	1440
Oceania					
Australia	476	418	465	455	401
Europe					
Albania	875	884	885	909	909
Austria	2917	3106	3731	4071	3843
Belgium–Luxembourg	3551	3708	3857	5286	4167
Bulgaria	1227	1377	1232	1456	1720
Czechoslovakia	2415	3119	3701	3701	3457
Denmark	3289	3780	4122	4988	4444
Finland	1929	2022	2486	2093	2324
France	2132	3042	2901	3422	3311
East Germany	2346	2756	2934	3495	3639
West Germany	3298	3723	3595	4394	4277
Greece	1201	1745	1857	1778	1667
Hungary	1242	1618	1909	2568	1922
Ireland	3086	3721	3500	3636	3846
Italy	1889	2330	2371	2614	2518
Netherlands	3292	3952	3867	4386	3800
Norway	3312	3322	3103	2000	1887
Poland	1897	2076	2547	2691	2465
Portugal	725	626	698	631	607
Romania	1153	1150	1286	1563	1429
Spain	912	1092	1162	1395	1329
Sweden	3058	3393	3850	4016	3489
Switzerland	4013	4525	4717	5370	5289
United Kingdom	2929	3856	3857	4667	4828
U.S.S.R.	1276	1183	1568	1115	1266
Yugoslavia	1147	1399	1627	1707	1711

n.a. = not available.
Source: Ref. 14.

TABLE 8 Major Exporters of Rye (1000 metric tons) for 1975–1977 and 1982–1984

	1975	1976	1977	1982	1983	1984
North America						
United States	25.3	4.2	0.9	6.2	23.2	12.9
Canada	268.3	184.2	217.3	451.4	617.0	542.6
South America						
Argentina	3.6	3.2	1.4	24.0	0.3	1.5
Brazil	—	—	—	11.0	—	—
Asia						
Turkey	—	—	20.8	0.1	0.1	—
Europe						
Austria	—	19.6	—	32.2	54.1	30.0
Belgium–Luxembourg	3.1	9.1	0.8	2.1	1.6	1.8
Czechoslovakia	7.0	3.0	—	—	2.6	—
Denmark	20.8	72.7	62.6	61.0	44.7	169.3
France	29.7	62.0	36.7	22.7	9.3	10.5
Germany, Fed. Rep.	18.5	80.2	38.2	54.7	21.7	34.2
Netherlands	16.6	24.6	7.0	5.7	6.0	6.0
Poland	70.4	20.9	20.9	11.3	99.4	11.0
Sweden	116.0	125.0	123.7	23.6	30.2	41.3
Eastern Europe plus U.S.S.R.	77.9	24.0	31.8	16.3	114.9	28.5
World	581.4	609.6	541.9	711.3	927.9	882.3

Source: Ref. 15.

TABLE 9 Major Importers of Rye (1000 metric tons) for 1975–1977 and 1982–1984

	1975	1976	1977	1982	1983	1984
America						
United States	18.1	18.2	3.5	57.8	66.0	10.4
Brazil	—	0.3	—	—	0.4	—
Oceania						
Australia	2.2	4.0	2.6	0.4	13.2	3.2
Asia						
Singapore	5.9	—	—	—	5.4	—
Japan	53.7	39.4	141.1	19.7	133.3	342.0
Korea, Rep. of	—	—	—	—	75.3	263.9
Europe						
Belgium–Luxembourg	11.3	16.6	13.4	13.0	12.5	7.1
Bulgaria	49.2	2.5	—	0.8	1.9	0.7
Denmark	—	3.8	3.2	7.9	8.9	0.2
Finland	29.7	—	—	93.6	19.9	0.8
France	2.4	3.1	2.5	2.8	3.6	2.7
East Germany	36.7	—	—	—	198.0	50.0
West Germany	34.1	137.2	66.9	37.1	20.7	44.5
Hungary	16.5	3.3	—	6.5	13.5	0.1
Italy	18.1	0.2	2.5	4.2	2.0	7.1
Netherlands	45.3	55.7	46.7	51.2	40.3	44.2
Norway	40.5	81.3	42.3	30.8	38.3	33.0
Poland	69.4	225.3	170.0	62.5	—	—
Romania	85.1	—	—	26.0	32.0	—
Switzerland	20.0	23.2	4.9	32.8	26.5	42.1
United Kingdom	33.2	25.0	33.0	9.2	5.9	11.0
Europe plus U.S.S.R.	258.8	231.1	170.0	395.4	248.7	50.8
World	576.0	661.1	542.1	788.8	827.1	866.3

Source: Ref. 15.

TABLE 10 Unit Weights for Rye

Country	Unit	Net Weight (lb)	Net Weight (kg)
Europe			
Finland	hectoliter	158.73	72.0
Netherlands	hectoliter	154.32	70.0
Norway	hectoliter	154.32	70.0
Sweden	hectoliter	160.94	73.0
Spain	fanega	95.33	43.2
Turkey	Kile	62.00	28.0
EEC[a]	hectoliter	156.53	71.0
North America			
United States	bushel	56.00	25.4
Canada	bushel	56.00	25.4
Asia			
Japan	Koku	311.29	141.2
Korea, Republic of	Suk	273.37	124.0
Africa			
South Africa	bag	200.00	90.7
Zimbabwe	bag	200.00	90.7
Oceania			
Australia	bushel	60.00	27.2
New Zealand	bushel	56.00	25.4

[a]European Economic Community
Source: Ref. 30.

TABLE 11 Grain Stocks (1000 bushels) of Rye in United States (Jan. 1 of each year)

Year	On–Farm	Off–Farm	Total
1970	14,437	26,367	40,804
1971	21,500	33,115	54,615
1972	15,471	38,058	53,529
1973	7,343	13,640	20,983
1974	6,082	5,509	11,591
1975	5,299	3,777	9,076
1976	4,727	4,151	8,878
1977	5,879	2,910	8,789
1978	9,990	5,178	15,168
1979	10,885	6,842	17,727
1980	5,878	3,468	9,346
1981	5,710	2,172	7,882
1982	7,661	3,207	10,868
1983	7,015	3,204	10,219

Source: Ref. 44.

TABLE 12 Rye: Supply and Disappearance, United States, 1969–1983

		Disappearance							
		Domestic use							
Year (beginning June)	Supply total (1000 bu)	Food (1000 bu)	Rye flour per capita (lb)	Seed (1000 bu)	Industry (1000 bu)	Feed[a] (1000 bu)	Total (1000 bu)	Exports (1000 bu)	Total disappearance (1000 bu)
1969	47,970	5,392	1.2	6,054	4,310	9,529	25,285	805	26,090
1970	59,767	5,430	1.2	6,873	3,522	11,407	27,232	3,204	30,436
1971	78,795	5,195	1.2	5,262	3,005	16,276	29,738	2,177	31,915
1972	75,290	5,122	1.1	5,321	3,038	16,458	29,929	6,535	36,474
1973	63,494	6,250	1.3	4,980	2,547	8,029	21,806	27,513	49,319
1974	31,958	5,459	1.3	5,040	1,386	6,986	18,871	6,465	25,336
1975	23,524	4,172	1.0	4,790	2,060	6,981	18,003	1,117	19,120
1976	19,542	3,696	0.9	4,114	1,930	5,350	15,090	38	15,128
1977	21,084	3,648	0.8	4,584	1,875	6,984	17,091	24	17,115
1978	28,179	3,690	0.8	4,644	2,383	8,130	18,847	359	19,206
1979	31,369	3,523	0.7	4,034	2,116	7,082	16,755	2,422	19,177
1980	28,685	3,515	0.7	4,150	2,050	7,331	17,046	7,494	24,540
1981	23,399	3,458	0.7	4,160	2,242	8,915	18,775	1,529	20,304
1982	25,500	3,300	0.7	4,400	2,300	9,600	19,600	200	19,800
1983[b]	34,500	3,500	0.6	4,800	2,100	15,000	25,400	1,000	26,400

[a]Residual item; roughly approximates total use as feed.
[b]Preliminary.
Source: Ref. 43.

TABLE 13 Disappearance of Rye in Various Countries (1000 metric tons)

Country	Period	Animal feed	Industrial Uses: Seed	Industrial Uses: Nonfood Products	Food: Gross	Food: Extraction rate (%)	Food: Net
Austria	1955–1959	40	41	1	369	74.2	274.0
	1965–1969	73	27	2	298	75.2	223.4
	1979	85	18	2	183	76.9	140.7
	1980	91	15	2	177	78.0	138.1
	1981	97	15	2	170	77.8	132.2
	1982	95	14	2	169	78.0	131.8
Belgium–Luxembourg	1955–1959	287	9	—	10	70.0	7.0
	1965–1969	98	4	—	8	70.0	5.6
	1979	49	1	—	7	70.0	4.9
	1980	34	1	—	8	70.0	5.6
	1981	34	1	—	8	70.0	5.6
	1982	34	1	—	10	70.0	7.0
Canada	1955–1959	89	16	29	10	70.0	7.0
	1965–1969	123	24	63	11	67.7	7.4
	1979	184	24	—	15	67.7	10.2
	1980	115	34	57	16	67.7	10.8
	1981	139	34	60	14	67.7	9.5
	1982	176	32	62	13	67.7	8.8
Denmark	1955–1959	150	23	—	138	97.8	135.0
	1965–1969	60	7	—	111	98.0	108.8
	1979	69	10	8	88	85.0	74.8
	1980	78	9	6	96	85.0	81.6
	1981	49	10	6	100	85.0	85.0
	1982	75	14	7	98	85.0	83.3
Finland	1955–1959	26	20	2	158	98.0	155.0
	1965–1969	39	15	3	117	98.0	114.7
	1979	6	9	2	103	98.0	100.9
	1980	3	10	6	101	98.0	99.0
	1981	4	2	5	103	98.0	100.9
	1982	3	7	5	103	98.0	100.9
France	1955–1959	358	68	2	76	67.1	51.0
	1965–1969	276	33	2	30	65.0	19.5
	1979	284	7	2	31	65.0	20.2
	1980	321	6	2	34	66.0	22.4
	1981	286	6	—	37	66.0	24.4
	1982	303	6	—	33	65.0	21.5
Germany	1955–1959	1562	241	49	1700	80.4	1368.0
	1965–1969	1713	157	70	1206	82.6	996.2
	1979	984	76	48	1017	85.0	864.5
	1980	878	75	34	1015	85.0	862.7
	1981	745	59	36	999	85.0	849.1
	1982	718	81	35	936	85.0	795.6

TABLE 13 (Continued)

Country	Period	Animal feed	Seed	Industrial Uses: Nonfood Products	Food: Gross	Food: Extraction rate (%)	Food: Net
Italy	1955–1959	45	18	—	131	75.6	99.0
	1965–1969	37	15	—	26	76.0	19.8
	1979	24	2	—	11	76.0	8.4
	1980	30	2	—	5	76.0	3.8
	1981	26	2	—	9	76.0	6.8
	1982	27	2	—	5	76.0	3.8
Netherlands	1955–1959	482	24	2	83	86.7	72.0
	1965–1969	176	10	1	64	83.8	53.6
	1979	19	1	1	71	86.0	61.1
	1980	22	1	1	73	84.0	61.3
	1981	5	1	1	69	85.0	58.7
	1982	1	1	—	68	85.0	57.8
Spain	1955–1959	87	64	—	356	69.9	249.0
	1965–1969	188	45	—	181	60.0	108.8
	1979	131	29	—	98	60.0	58.8
	1980	135	30	2	99	60.0	59.4
	1981	152	29	—	103	60.0	61.8
	1982	88	31	—	52	60.0	31.2
Sweden	1955–1959	82	21	—	141	78.0	110.0
	1965–1969	19	14	—	127	82.8	105.2
	1979	56	13	8	120	85.0	102.0
	1980	46	10	9	128	84.0	107.5
	1981	37	12	7	131	85.0	111.3
	1982	22	12	8	134	81.0	108.5
Switzerland	1955–1959	11	2	1	27	81.0	22.0
	1965–1969	11	2	1	42	77.0	32.3
	1979	23	1	1	26	77.0	20.0
	1980	25	1	1	26	77.0	20.0
	1981	35	1	1	24	77.0	18.5
	1982	34	1	1	21	76.0	16.0
United Kingdom	1955–1959	6	2	1	—	—	—
	1965–1969	4	1	—	20	57.0	11.4
	1979	14	1	—	20	70.0	14.0
	1980	13	1	—	15	73.3	11.0
	1981	11	1	—	21	71.4	15.0
	1982	16	1	—	13	69.2	9.0
United States	1955–1959	151	239	119	121	79.3	96.0
	1965–1969	238	145	103	138	77.4	106.8
	1979	227	124	—	91	82.0	74.6
	1980	218	105	—	142	82.0	116.4
	1981	249	106	—	92	82.0	75.4
	1982	250	108	—	174	82.0	142.7

Source: Ref. 29.

TABLE 14 Grades of Rye in North America

United States[a]

Grade No.	Test weight (bu) minimum (lb)	Maximum limits: Damaged kernels		Maximum limits: Foreign material	
		Total (%)	Heat–damaged (%)	Total (%)	Materials other than wheat (%)
U.S. No. 1	56.0	2.0	0.1	3.0	1.0
U.S. No. 2	54.0	4.0	0.2	6.0	2.0
U.S. No. 3	52.0	7.0	0.5	10.0	4.0
U.S. No. 4	49.0	15.0	3.0	10.0	6.0

Special Grades:
Tough rye—moisture content 14–16%
Smutty rye—14 or more smut balls in 250 g rye
Garlicky rye—2 or more garlic bulblets in 100 g rye
Weevily rye—live weevils or other insects in grain
Ergoty rye—more than 0.3% ergot in rye

Canada[b]

Grade No.	Minimum (lb) weight/bushel	Maximum Limits-Foreign Material: Other than cereals (%)	Cereals other than wheat (%)	Total foreign matter including wheat (%)	Other characteristics
No. 1—Canada Western	58.0	0	0	0	Sound—ergot free[c]
No. 2—Canada Western	56.0	0.5	1.5	2.0	Sound—ergot free[c]
No. 3—Canada Western	54.0	1.0	3.0	5.0	Reasonably sound, slightly damaged, <0.33% ergot[c]
No. 4—Canada Western	—	2.0	7.0	10.0	More than 5% heat damage, <0.33% ergot[c]
Special Grades:					
Canada Western Ergoty Rye	—	2.0	7.0	10.0	>0.33% ergot[c]
Canada Western Ergoty Rye and other grains[d]	—	2.0	—	—	>0.33% ergot[c]

[a]From Ref. 42.
[b]From Ref. 8.
[c]After dockage removal.
[d]Excluded from regular grades because of admixture of other grains: rye is predominating grain.

TABLE 15 Rye Classifications and Grading Factors of European Countries Including U.S.S.R.

U.S.S.R.

Classes:

Class I rye—winter rye, class divided into five subclasses
Class II rye—hard spring rye, class divided into 2 subclasses
Class III rye—other spring rye, no subclasses

Grading Factors:

Condition	*Milling value*	*Baking value*
color	vitreousness	sprouting
odor	kernel size uniformity	rheological properties
taste	ash content	baking test
moisture	test weight	
foreign matter	1000–kernel weight	
damaged kernels	laboratory milling test	

European Economic Community (EEC) and German Democratic Republic (DDR)

Grain Characteristic	DDR maximum (%)	EEC maximum (%)
Moisture content	18	16.0
Broken kernels	—	2.0
Grain Besatz (shrunken kernels, other grains, insect damage)	—	1.5
Sprouted kernels	3	1.0
Black Besatz (wheat seed, ergot, unsound grain, chaff, impurities)	2	0.5
Hectoliter weight	—	71 kg
Falling number	Min. 110 sec (9 g)	—

Sources: Refs. 1,33,40.

TABLE 16 Basic Grades of Rye Flour Produced and Product Specifications

United States

	Moisture (%)	Ash (%)	Protein (%)	Color
White rye flour	14.5 max.	0.58–0.78	7.0–9.1	white
Medium rye flour	14.5 max.	1.11–1.39	10.1–12.8	medium white
Dark rye flour	14.5 max.	2.05–2.83	13.7–16.2	dark
Rye meal	—	—	—	—

Size classifications for coarse and fine rye meal:

USBS sieves	Pumpernickel rye meal	Fine rye meal
On 8	30	10
On 20	46	40
On 40	14	20
On 60	5	5
Thru 60	5	25

U.S.S.R.

Grade	Ash (%)	Sieve (% over)	Sieve (% Thrus)	Color
Grade I	0.65	No. 5/2	No. IX/90	white
Grade II	1.25	450 μ/2	No. IX/60	gray
Straight grade	1.70	670 μ/2	No. IX/30	specky

West Germany

Rye flour type	Ash (dry basis) Min. (%)	Ash (dry basis) Max. (%)
610	0.58	0.65
815	0.79	0.87
997	0.95	1.07
1150	1.10	1.25
1370	1.30	1.45
1590	1.53	1.63
1740	1.64	1.84
1800 (meal)	1.65	2.00

EEC

Rye flour type	Ash (dry basis) (%)
1	max. 0.90
2	0.91–1.15
3	1.16–1.35
4	1.36–2.30

Sources: Refs. 11,33,37.

TABLE 17 Composition of Rye Grown in Different Countries (dry basis)

Country	1000–Kernel weight (g)	Ash (%)	Protein (%)	Pentosans (%)	Starch (%)
United States	15.7–24.1	1.62–2.10	12.6–14.5	7.0–9.6	59.3–61.4
Canada	18.2–22.1	1.61–18.6	12.4–14.0	7.6–8.8	59.0–61.3
Argentina	12.0	2.16	15.0	10.0	56.0
U.S.S.R.	15.6–20.1	1.67–1.87	12.3–15.4	7.8–8.5	58.6–61.8
West Germany	18.8–33.7	1.67–2.24	9.0–13.5	6.6–9.2	57.5–64.1

Source: Refs. 11,28,36.

TABLE 18 Composition of Rye Mill Streams (dry basis)

	Break streams						Reduction streams				
	I	II	III	IV	V	VI	1	2	3	4	5
Yield (%)	18	10	14	7	6	4	12	3	5	3	3
Ash (%)	0.51	0.55	0.80	1.04	1.40	2.19	0.60	0.83	1.03	3.10	3.75
Starch (%)	82.9	81.5	75.4	68.5	66.5	55.4	80.7	73.6	69.3	43.3	32.5
Protein[a] (%)	5.3	6.5	8.5	10.2	11.8	14.7	6.6	8.7	9.7	17.1	17.6
Pentosan (%)	3.0	3.4	4.2	5.7	5.3	8.1	3.3	4.7	5.6	10.3	14.0

[a]N x 6.25.
Source: Ref. 11.

TABLE 19 Component Fractions of Rye Proteins (%)

Fraction	Lasztity (21)	Chen and Bushuk (9)
Albumin	15.2	34.7
Globulin	18.5	10.7
Prolamin	40.2	19.0
Acetic acid soluble	n.d.	9.4
Residue	n.d.	20.6

n.d. = not determined.

TABLE 20 Amino Acid Composition of Rye (cv. Prolific) Endosperm Proteins (mol%)[a]

Amino acid	Albumins	Globulins	Gliadins	Glutenins
Lysine	2.1	4.3	0.7	2.0
Histidine	1.5	2.1	1.3	1.3
Arginine	2.6	4.8	1.5	1.7
Aspartic acid	4.1	7.5	2.1	2.7
Threonine	3.2	4.0	2.0	2.5
Serine	5.1	5.5	4.9	5.3
Glutamic acid	27.7	16.4	36.7	33.7
Proline	16.6	6.5	20.3	16.6
Glycine	4.3	8.1	2.3	7.2
Alanine	4.5	7.4	2.5	3.4
Valine	10.5	14.0	9.6	9.4
Methionine	1.3	1.7	1.1	1.0
Isoleucine	3.5	3.9	3.2	2.3
Leucine	6.6	7.6	5.9	5.2
Tyrosine	1.6	2.2	1.0	2.3
Phenylalanine	4.6	3.5	4.6	3.4

[a]Cystine and tryptophan not determined.
Source: From Ref. 31 with permission.

TABLE 21 Amino Acid Composition of Rye From Different Countries (g amino acid/100 g protein)

Amino Acid	Canada	Canada	U.S.S.R.	United States	England	Poland
Lysine	3.49	3.18	4.23	3.8	3.72	4.54
Histidine	2.14	2.30	2.09	2.3	2.38	2.66
Ammonia	3.40	3.19	n.d.	2.7	2.72	n.d.
Arginine	4.55	4.60	5.62	4.9	4.93	6.12
Aspartic acid	6.82	7.09	7.16	7.2	6.46	7.84
Threonine	3.26	3.67	3.11	3.7	3.97	3.54
Serine	4.11	4.70	4.54	4.2	6.12	4.45
Glutamic acid	30.51	30.18	29.91	23.7	33.46	23.81
Proline	15.29	11.40	5.20	8.9	14.23	9.45
Glycine	3.82	3.98	4.79	4.5	3.99	4.90
Alanine	4.06	4.07	5.13	4.4	4.54	4.91
Cystine	2.65	2.51	1.19	1.7	3.45	n.d.
Valine	5.22	5.37	5.56	3.3	5.15	4.92
Methionine	2.15	1.28	0.65	2.9	1.93	1.93
Isoleucine	4.21	4.00	3.88	3.3	4.21	3.72
Leucine	6.65	6.56	7.00	5.9	7.79	6.72
Tyrosine	2.16	2.07	2.86	1.5	2.38	1.56
Phenylalanine	5.16	4.91	5.25	4.0	5.53	3.56
Tryptophan	n.d.	1.33	n.d.	2.2	1.94	1.07

n.d. = not determined.
Source: Refs. 5,7,13,18,34,41.

TABLE 22 Amino Acid Composition of Rye Grain, Flour, and Bran (% of total protein)

Amino acid	Rye grain[a]	Flour[a]	Bran[a]	Flour[b]
Alanine	5.13	4.50	5.36	4.06
Arginine	5.62	4.76	6.31	4.55
Aspartic acid	7.16	6.64	7.47	6.82
Cystine	1.19	1.06	1.90	2.65
Glycine	4.79	4.29	5.44	3.82
Glutamic acid	29.91	34.70	27.93	30.51
Histidine	2.09	1.86	2.19	2.14
Isoleucine	3.88	3.60	3.69	4.21
Leucine	7.00	6.59	6.76	6.65
Lysine	4.23	3.38	4.06	3.49
Methionine	0.65	0.65	0.44	2.15
Phenylalanine	5.25	5.28	4.56	5.16
Proline	5.20	5.93	4.93	15.29
Serine	4.54	4.81	4.53	4.11
Threonine	3.11	2.92	3.34	3.26
Tyrosine	2.86	2.79	2.66	2.16
Valine	5.56	5.19	5.32	5.22

[a]From Ref. 34.
[b]From Ref. 7.

TABLE 23 Lysine Content of Rye Cultivars

Variety and/or species	Protein (%)	Lysine in protein (g/16 g N)
Antelope	11.4	3.33
Carsten	9.2	3.65
Secale dalmaticum CPI22755	16.5	3.23
Explorer	14.1	3.30
USDA PI 168178	14.6	3.22
USDA PI 227870	10.5	3.71
Rye Gator	17.3	3.21
Prolific Spring	14.9	3.25
Balbo	11.2	3.63
Detenicke	16.0	3.37
Secale cereale 5–SC–18	12.8	3.22
Rye Korean I	15.3	3.05
Dominant	11.0	3.80
Volyanko	11.9	2.91
Afganistan Winter Rye	10.4	3.35
Canadian Spring Rye	9.2	3.71
Secale Montanum 23–282	16.9	3.10
Secale segetale 23709	17.2	3.02
Abruzzi A	9.9	2.55
Maia barroso 5053	10.6	4.26

Source: Ref. 46.

TABLE 24 Chemical Scores for Rye and Wheat and WHO Standard Reference Pattern

		Rye		Wheat	
Amino acid	WHO (1973) Standard Reference (mg/g protein)	Amino acid content (mg/g protein)	Chemical score	Amino acid content (mg/g protein)	Chemical score
Lysine	55	37	67	31	56
Threonine	40	37	92	31	77
Methionine + cystine	35	37	105	43	123
Leucine	70	67	96	72	103
Isoleucine	40	39	97	35	88
Valine	50	52	104	47	94
Phenylalanine + tyrosine	60	69	115	81	135
Tryptophan	10	8	80	—	—

Source: Refs 17 and 38.

TABLE 25 Rye Starch Compared with Starch from Hard Red Spring Wheat (HRS)

	Rye	HRS
Starch (% of flour)	72.3	72.0
Nitrogen in starch (%)	0.04	0.04
Ash in starch (%)	0.37	0.38
Fat in starch (%)	0.98	0.75
Mean starch diameter (μ)	28.2	18.7
Amylose (%)	24.5–30.1	24.0–28.9
Water–binding capacity	86.5	85.0
Intrinsic viscosity	1.96	1.88
Birefringence endpoint temp. (°C)	59.6	62.9
Gelatinization temp. range (°C)	9.6	8.6
Absolute density (30°C)	1.4209	1.4832

Source: Refs. 4,20.

TABLE 26 Saccharides in Rye—Comparison with Hard Red Spring Wheat (HRS) (% dry basis)

	Rye	HRS
Monosaccharides	0	0
Sucrose	1.31	0.63
Maltose	0.03	0.04
Maltotriose	0.01	0
Trisaccharides[a]	0.80	0.64
Tetrasaccharides[b]	0.31	0.01

[a]Raffinose and kestose.
[b]Nystose and frustosylraffinose.
Source: Ref. 3.

TABLE 27 Lipid Content of the Cereals and Distribution of Mass and Lipids Among the Anatomical Parts[a]

		Distribution of mass(%)			Distribution of lipids (%)		
Cereal	Total lipids (%)	Bran	Endo-sperm	Germ	Bran	Endo-sperm	Germ
Wheat	3.31	4.01	92.0	4.02	3.39	62.7	33.8
Rye	2.99	4.48	91.1	4.36	3.20	62.3	34.5

[a]Calculated on dry-weight basis.
Source: Ref. 48.

TABLE 28 Lipid Composition of Rye and Wheat

	Composition (wt %)		Lipid composition of anatomical parts of cerals (wt%)[a]					
			Wheat			Rye		
Lipid	Wheat	Rye	Bran	Endosperm	Germ	Bran	Endosperm	Germ
Monoglycerides	5.70	3.71	6.60	6.60	—	10.0	2.66	—
Diglycerides	2.85	1.14	3.90	3.50	2.50	10.6	1.60	3.20
Triglycerides	20.1	23.1	14.3	11.7	19.0	6.8	11.2	21.3
Free fatty acids	3.50	5.10	14.5	3.53	8.33	9.87	3.67	9.20
Steryl esters	2.90	1.85	23.5	0.67	7.67	14.3	2.52	4.06
Monogalactosyl diglycerides	2.93	3.73	—	7.06	—	—	5.2	—
Digalactosyl diglycerides	12.00	11.10	—	15.3	—	—	13.3	—
Lysolecithin	4.80	4.00	—	1.28	—	—	6.44	—
Phosphatidylcholine	1.20	0.75	—	2.50	4.10	—	0.60	6.62
Phosphatidylethanolamine	0.48	0.12	—	20.1	1.8	—	16.0	1.00
Phosphatidylinositol	1.90	2.75	—	2.44	7.92	—	—	9.92

[a]Lipids analyzed only for components listed.
Source: Ref. 48.

TABLE 29 Fatty Acid Composition of Lipids from Whole-Grain Wheat and Rye

Fatty acid	Composition (wt %)[a]	
	Wheat	Rye
C14	0.25	0.12
C15	0.26	0.22
C16	17.8	14.8
C16:1	0.26	1.18
C18	1.23	0.78
C18:1	16.5	17.1
C18:2	56.5	57.7
C18:3	6.56	7.36
C20:1	0.70	0.71

[a]Calculated as methyl esters.
Source: Ref. 48.

TABLE 30 Mineral Composition of Whole-grain Rye (mg/100 g dry basis)

Mineral	Range
Potassium	412–520
Phosporus	359–380
Magnesium	92–130
Calcium	31–70
Sodium	20–26
Iron	2.7–10
Manganese	2.5–7.5
Zinc	3.4
Copper	0.6–0.9

Sources: Ref. 19 and 28.

TABLE 31 Vitamin Composition of Rye Grain, Bran, Shorts, and Flour (c.v. Prolific)[a]

	Grain	Flour	Shorts	Bran
Thiamin	7.7	0.8	4.0	9.3
Riboflavin	2.9	1.3	2.8	4.1
Niacin	15.3	8.8	18.0	27.4
Biotin	0.054	0.011	0.078	0.119
Folacin	0.49	0.13	0.72	0.83
Pantothenic acid	6.3	2.7	9.1	10.9
Vitamin B6	3.4	0.46	3.7	6.6
Tocols				
α–T	13.1	—	—	—
β–T	3.9	—	—	—
α–T–3	14.8	—	—	—
β–T–3	11.8	—	—	—

[a]μg/g, dry basis.
Source: Refs. 2,27.

TABLE 32 Vitamins and Minerals in Rye and Its Milling Fractions

Nutrient	Rye grain	Bran	Middlings	Low grade flour
Potassium (mg/100 g)	520	n.d.	590	510
Phosphorus (mg/100 g)	380	1220	590	320
Calcium (mg/100 g)	70	140	60	20
Manganese (mg/100 g)	7.5	112	415	n.d.
Thiamin (mg/kg)	4.6	2.9	3.1	2.3
Riboflavin (mg/kg)	1.8	0.2	2.3	0.8
Niacin (mg/kg)	15.0	26.3	15.9	8.0
Pantothenic acid (mg/kg)	10.4	15.9	21.8	9.8

n.d. = not determined.
Source: Refs. 10,24.

TABLE 33 Composition of Whole Rye and Wheat and Their Milled Products (15% moisture basis)

	% Extraction					
	English rye			Manitoba wheat		
Component	100	85	75	100	85	75
Total N (g/100 g)	1.40	1.28	1.17	2.39	2.38	2.29
Protein (N x 5.7) (g/100 g)	7.98	7.30	6.67	13.64	13.57	13.05
Fat (g/100 g)	1.98	1.64	1.33	2.49	1.70	1.32
Starch (g/100 g)	69.0	73.0	75.0	63.0	67.2	6
Fiber (g/100 g)	1.56	0.84	0.48	2.15	0.33	0.10
Thiamin (Iu/g)	1.45	0.98	0.80	1.18	0.92	0.29
Riboflavin (μg/g)	2.90	2.00	1.40	1.70	1.00	0.70
Ash (g/100 g)	1.72	1.04	0.72	1.53	0.75	0.44
Potassium (mg/100 g)	412	203	172	312	146	87
Calcium (mg/100 g)	31.5	26.1	19.5	27.6	18.5	13.1
Magnesium (mg/100 g)	92	45	26	141.0	61.8	30.4
Iron (mg/100 g)	2.70	1.97	1.72	3.81	—	—
Total phosphorus (mg/100 g)	359	193	129	350	188	109
Phytate phosphorus (mg/100 g)	258	104	57	242	96	37

Source: Refs. 26,38.

TABLE 34 Alkylresorcinols in Grains and Milling Fractions

Sample	Grain (%)	Bran: Milling fraction (%)	Bran: Alkyl–resorcinols (%)	Shorts: Milling fraction (%)	Shorts: Alkyl–resorcinols (%)	Flour: Milling fraction (%)	Flour: Alkyl–resorcinols (%)	Mill type
Triticale 6–TA–204	0.085	27.7	0.213	6.0	0.087	64.8	0.026	Quadrumat
Rye—Prolific	0.124	33.6	0.186	10.5	0.115	52.0	0.031	Sr.
Wheat—Chris	0.062	21.7	0.211	—	—	78.3	0.038	Quadrumat
Triticale 6–TA–204	0.085	30.0	0.271	—	—	70.0	0.039	Jr.
Rye—Prolific	0.124	43.8	0.187	—	—	56.2	0.029	

Source: Ref. 45.

TABLE 35 Uses of Basic Rye Flour Types

Flour type	Suggested Uses
White rye	Jewish and other white rye breads Dusting flour Light Swedish rye bread
Medium rye	Bohemian, Polish, and Russian rye breads For blending with other rye flours In American-type rye breads Medium Swedish rye breads Base for rye sours
Dark rye	American rye breads (smaller percentage than medium rye flour) Dark, heavy German rye breads Pumpernickel breads Swedish rye breads Base for stronger rye sours
Rye meal (various granulations)	In various types of pumpernickel bread For dusting (hearth rye breads)
Cracked rye	Used in either white bread dough or rye to make a specialty cracked rye bread (should be soaked before adding to dough)
Rye flakes (flat, rolled whole rye)	For blending with meal and rye flour in pumpernickel bread In crisp, flat Swedish rye bread In rye crisp wafers

Source: Ref. 16.

REFERENCES

1. *Amtsblatt der Europäischen Gemeinschaft.* Nr. L100, April 28, 1969, p. 8.
2. Barnes, P. J., Cereal tocopherols. In *Proceedings of the 7th World Cereal and Bread Congress.* J. Halas and J. Kratochvil (Eds.). Elsevier Amsterdam, 1982, p. 1095.
3. Becker, R., Lorenz, K., and Saunders, R. M., Saccharides in maturing triticale, wheat and rye. *J. Agric. Food Chem., 25*: 1115(1977).
4. Berry, C. P., D'Appolonia, B. L., and Gilles, K. A., The characterization of triticale starch and its comparison with starches of rye, durum and HRS wheat. *Cereal Chem., 48*: 415(1971).
5. Bixler, E., Schaible, P. J., and Bandemer, S., Preliminary studies on the nutritive value of triticale as chicken feed. *Quart. Bull. Michigan Agri. Exp. Stn., 50*: 276(1968).
6. Bushuk, W., History, world distribution, production and marketing. In *Rye: Production, Chemistry, and Technology.* Bushuk, W.(Ed.), American Assoc. of Cereal Chemists, St. Paul, Minnesota, 1976, chap. 1.
7. Bushuk, W., Proteins of triticale: Chemical and physical characteristics. In *Triticale: First man-made cereal.* Tsen, C. C.(Ed.). American Assoc. of Cereal Chemists, St. Paul, Minnesota, 1974, p. 128
8. Canadian Wheat Board, *Canadian Grain Handbook. Crop year 1974–75.* The Board., Winnipeg, Canada, 1974.

9. Chen, C. H., and Bushuk, W., Nature of proteins in triticale and its parental species. I. Solubility characteristics and amino acid composition of endosperm proteins. *Can. J. Plant Sci.*, *50*: 9(1970).
10. Committee on Animal Nutrition, *Joint United States—Canadian Tables of Feed Composition.* National Academy of Sciences—National Research Council, Publ. 1232, Washington, DC, 1964.
11. Drews, E., and Seibel, W., Bread-baking and other uses around the world. In *Rye, Production, Chemistry, and Technology*. Bushuk, W.(Ed.). American Assoc. of Cereal Chemists, St. Paul, Minnesota, 1976, chap. 6.
12. Evans, L. E., and Scoles, G. J., Cytogenetics, plant breeding and agronomy. In *Rye, Production, Chemistry, and Technology.* Bushuk, W.(Ed.). American Assoc. of Cereal Chemists, St. Paul, Minnesota, 1976, chap. 2.
13. Ewart, J. A. D., Amino Acid analyses of cereal flour proteins. *J. Sci. Food Agric.*, *18*: 548(1967).
14. Food and Agricultural Organization, *Production yearbooks*, Vol. 31 and Vol. 39, FAO, Rome, 1978 and 1985.
15. Food and Agricultural Organization, *Trade yearbooks*, Vol. 31 and Vol. 38, FAO, Rome, 1977 and 1984.
16. Gordon, J., Basic production aspects of rye bread. *Baker's Dig.*, *44*(5): 38,(1970).
17. Hulse, J. H., and Laing, E. M., *Nutritive Value of Triticale Protein*, International Development Ctr., Ottawa, Canada, 1974, chap. 4.
18. Janicki, J., and Kowalczyk, J., The biological values of rye proteins compared with wheat protein. *Vitalst Zivilisationskr.*, 10: 14(1965).
19. Kent, N. L., *Technology of Cereals.* Pergamon Press, New York, 1966, chap. 3.
20. Klassen, A. J., and Hill, R. D., Comparison of starch from triticale and its parental species. *Cereal Chem.*, *48*: 647(1971).
21. Lasztity, R., *The Chemistry of Cereal Proteins.* CRC Press, Boca Raton, FL, 1984, chap. 4.
22. Lorenz, K., Alkylresorcinols in cereal grains. *Proc., 10th Natl. Conf. on Wheat Utilization Research*, Tucson, 1977, p. 72.
23. Lorenz, K., Ergot on cereal grains. *Crit. Rev. Food Sci. and Nutrition, 11*(4): 311(1979).
24. Lorenz, K., Rye. In *Handbook of Processing and Utilization in Agriculture.* Wolff, I. A.(ed.). CRC Press, Boca Raton, FL, 1982. p. 243.
25. Lorenz, K., Rye bread formulation and processing. Technical Bulletin, Vol. IX, Issue 5. American Institute of Baking, Manhattan, Kansas, 1987.
26. McCance, R. A., Widdowson, E. M., Moran, T., Pringle, W. J. S., and Macrae, T. F., The chemical composition of wheat and rye and of flours derived therefrom. *Biochem. J.*, 39: 213(1945).
27. Michela, P., and Lorenz, K., The vitamins of triticale, wheat and rye. *Cereal Chem.*, 53: 853(1976).
28. Miller, D. F., *Composition of Cereal Grains and Forages.* Publ. 585. National Academy of Sciences—National Research Council, Washington, DC, 1958.
29. Organization for Economic Cooperation and Development, *Food Consumption Statistics, 1955–1973 and 1973–1982*, Paris, France, 1985.
30. Pomeranz, Y., and Shellenberger, J. A., *Bread Science and Technology.* AVI Publ. Co., Westport, CT, 1971, p. 221.
31. Preston, K. R., and Woodbury, W., Amino acid composition and subunit structure of rye gliadin proteins fractionated by gel filtration. *Cereal Chem., 52*: 719(1975).
32. Qualset, C. O., and Hoskinson, P. E., A germplasm source population of rye. *Crop Sci., 6*: 219(1966).
33. Rozsa, T. A., Rye milling. In *Rye: Production, Chemistry and Technology.* Bushuk, W.(Ed.). American Assoc. of Cereal Chemists, St. Paul, MN, 1976, chap. 5.
34. Rukosuev, A. N., and Silant'eva, A. G., Amino acid composition of rye grain, sieved rye flour, bread flour and bran. *Vop. Pitan., 31*: 142(1972).
35. Schery, R. W., *Plants for Man.* Prentice-Hall, Englewood Cliffs, NJ, 1963, p. 389.
36. Schopmeyer, H. H., Rye and rye milling. *Cereal Sci. Today., 7*: 138(1962).

37. Seibel, W., Bruemmer, J. M., and Stephan, H., West German bread. In *Advances in Cereal Science and Technology*, Vol. II. American Assoc. of Cereal Chemists, St Paul, MN, 1978, chap. 7.
38. Simmonds, D. H., and Campbell, W. P., Morphology and chemistry of the rye grain. In *Rye: Production, Chemistry, and Technology*. Bushuk, W.(Ed.). American Assoc. of Cereal Chemists, St. Paul, MN, 1976, chap. 4.
39. Smith, B. W., Feed manufacture. In *Cereal Technology*. Matz, S. A.(Ed.). AVI Publishing Co., Westport, CT, 1970, chap. 3.
40. Staatsverlag der DDR, *DDR Standard Roggen* (TEL 2755). Amt für Standardisierung der DDR, Staatsverlag der DDR, Berlin, 1973.
41. Tkachuk, R., and Irvine, G. N., Amino acid compositions of cereals and oilseed meals. *Cereal Chem., 46*: 206(1969).
42. U.S. Dept. Agriculture, Agricultural Marketing Service, Grain Division, *Official Grain Standards of the United States*, U.S. Government Printing Office, Washington, DC, 1970.
43. U.S. Dept. of Agriculture, Economic Research Service, *Agricultural Statistics Grain and Feed*, 1984.
44. U.S. Dept. of Agriculture, Statistical Reporting Service Crop Reporting Board, *Agricultural Statistics Grain and Feed*, Washington, DC, 1984.
45. Verdeal, K., and Lorenz, K., Alkylresorcinols in wheat, rye, and triticale. *Cereal Chem., 54*: 475(1977).
46. Villegas, E., McDonald, C. E., and Gilles, K.A., Variability in the lysine content of wheat, rye, and triticale proteins. *Cereal Chem., 47*: 746(1970).
47. World Health Organization, *Energy and Protein Requirement.* WHO Tech. Rep. Ser. 522, FAO Nutr. Meetings Ser. 52, Rep. Joint FAO/WHO Ad Hoc Expert Comm., Rome, 1973.
48. Zeringue, H. J., Jr., and Feuge, R. O., A comparison of the lipids of triticale, wheat and rye grown under similar ecological conditions. *J. Am. Oil Chem. Soc., 57*: 373–376(1980).

9

TRITICALE: PRODUCTION AND UTILIZATION

J. Perry Gustafson
U. S. Department of Agriculture and University of Missouri, Columbia, Missouri

Walter Bushuck
University of Manitoba, Winnipeg, Manitoba, Canada

Anna R. Dera
University of Missouri, Columbia, Missouri

I. INTRODUCTION

Triticale (*X Triticosecale* Wittmack), the first man-made cereal, is the product of a cross between wheat (*Triticum* ssp.) and rye (*Secale cereale* L.). Wilson reported on the first cross between wheat and rye, resulting in the production of sterile plants (113). Rimpau reported on the first fertile triticale from a cross between hexaploid wheat (*T. aestivum* L.—Thell.) and diploid rye (84). From the 1930s to the 1950s, research on triticale never really progressed past the botanical curiosity stage despite considerable effort (64). Early plant breeders looked on triticale as a potential means of combining the best from wheat (i.e., high yield and breadmaking quality) with the best from rye (i.e., winter hardiness, high lysine and protein, drought tolerance, disease resistance, tolerance to heavy metals), but until recently, the realization of many of these goals eluded breeders. Research efforts have resulted in triticale being grown commercially in several areas around the world. Triticale is mainly grown as a feed grain, but the amount of triticale utilized for human consumption is increasing.

II. PLANT BREEDING

A. World Production

World production is difficult to assess because most countries do not have a reporting service for triticale acreage. The estimates given in Table 1 are based solely on communications with breeders working on triticale within the stated country and are subject to some degree of error. Much of the triticale currently grown is utilized on the farm without being reported. Most of the present cultivation is located in developed countries (Table 1). There are two reasons for the acreage in developed countries. First, most of the acreage resides on land that is not conducive to high wheat yields (i.e., the sandy soils of Poland). Second, in many countries triticale offers farmers an alternative cash crop that is not controlled by any marketing board.

A few countries established grain standards in the early 1970s to be used in marketing triticale. However, with the recent improvements in grain characteristics (i.e., test and ker-

triticale to the production circumstances of farmers" (105). This is especially true when adapting triticale to conditions of the Third World, where agricultural practices from country to country vary tremendously. A farmer's decision-making process on whether or not to grow triticale in Kansas is quite different from a farmer in Australia, Poland, or Rwanda. At present, varieties are very environment-specific and need to be tailored for each particular environmental condition. Once triticale varieties become available that can be grown over a wide range of environments, while continuing to maintain high yields, triticale production will increase to an even higher level than at present.

B. Breeding

1. Methodology

The production of triticale begins with the crossing of either tetraploid (*T. durum* L.) (genomes AABB) or hexaploid (*T. aestivum* L.—Thell.) (genomes AABBDD) wheat x diploid rye (*S. cereal* L.) (genomes RR) and doubling the resulting haploid hybrid by treating with colchicine, thus creating a fertile amphidiploid. This procedure results in primary hexaploid (genomes AABBRR) and octoploid (genomes AABBDDRR) triticales, respectively. Intercrossing primary triticales within their own ploidy levels was the generally accepted method of breeding from the 1930s until the mid-1950s (64). This method of crossing resulted in some improvements in plant type; however, the yield and stability of the best lines were still significantly below that of wheat.

In the mid-1960s, Kiss and Pissarev started intercrossing hexaploid triticale and hexaploid wheat (Fig. 1) (43,75). This crossing procedure resulted in significant advances for the improvement of triticale. Dwarfing genes, genes that improve quality, and disease resistance genes from hexaploid wheat were added to the hexaploid triticale gene pool. These crosses resulted in the first triticale varieties being released in Hungary and the U.S.S.R., but the cultivars still showed some sterility and meiotic instability.

In the late 1960s, the spring triticale program at the International Maize and Wheat Improvement Center (CIMMYT), Mexico, began making triticale–wheat crosses that provided a major breakthrough in helping to establish triticale as a crop. An uncontrolled triticale–wheat cross resulted in the selection of a triticale "Armadillo" containing wheat chromosome 2D that had replaced rye chromosome 2R (35). This "substituted" triticale contained 15 pairs of wheat and 6 pairs of rye chromosomes. Much of the importance of this substitute came from a gene for daylength insensitivity present on wheat chromosome 2D, which allowed the CIMMYT breeding program to grow two generations of triticale per year as well as expand into day neutral production areas around the world. Major improvements in the agronomic quality of triticale were made in the late 1970s, and triticale has reached a level where it is beginning to be competitive with wheat in many locations around the world (97).

One problem with triticale–wheat crosses is that they can result in the removal of potential valuable rye gene complexes, even though new genes from the genomes of wheat are made available. An example would be the rapid spread of the Armadillo 2D(2R) substituted types throughout breeding programs all over the world, which narrowed the diversity of triticale instead of widening it, and removed rye chromosome 2R from the gene pool.

Krolow formulated a methodology which involved the crossing of octoploid triticales with hexaploid triticales, thus allowing the substitution of D-genome chromosomes and chromosome segments from wheat for those of the A and B genomes, leaving the rye genome intact (45). This methodology has become the main approach to triticale germplasm expansion in the 1980s. Lukaszewski et al. showed that chromosomes from the D genomes of wheat can replace those of the A and B genomes in almost any possible com-

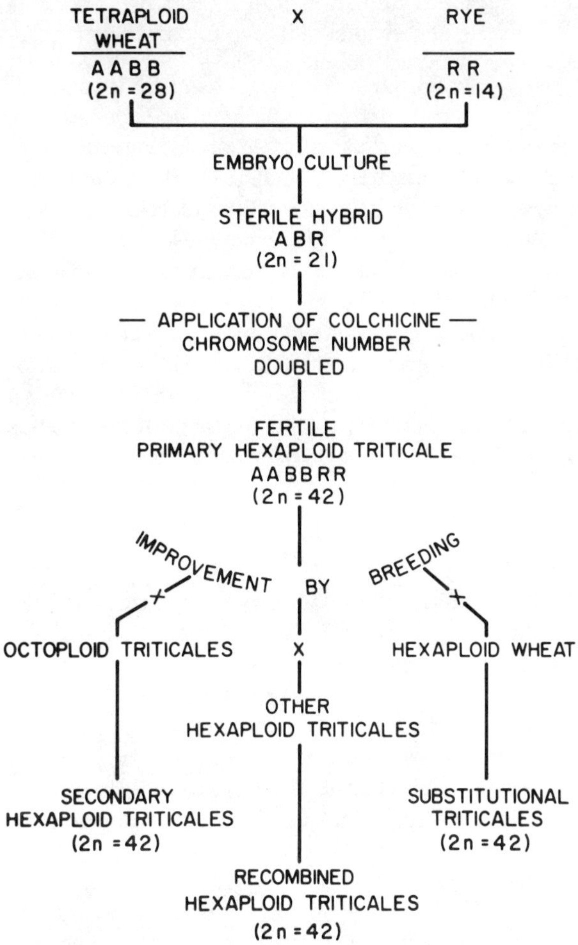

FIGURE 1 A schematic of the production of hexaploid triticales.

bination (53). An analysis of material from the CIMMYT breeding program by A. J. Lukaszewski (unpublished data) found that some D-genome chromosomes have replaced chromosomes from the A genome in approximately 30% of the triticales analyzed. The most common substitution involved chromosome 6D replacing chromosome 6A, resulting in a substituted triticale containing a complete rye genome.

In addition, breeders at CIMMYT have been intercrossing octoploid and hexaploid triticales and selecting triticales that are phenotypically octoploid in appearance. Analysis of several of these octoploid-appearing triticales showed that they were all hexaploid in their chromosome complement, but contained D-genome chromosomes in place of those from the A genome (34). Two interesting points are apparent from the above studies. First, all of the wheat genome substitutions involved chromosomes from the A and D genomes. No substitutions involving chromosomes from the B genome have been observed in either controlled studies or in a breeding program. Second, the largest majority of the substitution involved wheat chromosomes 6D and 6A. A potential drawback to this approach is that heavy selection pressure causing the 6D(6A) substitution could again narrow the gene base in much the same manner as the 2D(2R) substitution. However, it appears that hexaploid spring triticale is beginning to enter an exciting new chapter in its evolution toward becoming an established cereal crop.

2. Genetics and Cytogenetics

There have been several good reviews of triticale genetics and cytogenetics in recent years and the reader is referred to them for a detailed discussion (31,32,54). The genetics and cytogenetics of triticale are very closely associated with the interactions that occur between the genomes of wheat and rye. In most cases, genes from rye, when present in a triticale background, do not express themselves in a manner similar to that observed in diploid rye. This is seen in characters governed by single genes which do not often segregate in single gene ratios when in a triticale (e.g., hairy neck) (39). The rye chromosomes can even alter chromosome morphology when in a triticale background (74,95,102). In addition, rye chromosome pairing frequencies and heterochromatin content can also change when present in a wheat background (N. Jouve, unpublished data). All this points to the fact that the genetics and cytogenetics of triticale are very complex and continue to present challenges to the researcher.

C. Pathology

At the present time, any disease that attacks either wheat and/or rye also attacks triticale (115). It was originally thought that triticale would combine the best from both parental species. In many cases, triticale does just that by showing improved resistance to many wheat diseases, including *Puccinia* spp., *Ustilago* and *Urocystis* spp., *Tillitia* and *Neovossia* spp., and *Erysiphe* spp. (97). On the other hand, there are cases where very good genes for resistance present in either parental species are never fully expressed in triticale (89). In some instances, triticale is susceptible to diseases for which wheat or rye parents show resistance (63). The reasons for the erratic gene expression are not clear, but obviously there are many complex reactions occurring between the genes of wheat and rye.

The existing disease resistance available in triticale is barely adequate for present world needs. As world acreage increases, so will the natural selection pressure for pathogens that show new biotypes of virulence in high production locations. Because of the rather narrow gene base presently available in commercial triticale varieties, any disease epidemic would be disastrous to production. Such a situation occurred in Australia where a large majority of the varieties became susceptible to the new stem rust (*P. recondita* f. sp. *tritici*) race 34-2,12 (61,62). In addition, McIntosh et al. analyzed CIMMYT's twelfth International Triticale Screening Nursery and discovered that approximately 70% of all entries were susceptible to race 34- 2,12 (61). A quick change in commercial varieties and parents utilized in breeding programs has controlled the stem rust problem in Australia for the present time. Because problems of this type will continue to occur as acreage increases, breeders around the world need to continually expand their gene base for resistance to the pathogens occurring in major areas of production.

D. Agronomy

Triticale is similar in many respects to wheat and other cereals in its cultivation. The same machinery and chemicals utilized in the cultivation of wheat can also be used for triticale. This makes it easier to attract farmers to grow triticale because because they do not need additional equipment in order to maximize yields under most agronomic conditions. On the other hand, triticale does exhibit some differences from other cereals in its response to certain mineral conditions, and under some situations, triticale seedling emergence is different from that of wheat.

Triticale responds better than any other cereal, except rye, to acid soils, whether or not the soils are high in free aluminum. However, until recently, triticale had not been specifically adapted for cultivation on acid soils. Even though many triticales do well on acid soils, the problems involved with producing a highly acid-tolerant triticale are severalfold and can be associated with genes in wheat that inhibit rye gene expression. When a highly tolerant rye is crossed to a poorly tolerant wheat, the resulting triticale shows a tolerance level more characteristic of wheat than rye (7,8). A study by Aniol and Gustafson indicated that the incorporation of genes for a high level of tolerance from rye into wheat will not guarantee adequate expression (9). The aluminum tolerance expression level of rye genes appeared to depend on which wheat genome they were placed into. Recent work has shown that suppressers are present in wheat which decrease the effectiveness of the aluminum tolerance system from rye (J. P. Gustafson, and K. Ross, unpublished data). More research is needed in order to obtain acid soil tolerance levels in triticale that are stable and at as high a level as rye.

Evidence by Graham and Pearce suggested that triticale was superior to wheat when grown on soils low in copper (29). Two possible explanations for this superiority were suggested (30). First, triticale has a greater ability to absorb copper than wheat because of its better root growth, which allows it to reach and absorb low amounts of copper available in the soil. Secondly, the superiority of triticale over wheat may be due to its more efficient ability to translocate nutrients from the root to the leaves.

Preharvest sprouting associated with late maturity had been a major problem inhibiting the commercialization of triticale in several countries (91). Originally, this problem was reduced by utilizing sources of earlier maturity which led to the production of varieties capable of escaping the high moisture periods during harvest, thus decreasing the chances of preharvest sprouting. However, new sources of genetic resistance to preharvest sprouting present in several wheat parents are now being utilized in triticale-breeding programs (10). Salmon et al. determined that sprouting resistance can be extended in spring triticales by germination inhibitors found in chaff extracts as in wheat and barley (87). Selection for high levels of these inhibitors could become an efficient method of selecting for preharvest sprouting resistance in triticale.

Recent work has shown that there is no relationship between the rye parent and the resulting triticale for plant height and yield performance (67). Oettler did show that the genes from wheat responsible for governing plant height were related to the agronomic performance of the triticale. However, for grain yield, neither wheat or rye show any association with triticale performances (66,67).

In general, triticale combines the best of both parents when grown in poor soils, under marginal climatic conditions, and at high elevations (108). In Australia there is evidence that some spring triticales are superior to the local wheats under drought conditions (101). From rye, triticale obtained the ability to grow in marginal soil and climatic conditions, and from wheat it obtained yield potential and a more usable grain than rye. Evidence for suppressers which inhibit full expression of many rye genes indicates that triticale can be made even better agronomically if the suppressers can be inactivated.

III. KERNEL MORPHOLOGY AND DEVELOPMENT

Bushuk and Larter stated, “The physiology and biochemistry of kernel shriveling has received considerable attention: however, the cause of the phenomenon remains to be discovered” (17). Nearly a decade later, after having received considerably more attention, the causes for seed shriveling in triticale still need further study. Originally, four basic theories were introduced as possible explanations for this anomaly which is characterized by slow growth and/or cavities within the endosperm (104). High levels of α-amylase, a starch-degrading enzyme, was the first proposed explanation. Others followed, including insufficient levels of substrates, high activity of acid phosphatases, and the production of excessive endosperm aberrations. Although each idea has merit, only two still remain as possible primary determinants of seed shriveling (Fig. 2.).

A number of papers have been published on the apparent relationship between shriveled triticale grain and high levels of α-amylase activity, especially in later stages of kernel development (15,25,44,83,99). The assumption was, if a high level of starch-degrading enzyme is present, then there must be a high level of starch degradation resulting in shrunken endosperm. However, exceptions to this assumption began to appear. Agrawal reported low levels of α-amylase in shriveled-kernel triticale, as well as high levels of α-amylase in a plump-kernel triticale (1). Macri et al. compared α-amylase levels in secondary hexaploid triticales (4T, 11&, "Impala", and "Carman") by analyzing whole meal and milled flour (58). They determined that the whole meal contained much higher levels of enzyme activity than did the milled flour. Thus, the α-amylase must be principally limited to the aleurone layers of the triticale grain, since this fraction is missing from milled flour. Therefore, it is unlikely that aleurone-confined α-amylase activity is the primary cause of seed shriveling since it is shielded from the endosperm during kernel development.

Pena et al. also concluded there was no direct relationship between α-amylase activity and the degree of shriveling in triticale (73). They could not correlate α-amylase level with shriveling when comparing triticales with varying degrees of shriveling.

It is generally agreed that α-amylase activity does increase during the development of the triticale kernel. Ching et al. were able to ascertain that the α-amylase activity increase was not synchronized with shriveling and did not directly affect starch accumulation or degradation in a developing kernel (19). Other factors proposed as possible causes of seed shriveling and starch accumulation include water, photosynthate availability, and acid phosphatases.

Branlard et al. calculated fresh weight percentage values of developing shriveled-type triticale kernels, but could not find variation in water content that accounted for the shriveling in later stages of kernel development (15). They also found no abnormal decreases in water content for the triticales and concluded that water deficiency does not cause seed shriveling in triticale.

Several researchers analyzed triticale's ability to efficiently produce and transport photosynthates. Gas-exchange studies were executed to determine the photosynthetic rates for two triticales: IMJ–3, a shriveled-kernel type, and S71–142, a plump-kernel type (2). The triticale with the higher photosynthetic rate and CO_2 uptake was IMJ–3, the shriveled-kernel type triticale. IMJ–3 also had larger leaves and a 19% higher mean rate of dark respiration overall than did S71-142. This suggests that IMJ–3 fixes a greater amount of carbon and creates more photosynthate than S71–142, yet its seed remains shriveled. Later carbon partitioning experiments executed by allowing photosynthesis in the presence of $^{14}CO_2$ disclosed that transport of assimilates was not adversely hindered in shriveled-kernel type triticales (3).

There was no apparent relationship between soluble sugars or reducing sugars and seed shriveling, since no significant differences could be distinguished between reducing sugar content of plump and shriveled triticales (26,73). Because these sugars are used to synthesize starch, one can conclude that the level of starch precursors is not a restricting factor of starch synthesis in the triticale varieties analyzed. In addition, starch synthase and starch phosphorylase enzyme activity levels could not be closely correlated with starch accumulation in triticale (18,73).

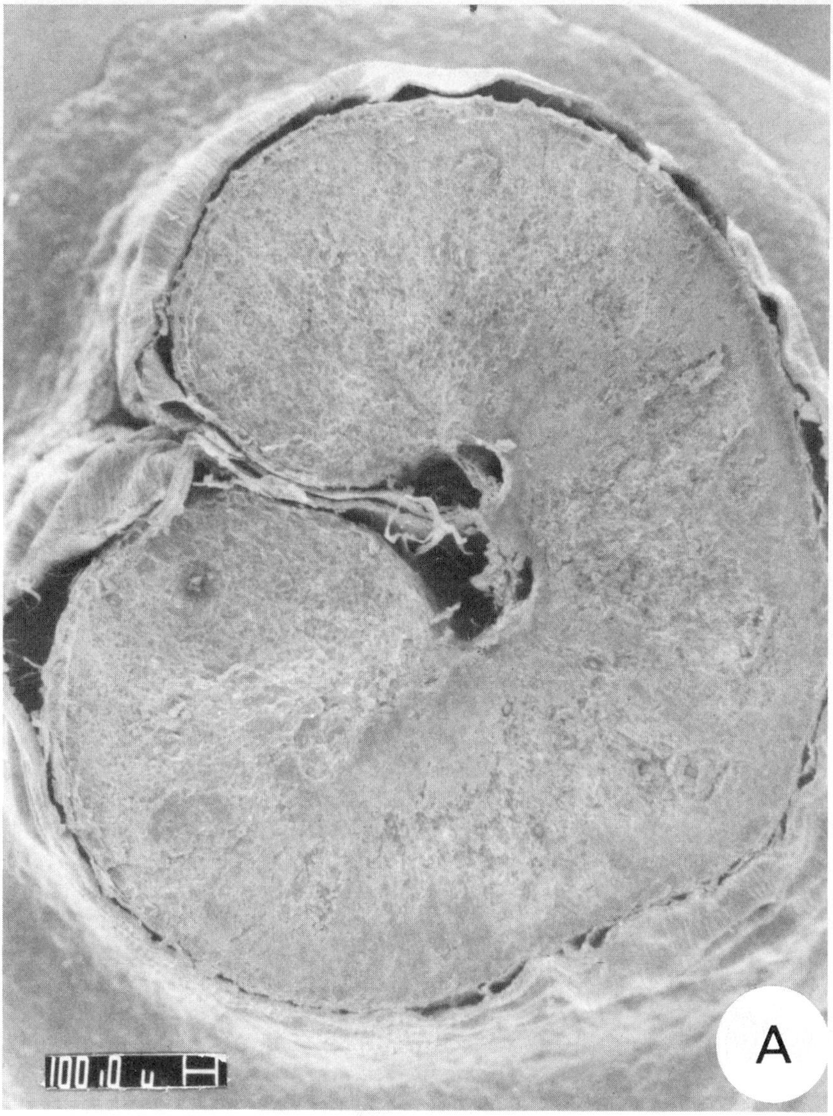

FIGURE 2 Scanning electron micrographs of P-5 triticale grain: (A) plump grain;

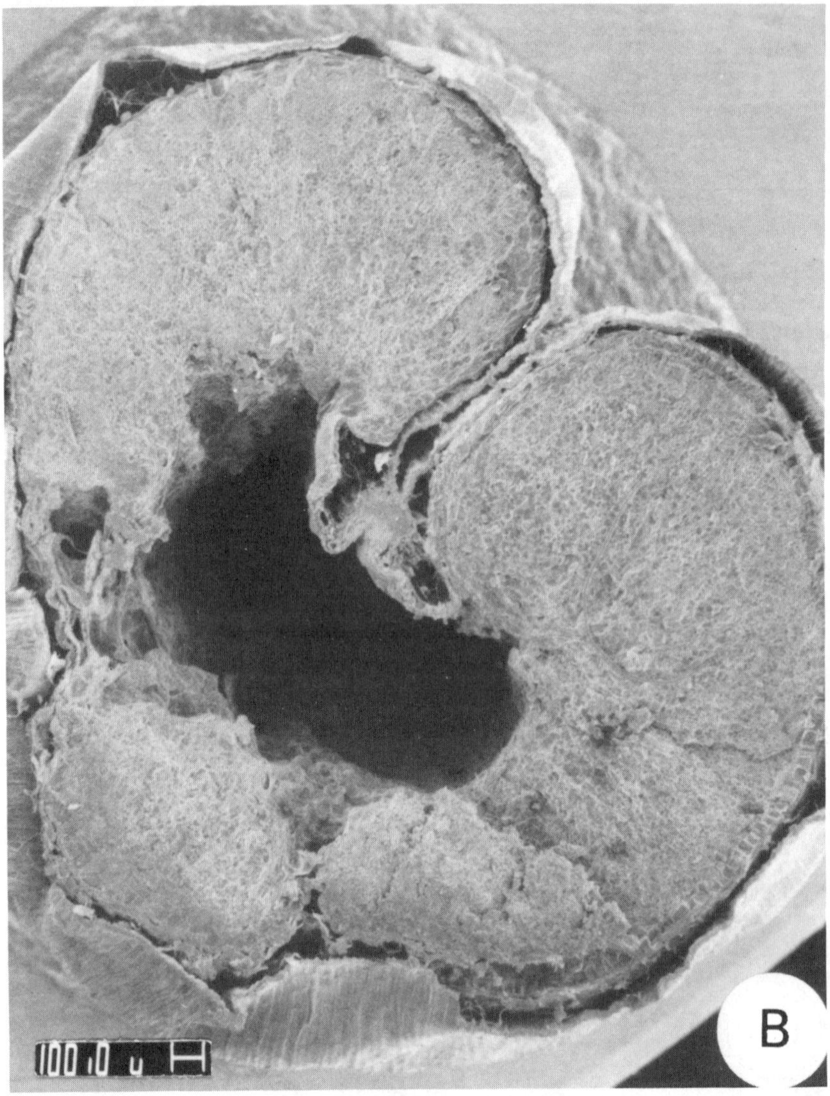

FIGURE 2 (B) medium shriveled grain;

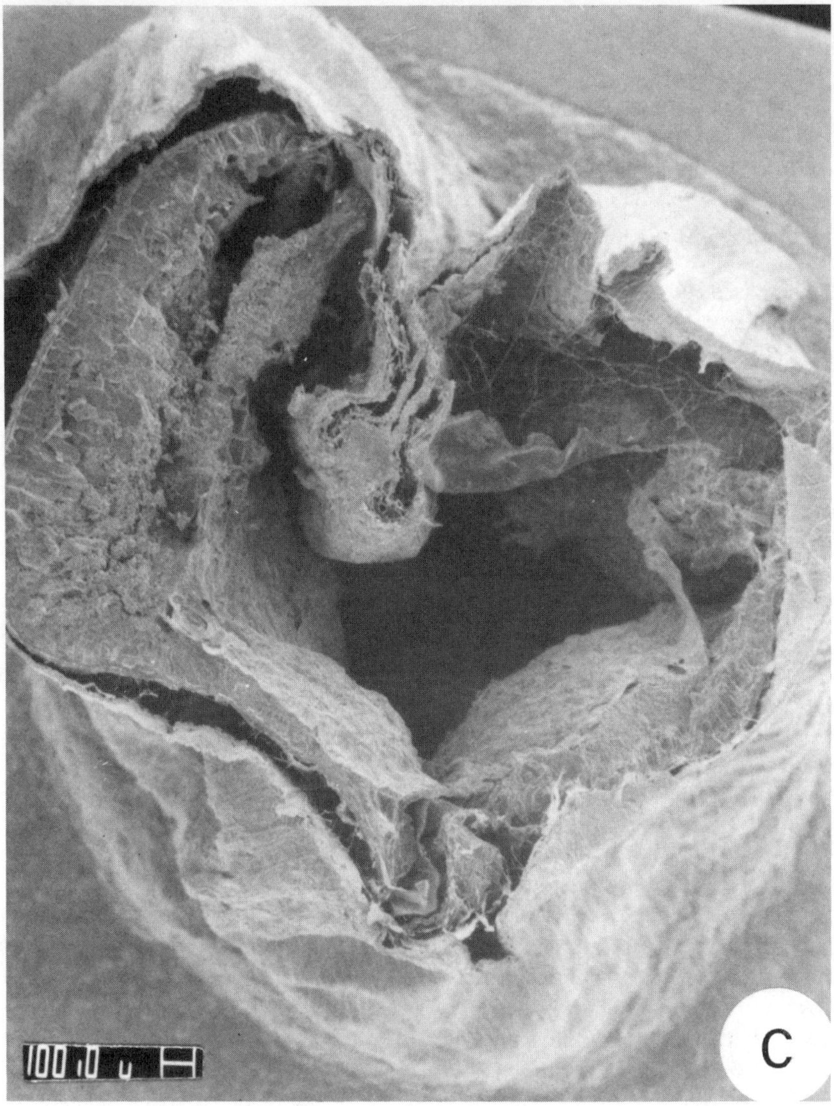

FIGURE 2 (C) highly shriveled grain. From Ref. 73.

So the possible causes for seed shriveling in triticales have been narrowed down. Photosynthesis is producing an adequate amount of sugars, which have no problem being transported. Due to the abundance of starch-synthesizing enzymes, there is no difficulty in converting these sugars into starch, which is safe from the elevated α-amylase and can accumulate within each triticale grain.

Shriveled triticale kernels analyzed in more biochemical detail have been shown to contain significantly higher levels of acid phosphatase activity as compared with plump types (19). This increased enzyme level was positively correlated with a decrease in dry seed weight and a premature stabilization of starch accumulation. In other words, as acid phosphatase increased, so did the degree of seed shriveling. The specific enzymes analyzed were para-nitrophenyl phosphatase, phospatidic phosphatase, glucose–1–phosphatase, ADPase, and ATPase. Normally, acid phosphatase is the major enzymatic component of lysosomes, necessary organelles responsible for vacuole formation and digestion of useless organelles (60). In shriveled-kernel type triticales, these enzyme levels are elevated, especially in later stages of seed maturation, and may create several biochemical imbalances. These enzymes may diminish available substrates necessary for starch synthesis. They might also consume the limited energy supply, which would slow down growth and hinder cell maintenance. Alternatively, acid phosphatases may degrade cellular membranes and other structures; this could directly create areas of necrosis (19).

The question still remains: why are the enzyme levels higher in triticale than either wheat or rye? Ching et al. theorized that because mitosis is unstable in many triticales (see Ref. 104), enzymes are needed to degrade the aberrant nuclei (18).

Kaltsikes et al. were the first to correlate the presence of rye heterochromatin in triticale with seed shriveling (12,42). Later, Gustafson and Bennett correlated the presence of rye telomeric heterochromatin with the production of aberrant nuclei in a ceonocytic endosperm, which in turn was correlated with test and kernel weight, and ultimately yield (33). Bennett hypothesized that since rye telomeric heterochromatin is late replicating, it is the cause of greater numbers of aberrant nuclei being produced in the triticale endosperm, which Gustafson and Bennett went on to show had a direct effect on test weight and thousand kernel weight (12,33). However, Varghese and Lelley found evidence that suggests that wheat, as well as rye chromosomes may be responsible for aberrant nuclei created by chromosome bridging (109). They concluded that disruption of spindle fibers was a more likely explanation for the production of aberrant nuclei.

Varghese et al. also analyzed mitosis of triticale embryos and found a correlation between slower nuclear divisions and a higher number of aberrant nuclei but determined that the cell division rate itself was not a direct cause of seed shriveling (110). Since the nuclear division rate of a cellular embryo is quite different from that in an early developing ceonocytic endosperm, it is not surprising that the embryo nuclear division rate was not related to seed shriveling. They went on to state that seed shriveling in triticale was mainly under genic control rather than heterochromatic control (110). Currently, it appears as though triticale's unique genome constitution (genic and heterochromatic) both act to determine the extent of aberrant nuclear production in the coenocytic endosperm, which indirectly creates an environment conducive to cellular necrosis and seed shriveling. It is still not known whether the genic or heterochromatic factors plays the major role in controlling seed shriveling in triticale.

In the past decade, many potential alternative causes for seed shriveling in triticale have been examined and either rejected or accepted for further study. The only definite correlations remaining are that shriveled seeds have a high production of aberrant nuclei in the ceonocytic endosperm and have higher levels of acid phosphatase, and that there may be an added genic component influencing seed shriveling. It still needs to be determined whether these symptoms, share a common cause or one is a result of the other.

IV. COMPOSITION AND QUALITY

A major attribute of triticale is its unique nutritional quality. It is generally accepted that triticale has high protein levels and an amino acid composition preferable to wheat (Tables 2 and 3). Such benefits clearly label triticale an excellent food grain. Additional information on triticale quality parameters can be found in an article by Lorenz (47). However, that data was derived from analyses of 20-year-old varieties and might not be comparable to present-day varieties.

Any antinutritive factors found in triticale are likely to be from its rye parent, which is known to reduce feed intake and diminish the growth rate of animals (27,37). Some compounds from rye that could impede triticale's nutritional performance include alkyl resorcinols, protease inhibitors, and nonstarchy polysaccharides (Table 4).

Rakowska et al. noted that the feed/grain ratio was significantly correlated ($r = +0.24$) with the pentosan (a component of nonstarchy polysaccharides) content in rye, but could not find a similar correlation for triticale (81). The amount of pentosans may only be a partial explanation for a lack of correlation, if triticale pentosans differ in molecular size and structure from wheat and rye (81). This would be possible if triticale contained hybrid molecules whose parts came from both wheat and rye or if the triticale had inherited a specific pentosan type from its wheat parent.

Rakowska et al. also obtained results contradicting trypsin inhibitors as major factors of triticale quality (81). They found no relationship between antitrypsin activity in triticale and protein digestibility in pigs or rats and concluded the trypsin inhibitor level was too low to create an effect. A decrease in digestibility is known to cause diarrhea in chicks; however, the 200 triticale lines tested did not show any indication of causing diarrhea or reduced digestibility. Also, in their experiments, triticale gave a better growth response overall than wheat or rye.

A nutritional analysis of triticale by Radcliffe et al. confirmed higher lysine values as compared to wheat, but none of the nine CIMMYT triticales studied contained significantly higher levels of protein than the local Australian wheat check, "Warimba" (80). Decreased protein values have accompanied the improvement of triticale's agronomic qualities (86). This is not surprising since improvements in grain size are accompanied by a decrease in percent protein in cereals. The decrease in protein is due to a disproportionate increase in starch content.

Since lysine is the first limiting amino acid in most cereals, because it contains a higher level of lysine than wheat, triticale would be nutritionally advantageous (Tables 2 and 5). Unfortunately, this has not always been observed (17). Growth trials with rats could not verify triticale's nutritional advantage since a significant difference between the wheat and triticales tested could not be determined (80). They proposed that either lysine cannot be made fully available to the animal or an antinutritional component is limiting growth. They specifically studied alkyl resorcinol levels but were unable to find a correlation between this compound and reduced seed conversion efficiency.

Although the protein contents of many triticale varieties are equivalent to wheats (Table 6 and 7), the percent and quality of each protein class remains dissimilar. Lupano and Anon separately characterized the protein components of the endosperm and germ of an Argentinian triticale (Line 47746) (55). Of the extracted endosperm proteins, 33% were albumins and globulins, the most likely lysine-containing fraction, and 29% were alcohol soluble. In contrast, the germ consisted of 5.9% albumin- and globulin-type proteins and only 10% alcohol-soluble type proteins. The triticales had a higher proportion of water-soluble protein than bread wheat and a lower percentage of alcohol-soluble, insoluble, and acid-soluble proteins, which are divided into two groups: pH 5.8 soluble and pH 5.8 insoluble. The acid (pH 5.8)-insoluble group is the protein fraction which contributes to good bread-making quality (55,70). This variation is critical since it is the interaction between appropriate concentrations of gluteninlike and gliadinlike proteins and their effect on other dough and flour components that produces optimal viscoelastic properties. However, this can be only a partial explanation for triticale's bread properties, because even triticales that produce a level of gluten comparable with bread wheat have weak dough strengths (69,70,97). Since a greater number of nonaggregated proteins are found in triticale glutenin, Pena suggested that the lack of aggregation contributed to triticale's inferior bread-making quality (70).

V. UTILIZATION

A. Food Grain

There is still considerable interest in the potential of triticale as a food grain. However, the predictions of the 1970s with regard to the anticipated rapid expansion of its use for baked products have not been realized. The suspected advantages of triticale over wheat and rye for products traditionally made from these grains have not been confirmed. Also, the indefinite supply of triticale grain of a uniform quality has mitigated against investment of funds for food product developmental research based on this cereal. However, there has been some potential useful research on the milling and baking properties of improved strains of triticale since the last major review (17).

1. Milling

Triticale grain can be milled into flour by essentially the same milling processes that have been perfected for milling wheat and rye. Because of the wide variation in kernel hardness and the degree of kernel shriveling, there is no standard milling procedure for triticale. Accordingly, a constant milling procedure applied to different triticale varieties will give a wide range of milling yields with the flours varying in ash content (Table 7) (17). However, for any one grain sample, the milling procedure can be optimized to produce a straight grade flour of acceptable ash content (e.g., below 0.55%). Such optimization generally produces yields in the range of 60–65% of flour with ash contents between 0.45 and 0.55%. The flour varieties of secondary hexaploid triticales used in the recent study by Macri et al. gave flour yields of 58.6–68.3% and ash content of 0.44–0.56% when milled on a Buhler experimental mill after tempering to a 14.5% moisture content (57). Comparative values for one sample of Canadian hard red spring wheat (cv. Marquis) were 71.7% yield and 0.45% ash content. Several international triticale breeding programs include kernel characteristics vis-a-vis milling properties among the breeding objectives. The CIMMYT program in Mexico continues to show improvement in kernel hardness and degree of shriveling (5,6). Similar improvements have been achieved in Australia and in Poland (38,92). Triticale breeders predict that with continued breeding it should be possible to develop varieties with complete plump kernels and varying degrees of kernel hardness as required for different milling uses.

Production of potentially useful fractions of triticale by wet milling and fractionation and by air classification has been reviewed previously (17). Fractions with interesting functional and nutritional properties by both types of processing have been prepared in the laboratory. However, such products have not yet found commercial application.

2. Dough and Baking Properties

Early interest in triticale flour for the production of baked products, catalyzed mainly by its demonstrated higher nutritional quality (due to higher protein and lysine contents), has not resulted in a continued growth of its use in the baking industry. Use of triticale flour for the production of wheat-type bread has been limited because of its higher α-amylase activity and tendency to produce weaker doughs. In spite of these deficiencies, satisfactory bread can be produced from triticale by the straight dough-baking method with the appropriate adjustments in water absorption, dough-mixing speed and time, and fermentation time (11,106,107). With appropriate adjustments in processing, most of the baked products commonly made from wheat flour can be produced from triticale (Table 8) (17,21,22).

Recent work on the baking properties of triticale has concentrated on the newly developed varieties of secondary hexaploid triticale (5,6,56,57,70,92). New intensive selection procedures have resulted in a substantial improvement in kernel characteristics and thereby a marked improvement in milling procedures. However, this effort has tended to bring triticale closer to wheat in terms of gluten strength and amylase activity (15,57,58,70,72). The overall bread making quality of the new triticale varieties is considerably better than that of the earlier ones, but it is still somewhat inferior to that of bread wheat of the same protein content. This inferiority appears to be due to a deficiency in protein "quality" as reflected by a lower percentage of gluten as compared to bread wheat.

Following the suggestion of Branlard and Barloy, research is currently in progress on the substitution of chromosome 1D from bread wheat into an isogenic line of hexaploid triticale in an attempt to improve the quality of the protein for bread making (Larter, personal communication) (14). If the appropriate segment of wheat chromosome 1D can indeed be incorporated into a stable hexaploid triticale background, it may be possible to achieve a substantial improvement in baking quality while maintaining the other attributes of triticale. Furthermore, there is evidence that the quality of the derived triticale depends on the quality of the rye and wheat parents (103). This factor should be taken into consideration in triticale variety improvement. However, one should also keep in mind that octoploid triticales, which contain the complete D genome from bread wheat, never had good quality. A complex interaction between wheat protein genes and those of rye could be causing the poor quality in triticale whether or not wheat chromosome 1D is present.

Even though triticale does not possess the protein composition of hexaploid wheat, it can still be utilized for making wheat bread products by blending triticale flour with wheat and modifying mixing and baking procedures. Blends of 25% or even 50% triticale flour with wheat flour produce a commercially acceptable product. Clercal, Lasko, and other triticales were tested for their ability to make acceptable French-style bread such as baguettes. Triticale flour yields were very promising at 68 to 71% compared with a minimum of 72% for the wheat check. The bread produced had an acceptable quality, and its taste was preferred over bread made with rye (11).

However, all triticale and triticale-wheat flours require alterations of the standard baking procedures. In order to create a dough with the desired viscoelastic properties, the mixing speeds, fermentation time and temperature, proofing time, and time between mixing and first punch must all be decreased to compensate for triticale's weak dough strength (97). To obtain optimal conditions for dough development, bulk fermentation should be eliminated and the levels of oxidizing agents and yeast need to be increased (70).

Triticale flour can also be used with minor modifications in the production of cookies, tortillas, cakes, and other soft wheat products since triticale basically performs like a soft wheat (Table 9). Crackers made from 25, 40, 55, and 70% triticale flour were tested for taste and overall acceptability. The lower triticale percentage mixtures were better accepted than higher percentage crackers (41). Yet even the 25% triticale crackers contained 11.8% protein, 0.7% more than whole wheat crackers, and the 40% triticale crackers had protein levels as high as 12.09%. So even small amounts of triticale added to everyday foods can significantly increase the product's nutritional value.

B. Feed Grain

Some of the early triticale feeding trials were encouraging. However, some varieties caused reductions in weight gain while others exhibited poor feeding efficiency. The possible reasons for such inconsistencies are as varied as the triticale varieties themselves. Variations in protein content, amino acid composition, nutritional availability, and antinutritional factors can all have an effect upon triticale's performance as a feed grain.

In trials using the common vole as a model, the Canadian triticale 6TA-131 gave the best body weight gain and possessed the highest protein efficiency ratio in a comparison with Grana wheat and Dankowskie Zlote rye (79). It seems that even though some triticales are inappropriate sources of animal feed, others can be utilized successfully.

"Beagle 82", a triticale released by the University of Georgia Coastal Plain Experiment Station, was utilized in a comprehensive study as a substitute feed for pigs. Three age groups of pigs were fed a total of nine diets in order to analyze the ability of triticale to replace soybeans or corn, the more expensive feed grains (36). Although a complete substitution of triticale for corn in diets of nursery pigs (5 weeks old) resulted in decreased rates of gain and feed efficiency, Hale and Utley suggested this might be due to a lower energy level in this diet as compared with corn/soybean meal mixes (36). Overall, they concluded that "Beagle 82" could adequately replace corn in pig diets. Pigs fed triticale found this grain palatable and were able to produce carcasses of comparable quality to corn-fed pigs.

Gatel et al. also obtained promising results with French triticale, "Clercal," as a feed for pigs (28). Using triticale in cereal/soybean meal mixtures enabled a reduction in the amount of the expensive soybean meal required. Compared with wheat/soybean meal diets, "Clercal" diets produced 4.3% higher growth rates, most likely due to an increased feed consumption with the triticale diet. The overall conclusion from tests on weaned piglets, growing-finishing pigs, and broiler chickens was that triticale rated equal to wheat and corn.

C. Forage

Triticale has been used as a forage from Palouse silt loam of the Pacific Northwest, to loam soils of Guelph, Ontario, Canada, to the fine sandy loam of Texas, and although not originally intended for this use, it has gained acceptance in most areas as an alternative forage. Both winter and spring varieties may be utilized as a nutritious food supply for livestock.

Performance trials in southern Kansas showed equivalent forage yields for triticale, wheat, rye, and barley (88). Trials in Georgia gave similar yields for triticale, wheat, and oats (16). However, Bishnoi and Hughes obtained more promising results in yield trials of winter triticales grown in Florida (13). These triticales were able to produce yields greater than wheat forages and similar to rye. Triticale forage yield trials in east Texas averaged about 6 t/ha dry matter over a season, which was higher than wheat forage yields of 4.8 to 5.6 t/ha for the same season. However, one noted disadvantage of triticale was its variability of yield from harvest to harvest as compared with wheat and rye (Table 10) (65). Another area of Texas gave very encouraging yields for their winter forage triticales, which were similar to rye and oats but were slightly higher than wheat under both dryland and irrigated conditions (40).

Shukla et al. evaluated triticale as a potential replacement for oats as the winter forage in northern India (96). Although the forage yields for triticale under irrigated conditions were relatively poor compared to oats and barley, triticale did much better under dryland conditions. Without irrigation, triticale could produce higher yields than oats, but still finished behind barley.

Two spring triticales, VT75229 and CF76, have also been tested on their forage capabilities. Ciha determined that for the area of eastern Washington, triticale could produce yields equivalent to other spring cereals if harvested at the soft-dough stage (113–116 days) but averaged lower yields if harvested earlier (20).

Although yields are adequate, there appears to be a slight advantage to using some triticales over other cereal forages when one analyzes its protein content. The spring triticales VT5229 and CF76 contained 0.80 and 0.87t/ha crude protein, respectively, at the soft dough stage. The local commercial soft spring wheat "Fielder" only managed to produce 0.75t/ha crude protein. Crude protein values of 0.57t/ha were also obtained for "Corgo" at a similar stage of development.

Poysa conducted an interesting study to determine the effect of harvesting a winter cereal as a forage and then allowing the regrowth to set seed so the grain could also be harvested (78). Although for most of the triticales tested, a forage harvest resulted in reduced grain yields, "OAC Wintri," a licensed winter triticale in Canada, actually increased its grain yield following an early forage harvest. This increased yield was attributed to a decrease in lodging of the cut plants. The only potential drawback was a delay of the heading date. However, this is a small concession for obtaining 1.5t/ha of forage with 30.5% protein, and an additional 0.5t/ha more grain than one could obtain without taking a forage harvest (78).

In the future, it may become more common to predict the quality of triticale forages by utilizing near infrared reflectance spectroscopy (NIRS) as is more commonly done for perennial forages. Marten et al. determined that the ability of NIRS to accurately assess such parameters as crude protein concentration and in vitro digestible dry matter is as accurate for triticale as for the perennial forage species (59).

D. Other Uses

1. Brewing

The utilization of triticale in brewing has been explored since the first cultivar was licensed. Pomeranz et al. indicated that the amount of nitrogenous material contained in the worts of the lines they analyzed was higher than on those from existing barleys (77). They were also higher in diastatic power and α- and β-amylase activities. However, beer made from triticale was naturally darker and had a higher pH value than beers made from barley and contained less alcohol (77).

Recently, Sobral analyzed some modern high-yielding triticales and found out that using 30% unmalted triticale did not affect the organoleptic characteristics of beer (98). When malted triticale was used, it seemed that optimization of the malting process was required. If the triticale varieties that Sobral used are to be utilized in making beer, then the brewing diagram must be adjusted in order to avoid having too much color and too long a period for proteolytic enzyme activity. In addition, the fermentation diagram needs to be adjusted to the high content of amino acids in the wort (Table 11).

VI. VULNERABILITY, PROBLEMS AND POTENTIALS

The vulnerability of the triticale as a commercial cereal crop stems from its limited gene base. The pathology section discussed the problems that can arise when large acreages of genetically closely related triticales are grown. A large production area of any cereal leaves it open to attack from the pathogens in that location, and the narrower the gene base, the easier it will be for the pathogens to break down any resistance present in the variety. This problem will be repeated over and over in locations growing large acreages of triticale until there are enough varieties in production from genetically diverse backgrounds.

The problem of a narrow gene base is further complicated, even though primary triticales are technically easy to make, by the fact that the agronomic characteristics of primary triticales never achieve the attributes present in the parents. Most new primary triticales are very disappointing, and a great deal of breeding work is needed before they can be useful as parents in a breeding program. This continues to slow down the expansion of triticale's gene base.

Finally, the acceptance of triticale grain by millers and bakers has been poor in many cases. Triticale does not mill and bake like wheat, and bakers are extremely reluctant to change procedures unless they can be guaranteed a constant supply of grain and a marketable product. On the other hand, farmers will not venture into a new crop unless they can be assured that there is a market for that grain.

A great deal has been written about the potential of triticale. Shebeski ended his presentation with the following: "*Triticale* will surely plateau at yields about 50 percent above those of wheat, should be grown whenever wheat is grown and beyond . . . " (94). Skovmand et al. concluded (conservatively) that triticale will eventually be grown on at least the same amount of acreage as durum (97). They then pointed out (realistically) that triticale will be grown on a much more acreage than durum because of its adaptability to marginal soils. These authors and others have written many glowing words about the future of triticale.

The current triticales are very sound agronomically and are able to compete with other cereals in many areas of the world. This is evident when one looks at the current world acreage figures (see Table 1). Acreage has been increasing annually for several years. This increase has been due mainly to increased production on the marginal soils that occur in many countries. Triticale yields will continue to increase because of the plant's potential for higher biomass production than other cereals.

Contributions from the U.S. Department of Agriculture, Agricultural Research Service, the University of Missouri Agricultural Experiment Station, Columbia, Missouri, U.S. A., Journal Series No. 10661, and the Department of Food Science, University of Manitoba, Winnipeg, Manitoba, Canada (Publication No. 11, p. 303).

TABLE 1 World Production of Triticale in 1986 (unless otherwise stated)

Country	Cultivation (ha)[a]	
Argentina	25,000[b]	
Australia	140,000	
Belgium	5,000	
Brazil	16,000	30,000 (est. 1987)
Bulgaria	10,000[b]	
Canada	15,000[b]	
China	25,000[b]	
Denmark	25[b]	
Federal Republic of Germany	30,000	
France	200,000	
German Democratic Republic	6,000[b]	
Greece	150[b]	
Hungary	5,000[b]	
India	500[b]	
Italy	15,000	
Kenya	5,000[b]	
Luxembourg	400[b]	
Mexico	8,000[b]	
New Zealand	150[b]	
Poland	309,000	400,000(est.1987–88)
Portugal	7,000	
Spain	35,000	
South Africa	15,000[b]	
Switzerland	5,000[b]	
Tanzania	380[b]	
The Netherlands	1,000	
Tunisia	5,000[b]	
United Kingdom	16,000	
United States	100,000[b]	
U.S.S.R.	250,000[b]	

[a]Countries where the acreage is unknown are assumed to have fewer than 1000 hectares.
[b]1984.

TABLE 2 Amino Acid Content of Triticale (g/100 g protein)

Amino acid	Triticale (Yoreme)	Wheat (INIA)	Rye (Snoopy)
Lysine	3.44	2.83	4.02
Threonine	3.55	2.98	4.06
Methionine[a]	1.28	1.42	1.35
Isoleucine	3.45	2.68	3.70
Leucine	7.20	7.22	7.75
Phenylalanine	4.94	3.77	4.74
Valine	4.48	3.73	5.10
Tryptophan	1.02	1.10	ND

[a]Partial destruction during hydrolysis.
Source: From General Laboratories, CIMMYT, 1982, with permission.

TABLE 3 Percent of True Protein Digestibility, Biological Value, and Net Protein Utilization of Some Varieties of Triticale Compared with a Wheat Variety, Using Male White Rats

	Triticale			Wheat
	Mapache	Beagle	PC–297	Hermosillo–77
True protein digestibility	92.7	91.0	91.5	92.0
Biological value	66.1	69.9	59.3	57.6
Net protein utilization	61.3	63.7	54.2	52.9

Source: Ref. 112.

TABLE 4 Nonstarchy Polysacharride Content (% air dry matter)

Cereal	Pentosans		Uroids	β-Glucans
	Total	Soluble		
Triticale (Lasko)	7.6	0.9	7.7	3.2
Rye (Dankowski Zlote)	8.6	1.7	5.4	4.1
Barley (Damazy)	6.9	0.5	4.4	4.3
Wheat (Grana)	4.4	0.5	4.5	3.9

Source: Ref. 81.

TABLE 5 Percent Amino Acid Content of Triticale, "Beagle 82"

Alanine	0.51	Lysine	0.43
Arginine	0.65	Methionine	0.22
Aspartic acid	0.71	Phenylalanine	0.58
Cystine	0.25	Proline	1.21
Glutaminc acid	3.12	Serine	0.60
Glycine	0.53	Threonine	0.41
Histidine	0.23	Tryptophan	0.11
Isoleucine	0.47	Tyrosine	0.33
Leucine	0.88	Valine	0.52

Values listed on 90% dry matter basis.
Source: Ref. 36.

TABLE 6 Percent Protein of Triticale Grains

Triticales:				Durum wheats:	
AM 3425	14.0	Okla F5	18.1	Ward	12.6
AM 3706	15.5	X2805	16.5	Cando	13.3
AM 2794	15.0	77 CFN	17.1	Rolette	13.2
AM 2871	14.2	6TB 227	15.9	Cocorit	14.1
AM 3695	16.0	Beagle	17.5		
AM 2853	14.5	CIMMYT (bulk)	15.7	Bread wheats:	
AM 3693	14.9	A2–31–3	15.4	Dannie	15.0
AM 2873	13.0	Mex F8	15.1	Calidad[a]	14.8
AM 2137	14.9	Rahum[a]	12.4	Hermosillo[a]	13.0
AM 1959	15.9	Mapache[a]	12.3		
AM 3685	16.6	Pm 28 Bulk-Cml "s"[a]	12.7	Rye:	
AM 2059	14.5	1A-M2A x Pi62/Bgl "s"[a]	11.8	Snoopy	13.7
AM 3948	15.3	Drira-Arm "s"[a]	12.0		
AM 3426	16.5	IRA x M2A[a]	12.6		
AM 2838	18.3	Bgc-Bulk e2[a]	13.7		

% protein = N × 5.7.
[a]% protein = N × 5.83 (72).
Source: Ref. 93.

TABLE 7 Chemical Composition of Triticale and Wheat Grown in the Yaqui Valley, Mexico, 1977–1978

	Triticale[a]			Wheat[a]
	Mapache	Beagle	PC–297	Hermosillo–77
Protein[b]	15.91	12.59	17.21	17.84
Fat	2.21	1.95	1.59	2.06
N-free extracts	77.81	80.81	77.38	76.21
Crude fiber	2.12	2.58	1.98	2.30
Ash	1.95	2.07	1.84	1.59
Available carbohydrate	68.55	70.86	68.66	68.08
Tannin	0.61	0.53	0.56	0.47

[a](% dry basis).
[b](% N x 6.25).
Source: Ref. 112.

TABLE 8 Changes in Bread-Making Technology That Improve the Product Obtained from Triticale Flour

Process	Compared with wheat	Ref.
Mixing speed	slower	17,49,50
Fermentation time	shorter	23,46,107
Proofing time	shorter	17
Fermentation temperature	lower	4
Yeast concentration	higher	4
Time from mixing to first punch	shorter	4

Source: Ref. 97.

TABLE 9 Unleavened Products from Triticale

Product	Ref.
Chapatis	82, 85, 93, 99
Pancakes	
Tortillas	24
Biscuits	24
Cookies	48, 83, 107
Cakes	71
Pizza crust	111
Sweet goods	71
Breakfast cereal	52
Pasta products	51, 68
Noodles	51
Distilled products	76, 77
Protein concentrates	90, 114
Waffle mix	85

Source: Ref. 97.

TABLE 10 Triticale Forage Yield (tt/ha)

Grazer Blend[a]	12.55	A–476	6.44
EXP A876–6[a]	10.48	A–313A–737	6.29
EXP A313A–16[a]	14.53	A–876–6	6.26
EXP A386A–12[a]	10.78	ARK 2014	5.95
B–227–8	7.81	B–227–14	5.92
876–J	7.35	ARK 2309	5.81
B–858	7.18	NF–55	5.69
A–313–A–16	7.15	ARK 2307	5.26
A–313–A–15	6.99	Coker 68–15 wheat	5.26
CO–61	5.99	NF–12	5.18
CO–53	6.52	ARK 2301	5.09
A–876–28	6.47	NF–185	4.98

Conversion to SI units by authors.
[a]From Ref. 40.
Source: Ref. 65.

TABLE 11 The Free Amino Acid Composition of Industrial Worts from Modern Portugese Triticales (mg/liter amino acid)

Asp	5.2[a]	Met	2.1	Thr	5.1
Leu	4.8	Ser + Asn	7.7	Ile	11.4
Glu	5.7	Tyr	6.9	Pro	29.4
Phe	8.9	Gly	2.8	Lys	8.8
Ala	9.4	His	5.3	Val	8.6

[a] mg/100 ml.
Source: Ref. 98.

REFERENCES

1. Agrawal, P. K., Changes in amylase, starch, and reducing sugars during grain development in triticale and their relation to grain shriveling. *Cereal Res. Comm.*, *5*:225–233 (1977).
2. Agrawal, P. K., and Fock, H. P., Carbon dioxide exchange in attached flag leaves and ears of hexaploid triticale in relation to seed shriveling. *Field Crops Res.*, *8*:315–321 (1984).
3. Agrawal, P. K., and Fock, H. P., Incorporation of $^{14}CO_2$ into carbohydrates in the developing kernels of triticale (x *Triticosecale* Wittmack) in relation to kernel shriveling. *Field Crops Res.*, *10*:143–149 (1985).
4. Amaya, A., and Pena, R. J., *CIMMYT Annual Report.* International Maize and Wheat Improvement Center, El Batan, Mexico, 1981.
5. Amaya, A., and Skovmand, B., Current status of hexaploid triticale quality. In *Genetics and Breeding of Triticale.* EUCARPIA Meeting, M. Bernard and S. Bernard (Eds.) Clermont–Ferrand, France. INRA, Paris, 1985, pp. 603–606.
6. Amaya, A., Pena, R. J., and Varughese, G., Influence of grain hardness in the milling and baking properties of recently developed triticales. In *Proceedings International Triticale Symposium*, N. L. Darvey (Ed.). The Australian Institute of Agricultural Science, Sydney, 1986, pp. 511–524.
7. Aniol, A., Aluminum uptake by roots of two winter wheat varieties of different tolerance to aluminum. *Biochem. Physiol. Pflanz.*, *178*:11–20 (1983).
8. Aniol, A. Inheritance of aluminum tolerance to triticale and parental species. In *Proceedings International Triticale Symposium*, N. L. Darvey (Ed.). The Australian Institute of Agricultural Science, Sydney, 1986, pp. 283–288.
9. Aniol, A., and Gustafson, J. P., Chromosome location of genes controlling aluminum tolerance in wheat, rye, and triticale. *Can J. Genet. Cytol.*, *26*:701–705 (1984).
10. Baier, A. C., and Nedel, J. L., Triticale in Brazil. In *Proceedings International Triticale Symposium*, N. L. Darvey (Ed.). The Australian Institute of Agricultural Science, Sydney, 1986 pp. 270–282.
11. Beaux, Y., and Martin, G., Bread–making aptitude in hexaploid triticale. In *Genetics and Breeding of Triticale.* EUCARPIA Meeting, M. Bernard and S. Bernard (Eds.) Clermont–Ferrand, France. INRA, Paris, 1985, pp. 651–655.
12. Bennett, M. D., Heterochromatin, aberrant endosperm nuclei and grain shriveling in wheat–rye genotypes. *Heredity*, *39*:411–419 (1977).
13. Bishnoi, U. R., and Hughes, J. L., Agronomic performance and protein content of fall–planted triticale, wheat and rye. *Agron. J.*, *71*:359–360 (1979).

14. Branlard, G., and Barloy, D., High molecular weight glutenin subunits in seed of triticale and its parental species. In *Genetics and Breeding of Triticale*. EUCARPIA Meeting, M. Bernard and S. Bernard (Eds.). Clermont–Ferrand, France. INRA, Paris, 1985, pp. 607–611.
15. Branlard, G., Bernard, M., Antraygue, C., and Barloy, D., Shriveling and alpha amylase activity in triticale and its parental species: preliminary studies. In *Genetics and Breeding of Triticale*. EUCARPIA Meeting, M. Bernard and S. Bernard (Eds.) Clermont–Ferrand France. INRA, Paris, 1985, pp. 617–632.
16. Brown, A. R., and Almodares, A., Quantity and quality of triticale forage compared to other small grains. *Agron. J.*, *68*:264–266 (1976).
17. Bushuk, W. and Larter, E. N., Triticale: Production, Chemistry and Technology. In *Advances in Cereal Science and Technology*, Vol. III, Y. Pomeranz (Ed.). Amer. Assoc. Cereal Chemists, St. Paul, MN, 1980, pp. 115–157.
18. Ching, T. M., Poklemba, C. J., and Metzger, R. J., Starch synthesis shriveled and plump triticale seeds. *Plant Physiol.*, *73*:652–657 (1983).
19. Ching, T. M., Thompson, D. M., and Metzger, R. J., Acid phosphatases and seed shriveling in triticale. *Plant Physiol.*, *76*:478–482 (1984).
20. Ciha, A. J., Forage production of triticale relative to other spring grains. *Agron. J.*, *75*:610–613 (1983).
21. Cooper, K., Triticale food use—Australia. In *Genetics and Breeding of Triticale*. EUCAR PIA Meeting, M. Bernard and S. Bernard (Eds.). Clermont–Ferrand France. INRA, Paris,1985, p. 657.
22. Cooper, K. V., Triticale food use in Australia. In *Proceedings Intl. Triticale Symp.*, N. L., Darvey (Ed.). Australian Institute of Agricultural Science, Sydney, 1986, pp. 538–539.
23. CIMMYT, *Report on Wheat Improvement*. International Maize and Wheat Improvement Center, El Batan, Mexico, 1977.
24. CIMMYT, *Report on Wheat Improvement*. International Maize and Wheat Improvement Center, El Batan, Mexico, 1980.
25. Dedio, W., Simmonds, D. H., Hill, R. D., and Shealy, H., Distribution of alpha–amylase in the triticale kernel during development. *Can. J. Plant Sci.*, *55*:29–36 (1975).
26. Dhaliwal, A. S., and Sharma, H. L., Changes in some carbohydrate fractions and some related enzymatic activities in plump and shriveled triticale and in wheat grains. *Aust. J. Plant Physiol.*, *13*:249–255 (1986).
27. Fernandez, R., Lucas, E., and McGinnis, J., Fractionation of chick growth depressing factor from rye. *Poultry Sci.*, *52*:2252 (1973).
28. Gatel, F., Lavorel, O., Fekete, J., Grosjean, F., and Castaing, J. Feeding value of triticale for monogastrics: weaned piglets, growing–finishing pigs and broilers. In *Genetics and Breeding of Triticale*, EUCARPIA Meeting, M. Bernard and S. Bernard (Eds.), Clermont–Ferrand, France. INRA, Paris, 1985, pp. 659–670.
29. Graham, R. D., and Pearce, D. T., The sensitivity of hexaploid and octoploid triticales and their parent species to copper deficiency. *Aust. J. Agric. Res.*, *30*:791–799 (1979).
30. Graham, R. D.., Anderson, G. D., and Aschv, J. S., Absorption of copper by wheat, rye, and some hybrid genotypes. *J. Plant Nutr.*, *3*:679–686 (1981).
31. Gupta, P. K., and Priyadarshan, P. M., Triticale: present status and future prospects. *Adv. Genet.*, *21*:255–345 (1982).
32. Gustafson, J. P., Cytogenetics of triticale. In *Cytogenetics of Crop Plants*, M. S. Swaminathan, P. K. Gupta, and U. Sinha (Eds.). Macmillan India Ltd., New Delhi, 1982, pp. 228–250.
33. Gustafson, J. P., and Bennett, M. D., The effect of telomeric heterochromatin on early endosperm development and kernel characteristics at maturity. *Can. J. Genet. Cytol.*, *24*: 83–92 (1982).
34. Gustafson, J. P., Dillé, J. E., and Skovmand, B., Wheat substitutions in hexaploid triticale. *Plant Breeding*, *102*:109–112. (1989).
35. Gustafson, J. P., and Zillinsky, F. J., Identification of D–genome chromosomes from hexaploid wheat in a 42–chromosome triticale. In *Proceedings of the 4th International*

Wheat Genetics Symposium, E. R. Sears and L. M. S. Sears (Eds.). Missouri Agr. Exp. Sta., Columbia, MO, 1973, pp. 225–232.

36. Hale, O. M., and Utley, P. R., Value of Beagle 82 triticale as a substitute for corn and soybean meal in the diet of pigs. *J. Animal Sci.*, *60*:1272–1279 (1985).
37. Hulse, J. H., and Laing, E. M., *Nutritive Value of Triticale*. International Development Research Centre, Ottawa, Canada, 1974.
38. Huskowska, T., Wolski, T., and Ceglinska, A., Breeding winter triticale for improvement of grain characters. In *Genetics and Breeding of Triticale*. EUCARPIA Meeting, M. Bernard and S. Bernard (Eds.), Clermont–Ferrand, France. INRA, Paris, 1985, pp. 641–649.
39. Jain, S. K., Cytogenetics of rye (*Secale* ssp.). *Bibliographia Genetica*, *XIX*:1–86 (1960).
40. Jones, R. M., Gardenhire, J. H., and Read, J. C., Small grain forage yields under irrigated and dryland conditions. *PR Tex. Agric. Exp. Stn.* (4141):25–28 (1983).
41. Kahn, C. B., and Penfield, M. P., Snack crackers containing whole–grain triticale flour: crispness, taste and acceptability. *J. Food Sci.*, *48*:266–267 (1983).
42. Kaltsikes, P. J., Roupakias, D. G., and Thomas, J. B., Endosperm abnormalities in *Triticum–Secale* combinations. I. X *Triticosecale* and its parental species. *Can. J. Bot.*, *53*:2050–2067 (1975).
43. Kiss, A., Neue Richtung in der Triticale–Zuchting. *Z Pflanzenzucht.*, *55*:309–329 (1966).
44. Klassen, A. J., Hill, R. D., and Larter, E. N., Alpha–amylase activity and carbohydrate content as related to kernel development in triticale. *Crop Sci.*, *11*:265–267 (1971).
45. Krolow, K. D., 4x triticale production and use in triticale breeding. In *Proceedings of the 4th Inernational Wheat Genetics Symposium*, E. R. Sears and L. M. S. Sears (Eds.). Missouri Ag. Exp. Sta., Columbia, MO, 1973, pp. 237–243.
46. Lorenz, K., Food uses of triticale, *Food Technol.*, *26*:66 (1972).
47. Lorenz, K., The history, development, and utilization of triticale. *Critical Reviews in Food Technology*, *5*:175–280 (1974).
48. Lorenz, K., Protein fortification of cookies. *Cereal Foods World*, *28*:449–452 (1983).
49. Lorenz, K., and Welsh, J., Agronomic and baking performance of semi–dwarf triticales.*Cereal Chem.*, *54*:1049–1056 (1977).
50. Lorenz, K., Disalver, W., and Lough, J., Evaluation of triticale in the manufacturing of noodles. *J. Food Sci.*, *37*:764 (1972).
51. Lorenz, K., Welsh, J., Normann, R., and Maga, J., Comparative mixing and baking properties of wheat and triticale flours. *Cereal Chem.*, *49*:187–193 (1972).
52. Lorenz, K., Reuter, F. W., and Sizer, C., The mineral composition of triticales and triticale milling fractions by x–ray fluorescence and atomic absorption.*Cereal Chem.*, *54*:534–542 (1974).
53. Lukaszewski, A. J., Apolinarska, B., and Gustafson, J. P., Introduction of the D–genome chromosomes from bread wheat into hexaploid triticale with complete rye genome. *Genome*, *29*:554–561 (1987).
54. Lukaszewski, A. J., Gustafson, J. P., Cytogenetics of triticale. *Plant Breeding Reviews*, *5*: 41–94 (1986).
55. Lupano, C. E., and Anon, M. C., Characterization of triticale protein. *Cereal Chem.*, *62*: 174–178 (1985).
56. Macri, L. J., The influence of alpha–amylase and protease activity on the baking quality of four secondary hexaploid triticales. M.Sc. thesis, University of Manitoba, Winnipeg, Manitoba, Canada, 1985
57. Macri, L. J., Ballance, G. M., and Larter, E. N., Factors affecting the breadmaking potential of four secondary hexaploid triticales. *Cereal Chem.*, *63*:263–267 (1986).
58. Macri, L. J., Ballance, G. M., and Larter, E. N., Changes in the alpha–amylase and protease activities of four secondary hexaploid triticales during kernel development. *Cereal Chem.*, *63*:267–270(1986).
59. Marten, G. C., Halgerson, J. L., and Cherney, J. H., Quality prediction of small grain forages by near infrared reflectance spectroscopy. *Crop Sci.*, *23*:94–96 (1983).

60. Marty, F., Branton, D., and Leigh, R. A., Plant vacuoles. In *The Biochemistry of Plants*, Vol.1. The Plant cell, Vol. 1, N. E. Tolbert (Ed.). Academic Press, New York, 1980, pp. 625–658.
61. McIntosh, R. A., Luig, N. H., Milne, D. L., and Cusick, J., Vulnerability of triticales to wheat stem rust. *Can. J. Plant Path.*, *5*:62–69 (1983).
62. McIntosh, R. A., and Singh, S. J., Rusts—real and potential problems for triticale. In *Proc. Intl. Triticale Symp.*, N. L. Darvey (Ed.). Australian Institute for Agricultural Science, Sydney, 1986, pp. 199–207.
63. Morrison, R. J., Larter, E. N., and Green, G. J., The genetics of resistance to *Puccinia graministritici* in hexaploid triticale. *Can. J. Genet. Cytol.*, *19*:683–693 (1977).
64. Muntzing, A., Triticale: results and problems. *Z. Pflanzenzucht.* (Suppl.) *10*:1–103 (1979).
65. Nelson, L. R., 1981–82 forage yields for oats, ryegrass, rye wheat and triticale. *PR Tex Agr. Exp. Sta.* (1441):29–35 (1983).
66. Oettler, G., The influence of the wheat and rye genome on the performance of primary triticale. In *Genetics and Breeding of Triticale.* EUCARPIA Meeting, M. Bernard and S. Bernard (Eds.), Clermont–Ferrand, France. INRA, Paris, 1985, pp. 125–134.
67. Oettler, G., Variation and covariation of agronomic characters in primary triticale and their wheat and rye parents. In *Proceedings of the International Triticale Symposium*, N. L. Darvey (Ed.). The Australian Institute of Agricultural Science, Sydney, 1986, pp. 120–123.
68. Pape, G., Beleia, A., Campos, J. E., Mazzari, M. R., Della Modesta, R. C., and NogueiraFernandez, V. L., EMBRAPA Bull. No. 004, 1982, pp. 1–44.
69. Payne, P. I., Corfield, K. G., Holt, L. M., and Blackman, J. A., Correlations between the inheritance of certain high–molecular–weight subunits of glutenin and breadmaking quality in progenies of six crosses of bread wheat. *J. Sci. Food Agric.*, *32*:51–60 (1981).
70. Pena, R. J., The influence of glutenin proteins on the mixing and baking properties of four secondary hexaploid triticales. Ph.D. thesis, University of Manitoba, Winnipeg, Manitoba, Canada, 1984.
71. Pena, R. J., and Amaya, A., Triticale; un nuevo cereal–desarollo. *Panagfa*, *8*:45–50 (1980).
72. Pena, R. J., and Bates, L. S., Grain shriveling in secondary hexaploid triticale. I. Alpha–amylase activity and carbohydrate content of mature and developing grains. *Cereal Chem.*, *59*:454–458 (1982).
73. Pena, R. J., Nagarajan, P., and Bates, L. S., Grain Shriveling in secondary hexaploid triticales. II. Morphology of mature and developing grains related to grain shriveling. *Cereal Chem.*, *59*:459–468 (1982).
74. Pieritz, W. J., Elimination von Chromosonem in amphidiploiden Weizen–roggen–basterden (triticale). *Z. Pflanzenzucht.*, *64*:90–109 (1970).
75. Pissarev, V., Different approaches to triticale breeding. *Hereditas Suppl.*, *2*:279–290 (1966).
76. Pomeranz, Y., Malting of triticale. In *Triticale: First Man–Made Cereal*, C. C. Tsen (Ed.). Am. Cereal Chem., St. Paul, MN, 1974, pp. 261–270.
77. Pomeranz, Y., Burkhart, B. A., and Moon, L. C., Triticale in malting and brewing. *Proc. Am. Soc. Brew. Chemists*:40 (1970).
78. Poysa, V. W., Effect of forage harvest on grain yield and agronomic performance of winter triticale, wheat and rye. *Can. J. Plant Sci.*, *65*:879–888 (1985).
79. Pronczuk, S., Biological value of triticale tested by common vole (*Microtus arvalis* Pall.) In *Genetics and Breeding of Triticale*, EUCARPIA Meeting, M. Bernard and S. Bernard (Eds.),Clermont–Ferrand France. INRA, Paris, 1985, pp. 671–674.
80. Radcliffe, B. C., Eagan, A. R., and Driscoll, C. J., Nutritional evaluation of triticale grain as an animal feed. *Aust. J. Exp. Agric. Anim. Husb.*, *23*:419–425 (1983).
81. Rakowska, M., Boros, D., and Raczynska–Bojanowska, K., Nutritive value of triticale grain as compared to wheat and rye. In *Genetics and Breeding of Triticale*, EUCARPIA Meeting, M. Bernard and S. Bernard (Eds.), Clermont–Ferrand, France. INRA, Paris, 1985, pp. 675–683.
82. Rao, G. V., Rao, G. C. P. R., Oatsala, C. N., Kermar, G. V., and Shurpaleker, S. R., Bread, biscuit and chapati making quality of Indian triticales. *J. Food Sci. Technol.*, *15*:11–15(1978).

83. Rao, V. R., Mehta, S. L., and Joshi, M. G., Peroxidase and amylase activity in developing grains of triticale, wheat and rye. *Phytochemistry, 15*:893–895 (1976).
84. Rimpau, W. Kreuzungprodukte landwirtschaftlicher kulturpflantzen. *Landwirtsch. Jahrb., 20*:335–371 (1891).
85. Rodgers, N., Triticale muscles into food markets. *Farm J., 97*:31 (1983).
86. Rodriguez Sanz, R., Sanchez–Monge, E., Carrillo, J. M. and Vasquez Muniz, J. F., Protein–lysine–yield–environment relationships in new triticale varieties. In *Genetics and Breeding of Triticale*, EUCARPIA Meeting, M. Bernard and S. Bernard (Eds.), Clermont–Ferrand France. INRA, Paris, 1985, pp. 633–640.
87. Salmon, D. F., Helm, J. H., Duggan, T. R., and Lakeman, D. M., The influence of chaff extracts on the germination of spring triticale. *Agron. J., 78*:863–867 (1986).
88. Sapra, V. T., Sharma, G. C., Hughes, J., and Bradford, R. R., Triticale, a wheat–rye hybrid. *J. Tenn. Acad. Sci., 48*:59–61 (1973).
89. Sarri, E. E., Varughese, G., and Abdalla, O. A., Triticale diseases: distribution and importance. In *Proc. Intl. Triticale Symp.*, N. L. Darvey (Ed.). The Australian Institute of Agricultural Science, Sydney, 1986, pp. 208–231.
90. Saunders, D. A., Betschart,A. A., Connor, M. A., Edwards, R. H., and Kohler, G. O., In *Triticale: First Man–Made Cereal*, C. C. Tsen (Ed.). Am. Cereal Chem., St. Paul, MN, 1974,pp. 280–291.
91. Scarth, R., Larter, E. N., Helm, J., Salmon, D. F., and Townley–Smith, T. F., The triticale breeding program in western Canada. In *Genetics and Breeding of Triticale*, M. Bernard and S. Bernard, (Eds.). EUCARPIA Meeting, Clermont–Ferrand, France. INRA, Paris, 1986, pp. 465–471.
92. Schuller, E. H., Valuation and milling for human consumption of bread. In *Proc. Int. Triticale Symp.*, N. L. Darvey (Ed.). The Australian Institute of Agricultural Science, Sydney, 1986, pp. 525–527.
93. Sekhon, K. S., Sharma, G. C., Paul, A. D., and Bietz, J. A., Nitrogen fertilization effects and anatomical, protein and amino acid characteristics of yellowberry in triticale. *Crop Sci., 23*:699–703 (1983).
94. Shebeski, L. H., Triticale in retrospect and prospect. *Hodowla Roslin Aklimatyzacja I Nasiennictwo, 24*:279–285 (1980).
95. Shigenaga, S., and Larter, E. N., Karyotype analysis of hexaploid triticale. *Can. J. Genet. Cytol., 13*:585–591 (1971).
96. Shukla, N. P., Lal, M., and Saxena, D. C., Effect of soil moisture regimes on water use and forage production by oats, barley and triticale. *Indian J. Agron., 29*:274–276 (1984).
97. Skovmand, B., Fox, P. N., and Villareal, R. L., Triticale in commercial agriculture: progress and promise. *Adv. Agron., 37*:1–45 (1984).
98. Sorbral, J., Triticale como potencial materia prima no farico de cerveja. In *Proc. 4th Reuniao Portuguesa Sobre Triticales.* Via Real, Portugal, June 15–17, 1987.
99. Srivastava, A. K., (1978). Psychology of kernel shriveling and its reduction in triticale. *Plant Cell Physiol., 19*:337–344 (1978).
100. Srivastava, J. P., Prospects of triticale as a commercial crop in India. *Triticale: Proc. Int. Symp. 1973*, Int. Develop. Res. Centre, Monogr. 024e, 1974, pp. 87–92.
101. Sweeny, G. C., Harria, H. C., and Jessop, R. S., Water use by triticale and wheat. In *Proc. Intl. Triticale Symp.*, N. L. Darvey (Ed.). The Australian Institute of Agricultural Science,Sydney, 1986, pp. 186–184.
102. Tarkowski, C., and Stefanowska, G., Chromosome morphology in the genome of rye, *Secale cereale* L. and in *Triticale* 6x and 8x. *Genet. Polon., 13*:83–89 (1972).
103. Tarkowski, C., and Stefanowska, G., Gruszecka, D., and Tochman, L., The effect of wheatand rye on the quality of protein in triticale. In *Genetics and Breeding of Triticale*, M. Bernard and S, Bernard (Eds.). EUCARPIA Meeting, Clermont–Ferrand, France. INRA, Paris, 1985, pp. 613–616.
104. Thomas, J. B., Kaltsikes, P. J., Gustafson, J. P., and Roupakias, D. G., Development of kernel shriveling in triticale. *Z. Pflanz., 85*:1–27 (1980).

105. Tripp, R., and Winkelmann, D., Socioeconomic factors influencing the adoption of triticale. In *Proc. Intl. Triticale Symp.*, N. L. Darvey (Ed.). The Australian Institute of Agricultural Science, Sydney, 1986, pp. 345–371.
106. Tsen, C. C., In *Triticale: First Man–Made Cereal*, Tsen, C. C. (Ed.), Am. Cereal Chem., St. Paul, MN, 1974, pp. 234–242.
107. Tsen, C. C., and Hoover, W. J., Baking quality of triticale flours. *Cereal Chem.*, *51*:365–375 (1974).
108. Varaghuse, G., Saari, E. E., and Abdalia, O. S., Two decades of triticale breeding and research at CIMMYT. In *Proc. Intl. Triticale Symp.*, N. L. Darvey (Ed.). The Australian Institute of Agricultural Science, Sydney, 1986, pp. 148–169.
109. Varghese, J. P., and Lelley, T., Origin of nuclear aberrations and seed shriveling in triticale: a re–evaluation of the role of C– heterochromatin. *Theor. Appl. Genet.*, *66*:159–167 (1983).
110. Varghese, J. P., Lelley, T., and Robbelen, G., Cell division rate, cell number and nuclear aberrations in relation to seed shriveling in triticale.*Z. Pflanz.*, *92*:102–116 (1984).
111. Velasquez, P., Master's thesis. Universidad Ibero Americana, 1983.
112. Villegas, E. M., Eggum, B. O., Vasal, S. K., and Kohli, M. M., Progress in nutritional improvement of maize and wheat. *Food Nutr. Bull.*, *2*:17–24 (1980).
113. Wilson, A. S., On the fertilization of cereals. *Trans. Proc. Bot. Soc.* (Edinburgh) *12*:237–242 (1875).
114. Wu, Y. Y., Stringfellow, A. C., Anderson, R. A., Sexson, K. R., and Wall, J. S., Triticale for food uses. *J. Agric. Food Chem.*, *26*:1039–1048 (1978).
115. Zillinsky, F. J., *Common Diseases of Small Grain Cereals: A Guide to Identification.* International Maize and Wheat Improvement Center, El Batan, Mexico, 1983.

10

Wild Rice: Processing and Utilization

Ervin A. Oelke
University of Minnesota, St. Paul, Minnesota

James J. Boedicker
University of Minnesota, Grand Rapids, Minnesota

I. INTRODUCTION

Wild rice (*Zizania*) is a native North American genus. The species, *Z. palustris* L., is found predominantly in the Great Lakes region of the United States and Canada. It grows in natural stands in the shallows of lakes and rivers and produces a grain that has been harvested for food for some 10,000 years (1).

The wild rice plant is 5–6 ft tall with 5 or 6 leaves on the main tiller and can develop 20–30 tillers (2). It is a cross-pollinated species and matures in about 120 days, requiring about 2600 growing degree days (40° F base temp.). Kernels of mature grain with lemma and palea (hulls) removed (caryopsis) vary from 0.3 to 0.6 in. in length and from 0.06 to 0.18 in. in diameter. The caryopsis consists of an impermeable pericarp, large endosperm, and small embryo. The immature caryopsis is green but turns purplish-black as it reaches maturity. Seeds on any given tiller mature at different times, and seed on secondary tillers matures later than that on the main tiller. Wild rice in natural stands has little shattering resistance and kernels shatter easily upon reaching maturity. When wild rice shatters or is harvested, the hulls and caryopsis remain intact as a unit. A more complete description of the wild rice plant, as well as its cultivation, is contained in a production guide from Minnesota (2).

Since about 1950, wild rice has been in the process of becoming a domesticated crop in the United States and is now being grown commercially in both the United States and Canada (2,3). Prior to that time, natural stands were the only source of the grain, and supplies were limited and varied greatly from year to year. With the advent and growth of commercial production, supplies of wild rice have increased tremendously over the last 25 years. Natural stands continue to be harvested, but the proportion of total supplies derived from natural stands has steadily declined. In some areas, including the entire state of Minnesota, natural stands of wild rice, by law, must be harvested only by traditional canoe-and-flail method (Fig. 1), whereas in some parts of Canada, mechanized harvest is permitted. Included in Table 1 are annual harvest estimates from natural stands in Minnesota since 1963. Of all the wild rice harvested by hand, Minnesota is likely to account for more than half in any given year.

In Canada, commercial production of wild rice takes place predominantly in lakes leased from the various provincial governments (4,5). Lease provisions vary by province, but generally lease holders are permitted to seed the lakes and, in some cases, to control water levels, and are granted exclusive harvesting rights. Much of the wild rice acreage in

FIGURE 1 Canoe and flail method of harvesting lake and river wild rice. One person propels the canoe with a pole through wild rice beds, while the other person bends the stalk over the side of the canoe with one flail and taps off the ripe grain from the panicle with the other flail.

these leased lakes is harvested with the use of airboats (3,5). Shown in Table 1 are annual harvest estimates from lakes and rivers in four Canadian provinces since 1963.

In the United States, wild rice is being produced commercially as a "domesticated" field crop in diked, flooded fields (2). Minnesota and California account for most of the acreage (20,000 and 8,000 acres, respectively, in 1988) with additional amounts in Idaho and Wisconsin. Table 2 shows production totals from cultivated fields in Minnesota and California since 1968.

Growing wild rice as a field crop was first attempted near Merrifield, Minnesota in 1950–1952 (2). James and Gerald Godward diked a one-acre area, planted it with seed collected from a nearby lake, and flooded the field. The field was drained before harvest and the crop was harvested by hand. An additional 40 acres were planted by them in 1953 and harvested with a small pull-type combine. They had good crops the first few years, but leaf blight (*Bipolaris oryzae* B. de Haam) caused serious losses thereafter. However, they con-

FIGURE 2 Harvesting cultivated wild rice with a modified grain combine. Modifications include oversized reel and a large track support system.

tinued their pioneering efforts, and today one of their sons has nearly 1000 acres in wild rice production.

Initially, the only seed available for planting in fields was of shattering types found in natural stands. These early fields were harvested several times over a 2-to 3-week grain-ripening period with specially designed, multiple-pass harvesters. In 1963, Dr. Paul Yagyu and Mr. Erwin Brooks with the University of Minnesota, Department of Agronomy and Plant Genetics, discovered plants in a grower's field that retained their seed longer than the rest of the plants (2). From these few plants, they and other breeders developed varieties with more resistance to shattering than types growing in lakes and rivers. Today, most of the wild rice being grown in fields is of the more shattering resistant varieties. Yields of unprocessed grain from shattering types grown in fields typically ranged from 150 to 200 lb/acre, whereas, with shattering resistant varieties, yields have been reported as high as 1500 lb/acre in Minnesota and twice that amount in California.

The development of more shattering resistant varieties of wild rice was largely responsible for a tremendous expansion in field production that occurred in the late 1960s and early 1970s. Practically all the expansion at that time was taking place in Minnesota where acreage increased from 900 acres in 1968 to 18,000 acres in 1973 (2). The finding of improved shattering resistance was the development that eventually made possible the shift to more efficient harvest with grain combines (Fig. 2). The improved harvest efficiency from the use of combines, together with greater harvested yields from shattering resistant varieties, were major contributing factors to expanded field production at that time.

Unlike most other cereal grains, which are already relatively dry at harvest, the moisture content of wild rice grain when harvested typically ranges from 35 to 45% (2). Because of its high moisture content, freshly harvested wild rice requires proper handling prior to being dried for long-term storage to prevent deterioration and loss from heating and mold development in the interim. Wild rice must be harvested at this high moisture content to avoid excessive grain loss from shattering that increases as the grain matures and dries. Generally, freshly harvested grain is transported directly to a processing plant. If it cannot be transported immediately, it must be kept cool by refrigeration or by stirring and adding water.

II. PROCESSING OF WILD RICE

The basic steps in processing of wild rice are drying, hulling, and separation of hulls from the hulled grain. Native Americans traditionally dried the grain on mats placed either in the sun or on racks over a fire. In more recent times, drying in a metal container over an open fire came to be the preferred method (Fig. 3). Hulling was accomplished by walking in place (jigging) on the dried grain in a container (Fig. 4) or skin-lined hole in the ground. Finally, hulls were separated from the grain by tossing the grain and hull mixture into the air (Fig. 5) or by dropping it over a mat or blanket, the wind blowing the lighter hulls away in the process.

Caucasians, upon arriving in the region, acquired a liking for wild rice and began buying and harvesting it themselves (1). They processed their grain in small processing plants located throughout much of northern Minnesota and Wisconsin and in adjacent areas in southern Canada. Equipment in these plants generally consisted of a parcher, a huller and a fanning mill. The parcher was a drum-like container, heated externally with a wood fire and ro-

FIGURE 3 Hand "parching" wild rice in an open kettle. The grain is constantly stirred for about 45 min. until dried to about 10% moisture content.

FIGURE 4 Hulling the grain by "jigging". The warm grain is placed into a wooden container and hulled by an individual walking in place in the container.

FIGURE 5 Separating the chaff (hulls) from the parched grain by "winnowing". The mixture of hulls and kernels is tossed up allowing the wind to blow the chaff away.

tated by hand. The huller was also a drum-like container with hand-rotated paddles or beaters inside. The fanning mill used to separate out the hulls was also turned manually.

The expanded commercial production of wild rice has led to dramatic changes in the processing sector of the industry (4). Although basic processing procedures remain essentially unchanged from those used in the early plants, many of today's processing plants are modern, high capacity facilities with modern equipment and are capable of producing a cleaner and more uniform product. Some older, smaller plants that handle basic processing still exist, but most larger plants now offer additional services including grading and packaging, warehousing, and, in some cases, even an instantizing process. Some of the newer plants are capable of processing over 6 million pounds of freshly harvested wild rice over a several-week harvest season.

A. Distribution of Processing Plants

Minnesota, California, and Canada had a combined total of 31 commercial processing plants in 1983 (Table 3), most of these being in Minnesota (4). Three additional plants have been built in California since 1983 to process that state's increased production. Much of the wild rice harvested in Canada (7 million lbs in 1983) is processed in Minnesota. Prior to 1985, a considerable portion of California's production was processed in Minnesota as well. The 22 commercial processing plants in Minnesota range in size from very large operations to small plants. This total does not include numerous very small plants for processing an individual's or small group's own lake harvest.

B. Custom Processing and Fees

Most wild rice is processed on a custom basis, especially that from growers (4). The methods of charging for processing and processing fees vary, some charging on a green (freshly harvested) basis and others on a finished (processed) basis. Charging on a green basis is potentially disadvantageous to the owner of the grain in that it provides processors little immediate incentive to maximize processed grain yield. On the other hand, charging on a finished basis can penalize processors if grain yield is lower than expected. Actual charges in 1987 ranged from 18.5 to 85 cents per pound of finished grain. Such a wide range is largely the result of variations in processing efficiency with large operations geared to handle large batches, whereas small plants will still process lots as small as 100 pounds. For most of the grain processed, however, charges range from 40 to 50 cents per finished pound under both charging systems.

C. Major Processing Operations

The major processing operations of a typical modern wild rice plant are shown in a flow chart in Figure 6. These operations are discussed in detail in the following sections.

1. Separation of Immature Kernels

An optional first step in processing wild rice performed at some Minnesota plants is separation of immature kernels. Much of the wild rice harvested by grain combine in Minnesota has a relatively high concentration of very immature kernels. Processors find it advantageous to remove this fraction before proceeding with subsequent processing steps. High concentrations of immature kernels in wild rice harvested by combine in Minnesota are due to several interrelated factors: (1) uneven grain ripening, (2) the higher risk of substantial loss of harvestable yield through wind, disease, and frost, when harvest is delayed, (3) the desire of growers to minimize that risk, and (4) the inability to adjust combines to eliminate very immature grain without causing excessive loss of more mature grain as well.

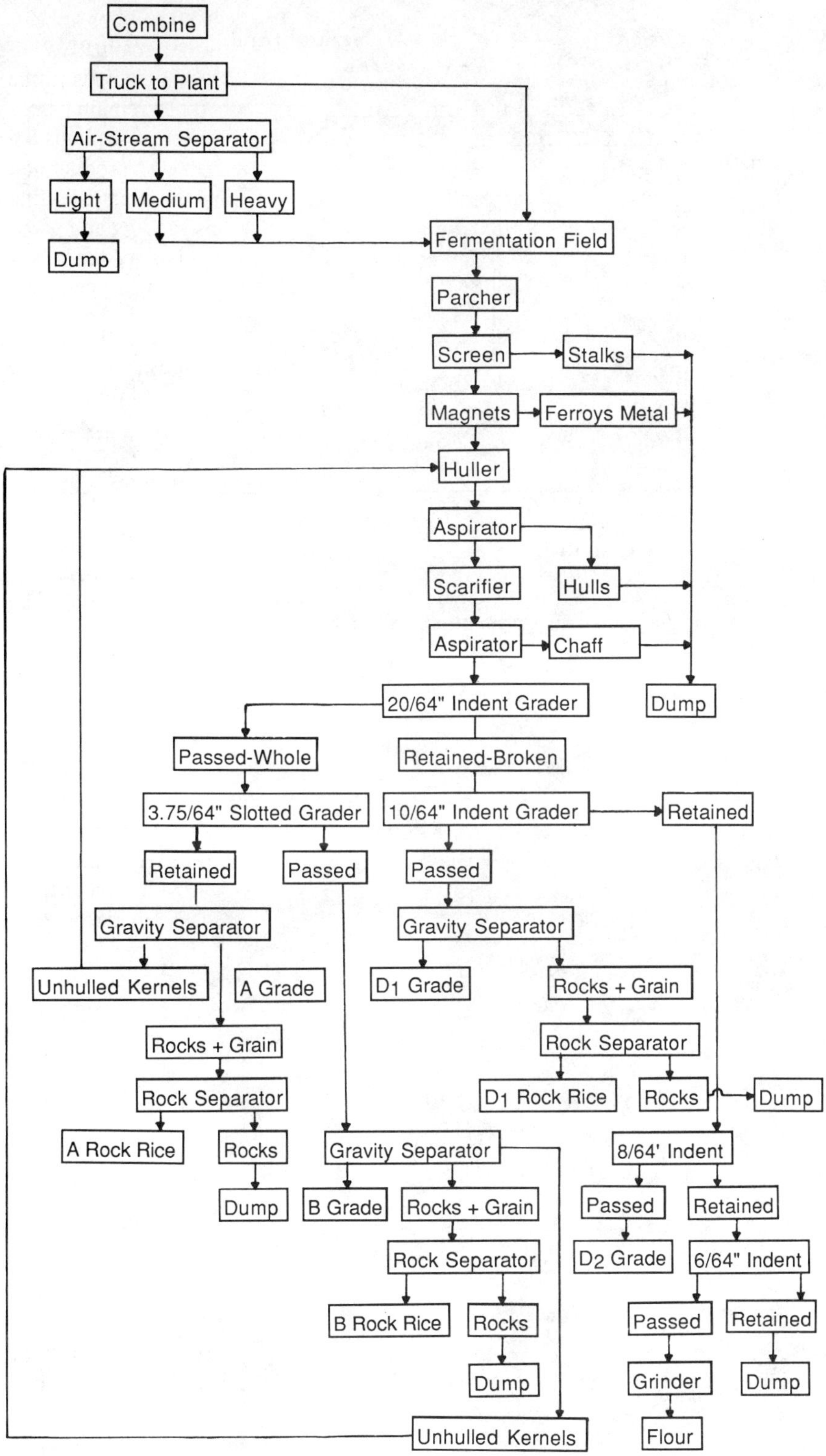

FIGURE 6 Operations flow chart for a typical processing plant.

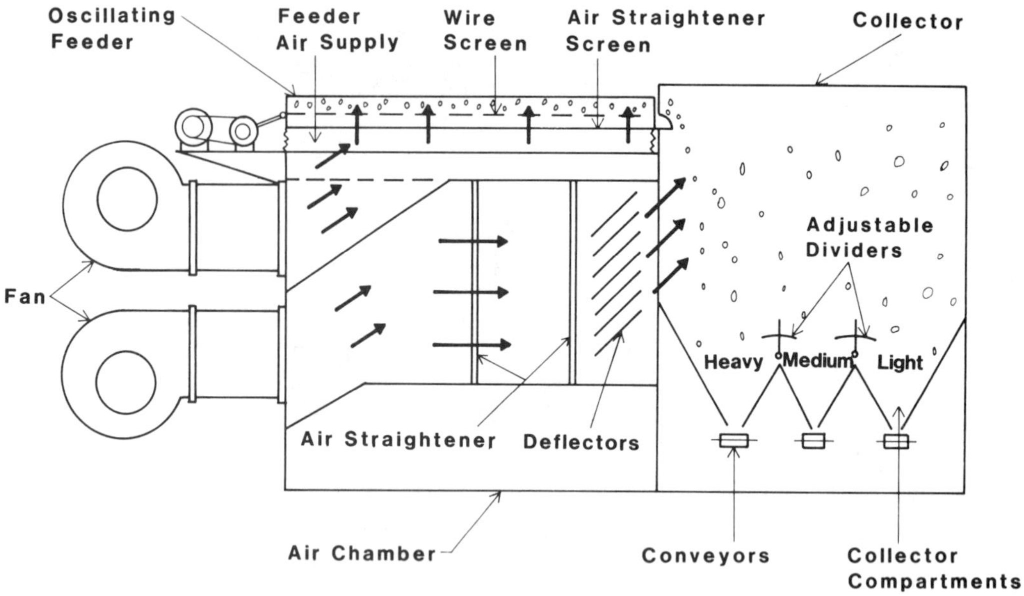

FIGURE 7 Schematic diagram of air stream separator, side view.

FIGURE 8 Air-stream separator installed at a processing plant.

Most large processing plants in Minnesota that process cultivated wild rice have an air-stream separator (Fig. 7 and 8) for this separation process (6,7). Wild rice harvested from fields planted to shattering types, natural stands harvested by multiple-pass (hand or machine) methods, or wild rice harvested by combine in California typically have low concentrations of very immature kernels; hence, this separation step is unnecessary for grain from these sources.

The air-stream separator shown in Figures 7 and 8 was built by the Agricultural Engineering Department of the University of Minnesota (7). It is designed to separate freshly harvested wild rice into three fractions designated as heavy, medium, and light. The heavy fraction consists mostly of plump, mature kernels; the medium fraction, kernels of intermediate stages of maturity; and the light fraction, predominantly small immature kernels and empty hulls and chaff. This separator is designed to permit the recycling of grain borderline in size between the medium and light fractions back through the air stream for more precise separation. Table 4 shows the characteristics of one lot of wild rice as delivered to a processing plant, and also its component maturity fractions, obtained both with and without the recycling (rerun) mode in operation. Light fractions, which usually yield only about 5%, generally cannot be processed profitably and are discarded by most processors. Processing plant capacity is essentially volume-limited. Separating and discarding the light fraction effectively increases plant capacity by a percentage at least equal to the percent reduction in volume that results from this separation. Some processors parch the heavy, mature fraction directly, thus bypassing fermentation which typically is the next step in processing.

2. Fermentation or Curing

Fermentation or curing of wild rice is a complex process involving respiration and numerous microorganisms (8,9). In the typical fermentation process, grain is formed into windrows on an impervious surface outdoors and periodically stirred and watered for several days (Figs. 9,10,11). Windrows range from 3 to 6 ft wide and 8 to 12 in. deep. Watering keeps moisture content high (which is desired for the next processing step) and, along with stirring, helps maintain aerobic conditions while limiting heat build-up to minimize dry matter loss and the development of undesirable molds. During normal weather conditions, windrows are stirred and watered once or twice daily. Fermentation periods normally range from 4 to 7 days. Over 20% of usable wild rice can be lost in 10 days through respiration and microbial action (Table 5) (8). In California, some processors ferment wild rice in 4 x 4 x 4-ft metal containers (Fig. 12), but the grain still requires watering and aeration to maintain its condition.

Although fermentation of wild rice is not completely understood, at least three important changes occur: (1) less mature kernels change color from green to brown, (2) hulls partially degrade, permitting easier hulling, and (3) some of the characteristic flavor is imparted to the grain. Wisconsin researchers (8) measured changes in color over 5 weeks of fermentation and found a significant color change after one week (Table 6). They (9) also identified microorganisms present on fermenting and parched wild rice (Table 7). They concluded that hull removal and final cooking virtually eliminate all viable microorganisms.

3. Drying or Parching

After fermentation, the grain is moved from the fermentation field (Fig. 13) to driers. The most commonly used drying procedure is referred to as parching. Parching at most plants is performed in batch-type, rotary drum parchers (Fig. 14), which typically are loaded by hand (Fig. 15). During parching, the 40–45% moisture content of the fermented grain is reduced to approximately 7%, and a slightly toasted flavor is imparted to the grain. In the process starch in the kernels is gelatinized, giving a glassy, translucent appearance to the inside of the kernel.

Parching drums are typically about 4 ft in diameter and 6–8 ft in length and are supported by rollers at the front and rear to permit rotation. The forward (open) end has vanes set at an angle to aid in retaining the grain during parching and in removing the grain (Fig. 14). Most parchers are fired with propane burners located below the drum. Typical parchers can dry from 250 to 400 lb/hr.

During parching the drum is rotated continuously, causing the grain to tumble, individual kernels intermittently coming in contact with the hot surface. Some parchers have small bars extending longitudinally through the parcher drum to provide better mixing as the grain tumbles. Some processors prefer to cover the front opening of the parcher during parching to help retain steam for quicker or more efficient drying with less temperature fluctuation within the drum. As parching of each batch nears completion, the grain is periodically sampled to monitor progress of the gelatinization process. Parching is complete when kernels reach a hard, flinty condition and the starch in the kernels is completely gelatinized. When parching is completed, the rear of the drum is raised slightly and the direction of rotation reversed, causing the grain to be discharged. Batches typically range from 500 to 800 lb. Parching time per batch is about 2 hr. During parching, temperature in the drum is controlled so as not to exceed 275°F.

Some continuous flow parchers are in use and consist of four long drums, each about 30 ft long, placed side by side. As the drums rotate, the grain is moved successively from one drum to the next, the process being controlled such that parching is complete upon discharge of the grain from the fourth drum.

FIGURE 9 Cleaned, unprocessed grain being formed into windrows on a hard-surfaced area adjacent to a processing plant.

FIGURE 10 Windrows of wild rice curing ("fermenting") before parching.

FIGURE 11 Machine used for periodically stirring and wetting windrows of curing wild rice.

FIGURE 12 In California, grain is cured in metal containers. The grain is aerated and watered periodically.

As an alternative to the rotary drum parcher, the University of Minnesota Agricultural Engineering Department built and tested a large prototype continuous-flow dryer utilizing superheated steam as the drying medium (Fig. 16) (10,11). Components of this dryer included a feeding mechanism to meter fermented grain into the top of the dryer, a three-tiered woven wire conveyor system to move it through the dryer, several stirrers to mix the grain on its travel on the conveyors, a fan and ductwork to recirculate steam up through the tiers of grain, and a propane-fired furnace and heat exchanger to add heat to the recirculated steam. The source of steam in the dryer was moisture given off by the grain itself, excess steam being allowed to continuously escape to the atmosphere through a duct from the top of the dryer. Best overall dryer performance was achieved with a steam supply temperature of 295°F (230°F return temperature). Capacity of the dryer was about 1000 lb/hr of fermented grain. Grain residence time in the dryer was approximately 40 min. Advantages of the system included more mechanized grain handling and less than half the fuel requirement of rotary drum parchers per quantity of water removed. One possible disadvantage was that grain dried in the unit lacked the toasted, nutlike flavor characteristic of parched wild rice that many consumers prefer.

A commercial continuous-flow dryer modeled after the prototype and rated at 3000 lb/hr was built and operated for several seasons. It performed satisfactorily although kernel breakage tended to exceed that from a well-operated rotary drum parcher.

Neither of these continuous-flow dryers has been operated for at least 2 years, however, the concept remains viable. In the future, as production expands and energy costs again increase as predicted, it is conceivable that continuous-flow dryers will find their place in the

FIGURE 13 Loading the cured grain for transport to parchers.

FIGURE 14 Typical rotary parcher.

FIGURE 15 Filling a parcher with cured wild rice.

processing industry, particularly for wild rice destined for the blend market for which a toasted flavor is less important.

A minor step following parching or drying is separation of any large pieces of stalk and fragments of ferrous metal from the grain. Stalk pieces are separated out by screening. Ferrous metal fragments are captured by permanent magnets positioned closely over conveyors transferring the dried, screened grain to hullers (Fig. 17).

4. Hulling

Hulls remain intact during drying or parching and must be detached from the edible portion of the kernel. Although some barrel-type hullers are still used, the most common type of huller is the roller huller consisting of two closely spaced, counter-rotating rollers with rubberlike outer surfaces (Fig. 18) (6). The grain is fed to pass between the rollers, and as it does, a rubbing action created by the rollers operating at different peripheral speeds causes the hulls to become detached. The roller surface material found to give the best compromise between levels of kernel breakage and unhulled kernels is neoprene of 66 durometer hardness (Table 8; Figs. 19 and 20) (11). To maintain good huller performance, processors must pay close attention to roller clearance and condition. Effective feeder and roller conditioning systems along with sensitive means for adjusting roller clearance are important features of any roller huller system.

5. Scarification

Scarification is the process of removing part of the outer impermeable layer of the kernel. Wild rice used in mixes with rice (*Oryza sativa* L.) must be scarified sufficiently to reduce its cooking time to that of rice. The typical scarifier consists of a long, inclined tube with rubber paddles inside mounted to a rotating shaft. The degree of scarification that occurs is dependent upon inclination of the scarifier, speed of the shaft, and clearance of the rubber

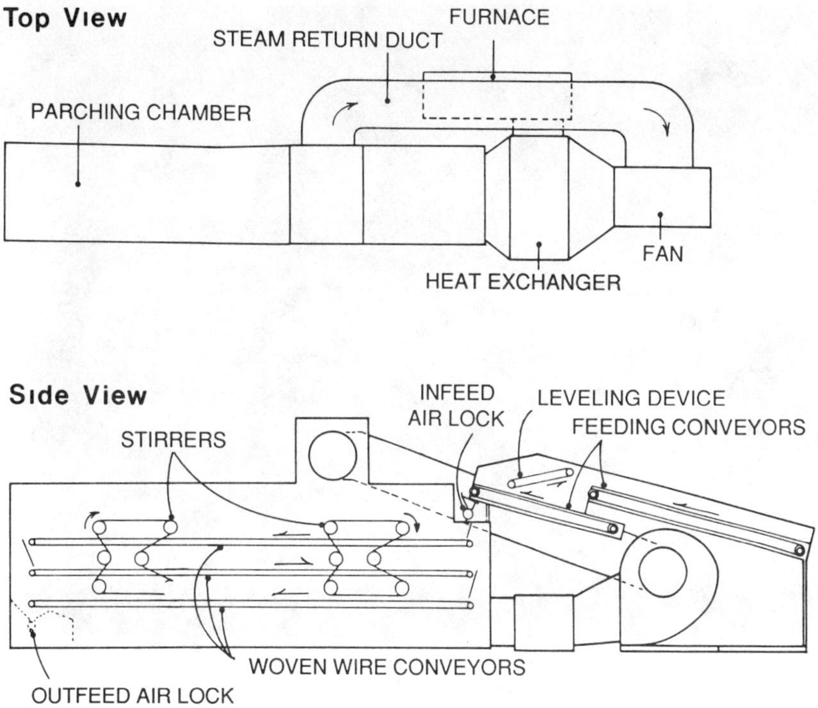

FIGURE 16 Schematic diagram of experimental continuous flow wild rice dryer.

FIGURE 17 Magnet that removes ferrous metal from grain before hulling.

FIGURE 18 Huller with two sets of rubber rollers which detach the hulls (lemmas and paleas) from the kernels.This huller handles two separate streams of parched grain.

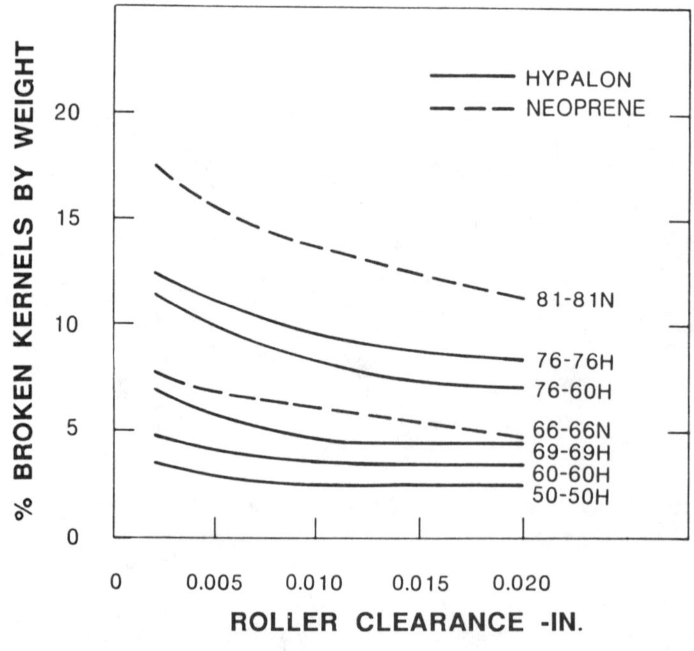

FIGURE 19 Percent broken kernels from heavy fraction from air-stream separator as influenced by roller clearance and surface hardness for 11 3/8–in. diameter hypalon and neoprene-covered rollers. The two numerical values at the end of each line indicate the hardness (durometer) of the covering on each of the paired rollers.

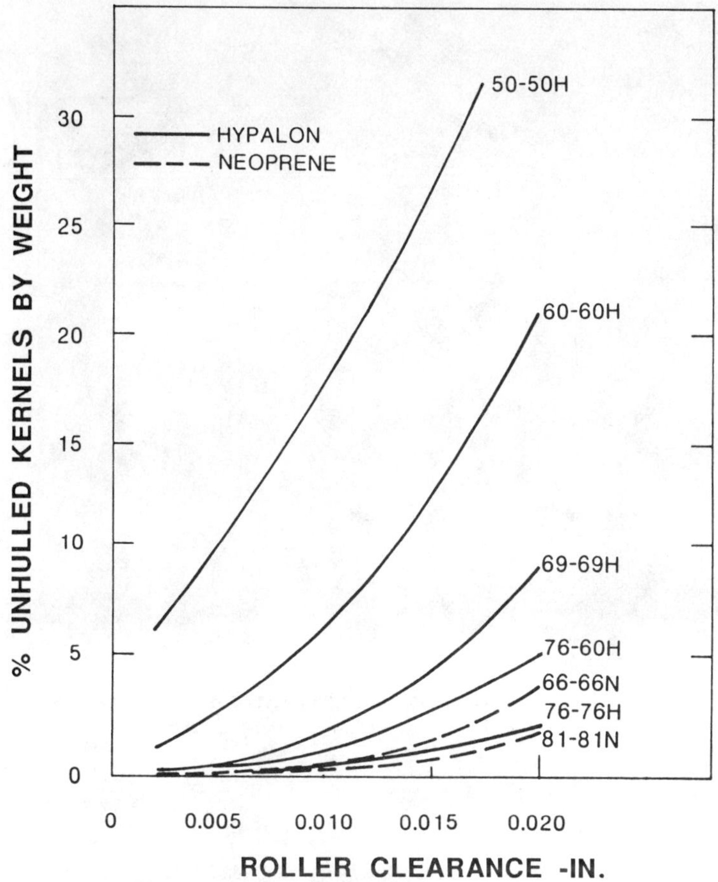

FIGURE 20 Percent unhulled kernels from heavy fraction from air-stream separator as influenced by roller clearance and surface hardness for 11 3/8–in. diameter hypalon and neoprene-covered rollers. The two numerical values at the end of each line indicate the hardness (durometer) of the covering on each roller of the paired set.

paddles. In some operations, grain is scarified after being size-graded for even better control of cooking time. Figure 21 shows two scarifying tubes.

6. Cleaning, Grading, and Packaging

In addition to screening and separation of ferrous particles as described above, grain is subjected to several other cleaning operations as it progresses through the plant. Aspirating devices are employed after hulling and scarifying operations to separate out hulls and chaff (Fig. 22). Aspirated material is conveyed outside the plant (Fig. 23). Gravity tables are used to remove unhulled grain as well as small pebbles and other debris (Fig. 24). Unhulled grain is either recycled directly to hullers or stored and rerun later.

Wild rice is graded principally on the basis of kernel size. Various types of length and width graders are used for the grading process (Fig. 25). No uniform grading standards for wild rice are yet in place across the industry, and plants sort processed wild rice differently, often depending on specifications of particular buyers. Figure 26 illustrates grading steps used by some processors. In this grading system, kernels are defined as being either whole or broken solely on the basis of length, regardless of whether the kernels are actually broken

FIGURE 21 Scarifying tubes are used to remove some of the outer layer from the parched kernels.

FIGURE 22 A vacuum aspirator to remove hulls from parched, hulled kernels.

FIGURE 23 Hulls and chaff are conveyed to the outside of the plant for disposal.

FIGURE 24 Gravity tables are utilized for debris removal and for size sorting of kernels.

or not. Whole grain (≥ 20/64 in. in length) is divided into A and B grades according to diameter. Grain classified as broken is further graded on the basis of length only. In 1984, the International Wild Rice Association proposed a somewhat different grading standard (Table 9) that some plants are now following to a limited extent (12). Some processing plants have developed their own grain specifications as shown in Table 10. These specifications differ from those depicted in Figure 26 and Table 9, and buyers need to be aware of actual grain specifications for the various grading systems in use. The industry, in cooperation with USDA, is presently developing a standard grading system for wild rice which should facilitate wild rice marketing.

Most plants bag processed wild rice into 100 lb sacks (Fig. 27), which are stored in clean, dry warehouses (Fig. 28). Many plants are equipped to package wild rice in small packages, and some will blend and package wild rice and rice according to customers' specifications (Fig. 29).

III. QUALITY

Even though industrywide quality standards have not been developed, several factors relating to quality are recognized by most marketers. Broken grain is undesirable because it detracts from appearance of the final product. Also undesirable is the presence of stress cracks, which can develop during parching due to excessive temperature and moisture gradients produced by drying the grain too rapidly. Stress cracks make kernels more susceptible to breaking in subsequent handling and processing operations. White centers in the kernel are another undesirable appearance-related characteristic that is quite evident in grain containing broken kernels. White centers are thought to result from gases becoming trapped in the

FIGURE 25 Width graders for final grading of rice.

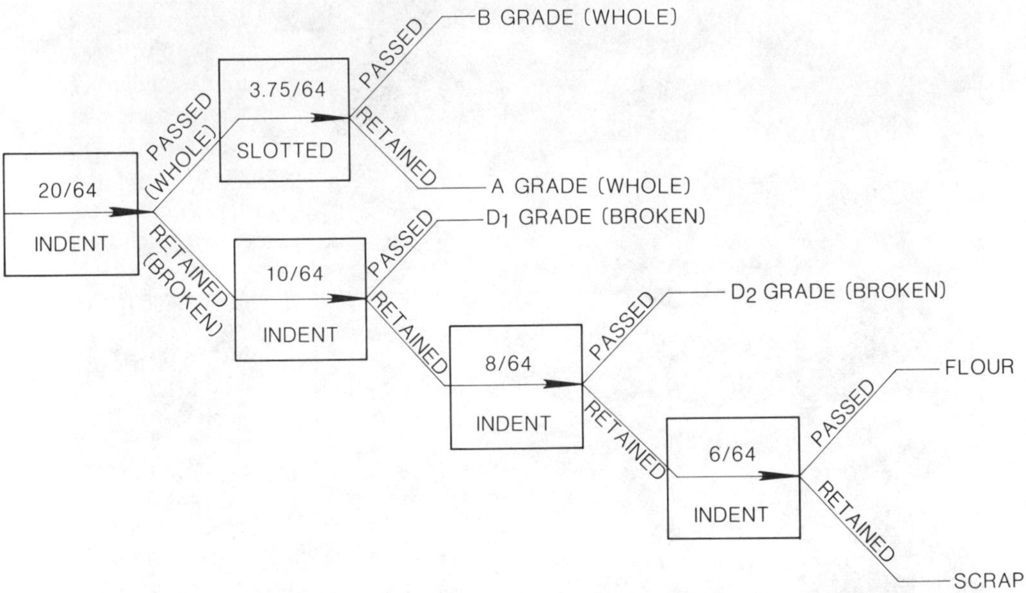

FIGURE 26 Size specifications utilized by some Minnesota processing plants for processed wild rice.(Numbers indicate fractions of an inch).

FIGURE 27 Processed wild rice is placed into poly-lined bags that can contain 100 pounds.

FIGURE 28 Processed wild rice is stored in clean, dry warehouses.

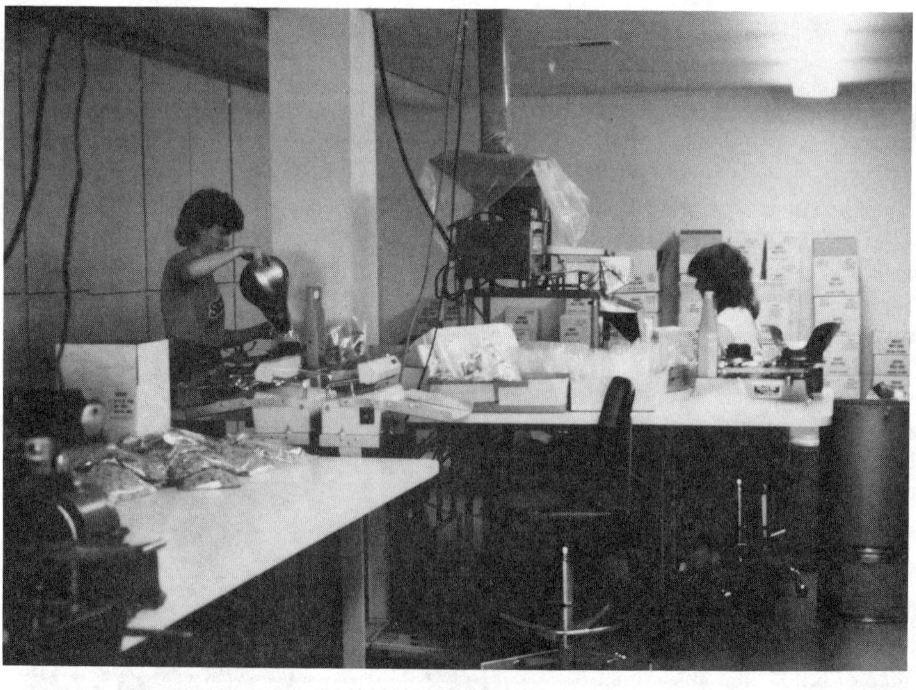

FIGURE 29 Blending rice and wild rice and packaging the blend in plastic bags. A typical packaging room in a processing plant.

center of the kernel during parching. Percentages of cracked, broken, and white-centered kernels vary considerably from plant to plant and even among batches processed at the same plant.

Two additional quality factors of importance are color and taste. Any green coloration, characteristic of inadequately fermented immature kernels, is generally undesirable. The preferred color is blackish-brown. Scarification, which results in removal of some of the seed coat, detracts from the desired dark color. However, for grain that must be scarified to reduce its cooking time, some of the color is removed during scarification. As for taste, many consumers prefer a slightly toasted flavor, as mentioned above. Any off-flavor, which can result from improper handling and fermentation, is highly undesirable (8).

Some in the industry claim that wild rice harvested from lakes and rivers is of better quality than cultivated wild rice. However, quality differences are due primarily to variations in how wild rice is handled and processed and not to any inherent differences related to how or where it is grown (8). The industry is continually striving to improve the quality and uniformity of processed wild rice, regardless of its source.

IV. NUTRITION

The wild rice kernel is similar in structure to other cereals, consisting of a pericarp, aleurone layer, endosperm, and embryo. The pericarp and embryo each represent about 5% of the caryopsis weight, with the endosperm and the aleurone layer making up the remaining 90% (13).

The nutritional quality of wild rice appears to be equal to or better than that of other cereals (14,15). Protein content is relatively high, and lysine and methionine comprise a higher percentage of the protein than in most other cereals (Tables 11,12). The SLTM value (the sum of lysine, threonine, and methionine contents), often used as a measure of nutritional quality of cereals, is slightly higher for wild rice than for oat groats, which is considered one of the better cereals for humans (8,16). The similarity in amino acid composition of processed and unprocessed wild rice indicates little reduction in nutritional quality through processing (8,14).

Researchers at the University of Wisconsin at Madison analyzed the different parts of the wild rice kernel for amino acids (Table 13) (8). They found that removal of the pericarp produced a slight reduction in the levels of histidine and arginine. However, total removal of the pericarp had little influence on either amino acid distribution or protein content of the kernel. Lysine concentration is highest in the embryo, and lysine content is reduced if the embryo is lost during processing or cooking.

The fatty acid composition of wild rice and several other cereal grains is given in Table 14 (8). Although wild rice contains less than 1% fat, linolenic acid comprises a much higher proportion of the fat than in rice, wheat, or oats. Linoleic and linolenic acids, which make up about 68% of the total fatty acids, are easily oxidized and make wild rice quite susceptible to the development of rancid odors. However, the high level of linoleic acid in the fat of wild rice makes the fat highly nutritious.

The mineral composition of wild rice compares favorably with oats, wheat, and corn (15). Wild rice, with the exception of calcium, has a higher mineral content than brown rice (Table 11). With rice, removing the pericarp (polishing) reduces mineral content considerably (15). With wild rice, however, scarifying causes no such reduction, according to Wisconsin researchers (8). Processed wild rice grain contains no vitamin A, but is an excellent source of the B vitamins—thiamine, riboflavin, and niacin (Table 15) (15).

V. MARKETS

Since 1968 the per-pound wholesale price of processed wild rice has ranged from a low of \$2.10 in 1987 to a high of \$5.15 in 1978 (Table 16) (17,18). Price variations between 1968 and 1977 reflected limited and erratic supplies due to variations in lake and early paddy harvests. The high prices of 1978 through 1980 reflected attempts by marketers to withhold supply, but these attempts were short-lived because the high prices simply encouraged increased production. Expansion in Minnesota was modest during 1978–1980, but California doubled production annually through 1981. By 1981, high inventory storage costs were forcing the sale of stocks and prices returned to market determined levels. Production increased 26% per year between 1982 and 1984, but markets were able to absorb this increase with only a slight drop in price. In 1985, California production more than doubled over 1984, and prices have declined sharply since then. The 1988 production in California is expected to be significantly lower than in 1987 and close to the 1984 level, and prices are expected to be about \$0.50 per pound higher than in 1987.

Wild rice markets have expanded at a vigorous rate over the last 12 years, especially during the 1982–1984 period when demand increased 52% (18). Most of this growth has been in the blend market, where sales have increased an average of 15% per year since 1961 when Uncle Ben's Inc. first introduced its packaged blend of wild rice, long grain rice, and herbs (18). Presently, other types of blends containing wild rice are also available such as soups, vegetable-based side dishes, and convenience foods such as dehydrated mixes and frozen entrees. Wild rice blends have helped to increase consumer awareness of wild rice in the United States from 8 to 30% in recent years (18).

Figure 30 shows the percent of wild rice sales by market outlet for the 1982 crop year (18). About one third of the wild rice sold that year was in the pure form. However, a large portion of even these sales eventually reaches the consumer in the form of blends prepared by restaurants and other users. Long grain wild rice is preferred for the pure market; however, a considerable amount of short or broken grain is also sold in pure form for use in soups and stuffings. The demand for wild rice, both in blends and in pure product, is expected to increase substantially as prices stabilize or decrease and production expands during the next few years.

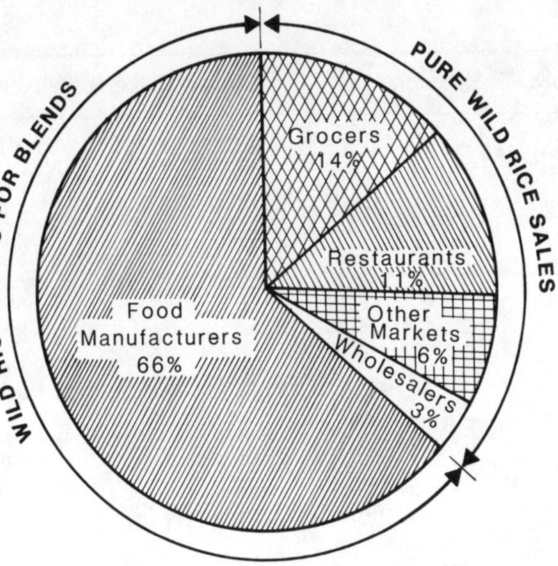

FIGURE 30 Percentage breakdown for wild rice sales by type of market outlet for the 1982 crop year.

TABLE 1 Wild Rice Harvested from Lakes and Rivers in Canada and Minnesota, 1963–1987[a]

Years	Minnesota	Manitoba	Saskatchewan	Ontario	Alberta	Total
1963	1,286	—	—	22	—	1,308
1964	514	—	—	23	—	537
1965	435	—	—	12	—	447
1966	429	—	—	18	—	447
1967	1,051	—	—	226	—	1,277
1968	524	—	—	126	—	650
1969	392	—	—	63	—	455
1970	489	60	1	26	—	576
1971	487	200	9	121	—	817
1972	414	273	22	436	—	1,145
1973	406	250	5	519	—	1,180
1974	400	55	9	4	—	468
1975	200	57	17	37	—	311
1976	800	140	39	450	—	1,429
1977	437	462	34	344	—	1,277
1978	220	190	25	62	—	497
1979	304	240	64	120	—	728
1980	1,000	560	128	388	—	2,076
1981	400	182	200	272	—	1,054
1982	440	166	208	75	—	889
1983	480	134	243	76	—	933
1984	540	249	450	160	—	1,399
1985	161	258	117	14	4	554
1986	179	342	304	40	7	872
1987	500	661	515	576	8	2,260

[a]1000 processed pounds, estimated using 40% yield of processed from unprocessed wild rice.
Source: 1963–1981 from Ref. 4; 1982–1987 from Ref. 2.

TABLE 2 Wild Rice Harvested from Cultivated Fields in Minnesota and California, 1968–1989

Year	Minnesota (1000 processed lb)	California[a] (1000 processed lb)
1968	36	0
1969	160	0
1970	364	0
1971	608	0
1972	1,496	0
1973	1,200	0
1974	1,036	0
1975	1,233	0
1976	1,809	0
1977	1,031	0
1978	1,761	100
1979	2,155	200
1980	2,320	400
1981	2,274	500
1982	2,697	880
1983	3,200	2,500
1984	3,600	3,800
1985	4,200	7,900
1986	5,100	9,000
1987	4,200	4,200
1988	4,200	3,500
1989	3,978	4,242

[a]Data from Marcum, Cooperative Extension Service, University of California.
Source: 1968–1982 values from Ref. 4; 1983–1989 from Minnesota Agriculture Statistics, Minnesota Department of Agriculture and USDA-NASS.

TABLE 3 Location of Commercial Wild Rice Processing Plants, 1983

Area[a]	Number of plants
Minnesota	22
California	2
Ontario	1
Manitoba	5
Saskatchewan	1
Total	31

[a]Data for Canadian provinces do not include smaller processing plants.
Source: Ref. 4.

TABLE 4 Characteristics of One Lot of Wild Rice as Received at a Processing Plant and of Component Fractions Produced by the Air-Stream Separator

Seed lot	Rerun feature (%)	Moisture content (%)	Processed grain obtained—fresh weight basis (%)	Distribution of processed kernels by diameter (%)						
				>4 1/2	4 1/2 x 4[c]	4 x 3 1/2	3 1/2 x 3	3 1/2 × 2 1/2	2 1/2 × 2	<2
As received	N.A.[a]	43.3	40.8	2.1	13.1	37.0	25.9	12.8	5.4	3.6
Heavy	W[b]	37.0	51.4	2.5	18.5	45.8	23.0	7.2	2.2	0.7
	W/O[b]	37.8	49.8	1.1	12.7	47.0	26.9	9.0	2.7	0.8
Medium	W	45.2	39.0	1.0	3.4	19.8	31.6	26.3	12.3	6.2
	W/O	45.5	36.9	0.1	1.6	16.8	32.1	29.1	13.5	7.0
Light	W	60.2	5.8	0.1	0.2	0.9	1.9	8.2	25.6	63.3
	W/O	61.7	5.0	0.0	0.2	0.9	1.9	9.2	27.2	60.3

[a]Not applicable.
[b]W = separator operated in rerun mode; W/O = rerun mode not in operation. With rerun mode medium and light gractions can be recycled for more precice separation.
[c]4 1/2 x 4 means kiernels passed a 4.5/64 in slotted sieve and were retained on a 4/64 in slotted sieve.
Source: Ref. 7.

TABLE 5 Processed Grain Recovered from Grain as Influenced by Storage Time at Ambient Temperature

Length of storage and turning[a] (days)	Processed grain recovered[b] (%)
0	100
1	100
2	97
3	98
4	93
5	92
6	93
7	89
8	89
10	79

[a]Wet wild rice (ca. 50% moisture) stored in 18- to 24- in deep bins with daily watering and turning.
[b]0 days storage used as base (100%).
Source: Ref. 8.

TABLE 6 Hunter Color-Color Difference L-Values (Lightness-Darkness) for Wild Rice as Influenced by Fermentation Time

Fermentation time (weeks)	Shattering lake type[a]	Nonshattering, MI variety[b]	Nonshattering, Johnson variety[b]
0	21.7[c]	21.0	19.8
1	20.0	17.8	19.0
2	20.3	18.8	19.2
3	19.1	19.4	19.1
4	20.5	18.1	19.1
5	19.5	18.0	19.6

[a]Fermented at 40°F.
[b]Fermented at 70°F.
[c]Higher numbers indicate lighter color; standard deviation = 0.5.
Source: Ref. 8.

TABLE 7 Kinds of Microorganisms and Frequency of Isolation from Fermenting and Parched Wild Rice

Source and kind of microorganism	Fermenting wild rice[a]	Parched wild rice[a]
From plate count agar (86°F)		
Gram-positive rods	5/27	17/25
Gram-negative rods	14/27	6/25
Gram-positive cocci	8/27	2/25
Bacillus species	0/27	15/25
Corynebacterium species	0/27	2/25
Lactobacillus species	2/27	0/27
Leuconostoc species	3/27	0/27
From plate count agar (45°F)		
Pseudomonas species	23[b]	3[b]
Pigmented Pseudomonas species	8/23	0/3
From violet red bile agar		
Escherichia coli	2/44	1/6
Enterobacter aerogenes	5/44	1/6
Intermediates	37/44	4/6
From KF agar		
Streptococcus faecium	14/35	7/11
Streptococcus faecalis	6/35	1/11
Other streptococci	15/35	3/11
From acidified potato dextrose agar		
Aspergillus species	5/16	2/10
Penicillium species	2/16	3/10
Mucor species	8/16	5/10
Rhizopus species	1/16	0/16

[a]No. specific kind/total isolates.
[b]A total of 70 colonies was isolated from fermenting and parched wild rice.
Source: Ref. 9.

TABLE 8 Influence of Hardness of Huller Roller Covering Material and Roller Clearance Upon the Percent of Broken and Unhulled Kernels

Run number of variety		Material and hardness durometer	Roller clearance (in.)	K–2[a] (%)		Netum (%)	
K–2	Netum			Broken[b]	Unhulled	Broken	Unhulled
H2–16	H3–16	Neoprene, 50	.002	4.02	5.67	1.81	4.65
H2–15	H3–15	Neoprene, 56	.002	4.84	1.01	2.05	0.84
H2–4	H3–4	Neoprene, 50	.005	2.80	20.74	1.47	15.68
H2–3	H3–3	Neoprene, 56	.005	3.40	11.13	1.71	7.11
H2–2	H3–2	Neoprene, 66	.005	7.69	0.31	3.52	0.80
H2–1	H3–1	Neoprene, 73	.005	10.55	0.29	4.83	0.44
H2–10	H3–10	Neoprene, 50	.010	3.76	26.96	2.39	20.30
H2–9	H3–9	Neoprene, 56	.010	3.80	23.37	1.51	18.46
H2–8	H3–8	Neoprene, 66	.010	6.56	1.48	3.59	1.32
H2–6	H3–6	Red Rubber, 66	.010	4.46	27.17	2.27	25.02
H2–7	H3–7	Neoprene, 73	.010	7.30	0.98	4.31	1.23
H2–5	H3–5	Neoprene, 86	.010	19.22	0.38	9.96	0.41
H2–13	H3–13	Neoprene, 66	.015	6.27	3.00	2.89	3.09
H2–12	H3–12	Neoprene, 73	.015	8.62	1.39	4.29	1.35
H2–11	H3–11	Neoprene, 86	.015	20.00	0.97	9.99	1.05
H2–14	H3–14	Neoprene, 86	.020	17.10	2.54	9.30	1.94

[a]Wild rice harvested in 1981, frozen, then parched in the laboratory on June 22, 1982, immediately prior to hulling.
[b]Kernel pieces 1/4 in. or less in length.
Source: Refs. 10, 11.

TABLE 9 Wild Rice Grades—Length and Diameter, Quality of Grades, and Definition of Grades*

A. GRADES[a]—LENGTH AND DIAMETER

Screens	Diameter measure[b]	Letter grade	Grades				
	425 & above	T	T1	T2	T3	T4	T5
425							
	400	O	O1	O2	O3	O4	O5
400							
	375	P	P1	P2	P3	P4	P5
375							
	350	I	I1	I2	I3	I4	I5
350							
	325	N	N1	N2	N3	N4	N5
325							
	300	G	G1	G2	G3	G4	G5
300							
	Below 300	S	—	S2	S3	S4	S5

Number grade:	1 (Long Gr)	2	(Med) 3	4 (Short Gr)	5	6 (Small Broken)	7
Length measure:[c]	30	26	20	12 10	8	6	
Riddle:[d]			#I	000			

[a]Although there is a potential of 49 grades, the smaller lengths cannot be used.
[b]Measurements expressed in 64th of a inch (ex. 425 = 4 1/4 /64).
[c]Measurements expressed in 64th of an inch (ex. 32 = 32/64).
[d]*Measurement of Riddle*: #1-20/64; #000-12/64.
Equation: Fraction of inch to 64ths of an inch: 1 = 64, 1/2 = 32, 1/3 = 21 1/3, 1/4 = 16, 1/5 = 12 4/5.

TABLE 9 (Continued)

B. QUALITY OF GRADES

Description	Foreign material: Solids	Foreign material: Dust and chaff	Foreign material: Other[a]	Chalky kernels	Damaged kernels[b]	Percent moisture	Other
Premium	< 0.0001	< 0.002	< 0.01	< 0.1	< 0.1	3–10	All FDA standards
Choice	< 0.0002	< 0.005	< 0.025	< 0.5	< 0.2	3–11	All FDA standards
Good	< 0.0005	< 0.01	< 0.05	< 2.0	< 1.0	3–12	All FDA standards
Standard	< 0.001	< 0.02	< 0.10	< 5.0	< 3.0	3–13	All FDA standards

[a]Other—unhulled, weed seed, other grain, etc.
[b]Damaged kernels—heat, frost, etc.

C. DEFINITION OF GRADES

Description	Percent long grains	Percent medium grains	Percent short grains	Percent brokens
Long grain	85 min.	10 max.	7 max.	3 max.
Medium grain	10 max.	80 min.	10 max.	7 max.
Short grain	5 max.	10 max.	80 min.	10 max.
Broken	1 max.	5 max	20 max	80 min.
Mill run	70 min.	20 max.	10 max.	5 min.

*Adopted by Processor Committee of the International Wild Rice Association, May 22, 1984.
Source: Ref. 12.

TABLE 10 Standard Specifications for Wild Rice for a Wild Rice-Processing Plant in Minnesota

Physical form:	
Variable	*Specifications*
Form:	Kernels
Color:	Greenish-brown to dark brown to charcoal black
Flavor/Odor:	Nutty flavor—faintly roasted odor
Analytical Specifications:	
Variable	*Specifications*
Moisture:	Less than 10% by A.O.A.C.
Length:	Greater than 20/64 in. = long grain
	Less than 20/64 in.–greater than 10/64 in. = medium grain
	Less than 10/64 in.–greater than 8/64[a] in. = short grain
Width:	2.5/64–5.5/64 in.
Water absorption:	Upon rehydration (cooking) product will typically increase in weight 250–375%
Microbiological standards:	Provided by an outside, reputable analytical laboratory and consulting service, 55 years in service
Total aerobic plate count	3,000,000/g *maximum*
E. coli	Less than 3/10 g
Staphylococcus	Negative/25 g
Salmonella	Negative/100 g
Aflatoxin	Less than 2 ppb

[a] Product less than 8/64 in. (chits and fines) discarded.
Source: Deer River Wild Rice, Inc., Deer River, Minnesota.

TABLE 11 Composition of Wild Rice, Cultivated Brown Rice, and Wheat

Nutrient	Wild rice[a]	Cultivated brown rice	Wheat
Protein (%)	13.8 (12.8–14.8)	8.1	14.3
Ash (%)	1.7 (1.4–1.9)	1.4	2.0
Fat (%)	0.6 (0.5–0.8)	1.9	1.8
Fiber (%)	1.2 (1.0–1.7)	1.0	2.9
Ether extract (%)	0.5 (0.3–1.0)	2.1	1.9
Phosphorus (%)	.28	.22	.41
Potassium (%)	.30	.22	.58
Magnesium (%)	.11	.12	.18
Calcium (ppm)	20	32	46
Iron (ppm)	17	10–17	60
Manganese (ppm)	14	30–39	55
Zinc (ppm)	5	24	–
Copper (ppm)	13	4–7	8
Nitrogen (free % extract)	82.4	87.4	78.9

[a]Numbers in parentheses indicate ranges in values.
Source: Refs. 6, 14, 15.

TABLE 12 Amino Acid Composition of Processed and Unprocessed Wild Rice, Two Varieties of Unprocessed Cultivated Rice, and Wheat[a]

Amino acid	Unprocessed wild rice	Processed wild rice	Cultivated rice (cv. Bray)	Cultivated rice (cv. Lebonnet)	Wheat
Lysine	5.23	4.05	3.78	4.28	2.79
Histidine	2.75	2.66	2.53	2.64	2.38
Ammonia	2.37	2.41	2.39	2.29	3.98
Arginine	8.18	7.31	8.85	8.77	4.67
Aspartic	10.40	10.26	9.61	9.81	5.40
Threonine	3.64	3.60	3.68	3.79	2.90
Serine	5.46	5.23	5.16	5.01	4.76
Glutamic	18.70	18.24	19.35	18.34	35.39
Proline	4.11	4.12	4.96	4.63	11.75
Glycine	4.84	4.76	4.96	5.16	4.33
Alanine	5.83	5.82	5.80	5.94	3.57
Cystine	2.78	—[b]	3.92	3.60	3.28
Valine	5.76	5.70	5.82	5.85	4.44
Methionine	2.79	3.01	2.19	2.06	1.42
Isoleucine	4.36	4.26	4.01	4.06	3.63
Leucine	7.39	7.33	8.32	8.30	7.19
Tyrosine	3.75	3.50	4.47	4.34	2.86
Phenylalanine	5.11	5.01	5.01	4.88	5.28
Tryptophan	1.80	1.64	1.78	1.72	1.57

All values are g amino acid/100 g protein.
[a] Wild rice = *Zizania palustris*, cultivated rice = *Oryza sativa*.
[b] Not determined.
Source: Refs. 6,14,15.

TABLE 13 Amino Acid Distribution in Kernel Fractions of Parched Wild Rice

Amino acid	Whole kernel	Endosperm plus embryo	Endosperm plus pericarp	Embryo alone
Alanine	5.7	5.9	6.0	7.8
(Ammonia)	2.3	2.1	2.3	1.7
Arginine	8.1	7.3	6.7	6.1
Aspartic acid	10.0	10.3	10.6	10.6
Cysteine	1.4	1.3	—[a]	—
Glutamic acid	19.9	20.5	20.8	13.6
Glycine	4.9	4.9	5.1	6.6
Histidine	3.5	2.7	—	—
Isoleucine	4.1	4.3	4.5	4.9
Leucine	7.3	7.5	7.7	8.8
Lysine	4.7	4.2	4.5	7.7
Methionine	2.2	2.4	2.8	1.9
Phenylalanine	5.0	5.2	5.4	5.2
Proline	3.6	4.5	3.5[b]	2.4[b]
Serine	5.5	5.7	5.8	5.4
Threonine	3.4	3.4	3.4	5.8
Tyrosine	2.7	2.5	3.7	3.8
Valine	5.6	5.5	6.1	6.7
SLTM	10.3	10.0	10.7	15.4
% Protein	13.3	13.6	—	—

All values are g/100 g recovered amino acids.
[a]Not determined.
[b]Estimated from the 570 nm curve.
Source: Ref. 8.

TABLE 14 Fatty Acid Composition (% of total) of Hexane Extracts of Wild Rice, Rice, Oats, and Wheat

Fatty acid[a]	Wild rice	Brown rice	Polished rice	Oat groats	Wheat
Palmitic (16.0)	14.5	20.4	33.8	16.2	24.5
Stearic (18.0)	1.1	1.6	2.7	1.8	1.0
Oleic (18.1)	15.9	41.3	43.3	41.2	11.5
Linoleic (18.2)	37.7	34.5	18.0	38.8	56.3
Linolenic (18.3)	30.0	1.0	0.6	1.9	3.7

[a]Number in parentheses indicates number of carbon atoms in molecule, with number of unsaturated carbons after decimal.
Source: Refs. 8,15.

TABLE 15 Vitamin Content of Wild Rice and Other Cereals

Vitamin	Wild rice	Brown rice	Polished rice	Oats	Hard red winter wheat	Corn
Thiamin (mg/100 g)	0.45	0.34	0.07	0.60	0.52	0.37
Riboflavin (mg/100 g)	0.63	0.05	0.03	0.14	0.12	0.12
Niacin (mg/100 g)	6.2	4.7	1.6	1.0	4.3	2.2
Vitamin C (mg/100 g)	0	0	0	0	0	0
Vitamin A (IU)	0	0	0	0	0	490

Source: Refs. 8,15.

TABLE 16 Wild Rice Wholesale Prices, 1968–1987

Year	Wholesale prices ($ per processed pound)
1968	3.27
1969	2.66
1970	2.88
1971	2.71
1972	2.34
1973	2.11
1974	2.37
1975	2.51
1976	2.68
1977	4.25
1978	5.15
1979	5.01
1980	4.47
1981	3.79
1982	3.40
1983	3.35
1984	3.30
1985	3.25
1986	2.75
1987	2.10

Source: Refs. 4, 17, 18.

REFERENCES

1. Steeves, T. A., Wild rice—Indian food and a modern delicacy. *Econ. Bot., 6*:107–142 (1952).
2. Oelke, E., Grava, J., Noetzel, D., Barron, D., Percich, J., Schertz, C., Strait, J., and Stucker R., *Wild Rice Production in Minnesota.* Agr. Ext. Serv., Univ. of Minnesota, AG-BU-0546, 1982.
3. Stevenson, S., *Wild Rice Report 1987—Northwestern Region of Ontario.* Ont. Min. Nat. Res., Kenora, Ont, 1988.
4. Winchell, E. H., and Dahl, R. P., *Wild Rice Production, Prices and Marketing.* Agr. Exp. Sta., Univ. of Minnesota, Misc. Pub. 29, 1984.
5. Horner, D., and Orcajada, P., *Wild Rice Production in Saskatchewan.* Saskatchewan Agr., La Ronge, Saskatchewan, 1985.
6. Oelke, E. A., and Strait, J., Wild rice production and processing, in I. A. Wolff (Ed.), *Handbook of Processing and Utilization in Agriculture*, Vol II, Part I—Plant Products. CRC Press, Inc., 1982, pp. 329–357.
7. Strait, J., Nordquist, D. W., and Boedicker, J. J., Separation of immature kernels from combined wild rice. In *Progress Report of 1976 Wild Rice Research Report.* Agr. Exp. Sta., Univ. of Minnesota, 1977, pp. 97–110.
8. Lund D., Lindsay, R., Marth, E., and Stuiber, D., *Methods to Extend the Storage Life of Green, Wet Wild Rice.* Dept. of Food Science, Univ. of Wisconsin, 1977.
9. Goel, M. C., Marth, E. H., Stuiber, D. A., Lund, D. B., and Lindsey, R. C., Changes in the microfilm of wild rice during curing by fermentation. *Milk Food Technol., 35*:385–391 (1972).
10. Strait, J., Donaldson, E., and Boedicker, J., Wild rice processing research. In *Minnesota Wild Rice Research, 1981*, Agr. Exp. Sta., Univ. of Minnesota, 1982, pp. 85–91.
11. Strait, J., Donaldson, E. J., Sigurdson, A., Voehl, M., and Boedicker, J. J., Wild rice parching and hulling research. In *Minnesota Wild Rice Research*, 1983, Agr. Exp. Sta., Univ. of Minnesota, 1984, pp. 91–106.
12. Processors Committee, Minutes of processors meeting, May 22, 1984. International Wild Rice Association, Grand Rapids, Minnesota, 1984.
13. Albrecht, K. A., Oelke, E. A. and Brenner, M. L., Abscisic acid levels in the grain of wild rice. *Crop Sci., 19*:671–676 (1979).
14. Candlish, B., Kneale, J., Peterson, R., Punter, D., and Stone, G., *A Guide to Wild Rice Production.* Agric. Canada, Manitoba Agric. Bull. Agdex 116, 1984.
15. Anderson, R. A., Wild rice: Nutritional review. *Cereal Chem., 53*:949–955 (1976).
16. Oelke, E. A., Amino acid content in wild rice (*Zizania acquatica* L.) grain. *Agron. J., 68*:146–148 (1976).
17. Nelson, R. N., and Dahl, R. P., *The Wild Rice Industry: Economic Analysis of Rapid Growth and Implications for Minnesota.* Dept. of Agric. and App. Econ., Univ. of Minnesota, Staff Paper Series P86-25, 1986.
18. Nelson, R. N., and Dahl, R. P., *Wild Rice Market Shows Vigorous Growth*, Agr. Ext. Ser., Univ. of Minnesota, Minnesota Agr. Econ. No. 649, 1985.

11

Cereal Proteins: Composition of Their Major Fractions and Methods for Identification

George L. Lookhart
U.S. Department of Agriculture
Manhattan, Kansas

I. INTRODUCTION

The ability to distinguish among different cultivars of particular crops is especially important to the agricultural sector. Differences exist among cultivars in their quality and other agronomic properties. Millers and bakers need to be certain the wheat cultivars they use are suitable for their intended useage, whether for bread, cake, pasta, crackers, or gravies. Brewers and maltsters want to be sure of using those cultivars that are suitable for malting. Farmers need to know the identification of the cultivars they are planting, since there may be a premium for a particular variety due to its quality or a yield benefit from a particular variety due to its genetic potential resistance or adaptability.

The methods traditionally used to assess cultivar identity and purity vary in detail from crop to crop, but generally they involve a lengthy and detailed morphological survey of the field–grown plants. Those methods are effective but consume much time and space, since large land areas and long time periods are required to grow the plants so their agronomic properties can be studied. For example, to identify barley varieties using the agronomic method, over 80 separate morphological characters are measured (17). As a result, laboratory–based (biochemical) methods for cereal cultivars have received considerable attention. Such analyses are much faster and less costly than traditional techniques and require less space and personnel.

The identification of cultivars is in effect an expression of genotype characterization. All cultivars must have differences within their genomes, and since the primary product of a structural gene is a protein, that protein may become a marker for the system or for a chromosome. For cultivar differentiation, it is necessary to study those proteins that exist in multiple forms. The most commonly used proteins for cultivar differentiation are the storage proteins, which are polymorphic in size, charge, or both of these parameters in nearly all species examined (17). (See appendix for abbreviations used throughout this chapter.)

Proteins in cereal grains can be divided in two broad groups based on their biological functions: biologically active enzymes and biologically inactive storage proteins. The storage proteins make up most (up to 80%) of the total proteins (Table 1). They are genotypic (dependent on the genetic background of the plant), and they are used most often for varietal identification.

Mention of firm names or trade products does not constitute endorsement by the U.S. Department of Agriculture over others not mentioned.

(dependent on the genetic background of the plant), and they are used most often for varietal identification.

The classical fractionation procedure of Osborne (73) has been used for years to divide cereal proteins into four major groups based on their solubilities. The albumins are soluble in water, the globulins are soluble in diluted salt solutions but not water, the prolamins are soluble in aqueous alcohols but not water or salt solutions, and the glutelins are soluble in acid or alkali but not alcohol, water, or salt solutions. Although much criticism has been leveled against this procedure, it has provided the basis for, and been most useful in, both structural and functional investigations of cereal proteins (10).

Large differences in estimating the relative proportion of proteins of the four major solubility fractions are found in different cereal species and even within a given species analyzed by different researchers (Table 1). The substantial variations in the proportions of each group within a species may be due to differences in the extraction methodology. However, protein content, growth location, and varietal differences may have larger predominant effects.

The amino acid compositions of each of the five major cereal grains are listed in Table 2 (15). The amounts of the essential amino acids (EAA) are listed at the top of the table and summed to give a total nutritional score (total EAA). The larger the score, the greater the amount of essential amino acid provided. Wheat has the lowest score (33), whereas corn has the highest (42). Compositions of the Osborne solubility fractions are listed in Tables 3–7 (15) for the cereals wheat, rice, corn, barley, and oats, respectively. Considerable variation exists for amino acid compositions both among cereal and among fractions with a given cereal.

The two storage protein fractions or groups, prolamins and glutelins, make up the bulk of the proteins in all cereals, except for oats. In general, storage proteins constitute 80–85% of the total wheat and barley proteins (in nearly equal proportions) and 85–90% of the rice proteins (mainly glutelins). Rice and oats have the lowest amount of prolamin of the cereals reviewed in this chapter, and those prolamins have the lowest proline values (Tables 4 and 7).

Electrophoresis is the process whereby charged species move under the influence of an external electric field. It is used to separate proteins by diffusing through a polymer matrix based on their apparent size and charge.Polyacrylamide gel electrophoresis (PAGE) is the major method used today. It uses crossed–linked and polymerized acrylamide to form the separation matrix. The three major variations of PAGE are (1) acidic buffers to ionize the proteins and keep them soluble., (2) the detergent sodium dodecylsulfate (SDS) to solubilize proteins and to ensure that all proteins have a uniform charge distribution, and (3) the use of pH gradients in a gel system to separate proteins by the difference in their surface charges.

The first method, acid PAGE, separates proteins by a combination of their apparent sizes and charges. Thereby, a small protein moves faster than a large protein with the same charge, but the higher the charge the faster the proteins move. This is the technique that has been used most often in "fingerprinting" or identifying cultivars for plant variety protection or for specific end–uses.

The SDS–PAGE procedure separates proteins on the basis of their apparent sizes as the surface charge is made uniformly negative by the SDS. It is envisioned that the SDS wraps itself around the proteins and thus gives a homogeneous charge to mass ratio. This technique is used for varietal identification and also for potential end–use quality (glutenin high molecular weight subunits) and molecular weight determinations.

The third technique, isoelectric focusing (IEF), separates proteins or polypeptides on the basis of their isoelectric points by setting up a pH gradient and moving the proteins through the gel to the point in the gel where the surface charge is the same as the charge in the gel (their corresponding isopotential points). It is a very sensitive and high resolution method. It is normally used when the other two techniques are not sufficiently powerful for resolving the proteins in question, for example, in identification of corn varieties. IEF is more expensive than the other methods as the ampholines that make the pH gradients are expensive to buy and difficult to make.

A relatively new method for separating cereal proteins and utilizing that separation as a fingerprint for identifying wheat varieties was reported by Bietz (5). That method utilized high–performance liquid chromatography (HPLC) to separate the proteins based on their relative hydrophobicity (the more polar proteins elute first, the more hydrophobic elute later). There are four components in any HPLC system: a pump, an injector, a column, and a detector. The pumping system consists of one or more high pressure pumps capable of producing accurate and reproducible flow rates of various solvents at pressures up to 5000 psi. The injector, usually a multiport valve, introduces sample into the high pressure solvent line without stopping the flow or changing the pressure substantially. The column or separation system is usually made of stainless steel and ranges from 10 to 30 cm in length and 4 to 4.6 mm in diameter. The packing for the column is the heart of the system in that it must not crush at high pressures (silica backbone) and flow rates but still must have high surface areas (small porous particles) for separations (~100,000 theoretical plates/m). The detector must be able to withstand moderate pressures without breaking. For UV operation, a quartz cell is required.

Most HPLC systems use variable wavelength UV–VIS detectors to be able to detect compounds at their maximum absorbance wavelength. Many other types of detectors are also used, including conductivity, refraction index, and fluorescence. Most protein separations for varietal identifications use wide pore (300°) C_{18} (reversed-phased) columns at elevated temperatures (45°C) and are eluted at 1 ml/min with water and acetonitrile gradients (25–50%) CH_3CN), each containing 0.1% TFA with detection at 210 nm.

In this chapter, methods for extracting various protein fractions from cereal grains (wheat, rice, maize, barley, and oats) and electrophoretic analysis of those fractions will be discussed. The HPLC method will only be mentioned when it is used in a complementary way, since HPLC procedures have been comprehensively reviewed by Wrigley and Bietz (118).

II. WHEAT

Wheat (*Triticum* species) ranks first in production among cereal grains grown in the world today. Wheat contributes one third of the total annual cereal protein production, and as such its importance as the major cereal grain can not be overstated. The characteristics of its protein fractions and the amino acid composition of the fractions were related to nutrition in a recent book by Lasztity (51).

Proteins in wheat have been characterized in three basic ways: by their solubilities (73), electrophoretic mobilities (42,116), and relative molecular weights (78,110). The classical Osborne fractionation procedure (73) separated proteins into four groups on the basis of their differential solubilities as previously described. Armed with those solubility definitions, Jones et al. (42) segregated the gliadin proteins into four subfractions on the basis of their electrophoretic mobilities. He referred to them as the α, β, γ, and ω gliadins.

Woychik et al. (116) used a more discriminating electrophoretic procedure and further subdivided the α and β fractions into two and four fractions, respectively. Subsequently, the γ and ω fractions were found to have three and eight components, respectively (13,37). At the same time, Cluskey et al. (16) and Graham (34) showed the uniqueness of gliadin electrophoresis in identifying wheat cultivars. Several reviews dealing with the use of electrophoresis in identifying wheat varieties by gliadin analysis are available (3,17,45,47,49,117).

The wheat–storage proteins, gliadins and glutenins, are the most often used in varietal identification. While the methods of separating the polypeptides are quite varied, the acid– and SDS–PAGE procedures have been used extensively. In the last 5 years, the use of HPLC has grown extensively. This technique was first used in characterizing wheat proteins by Bietz (5) and was recently reviewed by Wrigley and Bietz (118).

The protein fraction used most commonly for identifying cereal varieties is the prolamin fraction. Konarev et al. (49) reported that among cereal seed proteins, the alcohol–soluble fraction (prolamins and nonprolamins) was most suitable for identification of genomes and genotypes. For genomes, they found the nonprolamins (presumably albumins and globulins) most useful. For genotype identification, they found the prolamins most useful. Doekes (22), Bourdet and Autran (6), and Nierle (68) reported that the gliadins (prolamins) were best suited for wheat varietal identification.

Various procedures have been used to extract and analyze proteins for varietal identification, and a few of these are listed in Table 8. For the nonreducing extraction methods listed in the table, the solvent systems range from the classical aqueous ethanol (EtOH) and urea to the more modern solvents dimethylsulfoxide (DMSO) and dimethylformamide (DMF). Not only the solvents vary, but the ratios of solvent to flour or ground seed (meal) varies from 1:1 with the DMF procedure (procedure 9, Table 8) to 6:1 with the urea method (procedures 11 and 12, Table 1). Aluminum lactate (AL) is the buffer most often used. Jones et al. (42) found that AL buffers gave the most symmetrical electrophoretic patterns. Problems with obtaining pure AL have been stated as the reason that some researchers have switched to sodium lactate or other acidic buffers (25,36,46,60).

In most cases, 6% acrylamide gels are used, and the trend is to relatively thin gels, 1.5 or 3 mm. These gels give improved separations and reduce run times. The highest voltage system listed in Table 8 is 1000 V (106), but at such high electrical densities heat effects can cause the bands to bend (smile). The trend in the last few years is to use moderately high voltages (500 V) for shorter periods of time (2 hr) with glycine–acetic acid buffer (procedures 5 and 13, Table 8) or the AL (procedure 8, Table 8) buffer system at pH 3.1 in 6% polyacrylamide gels of 3mm thickness or less (Kasarda, private communication, 1988).

Three methods for identifying wheat varieties by the analysis of albumin and globulin fractions are also listed in Table 8. The water–to–flour extraction ratio was constant at 3:1, which is also common for gliadin extraction. The solvent extraction systems were chosen in accordance with the protein solubilities determined by the Osborne fractionation procedure: water for albumins and salt for globulins. The electrophoretic methods vary from the older AL–urea 13% SGE system at pH 3.1 to tri–borate pH 8.9 SDS–PAGE procedure. The paucity of methods for separating these protein fractions points to their limited usefulness in varietal identification.

The identification of wheat varieties by analysis of gliadin fractions can be complemented by analysis for their glutenin–subunit compositions, since syntheses of those two groups of proteins are under separate genetic control (56). The glutenin subunits are extracted with the aid of β–mercaptoethanol (ME) to break disulfide bonds (reduction), and the anionic detergent SDS to disrupt various noncovalent bonds. The ME concentration ranged from 0.5 to 2% and the solvents ranged from alcohol (55% 2–propanol) to tri–glycine or tri–borate with SDS. Protein extracts reduced and alkylated with 4–vinyl pyridine (4–VP) resolved more high molecular weight gliadin bands on SDS–PAGE analysis than when only reduced (18). That technique reportedly helped differentiate varieties that were indistinguishable by normal starch gel electrophoresis (SGE). The group of largest size glutenins (slowest moving) or high molecular weight (HMW) subunits have been correlated with the potential bread–baking quality (77), while total glutenin extracts have been used for varietal identification (24).

Hussein and Stegemann (39) reported that a total protein fraction could be extracted directly from flour or ground meal with 0.125 M tris–borate, pH 8.9, containing 5% SDS and 1% ME. Similar extraction procedures have been reported by many authors (24,77,96). However, the electrophoretic (separation) procedures used to separate the resultant extracts vary with respect to the total acrylamide (T) and the bis–acrylamide (cross–linker) concentration in the running and stacking gels. The comprehensive review of polyacrylamide gel electrophoresis recently published by Rothe and Maurer (88) and the classical paper by Weber and Osborne (113) on determining molecular weights of proteins by SDS–PAGE are indispensable for workers in this field.

III. RICE

Rice (*Oryza Sativa*) is second only to wheat as the leading food crop in the world. It is grown on every continent except Antarctica and is the food staple of over one half of the world's population. A review on the nutritive value and characteristics of the rice proteins has been reported by Lasztity (52).

Rice seed proteins can be grouped into four fractions according to their solubilities on Osborne fractionation: albumins, globulins, prolamins, and glutelins. The major storage protein fraction is the glutelin (43). Villareal and Juliano (109) indicated that little variation was found in the SDS–PAGE patterns of glutelins from various milled rice varieties. Damardjati et al. (20) found nearly identical electrophoretic patterns of the total protein extracts prepared from a limited number of Indonesian rices. However, Sarkar and Bose (90) reported differences in the electrophoretic patterns prepared with salt soluble proteins (globulins) of rice. It is the intent of this section to present several examples of extraction and analysis methods that have been successful in identifying rice varieties.

Table 9 lists some of the major extraction and analysis methods used for analysis of rice seed proteins. In the first three extraction procedures, both albumins and globulins are initially extracted. Pure globulins are analyzed only in the second procedure, where the globulins were precipitated with saturated ammonium sulfate, redissolved in 5% NcCl, dialyzed against water to reprecipitate the globulins, and then redissolved in 5% NaCl. The separation procedures vary from classical acid–PAGE and AL starch gels containing urea to the method of combining electrophoresis and immunochemical response to rabbit antisera. Kim and Jo (47) reported the use of acid (pH 3.2) and alkaline (pH 8.95) starch gel electrophoresis (SGE) to identify rice varieties from their albumin and globulin fractions. They also reported that SDS–PAGE of the rice glutelin fraction could be used for varietal identification.

There are few reports in the literature on the use of prolamin fractions for identifying rice varieties. Most authors have used other fractions because of the limited amount of prolamin (3–7%) in rice and its lack of importance in determining the quality of milled rice. However, Konarev et al. (49) reported that the cereal prolamins were most useful for genotypic (varietal) identification. In the methods listed in the prolamin extraction and analysis section of Table 3, the EtOH solvent extracts mainly prolamins, but also 6 M urea extracts glutelins, globulins, and albumins. The separation methods vary from the traditional acid–PAGE to the modern reversed–phase HPLC method.

The rice high molecular weight storage protein fraction (glutelin or oryzenin) can be extracted with acid (5 N HAc) or with urea and ME (procedures 7,8, and 9, Table 9). Glutelin is normally extracted with dilute NaOH, but du Cros et al. (26) reported that discreet electrophoretic bands could not be obtained from such extracts. They reported the most promising solvents were mild aqueous acids and strong (3 M) urea solutions.

Distributions among varieties were accentuated by the glutelin proteins (26) and by the prolamin proteins (59) extracted from the bran when brown rice was extracted instead of milled rice. All of the major U.S. rice varieties were identified by the HPLC patterns of their prolamin fractions extracted from brown rice (59). Electrophoretic methods ranged from AL PAGE to Na lactate gradient gels to SDS–PAGE. Isoelectric focusing (IEF) and two-dimensional electrophoresis utilizing IEF in one direction and PAGE in the second direction were helpful in identifying rice varieties (26).

The analysis of embryo proteins by SDS–PAGE was reported by Aliaga–Morell et al. (1) to be effective for identifying rice varieties. Rice is normally characterized as belonging to either the Japonica or Indica subspecies. Chen et al. (14) reported in a study of 32 cultivars that all Japonica types exhibited an SDS–PAGE band that migrated into the gel 56 units, whereas the Indica types all had a band at 57 units. Each accession also exhibited a specific separation pattern.

Researchers involved in identifying rice varieties have developed various methods with each group of proteins to solve the identification problems. It is obvious that no one system or protein fraction will be sufficient to identify all varieties.

IV. MAIZE

Maize (*Zea mays*) ranks third in production among cereal grains. Its rapid increase in production in the last 20 years has placed it, along with rice, as the cereal produced in greatest quantity after wheat. Most maize is used directly for feed, but a smaller amount is used for food and some is processed industrially into starch, oil, sugar syrup, and protein concentrates. Reviews of the nutritional aspects as a characterization of the maize protein fractions were recently published by Chung and Pomeranz (15) and Lasztity (53). Cooke (17) has published an excellent review of the general electrophoretic procedures, and Righetti and Bosisio (84) have reviewed the use of IEF procedures for identifying maize varieties.

The proteins in maize (corn) have been classified by the Osborne (73) fractionation procedure. The major storage protein fraction in maize, the prolamin fraction, is called zein. It constitutes more than 35% of the grain protein (111). Turner et al. (107) first demonstrated that zein was heterogeneous by using starch gel electrophoresis (SGE) and AL buffer containing 8 M urea. Since that time, each improvement in methods has brought with it an increase in the number of demonstrated polypeptides (better extraction and resolving methods), and these same methods have made it possible to elucidate the genetics of individual polypeptides.

Zein polypeptides do not vary greatly in their molecular sizes but do vary greatly in their isoelectric points (17), so the favored system for variety identification is IEF. Cooke (17) reported 28 zein components were separated on a 6–9 pH gradient. Wilson et al. (115) reported that the protein patterns were affected by the composition of the buffer in an SDS–PAGE system. Recently, Wilson (114) has shown that reduced and alkylated zeins can be separated into 41 different components, each belonging to one of four groups with molecular weights of 13.5, 18, 24, and 26.5 kDa.

Table 10 lists several of the various extraction and electrophoretic methods that have been utilized to identify corn genotypes. Nine methods are shown for the zeins and only one each for the globulins and allozymes, which closely reflects the amount of work done on each of the protein fractions.

The zeins are difficult to extract, so ground corn is usually defatted with petroleum ether, and the albumins and globulins are removed with a weak (0.5 M) NaCl solution and washed with water. The insoluble starchy material can then be extracted with a 55–70% alcohol solution containing a reducing agent.

Procedure 3 of Table 10 (33) resulted in the preparation of four zein fractions, each of which was then separated by IEF. The extraction method (procedure 9, Table 10) of Damerval et al. (21) eliminated protease activity during extraction. Only in procedure 7, Table 10 (112) were the zeins reduced and alkylated prior to electrophoresis.

The separation methods for zeins listed in Table 10 include the following: one AL starch gel with urea; one AL acrylamide gel with urea; two SDS acrylamide gels with different acrylamide concentrations (12 and 20%); four very similar IEF systems; one agarose–IEF systems; and one IEF rod system using urea and Triton X100, an anionic detergent.

Wilson (114) reported that the agarose–IEF system required only 1/5 to 1/10 of the zein material required for analysis with polyacrylamide gel (PAG) PAG–IEF method. In many cases more than one method was required to positively identify some genotypes (115).

The glutelin extraction method, also listed in Table 10, first extracted some of the albumins, globulins, and some zeins with the weak salt solution (0.5 M sodium phosphate pH 7.8) containing 0.5 M sodium chloride (NaCl) and dithiothreitol (DTT). The glutelins were then extracted with an alkaline buffer containing urea and a reducing agent. The polypeptides were alkylated before electrophoretic analysis with 4–vinyl pyridine. Wilson et al. (115) reported that the pyridylethylated (PE) glutelins alkylated in the presence of DTT gave much sharper bands than those alkylated in the presence of ME. The IEF and SDS PAGE procedures used for glutelin analysis were similar to those used to analyze the zeins.

Cardy and Kannnenberg (12) have published a comprehensive survey of allozymic variation in maize. Their extraction and analysis methods are listed in Table 10. They extracted 5–day–old coleoptile tissue with an alkaline sodium ascorbate buffer, separated the extracted proteins on starch gels, and stained for isoforms of 12 different enzymes. They were able to distinguish among 88 inbred lines and 146 hybrids of maize.

V. BARLEY

Barley (*Hordeum vulgare*) ranks fourth in production among cereal grains. Its major uses are in the brewing industry and in animal feeds. Only a small amount of it is used as food. Lasztity (54) recently published a review of the nutritional aspects and protein composition of barley proteins. A general review of the electrophoretic methods available for identifying barley and other cereal varieties has been published by Cooke (17).

The proteins of barley are classified by their solubilities, as in the other cereals: albumins, globulins, prolamins (hordeins), and glutelins. The amount of albumins is generally low (3–5%). The endosperm contains a considerable amount of globulin proteins (10–20%), but the main component proteins are the prolamins (hordeins) and glutelins, which each comprise from 40 to 45% of the total protein. Biochemical methods for identifying barley varieties have received less attention than in wheat, but several methods are available and a representative sampling of these are listed in Table 12. Most published methods utilize the hordeins. However, Jones (41) has reported the use of esterase patterns from 4–day germinated root tissue to identify U.S. barley varieties that were not differentiated by SDS–PAGE.

Shewry et al. (95) reported that in order to get reproducible patterns, the barley proteins had to be vigorously extracted with 50–60°C temperatures and a strong reducing agent (ME). In 8 of the 11 extraction procedures listed in Table 11, ME was used to reduce the proteins for extraction. Of the other three, one used monothioglycerol (MTG) as the reductant and the others either urea or 2–chloroethanol. The first extraction method (procedure 1, Table 11) was used on single or half seeds, the others on ground meal. The protein separation methods vary considerably, from SDS–PAGE (procedures 1, 2, 7, 10, 11 in Table 10) to IEF (procedures 2, 3, 8) to acid PAGE (procedures 1, 4, 5, 6, 9) to SGE (procedure 2). Within each procedure the buffers, pHs, % acrylamide concentrations, gel thickness, and analysis times varied.

For the different SDS–PAGE procedures, various tris buffers with and without urea were used. The acrylamide concentrations ranged from 12.5 to 17.5% but the pH was consistent, 8.8 to 8.9. The IEF systems used different amounts of urea but, more importantly, the ampholines used all covered the range from 5 to 9. The acid PAGE methods were most varied. The buffers ranged from AL to Na lactate to glycine–acetic acid. The acrylamide concentrations varied from 6 to 9%, one method used 3–27% acrylamide gradient gels and the pH values varied from 2.9 to 4.6.

The most recent method (procedure 11, Table 11) describes the use of silver staining, which is more sensitive than, but complements, the normal Coomassie blue staining procedure. The degree of discrimination with any of these methods for barley is less than that obtained with the wheat system. Cooke (17) listed two reasons for the restricted degree of barley variety identification by hordein electrophoresis: (1) the closely related ancestry of most modern barley varieties, and (2) the close proximity of the genes coding for the hordeins. A probably more important consideration to those listed by Cooke (17) is that barley is a diploid and thus has only two copies of the protein genes, whereas wheat is a hexaploid and has six copies of each gene, each of which is free to mutate to new forms at random. The occurrence of fewer polypeptide bands and the limited number of band combinations require the use of the various extraction and electrophoretic procedures in Table 10 and the Jones (41) esterase method.

VI. OATS

Oats (*Avena sativa*) rank sixth in production among cereal grains worldwide, after wheat rice, corn, barley, and the sorghum and millet group. Is is the most nutritionally balanced of the cereals, but lysine is still the first limiting of its amino acids (55).

The protein content of the oat groats (dehulled kernel) is the highest among the cereals, ranging from 12.4 to 24.5 %. Oat protein has some unique features, probably due to the fact that its proteins are distributed differently among the classical Osborne protein fractions than are those of other cereals. The proportion of albumins and globulins is much higher in oats, so the nutritional value is much better. In particular, lysine and aspartic acid contents are higher, whereas those of proline and glutamine are lower.

Oat cultivars have been identified by various electrophoretic (8, 9, 19, 25, 48, 57, 64, 72, 83, 85, 86, 98, 117) methods of characterizing oat storage proteins and isozymes. Lasztity (55) reviewed the distribution of oat proteins, the amino acid composition of the oat protein fractions, and methods used for varietal identification. This section lists some of the more recent studies, which have utilized different approaches.

Table 12 lists some examples of various methods that have been used for extracting and analyzing the oat protein fractions. The albumins and globulins section lists three major extraction methods that utilized buffer salts with varying ratios of solvent to solute; from 10:1 (procedure 1, Table 12) to 60:1 (procedure 3, Table 12). The electrophoretic analysis methods varied as well, from a tube gel system (procedure 1, Table 12) to SDS–PAGE slab procedures with either homogeneous acrylamide concentrations or linear acrylamide gradients (procedure 2, Table 12). The electrophoretic analysis times varied considerably, from the 1 hr (procedure 1, Table 12) of McDonald (64) to the 12 hr (procedure 3, Table 12) reported by Robert et al. (85,86).

The prolamins (avenins) were extracted with either 2–chloroethanol (2–Cl–EtOH) or 70% ethanol (EtOH). The solvent–to–solute ratios were 3:1 for the EtOH extracts to 6:1 for the 2–Cl–EtOH extracts. The EtOH system of Portyanko et al. (83) included Na lactate and urea, which would also extract the albumins and globulins. The analysis methods for the avenins were similar in that they all used lactate PAGE with an acid buffer. However, the concentrations of acrylamide varied from 7.5% (procedure 5 Table 12) to 13% (procedure 6, Table 12). Additionally, Lookhart (57) reported the use of reversed–phase HPLC to aid in differentiating those varieties whose electrophoretic patterns were very similar.

Singh et al. (98) reported the use of SGE to identify oat varieties by analysis of the iozyme forms of esterase, leucine aminopeptidase, and anodal and cathodal peroxidases that were extractable in aqueous solutions. McDonald (64), using the tube PAGE gel procedure (procedure 8, Table 12), identified varieties from their esterase and peroxidase electrophoretic patterns.

Cooke and Cliff (19) extracted nearly all the proteins from oats with a buffer containing SDS and ME. Their procedure utilized SDS–PAGE to separate all of the proteins on the basis of their relative molecular size and they used IEF to further characterize and identify the proteins.

VII. CONCLUSIONS

Prolamins and glutelins make up the bulk of the proteins in all cereals, except for oats, where the globulins play a larger role. The alcohol–soluble fraction (prolamin) is normally the most useful for varietal identification. However, the large numbers of closely related cultivars have led to the utilization of other protein components. The only consistencies in the cereal analyses described are that no one system or protein fraction has been found sufficient to identify all varieties of any cereal species and that Coomassie blue dye is commonly used for staining all the cereal protein bands.

A. Wheat

The methods for identification of wheat varieties by analysis of their storage proteins is varied. Gliadins and glutenins are the major storage proteins, and the amino acid compositions of those fractions are similar. The most commonly used identification procedure is to extract the gliadins with 70% ethanol and separate the resulting extract on a polyacrylamide gel (6%) of 3 mm thickness, with AL buffer system at pH 3.1 for 2 hr at 500 V. Staining of the resultant gel is done with Coomassie brilliant blue R250 (1%) in 6% trichloroacetic acid. Analysis of the glutenins by SDS–PAGE yields information on both varietal identification and genetic baking potential.

B. Rice

The major storage protein fraction is the glutelin (oryzenin). Among varieties, however, little variation in the glutelin fraction patterns have been found. Researchers have developed various methods within each group of proteins. The glutelins were characterized by SDS–PAGE and IEF, the prolamins by PAGE and HPLC, and the albumins and globulins by PAGE and SGE.

C. Maize

The major protein fraction in maize is the prolamin (zein) fraction. Zeins from different varieties have been found to be heterogeneous using SGE, but the most common procedure for varietal identification, IEF, makes use of their isoelectric point heterogeneities. In many cases more than one method is required to positively identify some varieties.

D. Barley

The major protein components in barley are the hordeins and glutelins. Most identification procedures utilize the hordeins. Wide variations in electrophoretic methods have been employed. Storage extraction solvents and more vigorous extraction conditions are required to get reproducible electrophoretic patterns from barley than required for wheat.

The degree of discrimination with any of the methods on barley is less than that obtained with the wheat system. However, various SDS–PAGE, IEF, and acid PAGE systems are available for identifying most barley varieties.

E. Oats

The major protein fractions in oats are albumins and globulins. The extraction solvents, solvent–to–solute ratios, and electrophoretic methods for varietal identification all vary widely. However, the most common procedure for globulins and total protein utilize SDS–PAGE, whereas that of prolamin analysis utilizes AL PAGE.

TABLE 1 Cereal Protein Distribution in the Four Major Solubility Fractions

	Protein fractions (% total proteins)				
Grain	Albumin	Globuling	Prolamin	Glutelin (+ residue)	Ref.
Wheat	5	10	69	16	10,82
	3–5	6–10	40–50	30–40	50
Rice	5	10	50	80	10,28,82
	trace	2–8	1–5	85–90	50
Corn	4	2	55	40	10,82
Barley	13	12	52	23	10,82
	3–5	10–20	35–45	35–45	28,50
Oats	1	78–80	10–16	5	10,50,82
	1	13	18	68	28

Source: With permission.

TABLE 2 Amino Acid Content of Five Cereal Grains (g/16 g N)

	Wheat		Rice		Corn		Barley		Oats	
Amino acid	1[a]	2[b]	1[a]	2[c]	1[a]	2[d]	1[a]	2[e]	1[a]	2[f]
Ile	3.3	3.7	3.8	3.8	3.7	3.7	3.6	3.6	3.8	3.9
Leu	6.7	6.7	8.2	8.7	12.5	13.6	6.7	6.5	7.3	7.4
Lys	2.9	2.4	3.8	4.0	2.7	2.6	3.5	3.4	3.7	4.2
Met	1.5	1.7	2.3	2.2	1.9	1.8	1.7	2.6	1.7	2.5
Cys	2.5	2.5	1.1	1.0	1.6	1.1	2.3	1.2	2.7	1.6
Phe	4.5	4.9	5.2	5.4	4.9	5.1	5.1	5.2	5.0	5.3
Tyr	3.0	2.8	3.5	3.6	3.8	4.4	3.1	2.5	3.3	3.1
Thr	2.9	2.8	3.9	3.9	3.6	3.6	3.3	3.1	3.3	3.3
Trp	1.1[g]	1.5	1.2[g]	1.3	0.6[g]	0.7	1.5[g]	—	1.3[g]	—
Val	4.4	4.5	5.5	5.5	4.8	5.3	5.0	5.0	5.1	5.3
Total EAAs[h]	32.8	33.5	38.5	39.4	40.1	41.9	35.8	33.1	37.2	36.6
Arg	4.6	4.0	8.3	9.4	4.2	3.8	4.7	4.4	6.3	6.9
His	2.3	2.3	2.5	2.8	2.7	2.8	2.1	2.1	2.1	2.2
Ala	3.6	3.4	6.0	6.4	7.5	7.9	4.0	4.1	4.5	5.0
Asp	4.9	4.8	10.3	10.8	6.3	—	5.7	6.3	7.7	8.9
Glu	29.9	33.1	20.6	22.6	18.9	—	23.6	27.4	20.9	23.9
Gly	3.9	3.8	5.0	5.3	3.7	3.4	3.9	3.8	4.7	4.9
Pro	9.9	11.0	4.7	4.7	8.9	8.3	10.9	12.3	5.2	4.7
Ser	4.6	4.9	5.4	6.0	5.0	4.8	4.0	3.5	4.7	4.2
N (g/100 g)[i]	2.38	2.90	1.45	2.12	1.52	—	1.89	2.18	2.23	2.74

[a]Calculated data (30).
[b]Calculated average of two varieties (105).
[c]Calculated average of six varieties (108).
[d]From Ref. 4.
[e]From Ref. 81. (An average of 113 cultivars, expressed as g/100 g recovered).
[f]From Ref. 87. (An average for oat groats of 289 cultivars of *Avena sativa*, expressed as g/100 g recovered).
[g]Determined by a microbiological method.
[h]EAAs = essential amino acids, including cysteine and tyrosine.
[i]On dry-weight basis.
Source: Ref. 15.

TABLE 3 Amino Acid Composition (% of Total amino acids) of Wheat Protein Fractions[a]

Amino acid or ammonia	Albumin 1[b]	Albumin 2[c]	Globulin, water[d]	Gluten[e]	Prolamin (gliadin)[f]	Glutelin (glutenin)[f]
Ile	2.6	1.8	1.3	3.7	4.1	3.2
Leu	8.2	7.9	9.1	6.5	6.8	6.5
Lys	3.3	5.9	12.4	1.1	0.6	1.7
Met	3.0	3.2	trace	1.5	1.5	1.6
Cys	6.4	7.6	13.6	1.4	1.0	1.0
Phe	2.2	0.2	3.1	4.6	5.4	4.0
Tyr	5.8	5.1	2.7	3.2	2.5	4.1
Thr	2.5	2.2	7.4	2.1	1.8	2.6
Trp	3.5	5.2	—	1.1	0.9	1.5
Val	6.9	10.3	2.2	4.4	4.1	4.1
Total EAAs[g]	44.4	49.4	51.8	29.6	28.7	30.3
Arg	8.5	8.1	14.6	3.1	2.3	3.1
His	1.8	trace	1.3	2.0	2.0	1.8
Ala	8.6	5.5	4.0	2.1	1.7	2.4
Asp	6.3	8.1	6.8	2.5	2.2	2.6
Glu	13.0	12.8	5.6	36.7	39.6	35.9
Gly	4.5	4.3	4.9	2.6	1.4	4.5
Pro	6.7	6.7	3.7	13.0	13.9	11.1
Ser	4.8	5.1	7.3	3.4	3.2	4.4
NH3	1.3	—	—	5.0	5.0	4.1

[a]Calculated from number of amino acid residues as reported by Kasarda et al. (45), except for water-soluble globulin.
[b]From Ref. 29.
[c]From Ref. 32.
[d]From Ref. 69.
[e]From Ref. 119.
[f]From Ref. 120.
[g]EAAs = essential amino acids, including cysteine and tyrosine.
Source: From Ref. 15 with permission.

TABLE 4 Amino Acid Composition of Rice Protein Fractions

	Amino acid composition (g/16 g N)						
	Brown rice[a]			Milled Rice[b]			
Amino acid	Albumin and globulin	Prolamin (oryzin)	Glutelin (oryzenin)	Albumin	Globulin	Prolamin (oryzin)	Glutelin (oryzenin)
Ile	3.1	4.9	3.7	3.9	2.9	4.5	4.5
Leu	6.3	11.8	7.7	7.5	6.3	10.8	7.5
Lys	4.6	0.3	3.0	4.7	2.4	0.5	3.3
Met	2.3	0.8	1.5	2.4	2.2	0.5	1.3
Cys	3.1	1.8	1.0	2.8	0.0	0.3	1.2
Phe	3.6	6.0	5.3	2.9	3.1	6.0	5.9
Tyr	4.0	8.4	5.2	3.7	4.8	8.3	4.9
Thr	3.8	2.4	3.4	4.4	2.8	2.3	3.6
Trp	1.4	0.9	1.5	1.8[c]	1.2[c]	0.9[c]	1.1[c]
Val	5.4	5.0	4.9	8.3	5.9	6.7	6.2
Total EAAs[d]	37.6	42.3	37.2	42.4	31.6	40.8	39.5
Arg	10.4	6.5	9.0	8.0	10.5	5.6	10.1
His	2.6	1.5	2.4	2.5	1.5	0.9	2.5
Ala	6.6	6.2	5.1	8.3	8.7	6.3	4.8
Asp	8.3	6.4	8.8	10.3	7.4	7.5	10.3
Glu	16.4	27.7	21.9	11.9	11.2	20.4	19.2
Gly	6.1	2.9	4.5	6.6	5.6	3.0	4.3
Pro	5.5	5.5	4.9	6.3	5.2	3.9	4.0
Ser	5.0	6.1	5.3	5.0	5.2	4.9	6.6
N (g/100 g)[e]	12.59	9.98	16.8	2.11	16.80	3.57	16.9

[a]From Ref. 74.
[b]From Ref. 104.
[c]From Ref. 43.
[d]EAAs = essential amino acids, including cysteine and tyrosine.
[e]On dry weight basis.

TABLE 5 Amino Acid Composition of Corn Protein Fractions[a]

Amino acid	Albumin			Globulin			Prolamin (zein)			Glutelin			
	1[b]	2[c]	3[d]	1[b]	2[c]	3[d]	1[b]	2[c]	3[e]	Whole[b]	G1[c]	G2[c]	G3[c]
Ile	4.6	3.3	3.7	4.3	4.0	3.0	4.3	3.2	4.4	3.6	2.7	2.4	3.5
Leu	9.8	6.0	6.4	8.5	8.0	5.9	20.1	19.7	23.0	12.0	12.7	8.4	10.5
Lys	6.3	5.5	6.4	5.9	5.2	6.0	0.1	0.1	0.0	3.0	0.1	1.7	4.2
Met	1.6	1.4	1.5	1.4	1.0	1.0	0.4	0.9	1.6	4.0	4.1	1.2	1.9
Cys	0.2	2.8	—	1.2	2.0	—	0.1	trace	1.7	1.4	trace	1.3	trace
Phe	4.4	2.2	3.5	4.8	4.9	4.6	6.2	6.5	7.1	4.4	5.6	2.7	4.4
Tyr	3.6	2.1	3.4	3.8	3.5	3.1	4.1	4.3	6.5	4.9	4.7	2.6	3.7
Thr	4.8	4.7	5.4	4.7	3.4	3.3	2.7	2.4	3.1	3.5	3.1	4.1	4.0
Trp	—	0.7	—	—	0.9	—	—	0.0	—	—	0.0	0.3	1.2
Val	6.7	4.8	5.9	6.6	5.9	5.7	4.4	3.2	3.7	5.4	3.6	5.8	5.9
Total EAAs[f]	42.0	33.5	36.2	41.2	38.8	32.6	42.4	40.3	51.1	42.2	36.6	30.5	39.3
Arg	7.4	6.6	7.5	7.7	9.9	12.5	1.4	1.3	1.7	4.1	2.2	4.1	5.9
His	2.4	2.8	2.5	2.5	3.4	3.9	0.9	0.9	1.2	3.3	2.0	7.3	3.5
Ala	7.9	10.3	7.6	7.4	6.6	5.3	10.0	13.0	10.4	7.2	10.9	4.1	7.0
Asp	8.8	9.4	9.7	9.2	6.6	7.7	5.7	5.3	5.6	6.0	4.1	3.1	7.7
Glu	14.5	15.0	12.7	13.7	17.2	16.8	24.3	26.8	29.0	19.0	23.7	25.4	19.1
Gly	6.1	9.0	7.0	6.8	5.4	5.5	1.2	0.8	1.9	4.1	2.8	5.2	4.9
Pro	6.4	6.9	5.2	6.8	6.8	3.8	9.2	8.6	14.7	9.9	11.8	15.9	7.7
Ser	4.8	5.6	5.0	4.8	4.9	5.6	4.8	5.0	5.9	4.3	5.7	4.1	4.9

[a]Amino acid compositions are expressed differently, depending on source.
[b]From Ref. 103. Amino acid composition is expressed as a percentage of total amino acids.
[c]From Ref. 92. Amino acid composition is expressed as g/100 g of amino acids recovered. C1, G2, and G3 = glutelin fractions 1, 2, and 3, respectively. Tryptophan was determined colorimetrically.
[d]From Ref. 75. Amino acid composition is expressed as g/100 g of protein.
[e]From Ref. 66. Amino acid composition is expressed as g/100 g of protein.
[f]EAAs = essential amino acids, including cysteine and tyrosine.

TABLE 6 Amino Acid Composition of Barley Protein Fractions[a]

			Prolamin (hordein)			Glutelin		
Amino acid	Albumin[b]	Globulin[b]	1[b]	2[c]	3[d]	1[b]	2[c]	3[e]
Ile	4.4	3.9	3.7	3.6	4.0	4.2	4.1	4.1
Leu	8.5	8.2	6.8	6.8	7.4	8.3	8.7	9.0
Lys	6.2	6.2	0.6	0.9	0.8	2.5	4.1	3.9
Met	2.4	2.2	1.0	0.6	0.8	1.5	1.7	2.1
Cys	0.5	0	1.4	1.3	2.7	0.5	1.8	2.2
Phe	5.2	4.8	7.5	4.7	8.3	5.7	5.2	5.7
Tyr	3.7	4.2	3.4	3.3	4.7	3.9	2.2	3.5
Thr	4.4	4.1	1.8	1.8	1.3	3.4	3.8	3.1
Trp	—	—	—	—	—	—	—	—
Val	6.8	6.9	4.0	3.9	3.7	5.8	5.6	6.0
Total EAAs[f]	42.1	40.5	30.2	26.9	33.7	35.8	37.2	39.6
Arg	5.8	8.0	2.7	2.8	2.8	4.5	4.5	5.4
His	2.5	2.1	1.2	1.5	1.9	2.3	2.7	2.3
Ala	7.2	5.1	1.5	1.7	2.5	3.2	4.3	5.7
Asp	11.4	8.5	1.6	1.5	1.6	4.3	6.0	5.6
Glu	17.8	18.3	38.9	41.2	36.4	29.9	25.4	21.6
Gly	4.8	5.0	1.2	1.4	2.3	3.2	4.4	5.0
Pro	6.7	8.2	19.4	19.9	18.8	12.2	10.4	10.6
Ser	4.5	4.3	3.2	3.2	2.6	4.4	4.8	4.5

[a]Data from Brandt (1976) and El-Negoumy et al (1977) are expressed as a percentage of total amino acids. All other data are expressed as g/16 g N.
[b]Calculated data (Chung and Pomeranz, 1985) from Brandt (1976) (Bombi variety).
[c]From Martens and Bach Knudsen (1980) (Bombi variety).
[d]From Singh and Sastry (1977b) (NP–113 variety).
[e]From Singh and Sastry (1978) (NP–113 variety).
[f]EAAs = essential amino acids, including cysteine and tyrosine.

TABLE 7 Amino Acid Composition of Oat Protein Fractions

Amino acid	Albumin		Globulin			Prolamin (avenin)			Glutelin		
	1[a]	2[b]	1[a]	2[b]	3[c]	1[a]	2[b]	3[c]	1[a]	2[b]	3[c]
Ile	4.8	3.1	4.3	4.5	4.4	3.7	2.8	3.3	5.0	4.1	4.8
Leu	8.6	6.4	6.8	7.5	7.5	10.6	11.2	10.5	8.1	9.4	7.4
Lys	8.2	8.1	5.5	4.2	5.0	3.3	0.7	1.1	5.0	2.9	3.3
Met	2.4	0.5	1.8	0.7	3.3	3.7	2.2	2.3	1.8	1.2	1.3
Cys	1.4	5.4	1.3	1.2	3.5	4.2	2.8	3.4	0.9	1.9	1.6
Phe	7.3	3.6	5.9	5.8	4.6	7.0	6.7	7.0	6.8	6.9	6.9
Tyr	3.1	4.0	2.4	4.2	4.2	1.7	2.1	2.4	4.9	3.9	4.6
Thr	5.6	4.7	3.6	3.5	4.1	2.3	1.4	1.5	4.4	2.8	3.5
Val	6.6	4.7	4.9	5.2	5.2	5.9	6.7	6.6	5.5	6.0	5.1
Total EAAs[d]	48.0	40.5	36.5	36.8	41.8	42.4	36.6	38.1	42.4	39.1	38.5
Arg	5.3	7.9	9.7	9.3	9.5	4.8	3.5	3.8	9.5	7.9	8.6
His	2.9	2.4	2.9	2.6	2.5	1.7	0.9	1.1	3.1	2.6	2.7
Ala	8.0	6.3	5.7	4.4	4.3	4.4	3.5	3.4	5.1	4.1	4.1
Asp	12.2	8.1	8.8	9.4	8.0	3.3	1.7	2.1	10.0	6.5	9.7
Glu	13.7	16.1	20.2	21.1	20.1	37.6	42.1	40.2	17.5	33.4	23.1
Gly	6.7	7.6	5.4	4.5	4.3	2.4	0.8	0.9	5.0	3.0	3.8
Pro	6.1	6.7	5.4	4.2	4.9	9.1	8.4	8.7	5.5	7.3	4.6
Ser	6.6	7.0	4.9	4.9	4.5	2.9	1.9	2.0	5.1	3.8	5.1

[a]From Ref. 23. Data are expressed as g/16 g N.
[b]From Ref. 121. Data are expressed as g/16 g N.
[c]Number of amino acid residues are reported in Ref. 79. Data are expressed as a percentage of total amino acids. Total protein amino acid data are for oatmeal defatted with water-saturated *n*-butanol.
[d]EAAs = essential amino acids, including cysteine and tyrosine. Tryptophan was not determined.

TABLE 8 Electrophoretic Methods for Wheat Variety Identification

Extraction	Methodology	Ref.
	NONREDUCING CONDITIONS	
A. *GLIADINS*		
ALCOHOL		
1. 3 × wt with 70% EtOH from gluten ball	AL SPINCO H SYSTEM pH 3.2, 8.7 V/cm, 3 hr	42
2. 3 × wt with 60% EtOH	SGE (9%), AL 0.5 M urea, pH 3.2, 8 hr	2
3. 3 × wt with 70% EtOH and flour 2 × vol buffer	AL PAGE (6%) 21° C 6 mm gels, 73 mA, 6.5 hr	11
4. 3 × wt with 70% EtOH	AL PAGE (6%) 7°C 1.5 mm gels, 1000 V, 40 min	106
5. 50 μL 70% EtOH with 100 μL Elect buffer	Glycine/Acetic PAGE (6%) 3 mm gels, pH 3.1, 400 V, 2.5 hr	60
6. 3 × wt with 70% EtOH	Na lactate PAGE (7.5%) 3 mm gel, pH 3.1, 20 mA, 4.5 hr	80
7. 3 × wt with 70% EtOH	Na lactate PAGE (6%) 15°C 6 mm gel, pH 3.1, 300 V, 5 hr	46
8. 3 × wt with 70% EtOH flour or ground meal	AL PAGE (6%) 20°C 3 mm gel, pH 3.1, 500 V, 2 hr	57
DIMETHYLFORMAMIDE		
9. 25 μL/mg seed	AL PAGE (6%) 20°C 50 mA, pH 3.1, 300 V, 6.5 hr	65
DIMETHYLSULFOXIDE		
10. 150 μL of 40% DMSO/seed	TRIS-lactate PAGE (6%) 1.5 mm gel, 400 V, 4.5 hr	36
11. 6 × wt with 2 M urea	AL urea SGE (12%) 20°C, pH 3.1, 200 V, 4 hr	63
12. 6 × wt with 1 M urea	Na lactate gradient PAGE (2–16%) 3 mm gel, pH 3.1, 4 hr	25
13. 100 μL urea (6%)/seed	Glycine/Acetic acid PAGE (6%) 2.7 mm gel, 20°C, pH 3.1, 400 V, 130 min	89
B. *ALBUMINS AND GLOBULINS*		
14. 3 × wt with water	AL 3 M urea SGE (13%) 3 mm gel, pH 3.1, 30 mA, 160 min	22
15. 3 × wt with water	TRIS-Borate PAGE (5%) CYANOGUM41 15°C , pH 8.9, 500 V, 2 hr	38
16. 3 × wt with 0.125 M TRIS-borate pH 8.9	TRIS-Borate SDS-PAGE CYANOGUM, (5–25%) 3 mm gel, pH 8.9,400 V 90 min	39

TABLE 8 (Continued)

Extraction	Methodology	Ref.
	REDUCING CONDITIONS	
GLUTENS		
17. TRIS-borate with SDS (5%) and ME (1%) 1 g meal/3 ml buffer	SDS-PAGE, SDS (0.1%) 3 mm gel, pH 7.1, 400 V, 90 min	38
18. 200 µL solvent/seed (55%), 2–propanol with ME (2%)	SDS-PAGE (17.5%) with & without 4–VP pH 8.9, 20 mA, 3 hr	96
19. 400 µL solvent/seed 0.1 M TRIS-glycine with ME (.05%)	TRIS-glycine PAGE (5%), pH 8.9	71

TABLE 9 Electrophoretic Methods for Rice Variety Identification

Extraction	Methodology	Ref.
ALBUMINS AND GLOBULINS		
1. 4 × wt with 0.1 M NaCl	Crossed immunoelectrophoresis pH 8.6, rabbit antisera	94
2. NaCl (5%) 1 hr PPT GLOB with ammonium sulf. (35%)	AL SGE with 7.5 M urea, pH 3.1 TRIS-Citrate SGE with 7.5 M urea, pH 8.9	40
3. 6 × wt with 0.5 M TRIS buffer pH 7.6	β-Alanine buffer PAGE (10%) 4°C, pH 4.5	90
4. Extract defatter flour with 0.5 M NaCl dialyzed vs. water	AL, 3 M urea and TRIS-borate SGE (12%), 25°C, pH 3.2 and pH 8.95	47
PROLAMINS		
5. 6 M urea	Acid-PAGE	67
6. 3 × wt with 70% EtOH	HPLC, C18 reversed phase	57
GLUTELINS (Oryzenin)		
7. 4 × wt with 3 M urea plus ME (1%)	Na Lact. PAGE gradient (3–27%) 23°C, 3 mm gel, 200 V, 90 min	25
8. 2 × wt with 5 N HAc	AL PAGE (6%) 20°C 6 mm gel, pH 3.1, 60 mA, 2 hr	DeJong and Bushuk, unpublished
9. 5 × wt with buffer 50 mM TRIS-HC1, SDS with 0.6 ME and 4 M urea pH 7.5	TRIS-glycine SDS-PAGE (14%), pH 8.3	14

TABLE 10 Electrophoretic Methods for Maize Variety Identification

Extraction	Methodology	Ref.
PROLAMINS (ZEINS)		
1. Defatted meal extract with EtOH (70%), dialyze vs. water, PPT SOL in AL-urea	AL, SGE (16%) 8 M urea 20 V/cm, 24 hr, AGAR gels (3%)	107
2. Defatted meal with extract EtOH (70%) with 0.6 M ME	TLIEF acrylamide (5%), 6 M urea pH 6–8 & 7–9 ampholytes (2% EACH) 2C, 13 W, 4 hr, SDS-PAGE (20%) 4°C, 50 V, 15 hr	102
3. 10 × wt with 0.5 M NaCl Extract insolubles Z1-55% IPA Z2=G1-55% IPA + ME G2-pH10 Borate + ME +NaCl G3-pH10 Borate + ME + NaCl + SDS	IEF acrylamide (5%), 6 M urea pH 6–8 & 7–9 ampholytes (2% each) 2°C, 13 W, 4 hr	33
4. Defatted meal extr with 0.5 M NaCl, insol. with EtOH (70%) and NaOAc (.5%)	AL with 8 M urea PAGE (5%)	76
5. Extract globulins-NaCl extract PPT-EtOH (70%) and ME (1.5%), lyoph, suspend solid in TRIS pH 8.2 with 6 M urea and ME (1%)	TLIEF acrylamide (5%), 6 M urea pH 6–8 & 7–9 ampholytes (2% each) 2°C, 13 W, 4 hr	70
6. Extract globulins-NaCl extract PPT with Nprop (50%) and ME (2%)	SDS-PAGE (12%), pH 8.9 20 mA, 3 hr	115
7. 35 × wt with EtOH (70%) dialyze, lyoph, reduce and alkylate	IEF acrylamide (5%) 6 M urea pH 6–8, ampholytes (2%) 13 W, 4 hr	112
8. IPA (55%) with ME (2%)	Agarose-IEF, 5 M urea with 2 mM DTT, pH 3.5–9.5 & 5–8 ampholytes (1% EACH), 1300 V-hr	114
9. TRIS buffer with SDS (4%) and ME (5%),heat 100°C, 3 min	IEF RODS (19.5 cm × 1 mm) acrylamide (2.84%) with 9.2 urea with Triton X100 (4%), ampholytes (4%), 900 V, 24 hr	21
GLUTELINS		
10. 6 × wt with Na_2HPO_4 (0.05 M) pH 7.8 with NaCl (0.5 M) and EDTA with DTT	IEF acrylamide (5%) with 6 M urea pH 5–7 & 7–9, ampholytes (1% each) 13 W, 3 hr	115
insolubles water wash Suspend starchy residue in TRIS-NO_3, pH 7.5 with 6 M urea and ME Alkylate with 4-VP 15 × wt with SDS buffer	SDS-PAGE (12.5%), pH 8.9, 3 hr	
ALLOZYMES		
11. 12 mm coleoptiles add 20 μL of pH 7.38 buffer Na ascorbate with sucrose	SGE (13%), 1 cm gels, 4°C	12

TABLE 11 Electrophoretic Methods for Barley Variety Identification

Extraction	Methodology	Ref.
PROLAMINS (HORDEINS)		
1. Single or half seeds 10 × wt with IPA (55%) and ME, dried,suspended in TRIS (pH 7.5) 8 M urea and ME	SDS-PAGE (17.5%), pH 8.9, 20 mA, 3 hr Urea PAGE (12.5%), 6 M urea pH 4.6, 15 mA, 18 hr	95–97
2. 20 × wt with IPA (55%) and ME, add equal vol NaCl (4%) PPT Hordeins, dry, reduce and alkylate	TRIS-borate SDS-PAGE (17.5%) pH 8.9, 3 hr SGE 3 M Urea, pH 3.3 IEF 6 M Urea, acrylamide (5%) pH 5–7 and 7–9 ampholytes (1% each) 13 W, 3 hr	95–97
3. IPA (50%) and ME (0.3%) dry and take up in 6 M urea	IEF PAGE (6.25%) 4.5 M urea pH 5–9 ampholytes 2 hr	93
4. 6 × wt with 1 M urea	Na lact PAGE gradient (2–37%) 3 mm gel, pH 3.1, 0.5–4 hr	25
5. IPA (55%) and MTG (2%)	AL PAGE (6%), pH 3.1 3 hr	61
6. Chloroethanol (50%)	HAc-glycine PAGE (7%) pH 2.9, 3.5 hr	91
7. EtOH (70%), dried then suspended in TRIS pH 6.7, SDS (2%) ME (1%) and 4 M urea	SDS-PAGE (12.5%) 4 M urea 0.125 M TRIS pH 8, 15 mA, 6 hr	44
8. 1.0 M Urea and ME (1%)	IEF PAGE (8.25%) 2 M urea pH 3.5–10 ampholytes, 3.5 hr	117
9. IPA (55%) and ME (2%)	AL PAGE (9%), pH 3.1 2 hr	18
10. TRIS SDS buffer (pH 6.8) and ME, DMF, water	SDS PAGE (17.5%) 1.5 mm gel, pH 8.8, 10 mA, 16 hr	101
11. IPA (55%) and ME (2%) diluted 8:1 with TRIS (pH 6.8) and DTT	SDS-PAGE (15%), 20 mA pH 8.8, 5 hr, Ag stain	35

TABLE 12 Electrophoretic Methods for Oat Variety Identification

Extraction	Methodology	Ref.
ALBUMINS & GLOBULINS		
1. 10 × wt with buffer 0.1 M NaAc and 0.1 N HAc pH 5	Disc-PAGE (7.5%), 7 cm tubes 3.3 mA, 1 hr	64
2. 25 × wt with buffer 50 mM TRIS-HCl (pH 8.0) with 1 M NaCl, dial	SDS-PAGE (9, 12, 15 or 5–20%) linear gradient, all with EDTA, with or without DTT	8
3. 60 × wt with buffer 1 M NaCl in 0.05 M TRIS (pH 8.5), PPT Glob, dial	TRIS-HCl SDS-PAGE (11%) 4 M urea SDS (0.1%), pH 8.8, 60 V, 12 hr	85,86
PROLAMINS		
4. 6 × wt with Solvent 2–Cl–EtOH (25%)	Na lact PAGE (12%) 480 V, 75 min	19
5. 3 × wt with EtOH (70%)	AL PAGE (7.5%) 6 mm gel, 10°C, 300 V, 5 hr	57
6. 3 × wt with EtOH (70%), add equal vol Na lact. buffer with 2 M urea	AL PAGE (13%) urea (7%) 1 mm gel, 32°C, pH 3.1, 580 V, 5 hr	83
ISOZYMES		
7. Esterase Leucine aminopeptidase Anodal peroxidase Cathodal peroxidase	TRIS-citrate SGE discontinuous buffer system, pH 8.0	98
8. Esterase Peroxidase	Disc-PAGE (7.5%), 3.3 mA, 1 hr	64
TOTAL PROTEIN		
9. 8 × wt with buffer SDS (2%), ME (5%), TRIS-HCl (0.0625 M) pH 6.8	SDS-PAGE (10 or 15%), TRIS-HCl (0.375 M), SDS (0.1%), pH 8.8	19

APPENDIX: ABBREVIATIONS

Abbreviations	Descriptions
2-Cl-EtOH	2-Chloroethanol
4-vp	4-Vinyl pyridine
Acryl	Acrylamide
Ag	Silver
Al	Aluminum lactate
C	Celsius
C18 RP	C18 reversed-phase column
cm	Centimeter
Defat	Defatted
Dial	Dialyzed
DMF	Dimethyl Formamide
DMSO	Dimethyl Sulfoxide
DTT	Dithiothreitol
EAA	Essential amino acids
EDTA	Ethylene diamine tetraacetic acid
Elect	Electrophoresis
EtOH	Ethanol
Extr	Extracted
g	Gram
Glob	Globulins
HAc	Acetic acid
HCl	Hydrochloric acid
HPLC	High-performance liquid chromatography
hr	Hour(s)
IEF	Isoelectric focusing
IPA	Isopropanol
L	Liter(s)
Lact	Lactate
Lyoph	Lyophilize
M	Molar
m	Milli
mA	Milliamp(s)
mm	Millimeter
ME	β-Mercaptoethanol
mg	Milligram(s)
min	Minutes
ml	Milliliter(s)
mM	Millimolar
MTG	Monothioglycerol
N	Normality (Concentration)
Na	Sodium
Na_2HPO_4	Sodium hydrogen phospate
NaAc	Sodium actetate
NaCl	Sodium choride
NaOAc	Sodium acetate
NO_3	Nitrate
Nprop	n-Propanol
PAGE	Polyacrylamide gel Electrophoresis
PPT	Precipitate
SDS	Sodium dodecyl Sulfate
SGE	Starch gel electrophoresis
SOL	Solubilized
Sulf	Sulfate

Abbreviation	Description
TFA	Trifluoroacetic acid
TLIEF	Thin layer isoelectric focusing
TRIS	Tris(hydroxymethyl) aminomethane
μ	Micro
V	Volt(s)
vol	Volume
W	Watts
wt	Weight

REFERENCES

1. Aliaga–Morell, J. R., Culianez–Macia, F. A., Clemente–Marin, G. and Primo–Yufera, E., Differentiation of rice varieties by electrophoresis of embryo protein. *Theor. Appl. Genet.,74*:224–232 (1987).
2. Autran, J. C. and Bourdet, A., L'identification des varietes de blé: Establissement d'un tableau general de determination fonde sur le diagramme electrophoretique des gliadines du grain. *Ann. Amelior. Plant., 25*:277–301 (1975).
3. Autran, J. C., Bushuk, W., Wrigley, C. W., and Zillmann, R. R., Wheat cultivar identification by gliadin electrophoregrams. IV. Comparison of international methods. *Cer. Foods World, 24*: (9)471–475 (1979).
4. Baudet, J., Mosse, J., Landry, J., and Moreaux, T., Etude sur les proteines du maise. I. Composition–acides amines des fractions azotees du grain. *Ann. Psysiol. Veg., 8*: 321–329(1966).
5. Bietz, J. A., Separation of cereal proteins be reversed–phase high–performance liquid chromatography. *J. Chrom., 255*:219–238 (1983).
6. Bourdet, A., and Autran, J. C., Electrophoretische Bestimmung des Gliadins zur Erkennung von Weizenmuschungen. *Getreide Mehl Brot., 30*: 203–207 (1976).
7. Brandt, A., Endosperm protein formation during kernel development of wild type and a high–lysine barley mutant. *Cereal Chem., 53*:890–901 (1976).
8. Brinegar, A. C., and Peterson, D. M., Separation and characterization of oat globulin polypeptides. *Arch. Biochem. Biophys., 219*(1):71–79 (1982).
9. Burgess, S. R., Shewry, P. R., Matlashewski, G. J., Altosaar, I, and B. J. Miflin., Characteristics of oat seed globulins. *Chem. Abstracts* 100, #20444, 1984, p. 302.
10. Bushuk, W. Utilization of cereal proteins. In *Utilization of Protein Resources.* (D. W. Stanley, E. D. Murray, D. H. Lee (Eds.). Food and Nutrition Press, Inc., Westport, CT, 1981.
11. Bushuk, W., and Zillmann, R. R., Wheat cultivar identification by gliadin electrophoregrams. I. Apparatus, method and nomenclature. *Can. J. Plant Sci., 58*:505–515 (1978).
12. Cardy, B. J., and Kannenberg, L. W., Allozymic variability among maize inbred lines and hybrids: Applications for cultivar identification. *Crop Science, 22*:1016–1020 (1982).

13. Charbonnier, L., Isolation and characterization of ω–gliadin fractions. *Biochem. Biophys. Acta, 359*:142–151 (1974).
14. Chen, L. F. O., Cheng, M. C., and Chen, S. C. G., Similarity and diversity of seed proteins in rice varieties. *Bot. Bull. Academia Sinica, 28*:169–183 (1987).
15. Chung, O. K., and Pomeranz, Y., Amino acids in cereal proteins and protein fractions. In *Digestibility and Amino Acid Availability in Cereals and Oil Seeds*., John Finley and Daniel Hopkins (Eds.). ACCS, St. Paul, MN 1985, Chap. 5.
16. Cluskey, J. E., Taylor, N. W., Charley H., and Senti, F. R., Electrophoretic composition and intrinsic viscosity of glutens from different varieties of wheat. *J. Sci. Fd. Agric., 38*: 325–335 (1961).
17. Cooke, R. J., The characterization and identification of crop cultivars by electrophoresis. *Electrophoresis, 5*:59–72 (1984).
18. Cooke, R. J., and Cliff, E. M., Barley cultivar characterization by Electrophoresis. I. A method for acid polyacrylamide gel electrophoresis of hordein proteins. *J. Nat. Inst. Agric. Bot., 16*: 189–195 (1983).
19. Cooke, R. J., and Cliff, E. M., The characterization of oat cultivars by electrophoresis. *J. Nat. Inst. Bot., 16*:415–429 (1984).
20. Damardjati, D. S., Soekarta, S. T., Nur, A., and Siwi, B. H., Evaluation of protein quality and properties on 6 varieties of indonesian rice. *Indonesian J. Crop Sci., 1*:1–20 (1985).
21. Damerval, C., Herbert, Y., and de Vienne, D., Is the polymorphism of protein amounts related to phenotypic variability? A comparison of two–dimensional electrophoresis data with morphological traits in maize. *Theor. Appl. Genet., 74*:194–202 (1987).
22. Doekes, G. J., Comparison of wheat varieties by starch–gel electrophoresis of their grain proteins. *J. Sci. Fd. Agric., 19*:169–176 (1968).
23. Draper, S. R., Amino acid profiles of chemical and anatomical fractions of oat grains. *J. Sci. Food Agric., 24*:1241–1250 (1973).
24. du Cross, D. L., Lawrence, G. J., Miskelly, D., and Wrigley, C. W., *Systematic Identification of Australian Wheat Varieties by Laboratory Methods*., Technical Publication No. 7, CSIRO Wheat Research Unit, North Ryde, Australia, 1980.
25. du Cross, D. L., Wrigley, C. W., Improved electrophoretic methods for identifying cereal varieties. *J. Sci. Fd. Agric., 30*:785–794 (1979).
26. du Cross, D. L., Wrigley, C. W., and Blakeney, B., Fractionation of rice–grain proteins by gradient gel electrophoresis and gel isoelectric focusing. Characterization of rice genotypes. *Il Riso, 28*(3): 275–284 (1979).
27. El–Negoumy, A. M., Newman, C. W., and Moss, B. R., Chromatographic fractionation and composition of the components of the salt–soluble proteins from Hiproly (CI 3974) and Hiproly Normal (CI 4362) barleys. *Cereal Chem., 54*:333–344 (1977).
28. Esen, A., Genetic mechanism and protein properties with special reference to plant proteins. In *Food Protein Deterioration: Mechanism and Functionality*., J. P. Cherry (Ed.). ACS Symp. Ser. 206, Am. Chem. Soc., Washington, D.C., 1982.
29. Ewart, J. A. O., Isolation and characterization of a wheat albumin. *J. Sci. Food Agric., 20*:730–733 (1969).
30. FAO, Amino acid content of foods and biological data on proteins. FAO Nutr. Stud. 24, Food Agric. Org., Rome, 1970.
31. FAO/WHO, *Energy and Protein Requirements*. Report of a joint FAO/WHO Ad Hoc Expert Committee. World Health Org. Tech. Rep. Ser. 522. World Health Org., Geneva. Also FAO Nutr. Rep. Ser. 52. Food Agric. Org., Rome, 1973.
32. Fish, W. W., and Abbott, D. C., Isolation and characterization of a water–soluble wheat flour protein. *J. Sci. Food Agric., 20*:723–730 (1969).
33. Gianazza, E., Righetti, P. G., Pioli, F., Galante, E., Soave, C., Size and charge heterogeneity of zein in normal and opaque–2 maize endosperms. *Maydica, 21*:1–17 (1976).
34. Graham, J. S. D., Starch gel electrophoresis of wheat flour proteins. *Aust. J. Biol. Sci., 16*:342 (1963).

35. Heisel, S. E., Peterson, D. M., and Jones, B. L., Characterization of hordein electrophoretic patterns of United States barley cultivars. *Cereal Chem.*, *63*:500–505 (1986).
36. Houwing, A., and Van Dreven, F., A proposed method of polyacrylamide gel electrophoresis in acid environments applied to gliadins of wheat grains. *Euphytica*, *36*:55–60 (1987).
37. Heubner, F. R., Rothfus, J. A., and Wall, J. S., Isolation and chemical comparison of different gamma–gliadins from hard red winter wheat flour. *Cereal Chem.*, *44*:221–229 (1967).
38. Hussein, K. R. F., Eustatiu, N., and Stegeman, H., Amount of wheat protein extracted by various solvents. *Rev. Roumaine Biochem.*, *14*:248–251 (1977).
39. Hussein, K. R. F., and Stegeman, H., Comparison of proteins from wheat. Kernels by various electrophoretic methods in polyacrylamide. *J. Argon. & Crop Sci.*, *146*:68–78 (1978).
40. Iwasaki, T., Shibuya, N., and Chikubu, S., Gel Filtration and electrophoresis of soluble rice proteins extracted from long, medium and short grain varieties. *Cereal Chem.*, *59*(3): 192–195 (1982).
41. Jones, B. L., Identifying United States malting barley varieties by electrophoresis of hordeins and esterase enzymes. *Proceedings of the III International Symposium on Biological Identification of Varieties*–Leningrad, U.S.S.R., 1987. (Abstract)
42. Jones, R. W., Taylor, N. W., and Senti, F. R., Electrophoresis and fractionation of wheat gluten. *Arch. Biochem. Biophys.*, *84*:363–376 (1959).
43. Juliano, B. O., The rice caryopsis and its composition. In *Rice Chemistry and Technology*, D. F. Houston (Ed.). American Association of Cereal Chemists, ST. Paul, MN, 19172, chap. 2.
44. Kapala, A., Variability of electrophoretic subunit patterns of hordein protein in spring barley (*Hordeum vulgare* L.). *Genetica Polonica*, 22(2):163–177 (1981).
45. Kasarda, D. D., Nimmo, C. C., and Kohler, G. O., Proteins and the amino acid composition of wheat fractions. In Y. Pomeranz (Ed.), *Wheat: Chemistry and Technology*, 1976, chap. 6.
46. Khan, K., Polyacrylamide gel electrophoresis of wheat gluten proteins. *Bakers Digest* (Oct.): 14–19 (1982).
47. Kim, S. I., and Jo, D. H., Fractionation and electrophoretic patterns of rice proteins. *J.Korean Agric. Chem. Soc.*, *26*(1):65–72 (1983).
48. Kim, S. I., and Moses, J., Electrophoretic patterns of oat prolamins and species relationships in Avena. *Can. J. Genet. Cytol.*, *21*:309 (1979).
49. Konorev, V. G., Gavrilyuk, I. P., Gubareva, N. K., and Peneva, T. I., Seed proteins in genome analysis, cultivar identification, and documentation of cereal genetic resources: A review. *Cereal Chem.*, *56*(4):272–278 (1979).
50. Konzak, C. F., Genetic control of the content, amino acid composition, and processing properties of proteins in wheat. *Adv. Genet.*, *19*:407–482 (1977).
51. Lasztity, R., Wheat proteins. In *The Chemistry of Cereal Proteins*. CRC Press, Boca Raton, FL, 1984, chap. 3.
52. Lasztity, R., Rice proteins. In *The Chemistry of Cereal Proteins*. CRC Press, Boca Raton, FL, 1984, chap. 8.
53. Lasztity, R., Maize proteins. In *The Chemistry of Cereal Proteins*. CRC Press, Boca Raton, FL, 1984, chap. 6.
54. Lasztity, R., Barley proteins. In *The Chemistry of Cereal Proteins*. CRC Press, Boca Raton, FL, 1984, chap. 5.
55. Lasztity, R., Oat proteins. In *The Chemistry of Cereal Proteins*. CRC Press, Boca Raton, FL, 1984, chap. 7.
56. Lawrence, G. J., and Shepherd, K. W., Variation in glutenin subunits of wheat.*Aust. J. Biol. Sci.*, *33*:221–233 (1980).

57. Lookhart, G. L., Identification of oat cultivars by combining polyacrylamide gel electrophoresis and reversed–phased high performance liquid chromatography. *Cereal Chem.*, *62*:345–350 (1985).
58. Lookhart, G. L., Albers, L. D., and Bietz, J. A., A comparison of PAGE and HPLC methods for analysis of gliadins polymorphism in the wheat cultivar Newton. *Cereal Chem.*, *63*(6):497–500 (1986).
59. Lookhart, G. L., Albers, L. D., Pomeranz, Y., and Webb, B. D., Identification of U.S. rice cultivars by high performance liquid chromatography. *Cereal Chem.*, *64*:199–206 (1987).
60. Maier, G., and Wagner, K., Routine method for identification of wheat varieties by polyacrylamide gel electrophoresis. *Z. Lebensm. Unters. Forsch.*, *170*:343–345 (1980).
61. Marchylo, B. A., and Laberge, D. E., Barley cultivar identification by electrophoretic analysis of hordein and environmental effects. *Can. J. Plant Sci.*, *60*:1343–1350 (1980).
62. Martens, H., and Bach Knudsen, K. E., Fractioning barley proteins by computer factor analysis. *Cereal Chem.*, *57*:97–105 (1980).
63. McCausland, J., and Wrigley, C. W., Electrophoretic analysis of wheat and rye mixtures in meal, flour and baked goods. *J. Sci. Fd. Agric.*, *27*:1197–1202 (1976).
64. McDonald, Jr., M. B., Oat cultivar characterization using electrophoresis. *J. Seed Technol.*, *5*(2):88–101 (1982).
65. Mecham, D. K., Kasarda, D. D., and Qualset, C. O., Identification of western U.S. wheat varieties by polyacrylamide gel electrophoresis. *Hilgardia*, *53*(7):1–32 (1985).
66. Misrao, P. S., Mertz, E. T., and Glover, D. V., Studies on corn proteins. IX. Comparison of the amino acid composition of Landry–Moreaux and Paulis–Wall endosperm fractions. *Cereal Chem.*, *53*:699–704 (1976).
67. Monod, M., Marie, R., and Feillet, P., Fractionation of the proteins from some rice varieties by polyacrylamide gel electrophoresis. *C. R. Acad. Sci.*, *274*(13):1957(1972).
68. Nierle, W., Experiments on the electrophoretic recognition of wheat varieties. *Ber. Getreidechem. Tagg. Detmold.* 35–40 (1976).
69. Nimmo, C. C., O'Sullivan, M. T., and Bernardin, J. E., The relation of a "globulin" component of wheat flour purothionin. *Cereal Chem.*, *45*:28–36 (1968).
70. Nucca, R., Soave, C., Motto, M., and Salamini, F., Taxonomic significance of the zein isoelectric focusing pattern. *Maydica XXIII* :239–249 (1978).
71. Ohms, J. P., Electrophoretic differentiation between varieties of grain cultivars. *Z. Lebensm. Unters. Forsh.*, *170*:27–30 (1980).
72. Ohms, V. J. P., Identification of German oat varieties by electrophoresis of endosperm proteins. *Landwirtsch Forsch.*, *34*(3):88–94 (1981).
73. Osborne, T. B., *The Proteins of the Wheat Kernel.* Carnegie Institute, Washington Publ., 1907. pp. 1–119.
74. Palmiano, E. P., Almazan, A. M., and Juliano, B. O., Physiochemical properties of protein of developing and mature rice grain. *Cereal Chem.*, *45*:1–12 (1968).
75. Paulis, J. W., and Wall, J. S., Albumins and globulins in extracts of corn grain parts. *Cereal Chem.*, *46*:263–273 (1969).
76. Paulis, J. W., and Wall, J. S., Fractionation and characterization of alcohol soluble reduced corn endosperm glutelin proteins. *Cereal Chem.*, *54*:1223–1228 (1977).
77. Payne, P. I., Identification of a high–molecular–weight subunit of glutenin whose presence correlates with bread–making quality in wheats of related pedigree. *Theor. Appl. Genet.*,*55*:153–159 (1979).
78. Pence, J. W., Nimmo, C. C., and Hepburn, F. N., *Proteins. Wheat: Chemistry and Technology*, I. Hylinka (Ed.). Monograph Series III. American Association of Cereal Chemists, St. Paul, MN, 1964, chap. 6.
79. Pernollet, J. C., Kim, S. I., and Moses, J., Characterization of storage proteins extracted from *Avina sativa* seed protein bodies. *J. Agric. Food Chem.*, *30*:32–36 (1982).
80. Peruffo, A. D. B., Bovo, G., and Pogna, N. E., Gliadin pattern heterogeneity within foundation seed samples of Italian durum wheat varieties. *Genet. Agr.*, *35*:191–194 (1981).

81. Pomeranz, Y., Robbins, G. S., Smith, R. T., Craddock, J. C., Gilbertson, J. T., and Moseman, J. G., Protein content and amino acid composition of barleys from the World Collection. *Cereal Chem.*, *53*:497–504 (1976).
82. Pomeranz, Y., Proteins and Amino Acids of Barley, Oats, and Buckwheat. In *Protein Nutritional Quality of Foods and Feeds*. Part 2. Quality—Plant Breeding, Composition, Processing, and Antinutrients. M. Friedman (Ed.). Marcel Dekker, New York, 1975.
83. Portyanko, V. A., Pomortsev, A. A., and Sozinov, A. A., Intravarietal and intervarietal polymorphism of oat prolamines. *Soviet Agricultural Sciences*, *1*:12–15 (1987).
84. Righetti, P. R., and Bosisio, A. B., Application of isoelectric focusing to the analysis of plant and food proteins. *Electrophoresis*, *2*:65–75 (1981).
85. Robert, L. S., Matlashewski, G. J., Adeli, K., Nozzolillo, C., and Altosaar, I., Electrophoretic and developmental characterization of oat globulins in cultivars of different protein content. *Cereal Chem.*, *60*(3):231–234 (1983).
86. Robert, L. S., Nozzolillo, C., and Altosaar, I., Molecular weight and charge heterogeneity of prolamins (avenins) from nine oat (*Avena sativa* L.). Cultivars of different protein content and from developing seeds. *Cereal Chem.)*, *60*(6):438–442 (1983).
87. Robbins, G. S., Pomeranz, Y., and Briggle, L. W., Amino acid composition of oat groats. *J. Agric. Food Chem.*, *19*:536–539 (1971).
88. Rothe, G. M., and Maurer, W. D., One dimensional PAA gel electrophoretic techniques to separate functional and denatured proteins. In *Gel Electrophoresis of Proteins*, Michael J. Dunn (Ed.), 1986, chap. 2.
89. Salmon, S. E., and Burbridge, K. M., Wheat variety identification by polyacrylamide gel electrophoresis. *FMBRA Bulletin*, *2*:78–88 (1985).
90. Sarkar, R., and Bose, S., Electrophoretic characterization of rice varieties using single seed (salt soluble) proteins. *Theor. Appl. Genet.*, *68*:415–419 (1984).
91. Schildback, R., *Kali–Briefe* (Buntehof) *15*:691–701 (1981).
92. Schonhaus, I., and Sgarbieri, V. C., Inherited characteristics of composition and protein nutritive value of a new cultivar of maize (Nutrimaize) in two stages of maturity. *J. Agric. Food Chem.*, *31*:1–7 (1983).
93. Scriban, R., Autran, J. C., Strobbel, B., Nicolaidis, M., synthesis of various analytical investigations on the chemotaxonomy of French barleys and malts. *Brewers Digest* (Dec.):42 (1979).
94. Shadi, A. I., and Djurtoft, R., Studies of rice proteins by crossed immunoelectrophoresis, gel electrophoresis, and isoelectric focusing. *Cereal Chem.*, *56*(5):402–406 (1979).
95. Shewry, P. R., Ellis, R. S., Pratt, H. M., and Miflin, B. J., A comparison of methods for the extraction and separation of hordein fractions from 29 barley varieties. *J. Sci. Fd. Agric.*, *29*:433–441 (1978).
96. Shewry, P. R., Faulks, A. J., Pratt, H. M., and Miflin, B. J., The varietal identification of single seeds of wheat by sodium dodecyl–sulfate polyacrylamide gel electrophoresis of gliadin. *J. Sci. Fd. Agric.*, *29*:847–849 (1978).
97. Shewry, P. R., Pratt, H. M., and Miflin, B. J., Varietal identification of single seeds of barley by analysis of hordein polypeptides. *J. Sci. Fd. Agric.*, *29*:587–596 (1978).
98. Singh, R. S., Jain, S. K., and Qualset, C. O., Protein electrophoresis as an aid to oat variety identification. *Euphytica*, *22*:98–105 (1973).
99. Singh, U, and Sastry, L. V. S., Studies on the proteins of the mutants of barley grain. 2. Fractionation and characterization of the alcohol–soluble proteins *J. Agric. Food Chem.*, *25*:912–917 (1977).
100. Singh, U., and Sastry, L. V. S., Studies on the proteins of the mutants of barley grain. 3. Fractionation and characterization of the glutelin fraction. *J. Agric. Food Chem.*, *26*: 689–691 (1978).
101. Smith, D. B., and Payne, P. I., A procedure for the routine determination of electrophoretic band patterns of barley and malt endosperm proteins. *J. Nat. Inst. Agric. Bot.*, *16*:487–498 (1984).

102. Soave, C., Righetti, P. G., Lorenzoni, C., Gentinetta, E., and Salamini, F., Expressivity of the opaque–2 gene at the level of zein molecular components. *Maydica XXII* (1976):61–75 (1976).
103. Sodek, L., and Wilson, C. M., Amino acis composition of proteins isolated from normal, *opaque*–2, and *floury*–2 corn endosperm by a modified Osborne procedure. *J. Agric. Food Chem.*, *19*:1144–1150 (1971).
104. Tecson, E. M. S., Esmana, B. V., Lontok, L. P., and Juliano, B. O., Studies on the extraction and composition of rice endosperm glutelin and prolamin. *Cereal Chem.*, *48*:168–181(1971).
105. Tkachuk, R., and Irvine, G. N., Amino acid compositions of cereals and oil seed meals. *Cereal Chem.*, *46*:206–218 (1969).
106. Tkachuk, R., and Mellish, V. J., Wheat cultivar identification by high voltage gel electrophoresis. *Ann. Technol. Agric.*, *39*(2):207–212 (1980).
107. Turner, J. E., Boundy, J. A., and Dimler, R. J., Zein: A heterogeneous protein containing disulfide linked aggregates. *Cereal Chem.*, *42*:452–461 (1965).
108. Vidal, A. J., and Juliano, B. O., Comparative composition of waxy and nonwaxy rice. *Cereal Chem.*, *44*:86–91 (1967).
109. Villareal, R. M., and Juliano, B. O., Properties of glutelin from mature and developing rice grain. *Phytochemistry*, *17*:177 (1978).
110. Wall. J. S., Cereal proteins. In *Symposium of Foods*, H. W. Schultz and A. F. Anglemeier (Eds.). AVI Publishing Co., N1, Westport, CT, 1964, chap. 14.
111. Wall, J. S., and Paulis, J. M., Corn and sorghum proteins. In *Advances in Cereal Sciences and Technology* II. Y. Pomeranz (Ed.). American Association of Cereal Chemists, St. Paul, MN,1978.
112. Wall, J. S., Fey, D. A., and Paulis, J. W., Improved two–dimensional electrophoretic-separation of zein proteins: Application to study of zien inheritance in corn genotypes.*Cereal Chem.*, *61*:141–146 (1984).
113. Weber, K., and Osborne, M., The reliability of molecular weight determination by sodium dodecyl sulfate–polyacrylamide gel electrophoresis. *J. Biol. Chem.*, *244*(16): 4406–4411 (1969).
114. Wilson, C. M., A nomenclature for zein polypeptides based on isoelectric focusing and sodium dodecyl sulfate polyacrylamide gel electrophoresis. *Cereal Chem.*, *62*(5): 361–365(1985).
115. Wilson, C. M., Shewry, P. R., and Miflin, B. J., Maize endosperm proteins compared by sodium dodecyl sulfate gel electrophoresis and isoelectric focusing. *Cereal Chem.*, *58*(4):275–281 (1981).
116. Woychik, J. W., Boundy, J. A., and Dimler, R. J., Starch gel electrophoresis of wheat gluten proteins with gluten proteins with concentrated urea. *Arch. of Biochem. and Biophys.*, *94*:477–482 (1961).
117. Wrigley, C. W., Autran, J. C., and Bushuk, W., Identification of cereal varieties by gel electrophoresis of the grain proteins. *Adv. Cer. Sci. Tech.*, *5*:211–259 (1982).
118. Wrigley, C. W., Bietz, J. A., Proteins and amino acids. In *Wheat Chemistry and Technology*, Y. Pomeranz (Ed.). American Association of Cereal Chemists, St. Paul, MN, 1968, vol. 1, chap. 5.
119. Wu, Y. V., and Dimler, R. J., Hydrogen ion equilibria of wheat gluten. *Arch. Biochem. Biophys.*, *102*:230–237 (1963).
120. Wu, Y. V., and Dimler, R. J., Hydrogen ion equilibria of wheat glutenin and gliadin. *Arch. Biochem. Biophys.*, *103*:310–318 (1963).
121. Wu. Y. V., Sexon, K. R., Cavins, J. F., and Inglett, G. E., Oats and their dry–milled fractions. *J. Agric. Food Chem.*, *20*:757–761 (1972).
122. Youngs, V. L., Peterson, D. M., and Brown, C. M., Oats. In *Advances in Cereal Science and Technology*, Vol. V, Y. Pomeranz, (Ed.). Am. Assoc. Cereal Chem., St. Paul, MN, 1982, pp.49–105.

12

CARBOHYDRATE COMPOSITION OF CEREAL GRAINS

Robert Becker and Grace D. Hanners
U.S. Department Of Agriculture, Albany, California

I. INTRODUCTION

Books describing the carbohydrate composition of cereal grains and their mill fractions are common (1–4). The carbohydrate components currently subject to the most attention are those nonstarch polysaccharides and digestion–resistant starch constituents broadly classified as total and soluble dietary fiber (TDF and TSDF). Isolated cereal grain starch is another common research topic, due to its economic importance.

The total dietary fiber (TDF) from cereal grains is mainly comprised of nonstarch polysaccharides. The accurate definition and measurement of TDF and its constituents, such as TSDF, have posed a difficult problem for food chemists. Recent research has been aimed at producing rapid, practical, reproducible, and accurate assays for TDF and its components (5–11). A new official method for dietary fiber determination in foodstuffs was published by the Association of Official Analytical Chemists in 1985 (12). However, this method does not appear to have been universally accepted (13), and considerable methods development has continued (14–18). The total and soluble dietary fiber values given in this discussion were obtained using recognized methods, but the reader should be aware that different methods may produce substantially varying results.

The interest in TDF has also led to a reevaluation of starch measurement techniques (see Sec. IV). Problems with the accurate measurement of starch have been associated with the gelatinization and enzymatic hydrolysis of the polysaccharide. Measurement of the amylose content of starch based on iodine binding has been shown to be affected by the iodine–binding capacity of amylopectin in certain rice starches. Developments in these areas should be followed closely for reports of anomalous amylopectins in other cereals.

Compositional data for all cereal grains are dependent upon the variety and degree of milling. Varietal compositional variations may be influenced by environmental conditions and are discussed when specific information is available.

This chapter is intended to provide the reader with a very brief introduction and update on cereal grain carbohydrates, particularly dietary fiber and isolated starches.

II. WHEAT

The carbohydrate composition of wheat and wheat products has been widely reported (2,19,20), and is shown in Table 1. These results are subject to variations due to cultivar, environmental effects, and processing/milling conditions.

The main milling fractions from wheat are composed of germ, bran, and endosperm material (21,22). The germ constitutes about 2% of the wheat kernel and consists of the embryonic axis and scutellum tissue. In the kernel, it functions as the storage, digestion, and absorbing organ during germination. It is composed of about 50% carbohydrates, which are approximately one third free sugars, one third starch, and one third cellulose and hemicellulose.

The bran is considered to be all of the outer structures inward through the aleurone layer and is about 70% total carbohydrates, primarily cellulose (35%) and hemicellulose (43%) with lesser amounts of starch (14%) and free sugars (8%) (23). The noncarbohydrate components are protein (17%), fat (5%), and ash (7%).

Starch is the principal carbohydrate constituent of the endosperm (96%) and serves as the potential food energy source for the germ upon germination (21). The remainder of the carbohydrate in the endosperm is made up of hemicellulose (2%), cellulose (0.3%), and free sugars (1.5%) (23). Depending on the degree of milling, the components typically of bran and germ occur in progressively smaller amounts in the endosperm flour as more of the exterior of the kernel is removed, so endosperm flour yield may vary from 65 to 84% of the kernel. In general, starch is concentrated in the endosperm, free sugars in the germ, and cellulose and hemicellulose in the bran.

Recent work on wheat carbohydrates has focused on total dietary fiber (TDF) and the water–soluble or insoluble components of dietary fiber, starch and its digestion–resistant components, and specific determinations of pentosans, β–glucans, and individual component sugars. A diet high in wheat bran as a source of fiber has been shown to be effective in the prevention and treatment of diverticular disease (24). Recent work (25) has shown a similar diet high in TDF to be associated with lower incidence of some kinds of cancer, particularly colon cancer.

The aleurone cell walls of wheat, like barley, contain large amounts of arabinoxylans and β–glucans and small amounts of cellulose and glucomannans. The major endosperm cell–wall polysaccharides of wheat are pentosans, comprised mainly of arabinose and xylose. The amounts of pentosans observed in wheat (6.6%) are in the range of levels found in barley, oats, triticale, and rye (5.7–8.5%) and significantly higher than amounts found in rice (1.2%) (26). The total β–glucan content in wheat (0.65%) is close to that of triticale (0.52%) while being higher than rice (0.10%) and lower than rye (1.89%), oat (3.37%), and barley (4.36%) (26).

Recent reviews have described the involvement of wheat flour pentosans in dough development, loaf volume, crumb and crust structure, and antistaling activity (27–29). Pentosans are thought to have antinutritional properties (30) and have been shown to adversely affect the separation of starch and gluten (31). They are thought to make a major contribution to dough properties by absorbing approximately 10 times their weight in water to form stable gels. The pentosan polymers are probably further strengthened by oxidative cross–linking with phenolic–type acids, e.g., ferulic acid (32), covalent attachment of the complex to protein, and general pentosan–protein–starch hydrogen bonding.

The free sugars in wheat kernels, which occur mainly in the aleurone layer and germ, are sucrose, raffinose, and neokestose (19). Other sugars such as stachyose, fructosylraffinose, glycerol, xylose, arabinose, glucose, fructose, and low molecular weight fructosans have also been reported (19).

Starch, the preponderant endosperm carbohydrate in wheat, is commercially produced as a by–product of gluten manufacture. Wheat starch (Figs. 1,2) has physiochemical properties similar to corn starch (Table 2), although viscosity and gel strength are usually lower. When used in baking, it can replace portions of wheat flour, increase cake volumes and tenderness, and reduce fat absorption in doughnuts, as well as having applications in the confection and canning industries (33).

III. CORN

There are several economically important types of corn in the United States: dent or field corn, which may be normal, waxy, or high amylose, popcorn, and sweet corn. Smaller amounts of flint and flour corn are grown in special climatic zones and in South America, Asia, and Europe. Field corn has well–defined grading standards which consider variety, color, test weight per bushel, moisture, and foreign material. Other corn varieties have more subjective standards; for example, sweet corn is graded on sweetness and tenderness and popcorn on kernel expansion. The carbohydrate compositions of these types vary considerable, in part due to end–use factors. For example, different varieties of hybrid field corn may be selected for use as feed or as raw material for corn oil, starch, or protein, while sweet corn is harvested at an earlier stage of maturity for vegetable consumption. The other types of corn vary in physical characteristics as well as in chemical composition.

Traditionally, corn grain has been used for livestock and poultry feed, primarily because it can be grown in a variety of geographic zones, is highly palatable, easily shelled and cracked, and has high digestibility. It is also processed to produce value–added products such as human food, starches, sweeteners, and ethanol by fermentation.

As in other cereal grains, starch is the preponderant carbohydrate in corn (Table 3, Figs. 3,4,5), varying from about 54% in sweet corn (34) to 68–74% in dent (35). The amylose content ranges from about 24% in normal dent to about 1% in waxy starch and as high as 80% in some of the starches from high amylose hybrid varieties (34). The free sugar content of sweet corn (12%) is undoubtedly responsible for its sweet taste (34), and varies with maturity and environmental conditions. The total dietary fiber of corn flour (15%) is concentrated in the bran fraction (82–92%), which has substantially more fiber than wheat bran (43–45%) (36). The amount of soluble dietary fiber in corn is less than in any other cereal except rice (37).

Corn starch has a wide variety of industrial applications, with uses ranging from thickening and gelling agents in puddings and fillings, to molding for confections, to use as a white colorant or coating. The gelatinized and cooled starch dispersions of normal corn starch have characteristics similar to wheat starch (Table 2), while the waxy maize starches have a higher hot–paste viscosity, softer texture, are clearer, have greater stability to retrogradation and higher freeze–thaw stability (38). In high amylose corn starch, the larger amount of linear starch rigidifies the micellar network and restricts swelling so the pastes show little breakdown during cooking.

Chemical modification of corn starches can be used to extend the functional characteristics to satisfy a wide range of industrial requirements.

IV. RICE

The chemistry and technology of rice have been ably reviewed by Juliano (3). Selvendran et al. (39) discuss a number of cereals including rice in their review of the chemistry, analysis, and properties of dietary fiber. Carbohydrate compositional information on rice is presented in Table 4, and the physiochemical properties of rice starch are shown in Table 2.

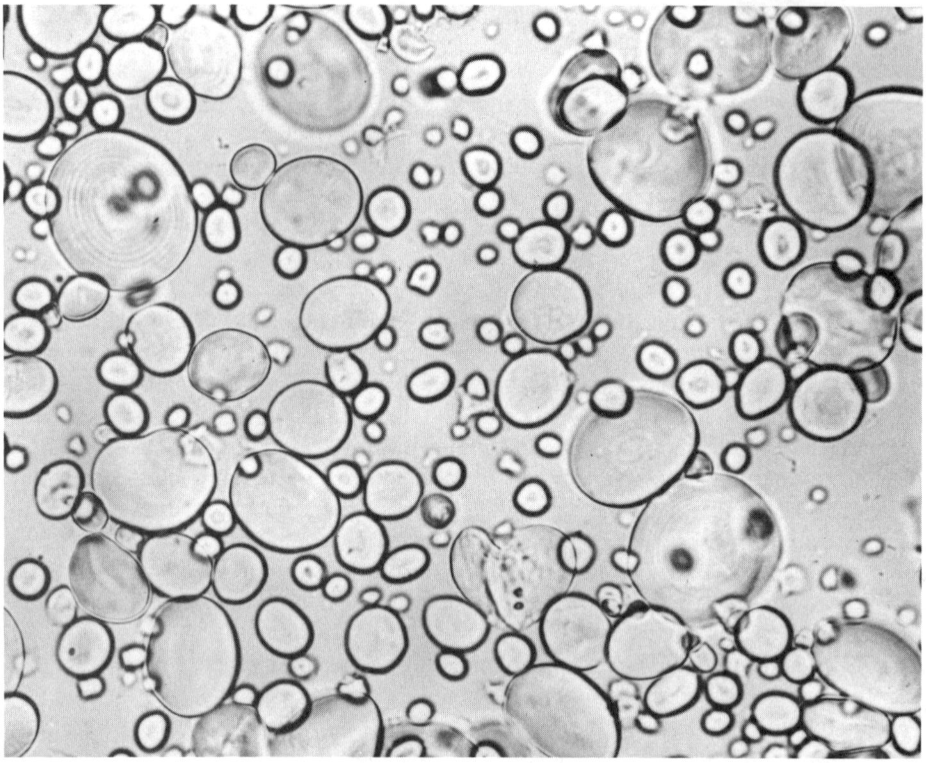

FIGURE 1 Wheat starch 700 ×. (From Ref. 154.)

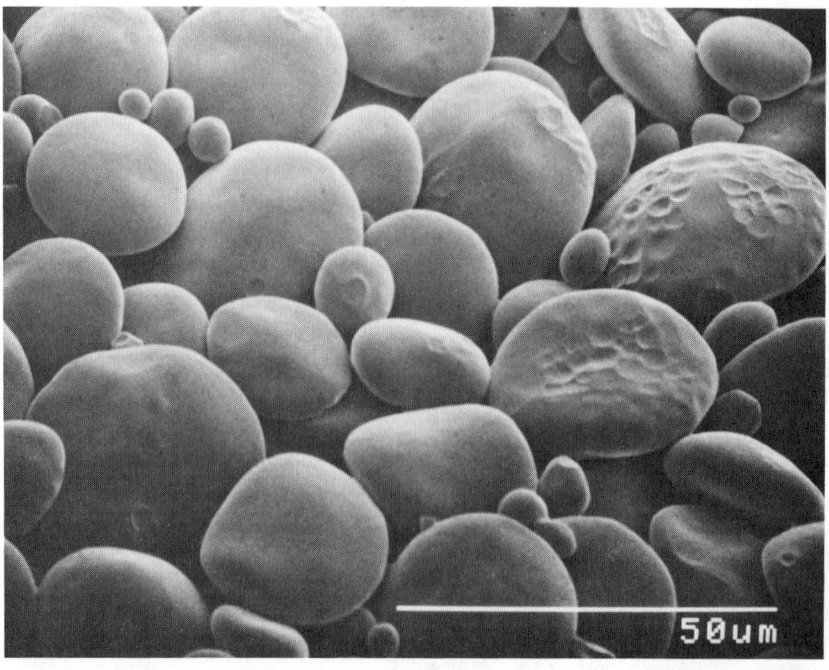

FIGURE 2 Scanning electron micrograph of wheat starch. (Photo by R. K. Johnston.)

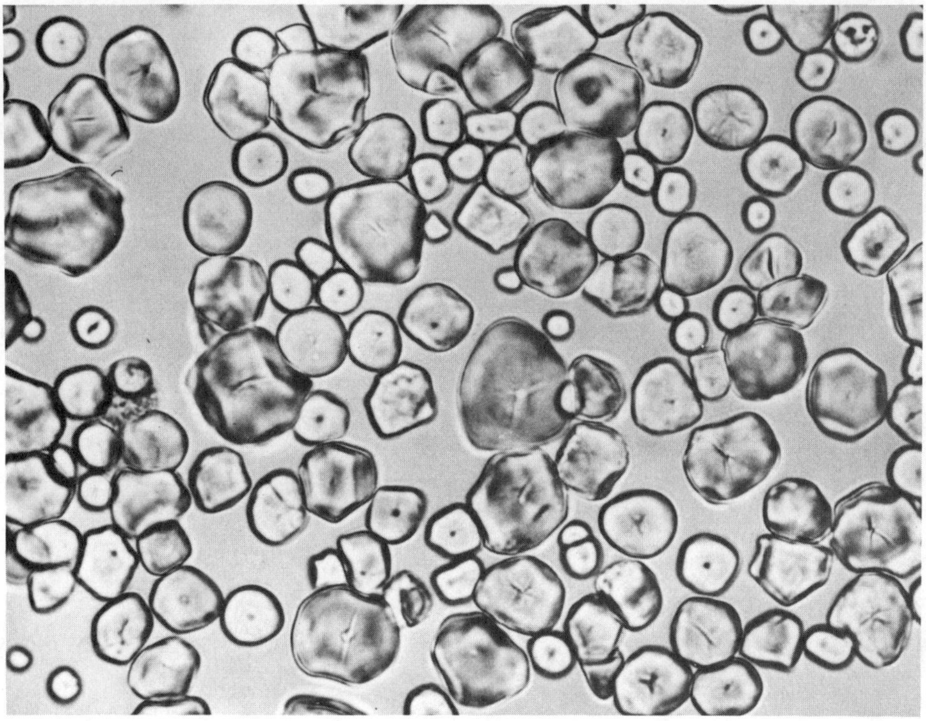

FIGURE 3 Corn starch, 700×. (From Ref. 154.)

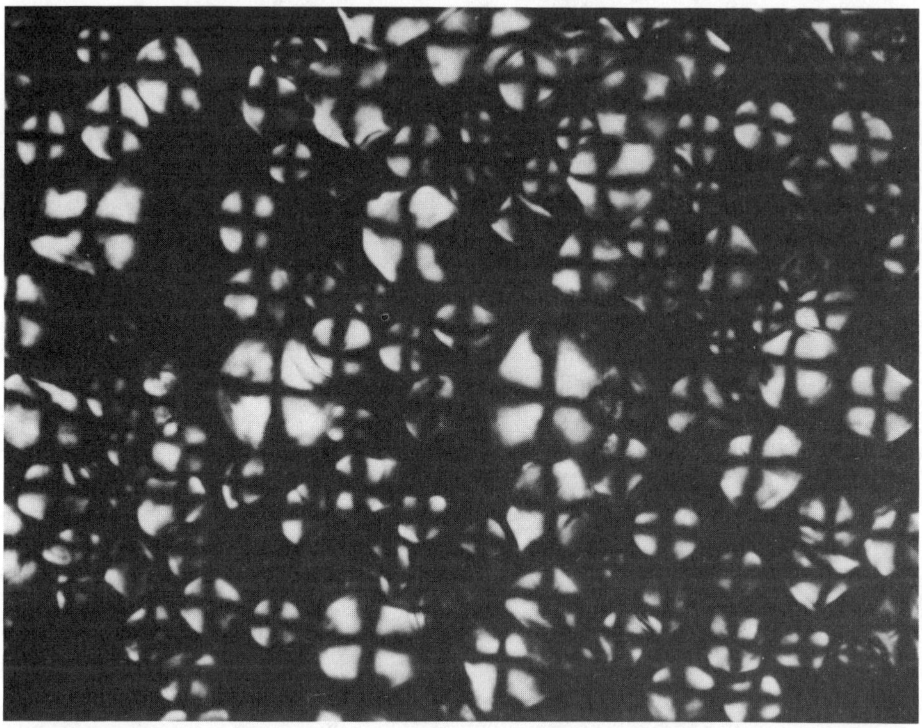

FIGURE 4 Corn starch photographed using polarized light, 700× . (From Ref. 154.)

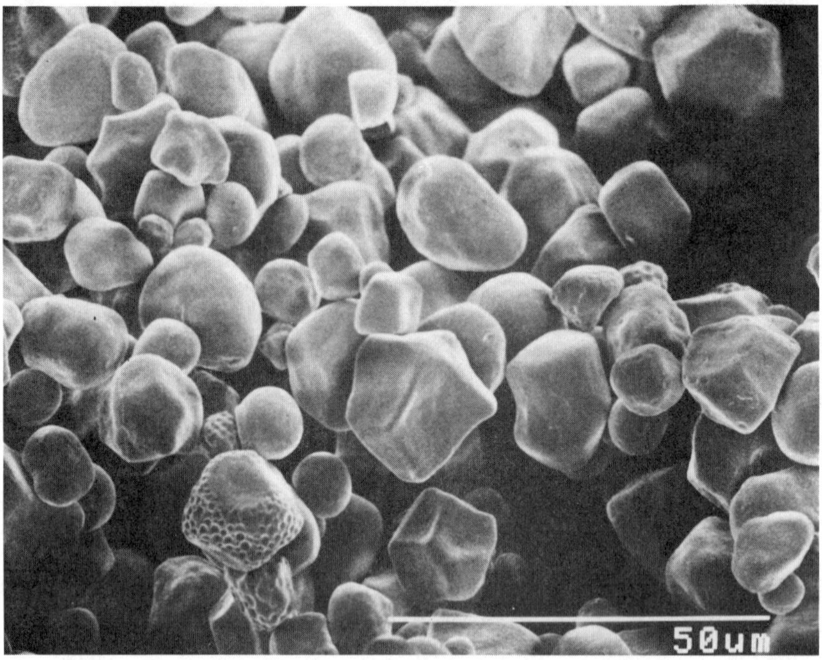

FIGURE 5 Scanning electron micrograph of corn starch. (Photo by R. K. Johnston.)

Starch comprises 90% of the carbohydrates of milled rice. The amylose content of the starch is thought to be the major factor determining the texture of cooked rice (40). During cooking, it influences water absorption, volume expansion, and dissolved solids. After cooking, it affects the color, gloss, stickiness, and softness of the cooked rice. Cooked rices having starches with amylose contents greater than 25% are flaky and dry, whereas those with less than 25% tend to be sticky and moist (41).

Since the amylose content is the most important factor affecting the cooking and eating quality of milled rice, considerable effort has been directed towards methodology to identify and quantify amylose. In general, rice starch contains either very low (0–9%), low (9–20%), intermediate (20–25%), or high (25–33%) amylose levels (42). Others classify starches only according to three amylose levels: waxy (0–9%), normal (9–30%), and high (+30%). Representative data are shown in Table 4. The amylose content of long grain rice ranges from 23 to 26 %, while medium grain ranges from 15 to 20% and short grain ranges from 18 to 20% (43). The reproducibility of a revised simplified amylose assay was evaluated by an international cooperative test (44). The revised method incorporated adaptations designed to eliminate the interference of lipids in amylose determinations due to competition between lipid and iodine in forming amylose complexes. Of the methods evaluated, procedures using amylose–amylopectin standards in place of amylose alone, and lower pH, gave truer and more reproducible amylose values in six samples which had amylose values ranging from 0.6 to 25.4%.

Villareal and Juliano (45) found 1.7% amylose in two waxy varieties and 16.2–32.0% amylose in eight nonwaxy varieties of rice starch granules, and demonstrated that amylose content was correlated to starch granule residual protein content. Asaoka et al. (46) examined the structure and properties of endosperm starch from high amylose mutants of rice.

Starch from indica or tropical rice has been characterized by several researchers (24–49). Another important quality test, especially in breeding programs, is the alkali digestibility test of Little et al. (50) as modified by Bhattacharya and Sowbhagya (51). The test indexes the extent of degradation in rice kernels caused by alkali and very significantly correlates with the starch gelatinization temperature (GT). Although the test of Little et al. (50) readily estimates the final GT of starch, the amylograph method of Halick et al. (52) remains the preferred objective measure of starch GT (53). An international cooperative test involving 11 laboratories showed comparable results for alkali digestibility of five widely different milled rice samples by both methods and suggested the use of 1.7% KOH in the assay (54).

Maher (55,56) evaluated the amount of NaOH needed to gelatinize various starches, including two rice samples, and described their distinctive viscosity patterns. The acid resistance of various starch granules, including rice, was compared by Nara et al. (57). Scanning electron microscopy was used to show that endosperm starches of wheat, barley, and rye have a common mode of swelling different from those of rice, potato, maize, and oat.

The amylograph viscosities of milled rice flours have been shown to be altered by cold water–soluble polysaccharides (58,59) while lipids, nonstarch polysaccharides, and amylose were shown to contribute to the hardening of waxy starch gel consistency (60). The cold water–soluble carbohydrates in waxy milled rices ranged from 1.2–3.4% soluble in 70% ethanol (mainly sucrose) to 1.9–3.5% insoluble in 70% ethanol (the high molecular weight fraction) (59). These high molecular weight carbohydrates suppressed amylograph peak viscosities as did rice bran hemicellulose (61). Perez et al. (60) found the cold water–soluble fractions of two waxy and six nonwaxy rice flour samples to be 1.74% and 1.79% of milled rice respectively, and to contain 0.39 and 0.26% of hot 80% ethanol solubles as percent of milled rice. The hot water–soluble fractions obtained after the cold water extraction for these same samples were 8.84 and 6.16% and contained 0.14 and 0.08% of hot 80% ethanol solubles as percent of milled rice. The water extract minus the ethanol extract is presumably a measure of the high molecular weight fraction. The water–soluble polysaccharides from sugary rice endosperm were determined by Asaoka et al. (62). Kuo et al. (63) found no raffinose sugars in dry rice seeds, and 0.56% of total sugar was identified as sucrose.

Changes in the cooking properties of rice grains during storage were shown to depend largely on changes in structure-maintaining properties, such as endosperm cell walls and endosperm proteins, rather than on changes in rice starch itself (64).

A cooperative test on amylography of milled rice flour for pasting viscosity and starch gelatinization temperature has been performed on rices with amylose contents of 1.2, 17.7, 27.8, and 28.2% dry basis (53). The test showed significant variation due to sample, laboratory, and interaction between sample and laboratory. The starches from seven intermediate amylose and seven waxy milled rices that differed in eating quality were found to have 21.8–25.55% and 1.2–2.7% amylose, respectively, by Merca and Juliano (50).

A convenient method to measure the chain length of amylopectin, without prior removal of amylose from starch, was developed by Hizukuri et al. (65). This method was used to assay 27 starch samples from different commodities, including rice. Takeda et al. (66) examined the iodine–binding capacity of amylopectin and found that rice starches with ordinary amylose contents (15.5–18.5%) had apparent high amylose contents (30–32%) due to a high affinity (1.62–2.57 g/100g) of long chain amylopectin for iodine. The amylopectin from the low and high amylose starches had average chain lengths of 19–20 and 21–22 glucose units and similar β–amylolysis limits (56–59%), but the iodine-binding amylopectins had more (up to 14.2%) long chains (85–180 glucose units) and appeared to be derived from long β–chains with side chains widely spaced and located far from the nonreducing terminus. The implications of this discovery for interpretation of amylose determinations have not been fully explored by the starch industry.

Although amyloses are defined as unbranched molecules, pure amyloses obtained from various sources have been shown to contain limited numbers of branched linkages and to be mixtures of branched and unbranched molecules (67). For example, a sample of rice amylose was found by Takeda et al. (67) to contain a mole fraction ratio of 0.69 unbranched to 0.31 branched amylose.

Juliano and Villareal (68) showed that the harder texture of cooked rice products from high gelatinization temperature waxy rices, compared to the texture of products from the low gelatinization temperature waxy rices, may be due to the higher molecular weight of their amylopectin.

A number of workers have used differential scanning calorimetry (DSC) to characterize starches by correlating their thermal and physiochemical characteristics. Russell and Juliano (69) examined eight milled rice starch samples using DSC and demonstrated that sodium dodecyl benzenesulfonate removed and replaced some of the starch lipid. Purified rice starches from eight varieties were analyzed by Biliaderis et al. (70) for amylose content, which ranged from 2.1% for waxy to 17–33% in nonwaxy starch, and correlated with thermal and physiochemical characteristics.

Rice endosperm cell walls contain about 48% cellulose, about 10% pectins (composed of galacturonic acid, arabinose, and xylose), smaller amounts of arabinoxylans, and a complex of xyloglucan, β–glucan, and arabinoxylan. The rice glucan contains 80% of β–(1,3) and 20% of β–(1,4) linkages (71,72).

Rice hull contains 37% cellulose and about 20% hemicellulose composed of xylan. The structures of an arabinoxylan (73) and a xyloglucan (74) from rice hull have been described.

Starches from the stem, pollen, and grain of rice contained 15.3–17.2 21.0–27.2 and 0–28.5% amylose, respectively (75).

Commercial rice starch is isolated from brewers rice, but its use is restricted by its high price relative to corn, wheat and potato starch. Rice starch has been used as a pudding starch, in baby foods, and as a laundry aid. The small granule size (Table 2, Figs. 6,7) makes it a useful cosmetic dusting powder (41). The physiochemical properties of rice starch are similar to other cereal grains, except for wide varietal differences and small granule size (Table 2).

V. OATS

The chemistry and technology of oats (Table 5) have been thoroughly reviewed in the recent AACC monograph edited by Webster (4).

Oat starch is more like rice starch than wheat starch in both granule size and shape. Its amylose content (16–27%) is similar to wheat starch (76), but oat amylose is more linear and oat amylopectin is more branched than that found in wheat (77). Recent studies using differential scanning calorimetry have shown a sizeable amylose–lipid complex in oat starch: 60% of the natural amylose is complexed by starch lipid during heating in an aqueous environment (78). This lipid is strongly bound within the starch granule causing a high difference between true and apparent amylose content (79). The amylose content of six lines of oats grown at the same location ranged from 25.2 to 29.4% (total) and 17.3 to 20.7% (apparent) (79). If lipid is removed, oat starch pasting behavior becomes similar to that of wheat rather than that of native oat starch (78).

The (1–3,1–4)–β–D–glucan content of three oat cultivars ranged from 2.5 to 6.6% (79) with one report of 73% of the glucan being water insoluble (80). Wood and Weisz (82) developed a nonenzymatic method for the determination of β–glucan using precipitation by dyes with subsequent high performance liquid chromatography. A fluorescence method for β–glucan determination was reported by Jensen and Aastrup (83) and Jorgensen (84). According to Wood (85), oat gum (mostly β–glucan) compares favorably with guar and locust bean gums and, especially considering its beneficial effects in the human diet, should have good potential as an industrial hydrocolloid for food use.

Total dietary fiber in oat bran (19.4%) is almost double that of oat flour (10.9%) (36); wild oat bran contains 20% total dietary fiber (86,87). The levels of free sugars are typical of other cereal grains (all less than 3% of the whole grain) according to Kuo et al. (63); however, McLeod and Preece (88) reported the free sugar content of oats (1.4%) to be small compared to most other grains (1.0–7.1%). Total free sugar content of whole oats was reported as 1.4%, with sucrose the predominant constituent (0.64%), followed by raffinose (0.19%), fructose (0.09%), glucose (0.05%), glucodifructose (0.04%), and a trace of maltose. The presence of stachyose (0.08%) in oat flour was reported by MacArthur and D'Appolonia (77).

Consumption of diets containing high levels of mucilaginous fibers, particularly the soluble fiber found in oat bran, has been shown to significantly lower blood lipid levels and increase fecal excretion of fat and bile acids (89,90). The decrease in serum lipids is pronounced in the lower density lipoproteins, while beneficial high density lipoproteins were shown to remain constant or increase (91).

Traditionally, oat starch has not generated widespread commercial interest due to its functional similarity to corn and wheat starch (Table 2). It can be used in composite breads, but imparts a crumb–softening effect. Recent studies have identified some oat cultivars whose starch pastes have unusual properties, such as the development of abnormally rapid cold viscosities. These oat starch gels are reportedly more elastic, adhesive, translucent, and

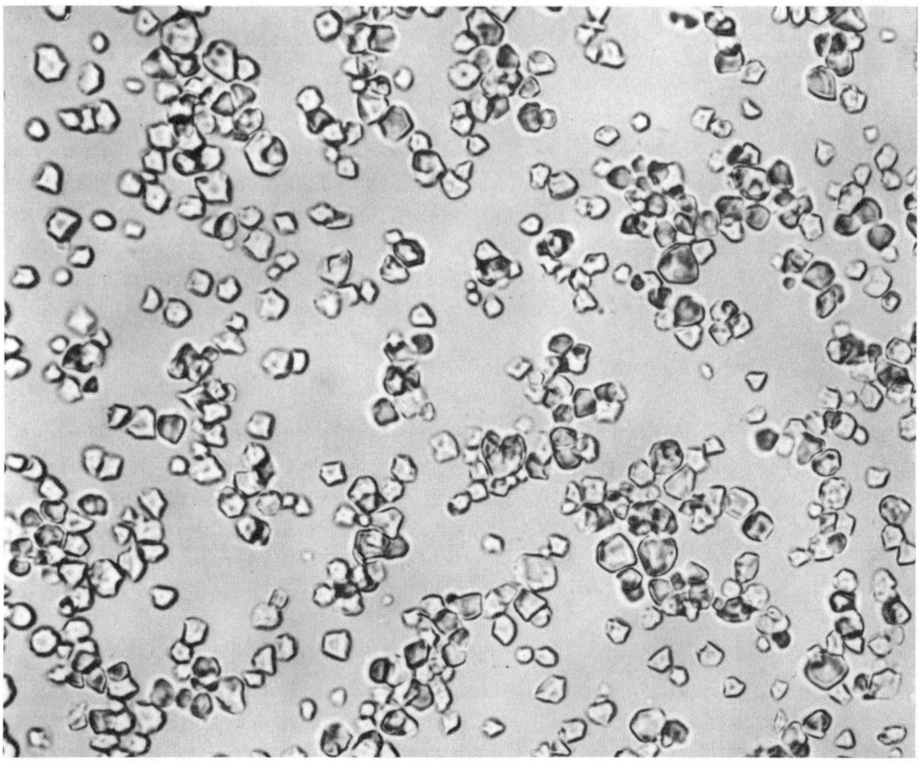

FIGURE 6 Rice starch, 700×. (From Ref. 154.)

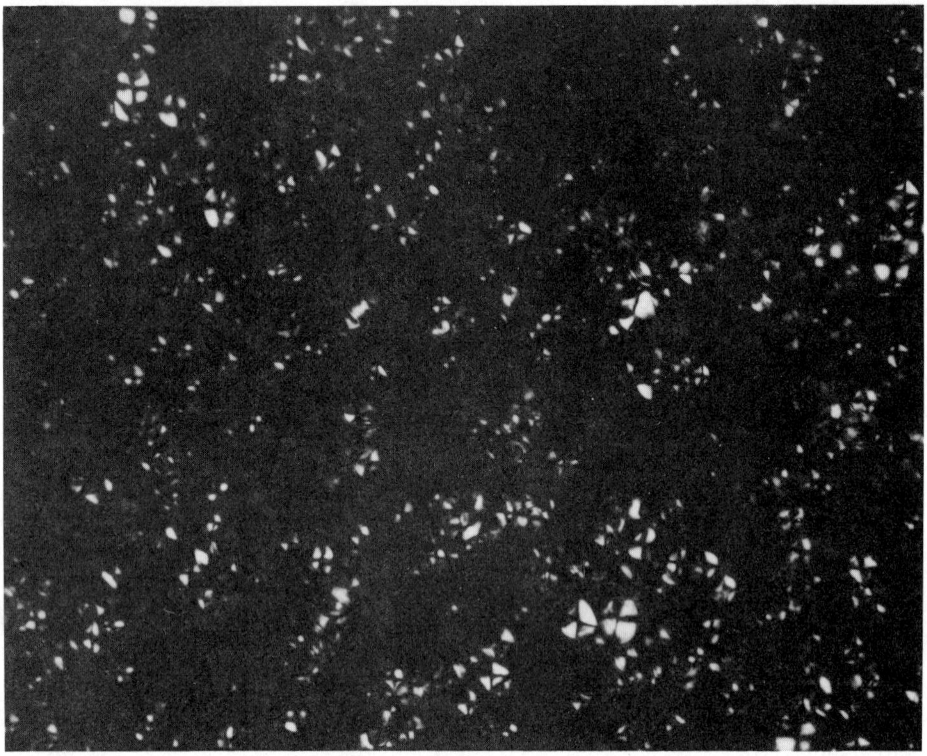

FIGURE 7 Rice starch photographed using polarized light, 700×. (Photo courtesy Corn Products Co., Summit-Argo, IL.)

show greater storage stability than corresponding wheat and corn starches (78). The pastes have higher cooling cycle viscosities than comparable concentrations of wheat or corn starches, and cooked oat starch granules are more sensitive to shear. The gels are influenced by salts, sucrose, gums, and mild acids. Industrial applications as ingredients in adhesives, paper products, and pharmaceuticals have been suggested (78,92,93).

VI. BARLEY

The use of barley in brewing has provided the impetus for detailed research into its chemical composition. The (1–3)(1–4)–β–D–glucans are important in brewing and have been the subject of many investigations (80). Residual, unfermented glucans form highly viscous aqueous solutions, which reduce the rate of wort and beer filtration and may lead to haze, precipitation and gel formation in stored beer. In chicken diets, barley β–glucans exhibited antinutritional properties by their "gumminess" and indigestibility, which affects food intake.

The reported content of β–glucans in barley ranges from 2.0 to 8.6% in malting and feed cultivars from Australia, Scandinavia, Britain, and North America (Table 6). The water–soluble glucans comprised 30–60% of the total. Henry (94) used a set of 50 samples, including 31 different varieties grown over six seasons in 13 different sites and containing 2.90–5.16% β–glucans, to calibrate near infrared reflectance to predict the β–glucan content of barley. The predictive capability of the calibration equation was demonstrated on a second set of 25 barley samples comprised of 22 varieties grown over five seasons at seven different locations. The β–glucans and pentosan contents of 50 six–row and 68 two–row cultivars were measured by Lehtonen and Aikasalo (95,96) while Aman and Newman (97) and Aman and Graham (16) found an average β–glucan content of 3.6% in Swedish barley and 4.5% (range 3.0–6.9) in 64 spring–sown barley samples. Total pentosan and β–glucan contents differ due to cultivar, but only total pentosans are affected by environment (98).

The conformation of barley (1–3)(1–4)–β–glucans in solution was studied by Buliga et al. (99). Ahluwalia and Fry (100) found that barley endosperm cell walls contained a feruloylated arabinoxylan and a nonferuloylated β–glucan. Kuo et al. (63) reported 0.63–0.79% raffinose and 1.18–1.42% sucrose as soluble oligosaccharides in two barley cultivars. Total dietary fiber in scotch and pearled barley was found to be 19.0% and 17.3%, respectively (36). The pentosan component of barley endosperm walls was shown by Ballance et al. (101) to consist of a family of arabinoxylans which vary in solubility, in the relative proportions of L–arabinose and D–xylose, and in the degree of substitution of (1–4)–linked xylosyl residues by L–arabinofuranosyl groups.

The amylose content of the barley starch is an important physiochemical parameter that has significant economic implications in the brewing and baking industries. Morrison et al. (79) determined amylose in starches from 31 barley cultivars. They distinguished between apparent amylose, which does not include lipid–complexed amylose, and total amylose as determined on defatted starch. Morrison et al. (102) found that environmental effects on the composition and properties of 16 Triumph barley starch samples were generally small and not related to weather statistics. These same authors also measured the amylose and lysophospholipid contents, swelling factor, and dimensions of large granules (A–type) and small granules (B–type) from 21 waxy, normal, and high amylose barley starches. The barley malt hydrolysis of large and small starch granules from normal and waxy barley cultivars was studied by MacGregor and Ballance (103). A procedure was developed by Makela et al. (104) to evaluate granule size distribution of starch in barley samples. The structure of amylopectins isolated from large and small starch granules of normal and waxy barley was studied by MacGregor and Morgan(105). MacGregor and Matsuo (106) studied starch degradation during the initial stages of germination of barley and wheat and suggested that the initial site of α-amylase synthesis is the embryo. Physiochemical properties of starch from two new varieties and two established barley varieties were compared by DeHaas and Goering (107). The effects of heat–moisture treatment (108,109) and steeping (110) on the physicochemical and functional baking properties of barley have been determined (Table 2). Barley starch has lower amylograph viscosities, swelling power, and water–binding capacity than wheat starch.

VII. SORGHUM

Sorghum is usually grown in semiarid areas as a dryland crop. Some cultures harvest the grain at an immature stage (about 21–31 days after flowering, [DAF]) for consumption as a sweet flavored vegetable, while most others grow the grain to maturity (usually about 60 DAF). When harvested at the earlier stage, the grain may contain 4 times the total sugar and 17 times the amount of sucrose found in the fully mature crop, which probably accounts for the sweet taste. Since the carbohydrate composition changes rapidly in the immature grain (111) and harvest dates could vary widely, only the composition of the mature seed is reported here.

Sorghum grain types have been classified as waxy, floury, corneous, sugary, and high lysine, each with a characteristic chemical composition (111). In addition, the terms "hard" and "soft" are used to designate vitreous and opaque endosperm and kernel appearance, although all of these descriptors are somewhat arbitrary (112).

Traditional foods from sorghum and their production, evaluation, and nutritional value have been reviewed by Rooney et al. (113). The textural characteristics of cooked sorghum pastes were shown to be correlated with the proportion of floury and corneous endosperm in the grain (114) and the chemical properties of the starches (crystalline structure of the granules and molecular weight of the components) from the corneous and floury endosperm (115).

The starch content of sorghum cultivars ranged from 32 to 79%, with the sugary and waxy sorghums generally having less starch than nonwaxy cultivars (Table 7). The amylose content ranged from 19 to 21% in starch from three sorghum cultivars with intermediate hardness (115), from 17 to 26% in two high lysine and two normal cultivars (111), and to 19% in *S. bicolor* and *S. vulgare* (116). Morrison et al. (79) found 27–30% amylose in a small random sample of sorghum. The physicochemical properties of sorghum starch were determined by Glennie (117), and the molecular properties of three sorghum starches from different origins were investigated by Craig and Stark (118). Normal sorghum starch has physicochemical properties similar to wheat starch (Table 2).

The free monosaccharides in sorghum grain range from 0.05 to 0.83% (116,119) and are mainly glucose and fructose. Sucrose varies from 0.20 to 1.68% and total sugars from 0.32 to 2.25% (116,120). The grain contains about 0.34% of the raffinose family of sugars (116,119,120). The β–glucans were found in amounts of about 1.0% and the pentosans range from 1.8 to 5.6% in mature cultivars, amounts similar to levels observed in oats. Phenolics and fiber content were found to be higher and fiber digestibility lower in bird–resistant versus non–bird–resistant varieties of sorghum (121).

VIII. MILLET

The chemistry of the four major millets has been reviewed by Hoseney et al. (112). The starch, pentosan, glucan, and total sugar content of several millets is shown in Table 8.

Monteiro et al. (122) analyzed the grain from 12 cultivars of Italian millet (*Setaria italica*) for soluble sugars, carbohydrates by difference, protein, ash, minerals, and phenolics and determined in vitro protein digestibilities using pepsin and trypsin. They found the overall composition to be similar to other millets. Muralikrishna et al. (123) examined three millets used in India, samai (*Panicum miliare*), sanwa (*Echinochloa frumentaceae*), and panivaragu (proso, *Panicum miliaceum*), for glucans and pentosans, as well as starch content and characteristics. Scanning electron microscopy and chemical analysis were used to study enzyme–digested starch granules from varagu (*Paspalum scrobiculatum*) by Paramahans and Tharanathan (124). The molecular weights and effects of acid modification of starches from potato, tapioca, chick pea, rice (waxy and nonwaxy), sorghum, maize, and finger millet were determined by Singh and Zakiuddin Ali (125). The composition of the pentosan fraction of pearl millet was determined by Bailey and Sumrell (126). Total dietary fiber in whole grain millet (unspecified variety) was 9.2% (36). Morrison et al. (79) found that random millet and sorghum starch preparations resemble maize starches in all respects. The physical properties of isolated pearl millet starch were determined by Abd Allah et al. (127) (Table 2).

In general, millets contain about 60% starch, which is 18–20% amylose (Table 8). They have amounts of free sugars typical of other cereal grains, generally 2–3% or less. The free sugars are usually comprised of sucrose, maltose, and maltotriose, small amounts of raffinose oligosaccharides, and small amounts of monosaccharides such as glucose and fructose. Small amounts of pentosan and glucan polysaccharides also occur in millet. The physicochemical properties of millet varieties differ somewhat, but are generally similar to wheat starch (Table 2).

IX. OTHER GRAINS

Rye (*Secale* ssp), Triticale (*Triticale* ssp), and wild rice (*Zizania aquatica* L.) are common alternate grains (Table 2,9). Rye contains about 5.4% total soluble fructose, which includes free fructose, fructose from sucrose, and fructose containing oligosaccharides, as compared to about 3% fructose in triticale, 2% in barley, 1% in oats. 0.8% in rice, and 2% in wheat (26). The grain production and utilization potential of a number of perennial grasses have been the focus of some recent interest (128–130).

Amaranth (*Amaranthus* ssp, Amaranthaceae) is a small–seeded grain that has become increasingly important in parts of South America, Asia, and India. Both glutinous and non-glutinous starch has been detected in the seeds of *Amaranthus hypochondriacus* (131–133). The differences in amylose content were attributed to both cultivation and environmental effects.

TABLE 1 Proximate and Carbohydrate Composition of Wheat[a]

Property (%)	Whole wheat	Flour	Bran
Moisture	11.5–14(20)	—	3.7–17.7(19)
Protein	13.3(20), 15.3(23)	14–18(23)	11.9–22.9(19)
Fat	2.0(20), 1.9(23)	0.9–2.4(23)	3.0–6.8(19)
Ash	1.7(20), 1.85(23)	0.4–1.4(23)	3.8–9.6(19)
Total dietary fiber	9.9–11.6(35), 11.8–12.1(8), 14.6 ± 1.1(36)	2.3–5.6(36)	42.6(36)
Total soluble fiber	2.07(37)	1.7(37)	—
β-Glucans	0.34–1.4(80), <0.7(29), 0.54(11), 1.4(5)	—	—
Pentosans	1.41–2.01(134), 2.3–2.4(95), 5.96(136), 6.71(18)	1.1–1.6(135) 1.6–2.6(23)	21.6–26.5(22)
Cellulose	—	0.3(23)	35.2(23)
Hemicellulose	—	2.4(23)	43.1(23)
Free sugars	2.1–2.6(63)	1.2–2.1(23)	7.6(23)
Starch	53.0(23)	65–74(2)	14.1(23)
Amylose (as % of starch)	17.32(23) 17–27(139)	—	—

[a]Sources shown in parentheses.

TABLE 2 Physicochemical Properties of Cereal Grain Starches[a]

			Corn (38)			Millet			
	Amaranth (137)	Barley	Normal	Waxy	High amylose	Proso (108)	Pearl (140)	Finger (140)	Oat
Granule size (μ range)	1–1.5	A: 35–40 (138) B: 2–3 (138)	5–25	—	—	—	4–22	4–22	3–12 (4)
Amylose content (%)	7.2	24–27 (139)	26	~1	up to 80	—	34	39	16–27 (4)
Amylograph viscosity peak (BU)	320	no peak (108)	—	—	—	490	—	—	760–855 (77)
Swelling power at T°C	3.5 T = 90	8.2 (108) T = 90	24 T = 95	64 T = 95	6 T = 95	11 T = 90	13 T = 85	13 T = 85	18–31 T = 95 (77,127)
Solubility % at T°C	37 T = 90	3.5 (108) T = 90	25 T = 95	23 T = 95	12 T = 95	3.1 T = 90	7.7 T = 85	7.5 T = 85	13–24 (77,127)
Water-binding capacity	127	49 (108)	—	—	—	49	—	—	86 (77)
Gelatinization temperature range, (°C)									
initial	62	51.5 (138)	62	63	85	65	56.5	62.5	—
midpoint	64	57	—	—	—	73	67	69	—
final	68	59.5 62 (108) 63 67	72	72	87	80	72	74	—

[a]Sources shown in parentheses.

TABLE 2 Continued

	Rice (3)				Sorghum		
	Long	Medium	Short	Rye(38)	Normal	Waxy	Wheat
Granule size (μ range)	1–9	—	—	2–35	3–27 (138)	—	5–35 (138)
Amylose content (%)	23–26	15–20	18–20	23	21–34 (112)	0.8–1 (112)	17–27 (139)
Amylograph viscosity, peak (BU)	765–940	890–980	820–870	—	330 (112)	980 (112)	350 (137)
Swelling power at T°C	—	—	—	—	26 (127) T = 90	—	12 (137) T = 90
Solubility, % at T°C	4–26 at 98°C			—	14 (127) T = 90	—	8.2 (137) T = 90
Water-binding capacity	—	—	—	—	68 (127)	—	72 (137)
Gelatinization temperature range, °C							
initial	71	65	65	57	63 (112)	63 (112)	58 (138)
midpoint	—	—	—	—	—	—	61
final	74	68	67	70	71.9	73.4	64

[a]Sources shown in parentheses.

TABLE 3 Proximate Chemical and Carbohydrate Composition of Corn[a]

Property (%)	Dent			Popcorn	Sweet corn	Bran
	Normal	Waxy	Amylomaize high amylose			
Protein	8.1–11.5(142) 9.5–10.9(34) 10.3(20), 9.9(143)	—	13.7(34)	11.9(34) 13.2(20)	12.1(143) 12.7(34) 12.8(20)	5(34)
Fat	1.3–1.5(142) 3.9–5.8(142) 4.3(34),4.5(20) 4.8(143)	—	5.9(34)	5.3(34) 5.2(20)	3.7(20) 8.4(34) 9.1(143)	3(34)
Ash	1.4 (34,20,143)	—	—	1.7(20) 1.9(34)	2.0(143) 2.1(34) 2.7(20)	—
Total dietary fiber	15 flour(36) 4 meal(36) 10.6 degermed meal(36)	—	—	—	9.4(37)	92(34) 82(36) 85(37)
Total soluble fiber	—	—	—	—	1.2(37)	1.16 (37)
Pentosans pentoglycans	5.8–6.6(141)	—	—	—	—	—
Cellulose	3.3–4.3(141)	—	—	—	—	—
Free sugars	1.61–2.22(142) 1.3(141), 2.6(34,143)	—	—	9.3(34)	12.0(34)	—
Starch	68–74(142)	—	57.4–62.9 (139)	62.3(34)	54.1(34)	0
Amylose	24(34)	~1(34)	50–80(34)	24(34)	24(34)	—

[a]Sources shown in parentheses.

TABLE 4 Proximate Chemical and Carbohydrate Composition of Rice[a]

Property (%)	Brown rice	Milled	Bran
Protein(3)	7.1–8.3	6.3–7.1	11.3–14.9
Fat(3)	1.6–2.8	0.3–0.5	15.0–19.7
Ash(3)	1.0–1.5	0.3–0.8	6.6–9.9
Total dietary fiber(36)	3.9 ± 0.2	2.3	20.3
Total soluble fiber (37)	0.89	0.92	—
Pentosans(3)	1.2–2.1	0.5–1.4	8.3
Cellulose	—	—	5.9–9.0(3)
β-Glucans	0.13(80)	0.11(81)	—
Starch(3)	66.4	77.6	13.8
Amylose			
Waxy	0.3–2.3(79) 1.7(49),1.4(60) <1(47) 1.2–2.7(59)		
Normal	12.2–28.6(79) 16.2–30.3(49) 16.0–33.0(47) 23.5(60), 21.8–25.5(59)		
High	30.2–32.0(144)		
Free sugars(3)	0.7–1.3	0.22–0.45	5.5–6.9

[a]Sources shown in parentheses.

TABLE 5 Proximate Chemical and Carbohydrate Composition of Oats[a]

Property (%)	Groats	Bran
Moisture	8.3 (20)	—
Protein (N × 5.85)	14.2 (20)	—
Fat	7.4 (20)	—
Ash	1.9 (20)0	—
Total dietary fiber	10.9 flour (36) 10.51 rolled (37)	19.4 (36) 15.72
Total soluble fiber	5.43 rolled (37)	7.84 (37)
β-Glucans	2.5–6.6 (80), 4.8–6.6 (5), 2.2–4.2 (16), 3.0 (11)	5.7 (85)
Pentosans	3.17 (18)	—
Total free sugars	1.14 (63), 1.4 (145)	2.6–3.4 (145)
Starch	43–64 (4)	—
Amylose	25–29 (79)	—

[a]Sources shown in parentheses.

TABLE 6 Proximate Chemical and Carbohydrate Composition of Barley[a]

Properties (%)	Whole grain
Moisture	10.8 (20)
Protein	9.6 (20)
Fat	1.1 (20)
Ash	1.3 (20)
Total dietary fiber	17.3 (36)
β-Glucans	3.9–8.1 (146), 2.0–8.6 (80), 3.0–6.9 (16), 5.1–7.2 (5), 3.5–5.3 (96), 2.8–5.6 (96)
Pentosans	4.4–7.8 (98), 6.7–9.8 (95), 7.3–11.0 (95)
Starch	54–63 (80)
Amylose	
Waxy	2.1–8.3 (102)
Normal	25.3–30.1 (102) 27.5–29.4 (80)
High	38.4–40.8 (102)

[a]Sources shown in parentheses.

TABLE 7 Proximate Chemical and Carbohydrate Composition of Sorghum[a]

Properties (%)	Whole grain	Flour (141)	Bran (141)
Moisture	11 (20)	—	—
Protein	11.0 (20) 12.3 (141)	11.2–13.0	5.2–7.6
Fat	3.3 (20) 3.6 (141)	0.4–0.8	3.7–6.0
Ash	1.7 (20)	0.30–0.44	2.02
Total dietary fiber	10.7 (35)	—	—
β-Glucans	1.0 (5)	—	—
Pentosans	2.5–5.6 (112) 2.09 (18)	—	—
Free sugars	0.32–2.25 (116)	—	—
Starch	72.3–75.1 (141) 32–79 (112)	81.3–83.0	34.6
Amylose	27–30 (79) 33–62 (112)	—	—

[a]Sources shown in parentheses.

TABLE 8 Proximate Chemical and Carbohydrate Composition of Millets[a]

Sample	Starch (%)	Amylose (%)	Pentosan(%)	Glucans(%)	Total sugar (%)
Pearl, *Pennisetum americanum*	56–65 (112)	17–22 (112)	2–3 (126)	—	2.6–5.4 (112)
Proso, *Panicum miliaceum*	—	12–20 (112)	—	—	—
Finger, (ragi) *Eleusine coracona*	57–59 (147) 60–61 (148)	6–18 (112)	6.2–7.2 (148)	—	0.6–0.7 (148)
Foxtail, *Setaria italica* 12 cv.	56 (147)	17.5 (149)	5.5 (148)	—	0.5 (148)
Varagu, *Paspalum scrobiculatum* (124)	72	—	0.2	1.1	—
Samai *Panicum miliare* (123)	57	18	0.42	0.65	—
Sanwa, *Echinochloa frumentaceae* (123)	62	20	0.16	0.28	—
Panivaragu Proso, *Panicum miliaceum* (123)	60	24	0.17	0.21	—
Italian, *Setaria italica*	—	—	—	0.42–0.69 (122)	—

[a]Sources shown in parentheses.

TABLE 9 Carbohydrate Composition of Some Cereal Grains and a Pseudocereal[a]

Sample	Starch (%)	Amylose (%)	Glucans (%)	Pentosan (%)	Total dietary fiber
Rye	52 (76)	27 (139)	1.9–2.9(5) 1.89 (26)	2.6 (26)	32 (36)
Triticale	53 (150)	23 (150)	1.2 (5), 0.65 (26) 0.34 (80)	1.5 (26)	—
Wild rice	60–65 (152)	—	—	—	5.2 (36)
Amaranth	48–69 (153)	0.14 (132) 0–21.8[b] (133) 0.1–11.1 (151)	—	—	10.2 (36)

[a]Sources shown in parentheses.
[b]Observed in the same species of *A. hypochondriacus* 10/89 (133).

REFERENCES

1. Hoseney, R. C., *Principles of Cereal Science and Technology*, American Association of Cereal Chemists, St. Paul, MN, 1986.
2. Pomeranz, Y. (Ed.), *Wheat: Chemistry and Technology*, AACC, St. Paul, MN, 1971.
3. Juliano, B. O. (Ed.), *Rice: Chemistry and Technology*, 2nd ed. AACC, St. Paul, MN, 1985.
4. Webster, F. H. (Ed.), *Oats: Chemistry and Technology*, American Association of Cereal Chemists, St. Paul, MN, 1986.
5. Prentice, N., Babler, S., and Faber, S., Enzymatic analysis of beta–D– glucans in cereal grains, *Cereal Chem.*, *57*:198–202 (1980).
6. James, W. P. T., and Theander, O., (Eds.), *The Analysis of Dietary Fiber in Food*, Vol 3. Marcel Dekker, Inc., New York, 1981.
7. Mongeau, R., and Brassard, R., Determination of neutral detergent fiber in breakfast cereals: pentose, hemicellulose, cellulose and lignin content, *J. Food Sci.*, *47*:550–555 (1982).
8. Asp, N. G., Johansson, C. G., Hallmer, H., and Siljestrom, M., Rapid enzymatic assay of insoluble and soluble dietary fiber, *J. Agric. Food Chem.*, *31*:476–482 (1983).
9. Prosky, L., Asp, N. G., Furda, I., Devries, J. W., Schweizer, T. F., and Harland, B. F., Determination of total dietary fiber in foods, food products, and total diets: interlaboratory study, *J. Assoc. Off. Anal. Chem.*, *67*:1044–1052 (1984).
10. Englyst, H. N., and Cummings, J. H., Simplified methods for the measurement of non–starch polysaccharides by gas–liquid chromatography of constituent sugars as alditol acetates, *Analyst*, *109*:937–942 (1984).
11. Aman, P., and Hesselman, K., An enzymatic method for analysis of total mixed–linkage β–glucans in cereal grains, *J. Cereal Sci.*, *3*:231–237 (1985).
12. Association of Official Analytical Chemists, Total dietary fiber in foods, enzymatic–gravimetric methods, first action, *J. Assoc. Off. Anal. Chem.*, *69*:399 (1986).
13. Englyst, H. N., Trowell, H., Southgate, D. A. T., and Cummings, J. H., Dietary fiber and resistant starch, *Am. J. Clin. Nutr.*, *46*:873–874 (1987).
14. Mongeau, R., and Brassard, R., A rapid method for the determination of soluble and insoluble dietary fiber: comparison with AOAC total dietary fiber procedure and Englyst's method, *J. Food Sci.*, *51*:1333–1336 (1986).

15. Englyst, H. N., and Hudson, G. J., Colorimetric method for routine measurements of dietary fiber as non–starch polysaccharides. A comparison with gas–liquid chromatography, *Food Chem.*, *24*:63–76 (1987).
16. Aman, P., and Graham, H., Analysis of total and insoluble mixed–linked (1–3), (1–4)–β–D–glucans in barley and oats, *J. Agric. Food Chem.*, *35*:704–709 (1987).
17. Hashimoto, S., Shogren, M. D., and Pomeranz, Y., Cereal pentosans: their estimation and significance. I. Pentosans in wheat and milled wheat products, *Cereal Chem.*, *64*:30–34 (1987).
18. Hashimoto, S., Shogren, M. D., Bolte, L. C., and Pomeranz, Y. Cereal pentosans: Their estimation and significance. III. Pentosans in abraded grains and milling by–products, *Cereal Chem.*, *64*:39–41 (1987).
19. Saunders, R. M., Wheat bran: composition and digestibility. In *Topics in Dietary Fiber Research*, G. A. Spiller and R. J. Amen (Eds.), Plenum Press, New York, 1978, pp. 43–58.
20. Watt, B. K., and Merrill, A. L., *Composition of Foods*, Agriculture Handbook No 8, Agricultural Research Service, U.S. Department of Agriculture, Washington, D.C., 1963.
21. Zeigler, E., and Greer, E. N., Principles of milling. In *Wheat: Chemistry and Technology*, Y. Pomeranz (Ed.), AACC, St. Paul, MN, 1971, pp. 115–200.
22. MacMasters, M. M., Hinton, J. J. C., and Bradbury, D., Microscopic structure and composition of the wheat kernel. In *Wheat: Chemistry and Technology*, Y. Pomeranz (Ed.), AACC, ST. Paul, MN, 1971, pp. 51–114.
23. D'Appolonia, B. L., Giles, K. A., Osman, E. M., and Pomeranz, Y., Carbohydrates. In *Wheat: Chemistry and Technology*, Y. Pomeranz (Ed.), AACC, St. Paul, MN, 1971, pp. 301–392.
24. Brodribb, A. J. M., Dietary fiber in diverticular disease of the colon. In *Medical Aspects of Dietary Fiber*, G. A. Spiller and R. Kay (Eds.). Plenum Press, New York, 1980.
25. Trowell, H., Dietary Fiber: A Paradigm. In *Dietary Fiber, Fiber Depleted Foods and Disease*, Trowell, Burkitt, and Heaton (Eds.), Academic Press, New York, 1985.
26. Henry, R. J., A comparison of the non–starch carbohydrates in cereal grains, *J. Sci. Food Agric.*, *36*:1243–1253 (1985a).
27. Hoseney, R. C., Functional properties of pentosans in baked foods, *Food Technol.*, *38*:114–117 (1984).
28. Shelton, D. R., and D'Appolonia, B. L., Carbohydrate functionality in the baking process, *Cereal Foods World*, *30*:437–442 (1985).
29. McCleary, B. V., Gibson, T. S., Allen, H., and Gams, T. C., Enzymic hydrolysis and industrial importance of barley beta–glucans and wheat flour pentosans, *Starch*, *38*:433–437 (1986).
30. Antoniou, T., Marquardt, r. R., and Cansfield, P. E., Isolation, partial characterization, and antinutritional activity of a factor (Pentosans) in rye grain, *J. Agric. Food Chem.*, *29*:1240–1247 (1981).
31. Wieg, A. J., Enzymatische Behandlung von Abwassern der Weizenstarkeindustrie, *Starch*, *36*:135–140 (1984).
32. Smith, M. M., and Hartley, R. D., Occurrence and nature of ferulic acid substitution of cell–wall polysaccharides in graminaceous plants, *Carb. Res.*, *118*:65–80 (1983).
33. Moore, C. D., Tuschhoff, J. V., Hastings, C. W., and Schanefelt, R. V., Applications of starches in foods. In *Starch: Chemistry and Technology*, 2nd ed., R. L. Whistler, J. N. BeMiller, and E. F. Paschall (Eds.), Academic Press, Orlando, FL, 1984, pp. 575–592.
34. Johnson, L. A., Corn: Its production, processing and utilization. In *The Handbook of Cereal Science and Technology*, K. Lorenz and K. Kulp (Eds.), Marcel Dekker, Inc., New York, (1990).
35. San Buenaventura, M. L., Dong, F. M., and Rasco, B. A., The total dietary fiber content of wheat, corn, barley, sorghum, and distillers' dried grains with solubles, *Cereal Chem.*, *64*:135–136 (1987).
36. Cardozo, M. S., and Eitenmiller, R. R., Total dietary fiber analysis of selected baked and cereal products, *Cereal Foods World*, *33*:414–418 (1988).

37. Anderson, J. W., and Bridges, S. R., Dietary fiber content of selected foods, *Am. J. Clin. Nutr.*, *47*:440–447 (1988).
38. Pomeranz, Y., *Functional Properties of Food Components*, Academic Press, Orlando, FL, 1985.
39. Selvendran, R. R., Stevens, B. J. H., and Dupont, M. S., Dietary fiber: Chemistry, analysis, and properties. In *Advances in Food Research*, C. O. Chichester, E. M. Mrak, and B. S. Schweigert (Eds.). Academic Press, Inc., New York, 1987, pp. 117–209.
40. Houston, D. F. (Ed.), *Rice: Chemistry and Technology*, American Association of Cereal Chemists, St. Paul, MN, 1972.
41. Juliano, B. O., Rice starch: Production, properties and uses. In *Starch: Chemistry and Technology*, 2nd ed., R. L. Whistler, J. N. BeMiller, and E. F. Paschall (Eds.), Academic Press, Orlando, FL, 1984, pp. 507–528.
42. Juliano, B. O., The chemical basis of rice grain quality. In *Proceedings of the Workshop on Chemical Aspects of Rice Grain Quality*, Int. Rice Res. Inst., Los Banos, Laguna, Philippines, 1979, pp. 69–90.
43. Webb, B. D., Bollich, C. N., Johnson, T. H., and McIlrath, W. O., Components of rice quality: Their identification, methodology, and stage of application in the United States breeding programs. In *Proceedings of the Workshop on Chemical Aspects of Rice Grain Quality*, Int. Rice Res. Inst., Los Banos, Laguna, Philippines, 1979, pp. 191–205.
44. Juliano, B. O., Perez, C. M., Blakeney, A. B., Castillo, T., Knogseree, N., Laignelet, B., Lapis, E. T., Murty, V. V. S., Paule, C. M., and Webb, B. D., International cooperative testing on the amylose content of milled rice, *Starch*, *33*:157–162 (1981).
45. Villareal, C. P., and Juliano, B. O., Waxy gene factor and residual protein of rice starch granules, *Starch*, *38*:118–119 (1986).
46. Asaoka, M., Okuno, K., Sugimoto, Y., Omura, T., and Fuwa, H., Characterization of endosperm starch from high–amylose mutants of rice (*Oryza sativa* L.), *Starch*, *31*:114–117 (1986).
47. Maningat, C. C., and Juliano, B. O., Properties of lintnerized starch granules from rices differing in amylose content and gelatinization temperature, *Starch*, *31*:5–10 (1979).
48. Maningat, C. C., and Juliano, B. O., Starch lipids and their effect on rice starch properties, *Starch*, *32*:76–82 (1980).
49. Hizukuri, S., Shirasaka, K., and Juliano, B. O., Phosphorus and amylose branching in rice starch granules, *Starch*, *35*:348–350 (1983).
50. Little, R. R., Hilder, G. B., and Dawson, E. H., Differential effect of dilute alkali on 25 varieties of milled white rice, *Cereal Chem.*, *35*:111–126 (1985).
51. Bhattacharya, K. R., and Sowbhagya, C. M., An improved alkali reaction test for rice quality, *J. Food Technol.*, *7*:323–331 (1972).
52. Halick, J. V., Beachell, H. M., Stansel, J. W., and Kramer, H. H., A note on the determination of gelatinization temperatures of rice varieties, *Cereal Chem.*, *37*:670–672 (1960).
53. Juliano, B. O., Perez, C. M., Alyoshin, E. P., Romanov, V. B., Bean, M. M., Nishita, K. D., Blakeney, A. B., Welsh, L. A., Delgado, L., El Baya, A. W., Fossati, G., Kongseree, N., Mendes, F. P., Brilhante, S. Suzuke, H., Tada, M., and Webb, B. D., Cooperative test on amylography of milled–rice flour for pasting viscosity and starch gelatinization temperatures, *Starch*, *37*:40–50 (1985).
54. Juliano, B. O., Blakeney, A. B., Butta, I., Castillo, D., Choudhury, N. H., Iwasaki, T., Shibuya, N., Kongseree, N., Lapis, E. T., Murty, V. V. S., Paule, C. M., Perez, C. M., and Webb, B. D., International cooperative testing of the alkali digestibility values for milled rice, *Starch*, *34*:21–26 (1982).
55. Maher, G. G., Alkali gelatinization of flours, *Starch*, *35*:271–276 (1983a).
56. Maher, G. G., Alkali gelatinization of starches, *Starch*, *35*:226–234 (1983b(.
57. Nara, S., Sakakura, M., and Komiya, T., On the acid resistance of starch granules, *Starch*, *35*:266–270 (1983).
58. Mod, R. R., Normand, F. L., Ory, R. L., and Conkerton, E. J., Effects of hemicellulose on viscosity of rice flour, *J. Food Sci.*, *46*:571–573 (1981).

59. Merca, F. E., and Juliano, B. O., Physicochemical properties of starch of intermediate–amylose and waxy rices differing in grain quality, *Starch, 33*:253–260 (1981).
60. Perez, C. M., Juliano, B. O., Pascual, C. G., and Novenario, V. G., Extracted lipids and carbohydrates during washing and boiling of milled rice, *Starch, 39*:386–390 (1987).
61. Mod, R. R., Normand, F. L., and Ory, R. L., *Proceedings of the 7th Rice Technical Working Group*, Texas, 1987, p. 54.
62. Asaoka, M., Okuno, K., Sugimoto, Y., Yano, M., Omura, T., and Fuwa, H., Structure and properties of endosperm starch and water soluble polysaccharides from sugary mutant of rice (*Oryza sativa* L.), *Starch, 37*:364–366 (1985).
63. Kuo, T. M., VanMiddlesworth, J. F., and Wolf, W. J., Contents of raffinose oligosaccharides and sucrose in various plant seeds, *J. Agric. Food Chem., 36*:32–36 (1988).
64. Shibuya, N., and Iwasaki, T., Effect of the enzymatic removal of endosperm cell wall on the gelatinization properties of aged and unaged rice flours, *Starch, 34*:300–303 (1982).
65. Hizukuri, S., Kaneko, T., and Takada, Y., Measurement of the chain length of amylopectin and its relevance to the origin of crystalline polymorphism of starch granules. *Biophys. Acta, 760*:188–191 (1983).
66. Takeda, Y., Hizukuri, S., and Juliano, B. O., Structures of rice amylopectins with low and high affinities for iodine, *Carb. Res., 168*:79–88 (1987a).
67. Takeda, Y., Hizukuri, S., Takeda, C., and Suzuki, A., Structures of branched molecules of amylose of various origins, and molar fractions of branched and unbranched molecules, *Carb. Res., 165*:139–145 (1987b).
68. Juliano, B. O., and Villareal, R. M., Varietal differences in physiochemical properties of waxy rice starch, *Starch, 39*:298–301 (1987).
69. Russell, P. L., and Juliano, B. O., Differential scanning calorimetry of rice starches, *Starch, 35*:382–386 (1983).
70. Biliaderis, C. G., Page, C. M., Maurice, T. J., and Juliano, B. O., Thermal characterization of rice starches: a polymeric approach to phase transitions of granular starch, *J. Agric. Food Chem., 34*:6–14 (1986).
71. Shibuya, N., and Iwasaki, T., Polysaccharides and glycoproteins in the rice endosperm cell wall, *Agric. Biol. Chem., 42*:2259–2266 (1978).
72. Shibuya, N., and Misaki, A., Structure of hemicellulose isolated from rice endosperm cell wall: Mode of linkages and sequences in xyloglucan, β–glucan and arabinoxylan, *Agric. Biol. Chem., 42*:2267–2274 (1978).
73. Watanabe, T., Shida, M., Furuyama, Y., Tsukamoto, K. I., Nakajima, T., Matsuda, K., and Kainuma, K., Structure of the arabinoxylan of rice hull, *Carb. Res., 123*:83–95 (1983).
74. Watanabe, T., Shida, M., Murayama, T., Furuyama, Y., Nakajima, T., Matsuda, K., and Kainuma, K., Xyloglucan in cell walls of rice hull, *Carb. Res., 129*:229–242 (1984).
75. Blakeney, A. B., and Matheson, N. K., Some properties of the stem and pollen starches of rice, *Starch, 36*:265–269 (1984).
76. MacMasters, M. M., Wolf, M. J., and Seckinger, H. L., The possible use of oat and other cereal grains for starch production, *Am. Miller Process., 75*:82–83 (1947).
77. MacArthur, L. A., and D'Appolonia, B. L., Composition of oat and wheat carbohydrates. II. Starch, *Cereal Chem., 56*:458–461 (1979).
78. Paton, D., Oat starch: Physical, chemical and structural properties, In *Oats: Chemistry and Technology*, F. H. Webster (ed.), AACC, St. Paul, MN, 1986, pp. 93–120.
79. Morrison, W. R., Milligan, T. P., and Azudin, M. N., A relationship between the amylose and lipid contents of starches from diploid cereals, *J. Cereal Sci., 2*:257–271 (1984).
80. Fincher, G. B., and Stone, B. A., Cell walls and their components in cereal grain technology. In *Advances in Cereal Science and Technology*, Vol. VIII, Y. Pomeranz (Ed.). AACC, St. Paul, MN, pp. 207–295.
81. Anderson, M. A., Cook, J. A., and Stone, B. A., Enzymatic determination of 1,3;1,4–β–glucans in barley grain and other cereals, *J. Inst. Brew., 84*:233–239 (1978).
82. Wood, P. J., and Weisz, J., Use of Calcofluor in analysis of oat beta–D–glucan, *Cereal Chem., 61*:73–75 (1984).

83. Jensen, S. A., and Aastrup, S., A fluorometric method for measuring 1,3;1,4–β–glucan in beer, wort, malt, and barley by use of Calcofluor, *Carlsberg Res. Commun.*, *46*:87–95 (1981).
84. Jorgensen, K. G., An improved method for determining β–glucan in wort and beer by use of Calcofluor, *Carlsberg Res. Commun.*, *48*:505–516 (1983).
85. Wood, P. J., Oat β–glucan: Structure, location and properties. In *Oats: Chemistry and Technology*, F. H. Webster (Ed.), AACC, St. Paul, MN, 1986, pp. 121–152.
86. Sosulski, F. W., and Wu, K. K., High–fiber breads containing field pea hulls, wheat, corn, and wild oat brans, *Cereal Chem.*, *48*:186–191 (1988).
87. Sosulski, F. W., Sosulski, K., and Olson, J. P., Nutritive value of wild oat groats and flakes, *Can. Inst. Food Sci. Technol. J.*, *18*:220–224 (1985).
88. McLeod, A. M., and Preece, I. A., Studies on the free sugars of the barley grain. V. Comparisons of sugars and fructosans with those of other cereals, *J. Inst. Brew. London*, *60*:46 3–55 (1954).
89. Kirby, R. W., Anderson, J. W., Sieling, B., Rees, E. D., Chen, W.–J.L, Miller, R. E., and Kay, R. M., Oat–bran intake selectively lowers serum low–density lipoprotein cholesterol concentrations of hypercholesterolemic men, *Am. J. Clin. Nutr.*, *34*:824–829 (1981).
90. Anderson, J. W., Story, L., Sieling, B., Chen, W.–J.L., Petro, M. S., and Story, J., Hypercholesterolemic effects of oat–bran or bean intake for hypercholesterolemic men, *Am J. Clinical Nutr.*, *40*:1146–1155 (1984).
91. Anderson, J. W., Dietary fiber in diabetes. In *Medical Aspects of Dietary Fiber*, G. A. Spiller and R. Kay (Eds.). Plenum Press, New York, 1980, pp. 193–221.
92. Paton, D., Oat starch. I. Extraction, purification and pasting properties, *Starch*, *29*:149–153 (1977).
93. Doublier, J. L., Paton, D., and Llamas, G., A rheological investigation of oat starch pastes, *Cereal Chem.*, *64*:21–26 (1987).
94. Henry, R. J., Near–infrared reflectance analysis of carbohydrates and its application to the determination of (1–>3), (1–>4)–beta–D–glucan in barley, *Carb. Res.*, *141*:13–19 (1985).
95. Lehtonen, M., and Aikasalo. R., Pentosans in barley varieties, *Cereal Chem.*, *64*:133–134 (1987).
96. Lehtonen, M., and Aikasalo, R., Beta–Glucans in two– and six–row barley, *Cereal Chem.* *64*:191–192 (1987).
97. Aman, P., and Newman, C. W., Chemical composition of some different types of barley grown in Montana, U.S.A., *J. Cereal Sci.*, *4*:133–141 (1986).
98. Henry, R. J., Genetic and environmental variation in the pentosan and β–glucan contents of barley, and their relation to malting quality, *J. Cereal Sci.*, *4*:269–277 (1986).
99. Buliga, G. S., Brant, D. A., and Fincher, G. B., The sequence statistics and solution conformation of a barley (1–>3, 1–>4)–beta–D–glucan, *Carb. Res.*, *157*:139–156 (1986).
100. Ahluwalia, B., and Fry, S. C., Barley endosperm cell walls contain a feruloylated arabinoxylan and an non–feruloylated β–glucan, *J. Cereal Sci.*, *4*:287–295 (1986).
101. Ballance, G. M., Hall, R. S., and Manners, D. J., Studies of some arabinoxylans from barley endosperm walls, *Carb. Res.*, *150*:290–294 (1986).
102. Morrison, W. R., Scott, D. C., and Karkalas, D. J., Variations in the composition and physical properties of barley starches, *Starch*, *38*:374–379 (1986).
103. MacGregor, A. W., and Ballance, D. L., Hydrolysis of large and small starch granules fromnormal and waxy barley cultivars by alpha–amylases from barley malt, *Cereal Chem.*, *57*:397–402 (1980).
104. Makela, M. J., Korpela, T., and Laakso, S., Studies of starch size and distribution in 33 barley varieties with a celloscope, *Starch*, *34*:329–334 (1982).
105. MacGregor, A. W., and Morgan, J. E., Structure of amylopectins isolated from large andsmall starch granules of normal and waxy barley, *Cereal Chem.*, *61*:222–228 (1984).
106. MacGregor, A. W., and Matsuo, R. R., Starch degradation in endosperms of barley and wheat kernels during initial stages of germination, *Cereal Chem.*, *59*:210–216 (1982).

107. DeHaas, B. W., and Goering, K. J., Barley starch. VII. New barley starches with fragmented granules, *Cereal Chem.*, *60*:327–329 (1983).
108. Lorenz, K., and Kulp, K., Cereal– and root starch modification by heat–moisture treatment. I. Physico–chemical properties, *Starch*, *34*:50–54 (1982).
109. Lorenz, K., and Kulp, K., Cereal– and root starch modification by heat moisture treatment. II. Functional properties and baking potential, *Starch*, *34*:76–81 (1982).
110. Lorenz, K., and Kulp, K., Steeping of barley starch. Effects on physiochemical properties and functional characteristics, *Starch*, *36*:116–121 (1984).
111. Fjeta, G., and Axtell, J. D., Dry–matter accumulation and carbohydrate composition in developing normal– and high–lysine sorghum grain, *J. Agric. Food Chem.*, *35*:981–985(1987).
112. Hoseney, R. C., Varriano–Marston, E., and Dendy, D. A. V., Sorghum and millets. In *Advances in Cereal Science and Technology*, Vol. IV. Y. Pomeranz (Ed.). AACC, St. Paul, MN, 1981, pp. 71–144.
113. Rooney, L. W., Kirleis, A. W., and Murty, D. S., Traditional foods from sorghum: theirproduction, evaluation, and nutritional value. In *Advances in Cereal Science and Technology*, Vol. VIII, Y. Pomeranz (Ed.), AACC, St. Paul, MN, 1986.
114. Cagampang, G. B., and Kirleis, A. W., Relationship of sorghum grain hardness to selected physical and chemical measurements of grain quality, *Cereal Chem.*, *61*:100–105 (1984).
115. Cagampang, G. B., and Kirleis, A. W., Properties of starches isolated from sorghum flouryand corneous endosperm, *Starch*, *37*:253–257 (1985).
116. Stark, J. R., Aisien, A. O., and Palmer, G. H., Studies on starches from Nigerian sorghum, *Starch*, *35*:73–76 (1983).
117. Glennie, C. W., Physico–chemical properties of sorghum starch thermally treated by different methods, *Starch*, *39*:273–276 (1987).
118. Craig, S. A. S., and Stark, J. R., A comparison of the molecular properties of sorghum starches of different origins, *Starch*, *36*:127–131 (1984).
119. Murthy, D. S., Singh, U., Suryaprakash, R., and Nicodemus, K. D., Soluble sugars in fiveendosperm types of sorghum, *Cereal Chem.*, *62*:150–152 (1985).
120. Subramanian, V., Jambunathan, ., and Suryaprakash, S., Note on the soluble sugars of sorghum, *Cereal Chem.*, *57*:440–441 (1980).
121. Reed, J. O., Phenolics, fiber, and fiber digestibility in bird resistant and non bird resistant sorghum grain, *J. Agric. Food Chem.*, *35*:461–464 (1987).
122. Monteiro, P. V., Gopal, D. H., Virupaksha, T. K., and Ramachandra, G., Chemical composition and *in vitro* protein digestibility of Italian millet (*Setaria italica*), *Food Chem.*, *29*:19–26 (1988).
123. Muralikrishna, G., Paramahans, S. V., and Tharanathan, R. N., Carbohydrate make–up of minor millets, *Starch*, *34*:397–401 (1982).
124. Paramahans, S. V., and Tharanathan, R. N., Carbohydrate composition of millet varagu,*Starch*, *32*:73–76 (1980).
125. Singh, V., and Zakuiddin Ali, S., Comparative acid modification of various starches, *Starch*, *39*:402–405 (1987).
126. Bailey, A. V., and Sumrell, G., Pentosans in pearl millet, *Cereal Chem.*, *56*:295–298 (1979).
127. Abd Allah, M. A., Mahmoud, R. M., El–Kalyoubi. M. H., and Abou Arab, A, A., Physicalproperties of starches isolated from yellow corn, sorghum, sordan and pearl millet, *Starch*, *39*:9–12 (1987).
128. Becker, R., Hanners, G. D., Irving, D. W., and Saunders, R. M., Chemical compositions and nutritional qualities of five potential perennial grains, *Lebens.–Wiss. u.–Technol.*,*19*:312–315 (1986).
129. Yensen, S. B., and Weber, S. B., Composition of *Distichlis palmeri* grain, a saltgrass, *J. Food Sci.*, *51*:1089–1090 (1986).
130. Fejer, S. O., and Fedak, G., Yield performance of *Secale cereale* X *S. montanum* progenies, *Cer. Res. Comm.*, *9*:331 (1981).

131. Okuno, K., and Sakaguchi, S., Glutinous and non–glutinous starches in perisperm of grain amaranths, *Cereal Res. Comm.*, *9*:305 (1981).
132. Sugimoto, Y., Yamada, K., Sakamoto, S., and Fuwa, H., Some properties of normal–andwaxy–type starches of *Amaranthus hypochondriachus* L., *Starch*, *33*:112–116 (1981).
133. Tomita, Y., Sugimoto, Y., Sakamoto, S., and Fuwa, H., Some properties of starches of grainamaranths and several millets, *J. Nutr. Sci. Vitaminol.*, *27*:471–484 (1981).
134. Shogren, M. D., Hashimoto, S., and Pomeranz, Y., Cereal pentosans: their estimation and significance IV. Pentosans in wheat flour varieties and fractions, *Cereal Chem.*, *65*:182–185(1988).
135. Ciacco, C. F., and D'Appolonia, B. L., Characterization of pentosans from different wheat flour classes and of their gelling capacity, *Cereal Chem.*, *59*:96–99 (1982).
136. Douglas, S. G., A rapid method for the determinations of pentosans in wheat flour, *FoodChem.*, *7*:139–145 (1981).
137. Lorenz, K, *Amaranthus hypochondriachus*—Characteristics of the starch and baking potential of the flour, *Starch*, *33*:149–153 (1981).
138. Snyder, E. M., Industrial microscopy of starches. In *Starch: Chemistry and Technology*, 2nd edition, R. L. Whistler, J. N. BeMiller, and E. F. Paschall (Eds.), Academic Press, Inc., Orlando, FL, 1984.
139. Zuber, M. S, Genic control of starch development. In *Starch: Chemistry and Technology*, Vol. I, R. L. Whistler and E. F. Paschall (Eds.), Academic Press, Inc., New York, 1965, pp. 43–64.
140. Malleshi, N. G., Desikachar, H. S. R., and Tharanathan, R. N., Physico–chemical propertiesof native and malted finger millet, pearl millet and foxtail millet starches, *Starch*,*38*:202–205 (1986).
141. Watson, S. A., Corn and sorghum starches: Production. In *Starch: Chemistry and Technology*, 2nd ed., R. L. Whistler, J. N. BeMiller, and E. F. Paschall (Eds.), Academic Press, Orlando, FL, 1984, pp. 417–464.
142. Watson, S. A., Manufacture of corn and milo starches. In *Starch: Chemistry and Technology*,Vol. II, R. L. Whistler and E. F. Paschall (Eds.), Academic Press, Inc., New York, 1967, pp. 1–52.
143. Matz, S. A., *Cereal Science*, The Avi Publishing Co., Inc., Westport, CT, 1969.
144. Juliano, B. O., Villareal, R. M., Perez, C. M., Villareal, C. P., Takeda, Y., and Hizukuri, S.,Varietal differences in properties among high amylose rice starches, *Starch*, *39*:390–393 (1987).
145. MacArthur–Grant, L. A., Sugars and non–starchy polysaccharides. In *Oats: Chemistry and Technology*, F. H. Webster (Ed.), AACC, St. Paul, MN, 1986, pp. 75–91.
146. Ullrich, S. E., Clancy, J. A., Eslick, R. F., and Lance R. C. M., β–Glucan content and viscosity of extracts from waxy barley. *J. Cereal Sci.*, *4*:279–285 (1986).
147. Wankhede, D. B., Shehnaz, A., and Rao, M. R. R., Preparation and Physicochemicalproperties of starches and their fractions from finger millet (*Eleusine coracana*) and foxtail millet (*Setaria italica*), *Starch*, *31*:153–159 (1979).
148. Wankhede, D. B., Shehnaz, A., and Rao, M. R. R., Carbohydrate composition of finger millet (*Eleusine coracana*) and foxtail millet (*Setaria italica*), *Qual. Plant*, *28*:293–303 (1979).
149. Wankhede, D. B., and Ramteke, R. S., Synergistic digestibility of several native starches by amylolytic enzymes, *Starch*, *34*:309–312 (1982).
150. Berry, C. P., D'Appolonia, B. L., and Giles, K. A., The characterization of triticale starch and its comparison with starches of rye, durum and HRS wheat, *Cereal Chem.*, *48*:415 (1971).
151. Stone, L. A., and Lorenz, K., The starch of *Amaranthus*—Physico–chemical properties and functional characteristics, *Starch*, *36*:232–237 (1984).
152. Hoseney, R. C., *Principles of Cereal Science and Technology*, The American Association of Cereal Chemists, St. Paul, MN, 1986.

153. Saunders, R. M., and Becker, R., *Amaranthus*: A potential food and feed resource. In *Advances in Cereal Science and Technology*, Volume IV, Y. Pomeranz (Ed.), The American Association of Cereal Chemists, St. Paul, MN, 1984.
154. Fitt, L. E., and Snyder, E. M., Photomicrographs of starches. In *Starch: Chemistry and Technology*, 2nd edition, R. L. Whistler, J. N. BeMiller, and E. F. Paschall (Eds.), Academic Press, Orlando, FL, 1984, pp. 675–689.

13

CEREAL LIPIDS

Okkyung Kim Chung
U.S. Department of Agriculture, Manhattan, Kansas

I. INTRODUCTION

Lipids are relatively minor constituents in cereal grains. However, they must be taken into consideration when discussing nutrition, grain storage, processing such as dry and wet milling, brewing, baking, cooking, and extrusion.

There have been extensive reviews on cereal lipids (1–5), barley lipids (6), corn lipids (7,8), oat lipids (9,10), rice lipids (11,14), sorghum and pearl millet lipids (15), and wheat lipids (16–23). The above reviews and other specific studies cited in this chapter are recommended for those who are interested in more detailed lipid chemistry of each cereal grain.

This chapter is intended to give a broad overview of comparative values of cereal lipids: how they differ in quantity and composition among various cereal grains. Thus, most data are presented in tables so that this chapter serves as a candid handbook. It includes data published mainly since the 1970s.

II. WHOLE GRAIN LIPIDS

A. Free and Bound Lipids of Nonstarch Lipids

Both lipid content and composition depend largely on extraction and purification procedures (extractant, extraction time and temperature, apparatus, the ratio of solvent to solute, purifying method, etc.) and to a lesser extent on samples (particle size, moisture content, varietal and class differences, the growing conditions of a given grain, etc.). Therefore, it is extremely difficult to compare lipid content or composition data reported by various researchers.

The terminology used to describe lipid preparations should be defined. For definition of abbreviations used throughout this chapter, see Table 1. Free lipids (FL) are the portion easily extractable with nonpolar solvents such as petroleum ether, hexane, diethylether, etc. by a Soxhlet, a Goldfish extractor, or shaking. Bound lipids (BL) are extracted from the FL–removed residue at room temperature with more polar solvents, generally alcohol alone or mixed with a small portion of another solvent, most commonly with water. Water–saturated

Mention of firm names or trade products does not constitute endorsement by the U.S. Department of Agriculture over others not mentioned.

n–butanol (WSB) is considered to be the most efficient solvent system, especially for wheat flour; a mixture of chloroform and methanol (2:1, 1:1, or 1:2 by volume) is also commonly used. The sum of FL and BL is termed nonstarch total lipids (NSTL). NSTL can also be obtained by the polar solvent extraction at room temperature without the FL extraction step. Therefore, BL or TL refer mainly to NSL for simplicity in this chapter. Lipids are generally reported in the literature as "crude fat," determined by ether extraction. These data are most available for whole grains, starchy endosperm, and other structural portions. They are equivalent to FL contents. FL contents are more readily comparable among the various laboratories than those of BL or NSTL. Any polar solvent, especially an alcohol–water mixture, extracts nonlipid materials in substantial amounts, and a purification procedure is required to remove lipids. Therefore, the reported quantities of BL or NSTL for a given grain are more difficult to compare between laboratories, since some are purified lipid contents and others are not, and the purified lipid content depends on the purification method.

Starch lipids (SL) are those bound to starch and most difficult to extract. Since true SL are present inside the starch granules, even a very polar solvent such as WSB cannot extract them at ambient temperature. Efficient extraction of SL requires mixtures of hot aqueous alcohols in proportions optimized for controlled swelling of starch granules and solubilization of lipids (23). The best solvents are *n*–propanol or isopropanol with water (3:1, by volume) used under nitrogen at 100 °C, but some *n*–butanol–water and methanol–water mixtures are also reasonably efficient at 100 °C (24). Recently, a third category was introduced: starch surface lipids (SSL) are some portions of NSL which become firmly absorbed onto or into starch granules during the separation of pure starch (23).

Data in Table 2 expressed on a dry basis were calculated from reported values (3,15,25–38). Also, some BL or TL contents were calculated by subtracting FL from TL or adding FL to BL, depending on availability of data. Lipids are a minor component of cereal grains (Table 2). FL contents range from 1.5 to 2% of kernel weights of barley, rice, rye, triticale, and wheat grains. They range from 3 to 7% of oats, millets, corn, and sorghum. However, BL contents are more uniform than FL contents in grains. Therefore, the FL/BL ratio is substantially higher for corn, millets, oats, and sorghum than for rye, triticale, and wheat grains. The FL/BL ratio was intermediate for barley and rice.

High oil–containing grains such as corn are continuously bred for high oil lines with improved production yield. Application of wide–line NMR spectroscopy for nondestructive analysis of oil content in single corn kernels made selection for higher oil content more efficient (39). Selections made for oil content and yield produced hybrids with 6–8.5% oil content and grain yields equal to those of good commercial hybrids (40).

Several kinds of millets exist, and the lipid data in literature are confusing. Rooney (15) compared FL (ether extraction) contents of several types of millets in a review paper. The average FL contents of pearl millet (*Pennisetum typhoides*) were 5.1% (4.1–5.6%, 14 samples), 5.4% (2.8–8.0%, 167 samples, Freeman and Bocan [41]), 5.6% (4.3–7.1%, 40 samples), and 6.2% (4.2–7.4%, 35 samples) (15). Other reported average FL contents were 4.8% (4.6–5.0%, 6 samples) for foxtail millet (*Setaria italica*), 5.8% (5.5–6.3%, 6 samples) for Japanese millet (*Echinochloa crusgalli*), and 4.2% (3.8–4.9%, 20 samples) for proso millet (*Panicum miliaceum*) (15). Taira (42) found slightly higher average FL (ether extraction) contents for glutinous foxtail millets (4.2–5.1%, average 4.7% of 21 samples) than for nonglutinous foxtail millets (4.0–4.7%, average 4.4% of 31 samples). Among the millets, pearl millet contains the most FL.

Lipid contents of rice in Table 2 were cited by Morrison (3) using the data of Nechaev and Sandler (2). Taira and Chang (43) reported the average nonglutinous brown rice FL (ether extraction) contents of 20 varieties of each Indica and Japonica types as 2.7% (2.38–2.91%) and 2.9% (2.54–3.58%), respectively. More recently, Taira et al. (44) reported the average FL contents of 15 nonglutinous varieties as 2.5% (2.24–2.97%) for Indica, 2.5% (2.12–2.94%) for Japonica, 2.7% (2.35–3.03%) for Sinica, and 2.6% (2.11–2.99%) for Japonica types.

B. Nonstarch Lipid Classes of Grains

Lipids can be separated into three broad classes by open column silicic acid chromatography. Nonpolar lipids (NL) are first eluted by chloroform, glycolipids (GL) by acetone, and last phospholipids (PL) by methanol. The mixture of GL and PL are polar lipids (PoL). After NL elution from a silicic acid column, PoL can be eluted with methanol without the GL elution step. Lipids can also be separated into various classes by thin–layer chromatography (TLC) using different developing solvent systems. Each individual lipid class migrates on the thin–layer plate differently, and the difference in migration rates makes it possible for complex lipids to separate into classes. The NL consists of SE, TG, DG, MG, and FFA (see Table 1). The total NL content is obtained by adding those NL class contents measured by densitometry. Thus, the NL content of a sample may differ, to some extent, depending on methodology used (column separation or TLC separation).

Data (1,12,26,28,33–35,37,45–53) in Table 3 may be compared for different grains only approximately, because some data were obtained by column chromatography and some by TLC. All cereal grain lipids are richer in NL than in other classes: 60–70% of TL are NL in wheat (hexaploid), triticale, and rye; 65–80% for barley and oat groats; 77–87% for sorghum and rice; and above 90% for corn and millet lipids (Table 3). Thus, the NL/PoL ratios are lowest for wheat, triticale, and rye; slightly higher for barley and oat groats; higher for sorghum and rice; and highest for corn and millet. There are significant varietal effects on the NL/PoL ratio for corn and especially for millet grains. Among wheats, tetraploid durum wheat NSTL shows the highest NL/PoL ratio.

Among all grains, wheat is richest in GL, followed by triticale, rye, and barley. Among wheat flour lipids, GL are the best as bread loaf volume improvers (17–19). Millet, corn, and sorghum lipids are poorest in GL. In general, PL are richest also in wheat, triticale, and rye lipids and slightly lower in barley, oat groats, sorghum, and rice. Although PL contents are higher than GL contents in both corn and millet, their lipids are poorest in PL among all grain lipids.

C. Fatty Acid Composition of Grain Lipids

All cereal grain lipids are rich in unsaturated fatty acids (FA) (Table 4). Palmitic acid (16:0) is a major saturated FA, and linoleic acid (18:2) is a major unsaturated FA for all cereals except for brown rice. For brown rice, oleic acid (18:1) is a major unsaturated FA. The presence of palmitoleic acid (16:1) and eicosenoic acid (20:1) is reported quite often but usually at levels below 1% of total FA composition.

Fatty acid compositions are generally similar for barley, rye, triticale, and wheat lipids. Rye lipids are somewhat higher in linolenic acid (18:3) than those of other cereals. Oat lipid FA composition is similar to that of brown rice, because oats and brown rice are rich in oleic acid. Millet lipids are generally higher in stearic acid (18:0) than all other cereal lipids. There are greater ranges in FA compositions of corn oils: Jellum (56) reported a range of 14–64% oleic acid and 19–71% linoleic acid for the world collection of 788 varieties of corn. The wide ranges in FA composition in corn were due to more lines having been examined in corn than in any of the other cereal grains (1).

Fatty acid composition differs depending on the lipid extractant. For example, FL were higher in both oleic and linoleic acids than BL of corn and pearl millet, whereas FL were lower in palmitic acid than BL of corn and oats (Table 4). Corn FA composition of NSTL is intermediate to those of FL and BL based on data compiled by Morrison (3).

Wheat lipid FA compositions are shown for wheat classes or subclasses in Table 5. The average of six HWW wheat and 14 SWS wheat lipids was lower in palmitic and stearic acids and higher in linoleic and linolenic acid than the overall average of 290 wheat lipids. The average FA compositions of 103 HRW and 46 HRS wheat lipids were very similar. According to Davis et al. (62), the average of 17 durum wheat lipids was higher in palmitic acid and lower in linoleic acid than overall wheat lipids. However, the range of four durum wheat lipid FA compositions reported by Youngs (63) is in agreement with the 290 wheat lipids reported by Davis et al. (62), except that durum wheat lipids were higher in linolenic acid.

III. NONSAPONIFIABLE LIPIDS IN GRAINS

Major cereal grain lipids are acyl lipids, which consists of FFA and FA esterified to alcohols, primarily glycerol or its derivatives. Other alcohols such as sterols, carotenoids, and tocopherols may be esterified with FA and, thus, these esters can be considered as acyl lipids (5). However, the nonacyl parts of these lipids often occur in the unesterified form. Nonsaponifiable lipids in cereals have been reviewed by Barnes (5), who included the sterols, triterpenols, carotenoids, and tocopherols together with the hydrocarbons.

A. Tocol Derivatives

Tocol derivatives (tocopherols and tocotrienols) are responsible for the vitamin E activity of plant tissues. Various combinations of all eight tocol derivatives are found among cereal grains (Table 6). In general, γ–T and γ–T–3 are not detectable or are present in low concentration in grains (e.g., wheat, rye, and triticale) which contain significant proportions of β–T and β–T–3. Conversely, the β–tocol derivatives are found in insignificant quantities when the corresponding γ–tocol derivatives are present (e.g., corn, millet, and rice). Barley grain is an exception and contains both β– and γ–tocol derivatives. It may be possible that β–tocopherol and β– tocotrienol (5,8–dimethyl derivatives) and γ–tocopherol and γ–tocotrienol (7,8–dimethyl derivatives) represent components of alternative biosynthetic routes to tocopherols and tocotrienols (5).

Total tocol derivative contents of grains and their compositions vary among grains. Barley and oat grains contain all four tocopherols and four tocotrienols, whereas rye, sorghum, and triticale contain nearly no γ–T–3, δ–T, and δ–T–3 (Table 6). Only the α– and β–tocol derivatives have been detected in wheats, but there are two reports concerning the presence of relatively large portions of γ– and δ–tocol derivatives (5).

Tocol derivative composition varies among maize inbreds (8) and wheat classes (62). Total tocol derivative content ranges from 2.6 to 10.2 mg/100 g corn, and α–T content ranges from 2.1 to 44.8% of total tocol derivative (Table 7). Among hard wheats, white wheats (HWW and HWS) contain more tocopherols than red wheats (HRW and HRS). White wheat tocopherols are richer in δ–T and poorer in α–T than red wheat tocopherols. Durum wheats contain the smallest amount of total tocopherols, and their tocopherol composition is most similar to that of HRW wheats.

Among the structural parts of grains, tocol derivatives are most abundant in the germ and least in the endosperm portions of both corn and wheat grains (Table 8).

B. Carotenoids

Carotenoids are polyisoprenoid compounds. Carotenoid hydrocarbons are known as carotenes and are the biosynthetic precursors of the oxygenated derivatives called "xanthophylls." The characteristic properties of the carotenoids reflect the presence of the conjugated polyene chain, which accounts for the color and sensitivity to light, heat, oxygen, and acid (5). Color due to carotenoids is an important factor in the use of cereal grains in food production, particularly in durum wheat used to make pasta. Certain carotenoids are precursors of vitamin A (retinol). Although they have no intrinsic activity, α–, β–, and γ–carotene are converted into vitamin A in the intestinal mucosa and liver (5).

Carotenoids are a very minor constituent in cereal grains (Table 9). They are most abundant in corn, yellow sorghum, and millet grains. There are generally more carotenoids in durum wheats than in common bread wheats. Lutein is the major carotenoid of wheat grain, although many carotenoids appear to be present. It has been detected in most studies of the carotenoids of other cereal grains. β–Carotene is also commonly found as a minor component, except in lipids extracted from barley grain, where it is present as a major component together with xanthophylls.

The carotenoid contents of five Orenburg barley varieties examined by Demchenko (71) were:

Carotenoid	*mg/kg Oil*
α –Carotene	2.8–4.0
β –Carotene	52–85
β –Carotene isomer	0.9–2.0
Xanthophylls	46–65

Ten carotenoid components are present in corn grain with zeaxanthin and lutein (xanthophylls) as major components (Table 10).

Carotenoids are not evenly distributed in corn kernel fractions (Table 11). They are abundant in endosperm, especially horny (vitreous) endosperm fractions obtained by hand–dissection: 74–86% of total carotenoids were distributed in horny endosperm (top portion, Table 11). Blessin et al. (74) further compared the carotenoid contents in the hand–dissected fractions with those in the dry–milled fractions. Fractionation was more crude using a sample from a commercial dry–milling operation than by hand–dissection. Thus, xanthophyll contents were substantially higher in bran and germ fractions from dry milling than in those obtained from hand–dissection (bottom portion, Table 11).

Chen and Geddes (75) determined chromatographically the carotene, free xanthophyll, and xanthophyll esters in various wheats (Table 12). They reported total carotenoid contents as being highest in soft white and lowest in HRS wheats. Though the total carotenoid content in durum wheat was comparable to those in other hexaploid wheats tested by Chen and Geddes (75), durum wheats are generally higher in carotenoid content than bread wheats. Lepage and Sims (76) reported that a durum (Mindum) flour and HRS (Thatcher) wheat flour contained total carotenoid contents of 3.7 and 2.8 mg/kg, respectively (Table 12). The carotenoid compositions in both durum wheat and flour were substantially different from those of three other classes of wheats; nearly 85% of the carotenoid was free xanthophyll (Table 12).

In Australian bread wheat flours, the carotenoid content varied from 1.5 to 4.3 mg/kg (77), whereas for the semolina from eight American durum wheats, it ranged from 4.4 to 6.9 mg/kg (78). Recently, the Canadian Grain Commission (79) reported that a concentration of

5.5 to 8.4 mg/kg, expressed as β – carotene on an as–is basis, was obtained from 10,608 durum samples examined by the Grain Research Laboratory in Winnipeg between 1972 and 1980 (80). Hard wheats have a much lower pigment content.

The carotenoids are not homogeneously distributed in the wheat kernel: the bran contains 0.93–2.22 mg/kg; the germ, 4.13–7.84 mg/kg; and the endosperm, 1.57–2.00 mg/kg (Table 13). The carotenoid composition differs among wheat classes and wheat fractions within a given class of wheat.

The variety and the growing location have highly significant effects on carotenoid contents of durum wheats (77,78). The carotenoid content varied from 3.3 to 10.1 mg/kg among 15 semolinas from durum wheats grown in France, and there was a 24 to 75% loss in carotenoid content during pasta–making (77). Also, during milling into semolina, large amounts of carotenoids are lost and can be found in the by–products (75,80).

Wall and Blessin (81,82) reported zeaxanthin and lutein as the major carotenoid components in yellow sorghum grain (Table 14).

C. Sterols

There are three classes of sterols: 4–demethylsterols, 4–monomethylsterols, and 4,4–dimethylsterols including triterpene alcohols. The major class is 4–demethylsterols, which are generally called sterols.

β –Sitosterol is the primary sterol in all cereal grains (Tables 15, 16) and their structural parts (Table 15, 17). The second major sterol is campesterol in corn and rice kernels and their parts (Table 15) and in rye and wheats (Table 16). However, in both common and wild oats the second dominant sterol is Δ^5–avenasterol (Table 15), and in sorghum it is stigmasterol (Table 16).

Compositions of sterols in dissected wheat and sorghum grain fractions are given in Table 17. Torres et al. (90) reported stigmasterol as the second major sterol in wheat endosperm and also a higher cholesterol content (Table 17) than reported by others (83,88,89) in Table 16. Palmer and Bowder (91) reported campesterol as the second major sterol in both endosperm and embryo fractions of sorghum (Table 17). Sorghum grain contained 394.0 mg 4–demethylsterols, 12.54 mg 4–monomethylsterols, and 65.6 mg triterpenes in 1 kg (on dry–weight basis) of grain (91). 4–Demethylsterols were found free (86% in S), esterified (13% in SE), and glycosidated (1% in SG). 4–Monomethylsterols were completely free (100% in S), whereas triterpenes were completely esterified (100% in SE). Three quarters of the total sorghum sterols were located in the embryo, whereas the triterpenes were equally distributed between the embryo and the endosperm of sorghum (91).

Sterol lipids can be found in four forms in cereal grains: free sterols (S), sterol esters (SE), sterol glycosides (SG), and steryl glycoside esters (ASG). β –Sitosterol is the major sterol in the sterol lipids of the wheat kernel (92), aleurone fraction (93), and milled flour (94) (Table 18). In milled wheat flour, 5 α –dihydrositosterol was reported to be high, as was campesterol (94).

Ohnishi et al. (92) reported that the relative proportions of wheat kernel sterols were 85% 4–demethylsterols, 3% 4–monomethylsterols, and 12% triterpenols including 4,4–dimethylsterols in SE and similar proportions in S. The composition of 4–monomethylsterols was 38.5 and 35.2% gramisterol in SE and S, respectively, 25.8 and 18.9% citrostadienol, 11.0 and 20.0 obtusifoliol, 2.8 and 0.47% lophenol, and 21.9 and 25.5% unknown sterol components (92). The composition of 4,4–dimethylsterols was 66.5 and 32.1% β –amyrin, respectively, in SE and S, 32.5 and 21.0% α –amyrin, 1.0 and 8.2% cyclobranol in SE and S, and 9.2% cycloartanol, 19.0% 24– methylenecycloartanol, and 10.5% unknown component in S (92).

Table 19 shows the sterol composition of rice bran oil (95,96) and its sterol lipids (97). The sterol composition of rice bran oil was reported by Itoh et al. (95) to be 53% 4–demethylsterols, 12% 4–monomethylsterols, and 35% triterpene alcohols including 4,4–dimethylsterols, which is substantially higher than those in corn oil and wheat germ oil (95,96,98).

Kuroda et al. (97) analyzed SE, S, SG, and ASG of bran separately (Table 19). 4–Methylsterols and triterpene alcohols with 4,4,–dimethylsterols were found along with the 4–demethylsterols in SE and S but not in SG or ASG. The principal FA components of SE were linoleic (58.3%), oleic (30.4%), and palmitic (7.4%) acids, whereas those of ASG were linoleic (42.5%), palmitic (29.9%), and oleic (22.7%) acids (97). The principal 4–demethylsterols of all four sterol lipids (SE, S, SG, and ASG) and bran oil were β–sitosterol, campesterol, and stigmasterol (Table 19). The principal 4–monomethylsterols of bran oil and sterol lipids (SE and S) were gramisterol and citrostadienol, and the principal 4,4–dimethylsterols were 24–methylenecycloartanol and cycloartenol.

Mahadevappa and Raina (98) reported the total sterol lipid content as 149 mg in 100 g finger millet including 13 mg SE, 91 mg S, 25 mg SG, and 20 mg ASG. The major FA, totaling 85–90%, were the same in both esterified sterols, but the proportions varied: palmitic, oleic, and linoleic acids comprised 24, 49, and 17% in SE, and 43, 36, and 7% in ASG. All four sterol lipids in finger millet contained 80–84% β–sitosterol, the remainder being stigmasterol (98).

In both corn oil and wheat germ oil, 4–demethylsterols are 87–98% of total sterols (Table 20). The 4–demethylsterol contents were 1441 and 1425 mg in 100 g of corn oil and wheat germ oil, respectively (99). The major 4– demethylsterols are β–sitosterol and campesterol in both corn oil and wheat germ oil. The major 4–monomethylsterols are gramisterol and citrostadienol. In addition, obtusifoliol is another major component in corn oil. The major 4,4– dimethylsterols are 24–methylenecycloartanol and cycloartenol in corn and wheat germ oils. A substantial amount of β–amyrin is present in wheat germ oil (Table 20).

IV. LIPIDS IN STRUCTURAL PARTS OF GRAINS

A. Lipid Contents in Various Grain Fractions

Endosperms are the major fractions of all cereal grains, and yet their lipid contents are significantly lower than germ and aleurone fractions (Table 21). Germs are the richest source of lipids among all cereal grain fractions, even though germs are relatively small fractions of grain kernels. The weight percentage of germ is 10–14% of corn, 8–12% of sorghum, 7% of oats, and 1–2% of rice and wheat kernel weights.

Lipids were unevenly distributed in grain fractions, and the lipid distribution differed among grains (Table 22). In corn kernels, 73–85% of corn lipids were distributed in the germ fractions (105,106), whereas in rye, triticale, and wheat kernels, 34–42% of the grain lipids were in the germ fractions (60). Price and Parsons (107) reported that the hulless barley (Prilar) and the hulless oat (James) lipids were distributed mainly in the bran–endosperm fractions (Table 22).

The lipid distribution is quite similar despite the differences in the four maize strains. H51 is an inbred; LG–11 is a three–way cross hybrid forage corn. Both the waxy maize and amylomaize are endosperm mutants, and the amylomaize is also a higher oil strain (8).

Among oat groat fractions, FL and TL were richest in the scutellum and BL were richest in embryonic axis (Table 23). Both red and white proso millet fractions contained similar lipid contents except for the bran FL contents being somewhat higher in white than those in red proso millets (30).

B. Lipid Compositions in Various Grain Fractions

Since the cereal lipid compositions are too complex to compare for all grains in one section, each will be discussed separately.

1. Barley

The average compositions of NL and PL for the two varieties, Kearney (winter type) and Prilar (spring type), are given in Table 24. In barley grain, NL are the major class of NSTL like other cereal grains (Table 3) and over one half of NL are TG (Table 24). The NL also contains 9.8% free sterols, 4.4% SE, and 5.7% HC (108). The two major classes of PL are PC and LPC (Table 24). The FA composition varies among lipid classes. The major FA is 18:2 for all classes except for PG and PA. The "others" in Table 24 include relatively small quantities of the other minor FA (12:0, 14:0, 16:1, and 20:0) (108,109).

The NSL contents and compositions in hulless barley (Prilar) fractions and their FA compositions of NL, GL, and PL are given in Table 25. Within a given lipid class, the FA composition differs depending on the structural parts of barley grain kernel: the hull lipids are more saturated than those in the other two fractions, and the NSL in bran–endosperm fraction are richer in 18:2 than those in the embryonic axis and hull fractions (Table 25). In addition to the major FA listed in Table 25, Price and Parsons (107) also reported the other minor FA (12:0, 14:0, 16:1, and 20:0).

Holmer et al. (110) studied the changes in lipid composition of germinating barley (Kenia) embryo (Table 26). Barley seeds were dissected before and after 5 days of germination, to distinguish between the scutellum, the coleoptile half of the embryo, and the coleorhiza half of the embryo. In tissues from the coleoptile and coleorhiza halves of the embryo, there was a concurrent decrease of TG with a marked increase in SE and ASG. In the scutellum, there was also a change in the TG, but the variations in both SE and ASG contents were much smaller (Table 26). MG and DG were virtually absent from embryonic tissue. The amounts of FA (18:2 and 18:3) increased after 5 days of germination in all the embryonic tissues, especially in the coleoptile half (110).

2. Oat Groats

The NSTL contents and compositions of hulless oat (James) grain fractions and their FA compositions are shown in Table 25: in addition to the FA listed, there were other minor FA (12:0, 14:0, 16:1, and 20:0) reported by Price and Parsons (107).

Youngs et al. (32) studied the NSL compositions of the two oat cultivars (Dal and Froker) grown in 2 years. The ranges of lipid compositions are given for groats, and the averages across years and cultivars are given for the groat fractions (Table 27). The most abundant class was TG, which averaged 41% of the NSTL in groats, and 39–58% in the four groat fractions. The net abundant component was DGDG: about 7% in the groats, and 8% in the bran and endosperm.

Sahasrabudhe (33) studied 12 oat cultivars grown at Ottawa, Canada, for the protein and lipid contents and their groat FA compositions. The NL compositional study was conducted for the six cultivars, and the GL and PL were characterized for only one cultivar (Hinoat) as shown in Table 28. Oats have shown a potential as a source of edible oil. There is a wide range (4–12%) of lipid contents in oats (Tables 3 and 28). The NSTL content extracted with chloroform–methanol (2:1 by volume) showed a significant correlation (r = 0.99) with the NL content. It indicates that the NSTL contents in oat groats are under genetic control as those in millet and corn.

3. Millet

The compositions of FL and BL of millet are shown in Table 29. FL contains approximately double TG and only half of PoL compared to BL, reported by Nechaev and Sandler (2).

NSTL are commonly extracted by a chloroform–methanol (C–M) mixture (2:1 by volume) or WSB. Salun and Kalugina (111) showed the average composition of two varieties. The WSB–extracted NSTL contain significantly more PoL. Osagie and Kates (28) extracted TL (NSL+SL) from millet seeds (*P. americanum*) grown in Zaria, northern Nigeria, with hot WSB for 1 hour. Thus, TL may contain all NSL and only a portion of SL, because a complete extraction of SL may take 5–6 hr at 100 °C (112). The similar PoL contents in FL by Nechaev and Sandler (2) and TL by Osagie and Kates (28) may be mainly due to the difference in millet varieties. With a given sample, hot WSB will extract the most PoL.

Among millet, GL, SG, and DGDG were the major classes, and LPC was the major PL class (Table 30). LPC content was 4.9% of TL and about 42% of PL; a substantial quantity of LPC, a major SL component, must have been extracted with hot WSB from the millet starch granules.

The FA profiles were similar among NL classes except for FFA. Cerebrosides (I and II) showed different FA compositions from the other GL classes. Substantially different FA compositions were shown for the PG and LPC from the rest of other PL classes. The "others" in Table 30 included other minor FA (14:0, 20:0, and 22:0). In general, millet PL contained less 18:0 and 18:3 than millet GL.

The lipid content of corn ranges very widely and is primarily determined by genetic selection like that of oats.

Tan and Morrison (106) studied the distribution of lipids in the germ, endosperm, pericarp, and tip cap of amylomaize, LG–11 hybrid maize, and waxy maize. The quantitative distribution of 23 acyl lipid classes and unsaponifiable matter in kernels of the three varieties were investigated. The lipid compositions of LG–11 hybrid corn are given in Table 31.

Corn lipids are extremely low in GL (Table 31) like millet and sorghum lipids (Table 3). The distribution of corn kernel lipids is about 80% in germ (Table 22). Germ lipids are mostly TG, with some SE and DG, and small amounts of GL and PL. The NSTL composition differed markedly in pericarp from those in the other kernel fractions: over half of pericarp lipids were unsaponifiable matters. Tip cap lipids had more TG, GL, and PL than pericarp lipids, but were otherwise similar.

The compositions of NL, GL, and PL were separately computed (Table 32). TG was over 90% NL in germ (106, 113), about 60% in endosperm NSL, but only 2.5% in endosperm SL. Over 90% NL was FFA in SL. Weber (113) detected substantial quantities of CB and sulfolipids (tentative identification) in GL of germ and endosperm NSL.

The major component in germ PL was PC, which was in good agreement between Tan and Morrison (106) and Weber (113). However, there were large differences in PL composition of the endosperm NSL; Tan and Morrison (106) reported 11.1% PC and 57.1% LPC, whereas Weber (113) reported 44.6% PC and 36.5% LPC plus unknown.

The FA compositions of NL, GL, and PL were higher levels of 18:0 and 18:3 (Table 33). For the LG–11 hybrid corn, germ lipids contained significantly more 18:2 and less 16:0 and 18:3 than other parts of kernel (106). For the H–51 inbred corn, germ lipids contained less 18:3 than other parts of kernel and more 18:3 than pericarp and tip cap. However, the 18:2 content was equal for both germ and endosperm lipids (39). The FA compositions in root and leaf lipids differ significantly from those of corn kernel or other kernel parts; corn leaf lipids contained a much higher level of 18:3 and lower levels of 18:1 and 18:2 (Table 33).

5. Rice

Rice hull lipid composition differs significantly from that in brown rice and its fractions (Table 34). Silicic acid fractionation of NSL of brown rice, bran, germ, and polish showed 86–91% NL, 2–5% GL, and 7–9% PL (13,51,114). Milled rice NSL had a lower NL fraction and a higher GL fraction. The ratios for the NL:GL:PL for milled rice are 82:8:10 by Choudhury and Juliano (51), 76:12:12 by Hirayama and Matsuda (50), and the range of (83–91):(2–4):(1–3) by Azudin and Morrison (115).

Azudin and Morrison (115) studied NSL and SL in milled rice of two waxy varieties (1.0–2.3% amylose) and 12 nonwaxy varieties (12.2–28.6% amylose). TL (NSL+SL) were extracted from rice flour and SL from purified rice starch; the composition of the NSL could be obtained by the difference, as shown in Table 34.

The major NL of NSL was TG, constituting 71–79% of NSTL (Table 34) and 83–87% of NL (51,114). The other important NL class was FFA, at 4–7% of NSTL and 13–17% of NL for brown rice, bran, germ, and polish. Unlike most other cereal NSL, the major GL of NSL of brown rice and its milling fractions were ASG and SG (Table 34). Major PL were PC and PE.

Choudhury and Juliano (51) reported that the distribution of brown rice NL was 14–18% in germ, 39–41% in bran, 15–21% in polish, and 25–33% in milled rice (12–14% in subaleurone layer and 12–19% in the endosperm). The distribution of the NSL of brown rice was 43% in bran, 19% in germ, 15% in polish, and 21% in milled rice. For GL, the distribution was 28% in bran, 7% in germ, 14% in polish, and 51% in milled rice; and for brown rice PL, 30% in bran, 14% each in germ and polish, and 42% in milled rice (51,114).

TL (NSL+SL) compositions are different between waxy and nonwaxy rice varieties (Table 35). Azudin and Morrison (115) reported that the two waxy rice (IR 29 and C441–4) starches prepared from the milled rice had very little amylose content (1.0–2.3%) and only traces of lipids (16–19 mg per 100 g starch), which were probably SSL, the NSL contaminants. The SSL were 100% FFA (Table 35). The TL in waxy rices were, therefore, NSL, and it is evident that they had suffered substantial lipolysis, judging by the high FFA values (115).

The nonwaxy starches contained 0.9–1.3% SL comprising, on average, 31.2% (29–45%) FFA, 61.5% (48–67%) LPL, and 3.2% GL (115), as shown in Table 35.

Choudhury and Juliano (51) extracted SL from milled rice after the NSL removal, using the one waxy variety (IR4445–63–1 with 2% amylose) and the two nonwaxy varieties (IR42 with 29% amylose and IR480–5–9 with 24% amylose). The SL composition of the milled rice of the waxy variety contained 41% PL and 7% GL, whereas the waxy starch by Azudin and Morrison (115) contained no GL and PL (Table 35). The differences in SL compositions of waxy rice and nonwaxy rice (both milled and brown) were different (51) but not to the extent shown by Azudin and Morrison (115).

The FA compositions of NSL and SL classes in the three brown rices and their milled fractions are given in Table 36 (13,51,114). The NSL in brown rice, bran, germ, and polish fractions showed similar FA compositions, which were substantially different from the SL in both brown and milled rices. The SL classes had similar FA compositions between brown and milled rices.

6. Rye

The NSL compositions of rye grain, its structural kernel fractions, and milled flour are shown in Table 37.

Zeringue and Feuge (60) dissected rye (Abruzzi) seeds into bran, germ, and endosperm fractions. NSTL were extracted from each fraction and whole kernel meals with 80% *n*–butanol. Eleven of the major lipid classes were measured densitometrically from the spots separated by TLC. The recovery % of those 11 classes ranged from 52 to 63%; the converted data on 100% recovery are shown in Table 37. The most abundant class was TG for rye grain and germ but not for bran and endosperm (60), whereas Chung and Tsen (34) reported that TG was the major class for milled rye flour lipids. Zeringue and Feuge (60) reported substantially higher DGDG levels in both grain and endosperm NSL than those in grains (Elbon and Balbo varieties) and their milled flours reported by Chung and Tsen (34). The endosperm NSL contained a much lower level of TG (17.7% of NSTL) and a higher level of PE than the rye flour NSL; such a large discrepancy may be, in part, due to the different varieties.

The NSL compositions of FL and BL from rye grains and their milled flour are given together with those from triticale grains and their milled triticale flours (Table 38). In rye grains as well as rye flours, FL contained higher levels of NL classes (except for SE) than BL, whereas BL contained higher levels of DGDG and PL. The amount of SE in BL seems very high. Since the lipids were measured by densitometry of charred spots on TLC plates, the SE band may have contained other material such as HC.

7. Triticale

Zeringue and Feuge (60) dissected triticale (AM 2147) seeds into bran, germ, and endosperm fractions. NSTL were extracted from those fractions in the same manner as done with the rye samples. Chung and Tsen (34) studied the four triticale varieties (Bronco, Armadillo, KS 385, and KS 419).

The NSL compositions of triticale grains reported by Zeringue and Feuge (60) and Chung and Tsen (34) were similar. Also, the NSL compositions were similar for the endosperm and milled triticale flours (Table 39). Triticale germ lipids were similar in compositions with rye germ lipids (Table 37). Triticale bran lipids contained a higher level of SE than rye bran lipids.

TG is a major component in FL of triticale grains and flours but only a minor component in BL (Table 38), which was the same for rye lipids. Triticale lipids are quantitatively similar to wheat and rye lipids analyzed under the same conditions (34).

8. Wheat

Among cereal grain lipids, the greatest research effort has been devoted to the study of lipids in wheat flour, but there are still some gaps in our knowledge, especially about the lipids in bran and germ (3). The most detailed and complete information was provided by Morrison (23), as shown in Table 40.

Wheat pericarp contains about 1.3% NSTL of its weight on dry basis; pericarp lipids consist of 86% NL, 7% GL, and 7% PL. Pericarp NL contain 40% DG, 25% FFA, 18% TG, 15% SE, and 1% MG.

Wheat does not exhibit the large range of lipid content and composition seen in some other cereals (3), although hexaploid wheat endosperm contained more NSL than tetraploid wheat endosperm (Table 40). Tetraploid wheat endosperm contained higher levels of TG, PG, and PE but lower levels of GL, especially DGDG, and APE than hexaploid wheat endosperm.

Wheat TL (NSL+SL) contains a substantially higher level of PL than NSL in endosperm and milled flour due to SL being very rich in PL (90–94% SL), especially LPL (Table 41).

Wheat germ oil contains very little GL (Table 41), which makes it similar to the amount of germ oils in barley (Table 26), corn (Tables 31 and 33), rice (Table 34), rye (Table 37), and triticale (Table 39). Aleurone lipids contain 7–10% GL with ASG, DGMG and DGDG as major classes.

The NSL composition of wheat endosperm differs substantially from that of milled flour mainly due to an increase in TG for milled flour. Morrison et al. (67) studied the distribution of acyl lipids in wheat flour millstreams and concluded that roughly three–fourths germ lipids and one–fourth aleurone lipids were transferred to endosperm material during milling process.

Endosperm SL contains over 90% PL, 4–6% NL, and 1–6% GL (Table 41). Therefore, SL composition is very different from that of NSL in any wheat fractions.

Whole wheat contains 2.5–3.3% TL (NSL+ SL), milled flour contains 1.7–2.0% NSTL, and wheat starch contains 0.8–1.2% SL (Table 42). The NSL contents in flours milled from high–grade spring, high–grade winter, and low–grade winter wheats are compared. The winter wheat flours contained higher levels of NSTL, TG, DGDG, and PC than the spring wheat flour.

SL in 100 g starch contained 35–62 mg NL, 9–54 mg GL, and 729–1047 mg PL (Table 42). Morrison (23) defined NL and GL of SL as nonstarch and starch surface lipids (SSL) and only LPL as true SL. Wheat starch LPL have a fairly uniform composition; average values are LPC = 86%, LPE = 10%, and LPG = 4% (23).

The lipid composition of starch is affected by the method used to isolate the starch and by its final purity. Crude starch contaminated with adhering proteins contains small amounts of nearly all the lipids classified as NSL. When the adhering proteins are removed, the starch contains very little TG or diacyl lipids such as DG, MGDG, DG DG, APE, PC, and PE (3), as shown in Table 42. Then, the SL consists of 90–94% LPL with 6–10% FFA and other monoacyl lipids (3,116–119).

The FA compositions of total TG, DG, and MG are very similar (Table 43). However, it is misleading when the stereospecific distributions of the FA are considered. The FA in SE and ASG are similar and generally segregate into more unsaturated or more saturated types (Table 43) according to the source of the lipid (3).

MGDG is the most unsaturated lipid in wheat, with only trace amounts of saturated acids at the 2–position. DGDG is similar in this respect but has much more saturated acids in the 1–position (Table 43). The *N*–acyl FA of APE and ALPE are less saturated than the *O*–acyl FA. The FA compositions of PE and PC are similar. The FA in LPE and LPC from the NSL are identical to the 1–position FA in PE and PC, respectively.

Starch LPE is much more unsaturated than nonstarch LPE and PE. Starch LPC is also slightly more unsaturated than nonstarch LPC (Table 43).

TABLE 1 Abbreviations

Grouping	Abbreviations	Definitions
General term	BL	Bound lipids
	CB	Cerebrosides
	db	Dry basis
	FA	Fatty acids
	FL	Free lipids
	GL	Glycolipids
	HC	Hydrocarbons
	LPL	Lysophospholipids
	NL	Nonpolar lipids
	NSL	Nonstarch lipids
	NSTL	Nonstarch total lipids
	PL	Phospholipids
	PoL	Polar lipids
	SL	Starch lipids
	SSL	Starch surface lipids
	TLC	Thin-layer chromatography
	TL	Total lipids
	WSB	Water-saturated *n*-butanol
	wt	Weight
Wheat class	HRS	Hard red spring
	HRW	Hard red winter
	HWS	Hard white spring
	HWW	Hard white winter
	SRW	Soft red winter
	SWS	Soft white spring
	SWW	Soft white winter
Lipid class		
Fatty acids	FFA	Free fatty acids
	12:0	Lauric acid
	14:0	Myristic acid
	16:0	Palmitic acid
	16:1	Palmitoleic acid
	18:0	Stearic acid
	18:1	Oleic acid
	18:2	Linoleic acid
	18:3	Linolenic acid
	20:0	Arachidic acid
	22:0	Behenic acid
Simple glycerides	MG	Monoglycerides
	DG	Diglycerides
	TG	Triglycerides

TABLE 1 (Continued)

Grouping	Abbreviations	Definitions
Glycosylglycerides	MGMG	Monogalactosylmonoglycerides
	DGMG	Digalactosylmonoglycerides
	MGDG	Monogalactosyldiglycerides
	DGDG	Digalactosyldiglycerides
	AMGMG	Acylmonogalactosylmonoglycerides
	AMGDG	Acylmonogalactosyldiglycerides
	ADGDG	Acyldigalactosyldiglycerides
Phosphoglycerides	PC	Phosphatidylcholine
	PE	Phosphatidylethanolamine
	PI	Phosphatidylinositol
	PG	Phosphatidylglycerol
	PS	Phosphatidylserine
	DPG	Diphsphatidylglycerol
	PA	Phosphatidic acid
	LPC	Lysophosphatidylcholine
	LPE	Lysophosphatidylethanolamine
	LPG	Lysophosphatidylglycerol
	LPI	Lysophosphatidylinositol
	LPS	Lysophosphatidylserine
	APE	*N*–Acylphosphatidylethanolamine
	ALPE	*N*–Acyllysophosphatidylethanolamine
Sterol lipids	S	Sterol
	SE	Steryl esters
	SG	Sterylglucoside
	ASG	Acylsterylglucoside
Tocol derivatives	α–T	α–Tocopherol
	α–T–3	α–Tocotrienol
	β–T	β–Tocopherol
	β–T–3	β–Tocotrienol
	γ–T	γ–Tocopherol
	γ–T–3	γ–Tocotrienol
	δ–T	δ–Tocopherol
	δ–T–3	δ–Tocotrienol

TABLE 2 Nonstarch Lipids of Cereal Grains

Grains	Nonstarch lipids[a] Free (%)	Bound (%)	Total (%)	Ratio of FL/BL[b]	Ref.
Barley	1.7–1.9 (1.8)	1.5–1.7 (1.6)	3.2–3.4 (3.3)	2.7–3.0 (2.9)	25
	1.8–3.9 (2.9)	0.7–0.8 (0.8)	2.5–4.7 (3.6)	2.6–4.9 (3.8)	26
Corn	4.9	0.2	5.1	24.5	15
	4.6–5.6 (4.8)	0.4–0.9 (0.6)	5.2–6.0 (5.6)	6.2–11.5 (8.0)	3
Millet	3.9–4.3 (4.1)	0.4–0.7 (0.5)	4.3–5.0 (4.6)	6.1–9.8 (8.2)	3
	5.0	0.5	5.5	10.0	27
	5.4	0.7	6.1	7.7	28
Pearl millet	5.6–7.1 (6.3)	0.6–0.9 (0.7)	6.2–7.9 (7.0)	7.2–11.0 (8.9)	29
Proso millet	3.3–3.9 (3.7)	0.5–1.3 (0.8)	3.9–4.9 (4.5)	2.8–7.1 (4.6)	30
Oats	4.4–7.0 (5.7)	0.5–0.9 (0.7)	4.9–7.9 (6.4)	7.7–8.8 (8.1)	3
	3.2–8.9 (6.3)	0.4–1.7 (1.2)	4.5–10.3 (7.5)	5.2–8.0 (5.3)	31
	5.5–8.0 (6.8)	1.4–1.6 (1.5)	6.9–9.6 (8.3)	3.9–5.0 (4.5)	32
	5.6	1.4	6.9	4.0	33
Rice	2.1–2.6 (2.3)	0.5–0.6 (0.5)	0.9–3.1 (2.9)	4.2–5.2 (4.8)	3
Rye	1.9–2.0 (1.9)	1.5–1.6 (1.6)	3.5–3.5 (3.5)	1.1–1.3 (1.2)	34
Sorghum	3.5–5.2 (4.2)	0.2–0.8 (0.4)	3.7–6.0 (4.6)	(10.5)	3
	2.8–4.4 (3.6)	0.6–0.8 (0.7)	3.6–5.0 (4.3)	(5.8)	35
	(3.7)	(0.2)	(3.9)	(16.7)	15
Triticale	1.7–2.2 (1.9)	1.5–2.6 (2.0)	3.2–4.6 (3.9)	0.8–1.2 (1.0)	34
	1.9	0.7	2.6	2.7	36
Wheat	1.4–2.6	0.7–1.2	2.1–3.8		3
HRW	1.3–1.7 (1.5)	0.8–1.0 (0.9)	2.2–2.6 (2.4)	1.4–2.1 (1.6)	37
	(2.0)	(1.5)	3.4–3.5 (3.5)	1.3–1.4 (1.4)	34
	1.6–2.6 (2.1)	1.0–1.2 (1.1)	2.7–3.9 (3.2)	1.5–2.3 (1.9)	38
HRS	1.6–1.9 (1.8)	0.8–1.0 (0.9)	2.4–2.9 (2.7)	1.8–2.1 (1.9)	37
	1.2–1.9 (1.6)	0.6–1.0 (0.8)	1.2–3.1 (2.2)	2.2–2.5 (2.4)	38
SRW	1.6	0.6	2.2	3.0	37
Soft White	1.7	0.9	2.6	1.8	37
Durum	2.1	1.2	3.3	1.8	37
	1.9–2.0 (2.0)	0.5–0.6 (0.6)	2.5–2.6 (2.5)	3.3–3.8 (3.6)	38

[a]Data in parentheses are the average values. For definitions of abbreviations, see Table 1.
[b]Calculated using the data reported in references.

TABLE 3 Nonstarch Lipid Classes of Cereal Grains

	Nonstarch lipids[a]						
Grains	TL (% Wt)	NL (% TL)	GL (% TL)	PL (% TL)	PoL (GL+PL) (% TL)	Ratio[b] of NL/PoL	Ref.
Barley	2.6–4.9 (3.8)	75–65 (70)	7–26 (17)	18–9 (13)	25–35 (30)	3.0–1.9 (2.5)	26
	3.1–4.6 (3.7)	65–78 (71)	7–13 (10)	14–26 (19)	22–35 (29)	1.8–3.6 (2.4)	45,46
Corn	2.6–13.8 (7.2)	88–96 (92)	—	—	4–12 (8)	7.3–24.6 (14.4)	47
	5.2	92	3	5	8	11.3	1,48
Millet	5.4	98	1	1	2	57.8	28
	6.9	94	2	4	6	16.2	28
	6.1	89	3	8	11	7.8	28
Oat groats	4.6–11.6 (8.3)	66–80 (72)	6–10 (8)	12–26 (20)	20–34 (28)	1.9–3.9 (2.6)	33
Brown rice	2.4–2.6 (2.5)	78–87 (83)	4–12 (8)	9–10 (10)	13–22 (17)	3.6–6.5 (5.1)	12,49,50
	2.9–3.4	85–87	4–6	8–9	13–15		51
Rye	3.6	71	10.7	18.3	29	2.4	52
	3.5	63–68 (65)	10–12 (11)	22–25 (24)	32–37 (35)	1.7–2.1 (1.9)	34
Sorghum	3.3	86	3	11	14	6.2	52
	3.6–5.0 (4.3)	77–85 (81)	2–6 (4)	13–17 (15)	15–23 (19)	3.4–5.8 (4.6)	35
Triticale	2.4	67	16	17	33	2.0	52
	3.2–4.6 (3.9)	53–63 (59)	10–18 (14)	26–29 (27)	37–47 (41)	1.1–1.7 (1.5)	34
	2.6	68	14	18	32	2.1	35
Wheat							
HRW	3.4–3.5 (3.5)	60–61 (60)	17–18 (17)	24–26 (25)	39–40 (40)	1.5	34
	2.2–2.6 (2.4)	65–70 (67)	15–17 (16)	15–19 (17)	30–36 (33)	1.8–2.3 (2.1)	37
HRS	2.3	62	22	16	38	1.6	52
	2.2–2.9 (2.5)	67–72 (69)	13–16 (15)	15–17 (16)	28–33 (31)	2.0–2.6 (2.3)	37
	1.4–1.9 (1.6)	69–71 (70)	—	—	29–31 (30)	(2.3)	53
SRW	2.1	69	14	17	31	2.2	37
Soft white	2.4	68	16	16	32	2.1	37
Durum	3.3	74	12	14	26	2.8	37

[a]Data in parentheses are the average values. (—) denotes data not reported. For definitions of abbreviations, see Table 1.
[b]Calculated using the data reported in references.

TABLE 4 Fatty Acid Composition of Grain Nonstarch Lipids

Grains	Fatty acid (wt %)[a]						
	14:0	16:0	18:0	18:1	18:2	18:3	Ref.[b]
Barley	2	21–24	< 2	9–14	56–59	4–7	3,54
	0.3–0.6 (0.4)	19–26 (23)	1.0–1.3 (1.1)	13–18 (16)	51–58 (54)	6–7 (6)	46
	—	19–22	1.0–1.7	14–21	51–60	4–6	26,55
Corn	—	11–17	< 3	33–39	43–49	< 2	3*
	—	19–22	2	18–28	31–39	2–6	3**
	—	7–19	< 4	23–46	35–66	< 3	3***
	—	6–22	0.6–15.0	14–64	19–71	0.5–2.0	56
	—	11–12	1.6–1.8	24–27	58–62	1.0–1.1	8
Millet—	16–17	5–7	25–27	47–48	3–13	28	
Pearl millet	tr–0.2 (0.1)	20–22 (21)	6.1–10.1 (7.9)	27–28 (28)	37–40 (38)	2.2–4.2 (3.2)	29*
	0.4–0.8 (0.6)	20–23 (22)	4.8–7.9 (6.1)	17–20 (19)	27–32 (29)	1.9–3.3 (2.8)	29**
	—	19	5.0	25	46	3.2	15
Foxtail millet							
Type A	—	7–10 (8)	3.4–7.2 (5.2)	11–15 (13)	66–72 (68)	2.2–4.1 (2.9)	42
Type B	—	7–9 (8)	1.0–1.9 (1.5)	14–17 (16)	69–72 (70)	2.1–3.3 (2.4)	42
Oats	0.4	19	2.2	39	38	1.3	32*
	0.9	26	2.0	29	41	1.4	32**
	tr–4.9 (1.4)	15–26 (20)	1.6–3.9 (2.4)	26–41 (35)	31–46 (38)	1.7–3.7 (2.4)	1*,33*
	tr–0.8 (0.6)	16–22 (19)	0.8–2.0 (1.5)	27–48 (38)	34–47 (40)	0.9–2.5 (1.7)	31,57
	—	16–22 (19)	1.1–2.1 (1.5)	22–33 (30)	44–52 (47)	1.9–3.4 (2.6)	55
Brown Rice	0.4–1.0	18–26	1.6–2.7	36–44	25–35	1.0–3.7	12
	0.2–0.4 (0.3)	14–21 (18)	1.3–2.5 (1.9)	37–52 (44)	29–39 (33)	1.0–2.0 (1.4)	43*,44*
	0.2–0.3 (0.2)	15–16 (16)	1.5–2.0 (1.7)	41–46 (44)	33–38 (35)	1.0–1.4 (1.2)	58*
Rye	—	12–19 (16)	< 1–2 (1)	12–16 (14)	57–65 (59)	3–12 (9)	59
	0.1	15	0.8	17	58	7	60
Sorghum	—	10–14 (12)	0.2–1.3 (0.8)	28–42 (34)	42–56 (50)	1.1–4.7 (3)	41
	—	15–17 (16)	0.9–1.2 (1.1)	21–27 (24)	55–58 (56)	2.1–3.0 (2.6)	35
Triticale	0.4	20	0.9	11	59	5.3	60
	0.9–1.8	16–18	0.7–1.9	8–9	57–58	3.5–4.2	61
Wheat	—	17–24	1–2	8–21	55–60	3–5	3
	—	24	1.3	18	55	2.3	62

[a]Data in parentheses are the average values. (—) denotes data not reported.
[b]Fatty acid composition of free lipids (*), bound lipids (**), and total lipids (***).

TABLE 5 Fatty Acid Composition of Wheat Nonstarch Lipids

Class or subclass[a]	Number of samples	Fatty acid (wt %) 16:0	18:0	18:1	18:2	18:3
HRW	103	22	1.3	19	55	2.3
HRS	46	24	1.3	19	54	1.9
HWW	6	21	0.7	16	59	3.0
HWS	2	24	—	20	54	2.2
SRW	14	22	1.2	20	54	2.5
SWW	88	25	1.4	15	55	2.5
SWS	14	21	0.7	16	59	2.8
Durum	17	26	1.4	21	50	1.8
	4[b]	20–24	0.9–1.3	14–17	56–57	4.2–5.6
All	290	24	1.3	18	55	2.3

[a]For definitions of abbreviations, see Table 1. (—) denotes data not reported.
[b]From Ref. 63.
Source: Ref. 62.

TABLE 6 Tocol Derivative Content in Cereal Grains

Grains	Tocol derivatives (mg/100 g)[a]									
	α–T	α–T–3	β–T	β–T–3	γ–T	γ–T–3	δ–T	δ–T–3	Total	Ref.
Barley	0.2–0.4	1.1–1.3	0.04–0.4	0.3—0.7	0.03–0.5	0.2	0.01–0.04	0.16	<5.0	3
	0.4	1.3	0.3	0.7	0.05	0.2	0.1	—		64
	0.5	1.3	0.02	0.7	0.07	0.8	0.02	0.07		5
Corn	0.6–2.1	0.2–0.5	—	—	0.5–1.1	—	—	—		3
	0.6	0.2	—	0.4	3.8	0.5	tr	—	—	64
	0.2–4.5	0.4–1.4			3.6–8.8	0.2–1.8	—	—	2.6–10.2	8
Millet	0.05	—	tr	—	1.3	—	0.4	—		64
Bulrush millet	0.13	—	—	—	5.54	0.53	0.08	0.2		5
Foxtail millet	0.19	0.04	0.04	—	2.78	0.06	—	—		5
Finger millet	0.32	0.05	0.03	—	1.76	—	—	—		5
Pearl millet	—	—	0.04	1.50	—	—	0.35	—		5
Oats	0.3–1.7	0.7–1.1	0.1–0.2	0.1–0.3	0.3	—	—	—		3
	0.7	0.7	0.2	0.1	0.3	—	—	—		64
	1.3–4.0	0.2–6.3	0.3–0.5	0.3–1.1	0.7–6.1	0.91	0.14	0.32	1.3–3.0	10
Milled rice	0.05–0.3	0.2–0.5	—	—	0.1–0.3	—	<0.04	—		3
	0.3	tr	—	—	0.3	0.5	0.04	—		64
	0.06	—	—	—	0.08	0.26	0.02	0.02		5
Rye	0.5–1.8	0.7–1.5	0.3–0.7	0.8–0.9	0.6	—	—	—		3
	0.8	1.3	0.4	0.9	0.6	—	—	—		64
Sorghum	0.08	—	—	—	1.15	—	—	—		5
Triticale	0.91	1.03	0.30	1.51	—	—	—	—	5	
Winter	0.7–0.95	—	—	—	—	—	—	—		65
Spring	1.35–1.45	—	—	—	—	—	—	—		65
Wheat	1.0	0.2	0.4	1.9	—	—	—	—		5
	0.9–1.8	0.3–0.7	—	2.5–3.6	—	—	—	—	4.9–5.8	3
	1.0	0.4	0.9	2.5	—	—	0.08	—		64

[a]For definition of abbreviations, see Table 1. Tocopherols include α–T, β–T, γ-T, and δ–T and tocotrienols include α–T–3, β–T–3, γ–T–2, and δ–T–3. (—) denotes data not reported and (tr) denotes trace amount.

TABLE 7 Tocol Derivative Composition of Corn Inbreds and Wheat Classes

Grains	Composition (% total)[a]						Content (mg/ 100 g)
	α–T	α–T–3	β–T	γ–T	γ–T–3	δ–T	
Corn inbreds[b]							
B37	44.8	13.2	—	36.1	5.9	—	5.2
C103D	25.3	14.3	—	43.3	17.1	—	2.6
B84	23.3	9.3	—	65.1	2.3	—	5.4
A619	8.1	4.0	—	69.7	18.2	—	10.2
W64A	2.1	5.1	—	87.8	5.0	—	6.7
Wheat[c]							
HRW (103)	51.9	—		20.8[d]	—	27.3	3.1
HRS (46)	47.6	—		24.4	—	27.9	3.2
HWW (6)	32.8	—		23.3	—	43.9	5.2
HWS (2)	26.4	—		24.1	—	49.5	9.9
SRW (14)	46.6	—		19.9	—	33.4	3.3
SWW (88)	43.0	—		23.7	—	33.3	3.3
SWS (14)	37.0	—		25.7	—	37.3	4.1
Durum (17)	55.6	—		21.1	—	23.3	2.8
All (290)	46.6	—		22.6	—	30.8	3.3

[a]For definitions of abbreviations, see Table 1. (—) denotes data not reported.
[b]From Ref. 8.
[c]Calculated data from Ref. 62. Number of samples are in parentheses.
[d]Sum of β–T and γ–T.

TABLE 8 Tocol Derivatives in Dissected Corn and Wheat Grain Fractions

Grain fractions	Tocol derivatives (mg/100 g)[a]							
	α–T	β–T–3	β–T	β–T–3	γ–T	γ–T–3	δ–T	Total
Corn[b]								
Germ	12.9–19.4 (95.8–95.9)c	—	—	—	33.2–38.8 (94.8–94.6)	—	0.73–1.04 (100)	47–58.9
Pericarp & tip cap	0.51–0.75 (3.7–3.8)	tr	—	—	1.72–2.04 (4.9–5.0)	tr	—	2.55–3.23
Endosperm	0.05–0.07 (0.4)	0.45–0.87 (H100)	—	—	0.10–0.19 (0.2–0.4)	0.38–1.89 (H100)	—	1.12–3.02
Wheat[d]								
Germ	25.56 (98.0)	< 0.2	11.44 (97.3)	<0.2	—	—	—	37.0
Pericarp, testa & aleurone	0.05 (1.2)	1.00 (81.2)	<0.04	6.86 (49.8)	—	—	—	7.91
Endosperm	0.007 (.08)	0.045 (18.8)	0.01 (2.7)	1.35 (50.2)	—	—	—	1.41
Wheat[e]								
Whole	0.9–1.8	0.3–0.7	—	2.5–3.6	—	—	—	4.9–5.8
Bran	1.6–3.3	1.1	1.0–1.3	2.9–5.4	—	—	—	
Patent flour	0.26–0.34	0.14	0.20–0.22	1.04–2.18	—	—	—	1.5–2.8

[a]For definitions of abbreviations, see Table 1. (—) denotes data not reported and (tr) denotes trace amount.
[b]From Ref. 66.
[c]Data in parentheses represent % distribution among the fractions of the whole grain.
[d]From Ref. 67.
[e]From Ref. 3.

TABLE 9 Carotenoid Content in Cereal Grains

Grains	Carotenoid Content (mg/kg)[a] Grain	Lipids	Ref.
Barley	1	100–150	5
Corn	1–58	—	68
Millet	6.7–6.9	—	69
Oats	—	100–310	2
Rye	—	15–28	2
Sorghum			
Common	1.5	—	5
Yellow	25–30	—	5
Wheat	1.8–5.8	43–99	2,70

[a](—) denotes data not reported.

TABLE 10 Carotenoid Content in Corn Grain

Carotenoids	Carotenoid Content (mg/kg)[a] b	c	d
Zeaxanthin	0.8–27.4	6.1–15.8	0.1–33.1
Lutein (xanthophyll)	2.0–33.1	7.0–14.9	0.1–16.1
Zeaxanthin and lutein esters	0.3–4.1	—	—
Cryptoxanthin	0.3–5.1	0.9–2.3	0.0–6.9
Fisoxanthin	—	0.5–2.0	—
Zeinoxanthin	0.0–7.8	—	0.1–5.1
Polyoxy compounds	0.4–3.0	—	0.1–7.5
α–Carotene		0.1–0.6	} 1.5–4.8
β–Carotene	0.1–5.4	0.4–1.7	
β–Zeacarotene	0.1–4.7	<0.6	

[a](—) denotes data not reported.
[b]From Ref. 68.
[c]From Ref. 72.
[d]From Ref. 73.
Source: Refs. 3,5.

TABLE 11 Carotenoid Content in Corn Fractions

Fractions	Weight[a] (%)	Carotenoids (mg/kg) Carotene		Xanthophyll		Total		Distribution[a] (% Total)
Whole[b]	100	1.7–2.0		13.9–19.1		15.9–20.8		100
Bran	7–8	0.2–0.3		1.1–1.8		1.4–2.1		1
Germ	10	0.5–1.3		3.5–4.1		4.0–5.4		2–4
Endosperm	82–83	3.4–4.8		23.8–40.1		27.3–40.3		95–97
Floury	(28–36)	(0.3–1.1)		(3.0–12.8)		(3.3–13.9)		(9–23)
Horny	(46–54)	(2.3–3.7)		(20.8–27.3)		(24.0–29.6)		(74–86)
		Hand-Diss.[d]	Dry-Milled	Hand-Diss.	Dry-Milled	Hand-Diss.	Dry-Milled	
Whole	100	1.7	1.7	19.1	19.1	20.8	20.8	100
Bran	—	0.2	0.8	1.4	11.4	1.6	12.2	—
Germ	—	1.3	1.8	4.1	7.7	5.4	9.5	—
Endosperm[c]	—	4.8	4.4	35.5	41.5	40.3	45.9	—
Floury	—	(1.1)	(1.5)	(10.0)	(15.6)	(11.7)	(17.1)	—
Horny	—	(3.7)	(2.9)	(25.5)	(25.9)	(29.2)	(28.8)	—

[a](—) denotes data not reported.
[b]Hand-dissected fractions of three yellow dent corns.
[c]For commercial dry-milled samples, corn break flour and corn grits represent, respectively, floury and horny endosperm fractions.
[d]Hand-dissected.
Source: Adapted from Ref. 74.

TABLE 12 Carotenoid Content and Composition in Wheat Grains[a]

Carotenoids	Wheat Class					
	HRW[b]	HRS		Soft white[b]	Durum	
		Wheat[b]	Flour[c]		Wheat[b]	Flour[c]
		Content (mg/kg)				
Carotene	0.21	0.18	0	0.25	0.15	0
Xanthophyll	0.79	0.99	0.60	1.33	1.67	3.14
Xanthophyll ester	1.04	0.63	2.20	0.72	0.15	0.56
Monoester	—	—	(1.30)	—	—	(0.36)
Diester	—	—	(0.90)	—	—	(0.20)
Total	2.04	1.80	2.80	2.30	1.97	3.70
		Composition (%)				
Carotene	10.3	10.0	0	10.8	7.6	0
Xanthophyll	38.7	55.0	21.6	57.8	84.8	84.8
Xanthophyll ester	51.0	35.0	78.4	31.4	7.6	15.1
Monoester	—	—	(46.5)	—	—	(9.8)
Diester	—	—	(31.9)	—	—	(5.3)

[a]For definitions of abbreviations, see Table 1. (—) denotes data not reported.
[b]Adapted from Ref. 75.
[c]Adapted from Ref. 76.

TABLE 13 Carotenoid Content and Composition in Wheat Fractions

Fractions	Carotenoids	HRW	HRS	Soft white	Durum
		Wheat class			
		Content (mg/kg)			
Bran	Carotene	0.02	0.04	0.03	0.10
	Xanthophyll	0.33	0.42	0.32	1.31
	Xanthophyll ester	0.60	0.47	0.53	0.81
	Total	0.95	0.93	0.88	2.22
Germ	Carotene	0.80	0.72	1.13	0.50
	Xanthophyll	5.98	5.78	9.70	2.93
	Xanthophyll ester	1.04	0.69	0.21	0.70
	Total	7.84	7.19	11.04	4.13
Endosperm	Carotene	0.11	0.09	0.21	0.08
	Xanthophyll	0.77	0.84	1.18	1.78
	Xanthophyll ester	1.12	0.64	0.79	0.10
	Total	2.00	1.57	2.18	1.96
		Composition (%)			
Bran	Carotene	2.1	4.3	3.4	4.5
	Xanthophyll	34.8	45.1	36.4	59.0
	Xanthophyll ester	63.2	50.5	60.2	36.5
Germ	Carotene	10.2	10.0	10.2	12.2
	Xanthophyll	76.2	80.3	87.8	70.8
	Xanthophyll ester	13.5	9.6	1.9	17.0
Endosperm	Carotene	5.5	5.7	9.6	4.1
	Xanthophyll	38.5	53.5	54.2	90.8
	Xanthophyll ester	56.0	40.8	36.3	5.1

Source: Adapted from Ref. 75.

TABLE 14 Carotenoid Content and Composition in Yellow Sorghum Grain

Carotenoids	Content (mg/kg)	Composition (%)
Zeaxanthin	2.8	36.3
Lutein	2.2	28.6
β–Carotene	0.8	10.4
Xanthophyll I	1.5	19.5
Xanthophyll II	0.4	5.2
Total	7.7	100

Source: Refs. 81,82.

TABLE 15 Composition of 4–Demethylsterols in Corn, Oats, and Rice[a]

	Composition (%)[b]							
	Corn			Oats			Rice	
	Kernel		Germ	Common		Wild	Kernel	Bran
Sterols	A	B	C	D	E	D	A	F
β–Sitosterol	65	66–70	66–76	39	69	57	56	54
Campesterol	21	16–21	16–23	6	10	8	22	20
Stigmasterol	5	4–10	6–8	5	8	4	11	8
Δ^5–Avenasterol	—	2–9	1–4	21	7	19	—	11
Δ^7–Avenasterol	—	—	—	14	—	6	—	4
Δ^7–Stigmasterol	—	0.2–1.0	1–5	6	—	2	—	2
Cholesterol	tr	0.2–0.4	—	6	2	4	3	—
Brassicasterol	—	—	—	—	—	—	—	—
5α–Stanols	9	—	—	—	—	—	—	—
Others	tr	—	1	3	—	1	5	—

[a](—) denotes data not reported and (tr) denotes trace amount.
[b]Composition as determined by: A, Knights (83); B, Miric et al. (84) cited by Weber (8); C, Morrison (3); D, Knights (85); E, Youngs et al. (32); F, Weihrauch and Gardner (86).

TABLE 16 Composition of 4–Demethylsterols in Rye, Sorghum, and Wheats[a]

	Composition (%)[b]						
	Rye	Sorghum		Wheat			
						Durum	Soft
Sterols	A	B	C	A	D	E	E
β–Sitosterol	59	55	54	55	41–53	56–62	71–73
Campesterol	20	20	9	21	14–27	25–39	20–24
Stigmasterol	tr	20	23	—	—	1–3	1–2
Δ^5–Avenasterol	8	—	—	—	—	<5	1–2
Δ^7–Avenasterol	—	—	—	—	—	<2	<1
Δ^7–Stigmasterol	—	—	—	—	—	<2	<1
Cholesterol	0.5	—	1	tr	0–4	1–3	1–2
Brassicasterol	—	—	—	—	—	—	—
5α–Stanols	9	—	—	18	—	—	—
Dihydrositosterol	—	—	—	—	6–23	—	—
Dihydrocampesterol	—	—	—	—	13–23	—	—
Dihidrobrassicasterol	—	—	13	—	—	—	—
Others	5	6	—	5	—	—	—

[a](—) denotes data not reported and (tr) denotes trace amount.
[b]Composition as determined by: A, Knights (83); B, Weihrauch and Gardner (86); C, Heupel et al. (87); D, Berrie and Knights (88); E, Artaud et al. (89).

TABLE 17 Composition of 4–Demethylsterols in Dissected Wheat and Sorghum Grain Fractions

Fractions	Composition (%)			
	Sitosterol	Campesterol	Stigmasterol	Cholesterol
Triticum durum[a]				
Endosperm	75.0	3.8	13.6	7.6
Bran	65.3	10.9	10.2	13.6
Embryo	55.2	12.8	16.2	15.8
Scutellum	69.6	17.4	10.4	2.6
T. aestivum[a]				
Endosperm	67.0	4.6	14.6	13.3
Bran	67.2	10.6	10.6	11.6
Embryo	63.3	16.1	9.4	11.2
Scutellum	81.9	14.5	2.3	1.3
Sorghum vulgare[b]				
Endosperm	47.3	26.7	17.7	1.0
Embryo	55.2	23.7	12.9	0.6

[a]From Ref. 90.
[b]From Ref. 91. They also reported 5.6 and 4.7% Δ^5–avenasterol and 1.6 and 2.9% Δ^7– avenasterol in sorghum endosperm and embryo, respectively.

TABLE 18 Composition of 4–Demethylsterols in Sterol Lipids of Wheat Kernel, Aleurone, and Flour[a]

Sterols	Composition (%)								
	Kernel[b]		Aleurone[c]			Milled flour[d]			
	SE	S	SE	S	SG	SE	S	SG	ASG
β–Sitosterol	71.8	65.6	55.8	63.0	50.2	75.2	63.2	65.6	63.8
5α–Dihydrositosterol	—	—	—	—	—	14.1	13.6	14.0	14.3
Campesterol	22.8	26.0	32.5	22.6	34.8	4.8	16.8	14.4	17.2
5α–Dihydrocampestrol	—	—	—	—	—	5.9	6.4	6.0	5.7
Stigmasterol	2.0	3.0	7.6	5.8	10.0	—	—	—	—
Δ^5–Avenasterol	2.9	3.6	—	—	—	—	—	—	—
Δ^7–Avenasterol	0.5	0.5	—	—	—	—	—	—	—
Δ^7–Stigmasterol	<0.1	1.0	—	—	—	—	—	—	—
Cholesterol	<0.1	0.3	4.1	8.6	5.0	—	—	—	—

[a]For definition of abbreviations, see Table 1. (—) denotes data not reported.
[b]From Ref. 92.
[c]From Ref. 93.
[d]From Ref. 94.

TABLE 19 Composition of Rice Bran Oil and Its Sterol Lipids

Sterols	Composition (%)[a]				
	Bran oil[b]	SE[c]	S[c]	SG[c]	ASG[c]
4–Demethylsterols	(53)[d]				
β–Sitosterol	49	51.6	57.2	77.2	80.2
Campesterol	28	20.2	16.2	9.3	8.0
Stigmasterol	15	9.2	18.8	11.8	11.3
Δ^5–Avenasterol	5	11.4	6.0	—	—
Δ^7–Avenasterol	1	2.7	—	—	—
Δ^7–Stigmasterol	2	3.6	tr	1.0	0.5
Cholesterol	—	1.3	1.8	—	—
4–Monomethylsterols	(12)				
Obtusifoliol	7	9.5	7.7	—	—
Unknown	6	1.5	1.4	—	—
Gramisterol	39	55.4	60.5	—	—
Unknown	8	5.7	6.6	—	—
Citrostadienol	40	27.9	23.6	—	—
4,4–Dimethylsterols	(35)				
Cycloartanol	9	2.5	3.9	—	—
Cycloartenol	41	26.6	19.6	—	—
24–Methylenecycloartanol	42	67.9	67.0	—	—
Unknown	tr	2.1	6.6	—	—
Cyclobranol	tr	0.9	2.9	—	—
Unknown	7	—	—	—	—

[a]For definitions of abbreviations, see Table 1. (—) denotes data not reported and (tr) denotes trace amount.
[b]From Refs. 95,96.
[c]From Ref. 97.
[d]Data in parentheses are % total sterols.

TABLE 20 Composition of Sterols in Corn and Wheat Germ Oils[a]

Sterols	Composition (%)[b]				
	Corn		Wheat Germ		
	A	B	A	B	C
4–Demethylsterols	(98.0)[c]	(92.5)	(87.1)	(92.4)	—
β–Sitosterol	66	60	67	60	64
Campesterol	23	17	22	19	29
Stigmasterol	6	6	tr	4	1
Δ^5–Avenasterol	4	10	6	7	2
Δ^7–Avenasterol	tr	1	2	2	2
Δ^7–Stigmasterol	1	tr	3	2	1
Cholesterol	tr	tr	tr	tr	—
Brassicasterol	tr	—	tr	—	<1
Others	—	6	—	5	—
4–Monomethylsterols	(1.0)	(4.0)	(5.4)	(3.8)	
Obtusifoliol	30	21	6	14	7
Δ^7–Sterol	—	6	—	6	—
Gramisterol	34	26	41	25	41
Cycloeucalenol	—	6	—	6	—
24–Ethyllophenol	—	4	—	5	—
Citrostadienol	30	29	46	30	38
Unknown	5	3	7	3	14
Others	1	5	—	11	—
4,4 Dimethylsterols	(1.0)	(3.5)	(7.5)	(3.8)	
Cycloartanol	2	—	3	2	5
β–Amyrin	4	1	18	12	22
α–Amyrin	—	2	8	7	—
Cycloartenol	38	43	17	25	17
24–Methylenecycloartanol	53	43	44	33	27
Cyclobranol	2	—	2	2	6
Δ^7–Sterol	—	5	—	6	—
Δ^8–Sterol	—	5	—	11	—
Unknown	1	—	2	—	10
Others	1	3	—	2	—

[a](—) denotes data not reported and (tr) denotes trace amount.
[b]Composition as determined by: A, Itoh et al. (95,96); B, Kornfeldt and Croon (99); and C, Lercker et al. (100), cited by Morrison (3).
[c]Values in parentheses are % total sterols.

TABLE 21 Weight Distribution of Grain Fractions and Their Free Lipid Contents in Grain Fractions[a]

Grains	Fractions	Weight (%)	Free lipid content (% fraction)	Ref.
Corn	Whole kernel	100	3.9–5.8	101
	Endosperm	79.7–83.5	0.7–1.1	
	Germ	10.2–14.1	31.1–38.9	
	Bran	4.4–6.2	0.7–1.2	
	Tip cap	0.8	3.7–3.9	
Oats	Whole kernel	100	5.4	102
	Groats	64–81	7.6	
	Hulls	19–36	0.6	
	Groats	100	—	
	Endosperm	69	6.2–6.7	
	Bran	24	—	
	Germ	7	11.2	
Rice	Whole kernel	100	1.9–3.1	3
	Brown rice	71–78	2.4–3.9	
	Hulls	23–29	0.2–0.4	
	Brown rice	100	2.9–3.4	51
	Inner endosperm	82.5–83.8	0.4–0.8	
	Subaleurone layer	4.9–5.2	5.6–8.5	
	Polish	4.1–4.4	10.2–15.0	
	Bran	5.9–6.4	19.4–25.5	
	Germ	1.3–1.5	34.1–36.5	
Sorghum	Whole kernel	100	3.2–3.9	103
	Endosperm	80.0–84.6	0.4–0.8	
	Germ	7.8–12.1	26.9–30.6	
	Bran	7.3–9.3	3.7–6.0	
Wheat	Whole kernel	100	1.8	104
	Bran	—	5.1–5.8	
	Pericarp	5.0–8.9	0.7–1.0	
	Testa (hyaline)	0.2–1.1	0.2–0.5	
	Aleurone	4.6–8.9	6.0–9.9	
	Endosperm	74.9–86.5	0.8–2.2	
	Outer endosperm	—	2.2–2.4	
	Inner endosperm	—	1.2–1.6	
	Germ	2.0–3.9	28.5	
	Embryonic axis	1.0–1.6	10.0–16.3	
	Scutellum	1.1–2.0	12.6–32.1	

[a](—) denotes data not reported.

TABLE 22 Nonstarch Lipid Content and Its Distribution in Grain Fractions

Grain	Type	Fraction	Weight (%)	Nonstarch lipid Content (% fraction)	Nonstarch lipid Distribution[a] (%)	Ref.
Corn	H51	Whole	100	4.9	100	105
		Germ	10.9	35.5	82.2	
		Endosperm	82.6	1.0	17.5	
		Pericarp	5.8	0.2	0.3	
		Tip cap	0.7	0.3	<0.1	
	LG–11	Whole	100	4.9	100	106
		Germ	8.4	38.7	78.4	
		Endosperm	86.0	1.0	20.7	
		Pericarp	4.2	0.3	0.3	
		Tip cap	1.4	1.6	0.6	
	Waxy maize	Whole	100	5.1	100	106
		Germ	10.8	35.6	84.8	
		Endosperm	81.0	0.8	14.3	
		Pericarp	6.6	0.2	0.3	
		Tip cap	1.6	1.7	0.6	
	Amylomaize	Whole	100	9.3	100	106
		Germ	15.0	33.7	73.2	
		Endosperm	74.7	2.4	26.0	
		Pericarp	7.8	0.2	0.2	
		Tip cap	2.5	1.8	0.6	
Barley	Prilar Hulless					
		Whole	100	3.2	100	107
		Bran-endosperm	90.2	2.8	77.1	
		Emryonic axis	3.0	19.6	17.9	
		Hull	6.8	2.4	50	
Oats	James Hulless					
		Groats	100	7.2	100	107
		Bran-endosperm	84.5	7.1	84.7	
		Embryonic axis	2.4	21.2	7.2	
		Hull	13.1	4.4	8.1	
Rye[b]		Whole	100	3.0	100	60
		Endosperm	91.1	2.1	62.3	
		Germ	4.4	23.6	34.5	
		Bran	4.5	2.1	3.2	
Triticale[b]		Whole	100	3.0	100	60
		Endosperm	91.2	1.8	55.2	
		Germ	4.2	30.0	41.9	
		Bran	4.6	1.9	2.9	
Wheat[b]		Whole	100	3.3	100	60
		Endosperm	92.0	2.2	62.7	
		Germ	4.0	27.8	33.8	
		Bran	4.0	2.8	3.4	

[a]Distribution in corn kernels was calculated by using data of Weber (105) and Tan and Morrison (106).
[b]Fractions were the milling products.

TABLE 23 Free, Bound, and Total Nonstarch Lipids in Oat and Proso Millet Grain Fractions

Grains	Fractions	Lipid content (% fraction)		
		Free	Bound	Total
Oats[a]	Groats	5.5–8.0	1.4–1.6	6.9–9.6
	Hull	2.0–2.3	0.6	2.6–2.9
	Scutellum	20.4–20.6	2.8–4.2	23.4–24.6
	Embryonic axis	10.6–12.6	3.3–4.1	14.7–15.9
	Bran	6.4–9.5	1.2–1.3	7.7–10.7
	Endosperm	5.2–6.8	1.0	6.2–7.8
Red proso millet[b]	Whole	3.5–3.9	0.5–0.9	4.3–4.9
	Bran	3.5–4.3	0.3–0.7	4.1–5.0
	Flour	3.6–4.1	0.5–0.9	4.2–4.9
White proso millet[b]	Whole	3.3–3.9	0.6–1.3	3.9–4.9
	Bran	4.0–6.8	0.3–0.5	4.3–7.3
	Flour	3.2–4.0	0.5–0.7	3.8–4.5

[a]From Ref. 32. Total lipids were the sum of FL and BL.
[b]From Ref. 30. The flour yields were 79.2–82.4% and 82.1–87.1% for red and white proso millets, respectively.

TABLE 24 Average Compositions of Nonpolar Lipids and Phospholipids of Two Barley Grains and Fatty Acid Composition of Each Lipid Class

Lipid[a] Class	Composition (%)	Fatty acid (%) 16:0	18:0	18:1	18:2	18:3	Others
Nonpolar Lipids[b]							
TG	51.9	19.0	0.8	20.9	38.6	4.8	15.9
1,3–DG	7.3	20.9	2.3	12.9	59.5	3.9	0.5
1,2–DG	8.4	20.7	2.8	12.8	57.9	5.0	0.8
FFA	9.0	34.8	3.8	8.5	49.1	2.9	0.9
Unknown 1	0.8	23.0	4.2	18.6	46.1	4.2	3.9
2	1.1	19.2	6.9	18.3	40.3	3.9	11.4
3	1.5	20.2	6.3	18.7	46.9	5.7	2.2
4	tr	25.6	18.1	27.3	8.4	5.6	15.0
Phospholipids[c]							
DPG	1.6	18.9	4.6	12.9	53.6	1.9	8.1
PG	0.5	35.1	11.5	26.4	17.3	1.7	8.0
PE	8.2	20.8	6.3	16.5	52.0	2.1	2.3
PC	44.4	18.7	2.2	18.9	55.3	3.9	1.0
PI	1.2	22.9	4.4	23.8	44.1	1.3	3.5
PS	4.9	30.6	3.6	6.3	48.6	2.8	8.1
PA	0.1	35.7	15.2	20.2	12.1	2.5	14.3
LPC	37.0	27.3	1.8	12.7	53.3	3.5	1.4
Unknown 1	0.4	37.2	24.0	11.2	17.3	0.8	9.5
2	1.7	27.0	13.8	9.8	42.9	2.8	3.7

[a]For definitions of abbreviations, see Table 1. Lipid composition is expressed as % nonpolar lipids or phospholipids.
[b]From Ref. 108.
[c]From Ref. 109.

TABLE 25 Nonstarch Lipid and Fatty Acid Compositions of Hulless Barley and Hulless Oat Grain Fractions

Grain fractions	Lipid[a] Content (%)	Class	Composition (%)	Fatty acid (%) 16:0	18:0	18:1	18:2	18:3	Others
Hulless Barley (Prilar)									
Hull	6.8	NL	75.9	26.1	5.4	21.5	30.6	5.0	9.7
		GL	18.2	41.9	7.3	15.3	17.1	9.3	6.2
		PL	5.9	37.9	2.2	10.0	22.2	10.5	3.0
Embryonic-axis	19.6	NL	75.8	19.5	0.6	20.4	49.7	9.6	0.2
		GL	6.4	17.4	3.9	14.7	40.6	23.2	0.2
		PL	17.8	17.5	0.2	19.1	45.2	9.4	0.2
Bran-endosperm	2.8	NL	64.4	20.6	1.3	16.6	56.8	4.4	0.3
		GL	12.5	17.9	2.1	5.9	68.3	5.6	0.2
		PL	23.1	30.4	1.5	17.4	56.4	2.0	0.6
Hulless Oat (James)									
Hull	4.4	NL	66.9	32.4	4.7	18.2	27.6	8.8	7.7
		GL	27.6	36.7	4.4	10.0	22.2	23.1	2.3
		PL	5.5	32.3	2.2	14.8	36.6	12.3	1.3
Embryonic-axis	21.2	NL	87.4	17.1	0.6	32.1	45.8	3.7	0.7
		GL	3.8	19.9	3.7	19.1	45.2	10.0	2.1
		PL	8.8	19.0	0.3	25.2	51.0	3.1	1.4
Bran-endosperm	7.1	NL	56.9	16.4	1.5	36.1	43.9	1.7	0.4
		GL	21.4	20.1	2.0	17.4	56.4	3.3	0.3
		PL	21.7	28.8	1.7	18.5	48.6	1.2	1.2

[a]For definitions of abbreviations, see Table 1. Lipid content is expressed as % of weight (dry basis) of fraction and lipid composition is expressed as % lipids.
Source: Ref. 107.

TABLE 26 Nonstarch Lipid Composition in Two-Row Barley Embryonic Tissues at 0 and 5-Day Germination

Lipid[a] class	Lipid composition (% lipids)					
	Coleorhiza		Coleoptile		Scutellum	
	0	5	0	5	0	5
SE	7.1	15.4	5.3	11.8	1.1	3.7
TG	65.2	55.1	58.1	39.7	86.3	71.4
FFA	2.0	8.4	3.6	12.1	0.5	2.8
ASG	4.1	12.3	5.8	17.1	3.4	8.6
PE	5.2	3.2	4.7	9.6	2.5	4.7
PC	16.4	5.5	22.5	9.6	6.2	8.8

[a]For definitions of abbreviations, see Table 1.
Source: Ref. 110.

TABLE 27 Nonstarch Lipid Compositions of Oat Groats and Groat Fractions

Lipid class[a]	Lipid composition (% total lipids)				
		Groat fractions			
	Groats	Bran	Scutellum	Embryonic axis	Endosperm
TG	32.4–50.6	38.6	50.1	57.7	41.0
DG	1.8–3.0	3.0	3.4	4.0	2.9
FFA	2.0–11.0	2.8	2.1	2.0	3.0
S	0.5–1.7	0.9	0.7	0.8	0.9
SG	0.9–1.4	0.9	0.5	0.8	1.2
MGMG	4.0–4.3	4.0	*	*	4.5
DGDG	6.9–7.6	8.0	*	*	8.0
PE	0.8–2.2	2.5	0.9	1.1	2.3
PC	3.2–6.1	3.6	2.6	2.8	3.5
LPE	0.4–1.7	1.7	*	*	1.7
LPC	1.8–2.7	2.9	*	*	3.0
Others[b]	16.0–34.2	31.0	39.7	30.8	28.0

[a]For definitions of abbreviations, see Table 1. (*) denotes those not measured.
[b]Lipids not measured, values obtained by differences.
Source: Ref. 32.

TABLE 28 Nonstarch Lipid Content and Composition of Oat Groats of Hinoat and Other Five Strains[a]

Lipid class[b]	Lipid content (mg/100 g, db)		Lipid composition (% TL)	
	Hinoat	Five strains	Hinoat	Five strains
Total NL	4,150	3,150–9,270	65.8	66.9–79.8
SE	130	130–460	2.1	2.8–4.0
TG	3,390	2,320–6,570	53.7	43.1–56.4
DG+MG	270	260–1,010	4.3	5.7–8.9
FFA	250	300–1,220	4.0	6.4–10.5
S	110	140–670	1.7	2.3–5.8
Total GL	606	260–1,040	9.6	5.8–9.6
SG	44	*	0.7	*
MGDG	114	*	1.8	*
MGMG	38	*	0.6	*
DGDG	246	*	3.9	*
DGMG	32	*	0.5	*
Others	132	*	2.1	*
Total PL	1,550	1,200–1,830	24.6	11.6–26.0
PG	145	*	2.3	*
PE	227	*	3.6	*
PC	467	*	7.4	*
PA+PI	56	*	0.9	*
PS	50	*	0.8	*
LPL	315	*	5.0	*
Others	290	*	4.6	*
Total NSL	6,310	4,620–11,660	100.0	100.0

[a]Some data are calculated. Five oat strains included Exeter, OA–290–5, Terra, Lodi, and CI–4492.

[b]For definitions of abbreviations, see Table 1. (*) denotes those not measured.

Source: Ref. 33.

TABLE 29 Composition of Free Lipids, Bound Lipids, Total Nonstarch Lipids, and Total Lipids (Nonstarch + Starch) in Millet Grains

Lipid class[a]	Lipid composition (% lipids)				
	Free[b]	Bound[b]	Total NSL[c] C–M	Total NSL[c] WSB	Total (NSL+ SL)[d] Hot WSB
Nonpolar lipids					
SE + HC	8.7	12.1	6.0	4.8	3.0
TG	66.4	34.7	44.7	38.6	73.5
FFA	3.1	9.2	16.7	14.5	1.8
S	4.5	11.4	9.5	5.4	5.4
DG	2.3	4.3	6.9	7.5	2.8
MG	15.2	33.4	3.3	3.5	2.8
Polar lipids	15.2	33.4	13.1	25.9	14.3

[a]For definitions of abbreviations, see Table 1.
[b]Average data of two varieties by Nechaev and Sandler (2).
[c]Average data of two varieties by Salun and Kalugina (111); C–M = $CHCl_3$–CH_3OH (2:1, V/V); WSB = water-saturated 1–butanol.
[d]From Ref. 28.

TABLE 30 Compositions of Total Lipids (Nonstarch + Starch) in Millet Grains and of Fatty Acids in Lipid Classes

Lipid[a]		Fatty acid (%)					
Class	Composition (% TL)	16:0	18:0	18:1	18:2	18:3	Others
Total NL	85.7	16.7	8.2	24.7	45.7	3.2	1.4
SE	3.0	18.0	8.7	30.8	42.5	tr	—
TG	73.5	15.9	5.3	28.8	46.2	2.5	0.8
DG, MG	2.8	24.1	4.8	22.7	40.8	7.6	tr
FFA	1.8	21.1	8.8	27.2	37.8	4.1	0.8
S	5.4	—	—	—	—	—	—
Total GL	2.6	19.4	4.3	24.4	39.5	10.9	1.5
SG	0.4	—	—	—	—	—	—
ASG	1.0	27.9	4.4	26.9	37.3	3.4	—
MGDG	0.3	20.2	5.9	26.0	34.3	13.7	—
MGMG	0.04	20.9	8.1	22.5	42.1	6.3	—
DGDG	0.4	16.6	5.6	20.4	38.6	18.7	—
DGMG	0.06	13.9	10.8	29.8	35.0	10.5	—
AMGDG	0.05	27.5	8.0	22.1	36.2	6.2	—
CB-I	0.2	32.5	11.9	23.1	23.2	9.3	—
CB-II	0.04	25.9	18.0	19.9	29.2	7.0	—
Unknown	0.1	21.0	10.4	23.0	35.5	10.1	—
Total PL	11.7	29.6	2.6	26.4	37.3	4.1	tr
PG	0.1	8.2	2.4	10.5	70.4	1.8	6.6
PE	0.8	20.0	2.3	21.9	52.4	1.8	1.6
PC (+ PI)	2.9	18.8	2.8	28.2	48.1	2.2	tr
LPE (+ PS)	2.5	41.8	2.4	20.2	33.0	2.6	tr
LPC	4.9	46.0	3.2	28.9	18.3	3.7	tr
PA	0.2	26.8	4.8	21.4	37.3	1.9	7.9
X4[b]	0.2	31.8	6.1	20.4	36.0	tr	5.7

[a]For definitions of abbreviations, see Table 1. (—) denotes data not reported and (tr) denotes trace amount. Lipid compositions were calculated using data by the hot WSB-extraction (28) to express % total lipids (TL).
[b]x4 = biphosphatidic acid.
Source: Ref. 28.

TABLE 31 Compositions of LG–11 Hybrid Corn Total Lipids (Nonstarch and Starch), Nonstarch Lipids in Corn Fractions, and Starch Lipids

Lipid class[a]	Corn lipids (% TL)	Nonstarch lipids (% NSL)				Starch lipids[b] (% SL)
		Pericarp	Germ	Tip cap	Endosperm	
Total NL	84.6	39.1	90.4	73.3	79.8	52.8
SE	2.5	14.9	1.9	10.0	6.9	0.6
TG	70.8	19.5	84.9	44.4	48.5	1.3
DG	2.8	3.4	2.9	3.3	4.6	0.5
FFA	8.1	1.1	0.7	13.3	18.8	48.6
MG	0.4	tr	0.1	2.2	1.0	1.8
Total GL	2.8	3.4	1.8	7.8	5.9	5.7
ASG	0.4	1.1	0.2	2.2	0.9	1.1
MGDG	0.6	1.1	0.3	1.1	1.7	1.8
MGMG		tr		1.1		
DGDG	1.1	tr	1.0	2.2	1.6	1.5
DGMG	0.7	1.1	0.4	1.1	1.6	1.3
Total PL	7.5	2.3	3.3	7.8	4.1	41.5
APE	0.3	1.1	0.3	1.1	0.4	—
ALPE	< 0.1	tr	0.1	—	0.1	—
DPG				—		—
PG	0.1	—	0.1	—	0.1	—
PE	0.4	—	0.4	1.1	0.1	—
PC	1.4	1.1	1.7	1.1	0.5	—
PI	0.5	—	0.6	1.1	0.4	—
PA	0.2		0.2	1.1	0.1	—
LPG	0.1	tr	—	—	—	0.7
LPE	0.3		—	tr	0.1	2.7
LPC	0.4		< 0.1	1.1	2.4	36.8
LPI, LPS	0.1	—	—	1.1	—	1.3
Total unsaponifiables	5.1	55.2	4.4	11.1	10.2	—

[a]For definitions of abbreviations, see Table 1. (—) denotes data not reported and (tr) denotes trace amount.

[b]Starch lipids (SL) in endosperm fraction.

Source: Ref. 106.

TABLE 32 Compositions of Nonpolar Lipids, Glycolipids, and Phospholipids in Whole Kernel, Germ, and Endosperm of Corn

	Composition (% lipids)					
				Endosperm		
Lipid class[a]	Whole corn[b]	Germ[b]	Germ[c]	NSL[b]	NSL[c]	SL[b]
Nonpolar lipids	100.0	100.0	100.0	100.0	100.0	100.0
SE	3.0	2.1	—	8.6	—	1.2
TG	83.6	93.8	99.4	60.8	58.9	2.5
DG	3.3	3.2	—	5.8	—	0.9
FFA	9.6	0.8	0.6	23.6	41.1	92.0
MG	0.5	0.1	—	1.2	—	3.4
Glycolipids	100.0	100.0	100.0	100.0	100.0	100.0
ASG	13.4	8.8	1.6	15.6	4.1	19.2
MGDG	23.0	15.9	9.3	28.8	8.1	31.5
MGMG	23.0	15.9	—	28.8	12.5	31.5
DGDG	40.3	54.1	21.0	27.8	15.5	26.0
DGMG	23.3	21.2	—	27.8	—	23.3
SG+CB	—	em	32.8	—	13.1	—
CB	—	—	22.0	—	8.6	—
Sulfolipid	—	—	7.0	—	19.7	—
Others[d]	—	—	6.3	—	18.4	—
Phospholipids	100.0	100.0	100.0	100.0	100.0	100.0
APE	3.4	7.5	1.1	9.5	1.2	—
ALPE	0.5	1.3	1.8	1.6	1.1	—
DPG	0.5	1.3	2.1	1.6	0.9	—
PG	1.0	2.6	4.3	1.6	1.9	—
PE	4.8	13.4	11.7	3.2	6.1	—
PC	18.3	51.5	54.5	11.1	44.6	—
PI	6.5	16.9	18.7	9.5	3.6	—
PS	—	—	0.9	—	—	—
PA	2.1	5.2	1.2	3.2	0.7	—
LPG	1.0	—	—	—	—	1.7
LPE	4.1	—	0.4	3.2	3.4	6.6
LPI, LPS	2.0	—	—	—	—	3.2
LPC	56.3	1.6	0.8	57.1	36.5	88.5
Unknown	—	—	2.5	—	36.5	—

[a]For definitions of abbreviations, see Table 1. (—) denotes data not reported.
[b]Calculated data for LG–11 hybrid corn from Tan and Morrison (106).
[c]From Ref. 113.
[d]Includes spots on the front and origin of thin-layer plates.

TABLE 33 Fatty Acid Composition of Lipids in Corn Grain Fractions

Grain fractions	Lipid Class[a]	Lipid Composition (%)	Fatty acid (%) 16:0	18:0	18:1	18:2	18:3
Germ[b]	NL	92.4	16.0	1.3	31.8	49.5	0.8
	GL	2.0	24.9	1.7	28.3	42.8	2.3
	PL	5.6	24.9	1.3	31.1	42.0	0.7
Endosperm[b]	NL	86.9	20.8	2.4	23.6	49.4	3.8
	GL	5.4	28.1	2.8	20.0	44.6	4.5
	PL	7.6	21.3	2.4	35.1	39.6	1.6
LG–11 hybrid corn[c]							
Pericarp			24.7	5.2	21.1	45.8	3.1
Germ			11.0	1.4	24.1	62.8	0.7
Tip cap			22.1	2.6	20.0	49.2	6.0
Endosperm, Total			24.7	1.9	15.6	53.5	4.3
Endosperm, SL			31.4	1.1	11.1	53.8	2.7
Starch, SL			36.8	1.8	10.1	47.9	3.5
H–51 inbred corn[d]							
Kernel			17.7	1.2	29.9	50.0	1.2
Pericarp			31.9	4.4	21.9	39.0	2.8
Germ			17.1	0.9	31.4	50.0	0.6
Tip cap			38.2	5.7	22.0	32.5	1.6
Endosperm			20.4	1.9	24.2	50.0	3.5
Root			24.9	7.9	7.6	51.9	7.7
Leaf			15.8	0.8	1.1	13.1	69.2

[a]For definitions of abbreviations, see Table 1.
[b]From Ref. 113.
[c]From Ref. 106.
[d]From Ref. 39.

TABLE 34 Composition of Nonstarch Lipids in Hull, Brown Rice, and Milling Fractions[a]

Lipid class[b]	Composition (% NSL)						
	Hull	Brown	Bran	Germ	Polish	Milled	Milled[c]
Total NL	64	86	89	91	87	82	83–91
SE	—	—	—	—	—	—	2–4
TG	—	71	76	79	72	58	55–83
FFA	—	7	4	4	5	15	3–29
MG	—	—	—	—	—	—	1–8
Total GL	25	5	4	2	5	8	2–4
SG	—	<1	<1	<1	1	2	—
ASG	—	3	2	1	3	4	—
MGMG	—	tr	tr	tr	tr	tr	1–3
DGDG	—	<1	<1	<1	<1	1	1–2
Total PL	11	9	7	7	8	10	7–13
PE	—	4	3	3	3	4	—
PC	—	4	3	3	3	4	—
LPE	—	—	—	—	—	1	—
LPC	—	<1	<1	<1	<1	2	—
Total NSL (% Fraction, db)	0.4	2.7	18.3	30.2	10.8	0.8	0.3–1.0

[a]Average composition of three varieties.
[b]For definitions of abbreviations, see Table 1. (—) denotes data not reported and (tr) denotes trace amount.
[c]Calculated data from Azudin and Morrison (115) by subtracting SL content from TL (NSL+SL). The range shows the average of two waxy rice varieties, of six normal rice varieties grown in the Philippines, and of six normal rice varieties grown in Portugal. MGMG includes MGDG, and DGDG includes DGMG.
Source: Refs. 13,51,114.

TABLE 35 Compositions of Total Lipids (Nonstarch + Starch) and Starch Lipids in Waxy and Nonwaxy Brown Rice and Milled Rice Samples

	TL (NSL + SL) composition (% TL)		Starch lipid composition (% SL)					
			Starch		Milled rice		Brown Rice	
Lipid class[a]	Waxy[b]	Non–waxy[b]	Waxy[b]	Non–Waxy[b]	Waxy[c]	Non–waxy[c]	Waxy[c]	Non–waxy[c]
Total NL	91.3	51.7	100	35.3	33	25	43	23
SE	3.3	1.2	0	0.7	—	—	—	—
TG	55.3	23.4	0	0.3	5	4	5	2
FFA	30.0	23.9	100	31.2	28	21	37	21
MG	2.7	3.2	0	3.1	—	—	—	—
Total GL	2.2	3.2	—	3.2	7	5	9	6
SG	—	—	—	—	1	1	2	1
ASG	—	—	—	—	3	2	3	2
MGMG	1.1	1.2	—	1.1	3	2	3	3
DGDG	1.1	2.0	—	2.1	tr	tr	tr	tr
Total PL	6.5	45.1	—	61.5	41	52	32	54
PE	—	—	—	—	3	4	3	4
PC	—	—	—	—	5	5	3	4
LPE	—	—	—	6.15	15	21	12	22
LPC	—	—	—	6.15	18	22	13	24
Others	—	—	—	—	18	18	16	17

[a]For definitions of abbreviations, see Table 1. (—) denotes data not reported and (tr) denotes trace amount.
[b]Average composition of 2 waxy rice varieties and of 12 nonwaxy varieties using data from Azudin and Morrison (115). MGMG contains MGDG and DGDG contains DGMG.
[c]From Ref. 51.

TABLE 36 Fatty Acid Compositions of Nonstarch Lipid and Starch Lipid Classes in Brown Rice and Milled Fractions

Rice fractions	Lipid class[a]	Fatty acid (%) 16:0	18:1	18:2
Nonstarch lipids				
Brown rice	TL	23–24	32–37	36–40
	NL	22–24	35–37	35–39
	GL	30–33	24–26	36–38
	PL	25–29	30–34	34–38
Bran	TL	22–25	36–37	36–37
	NL	22–23	38–41	37–41
	GL	25–27	30–31	35–39
	PL	20–25	36–43	32–36
Germ	TL	23–25	36–38	35–38
	NL	23–24	34–36	37–38
	GL	28–29	32–34	35–36
	PL	22–24	36–37	36–37
Polish	TL	23–25	36–39	37–39
	NL	22–25	35–37	37–39
	GL	26–30	26–28	35–39
	PL	25–26	34–37	33–35
Milled rice	TL	32–34	20–21	38–41
	NL	21–27	17–22	45–56
	GL	47–49	10–12	34–38
	PL	44–51	10–12	34–40
Starch lipids				
Brown rice	TL	43–48	12–17	29–40
	NL	23–38	20–23	30–53
	GL	53–55	7–12	26–32
	PL	45–54	9–14	30–39
Milled rice	TL	44–46	10–18	30–42
	NL	21–35	20–22	38–54
	GL	53–56	7–16	23–31
	PL	46–56	15–16	23–31

[a]For definitions of abbreviations, see Table 1.
Source: Refs. 13,51,114.

TABLE 37 Nonstarch Lipid Compositions of Rye Grain, Bran, Germ, and Endosperm Fractions and Milled Rye Flour

Lipid class[a]	Lipid composition (% lipids)					
	Grain[b]	Grain[c]	Bran[b]	Germ[b]	Endosperm[b]	Flour[c]
SE	3.2	8.1–9.2	27.7	7.4	4.0	17.3–17.6
TG	40.3	31.8–32.9	13.2	38.5	17.7	32.1–32.7
DG	2.0	8.3–10.1	20.6	5.8	2.5	8.5–12.4
MG	6.5	5.2–6.5	19.4	—	4.2	2.8–4.7
FFA	8.9	8.0–10.1	19.1	16.6	5.8	9.3–10.0
MGDG	6.5	8.0–10.1	—	—	8.2	9.3–10.0
DGDG	19.3	6.5–7.2	—	—	21.1	9.6–10.4
Other GL	—	3.3–5.0	—	—	—	1.3–2.1
PE	0.2	2.4–3.4	—	1.8	25.3	3.1–5.8
PC	1.3	8.3–9.2	—	12.0	1.0	2.6–3.2
PI	4.8	—	—	17.9	—	—
PS	—	5.0–6.8	—	—	—	2.8–3.1
LPC	7.0	5.9–6.8	—	—	10.2	3.5–5.1

[a]For definitions of abbreviations, see Table 1. (—) denotes data not reported.
[b]Data from Zeringue and Geuge (60) were converted on 100% recovery of TLC analyses by densitometry.
[c]Data from Chung and Tsen (34) of FL and BL were combined and recalculated to obtain TL composition.

TABLE 38 Nonstarch Lipid Compositions of Free Lipids and Bound Lipids in Rye and Triticale Grains and Their Milled Flours

Lipid class[a]	Lipid composition (% lipids)							
	Rye				Triticale			
	Grain		Flour		Grain		Flour	
	FL	BL	FL	BL	FL	BL	FL	BL
SE	4.9–6.4	11.8–13.0	10.8–11.6	23.1–25.0	4.9–6.5	7.0–16.5	6.4–8.5	7.1–16.2
TG	51.1–56.7	5.5–6.2	48.0–55.1	10.3–14.4	46.0–54.4	3.7–5.1	43.4–49.5	2.0–3.6
DG	10.8–8.6	7.9–9.1	11.3–16.4	5.7–8.0	4.6–15.2	3.4–11.4	4.9–8.3	3.3–6.5
MG	6.3–7.0	4.0–5.9	4.5–5.1	1.0–4.2	3.0–8.6	2.4–3.2	4.8–5.8	2.2–3.1
FFA	10.8–13.8	4.8–5.2	9.5–9.7	8.9–10.5	10.4–21.6	3.6–8.1	16.2–20.9	6.2–9.9
MGDG	10.8–13.8	4.8–5.2	9.5–9.7	8.9–10.5	10.4–21.6	3.6–8.1	16.2–20.9	6.2–9.9
DGDG	4.3–4.6	9.5–10.1	3.9–4.2	16.0–16.5	3.0–4.5	11.8–12.4	3.4–5.5	15.0–18.6
Other GL	0.2–0.7	7.5–10.0	—	2.8–4.2	1.3–2.4	5.4–13.9	—	4.3–12.4
PE	0.8–1.6	4.4–5.5	2.1–2.3	4.2–9.6	2.3–3.2	7.1–9.5	3.4–6.3	7.1–10.8
PC	4.1–4.5	13.2–15.4	1.4	3.8–6.3	33.3–4.8	13.9–15.8	2.6–4.2	8.8–13.5
LPC	—	13.6–14.5	—	7.4–10.3	—	11.5–14.8	—	13.5–17.8
Other PL	1.1–1.7	10.2–12.7	1.5–2.6	3.7–4.1	1.4–1.8	8.7–12.2	1.6–2.7	5.5–7.5

For definitions of abbreviations, see Table 1. (—) denotes data not reported.
Source: Ref. 34.

TABLE 39 Nonstarch Lipid Compositions of Triticale Grain, Bran, Germ, and Endosperm Fractions and Milled Triticale Flour

Lipid class[a]	Lipid composition (% lipids)					
	Grain[b]	Grain[c]	Bran[b]	Germ[b]	Endosperm[b]	Flour[c]
SE	3.9	6.7–11.1	42.1	10.8	3.6	7.1–12.7
TG	35.8	23.9–29.0	20.6	37.2	20.0	18.9–23.6
DG	8.7	5.6–12.7	12.2	2.8	7.8	4.4–7.1
MG	3.8	2.7–6.1	9.3	—	6.8	3.3–3.9
FFA	7.9	7.3–15.3	15.8	14.3	8.1	11.1–15.3
MGDG	3.9	7.3–15.3	—	—	6.7	11.1–15.3
DGDG	12.7	7.1–8.8	—	—	13.6	10.2–13.4
Other GL	—	3.3–8.8	—	—	—	2.5–7.8
PE	9.3	4.6–6.7	—	4.7	20.7	6.6–8.2
PC	6.8	8.8–9.8	—	11.2	5.0	6.5–9.0
PI	2.7	—	—	19.0	4.4	—
PS	—	5.4–6.5	—	—	—	3.9–5.5
LPC	4.5	6.4–6.8	—	—	3.3	7.8–10.7

[a]For definitions of abbreviations, see Table 1. (—) denotes data not reported.

[b]Data from Zeringue and Feuge (60) were converted on 100% recovery of TLC analyses by densitometry.

[c]Data from Chung and Tsen (34) of FL and BL were combined and reevaluated to obtain TL composition.

TABLE 40 Total Nonstarch Lipids in Wheat Kernel Fractions

	Lipid content (mg/100 g db of fraction)				
				Aleurone-free endosperm	
Lipids[a]	Pericarp	Germ	Aleurone	Tetraploid	Hexaploid
Total NL	1,137	15,264–22,905	6,288–8,821	365–402	354–391
SE	174	435–572	110–316	7–15	15–45
TG	205	14,306–21,958	5,245–7,978	263–272	140–222
DG	460	104–438	75–324	40–50	40–56
FFA	283	33–280	170–399	27–45	38–62
MG	15	107–198	47–83	10–36	22–75
Total GL	90	191–525	230–853	154–173	261–408
ASG	50	—	156–245	13–16	11–42
MGDG	*	—	49–87	22–24	27–79
MGMG	*	191–525	11–62	3–10	10–21
DGDG	*	—	43–211	113–115	145–283
DGMG	*	—	27–214	2–11	15–16
Total PL	93	1,979–2,646	1,478–1,575	234–273	190–386
APE	*	—	0–58	33–52	71–141
ALPE	*	—	0–47	42–61	21–84
DPG	*	82	10–91	7	—
PG	*	46–222	70–82	} 31–34	} 4–8
PE	*	359–378	128–152		
PC	*	877–2,378	783–885	18	16–67
PI	*	257–1,010	245–273	—	—
LPG		—	—	} 2.7	} 5–17
LPE	*	7–194	—		
LPC	*	81–359	60–206	87–96	39–95
LPI, LPS	*	—	—	—	—
PA	*	70–135	—	} 6	} 0–11
Others	*	—	8–74		
Total NSL	1,320	17,454–25,877	8,656–10,626	772–829	848–1093

[a]For definitions of abbreviations, see Table 1. (—) denotes data not reported and (*) denotes those not measured.
Source: Ref. 23.

TABLE 41 Compositions of Wheat Total Lipids (Nonstarch and Starch), Nonstarch Lipids in Wheat Fractions, and Starch Lipids[a]

Lipid class[b]	Wheat lipids (% TL)	Nonstarch lipids (% NSL)				Starch lipids (% SL)
		Germ	Aleurone	Endosperm	Milled flour	
Total NL	44.4–56.9	79.3–84.8	72.3–79.5	33.2–45.7	58.2–60.9	4.4–5.9
SE	2.1–2.5	2.7–3.1	1.3–3.6	1.8–4.7	1.9–4.2	0.6–1.0
TG	29.5–47.2	72.6–79.3	60.3–71.9	13.7–26.2	39.6–49.3	0.4–0.5
DG	1.5–2.8	1.7–2.2	0.9–3.4	3.8–5.5	3.3–5.0	0–0.5
FFA	2.4–8.9	0.9–2.0	2.1–6.5	3.6–7.3	3.3–6.5	2.4–3.5
MG	1.5–2.5	0.6–0.8	0.8–1.0	4.9–7.3	2.7–3.9	0.5–0.6
Total GL	12.5–14.4	0–3.5	6.7–9.8	30.7–38.3	23.3–26.8	1.2–5.5
ASG	1.0–1.2	0–1.2	1.8–2.8	1.1–3.4	0.8–4.2	0–1.2
MGDG	0.9–2.6	0–2.4 (MGDG, MGMG, DGDG, DGMG combined)	0.6–1.2	3.3–7.4	5.1–5.9	0.3–0.6
MGMG	0.5–1.1		0.5–1.1	0.9–1.9	0.9–1.4	0–1.1
DGDG	5.4–9.5		0.5–2.4	20.4–26.6	12.6–16.5	0–0.7
DGMG	2.7–4.5		1.5–2.5	1.5–1.7	0.6–3.4	0.6–1.8
Total PL	30.6–41.5	14.3–17.2	13.8–17.9	23.6–34.4	14.2–15.8	90.1–94.4
APE	1.0–7.9	0–0.1	—	8.4–13.4	4.2–4.9	—
ALPE	0.7–2.9	0–0.3	—	4.6–6.9	1.7–2.3	—
DPG	*	0.2–0.6	0.1–1.0	—	—	—
PG	0.9–2.2 (PG, PE combined)	0.1–0.9	1.1–2.7 (PG, PE combined)	0.5–0.7 (PG, PE combined)	0.8–1.1 (PG, PE combined)	—
PE		1.3–2.6				—
PC	2.2–6.8	6.4–10.4	9.0–9.6	3.0–6.3	3.9–4.9	0.5
PI	0–2.6	2.6–2.9	1.9–3.2	—	0.4–0.6	—
LPG	0.7–1.9	—	—	—	0.2	2.8–6.6
LPE	2.1–2.5	0.4	—	0.6–0.8	0.3–0.5	10.3–12.2
LPC	16.0–21.8	0.4–1.1	0.9–2.4	4.6–7.7	1.5–2.1	60.9–76.9
LPI, LPS	0–1.4	1.3	0.1–0.4	—	—	0.5–10.4
PA	0.4–0.8	—	—	0.3–0.7	—	1.7–2.2

[a]Calculated data from Hargin and Morrison (116) for wheat lipids, nonstarch lipids in germ, aleurone, and endosperm and starch lipids in endosperm and from Morrison et al. (112) for milled flours.

[b]For definitions of abbreviations, see Table 1. (—) denotes data not reported and (*) denotes those not measured.

TABLE 42 Total Lipids (Nonstarch and Starch) in Whole Wheats, Nonstarch Lipids in Milled Flours, and Starch Lipids in Wheat Starches

	Lipid content (mg/100 g db of sample)				
		Flour nonstarch lipids[c]			
Lipid class[a]	Whole wheat total lipids[b]	Spring (H)	Winter (H)	Winter (L)	Starch lipids in starch[b]
Total NL	1,304–1,889	1,008	1,136	1,102	35–62
SE	55–68	72	43	35	2–8
TG	867–1,566	674	909	892	2–5
DG	43–201	86	67	60	0–4
FFA	10–261	110	64	61	22–34
MG	37–72	66[d]	53[d]	54[d]	3–5
Total GL	265–436	453	524	422	9–54
ASG	28–45	71	18	15	0–12
MGDG	23–77	87	115	92	2–6
MGMG	13–33	23	17	21	7–15
DGDG	124–278	214	322	283	2–24
DGMG	36–114	58	52	11	—
Total PL	776–1,219	242	293	285	729–1,047
APE	24–232	72	95	85	—
ALPE	18–84	34	33	42	—
DPG	<21	—	—	—	—
PG, PE	22–64	13	19	20	—
PC	57–206	66	96	84	—
PI	<79	11[e]	9[e]	8[e]	—
LPG	18–56	10	5	4	23–54
LPE	62–74	10	7	9	79–104
LPC	469–627	36	29	33	499–864
LPI, LPS	<41	—	—	—	—
PA	11–21	—	—	—	—
Others	—	—	—	—	5–41
Total lipids	2,540–3,328	1,703	1,953	1,809	773–1,171

[a]For definitions of abbreviations, see Table 1. (—) denotes data not reported.
[b]From Ref. 23.
[c]From Ref. 112. (H) denotes high-grade and (L) low-grade wheats.
[d]Including AMGDG.
[e]Including PS.

TABLE 43 Fatty Acid Composition of Wheat Flour Nonstarch and Starch Lipid Classes

Lipid class[a]	Fatty acid (wt %)				
	16:0	18:0	18:1	18:2	18:3
Nonpolar lipids					
Triglyceride, total	14–20	1–2	14–17	56–64	4–6
1-position	24	2	14	54	5
2-position	4	tr	15	76	5
3-position	21	2	12	60	5
Diglyceride, total	15–22	1–2	12–14	58–64	3–6
1-position[b]	36–39	2–3	9–12	42–48	3–5
2-position[c]	1–4	tr	12–17	73–83	5
Monoglyceride	16–22	1–2	9–14	56–65	3–4
FFA	14–20	1–2	9–12	64–68	4–7
SE[d]	42–46	2–4	9–10	40–45	2–4
SE[e]	54–63	2–4	6–7	27–30	2–3
Glycolipids					
ASG[d]	35–36	4	9–10	41–49	3–4
ASG[e]	52–56	2–5	6–9	28–35	2–3
AMGDG, total	24–28	1–2	6–9	59–63	2–5
6-position	44	tr	3	44	6
1 + 2 positions	12	2	7	73	5
AMGMG	32	2	6	58	1
MGDG, total	5–7	tr	7–8	78–82	4–6
1-position	11	1	5	81	1
2-position	tr	tr	9	83	7
MGMG	7–11	1	5–8	77–79	4–5
ADGDG	17	2	13	64	2
DGDG, total	11–16	1–2	6–7	69–74	5–7
1-position	26	2	4	63	4
2-position	2	tr	7	83–84	7
DGMG	11–14	2	5–8	70–75	5–7
Phospholipids					
APE, total	17–20	1	8–10	53–65	3–4
N-acyl	9–17	1–2	10–12	65–74	3–8
O-acyl	21–28	1–2	8–10	56–64	2–3
ALPE, total	16–20	1–2	6–9	52–66	2–3
N-acyl	7–12	1	12–14	68–75	3–7
O-acyl	24–29	1	6–7	60–63	2–3
PE, total	19–22	1–2	9–10	61–67	3–4
1-position	29	3	9	56	3
2-position	4	tr	11	79	5
PC, total	18–23	1–2	12–15	60–63	2–3
1-position	37–38	3	8–11	47–49	2–3
2-position	2–4	1	16–18	74–77	2–4
Nonstarch, LPE	30–35	1–2	6–8	52–57	3–4
Nonstarch, LPC	39–45	1–2	5–7	45–49	2–3
Starch, LPG	19	4	2	74	1
Starch, LPE	20	2	3	74	1
Starch, LPC	35	2	7	54	2

[a]For definitions of abbreviations, see Table 1. (tr) denotes trace amount.
[b]1-position of 1,2-diglyceride (DG) and 1,3-DG and 2-position of 2,3-DG.
[c]2-position of 1,2-DG and 3-position of 1,3-DG and 2,3-DG.
[d]More unsaturated type.
[e]More saturated type.
Source: Ref. 23.

REFERENCES

1. Weber, E. J., Structure and composition of cereal components as related to their potential industrial utilization. IV. Lipids. In: *Industrial Uses of Cereals*, Pomeranz, Y. (Ed.). Am. Assoc. Cereal Chem., St. Paul, MN, 1973, pp. 161–206.
2. Nechaev, A. P., and Sandler, Zh. Ya., *Grain Lipids* (Lipidy Zerna) (Book), Kolos (Moscow), 1975. (Cited by Morrison [3]).
3. Morrison, W. R., Cereal lipids, In: *Advances in Cereal Science and Technology*, Vol. II, Pomeranz, Y. (Ed.). Am. Assoc. Cereal Chem., St. Paul, MN, 1978, pp. 221–348.
4. Morrison, W. R., Acyl lipids in cereals. In: *Lipids in Cereal Technology*, Barnes, P. J., (Ed.). Academic Press, New York, 1983, pp. 11–32.
5. Barnes, P. J., Non–saponifiable lipids in cereals. In: *Lipids in Cereal Technology*, Barnes, P. J. (Ed.). Academic Press, New York, 1983, pp. 33–55.
6. Baisted, D. J., Starch lipids in barley and malt. In: *Lipids in Cereal Technology*, Barnes, P. J. (Ed.). Academic Press, New York, 1983, pp. 93–110.
7. Inglett, G. E., Kernel structure, composition, and quality. In: *Corn: Culture, Processing, and Products*, Inglett, G. E. (Ed.). AVI, Westport, CT, 1970, pp. 123–137.
8. Weber, E. J., Lipids in maize technology. In: *Lipids in Cereal Technology*, Barnes, P. J. (Ed.). Academic Press, New York, 1983, pp. 353–372.
9. Youngs, V. L., Oat lipids. *Cereal Chem.*, *55*:591–597 (1978).
10. Hammond, E. J., Oat lipids. In: *Lipids in Cereal Technology*, Barnes, P. J. (Ed.). Academic Press, New York, 1983, pp. 331–352.
11. Juliano, B. O., The rice caryopsis and its composition. In: *Rice: Chemistry and Technology*, Houston, D. H. (Ed.). Am. Assoc. Cereal Chem., St. Paul, MN, 1972, pp. 16–74.
12. Juliano, B. O., Rice lipids. *Il Riso*, *25*:3–21 (1977).
13. Juliano, B. O., Lipids in rice and rice processing. In: *Lipids in Cereal Technology*, Barnes, P. J. (Ed.). Academic Press, New York, 1983, pp. 305–330.
14. Fujino, Y., Rice lipids. *Cereal Chem.*, *55*:559–571 (1978).
15. Rooney, L. W., Sorghum and pearl millet lipids. *Cereal Chem.*, *55*:584–590 (1978).
16. Mecham, D, K., Lipids. In: *Wheat Chemistry and Technology*, 2nd ed., Pomeranz, Y. (Ed.). Am. Assoc. Cereal Chem., St. Paul, MN, 1971, pp. 393–451.
17. Pomeranz, Y., Chemical composition of kernel structures. In: *Wheat: Chemistry and Technology*, Vol. I, Pomeranz, Y. (Ed.). Am. Assoc. Cereal Chem., St. Paul, MN, 1988, pp. 97–158.
18. Chung, O. K., and Pomeranz, Y., Wheat flour lipids, shortening and surfactants—A three way contribution to breadmaking. *Bakers Dig.*, *51*(4):32–44,153 (1977).
19. Chung, O. K., and Pomeranz, Y., Recent research on wheat lipids. *Bakers Dig.*, *55*(5):38–50,55,96,97 (1981).
20. Chung, O. K., Pomeranz, Y., and Finney, K. F., Wheat flour lipids in breadmaking. *Cereal Chem.*, *55*:598–618 (1978).
21. Morrison, W. R., Wheat lipid composition. *Cereal Chem.*, *55*:548–558 (1978).
22. Morrison, W. R., Lipids in wheat and their importance in wheat products. In: *Recent Advances in the Biochemistry of Cereals*, Laidman, D. L., and Wyn–Jones, R. G. (Eds.). Academic Press, London, 1979, pp. 313–335.
23. Morrison, W. R., Lipids, In: *Wheat Chemistry and Technology*, Vol. I. Pomeranz, Y. (Ed.). Am. Assoc. Cereal Chem., St. Paul, MN, 1988, pp. 373–439.
24. Morrison, W. R., and Coventry, A. M., Extraction of lipids from cereal starches with hot aqueous alcohols. *Staerke*, *37*:83–87 (1985).
25. Pomeranz, Y., Ke, H., and Ward, A. B., Composition and utilization of milled barley products. I. Gross composition of roller–milled and air–separated fractions. *Cereal Chem.*, *48*:47–58 (1971).
26. Bhatty, R. S., and Rossnagel, B. G., Lipids and fatty acid composition of Riso 1508 and normal barley. *Cereal Chem.*, *57*:382–386 (1980).
27. Prudthi, T. D., and Bhatia, I. S., Lipids in cereals. I. *Pennisetum typhoideum.*, *J. Sci. Food Agric.*, *21*:419–422 (1970).

28. Osagie, A. U., and Kates, M., Lipid composition of millet *Pennisetum americanum* seeds. *Lipids*, *19*:958–965 (1984).
29. Lai, C. C., and Varriano–Marston, E., Lipid content and fatty acid composition of free and bound lipids in pearl millets. *Cereal Chem.*, *57*:271–274 (1980).
30. Lorenz, K., and Hwang, Y. S., Lipids in proso millet (*Panicum miliaceum*) flours and brans. *Cereal Chem.*, *63*:387–390 (1986).
31. de la Roche, I. A., Burrows, V. D., and McKenzie, R. I. H., Variation in lipid composition among strains of oats. *Crop Sci.*, *17*:145–148 (1977).
32. Youngs, V. L., Puskulcu, M., and Smith, R. R., Oat lipids. I. Composition and distribution of lipid components in two oat cultivars. *Cereal Chem.*, *54*:803–812 (1977).
33. Sahasrabudhe, M. R., Lipid composition of oats (*Avena sativa* L.). *J. Am. Oil Chem. Soc.*, *54*:80–84 (1979).
34. Chung, O. K., Tsen, C. C., Triticale lipids. In: *Triticale: First Man–Made Cereal*, Tsen, C. C. (Ed.). Am. Assoc Cereal Chem., St. Paul, MN, 1974, pp. 191–200.
35. Osagie, A. U., Total lipids of sorghum grain. *J. Agric. Food Chem.*, *35*:601–604 (1987).
36. Zeringue, H. J., Jr., Singh, B., and Feuge, R. O., Triticale lipids: composition and breadmaking characteristics of triticale flours. *Cereal Chem.*, *58*:351–354 (1981).
37. Pomeranz, Y., Chung, O. K., and Robinson, R. J., Lipids in wheat from various classes and varieties. *J. Am. Oil Chem. Soc.*, *43*:511–514 (1966).
38. Skarsaune, S. K., Youngs, V. L., and Gilles, K. A., Changes in wheat lipids during seed maturation. I. Physical and chemical changes in the kernel. *Cereal Chem.*, *47*:522–532 (1970).
39. Weber, E. J., Corn lipids. *Cereal Chem.*, *55*:572–584 (1978).
40. Weber, E. J., and Alexander, D. E., Breeding for lipid composition in corn. *J. Am. Oil Chem. Soc.*, *52*:370–373 (1975).
41. Freeman, J. E., and Bocan, B. J., Pearl millet: A potential crop for wet milling. *Cereal Sci. Today*, *18*:69–73 (1973).
42. Taira, H., Lipid content and fatty acid composition of nonglutinous and glutinous varieties of foxtail millet. *J. Agric. Food Chem.*, *32*:369–371 (1984).
43. Taira, H., and Chang, W. L., Lipid content and fatty acid composition of Indica and Japonica types of nonglutinous brown rice. *J. Agric. Food Chem.*, *34*:542–545 (1986).
44. Taira, H., Nakagahra, M., and Nagamine, T., Fatty acid composition of Indica, Sinica, Javanica, and Japonica groups of nonglutinous brown rice. *J. Agric. Food Chem.*, *36*:45–47 (1988).
45. Parsons, J. G., and Price, P. B., Search for barley (*Hordeum vulgare* L.) with higher lipid content. *Lipids*, *9*:804–808 (1974).
46. Price, P. B., and Parsons, J. G., Lipids of six cultivated barley (*Hordeum vulgare* L.) varieties. *Lipids*, *9*:560–565 (1974).
47. Weber, E. J., Lipids of maturing grain of corn (*Zea mays* L.): I. Changes in lipid classes and fatty acid composition. *J. Am. Oil Chem.*, *46*:485–488 (1969).
48. Weber, E. J., Lipids of maturing grain of corn (*Zea mays* L.): II. Changes in polar lipids. *J. Am Oil Chem. Soc.*, *47*:340–343 (1970).
49. Fujino, Y., and Mano, Y., Classification of lipids and composition of fatty acids in brown rice. *J. Jap. Soc. Food Nutr.* (Eiyo To Shokuryo), *25*:472–474 (1972).
50. Hirayama, O., and Matsuda, H., Lipid components and distribution in brown rice. Nippon Nogei Kagaku Kaishi, *47*:371–377 (1973).
51. Choudhury, N. H., and Juliano, B. O., Effect of amylose content on the lipids of mature rice grain. *Phytochem.*, *19*:1385–1389 (1980).
52. Price, P. B., and Parsons, J. G., Lipids of seven cereal grains. *J. Am. Oil Chem. Soc.*, *52*:490–493 (1975).
53. Lin, M. J. Y., Youngs, V. L., and D'Appolonia, B. L., Hard red spring and durum wheat polar lipids. I. Isolation and quantitative determinations. *Cereal Chem.*, *51*:17–33 (1974).
54. Welch, R. W., Fatty acid composition of grains from winter and spring sown oats, barley, and wheat. *J. Sci. Food Agric.*, *26*:429–435 (1975).

55. Bhatty, R. S., Note: Distribution of lipids in embryo and bran–endosperm fractions of Riso 1508 and Hiproly barley grains. *Cereal Chem.*, *59*:154–156 (1982).
56. Jellum, M. D., Plant introductions of maize as a source of oil with unusual fatty acid composition. *J. Agric. Food Chem.*, *18*:365–370 (1970).
57. Youngs, V. L., and Puskulcu, H., Variation in fatty acid composition of oat groats from different cultivars. *Crop Sci.*, *16*:881–883 (1976).
58. Taira, H., and Itani, T., Lipid content and fatty acid composition of brown rice of cultivars of the United States. *J. Agric. Food Chem.*, *36*:460–462 (1980).
59. Weihrauch, J. L., and Matthews, R. H., Lipid content of selected cereal grains and their milled and baked products. *Cereal Chem.*, *54*:444–453 (1977).
60. Zeringue, H. J., Jr., and Feuge, R. O., A comparison of the lipids of triticale, wheat and rye grown under similar ecological conditions. *J. Am. Oil Chem. Soc.*, *57*:373–376 (1980).
61. Lorenz, K., Maga, J., Size, C., and Welsh, J., Variability on the limiting amino acid and fatty acid composition of winter wheats and triticales. *J. Agric Food Chem.*, *23*:932–938 (1975).
62. Davis, K. R., Litleneker, N., Le Tourneau. D, Cain, R. F., Peters, L. J., and McGinnis, J., Evaluation of the nutrient composition of wheat. I. Lipid constituents. *Cereal Chem.*, *57*:178–184 (1980).
63. Youngs, V. L., Durum lipids. In: *Durum Wheat: Chemistry and Technology*, Fabriani, G., and Lintas, C. (Eds.). Am. Assoc. Cereal Chem., St. Paul, MN, 1988, pp. 139–148.
64. Slover, H. T., Tocopherols in foods and fats. *Lipids*, *6*:291–296 (1971).
65. Lorenz, K., and Limjaroenrat, P., α–Tocopherol content of triticales and triticale milling fractions. *Lebensm. Wiss. Tech.*, *71*:86–88 (1974).
66. Grams, G. W., Blessin, C. W., and Inglett, G. E., Distribution of tocopherols within the corn kernel. *J. Am. Oil Chem. Soc.*, *47*:337–339 (1970).
67. Morrison, W. R., Coventry, A. M., and Barnes, P. J., The distribution of acyl lipids and tocopherols in flour millstreams. *J. Sci. Food Agric.*, *33*:925–933 (1982).
68. Quackenbush, F. W., Firch, J. G., Brunson, A. M., and House, L. R., Carotenoid, oil, and tocopherol content of corn inbreds. *Cereal Chem.*, *40*:250–259 (1963).
69. Salomatina, L. G., and Olifson, L. E., Chemical composition and physical properties of millet oil. *Maslo–Zhir. Prom.*, *35*:9–11 (1969).
70. Fortman, K. L., and Joiner, R. R., Wheat pigments and flour color. In: *Wheat Chemistry and Technology*, 2nd ed., Pomeranz, Y. (Ed.). Am. Assoc. Cereal Chem., St. Paul, MN, 1971, pp. 493–522.
71. Demchenko, A. I., On the content of carotenoids and tocopherols in barley seed oil. *Izv. Vyssh. Ucheb. Zaved. Pishch. Tekhnol.*, *5*:18–20 (1969).
72. Cabulea, I., Contribution to the study of carotenoid metabolism in the maize grain. *Proc. Meet. Maize and Sorghum Sect.* EUCARPIA (Eur. Ass. Res. Plant Breed.) *5*:85–91 (1971).
73. Grogan, C. O., and Blessin, C. W., Characterization of major carotenoids in yellow maize lines of differing pigment concentration. *Crop Sci.*, *8*(6):730–732 (1968).
74. Blessin, C. W., Brecher, J. D., and Dimler, R. J., Carotenoids of corn and sorghum. V. Distribution of tanthophylls and carotenes in hand–dissected and dry–milled fractions of yellow dent corn. *Cereal Chem.*, *40*:582–586 (1963).
75. Chen, K. T., and Geddes, W. F., Studies on the wheat pigments. MS thesis, University of Minnesota, St. Paul, MN, 1945.
76. Lepage, M., and Sims, R. P. A., Carotenoids of wheat flour: Their identification and composition. *Cereal Chem.*, *45*:600–604 (1968).
77. Laignelet, B., Lipids in pasta and pasta processing. In: *Lipids in Cereal Technology*, Barnes, P. J. (Ed.). Academic Press, New York, 1983, pp. 269–286.
78. Irvine, G. N., and Anderson, J. A., Variation in principal quality factors of durum wheats with a quality prediction test for wheat or semolina. *Cereal Chem.*, *30*:334–342 (1953).
79. Canadian Grain Commission, Canadian Amber Durum Wheat 1979 Crop. Crop Bull. 143. Grain Res. Lab., Winnipeg, Manitoba.
80. Quaglia, G. B., Other durum wheat products. In: *Durum Chemistry and Technology*, Fabriani, G., and Lintas, C. (Ed.). Am. Assoc. Cereal Chem., St. Paul, MN, 1988, pp. 263–282.

81. Wall, J. S., and Blessin, C. W., Composition and structure of sorghum grain. *Cereal Sci. Today*, *14*:264–266, 268–270, 276 (1969).
82. Wall, J. S., and Blessin, C. W., Composition of sorghum plant and grain. In: *Sorghum Production and Utilization*, Wall, J. S., and Ross, W. M. (Eds.). AVI Publ. Co., Westport, CT, 1970, pp. 118–166.
83. Knights, B. A., *The Gas Chromatography of Steroids*, Grant, J. K. (Ed.), Cambridge Univ. Press, Cambridge, 1967, pp. 211–221.
84. Miric, M., Lalic, Z., and Milhajlovic, M., Hrana Ishrana *22*:125–128 (1981). (Cited by Weber [8]).
85. Knights, B. A., Comparison of the grain sterol fractions of cultivated and wild oat species. *Phytochemistry*, *7*:2067–2068 (1968).
86. Weihrauch, J. L., and Gardner, J. M., Sterol content of foods of plant origin. *J. Am. Diet. Ass.*, *73*(1):39–47 (1978).
87. Heupel, R. C., Sauvaire, Y., Le, P. H., Parish, E. J., and Nes, W. D., Sterol composition and biosynthesis in sorghum: Importance to developmental regulation. *Lipids*, *21*:69–75 (1986).
88. Berrie, A. M. M., and Knights, B. A., Sterols of genus *Triticum*. *Phytochemistry*, *11*:2363–2365 (1972).
89. Artaud, J., Iatrides, M.–C., and Estienne, J., Application of high pressure liquid chromatography to the determination of soft wheat in pasta. *Ann. Falsif. Expert. Chim.*, *72*:153–157 (1979).
90. Torres, J. A., Carbonero, P., and Garcia–Olmedo, F., Endosperm sterol phenotype and germination in wheat. *Phytochemistry*, *15*:677 680 (1976).
91. Palmer, M. A., and Bowden, B. N., The pentacyclic triterpene esters and the free, esterified and glycosylated sterols of sorghum *vulgare* grain. *Phytochemistry*, *14*:1813–1815 (1975).
92. Ohnishi, M., Ito, S., and Fujino, Y., Composition and molecular species of waxy lipids in wheat grain. *Cereal Chem.*, *63*:193–196 (1986).
93. Singh, S. P., and Paleg, L. G., Low–temperature–induced GA_3 – sensitivity of wheat. V. Sterol conversions in the wheat aleurone tissue during imbibition. *Aust. J. Plant Physiol.*, *12*:549–555 (1985).
94. MacMurray, T. A., and Morrison, W. R., Composition of wheat flour lipids. *Sci. Food Agric.*, *21*:520–528 (1970).
95. Itoh, T., Tamura, T., and Matsumoto, T., Sterol composition of 19 vegetable oils. *J. Am. Oil Chem. Soc.*, *50*:122–125 (1973).
96. Itoh, T., Tamura, T., and Matsumoto, T., Methylsterol composition of 19 vegetable oils. *J. Am. Oil Chem. Soc.*, *50*:300–303 (1973).
97. Kuroda, N., Ohnishi, M., and Fujino, Y., Sterol lipids in rice bran. *Cereal Chem.*, *54*:997–1006 (1977).
98. Mahadevappa, V. G., and Raina, P. L., Sterol lipids in finger millet (*Eleusine coracana*). *J. Am. Oil Chem. Soc.*, *55*:647–648 (1978).
99. Kornfeldt, A., and Croon, L–B., 4–demethyl–, 4–monomethyl– and 4,4– dimethylsterols in some vegetable oils. *Lipids*, *16*:306–314 (1981).
100. Lercker, G., Capella, P., Conte, L. S., and Falli, B., Composition of wheat germ oil. *Riv. Ital. Sostanze Grasse*, *54*:177–182 (1977). (Cited by Morrison [3]).
101. Curtis, P. E., Leng, E. R., and Hageman, R. H., Developmental changes in oil and fatty acid content of maize strains varying in oil content. *Crop Sci.*, *8*:689–693 (1968).
102. Western, D. E., and Graham. W. R., Jr., Marketing, processing, uses and composition of oats and oat products. In: *Oats and Oat Improvement*, Coffman, F. A. (Ed.). Am. Soc. Agronomy, Madison, WI, 1961, p. 552. (Cited by Weber [1]).
103. Hubbard, J. E., Hall, H. H., and Earle, F. R., Composition of the component parts of the sorghum kernel. *Cereal Chem.*, *27*:415–420 (1950).
104. MacMasters, M. M., Hinton, J. J. C., and Bradbury, D., Microscopic structure and composition of the wheat kernel. In: *Wheat Chemistry and Technology*, Pomeranz, Y. (Ed.). Am. Assoc. Cereal Chem., St. Paul, MN, 1971, pp. 51–113.
105. Weber, E. J., The lipids of corn germ and endosperm. *J. Am. Oil Chem. Soc.*, *56*:637–641 (1979).

106. Tan, S. L., and Morrison, W. R., The distribution of lipids in the germ, endosperm, pericarp and tip cap of amylomaize, LG–11 hybrid–maize and waxy maize. *J. Am. Oil Chem. Soc.*, *56*:531–535 (1979).
107. Price, P. B., and Parsons, J., Distribution of lipids in embryonic axis, bran–endosperm, and hull fractions of hulless barley and hulless oat grain. *J. Agric. Food Chem.*, *27*:813–815 (1979).
108. Price, P. B., and Parsons, J. G., Neutral lipids of barley grain. *J. Agric. Food Chem.*, *28*:875–877 (1980).
109. Parsons, J. G., and Price, P. B., Phospholipids of barley grain. *J. Agric. Food Chem.*, *27*:913–915 (1979).
110. Holmer, G., Ory, R. L., and Hoy, C.–E., Changes in lipid composition of germinating barley embryo. *Lipids*, *8*:277–283 (1973).
111. Salun, I. P., and Kalugina, S. A., Changes in millet lipids during storage. *Izv. Vyssh. Ucheb. Zaved. Pishch. Teknol.*, *5*:21–24 (1973). (Cited by Morrison [3]).
112. Morrison, W. R., Mann, D. L., Wong Soon, and Coventry, A. M., Selective extraction and quantitative analysis of non–starch and starch lipids from wheat flour. *J. Sci. Food Agric.*, *26*:507–521 (1975).
113. Weber, E. J., The lipids of corn germ and endosperm. *J. Am. Oil Chem. Soc.*, *56*:637–641 (1979).
114. Juliano, B. O., Polysaccharides, proteins, and lipids of rice. In: *Rice: Chemistry and Technology*, 2nd ed., Juliano, B. O., (Ed.). Am. Assoc. Cereal Chem., St. Paul, MN, 1985, pp. 59–174.
115. Azudin, M. N., and Morrison, W. R., Non–starch lipids and starch lipids in milled rice. *J. Cereal Sci.*, *4*:23–31 (1986).
116. Hargin, K. D., and Morrison, W. R., The distribution of acyl lipids in the germ, aleurone, starch and non–starch endosperm of four wheat varieties. *J. Sci. Food Agric.*, *31*:877–888 (1980).
117. Soulaka, A. B., and Morrison, W. R., The amylose and lipid contents, dimensions and gelatinization characteristics of some wheat starches and their A– and B– granule fractions. *J. Sci. Food Agric.*, *36*: 708–718 (1985).
118. Soulaka, A. B., and Morrison, W. R., The bread baking quality of six wheat starches differing in composition and physical properties. *J. Sci. Food Agric.*, *36*:719–727 (1985).
119. Morrison, W. R., and Gadan, H., The amylose and lipid contents of starch granules in developing wheat endosperm. *J. Cereal Sci.*, *5*:263–275 (1987).

14

Minor Constituents of Cereals

Margaret Ann Bock
New Mexico State University, Las Cruces, New Mexico

I. VITAMINS

Cereal grains play an important role in meeting the nutrient needs of the human population. Like any food, they are good to excellent sources of some nutrients and low or void in other nutrients. In a given grain, the vitamins content varies from one part of the grain to another. This explains why the removal of certain components during the milling process results in loss of vitamins.

A. Fat-Soluble Vitamins

When considered as a whole, cereals are naturally low in lipids, therefore, they tend to be low in the fat-soluble vitamin A, which is present as the precursor carotenoids, and vitamins D, E, and K. The notable exception is corn. Corn is considered to be a low source (100–1000 IU of Vitamin A/100 g of grain) because of the presence of the carotenoids (1–8). In 1974, enrichment of wheat flour with vitamin A at the rate of 1.3 mg/lb of flour was suggested (9). However, at this point such enrichment is not permitted. Distribution of carotenoids in yellow dent corn fractions is given in Table 1. Aside from corn, minor amounts of carotene are found in barley (0.04 mg/100 g) and brown rice (0.013 mg/100 g).

Oil extracted from the corn and wheat germ provides vitamin E in the form of various tocopherols. Levels of tocopherols found in hard red winter wheat and durum wheat are shown in Table 2. Kutsky (1) and others (2–5,7–8) classify corn and wheat germ oils as high (50–300 mg of vitamin E/100 g) contributors and whole wheat flour as a low (0.5–5 mg/100 g) source of this vitamin. From a stability standpoint, the tocopherols are stable to acid in the absence of oxygen and visible light. They are labile at room temperature when exposed to oxygen, alkalies, and ferric salts. They are also unstable upon exposure to ultraviolet(UV) light. When unsaturated fatty acids such as those in corn oil are oxidized, considerable amounts of tocopherols are lost (12) (Table 3).

Although classified as low (10–100 IU/100 g) sources, various grain oils contribute vitamin D (1–8). In general, this vitamin is stable to heat, acids, and oxygen. In foods which are alkaline and exposed to air and light, vitamin D is slowly destroyed (12) (Table 3).

Vitamin K sources categorized as modest contributors (10–100 μg/100 g) include whole wheat and the wheat germ and bran. Corn is a poor source (0–10 μg/100 g) of vitamin K (1–8). When processing, this vitamin is stable to heat and reducing agents. However, it is labile in the presence of alcoholic alkali, oxidizing agents, strong acids, and light (12) (Table 3).

B. Water-Soluble Vitamins

Cereal grains are important sources of B vitamins, particularly thiamin, riboflavin, niacin, and pyridoxine (B_6), and cereal products are good sources of B vitamins, especially when enriched with thiamin, riboflavin, and niacin.

1. Thiamin

Dietary thiamin contribution of the various cereal grains and their components is outlined in Table 4. In the United States, flour made from wheat can be enriched with thiamin at the rate of 0.44–0.55 mg thiamin/100 g flour (13). In the mid-1970s a proposal for raising enrichment levels to 2.9 mg/100 g flour was suggested (9,14). According to Kutsky (1), barley, corn, oats, and brown rice are classified as moderate sources of thiamin, providing 100–1000 μg thiamin/100 g of grain. Rice bran and wheat germ are classified as high sources of thiamin, providing 1000–10000 μg/100 g of grain component.

Thiamin content of the selected cereal grains is outlined on Table 5. According to this compilation, thiamin content is highest in millet and lowest in rice. Toepfer et al. (11) indicated that durum wheat had a higher average thiamin content than hard red wheat (Table 6). Tables 7 and 8 are good illustrations of the differences in vitamin content of a grain according to fractions. As evident from Table 7, rice germ has the highest thiamin content; milled rice has the lowest. Oat and hulls are low in thiamin, whereas the finished groats, flour, chips, and meal have the highest portions of this vitamin (Table 8).

Thiamin is the most labile of the various B vitamins found in or added to cereal grain products. It is stable under acid conditions but is destroyed in large percentages in air (especially at higher pH), under autoclaving, and during exposure to sulfites and alkalies (12) (Table 3). The stability at acid pH is demonstrated by the small losses noted in fermented baked products (22–23). Lability under alkaline conditions is proved by the significant losses of thiamin in cookies made with baking soda (22) (Figure 1).

2. Riboflavin

Kutsky (1) and others (2–8) classify barley and rice as low (10–100 μg/100 g) dietary sources of riboflavin and corn, oats, rice bran, and wheat germ as moderate sources (100–1000 μg/100 g) (Table 4). Like thiamin, riboflavin enrichment of wheat flour at the rate of 0.26–0.32 mg/100 g flour is allowed in the United States (13). When grains are compared on a mg/100 g basis, rice contains the lowest amount of riboflavin, while millet contains the most (Table 5). Hard red wheat and durum wheat both contain similar amounts of this vitamin (Table 6). Unlike thiamin, riboflavin content is less variable from one rice component to another (Table 7). With the exception of oat shorts, riboflavin is fairly consistent among the various oat products (Table 8).

When exposed to heat in a dry form or in an acid medium, riboflavin is stable. However, this vitamin is very sensitive to light with the of destruction increasing as pH and temperature rise (12). Maleki and Dajhir (24) found that the percent lost from enriched baked Arabic bread decreased from 19% when no riboflavin was added to 9% when flour was enriched at 0.60 mg/100 g flour. Ranhotra and Gelroth (22) found about an 8% loss in fresh baked bread made from enriched flour (Figure 1).

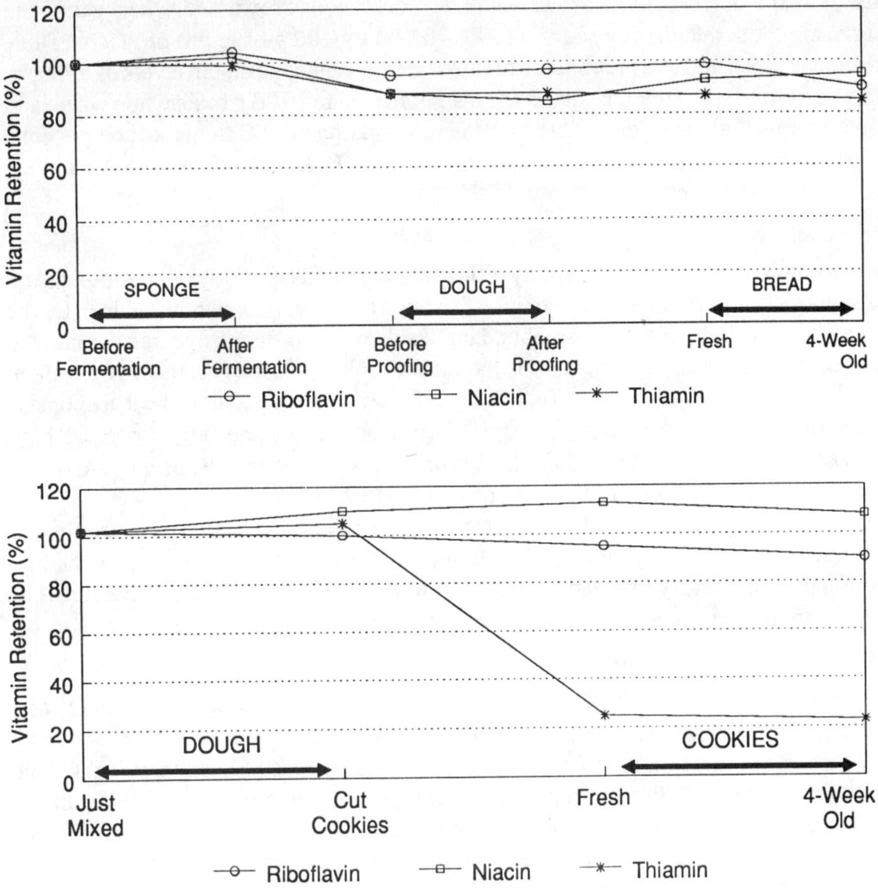

FIGURE 1 Retention of enrichment vitamins at different stages of bread and cookie making.

3. Niacin

Niacin is found in cereal grains in bound and free forms. Since the bound form is poorly utilized by humans (25–26), the niacin added in the enrichment process becomes an important contributor to dietary niacin. The amounts of niacin present in wheat in bound versus free forms are shown in Table 6. One half or less of the total niacin is in the free form. Bound niacin in sorghum constitutes 40% or more of the total niacin (27). Niacin in the form of nicotinic acid can be added to flour at the rate of 3.56–4.45 mg/100 g flour (13) according to enrichment standards. An increase to 24 mg/lb flour was proposed in the mid-1970s (9). In general, niacin is considered to be stable to air, light, heat, acids, and alkalies (12). When wheat flour enriched with niacin is baked into bread, losses in the baking process are minimal (1–2% in Arabic bread) regardless of the level of enrichment (22–24). Not only are losses in the baking process minimal, but Hepburn (28) demonstrated that the proportion of free niacin was higher in baked products like bread, cakes, and crackers than in the flour from which they were made. Alkali treatment of bound niacin, such as that seen in the process of making corn tortillas, improves niacin absorption probably through hydrolysis of bound niacin at the baking stage (23).

Kutsky (1) and others (2–8) indicate that barley, corn, oats, brown rice, wheat germ, and wheat bran are moderate dietary sources (1000–10000 μg/100 g of grain) of niacin (Table 4). Rice bran is classified as a high source of this vitamin. When the niacin levels of different grains are compared, rye contains the smallest amount on a mg/100 g basis while wheat and barley average the highest levels (Table 5). When comprising the different rice components, one can see that niacin is concentrated in the bran (Table 7). In oats, niacin is lowest in the finished groats and highest in the shorts (Table 8).

4. Pyridoxine (B_6)

Vitamin B_6 is composed of three distinctly different compounds: pyridoxine, pyridoxal, and pyridoxamine. The distribution of these three fractions in wheat is shown in Table 6. Of the three fractions, pyridoxine is the most abundant (Table 6). Oats on the average contain the least amount of this vitamin; rice contains the most (Table 5). However, most of the B_6 in oats is highest in the dry-milled oats (Table 8). Barley, corn, oats, rye, and wheat are considered to be moderate sources (100–1000 μg/100 g); wheat germ and brown rice are high (1000–10000 μg/100 g) (Table 4). Enrichment of flour with vitamin B_6 at the rate of 2.0 mg/lb flour has been proposed but never implemented (9,14).

Vitamin B_6, in the form of pyridoxine, is stable when exposed to heat, strong alkaline, or acid but is sensitive to light, especially UV light, in the presence of alkali. Pyridoxal and pyridoxamine are adversely affected by exposure to air, heat, and light. All three forms of this vitamin are destroyed at neutral pH when exposed to UV light (12).

5. Other Vitamins

Information about other vitamins found in cereals is sparse. According to Kutsky (1) and others (2–8), wheat bran is classified as a high source (90–300 μg/100 g) of folic acid; barley, corn, oats, rye, and wheat are categorized as moderate sources (30–90 μg/100 g) (Table 4). In the rice fractions, folic acid tends to be concentrated in the germ(Table 4). In the mid-1970s enrichment of flour with folic acid at the rate of 0.3 mg/lb was suggested (9,14). As with vitamins A and B_6, this proposal was not implemented. Large losses of folic acid are noted at autoclaving temperatures in the presence of acids and alkalies. This destruction is enhanced by oxygen and light (12).

Corn, barley, and rye have similar but low concentrations of pantothenic acid; wheat and oats have comparable but higher amounts of this vitamin (Table 5). In rice, a significant amount of this vitamin is lost in the milling process (Table 7). Oats, wheat, and rice are classified as moderate sources (0.5–2.0 mg/100 g) of pantothenic acid; wheat germ and bran are both considered by Kutsky (1) as high sources (2.0–10.0 mg/100 g) (Table 4).

From a dietary source standpoint, barley, corn, oats, rice, and wheat are considered to be medium sources (10–100 μg/100 g) of biotin (Table 4) (1–8). This vitamin is relatively stable when exposed to air, oxygen, and UV light (12).

Mature, dry cereal grains contain no detectable ascorbic acid (vitamin C). However, this vitamin can be detected in germinated cereal seeds.

II. MINOR CONSTITUENTS OF CEREALS

A. Minerals

Minerals are inorganic elements found in small or trace amounts in various dietary constituents. Based on the amount present in the human body, minerals are classified as major and trace minerals (elements) (29). The major minerals are those that are present in the body in amounts greater than 5 g. Trace minerals are those present in amounts less than 5 g.

The average mineral content of a given grain varies significantly from one part of the world to another. This is a function of factors such as the type of grain, the variety, growing conditions, and fertilizer application.

1. Calcium

About 95% of all mineral matter in cereal grains consists of phytates, phosphates, and sulfates of calcium, magnesium, and potassium (30). The aleurone layer contains 53% of the calcium present in wheat. Because 87% of the phytic acid also resides in this layer, calcium probably exists as a mixed salt of calcium–magnesium phytate (Ca_5Mg phytate) (30).

The mineral composition of various cereals is presented in Tables 9 and 10. The level of calcium is lowest in sorghum and highest in oats. Wheat germ and wheat bran are considered to be low sources of calcium (50–100 mg/100 g): buckwheat, oats, and wheat are classified as moderate sources (100–200 mg/100 g) (1–3,5–6,8,33) (Tables 11,12). One should note that certain grains can be enriched with calcium, thereby making them better sources of this mineral (13). In addition, grains (e.g., corn) that in some countries are traditionally soaked in lime (CaO) may show an increase in calcium content of several hundred milligrams depending on the concentration of the lime solution, the amount of time the grain has been soaked, and how thoroughly or how much the grain had been washed following the soaking process (34). A comparison of varietal mineral content of wheat, rye, barley, and oats is presented in Tables 13–16.

As can be seen from the Koivistoinen et al. study (35), calcium levels in wheat and barley tend to vary from one variety to another. Lorenz and Loewe (36) found variations in calcium content in various classes of hard wheats and also in various classes of U.S. and Canadian soft wheats. However, no differences between hard and soft wheats from either country were observed. According to Koivistoinen et al. (35), calcium levels in various rye and oat varieties tend to be reasonably consistent (Table 13–16).

The effects of various fertilizer treatments on mineral content of spring and winter wheat, barley, oats and rye grown in 10 localities in Finland are shown in Tables 17–21. These data (37) show that fertilizer treatment did not result in a variation in calcium content in the grains studied.

2. Phosphorus

Compared to other minerals, phosphorus is found in large quantities in cereal grains. It is mostly associated with the phytic acid (myoinositol hexaphosphoric acid) and its salts. In wheat, rice, and maize, 80% or more of the total phosphorus is accounted for by the phytate (30). Over 80% of the phytate is located in the aleurone portion of wheat and the pericarp of rice; in corn, over 80% is found in the germ (30). In wheat, phosphorus becomes incorporated into phytic acid during maturation (30). As seen in Table 9, barley contains the highest average levels of phosphorus and whole grain rice, the lowest (285 mg/100 g). From a dietary standpoint, barley, corn, and rice are considered moderate sources of phosphorus (100–200 mg/100 g); buckwheat, millet, oats, brown rice, rice bran, rye, wheat, wheat germ, wheat bran, and wild rice are classified as high sources (200–1200 mg/100 g) (Tables 11,12).

The data in Tables 13–16 indicate that quantities of phosphorus vary significantly from one wheat variety to another. This variation can also be seen in barley. In contrast, phosphorus content from one variety of rye or oats to another does not vary significantly. In the Syvalahti and Korkman study (37), phosphorus content of the grain was not affected by the fertilizer treatments of spring wheat, barley, and rye. Significant differences in phosphorus content were seen in the winter wheat and oats when different fertilizer treatments were used (Tables 17–21).

3. Magnesium

Eighty-seven percent of the magnesium in cereal grains is located in the aleurone layer (30). Because magnesium binds with phytic acid, much of the magnesium is probably present as calcium-magnesium phytate (Ca_5Mg phytate) or as potassium–magnesium phytate (30). The remainder is likely to be present in phosphates and sulfates (30).

From a dietary standpoint, brown rice is considered to be a poor source of magnesium (50–100 mg/100 g). Moderately good sources (100–200 mg/100 g) include barley, millet, oats, rye, wheat, and wild rice. Buckwheat, wheat bran, and wheat germ are considered to be high sources of this mineral (200–400 mg/100 g) (1–3,6,8,33,38) (Tables 11,12). In the mid-1970s, the Food and Nutrition Board proposed that wheat flour be enriched with magnesium at the rate of 200 mg/lb flour (9,14). However, this proposal was never implemented.

According to Tables 9 and 10, magnesium content is highest in wheat and lowest in rice. Magnesium levels vary significantly among wheat, barley, and oat varieties (Tables 13–16). There is little variation in magnesium from one rye variety to another (Tables 13–16). When fertilizer treatments differ, the quantity of magnesium varies significantly in spring wheat and barley. The different fertilizer treatments used by Syvalahti and Korkman (37) did not affect magnesium content in winter wheat, oats, and rye (Tables 17–21).

4. Iron

In corn, iron is concentrated in the outer cells of the scutellum and in the aleurone (30). Iron in wheat is located in the outer endosperm and bran. Patent flour contains about 5.4 μg/g compared to the 124 μg/g in the bran (30). Morris and Ellis (39) have identified a low molecular weight compound in wheat, which is believed to be monoferric phytate. This compound is thought to account for 60% of the iron in wheat bran.

According to Kutsky (1) and others (6,34,40–42), only rice bran, wheat germ, and wheat bran are considered to be good sources of dietary iron (5–18 mg Fe/100 g). Corn and rice are classified as low sources (0.1–1.0 mg Fe/100 g) (Tables 11,12). Although barley, buckwheat, oats, brown rice, rye, and wheat are moderately good sources, enrichment of wheat flour with iron at the rate of 2.89–3.6 mg/100 g of flour has been allowed since the late 1940s (13). In the mid–1970s, the Food and Nutrition Board suggested an increase in the allowed iron enrichment to 40 mg/lb flour (9,14). However, for various reasons related to health and safety, this proposed increase has never become official.

As shown in Tables 9 and 10, iron content is lowest in corn and highest in rye. The amount of iron present in various varieties of rye, barley, and oats dose not tend to vary significantly (Tables 13–16). This is not the case with wheat (35–36). Iron content of barley, oats, and rye is significantly affected by fertilizer treatments (33) (Tables 17–21).

5. Zinc

Information related to zinc content of cereal grains is somewhat limited. Based on the information that is available, zinc content is highest in wheat and lowest in rice (Tables 9,10). According to Koivistoinen et al. (35) and Lorenz and Loewe (36), zinc levels vary significantly from one wheat variety to another; similar variation has been seen among barley varieties. In contrast, there is little difference in zinc content between rye varieties and oat varieties studied by Koivistoinen et al. (35) (Tables 13–16). With the exception of oats and, to a lesser extent, spring wheat, differences in fertilizer treatments had little impact on zinc in the study conducted by Syvalahti and Korkman (37) (Tables 17–21).

From a dietary standpoint, the best sources of zinc in cereals are wheat germ and wheat bran (4–10 mg/100 g). Moderate sources (0.4–4 mg/100 g) include barley, corn, oats, brown rice, rye, and wheat. Corn oil and polished rice are poor sources of zinc (1–3,6,33,38,41–42) (Tables 11,12).

Research coming out of the Middle East has raised some concerns about the bioavailability of zinc from plant products. When the diet is high in cereal grains of high extraction and is consumed in the form of unleavened flour products, this concern may be very significant because zinc deficiency affects growth. However, research on products that are leavened indicates that zinc becomes more physiologically available to a degree greater than can be accounted for by the action of yeast on the phytate. Therefore, the issue of bioavailability appears to be complex, including factors like the impact of yeast, the impact of pH changes, and the phytate-to-zinc ratio (43).

6. Copper

The refining of cereals results in significant loss of copper. However, this loss is not as extensive as the loss of iron, manganese, or zinc (30). Copper in cereal bran is soluble and released at pH 1 but is bound at pH 6.8 (30). The copper content is lowest in corn and highest in oats (Tables 9,10). Wheat is the only grain with a significant variation in copper content among the varieties (35) (Tables 13–16). The copper content of spring wheat, barley, and rye is affected by different fertilizer treatments (Tables 17–21).

From a dietary standpoint, wheat germ and wheat bran are the only cereal grains that serve as good sources of copper (1–10 mg/100 g) (Tables 11,12). Barley, corn, oats, rice, rye, and wheat are considered to be moderately good sources (0.1–1 mg/100 g) of this mineral.

7. Sodium and Potassium

Sodium and potassium are minerals of concern in health care. Therefore, the amounts of these minerals in dietary constituents are of interest. Potassium levels are high in most cereal grains (Tables 9,10). The one notable exception is milled rice (Tables 9,10). Sodium levels are highest in barley and lowest in rice, especially in milled rice (Tables 9,10). According to the work by Koivistoinen et al. (35), potassium content is relatively stable among the different varieties of wheat, rye, barley, and oats studied (Tables 13–16). Their potassium levels found for wheat differ from those of Lorenz and Loewe (36).

Sodium levels vary significantly between the barley and oat varieties but not the wheat and rye varieties studied (Tables 13–16) (31). When comparing hard versus soft U.S. and Canadian wheat, Lorenz and Loewe (36) did not find any significant differences in sodium content. They did note significant differences in sodium quantities in soft and hard wheat classes.

From a dietary source standpoint, no cereal grain is considered to be a high or even moderate contributor of sodium to the diet (Tables 11,12). Buckwheat, rye, and wheat bran are considered to be good sources (400–1000 mg/100 g) of potassium (Tables 11,12).

8. Other Minerals

As their name suggest, trace elements are required only in very small amounts of a few micrograms to a few milligrams per day; nevertheless, they are necessary for good health. Until 1960, nine trace elements were considered essential. These included iron, iodine, copper, manganese, zinc, cobalt, molybdenum, selenium, and chromium (30). Because there is more information about iron, copper, and zinc, these minerals have been discussed previously in this chapter. Since 1960, six additional trace elements have been recognized. These include tin, vanadium, fluorine, silicon, nickel, and arsenic (30). Although the most recently recognized elements are considered to be essential, nutritional deficiency problems are unlikely to occur in man because of the extremely small amounts needed (30).

Information about these trace elements in cereal grains is sketchy. Syvalahti and Korkman (37) measured numerous elements when they were investigating the impact of different fertilizer treatments. They found that the amounts of elements like sulfur, aluminum, boron, bromine, cobalt, fluorine, manganese, molybdenum, selenium, cadmium, and lead tended to vary in the grains studied when fertilized treatments were changed (Tables 17–21). In addition, in some of their analyses, levels of elements like arsenic and mercury were consistently below analytically detectable levels. Lag and Steinnes (44) found that the locality where wheat or barley was grown affected the selenium and molybdenum content of these grains. They even observed large variations in molybdenum levels in these grains within a given area. When comparing their work to others, Lag and Steinnes (44) found some agreement with their own findings but also found variation from their study results. Cereal grain contributors of chromium, cobalt, fluorine, iodine, manganese, molybdenum, and selenium are listed in Tables 11 and 12. Levels of trace elements not previously discussed are shown in Tables 22 and 23. One should note that although buckwheat, corn oil, wheat bran, and wheat germ are considered to be high sources of cobalt, the only known need for cobalt by humans is that associated with the cobalt that comprises the center of the vitamin B_{12} compound, which is only supplied by animal products. Therefore, there is reason to question the value of cereal grains as a source of cobalt.

B. Enzymes

Cereal grains are complex biological systems involving many enzymes that go through several development stages from the planted seed to the ripe grain.

1. Amylases

Because cereals store their excess energy as starch, they contain two types of amylase. α-Amylase is an endoenzyme that randomly cleaves α(1,4)-glucosidic linkages. Such cleavage results in degradation of large molecules (amylose and amylopectin) into fragments with shorter and less complex glucose chains. When the grain is intact, only low levels of this enzyme are detectable. An increase in the activity of α-amylase is an indicator that germination is in progress.

β-Amylase is the second amylase commonly present in cereal grains. It is an exoenzyme that cleaves starch from the nonreducing end of the molecule. This enzyme cleaves alternate α(1,4)-glucosidic linkages, releasing the disaccharide maltose. This enzyme is found in intact grain and does not appear to increase in activity during germination.

Because there are also α(1,6) linkages in amylopectin at the branch points of the molecule, neither α-nor β-amylase is capable of completely degrading starch to simple sugars. When both amylases are functioning in combination, about 85% of the starch is degraded (15).

Wheat, barley, and rye appear to have much higher α-and β-amylase activities than other cereal grains. One should note that a large number of amylase enzymes are found in cereal grains (15). They vary based on the anatomical part and maturity of the grain. In addition, these enzymes appear to be different in the sound, mature grain as compared to the grain after germination.

2. Proteases

Proteases, referred to as proteinases and peptidases, are found in intact cereal grains. However, their levels of activity are low.

Most assays for determining proteolytic activity are based on the measurement of soluble nitrogen caused by proteolysis. Cereal proteins are large and insoluble, therefore, even though a significant amount of enzyme activity may be in progress, soluble nitrogen may not be produced to any great extent (15).

3. Lipase

All cereals have lipase activity. However, this activity varies widely from one cereal to another. Compared to lipase activity observed in wheat and barley, oats and pearl millet have a relatively high activity level. Interest in lipase activity is related to the fact that free fatty acids are more susceptible to oxidative rancidity. In addition, free fatty acids in sufficient quantity can give a product a soapy taste (15).

4. Lipoxygenase

Peroxidation of polyunsaturated fatty acids by oxygen is catalyzed by lipoxygenase. This enzyme occurs commonly in nature and attacks preferentially fatty acids and lipids which contain a *cis*-penta-1,4-diene unit (–CH=CH-CH_2-CH=CH-). A large number of isoenzymes having different activities are thought to exist.

With the exception of pearl millet, in which this enzyme is thought to be absent, lipoxygenase is found in most cereal grains. The reason that lipoxygenase is of concern centers around the fact that this enzyme promote the oxidation of many products (15).

5. Phytase

Phytase is an esterase that is responsible for the hydrolysis of phytic acid (myoinositol hexaphosphoric acid), converting it to inositol and free phosphoric acid. Because phytic acid is thought to bind calcium, magnesium, iron, and zinc, keeping them from being absorbed from food, the activity of this enzyme is very important (15).

6. Other enzymes

Cereals either contain or produce during sprouting enzymes that are involved in oxidizing o-phenylenediamine. Most of that activity appears to be associated with the bran but the significance of the enzyme in cereal grains is unclear. Peroxidase and catalase are also found in cereals. Their role is not clear (15).

7. Enzymes in Barley

In barley, hydrolysis of endosperm components during germination is catalyzed by enzymes like α-amylase. The embryo controls the release of the endosperm components. Activity of these hydrolytic enzymes is stimulated by gibberellic acid from the embryo which moves to the aleurone inducing de novo synthesis of α-amylase. Gibberellic acid appears to promote de novo synthesis and/or release of other enzymes that function in the endosperm.

Barley contains protease inhibitors including two main trypsin inhibitors and two main chymotrypsin inhibitors. The levels of trypsin inhibitors are relatively constant from one genotype to another; levels of chymotrypsin inhibitors vary with the cultivars. The two main chymotrypsin inhibitors and one of the main trypsin inhibitors are degraded by pepsin and, therefore, are not considered to have adverse nutritional effects. However, the other trypsin inhibitor is resistant to the action of pepsin, resulting in lower protein digestibility in cultivars where it is present (47). Research reviewed by Newman and McGuire (47) indicated that the level of this inhibitor is so low that it is unlikely to have a negative impact on the nutritive value of proteins.

8. Enzymes in Corn

Mature corn (dry) has a low level of α-amylase activity, which increases during germination in the scutellum as well as in the kernel with the embryo removed.

Lipase from scutella of corn seedlings has been isolated and characterized by Lin et al. (48) and Lin and Huang (49). Corn lipase is present in the lipid bodies of germinated seeds and is active only on triglycerides containing oleic or linoleic acids. This lipase hydrolyzes triglycerides to three fatty acids for β-oxidation in the glyoxysomes.

The role of lipoxygenase in corn is unclear. Lipoxygenase activity results in fatty acid hydroperoxides formed by oxidation of polyunsaturated fatty acids. These hydroperoxides are highly reactive and potentially damaging to the cellular components. Regulation of plant growth is one of the proposed roles of lipoxygenase (50). The metabolites resulting from lipoxygenase activity resemble prostaglandins and leukotrienes, which are synthesized in mammals from oxygenated fatty acids. These metabolites may also act as metabolic regulators.

9. Enzymes in Millet and Sorghum

Sorghum protein bodies contain protease, phytase, β-and α-glucosidase, β-galactosidase, and several acid phosphatases. Activity of the protease and phytase is outlined in Table 24. In these cereals, phytase and phytic acid are unequally distributed throughout the grain. Phytase activity of sorghum is lower than that in rye and wheat.

As in other grains, amylase activity increases rapidly with germination. Some malted cultivars of finger millet show higher α-amylase activity than comparable wheat and barley malts. In contrast to barley malt, where β-amylase is the major amylase, in sorghum α-amylase predominates at a ratio of 2 or 3 to 1 (27).

10. Enzymes in Rice

Akazawa (52) has presented an extensive review of the enzymes of rice in Rice Chemistry and Technology. The information in Figure 2 summarizes the proposed activities involved in starch synthesis and degradation in rice seeds.

Sucrose metabolism in rice appears to be under the control of two enzymes: sucrose synthetase and sucrose-6-phosphate synthetase. Questions have arisen regarding which of the enzymes reactions is most important in sucrose synthesis and whether reversal of the sucrose synthetase reaction is involved in utilization of sucrose for starch synthesis. Another set of enzymes thought to be involved in the production of UDP-glucose and ADP-glucose are UDP-glucose and ADP-glucose pyrophosphorylases. Although no such enzymes have

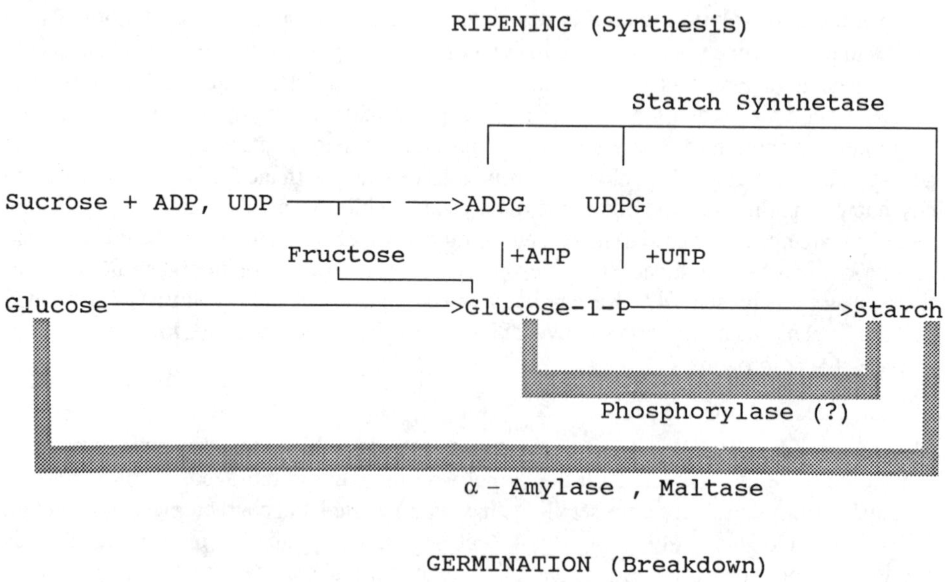

FIGURE 2 Enzymatic activity during starch synthesis and breakdown in rice seeds (From Ref. 52.)

been isolated, enzymes like hexokinase, phosphoglucoisomerase, and phosphoglucomutase are thought to be involved in the overall transformation system of sucrose to starch. As can be seen in Figure 2, questions exist about the presence and/or role of phosphorylase in the formation and degradation of starch. An enzyme (amylo-1,4–1,6-transglucosidase/Q enzyme/branching enzyme) is thought to play a role in the branching of amylopectin and is also thought to be present in the developing rice seed.

From Figure 2, one can see that α-amylase may be involved in the breakdown of starch to glucose. Research into the stimulating effect of gibberellic acid on de novo synthesis of α-amylase in aleurone cells has led to the hypothesis that a similar mechanism is operative in germinating rice seeds.

As in other cereals, phytin is also present in rice. However, there is only limited information about the formation of this compound. Phytase isolated from rice has been used for investigations of plant phosphatase. However, little investigation related to the role of phytase in the physiology of rice seed germination has been conducted.

During storage of polished rice seeds, an increase in free fatty acids occurs. Studies by Funatsu et al. (53) have detected two major isozymes of lipase in rice bran.

11. Enzymes in Rye

When rye kernels are dissected into pericarp, aleurone, and endosperm, at least three different α-amylase activities can be noted at various points of development. Activity in the pericarp declines as the grain matures. α-amylase activity fluctuates in the aleurone during development but diminishes at or close to maturity. This activity increases in the endosperm, and at maturity this is the major activity associated with this enzyme in the grain. As in other grains, β-amylase activity is present in low concentrations at maturity. There has been little investigation of this enzyme in rye (54).

Proteolytic enzymes similar to those observed in wheat are present in rye. Concentrations of these enzymes tend to be higher in the bran layers. Protease activities observed in some rye samples appear to be comparable with those observed in triticale but higher than those of wheat (54).

12. Enzymes in Wheat

Two α-amylases can be detected in ungerminated wheat. One fraction is heat and inhibitor sensitive; the second resembles the α-amylase of germinated wheat (43). As in other cereal, α-amylase of wheat acts on amylose in a random manner.

β-Amylase has also been identified. It consists of three fractions, differing in electrophoretic behavior. β-Amylase splits maltose from the starch molecule starting at the nonreducing end. Almost complete hydrolysis to maltose is observed when amylose is the substrate. In amylopectin, the action stops short of the branch points. A comparison of total amylase activity in selected varieties of cereal grains is outlined in Table 25. The impact of pH, temperature, substrate, and inhibitors on the amylase of wheat is discussed by Kruger and Reed in Wheat Chemistry and Technology (55).

Proteinases and peptidases occur in wheat. The hull and endosperm have little or no proteolytic activity even after germination. Both proteinase and dipeptidase activities are observed in the scutellum and embryo; rapid increases in activity of these enzymes are observed on germination (55).

Lipases are enzymes that catalyze the hydrolysis of esters, regardless of the compound hydrolyzed. Low-grade flours have higher lipase activity than clear or patent flours, bran, or wheat germ. The scutellum of the wheat berry has 3 1/2–4 times as much lipase activity as the cotyledon and endosperm, 4–5 times as much as the hull, and 15 times as much as the radicles. As in other grains, the possible deleterious effects of fatty acids released during lipolytic activity may have on storage and baking quality are of concern (55).

Phosphatases like phytase also occur in wheat and other cereal grains. Phytase activity increases sixfold on germination in soft wheat and is about 20% greater on germination in hard wheat. When bread products using high extraction flours constitute a significant part of the diet, the activity of phytase may be important releasing minerals like calcium, magnesium, iron, and zinc which have complexed with phytic acid (55).

Aside from phytase, other phosphatases also occur in wheat (e.g., hexosediphosphatase and pyrophosphatase). Adenosine-triphosphatase, glycerophosphatase, and thiamine pyrophosphatase have been detected histochemically from wheat and sprouts (55).

Oxidases present in wheat include lipoxidase, phenol and aromatic amine oxidases, catalase, and peroxidase. Although lipoxidase activity, which is responsible for catalyzing peroxidation of polyunsaturated fatty acids, is only one fourth of that in soybeans, activity of this enzyme is highest in red wheats and lowest in white and durum wheats. In the milled fractions its activity is highest in the germ and lowest in the patent flour (germ > shorts > clear > flour > bran > patent flour) (55).

The phenol and aromatic amine oxidases catalyze the oxidation of a variety of aromatic substances using molecular oxygen. Oxidases that include a compound capable of oxidizing *o*-phenylenediamine, have been detected in wheat and barley homogenates but not in corn or rice. Pyrogallo oxidase and catechol oxidase activity have also been reported in wheat (55).

Catalase, a hemoprotein that catalyzes the degradation of hydrogen peroxide to water and oxygen, is present in wheat. Its physiological function is not well understood. Higher activity of the enzyme was found in rust-suspectable varieties than in rust-resistant wheat varieties (55).

Peroxidase is also a hemoprotein. It is involved in the degradation of a number of aromatic amines and phenols by hydrogen peroxide. The activity of this enzyme is greatest in wheat, followed by barley, corn, and rice (55).

C. Pigments

Color of a given plant is most likely due to the pigments present. Most pigments are located in plastids, which are specialized bodies in the protoplasm of the cell. The chief plant pigment are classified as the carotenoids, the chlorophylls, the anthoxanthins, and the anthocyanins.

1. Pigments in Barley

Anthocyanins of the cyanidin types can be found in the green organs of barley cultivars; catechins are located in the seed coat and mature kernel. These are phenolic compounds; they form insoluble complexes with protein that are thought to cause the haze in beer. There is also some possibility that these complexes reduce the digestibility of the proteins in animals. In addition, tannins, which are polyphenolic compounds, complex with proteins and often account for off-colors that are noted, especially during processing. Tannins also are present in barley. Their content varies depending on the variety (47).

2. Pigments in Corn

Corn is one of the few cereal grains that contain carotenoids to any degree. The carotenoids in corn are classified as carotenes and xanthophylls. Carotenes are important constituents of corn because they are the precursors of vitamin A. Xanthophylls are responsible for the yellow of egg yolk and skin color of chickens fed corn-based diets (56). Corn kernels can differ in color. Colors range from white to yellow, orange, red, purple, and brown. Differences in color may be due to genetic differences in the pericarp, aleurone, germ, and endosperm (57). In addition, carotenoids also function as antioxidants. The distribution of carotenoids in yellow dent corn is outlined in Table 1. The predominant xanthophylls in corn are lutein and zeaxanthin; β-carotene is the principal carotene. One should note that these pigments are subject to oxidation and, therefore, can be destroyed during storage if not handled properly.

3. Pigments in Rice

Anthocyanin pigments have been found in the red and purple cultivars of rice. Phenolic compounds have been reported present in the hulls. It is thought that the seed coats and pericarp contain the same phenolic compounds observed in other tissues of the rice plant. The anthocyanin pigments account for the yellowing in rice observed during cooking in alkaline waters (31).

Chlorophyll is responsible for the greenish color of crude oil extracted from rice bran. However, this color is easily bleached during conventional refining of rice (31).

4. Pigments in Sorghum and Millet

The presence of tannins in some genotypes of sorghum is very important. These polyphenols are thought to cause the deep red to brown colors seen in some grains as they ripen. They also may be responsible for the off-colors noted in some products made from sorghum. Seed color, tannin content, and protein content tend to be correlated. Bird damage appears to be negatively correlated with the amount of tannin and the seed color (27).

Although of questionable nutritional significance, β-carotene is higher in yellow than in white endosperm sorghums (27).

In sorghum hulls, research indicates that the pigment consists of apigenin, phlobaphene, and durasantalin. Color in the seed coat is attributed to durasantalin and phlobaphene. Various other pigments have been observed with selected extracting procedures (27).

The major pigments in pearl millet are flavonoids, which are pH sensitive. Glucosylvietexin is the major c-glycosylflavnoid in millet. Alkaline-labile phenolic acid in the form of ferulic acid has also been detected in the flour from millet. These phenolics are concentrated in the peripheral endosperm, and their content is, therefore, reduced upon dehulling (27).

5. Pigments in Wheat

In wheat, origin, growing conditions, and variety influence the degree and/or content of yellow pigment. This presents a real problem because flours for bread and cakes need to have a minimum of yellow color, whereas flours for pasta products need to retain yellowness. Research indicates that the principal pigments in wheat associated with yellowness are free xanthophyll and xanthophyll esters. Aside from xanthophyll and its esters, other pigments reported in flours include carotene, flavones (aricin), kryptoxanthin, and degradation products of chlorophyll (55). One should note that xanthophyll has no potential to form vitamin A and that carotene is probably of no nutritional importance. This is likely a function of low amounts present and high susceptibility of carotene to oxidation.

TABLE 1 Distribution of Carotenoids in Yellow Dent Corn Fractions

Fraction	Variety	Carotenes (ppm)
Whole corn	S X 20	1.8
	Pioneer 329	2.0
	Mix[a]	1.7
Bran	S X 20	0.3
	Pioneer 329	0.3
	Mix[a]	0.2
Germ	S X 20	0.5
	Pioneer 329	1.0
	Mix[a]	1.3
Floury endosperm	S X 20	1.1
	Pioneer 329	0.3
	Mix[a]	1.1

[a]Commercial mix.
Source: Ref. 10.

TABLE 2 Tocopherol Content of Hard Red and Durum Wheat

Tocopherols	Hard red wheat (μg/g[a])	Durum wheat
Total	58	58
α	13	10
β	7	5
γ	—	—
δ	—	—
α–3	5	7
β–3	33	36

[a]Dry wt. basis.
Source: Ref. 11.

TABLE 3 Stability of Nutrients Under Selected Conditions

Vitamin	Neutral pH 7	Acid pH <7	Alkaline pH >7	Air or oxygen	Light	Heat	Max. cooking loss (%)
A	S	U	S	U	U	U	40
Carotene	S	U	S	U	U	U	30
D	S		U	U	U	U	40
E	S	S	S	U	U	U	55
K	S	U	U	S	U	S	5
Thiamin	U	S	U	U	S	U	80
Riboflavin	S	S	U	S	U	U	75
Niacin	S	S	S	S	S	S	75
B6	S	S	S	S	U	U	40
Folic acid	U	U	S	U	U	U	100
Panothenic Acid	S	U	U	S	S	U	50
Biotin	S	S	S	S	S	U	60

S = stable, U = unstable.
Source: Ref. 12.

TABLE 4 Levels of Water-Soluble Vitamins Contributed by Cereal Grains or Grain Components

Grain	Level[f]	Thiamin	Riboflavin	Niacin	Vitamin B_6	Folic Acid	Pantothenic Acid	Biotin
	Low		X					
Barley	Medium	X		X	X	X		X
	High							
	Low							
Buckwheat	Medium							
	High							
	Low							
Corn	Medium	X	X	X	X	X		X
	High							
	Low							
Millet	Medium							
	High							
	Low							
Oats	Medium	X	X	X	X	X	X	X
	High							
	Low		X			X[a]		
Rice	Medium	X[a]	X[b]	X[a]			X	X
	High	X[b]		X[b]	X[a]			

	Low							
Rye	Medium		X	X	X			
	High							
	Low							
Sorghum	Medium							
	High							
	Low							
Triticale	Medium							
	High							
	Low							
Wheat	Medium		X[c]	X[c,d]	X	X	X	X
	High	X[c]			X[c]	X[d,e]	X[c,e]	
	Low							
Wild Rice	Medium							
	High							

[a]Brown rice.
[b]Rice bran.
[c]Wheat germ.
[d]Wheat.
[e]Wheat bran.
[f]Denotes quantity, not quality.
Source: Refs. 1, 2–5, 7–8.

TABLE 5 Vitamin Composition of Selected Cereal Grains (mg/100 g)

Vitamin	Wheat	Rye	Barley	Oats	Rice	Corn	Sorghum	Millet	Triticale	Wild rice
B_1	0.55	0.44	0.57	0.77	0.33	0.44	0.58	0.73	0.65	0.45
B_2	0.13	0.18	0.22	0.18	0.09	0.13	0.17	0.38	0.25	0.63
Niacin	6.40	1.50	6.4	1.8	4.9	2.6	4.8	2.3	3.5	6.2
Pantothenic acid	1.36	0.77	0.73	1.4	1.2	0.70	1.0			
B_6	0.53	0.33	0.33	0.13	0.79	0.57	0.60			

Source: Refs. 14–19.

TABLE 6 B-Vitamin Concentration in Hard Red versus Durum Wheat (mg/100 g, dry wt. basis)

Vitamin	Hard red wheat	Durum wheat
Thiamin	0.57	0.67
Riboflavin	0.12	0.11
Niacin		
Total	7.4	11.1
Free	3.6	4.7
Vitamin B_6	0.35	0.43
Pyridoxine	0.26	0.33
Pyridoxal	0.05	0.06
Pyridoxamine	0.04	0.04

Source: Ref. 11.

TABLE 7 Vitamin Content of Rice and Rice Components (mg/100 g)

Vitamin	Brown rice	Milled rice	Rice bran	Rice polish	Rice germ
Thiamin	0.34	0.07	2.26	1.84	6.5
Riboflavin	0.05	0.03	0.25	0.18	0.5
Niacin	4.7	1.6	29.8	28.2	3.3
Pyridoxine	1.03	0.45	2.5	2.0	1.6
Pantothenic acid	1.5	0.75	2.8	3.3	3.0
Folic acid	0.02	0.02	0.15	0.19	0.43

Source: Refs. 16, 20.

TABLE 8 Selected B-Vitamin Content of Oats and Oat Products (mg/100 g)

Product[a]	Thiamin	Riboflavin	Niacin	Pyridoxine
Dry milled oats	0.65	0.14	1.15	0.20
Finished groats	0.77	0.14	0.97	0.12
Hulls	0.15	0.16	1.04	—
Oat shorts	0.44	0.35	1.62	—
Oat flour, chips, and meal	0.78	0.17	1.25	—

[a]Products from same mill fraction.
Source: Refs. 16, 21.

TABLE 9 Mineral Composition of Rye, Wheat, Barley, Corn, Oats, and Rice (mg/100 g, dry wt.)

			Barley			Oats		Rice	
	Rye	Wheat	Whole grain	Kernel only	Corn	Whole grain	Kernel only	Whole grain	Kernel only
Phosphorus	380	410	470	400	310	340	400	285	290
Potassium	520	580	630	600	330	460	380	340	120
Calcium	70	60	90	80	30	95	66	68	67
Magnesium	130	180	140	130	140	140	120	90	47
Iron	9	6	6	—	2	7	4	—	6
Copper	0.9	0.8	0.9	—	0.2	4	5	0.3	0.4
Mangenese	7.5	5.5	1.8	—	0.6	5	4	6	2
Zinc	3.4	4.4	4.0	—	—	3.9	—	1.5–2.2	1.2–2.1
Sodium	3.1	4.6	11.8	—	—	8.6	—	3.1–6.9	2.2–5.1

Source: Refs. 15, 17, 31–32.

TABLE 10 Mineral Composition of Sorghum, Triticale, and Wild Rice[a]

	Sorghum	Triticale	Wild rice
Phosphorus	405	0.19%	0.4–0.5%
Potassium	400	1.21%	0.4–0.6%
Calcium	20	0.21%	0.01–0.03%
Magnesium	150	0.16%	0.1–0.2%
Iron	6	—	12–51 ppm
Copper	0.5	3.9 ppm	1.8–14.5 ppm
Manganese	1.5	37 ppm	
Zinc	0.0008%	36 ppm	40–121 ppm
Sodium	0.00008%	—	—

[a]mg/100 g (dry wt) unless otherwise noted.
Source: Refs. 15, 17, 31–32.

TABLE 11 Levels of Minerals Contributed by Cereal Grains or Grain Components

Grain	Level[1]	Ca	P	Mg	Fe	Zn	Cu	Na	K
Barley	Low							X	X
	Medium		X	X	X	X	X		
	High								
Buckwheat	Low								
	Medium	X			X				
	High		X	X					X
Corn	Low				X	X[e]	X		X
	Medium		X			X			
	High								
Millet	Low								
	Medium			X					
	High		X						
Oats	Low								
	Medium	X		X	X	X	X		
	High		X						
Rice	Low			X[c]	X	X[f]		X[c]	
	Medium		X		X[c]	X[c]	X		X[c]
	High		X[c,d]		X[d]				

Rye	Low								
	Medium			X	X	X	X		
	High		X						X
Sorghum	Low								
	Medium								
	High								
Triticale	Low								
	Medium								
	High								
Wheat	Low	X[a,b]						X	
	Medium	X		X	X	X	X		X
	High		X	X[a,b]	X[a,b]	X[a,b]	X[a,b]		X[b]
Wild rice	Low								
	Medium			X					
	High		X						

[a]Wheat germ.
[b]Wheat Bran.
[c]Brown rice.
[d]Rice bran.
[e]Corn oil.
[f]Polished rice.
[g]Corn starch.
[h]Corn meal.
[i]Oats and oat bran.
[j]Wheat.
[k] Wheat flour.
[l] Denotes quantity, not quality.

TABLE 12 Levels of Trace Minerals Contributed by Cereal Grains or Grain Components

Grain	Level[1]	Cr	Co	F	I	Mn	Mo	Se
Barley	Low	X	X		X			
	Medium							
	High					X	X	X
Buckwheat	Low							
	Medium							
	High		X			X	X	
Corn	Low	X			X	X[e]		X
	Medium		X[e]					X
	High		X[g,h]					
Millet	Low							
	Medium							
	High							
Oats	Low				X			
	Medium							
	High					X[i]	X	
Rice	Low	X[f]	X[f]		X			
	Medium	X[c]	X[c]			X	X	X[f]
	High							X[c]

Rye	Low				X			
	Medium							
	High					X		
Sorghum	Low							
	Medium							
	High						X	
Triticale	Low							
	Medium							
	High							
Wheat	Low				X[a,b]			
	Medium		X	X[a,b]			X[a]	
	High	X[a,b]	X[a,b]			X[a,b]	X[b,j]	X[a,b,k]
Wild rice	Low							
	Medium							
	High							

[a]Wheat germ.
[b]Wheat Bran.
[c]Brown rice.
[d]Rice bran.
[e]Corn oil.
[f]Polished rice.
[g]Corn starch.
[h]Corn meal.
[i]Oats and oat bran.
[j]Wheat.
[k]Wheat flour.
[l]Denotes quantity, not quality.

TABLE 13 Varietal Differences in Mineral Content of Wheat (mg/100 g, dry wt.)

Wheat variety	Fe	Cu	Mn	Zn	Mg	K	Na	Ca	P
Apu	7.45	0.90	7.06	4.63	173	381	5.61	49.2	421
Norrona	4.76	0.63	5.56	4.03	164	363	3.72	35.6	378
Svenno	5.41	0.62	5.48	4.29	171	378	4.66	35.6	416
Timantti	6.07	0.70	6.06	4.59	162	373	5.12	40.4	434
Selkirk	5.70	0.73	6.20	5.54	200	347	4.14	43.7	465
Timanti II	6.16	0.90	5.16	4.23	176	399	5.79	80.7	464
Erli	4.36	0.57	5.01	3.59	165	371	3.67	31.7	378
Touko	5.07	0.59	5.88	4.60	192	377	3.08	37.1	470

Source: Ref. 35.

TABLE 14 Varietal Differences in Mineral Content of Rye (mg/100 g, dry wt.)

Rye variety	Fe	Cu	Mn	Zn	Mg	K	Na	Ca	P
Pekka	4.43	0.71	4.14	3.41	124	480	2.18	37.5	427
Maatiaisruis	4.50	0.56	3.88	3.78	125	519	2.41	35.2	406
Toivo	4.50	0.63	4.14	3.40	124	521	3.30	36.9	456
Ensi	4.58	0.58	3.75	3.65	124	485	2.96	39.8	377
Oiva	4.42	0.67	4.08	3.63	125	523	3.00	36.1	449
Sangaste	4.17	0.72	4.13	3.17	124	508	2.70	33.0	381

Source: Ref. 35.

TABLE 15 Varietal Differences in Mineral Content of Barley (mg/100 g, dry wt.)

Barley variety	Fe	Cu	Mn	Zn	Mg	K	Na	Ca	P
Tamini	4.75	1.16	2.50	3.63	146	521	13.55	43.3	425
Paavo	4.92	1.00	1.83	3.40	125	566	10.00	38.4	365
Balden	5.21	1.08	1.91	3.89	135	523	8.70	34.1	411
Pirkka	5.48	1.25	2.25	4.32	148	518	12.81	38.5	449
Otra	5.04	1.13	2.23	3.80	148	530	10.93	40.5	429

Source: Ref. 35.

TABLE 16 Varietal Differences in Mineral Content of Oats (m/100 g, dry wt.)

Oats variety	Fe	Cu	Mn	Zn	Mg	K	Na	Ca	P
Siste	5.00	1.61	5.79	3.82	151	421	8.47	68.6	358
Pendek	4.55	1.16	7.04	3.47	139	406	10.16	62.3	364
Sol II	4.91	1.23	7.98	3.93	142	392	7.22	65.1	357
Kyro	5.47	1.30	7.17	4.05	148	414	6.79	75.0	376
Elo	5.19	1.70	8.08	4.17	150	447	8.55	67.9	382
Jammi	4.57	1.22	7.68	4.05	144	422	7.10	66.8	392
Kultasade II	5.14	1.20	5.59	3.64	142	409	10.71	69.8	355
Nip	5.90	0.97	8.37	3.90	170	410	10.85	71.6	398

Source: Ref. 35.

TABLE 17 Effects of First Year Fertilization on Mineral Content of Spring Wheat[a]

Treatment[b]	Ca (g)	Mg (g)	P (g)	K (g)	S (g)	Al (mg)	B (g)	Br (mg)	Cr (mg)
1	0.34	1.46	4.34	5.2	1.48	36.2	1.5	4.3	0.04
2	0.34	1.48	4.33	5.2	1.53	12.4	1.2	2.1	0.03
3	0.34	1.48	1.33	5.0	1.58	12.5	1.3	2.6	0.02
4	0.34	1.44	4.35	5.1	1.55	9.1	1.8	2.5	0.03
5	0.34	1.43	4.37	5.1	1.59	16.8	1.6	2.3	0.03
	Co (mg)	Cu (mg)	F (mg)	Fe (mg)	Mn (mg)	Mo (mg)	Ni (mg)	Rb (mg)	Se (mg)
1	0.03	6.2	0.92	76	50	0.09	0.3	4.9	0.01
2	0.02	6.5	0.78	64	55	0.10	0.3	4.5	0.01
3	0.02	6.7	0.68	62	55	0.10	0.3	4.2	0.01
4	0.02	6.1	0.80	62	57	0.10	0.3	3.9	0.03
5	0.02	6.2	0.77	67	54	0.07	0.3	3.8	0.01
	Si (mg)	Zn (mg)	As (mg)	Cd (mg)	Pb (mg)	Hg[c] (mg)			
1	760	54	<0.05	0.05	0.07	0.004			
2	150	55	<0.05	0.06	0.04	0.004			
3	—	54	<0.05	0.06	0.04	0.004			
4	—	55	<0.05	0.06	0.04	0.004			
5	—	55	<0.05	0.08	0.04	0.004			

[a]Per kg dry matter.
[b]1 = No fertilizer;
2 = NPK (N, P, K), Ca, S, Cl;
3 = NPK, Ca, S, Cl, Mg, B Cu, Mn;
4 = NPK, Ca, S, Cl, Mg, B, Cu, Mn, Zn, Fe, Co, Mo, Se, I, Cr, Sn, and F;
5 = NPK, Ca, S, Cl, Hg, Cd, As, and Pb.
[c]Less than 0.004.

Mg (95 kg/ha);
B (2.1 kg/ha);
Cu (12.5 kg/ha);
Mn (13 (kg/ha);
Zn (1.75 kg/ha);
Fe (10 kg/ha);
Co (0.21 kg/ha);
Mo (0.39 kg/ha);

Se (0.21 kg/ha);
I (0.84 kg/ha);
Cr (0.13 kg/ha);
F (0.13 kg/ha);
Hg (0.17 kg/ha);
Cd (0.17 kg/ha);
As (0.45 kg/ha);
Pb (0.35 kg/ha).

Experiment conducted in 10 locations in Finland over 3 years.
Source: Ref. 37.

TABLE 18 Effects of First Year Fertilization on Mineral Content of Winter Wheat[a]

Treatment[b]	Ca (g)	Mg (g)	P (g)	K (g)	S (g)	Al (mg)	B (g)	Br (mg)	Cr (mg)
1	0.30	1.39	3.98	5.3	1.33	5.7	1.0	2.6	< 0.02
2	0.30	1.39	3.88	5.2	1.38	6.3	1.0	3.5	< 0.02
3	0.30	1.36	3.83	5.4	1.40	4.4	1.2	1.5	< 0.02
4	0.30	1.35	3.80	5.2	1.45	4.3	1.3	1.6	< 0.02
5	0.30	1.38	3.75	5.2	1.53	5.2	1.0	1.4	0.03
	Co (mg)	Cu (mg)	F (mg)	Fe (mg)	Mn (mg)	Mo (mg)	Ni (mg)	Rb (mg)	Se (mg)
1	0.02	5.1	0.64	64	43	0.19	0.3	4.0	0.01
2	0.02	5.3	0.58	65	46	0.19	0.4	4.4	0.01
3	0.02	5.3	0.58	66	43	0.17	0.3	4.1	0.01
4	0.02	5.4	0.75	67	43	0.13	0.3	4.2	0.03
5	0.03	5.7	0.69	71	47	0.13	0.3	4.3	0.01
	Si (mg)	Zn (mg)	As (mg)	Cd (mg)	Pb (mg)	Hg[c] (mg)			
1	360	41	< 0.05	0.13	0.06	0.004			
2	310	42	< 0.05	0.12	0.08	0.004			
3	—	41	< 0.05	0.10	0.05	0.004			
4	—	42	< 0.05	0.09	0.05	0.004			
5	—	43	< 0.05	0.11	0.05	0.004			

[a]Per kg dry matter.
[b]1 = No fertilizer;
2 = NPK (N, P, K), Ca, S, Cl;
3 = NPK, Ca, S, Cl, Mg, B Cu, Mn;
4 = NPK, Ca, S, Cl, Mg, B, Cu, Mn, Zn, Fe, Co, Mo, Se, I, Cr, Sn, and F;
5 = NPK, Ca, S, Cl, Hg, Cd, As, and Pb.
[c]Less than 0.004.

Mg (95 kg/ha); B (2.1 kg/ha); Cu (12.5 kg/ha); Mn (13 (kg/ha); Zn (1.75 kg/ha); Fe (10 kg/ha); Co (0.21 kg/ha); Mo (0.39 kg/ha); Se (0.21 kg/ha); I (0.84 kg/ha); Cr (0.13 kg/ha); F (0.13 kg/ha); Hg (0.17 kg/ha); Cd (0.17 kg/ha); As (0.45 kg/ha); Pb (0.35 kg/ha).

Experiment conducted in 10 locations in Finland over 3 years.
Source: Ref. 37.

TABLE 19 Effects of First Year Fertilization on Mineral Content of Rye[a]

Treatment[b]	Ca (g)	Mg (g)	P (g)	K (g)	S (g)	Al (mg)	B (g)	Br (mg)	Cr (mg)
1	0.36	1.14	3.72	6.0	1.33	9.7	1.5	1.9	0.02
2	0.35	1.13	3.62	6.0	1.30	20.5	1.5	1.8	<0.02
3	0.37	1.16	3.59	6.0	1.33	20.5	1.6	1.5	<0.02
4	0.35	1.15	3.60	6.0	1.31	10.5	1.6	1.2	<0.02
5	0.35	1.14	3.56	5.7	1.32	9.9	1.5	1.1	<0.02
	Co (mg)	**Cu (mg)**	**F (mg)**	**Fe (mg)**	**Mn (mg)**	**Mo (mg)**	**Ni (mg)**	**Rb (mg)**	**Se (mg)**
1	0.03	6.4	0.62	49	41	0.15	0.2	3.9	0.01
2	0.03	6.1	0.64	51	41	0.15	0.2	3.8	0.01
3	0.03	6.5	0.63	55	39	0.15	0.2	3.9	0.01
4	0.03	6.6	0.70	52	40	0.18	0.2	3.8	0.04
5	0.03	6.6	0.91	53	39	0.13	0.2	3.9	0.01
	Si (mg)	**Zn (mg)**	**As (mg)**	**Cd (mg)**	**Pb (mg)**	**Hg[c] (mg)**			
1	100	41	<0.05	0.03	0.11	0.004			
2	800	40	<0.05	0.02	0.12	0.004			
3	—	39	<0.05	0.02	0.14	0.004			
4	—	40	<0.05	0.02	0.10	0.004			
5	—	39	<0.05	0.03	0.10	0.004			

[a]Per kg dry matter.
[b]1 = No fertilizer;
2 = NPK (N, P, K), Ca, S, Cl;
3 = NPK, Ca, S, Cl, Mg, B Cu, Mn;
4 = NPK, Ca, S, Cl, Mg, B, Cu, Mn, Zn, Fe, Co, Mo, Se, I, Cr, Sn, and F;
5 = NPK, Ca, S, Cl, Hg, Cd, As, and Pb.
[c]Less than 0.004.

Mg (95 kg/ha); B (2.1 kg/ha); Cu (12.5 kg/ha); Mn (13 (kg/ha); Zn (1.75 kg/ha); Fe (10 kg/ha); Co (0.21 kg/ha); Mo (0.39 kg/ha); Se (0.21 kg/ha); I (0.84 kg/ha); Cr (0.13 kg/ha); F (0.13 kg/ha); Hg (0.17 kg/ha); Cd (0.17 kg/ha); As (0.45 kg/ha); Pb (0.35 kg/ha).

Experiment conducted in 10 locations in Finland over 3 years.
Source: Ref. 37.

TABLE 20 Effects of First Year Fertilization on Mineral Content of Barley[a]

Treatment[b]	Ca (g)	Mg (g)	P (g)	K (g)	S (g)	Al (mg)	B (g)	Br (mg)	Cr (mg)
1	0.41	1.35	4.12	6.9	1.39	44.6	1.0	2.2	0.03
2	0.42	1.29	3.99	6.6	1.48	24.2	1.1	1.2	0.03
3	0.41	1.32	3.95	6.4	1.54	25.5	1.3	1.2	0.02
4	0.42	1.30	3.97	6.7	1.51	24.6	1.3	1.3	0.03
5	0.42	1.30	4.04	6.9	1.52	21.0	1.1	1.4	0.02
	Co (mg)	Cu (mg)	F (mg)	Fe (mg)	Mn (mg)	Mo (mg)	Ni (mg)	Rb (mg)	Se (mg)
1	0.03	8.6	1.09	66	26	0.16	0.2	5.9	0.01
2	0.03	8.0	0.88	56	27	0.12	0.2	5.5	0.01
3	0.03	8.6	0.87	58	26	0.10	0.2	5.2	0.01
4	0.03	8.4	0.90	61	27	0.14	0.2	5.1	0.06
5	0.03	7.9	0.85	54	27	0.13	0.2	5.2	0.01
	Si (mg)	Zn (mg)	As (mg)	Cd (mg)	Pb (mg)	Hg[c] (mg)			
1	500	42	< 0.05	0.03	0.37	0.004			
2	200	43	< 0.05	0.03	0.36	0.004			
3	—	44	< 0.05	0.03	0.36	0.004			
4	—	43	< 0.05	0.03	0.36	0.004			
5	—	43	< 0.05	0.04	0.36	0.004			

[a]Per kg dry matter.
[b]1 = No fertilizer;
2 = NPK (N, P, K), Ca, S, Cl;
3 = NPK, Ca, S, Cl, Mg, B Cu, Mn;
4 = NPK, Ca, S, Cl, Mg, B, Cu, Mn, Zn, Fe, Co, Mo, Se, I, Cr, Sn, and F;
5 = NPK, Ca, S, Cl, Hg, Cd, As, and Pb.
[c]Less than 0.004.

Mg (95 kg/ha);
B (2.1 kg/ha);
Cu (12.5 kg/ha);
Mn (13 (kg/ha);
Zn (1.75 kg/ha);
Fe (10 kg/ha);
Co (0.21 kg/ha);
Mo (0.39 kg/ha);

Se (0.21 kg/ha);
I (0.84 kg/ha);
Cr (0.13 kg/ha);
F (0.13 kg/ha);
Hg (0.17 kg/ha);
Cd (0.17 kg/ha);
As (0.45 kg/ha);
Pb (0.35 kg/ha).

Experiment conducted in 10 locations in Finland over 3 years.
Source: Ref. 37.

TABLE 21 Effects of First Year Fertilization on Mineral Content of Oats[a]

Treatment[b]	Ca (g)	Mg (g)	P (g)	K (g)	S (g)	Al (mg)	B (g)	Br (mg)	Cr (mg)
1	0.91	1.45	4.26	6.21	1.78	22.0	1.5	4.8	0.03
2	0.94	1.43	4.31	6.46	1.83	93.0	1.7	3.3	0.05
3	0.90	1.46	4.33	6.33	1.86	26.3	2.1	3.1	0.04
4	0.88	1.45	4.39	6.38	1.86	25.4	2.3	3.5	0.04
5	0.88	1.42	4.34	6.38	1.83	27.4	1.6	3.1	0.04
	Co (mg)	Cu (mg)	F (mg)	Fe (mg)	Mn (mg)	Mo (mg)	Ni (mg)	Rb (mg)	Se (mg)
1	0.06	6.7	0.99	61	70	0.24	2.4	5.5	0.01
2	0.06	6.5	1.16	89	69	0.17	2.5	5.4	0.01
3	0.06	6.8	1.11	116	69	0.16	2.4	5.1	0.01
4	0.06	6.4	0.91	64	57	0.19	2.5	5.5	0.05
5	0.06	6.7	1.08	64	70	0.19	2.5	5.3	0.01
	Si (mg)	Zn (mg)	As (mg)	Cd (mg)	Pb (mg)	Hg[c] (mg)			
1	600	53	< 0.05	0.04	0.12	0.004			
2	800	53	< 0.05	0.04	0.16	0.004			
3	—	52	< 0.05	0.04	0.13	0.004			
4	—	55	< 0.05	0.04	0.12	0.004			
5	—	53	< 0.05	0.05	0.13	0.004			

[a]Per kg dry matter.
[b]1 = No fertilizer;
2 = NPK (N, P, K), Ca, S, Cl;
3 = NPK, Ca, S, Cl, Mg, B Cu, Mn;
4 = NPK, Ca, S, Cl, Mg, B, Cu, Mn, Zn, Fe, Co, Mo, Se, I, Cr, Sn, and F;
5 = NPK, Ca, S, Cl, Hg, Cd, As, and Pb.
[c]Less than 0.004.

Mg (95 kg/ha); B (2.1 kg/ha); Cu (12.5 kg/ha); Mn (13 (kg/ha); Zn (1.75 kg/ha); Fe (10 kg/ha); Co (0.21 kg/ha); Mo (0.39 kg/ha); Se (0.21 kg/ha); I (0.84 kg/ha); Cr (0.13 kg/ha); F (0.13 kg/ha); Hg (0.17 kg/ha); Cd (0.17 kg/ha); As (0.45 kg/ha); Pb (0.35 kg/ha).

Experiment conducted in 10 locations in Finland over 3 years.
Source: Ref. 37.

TABLE 22 Selected Trace Elements in Various Cereal Grains[a]

Element	Barley	Corn	Millet	Oats	Sorghum	Triticale	Wheat	Wild rice
Aluminum		10.7–21.7			0.0004%			
Boron					0.001%			
Cobalt	< 1–2.8	0.022	< 1	0.064	0.304		0.108[b]	
Chromium					0.00004			< 1
Chlorine	0.13	0.06		0.11	0.10			
Iodine		0.344						
Nickel	< 1	< 1–1.74	< 1		< 1		< 1	
Sulfur	0.19	0.14		0.23	0.18		0.25[b]	
Lead					0.001%			
Silicon					0.004%			
Molybdenum					0.0001%			
Tin					0.00004%			
Titanium					0.00004%			

[a]ppm unless otherwise noted.
[b]Wheat bran.
Source: Refs. 18–19, 45–46.

TABLE 23 Trace Element Inorganic Constituents of Brown Rice and Its Fractions (μg/g, dry wt.)

	Brown rice	Milled rice	Rice bran	Rice embryo	Rice polish
Aluminum	n.d.	0.73–7.23	53.5–369	n.d.	n.d.
Chlorine	203–275	163–239	510–970	1,520	n.d.
		205–372			
Iron	26–46	1.8–13.6	190	130	280
	15.3–35.7	3.3–3.6	140–316	110; 130	102
	6.8–27.8	4.6–26.8	130–200	489	
	24	4.9–7.2	530		
Manganese	13–42	9.9–13.6	406–877	120; 140	65
		22–33	110; 110		
Silicon	280–1,900	140–370	1,700–4,400	560–1,900	560–1,200
		107–180	12,200–16,300		
Zinc	15–22	12–21	80	100;300	50; 80

Source: Ref. 31.

TABLE 24 Distribution of Protease and Phytase Activity in Soluble and Insoluble Protein Fractions of Protein Body Preparations of Sorghum, Pearl Millet, and Finger Millet

	Protease (μmol Glycine equivalent released/hr/ml)	Phytase (μmol P released/hr/ml)
Sorghum		
Soluble	0.38	0
Insoluble	2.72	0.10
Pearl millet		
Soluble	2.68	0
Insoluble	5.12	0.12
Finger millet		
Soluble	1.72	0
Insoluble	3.04	0

Source: Ref. 51.

TABLE 25 Comparison of Saccharifying Activity in Selected Cereal Grains

Cereal species	Ungerminated[a]		Germinated[a]	
	Total	Free	Total	Free
Barley	49.6–77.2	20.1–43.0	58.8–92.9	48.1–83.0
Wheat	90.1–116.2	44.8–59.2	93.0–170.9	77.8–138.9
Rye	85.0–110.5	65.8–89.7	104.2–137.9	99.0–129.0
Oats	0.7–3.7	0.7–4.5	16.6–24.2	18.1–25.6

[a]Activity expressed in Lintner.
Source: Ref. 55.

REFERENCES

1. Kutsky, R. J., *Handbook of Vitamin, Minerals and Hormones*. Van Nostrand Reinhold Company, New York, 1981.
2. Altman, P. L., and Dittmer, D. S., *Metabolism*, 1st ed. Am. Soc. Exp. Biol., Washington, D.C., 1968.
3. Diem, K., *Documenta Geigy Scientific Tables*, 6th ed. Geigy Pharmaceuticals, Ardsley, NY, 1962.
4. Burton, B. T., *The Heinz Handbook of Nutrition*, 3rd ed. McGraw-Hill, New York, 1976.
5. Kirschmann, J. D., *Nutrition Almanac*. McGraw-Hill, New York, 1975.
6. Pfeiffer, C., *Metal and Elemental Nutrients*. Keats, New Canaan, CT, 1975.
7. Sebrell, W. H., Harris, R. S., Gyorgy, P., and Pearson, W. N., *The Vitamins*, 2nd ed., Vols. I–VII. Academic Press, New York, 1967.
8. Watt, B. L., and Merrill, A. L., *Composition of Foods*. Agriculture Handbook #8. U.S. Government Printing Office, Washington, D. C., 1975.
9. Hepburn, F. N., Status report on FNB proposals for cereal fortification. *Cereal Fds. World*, 28(8): 360–362 (1976).
10. Blessin, C. W., Brecher, J. D., and Dimler, R. J., Carotenoids of corn and sorghum. III. Variations in xanthophylls and carotenes in hybrid, inbred and exotic corn lines. *Cereal Chem.*, 40: 436–442 (1963).
11. Toepfer, E. W., Polansky, M. M., Eheart, J. F., Slover, H. T., and Morris, E. R., Nutrient composition of selected wheat products. XI. Summary. *Cereal Chem.*, 49: 173–186 (1972).
12. Harris, R. S., General discussion on the stability of nutrients. In *Nutritional Evaluation of Food Processing*, E. Karmas and R. Harris (Eds.). AVI, New York, 1988.
13. National Academy of Science, *Flour and Bread Enrichment 1945–1950*. NAS, Washington, D.C., 1950.
14. National Academy of Science, *Proposed Fortification Policy for Cereal-Grain Products*. NAS, Washington, D.C., 1974.
15. Hoseney, R. C., Minor constituents of cereals. In *Principals of Cereal Science and Technology*. Am. Assoc. of Cereal Chem., St. Paul, MN, 1986.
16. Wu, Y. V., and Inglett, G. E., Effects of agricultural practices, handling, processing and storage on cereals. In *Nutritional Evaluation of Food Processing*, 3rd ed, E. Karmas and R. Harris (Eds.). Van Nostrand Reinhold Company, New York, 1988.
17. Simmonds, D. H., and Campbell, W. P., Morphology and chemistry of rye grain. In *Rye: Production, Chemistry and Technology*, W. Bushuk (Ed.). Am. Assoc. Cereal Chemistry, St. Paul, MN, 1976.
18. Anderson, R. A., Wild Rice: Nutritional review. *Cereal Chem.*, 53(6): 949–955 (1976).
19. Anderson, R. A., Wild Rice: its history, current production, use. *The Rice Journal*, (July-August): 34–38 (1978).
20. Houston, D. F., and Kohler, G. O., *Nutritional Properties of Rice*. FNB-NRC, NAS, Washington, D.C., 1970.
21. Geddes, W. F., Oats and oat products. In *Nutritional Evaluation of Food Processing*, R. S. Harris and H. von Loesecke (Eds.). John Wiley, New York, 1960.
22. Ranhotra, G. S., and Gelroth, J. A., Stability of enrichment vitamins in bread and cookies. *Cereal Chem., 63*(5): 401–403 (1986).
23. Ranhotra, G. S., and Bock, M. A., Effects of baking on nutrients. In *Nutritional Evaluation of Food Processing*, E. Karmas and R. Harris (Eds.). AVI, New York, 1988.
24. Maleki, M., and Daghir, S., Effect of baking on retention of thiamine, riboflavin and niacin in Arabic bread. *Cereal Chem., 44*: 483–487 (1967).
25. Committee on Dietary Allowances, *Recommended Dietary Allowances*. FNB-NRC, NAS, Washington, D.C., 1980.
26. Rao, B. S. N., and Gropalan, C., Niacin. In *Present Knowledge in Nutrition*, 5th ed., Olson et al. (Eds.). The Nutrition Foundation, Washington, D.C., 1984.
27. Hulse, J. H., Laing, E. M., and Pearson, O. E., *Sorghum and the Millets: Their composition and Nutritive Value*. Academic Press, New York, 1980.

28. Hepburn, F. N., Nutrient composition of selected wheats and wheat products. VII. Total and free niacin. *Cereal Chem., 48*: 369–372 (1971).
29. Shils, M., and Young, V., *Modern Nutrition in Health and Disease*, 7th ed. Lea and Febriger, Philadelphia, PA, 1988.
30. Hazell, T., Minerals in foods: Dietary sources, chemical forms, interactions, bioavailability. *Wld. Rev. Nutr. Diet., 46*: 1–123 (1985).
31. Juliano, B., The rice caryopsis and its composition. In *Rice Chemistry and Technology*, D. F. Houston (ed.). Am. Assoc. Cereal Chem., St. Paul, MN, 1972.
32. Tingle, J. N., and Dawley, W. K., Mineral composition of whole plant cereals for silage in Central British Columbia. *Can. J. Plant Sci., 52*: 805–809 (1972).
33. Williams, R.J., *Nutrition Against Disease*. Pitman Publishing Co., New York, 1971.
34. Bressani, R., Paz, R., Paz, Y., and Scrimshaw, N., Chemical changes in corn during preparation of tortillas. *J. Agri. and Fd. Chem., 6*(10):770–774 (1958).
35. Koivistoinen, P., Nissinen, H., Varo, P., and Ahlstrom, A., Mineral element composition of cereal grains from different growing areas in Finland. *Acta Agri. Scandinavica, 24*: 327–334 (1974).
36. Lorenz, K., and Loewe, R., Mineral composition of U.S. and Canadian wheats and wheat blends. *J. Agri. and Fd. Chem., 25*(4): 806–809 (1977).
37. Syvalahti, J., and Korkman, J., The effect of applied mineral elements on the mineral content and yield of cereals and potatoes in Finland. *Acta Agri. Scandinavica* Suppl., *20*: 80–88 (1978).
38. Johnson, A. H., and Peterson, M. S., *Encyclopedia of Food Technology*. AVI Publishing Co., Westport, CT, 1974.
39. Morris, E., and Ellis, E., Isolation of monoferric phytate from wheat bran and it biological value as an iron source to the rat. *J. Nutr., 106*: 753–760 (1976).
40. Davies, K., *Clinical Significance of Essential Biological Metals*. Charles C. Thomas, Springfield, IL, 1972.
41. Schroeder, H. A., *The Trace Elements and Man*. Devin-Adair, Greenwich, CT, 1977.
42. Underwood, E. J., *Trace Elements in Human and Animal Nutrition*. Academic Press, New York, 1977.
43. Reinhold, J., Parsa, M., Karimian, N., Hammick, J., and Ismail-Beigi, F., Availability of zinc in leavened and unleavened wholemeal breads as measured by solubility and uptake by rat intestine *in vitro*. *J. Nutr., 104*: 976–982 (1974).
44. Lag, J., and Steinnes, E., Content of some trace elements in barley and wheat grown in Norway. *Scientific Reports of the Agri. Univ. of Norway, 57*: 2–11 (1978).
45. Adams, R., Symposium: New concepts and developments in trace element nutrition. Variability in mineral and trace element content of dairy cattle feeds. *J. Dairy Sci., 58*(10): 1538–1548 (1975).
46. Ara, J., Kahn, H. H., and Nadeem, M., Trace element content of some varieties of cereals. *Pakistan J. Biochem., 14*(2): 92–97 (1981).
47. Newman, C. W., and McGuire, C. F., Nutritional quality of barley In *Barley*, D. C. Rasmusson (Ed.). Am. Soc. of Agronomy, Crop Sci. Soc. of Am., Soil Sci. of Am., Madison, WI, 1985.
48. Lin, Y. H., Wimer, L. T., and Huang, A. H., Lipase in the lipid bodies of corn scutella during seedling growth. *Plant Physiol. 76*:719–722 (1983).
49. Lin, Y. H. and Huang, A. H., Purification and initial characterization of lipase from the scutella of corn seedlings. *Plant Physiol., 76*: 719–722 (1984).
50. Vick, B. A., and Zimmerman, D. C., Biosynthesis of jasmonic acid by several plant species. *Plant Physiol., 75*: 458–461 (1984).
51. Adams, C. A., Novellie, L., and Liebenberg, N., Biochemical properties and ultrastructure of protein bodies isolated from selected cereals. *Cereal Chem., 53*: 1–12 (1976).
52. Akazawa, T., Enzymes of rice. In *Rice Chemistry and Technology*, D. F. Houston (Ed.). Am. Association of Cereal Chemists, St. Paul, MN, 1972.
53. Funatsu, M., Aizono, Y., Hayashi, K., Watanabe, M., and Eto, M., Biochemical studies on bran lipase. Part 1. Purification and physical properties. *Agr. Biol. Chem., 35*(5): 734 (1971).

54. Simmonds, D. H., and Campbell, W. P., Morphology and chemistry of rye grain. In *Rye: Production, Chemistry and Technology*, W. Bushuk (Ed.). Am. Assoc. Cereal Chemists, St. Paul, MN, 1976.
55. Kruger, J. E., and Reed, G., Enzymes and color. In *Wheat Chemistry and Technology*, 3rd ed., Vol. 2, Y. Pomeranz (Ed.). Am. Assoc. of Cereal Chemists, St. Paul, MN, 1988.
56. Wright, K. N., Nutritional properties and feeding value of corn and its by-products. In *Corn Chemistry and Technology*, S. A. Watson and P. E. Ramstad (Eds.). Am. Assoc. of Cereal Chemists, St. Paul, MN, 1987.
57. Watson, S. A., Structure and composition. In *Corn Chemistry and Technology*, S. A. Watson and P. E. Ramstad (Eds.). Am. Assoc. of Cereal Chemists, St. Paul, MN, 1987.

15

QUALITY EVALUATION OF CEREALS AND CEREAL PRODUCTS

Vladimir F. Rasper
University of Guelph, Guelph, Ontario, Canada

The quality of cereals and various cereal products is determined by a variety of characteristics which may be assigned different significance depending on the desired end use or type of product. These characteristics can be divided into chemical, enzymatic, and physical. Likewise, the individual quality testing methodologies can be classified into those concerned with the chemical components of the tested material, assays for a quantitative estimation of enzymatic activity, and tests dealing with various physical or physicochemical properties. Procedures based on small-scale preparation of products, e.g., laboratory milling and experimental baking tests, are also part of the methodology. Tables 1 and 2 present a list of procedures most commonly employed as part of the Approved Methods of the American Association of Cereal Chemists (1) or Standards of the International Association for Cereal Chemistry and Technology (ICC). The expected ranges of values measured by these procedures for different types of wheat flour are given in Tables 3–5.

I. CHEMICAL TESTING

A. Moisture

Determination of moisture is an essential step in quality evaluation of cereal grains and products of their processing. The behavior of the grains in both storage and milling depends to a great extent on their moisture content. Moisture content also influences the keeping quality of flour and bakery products. The knowledge of moisture content is required for comparing production data at a uniform level of dry solids and for compliance with government regulations. Similarly, since compositional percentages are inversely related to moisture content in the analyzed material, results of chemical analyses have to be reported on a fixed moisture basis. In North America, it has become a common practice to report the results on the 14% basis, whereas in Europe the dry solids basis is often preferred.

There are many methods of testing moisture content, and their results may vary considerably. It is, therefore, important that all tests are made by the same method if the results are intended for the same purpose. In reporting the results, the method used should be indicated so that in comparing results of different methods, a proper correction factor can be applied.

When moisture is determined by drying, the measured value is defined as a loss in weight under the conditions of the specific procedure. The loss includes both water and other volatiles. If the original moisture content in the sample exceeds 16%, drying is usually carried out in two stages. The moisture content is first reduced to 16% or less (13% for rice) by keeping the sample in a well-ventilated and warm place. Then the sample is dried at an elevated temperature: 103°C for 72 hr (1) (AACC Method 44–15A) or at 135°C for 2 hr (AACC Method 44–19). An alternate procedure uses a partial vacuum equivalent to 25 mm mercury or less at 98–100°C for approximately 5 hr (AACC Method 44–32). Other techniques such as measurement of the dielectric constant (AACC Method 44–11), azeotropic distillation with toluene (AACC Method 44–51), and near-infrared reflectance (NIR) spectroscopy are also used. According to the ICC Basic Reference Method (17,18), moisture content in the sample is defined as the loss in weight sustained by the material when equilibrated in an anhydrous atmosphere at a temperature between 45 and 56°C and a pressure of 1.3–2.7 kPa (10–20 mmHg).

B. Protein

The quantity of protein in the tested sample is generally determined by the classical Kjeldahl analysis or some of its more recent modifications (1) (AACC Methods 46–10, 46–11, 46–12, 46–13). Protein nitrogen is reduced and transformed to ammonium sulfate by hot digestion of the dry sample with concentrated sulfuric acid in the presence of a catalyst. Ammonium is then liberated from the sulfate by distillation in the presence of sodium hydroxide and driven into a known volume of standard acid solution. From the quantity of unreacted acid determined by titration, the quantity of released nitrogen is established and converted to protein by multiplying the percentage of nitrogen with the appropriate conversion factor (5.7 for wheat and wheat flour, 6.25 for most foods and feeds). Another method is based on the dye-binding capacity of specific groups of amino acids in protein (1,37) (AACC Method 46–14A). More recently, methods have been developed to determine protein by near-infrared reflectance spectroscopy (1,25,41) (AACC Method 39–10).

Protein determination is of particular significance in quality testing of wheat and wheat flour. The quantity and quality of wheat protein have been recognized as factors having a decisive effect on the physicochemical properties of wheat flour dough and, consequently, on its handling properties and baking potential. Since these relevant functions of wheat protein have been attributed primarily to gluten-forming proteins, protein determinations in wheat or wheat flour is very often supplemented by a quantitative estimation of wet and dry gluten. Gluten is prepared by washing out most of the starch and solubles from a piece of dough developed from tested flour and water (1) (AACC Method 38–10). This tedious and poorly reproducible procedure has been made easier by the introduction of automatic gluten washers (Fig. 1) (1) (AACC Method 38–11). The washed-out gluten can be examined visually for its color and elasticity. The difference between gluten weight before and after its drying can be taken as a rough estimate of its hydration capacity. Some other tests used for the evaluation of the gluten quality of wheats and wheat flours are mentioned in Section IV.

C. Mineral Content

Mineral content, more commonly known as ash content, has always been considered an important criterion in flour quality evaluation. Although not directly related to the technological performance of flour, it serves as an indicator of the degree of separation of starchy endosperm from the bran during the milling process. Thus, the ash content determination not only provides an easy criterion for classifying flours into different milling types but becomes a tool for controlling the entire milling operation. The conditions of the determina-

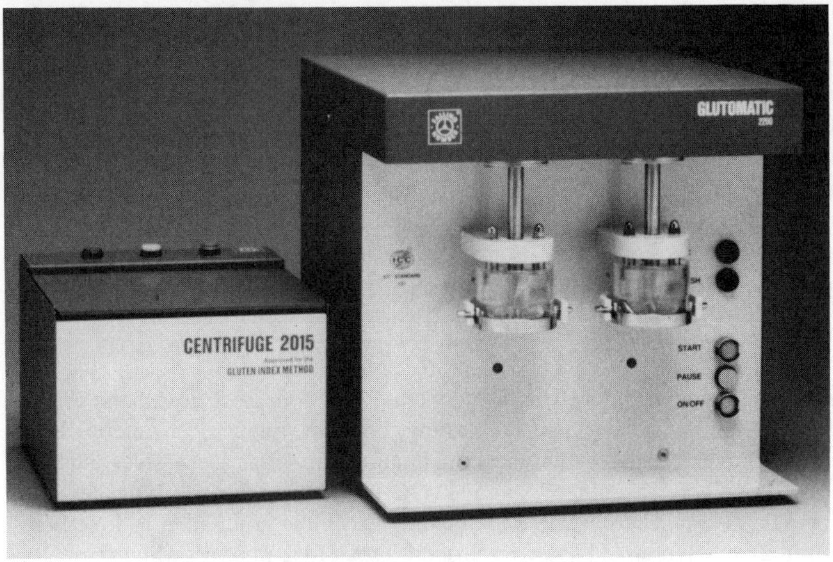

(A)

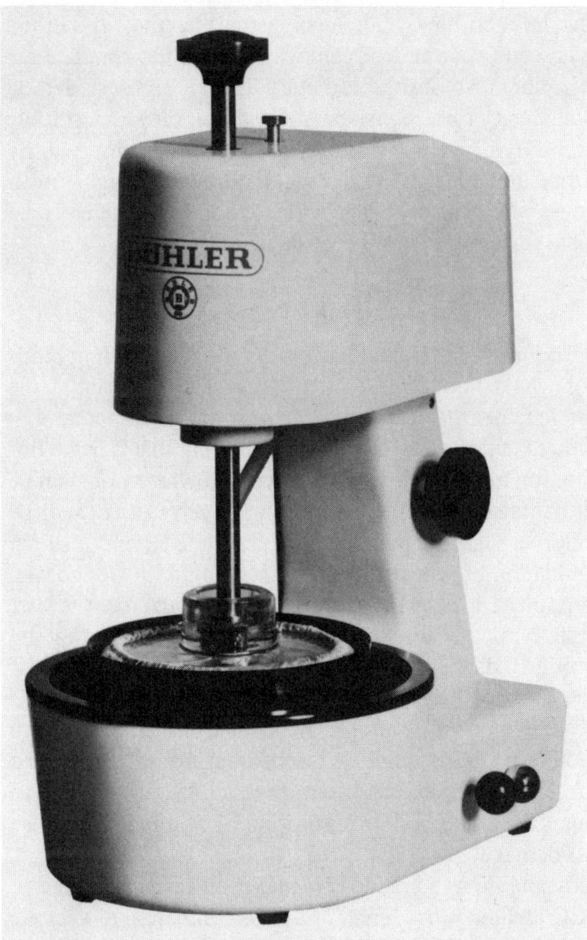

(B)

FIGURE 1 Gluten washers: (A) Glutomatic 2200. (Courtesy of Falling Number A.B. Stockholm, Sweden.) (B) Glutinex MLI 500. (Courtesy of Bühler Brothers Ltd. Uzwil, Switzerland.)

tion may vary with the type of the tested material and procedure used. According to the AACC Method 08–01, ash is determined as a residue after incinerating the sample at 550°C (for soft wheat flour) or 575–590°C (for hard wheat flour) until light grey ash is obtained (1). Using an accelerated method, applicable only to flour, the sample is wetted with magnesium oxalate solution and incinerated for 30–45 min at 700°C (1) (AACC Method 08–02). The ICC Standard No. 104 (13) for ash determination in cereal materials prescribes incineration at 900°C until the residue is white or nearly white. For a rapid determination of ash without any incineration involved, NIR spectroscopy can be applied (40).

D. Fiber

Mineral content correlates very closely to *fiber* content. They both are related to the amount of bran in the kernel or in the flour. In cereal testing, fiber has usually been determined as crude fiber in the form of a residue remaining after treating the sample successively with hot concentrated acid and hot concentrated alkali (1) (AACC Method 32–10). In wheat grain the crude fiber content ranges from 2 to 2.5% (14% m.b.). Flour crude fiber is directly related to the milling extraction and averages around 0.5% (14% m.b.) in white flour. Recently, fiber content in cereals and cereal products is more frequently reported as "dietary fiber." This term designates a residue indigestible by human digestive enzymes. A method based on a combined action of amylase and neutral detergent was designed to simulate the digestive process in vitro for the determination of insoluble dietary fiber (1) (AACC Method 32–20). For the determination of total dietary fiber, consisting of both soluble and insoluble fractions, the enzymatic method uses amylolytic and proteolytic enzymes as proposed by Prosky et al. (29). Like crude fiber, total dietary fiber in flour will increase with higher milling extraction rates. It may range from approximately 3% in white wheat flour to over 9% in wholemeal. In wheat bran it accounts for more than 40% of the solids.

E. Starch

Starch constitutes the main chemical component of all cereal grains and their milling products. However, in quality testing of wheat flours, more attention is usually given to the physical condition of the starch granule rather than the actual quantity of this component. It is the degree of physical damage inflicted upon the granule during milling that is more important. While some positive contribution to the quality of some types of wheat flour can be expected from a limited amount of damaged granules, a distinctly adverse effect will be noticed if the amount becomes excessive. The first attempt to measure the extent of starch damage involved microscopic examination of wheat flour after staining with Congo red. Physical damage enhances the absorption of dyes by the starch granule. Congo red was later replaced with 0.2% solution of crystal violet or solution of chlorzinciodide (43). Several methods exploit the higher susceptibility of damaged starch granules to amylolytic degradation and quantitatively estimate the level of products of this degradation (1,8) (AACC Method 76–30A). Other methods depend upon the development of color formed on addition of iodine reagent to an extract of flour containing amylose leached out from the damaged granules by a solution of ammonium sulfate, formamide, and sulfosalycilic acid (12,42). An amperometric technique (23) measures the rate of iodine absorption by starch granule. The rate increases with higher degree of damage.

Specific methods are used in analyzing flours for added *bleaching* and *maturing agents*. The Approved Methods of the AACC include procedures for the determination of acetone peroxide (Method 48–05), benzoyl peroxide (Method 48–07), chlorine dioxide (Method 48–30), potassium bromate (Method 48–42), ammonium persulfate (Method 48–62), and azodicarbonamide (Method 48–71A).

II. TESTING ENZYMATIC ACTIVITY

In testing cereal grains and their milling products for enzymatic activity, *amylolytic activity*, i.e., the activity of starch-hydrolyzing enzymes, is considered of primary importance. While only small amounts of α-amylase are present in sound grain, the activity of this enzyme increases markedly during pre-or postharvest germination. Some level of this activity is desired in wheat flour for a sufficient development of gas during fermentation and early stages of baking of yeasted doughs. If the concentration of α-amylase is not sufficient, flour is supplemented using preparations from malt, fungi, or bacteria. Excessively high levels, however, impair the quality of both dough and final baked product. Unlike α-amylase, β-amylase is present in higher quantity even in sound grain, but the quantity remains practically unchanged during germination. For that reason, the assays of amylolytic activity deal primarily with the activity of α-amylase.

The AACC Method 22–07 (1) is a colorimetric assay based on the reaction of α-amylase with a buffered solution of potato amylose dyed with Cibachrom blue F3GA. Another AACC method (Method 22–07) measures the rate of decrease in light-scattering ability of a dilute suspension of a β-limit dextrin substrate treated with enzyme extracted from the tested flour. The ICC Standard No. 108 (14) is a similar colorimetric method. Other procedures are based on quantitative chemical estimation of reducing sugars produced in situ in a buffered flour suspension after a standard incubation period (1) (AACC Method 22–15). Such methods measure the effects of the combined activity of α-amylase and β-amylase, as well as susceptibility of damaged starch granules to amylolytic activity. In that respect, the results are more a reflection of the total gas production ability of the tested flour than the actual amylolytic activity. A number of methods evaluate this combined effect by measuring the quantity of gas produced by the flour under the specific test conditions (see Sec. IV). Methods for the estimation of amylolytic activity from viscometric measurements on starch pastes are found in Section III.

Some of the amylase preparations used for the adjustment of amylolytic activity of wheat flour contain appreciable quantities of *proteolytic enzymes*. In flour unsupplemented with such preparations, higher levels of proteolytic activity indicate sprouted grain. In order to assay this activity, nitrogen released from a buffered hemoglobin substrate by incubating it with the enzyme-containing flour extract is estimated quantitatively by Kjeldahl analysis (1) (AACC Method 22–60). Solubilized hemoglobin can also be measured spectrophotometrically (Method 22–62) or colorimetrically (Method 62–63).

Unfavorable storage conditions may enhance the action of *lipases*, especially in materials high in fat. Since higher lipase activity leads to an increase in fat acidity, the extent of the lipase action can be expressed in terms of milligrams of potassium hydroxide required to neutralize the free fatty acids extracted from a 100-g sample (1) (AACC Methods 62–01A, 62–02A, and 62– 03A). Fat acidity may range from about 20 for sound wheat to over 100 if the grain was exposed to adverse storage conditions.

III. PHYSICAL TESTING

A. Physical Grain Testing

1. Test Weight

Test weight, measured as weight of grain per unit volume, is the simplest and most widely used criterion of milling quality of grain. Although many factors may influence the relationship between test weight and milling yield, the latter usually increases with increasing test weight unless the weight value exceeds 75 lb per bushel (10).

In countries using Imperial units, test weight is expressed in terms of lb per bu. In the United States, the Winchester bushel (2,150.42 in.^{3}) is used, which slightly differs from the Imperial bushel (2,219.36 in.^{3}). In countries using the metric system, test weight is defined in terms of kilograms per hectoliter. For conversion, the following factors can be used:

From	To	Multiply by
lb/Winchester bu	lb/Imperial bu	1.032
lb/Winchester bu	kg/hectoliter	1.282
lb/Imperial bu	kg/hectoliter	1.247

2. Hardness of grain

Hardness of grain constitutes a comparative measure for distinguishing between soft and hard wheats. The criterion is also widely used in testing corn for hardness and breakage susceptibility (26). Methods used for testing hardness employ either a single grain or a bulk sample (27). The single grain tests involve abrasion, cutting, crushing, or penetration. In bulk tests hardness is estimated from the power or time required to grind a given quantity of grain, from the quantity of abraded material, or from the particle size of the ground material. The latter is determine by sieving (39), sedimentation (17,18) (ICC Standard No. 127), centrifugation (1) (AACC Method 50–10), light-diffraction technique (2), Coulter counter, or NIR spectroscopy (41).

In grading of wheat, other physical criteria are considered. *Kernel weight*, which is usually expressed in grams per 1000 kernels, is the function of kernel size and density. It may be a reliable predictor of milling yields as large, dense kernels usually have a higher ratio of endosperm to nonendosperm components than smaller, less dense kernels. U.S. red winter and hard red spring wheats range in kernel weight from 20 to 32 g/1000 kernels. Soft red winter, white, and durum wheat average approximately 35 g/1000 kernels (10).

Milling yield can also be predicted from *kernel size and shape* (32). Based on the *pigmentation of bran*, wheats are classed as either red or white. These two colors are varietal characteristics but may be influenced by environmental factors. Inspection of grain for *damaged kernels* and impurities is also part of the grading procedure. The ICC system adopted for the evaluation of wheat intended for milling includes testing for extraneous material referred to as "Besatz" (15). "Besatz" includes dockage, foreign material, damaged kernels, shrunken and broken kernels, and wheat of other classes.

3. Experimental milling

Experimental milling provides the miller with advanced information on the performance of the tested grain in the commercial scale process. In breeding programs it serves as a useful tool which not only enables the evaluation of the milling potential of the tested lines but helps in predicting their end-use quality.

Prior to any experimental milling, the grain must be cleaned and conditioned (tempered) (1) (AACC Method 26–10). Numerous milling devices exist for the actual milling operation. They all work on the principle of grinding and sifting and only differ in the extent of these operations. The Bühler Laboratory Mill (Bühler Bros, Inc. Uzwil, Switzerland) is a continuous automatic mill with pneumatic conveying and a milling system consisting of three breaks and three reductions (Fig. 2). It can produce straight-grade flour comparable in quality to commercially milled flour. Some modifications, however, are required, such as the addition of a bran finisher, in order to achieve extraction rates characterizing commercial products (5).

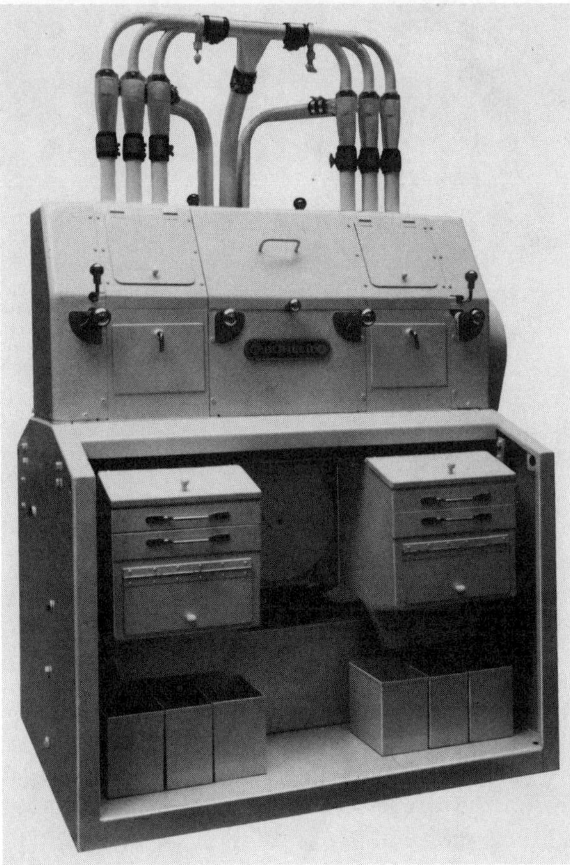

FIGURE 2 The Bühler experimental mill MLU 202. (Courtesy of Bühler Brothers Ltd., Uzwil, Switzerland.)

Another automatic mill is the Brabender Quadrumat Senior (Fig. 3) with two four-roll system units, one for breaking and one for reduction, and a two-section plansifter. The Brabender Quadrumat Junior (Fig. 4), a compact unit with four 3-inch diameter rolls, is well suited for milling small quantities of wheat (down to 20 g).

Some millers prefer batch-operated experimental mills such as the Allis-Chalmers Mill, because of the ease in adjusting the milling procedure on the basis of a visual examination of the yields and quality of stocks throughout the millflow.

In evaluating the results of experimental milling, two factors are usually considered: flour extraction and flour ash. The lower the flour ash and the brighter the flour color, the more desirable the wheat for milling. The following two formulas are used in the evaluation of wheat milling quality from the experimental milling data (31):

Milling rating = % extraction of straight grade flour – (ash × 100)

Milling value = % extraction of straight grade flour – Kent Jones flour color

Higher milling ratings and milling values are preferred. Milling quality of different wheats can also be ascertained by comparing their cumulative ash curves (22). Cumulative

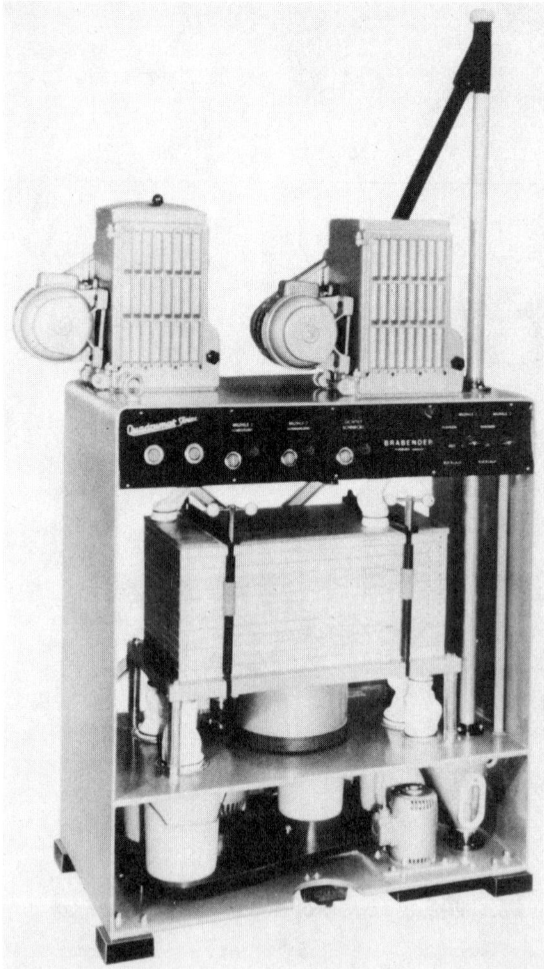

FIGURE 3 The Brabender Quadrumat Senior laboratory mill. (Courtesy of C.W. Brabender Instruments Co., South Hackensack, N.J.)

ash curves are constructed by arranging mill streams in ascending order of ash on a constant moisture basis and by plotting cumulative ash against cumulative extraction for each successive blend of millstreams. Wheats that exhibit the lowest initial flour ash and the slowest rate of ash increase with increasing flour extraction are preferred. The results of this comparison can be expressed in terms of a single numerical score—the curve index. A line is drawn from the 30% extraction point on the cumulative curve to the 70% extraction point (Fig. 5). The distance on the 50% extraction level from the curve to the drawn line, when measured at right angle to the line, is called depth D. It is used in the calculation of the curve index:

$$\text{Curve index} = L - 2D$$

where L is the length of the line between the 30% and 70% extraction points. A lower curve index indicates better milling quality.

FIGURE 4 The Brabender Quadrumat Junior laboratory mill. (Courtesy of C.W. Brabender Instruments Co., South Hackensack, N.J.)

B. Flour Color Testing

Color testing of wheat flour is done for the purpose of evaluating either its whiteness, which primarily determines the extent of the oxidation of carotenoid pigments by bleaching compounds, or the presence of bran particles. Testing of flour for whiteness is based on measuring light reflectance of the sample within the blue range of light spectrum. Since improvement in flour color results from the oxidation of the pigments by the bleaching agents as well as natural oxidation during storage, the measured values vary not only with the extent of bleaching but the age of the flour. Correlation between color measured in the blue range of light spectrum and ash content in flour is rather limited. If color is to be taken as a measure of contamination with bran particles, reflectance meters with a light source in the green band of the light spectrum should be used. Among them, the Kent-Jones Grader (20) and the Agtron Color Meter set on "green mode" (1,9,33) (AACC Method 14–30) are the most common. More recently, NIR analyzers have also been employed for this purpose.

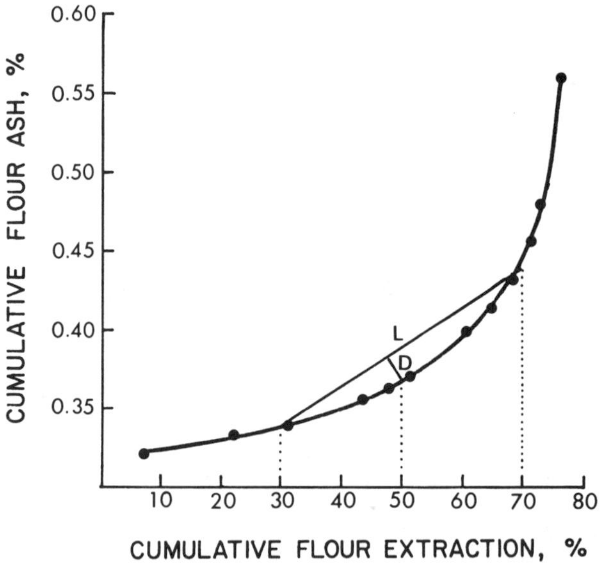

FIGURE 5 Measurement of the curve index. (Reprinted, with permission, from Ref. 22.)

	PHASE I	PHASE II	PHASE III
FUNCTION	Dough Mixing	Structural transformation of dough by fermentation and maturing	Gelatinization of crumb in oven
MEASURING METHOD	FARINOGRAPH	EXTENSOGRAPH	AMYLOGRAPH
TYPE OF DIAGRAM	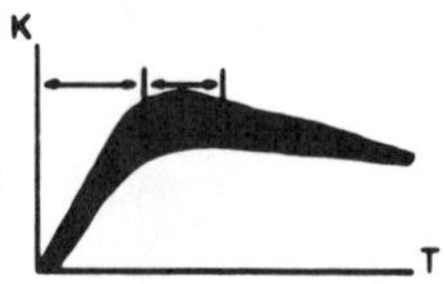 Farinogram	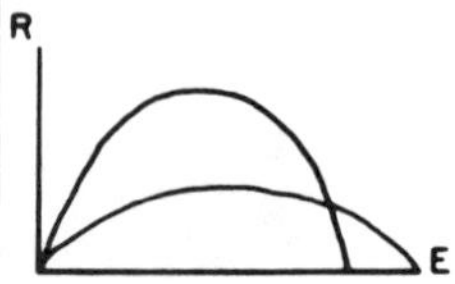 Extensogram	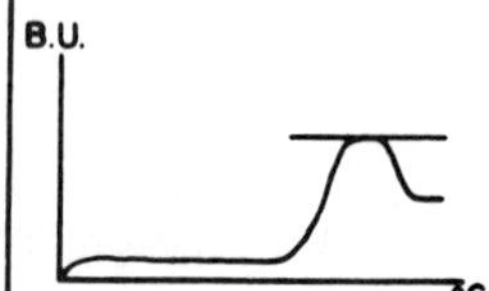 Amylogram
FLOUR SPECIFICATION DATA OBTAINED	Water Absorption Mixing Time Mixing Tolerance General Strength (dynamic, under machining abuse)	Relation Strain/Stress indicating degree of maturing action present or needed. General Strength (static)	Gelatinization characteristics in oven
CORRECTION POSSIBILITIES	Change of Wheat Mix	Aging by storage Chemical Oxidizers and Heat Conditioning	Diastatic Malt

FIGURE 6 Three-phase system of physical flour testing. (Courtesy of C.W. Brabender Instruments Co., South Hackensack, N.J.)

The "slick" (Pekar) test is a rough but simple guide in assessing flour color visually (1) (AACC Method 14–10). Flour is placed on a flat piece of wood or metal, pressed down, trimmed, and immersed in water. The color may be judged at several stages: (1) before immersion in water (dry), (2) shortly after immersion, and (3) after the flour has been dried.

C. Physical Dough Testing

Physical properties of dough not only determine performance at the various stages of the technological process but have a pronounced effect on the quality of the finished product. Since the physical condition of a dough depends, to a large extent, on the quality of flour used, testing doughs for their physical properties has become an essential part of methodology for quality evaluation of wheats and wheat flours. The methodology was designed with a "three-phase" concept, which reflects the relevance of the individual physical qualities of dough at the different stages of the baking process (Fig. 6).

The first phase of this testing system is concerned with the behavior of dough when it is developed from flour and water and subsequently subjected to overmixing. Recording mixers are used to measure and record the changes in the resistance of dough to mixing with time. The recorded mixing curves are characterized by an ascending part which indicates changes during the dough development process, while the subsequent decline in resistance is taken as a sign of a steady breakdown of the dough structure upon mixing beyond the point of optimum development.

The Brabender farinograph (Fig. 7) is one of the most widely used recording dough mixers (6). The two Z-shaped blades of the farinograph mixer rotate at constant but different speeds and subject the dough to a relatively gentle mixing at constant temperature. A representative farinograph record, or farinogram, with the commonly measured indices as defined by the ICC method (16) is shown in Figure 8. Dough development time and stability increase with increasing strength of flour, whereas mixing tolerance index and degree of softening relate to increasing strength inversely. The AACC method (1) (AACC Method 54–21) uses additional indices. Arrival time is defined as time required for the top of the curve to reach to 500 FU (farinograph units) line on the recording chart and serves as a measurement of the rate at which the water is taken up by flour. Departure time equals the sum of the arrival time plus the stability. Longer arrival and departure times indicate stronger flour. Twenty-minute drop is measured as the difference between the height of the center of the curve at its peak and 20 minutes after the first addition of water. The higher the value, the weaker the flour. To express the strength of the tested flour as a single score, the valorimeter value may be determined from the dough development time and the descending slope of the curve by means of a special nomograph known as the valorimeter; the higher the value, the stronger the flour.

Water absorption of flour is another relevant property that can be determined by means of a farinograph. It is defined as the amount of water required for a dough to reach a definite consistency (normally 500 FU) at the point of optimum development. Stronger flours with higher protein content and better gluten quality are characterized by higher absorptions.

Figure 9 presents an example of how the farinograms and water absorption values change with the increasing strength of flour.

More information about the dough-mixing characteristics may be obtained if the standard farinograph mixing bowl is replaced with the resistograph mixer. The resistograph combines mixing with stretching, pressing, and kneading, thus imparting high shear and high work input to the dough. The resistograms have two maxima, which become more pronounced with medium and weak flours (Fig. 10). The first maximum is related to binding

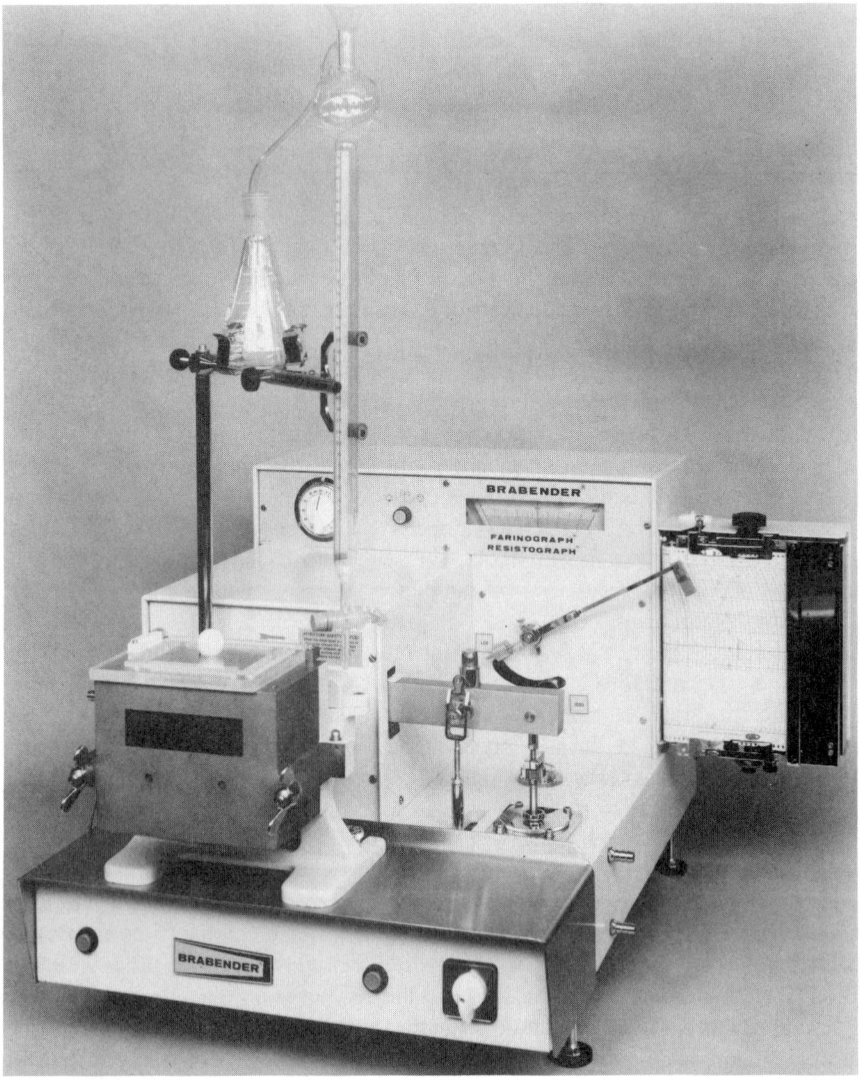

FIGURE 7 The Brabender farinograph/resistograph. (Courtesy of C.W. Brabender Instruments Co., South Hackensack, N.J.)

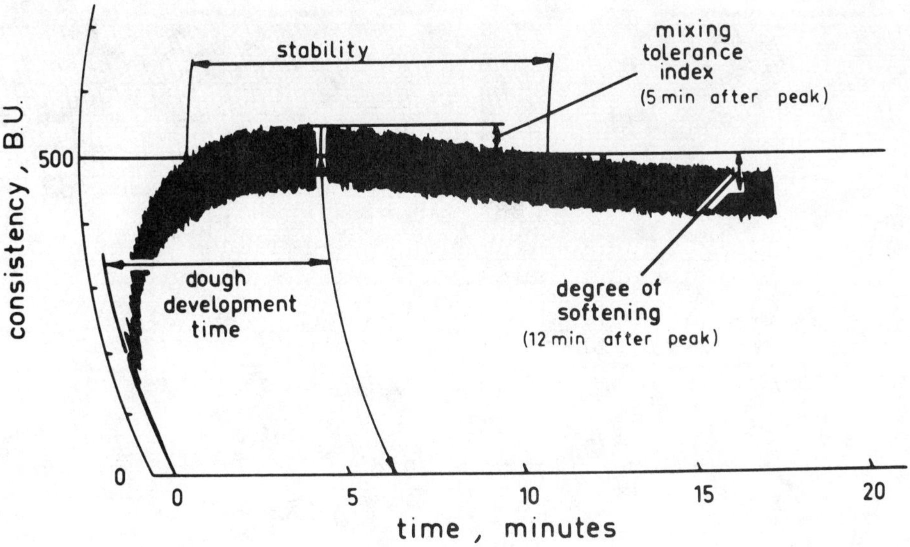

FIGURE 8 Representative farinogram showing some commonly measured indices. (Reprinted, with permission, from Ref. 3.)

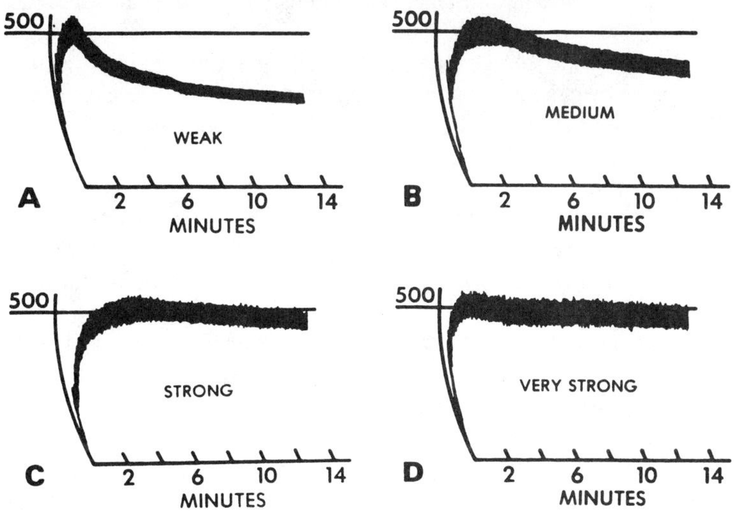

FIGURE 9 Farinograms of flours of different strength. Farinograph absorptions: flour A—54%, flour B—57%, flour C—64.5%, flour D—62.7%. (Reprinted, with permission, from Ref. 28.)

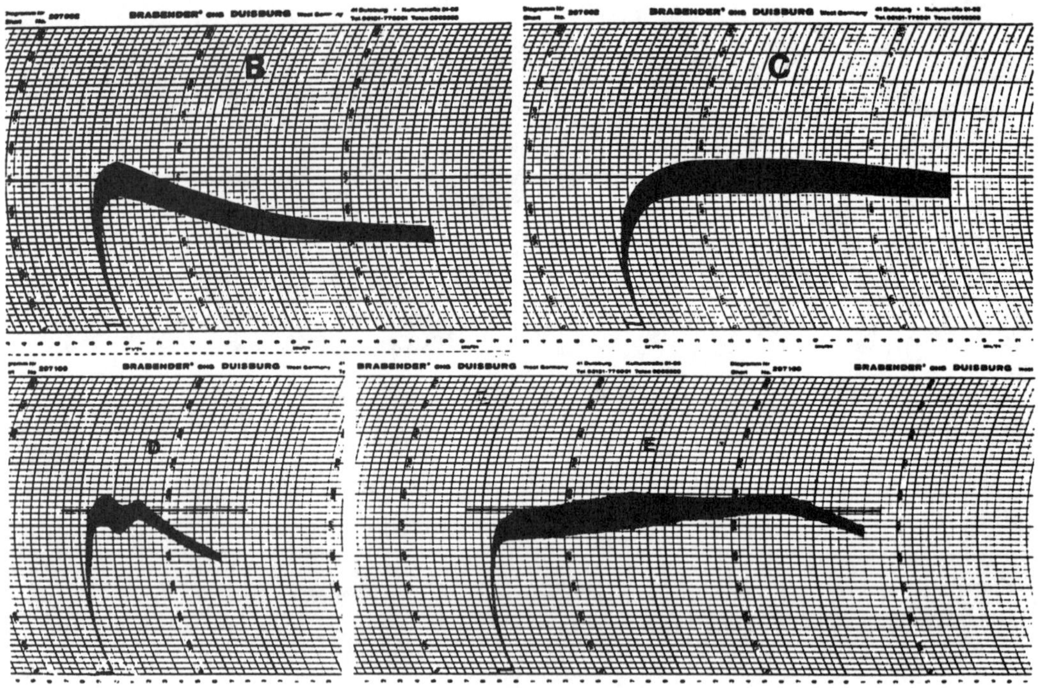

FIGURE 10 Comparison of farinograms and resistograms for identical flours: (B) farinogram of soft flour, (C) farinogram of strong flour, (D) resistogram of soft flour, (E) resistogram of strong flour. (Courtesy C.W. Brabender Instruments Co., South Hackensack, N.J.)

water by flour; the second one measures the stickiness and extensibility at breakdown of the dough.

Another variant of the farinograph, the Brabender Do-corder mixer (Fig. 11), was designed to simulate conditions of mechanical dough development (24). It has a nearly closed mixer in which dough can be subjected to mixing at variable speeds and work-input levels higher than in an ordinary farinograph mixer (Fig. 12).

The mixograph (TMCO, Lincoln, Nebraska), originally designed by Swanson and Working (36), is another widely used recording mixer (Fig. 13). The mixing action is provided by four planetary pins revolving about three stationary pins on the bottom of the mixing bowl. The mixing can be described as a pull, fold, and repull action, which is more severe than that produced by the farinograph. As a result, the main advantage of the mixograph is the speed with which a test can be conducted (21). However, in comparison with the farinograph, the mixograph is more difficult to standardize and has a limited applicability for the determination of water absorption of the tested flour. The shape of a mixogram (Fig. 14) can be characterized by indices similar to those defined for the farinogram (1) (AACC Method 54–40). Peak time is similar to farinograph dough development time. Peak height provides information about flour strength and absorption. Height of curve at a specified time past the peak is similar to the farinograph tolerance index. Higher values indicate a greater tolerance to overmixing. The same applies for the values of the angle between the ascending and descending portions of the curve at the peak. A higher toler-

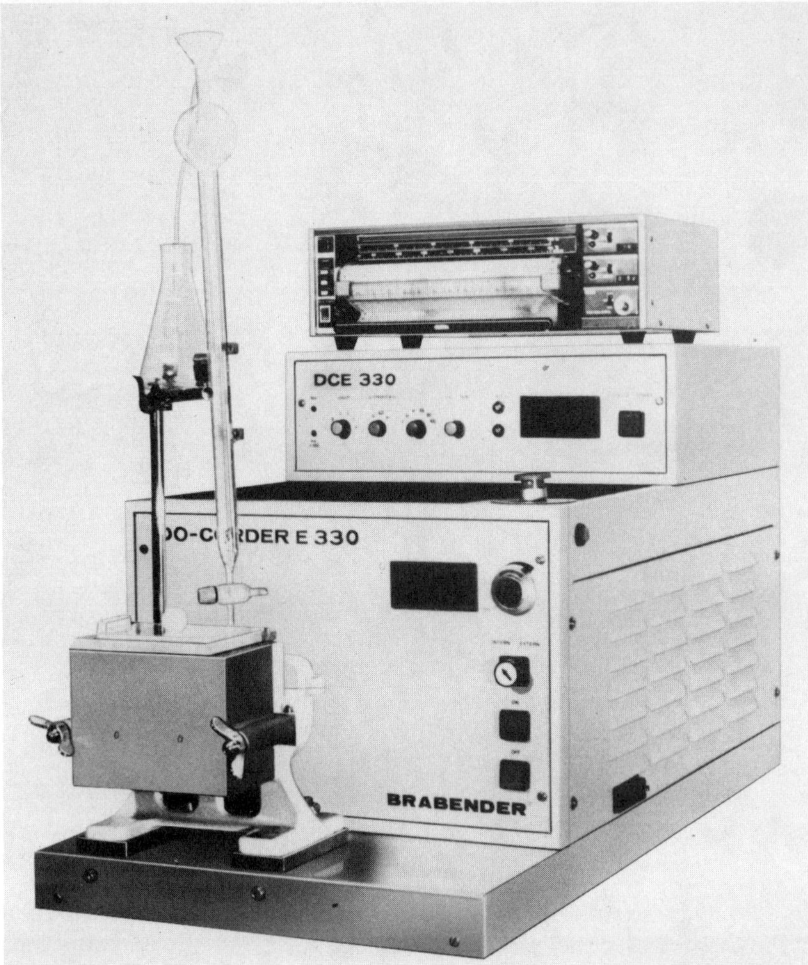

FIGURE 11 The Brabender Do-corder. (Courtesy C.W. Brabender Instruments Co., South Hackensack, N.J.)

ance to overmixing and overall strength of flour can also be judged from the area under the curve. Examples of mixograms milled from different classes of wheat are shown in Figure 15.

Techniques constituting the second phase of the three-phase testing system provide information on the potential behavior of dough during its rise due to the development and expansion of gas at the fermentation and early baking stages. Load-extension instruments are applied to measure the resistance of dough to extension. The Brabender extensigraph (Fig. 16) stretches a cylindrically shaped dough piece until it ruptures while the resulting force on the test piece is transmitted through a balanced lever system to a recorder. The dough (with 2% salt based on flour weight) is prepared in a farinograph mixer at its optimum absorption. According to the AACC procedure (1) (AACC Method 54–10) the dough is developed until its consistency reaches its maximum, whereas the ICC method (ICC Standard No. 114) uses

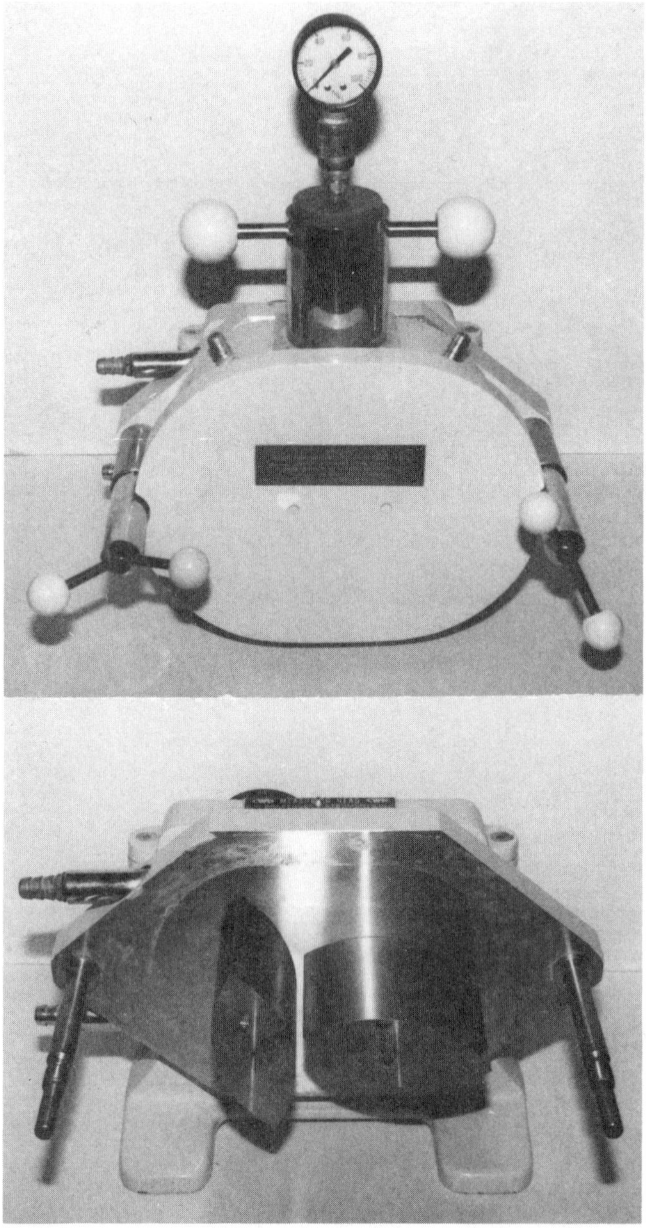

FIGURE 12 Two views fo the Developer Head for the Brabender Do-Coder. (Courtesy C.W. Brabender Insutrments Co., South Hackensack, N.J.)

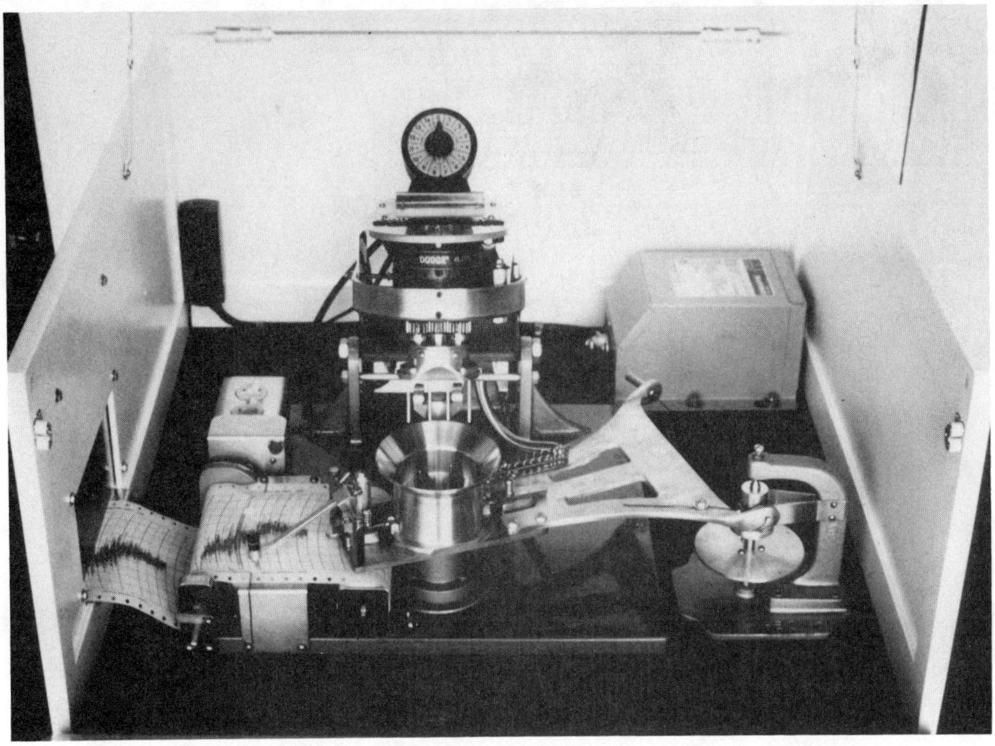

FIGURE 13 The mixograph. (Courtesy TMCO, Lincoln, NB.)

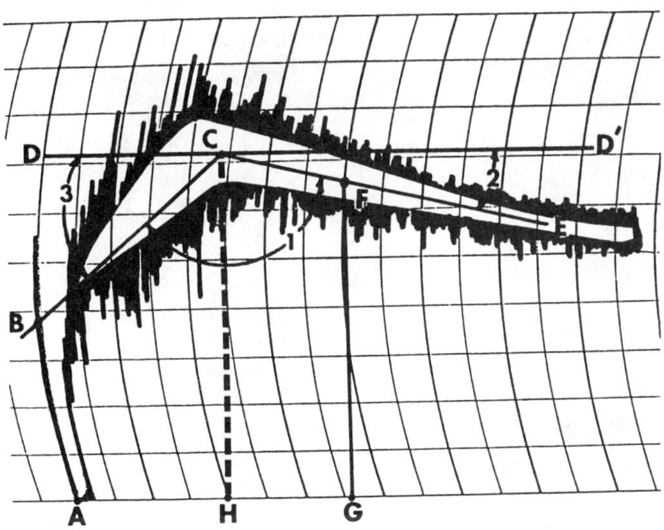

FIGURE 14 Representative mixogram showing the commonly measured indices: peak time = line CD. Maximum height = line CH. Angle between ascending and descending portions of curve at peak (lines BC and CE, angle 1). Alternately, angles formed by lines with horizontal lines (angles 2 and 3) have been used to describe curves. (Reprinted, with permission, from Ref. 21.)

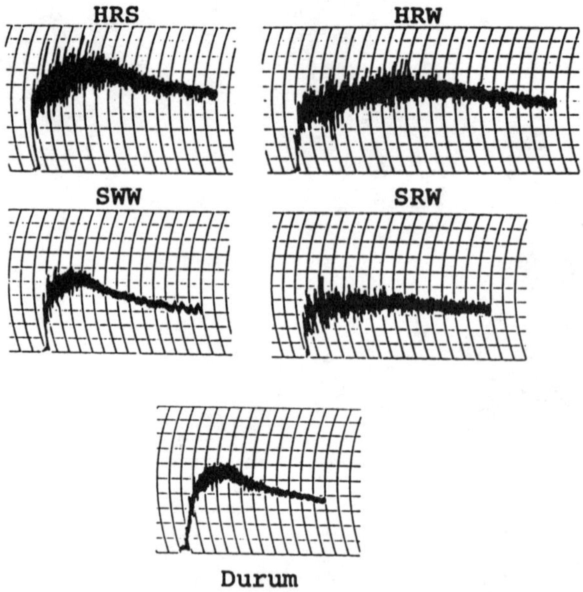

FIGURE 15 Mixograms of flours milled from different wheat classes (obtained with optimized absorption farinography). (Reprinted with permission from Ref. 21.)

FIGURE 16 The Brabender extensigraph. (Courtesy C.W. Brabender Instruments Co., South Hackensack, NJ.)

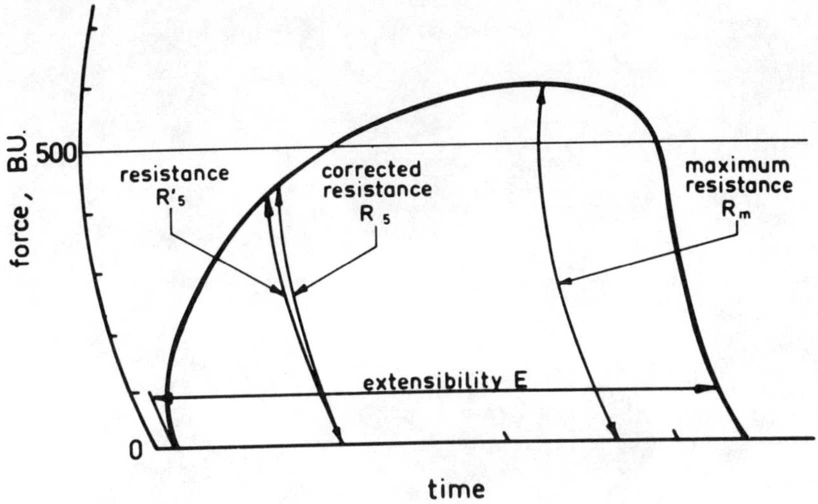

FIGURE 17 Representative extensigram with the most commonly measured indices. (Reprinted, with permission, from Ref. 3).

mixing for a fixed 5-min period. While the dough test piece is stretched, a curve of force versus time, an extensigram, is recorded (Fig. 17). Several extensigram indices provide a practical guide to general strength of the dough. They include:

1. The maximum resistance Rm or the resistance at fixed extension, usually corresponding to 50 mm transposition on the chart paper, R_5. The latter has the advantage of measuring the resistance at the same extension for all tested doughs.
2. The dough extensibility E.
3. The ratio of resistance and extensibility.
4. The area under the curve, which is proportional to the energy that is required to stretch the test piece to its rupture point. This index is a convenient single figure for characterizing flour strength. The stronger the flour, the more energy is required to stretch the dough.

The extensigraph response to different types of flour and the ways in which the shape of the extensigram may relate to the behavior of dough can be seen from Figure 18.

Apart from differentiation among flours, the extensigraph is also used to evaluate the effects of oxidizing agents on the physical qualities of dough. For example, Figure 19 shows how the type of extensigram changes after the addition of ascorbic acid when it is allowed to react in the dough for different times.

An apparatus similar to the Brabender extensigraph is the Halton (or Simon "Research") extensometer of the Association of British Flour Millers (11). The extensometer is part of a three-unit device that also includes a water absorption meter and a mixer-shaper unit. The absorption meter determines the optimum absorption of the dough (generally yeasted) from the extrusion time values measured on several doughs prepared from the same flour sample with varying amounts of water. Optimum absorption has been empirically linked to an extrusion time of 50 seconds. Doughs, after shaping in the mixer-shaper unit, are stretched between two pegs. The force exerted on the stationary peg is transmitted and recorded in the form of a curve that resembles the Brabender extensigram.

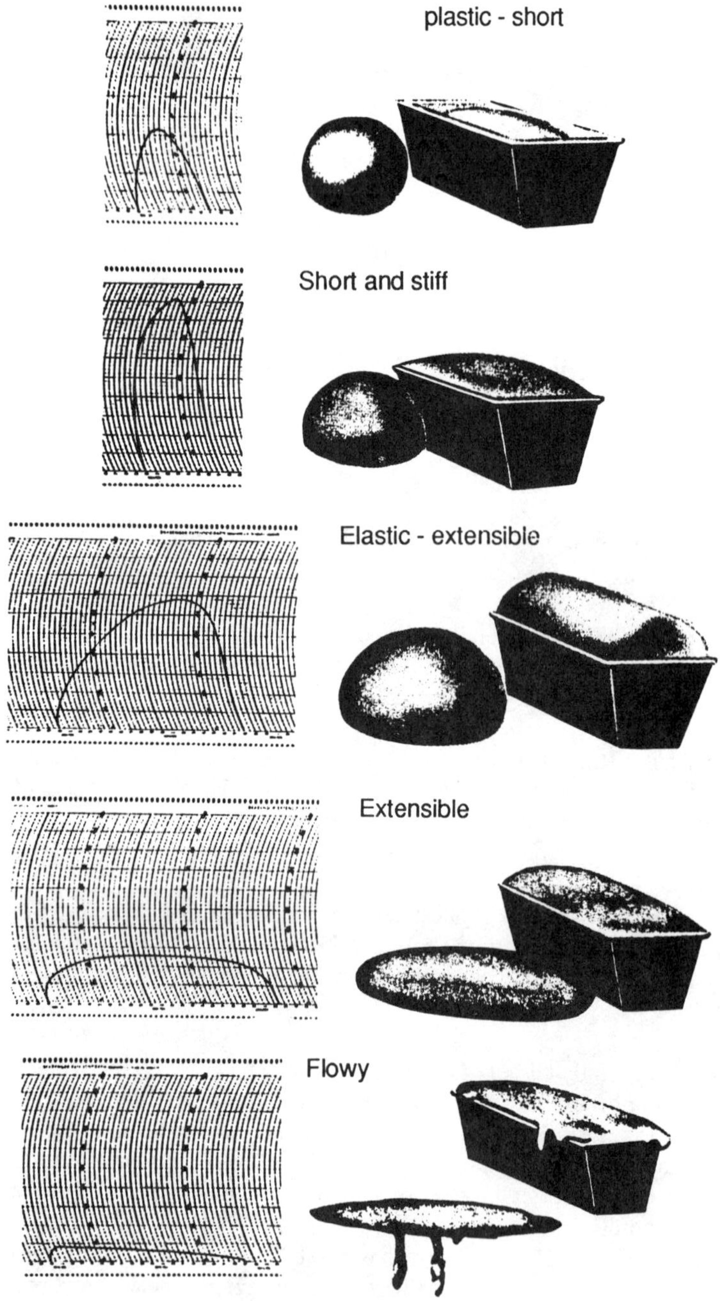

FIGURE 18 Extensigrams of different types of flour and their relation to the baking behavior of their doughs. (Courtesy C.W. Brabender Instruments Co., South Hackensack, NJ.)

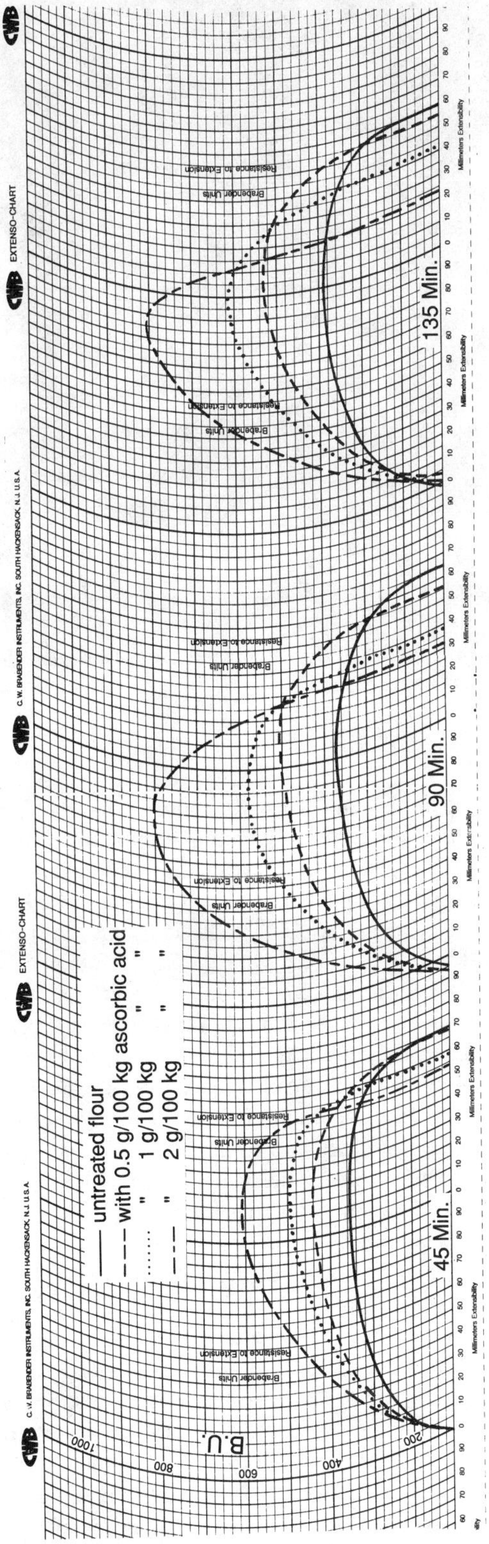

FIGURE 19 The effect of ascorbic acid on extensigrams recorded after different resting periods. (Courtesy Brabender Instruments Co., South Hackensack, NJ.)

FIGURE 20 The Chopin alveograph MA82 with computerized Relaxocalculator RCV4. (Courtesy of Chopin SA, Villeneuve la Garenne, France.)

Another load-extension apparatus, until recently more popular in several European countries than in North America, is the Chopin alveograph (Fig. 20). Unlike the Brabender extensigraph or Halton extensometer, which both stretch the test dough piece in only one direction, the alveograph subjects dough to extension in all directions by blowing a molded and rested sheet into a bubble (1,7,19) (AACC Method 54–30). From the physical viewpoint, such a mode of extension is well linked with the gas cell expansion in the rising dough. The instrument records the air pressure in the bubble as a function of inflation time. A typical alveograph record, an alveogram, is shown in Figure 21. Its interpretation is similar to that of the extensigram. The maximum height of the curve is taken as a measure of resistance to extension, and its length as a measure of extensibility. The area under the curve is usually converted into a W value, referred to as deformation energy, which represents the total work input in blowing up the test piece into a bubble. The W value is the most commonly used alveogram index and resembles, in its use, the area under the extensigram. Alveograms of flours with distinctly different properties are shown in Figure 22. In interpreting the alveograms, it has to be kept in mind that, unlike doughs for extensigraph testing, doughs intended for the alveograph are prepared with constant water addition to flour (51.4%), irrespective of its true absorption.

The third phase of the three-phase testing system concentrates on changes in physical properties taking place during the baking stage, the most striking of which result from the gelatinizing of starch and its degradation by amylolytic enzymes either inherent in or added to flour. Viscometry under controlled conditions of heat supply to the tested sample is well suited for this purpose. The Brabender amylograph (Fig. 23) is a torsion viscometer which provides a continuous record of the changes in viscosity of a buffered flour suspension at a uniform rate of temperature increase (1.5°C per minute) and constant stirring (1,34) (AACC

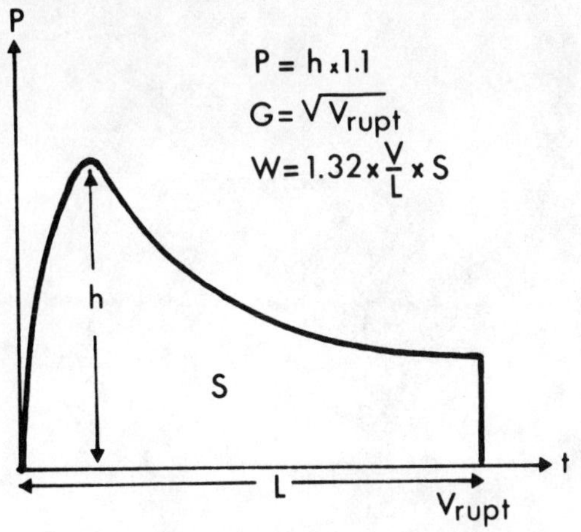

FIGURE 21 Representative alveogram showing the commonly used indices, where P is overpressure (mm), L is abscise at rupture (mm), G is swelling index (ml), V is volume of air (ml), and W is deformation energy (10^{-4}) (J). (Courtesy of Chopin SA, Villeneuve la Garenne, France.)

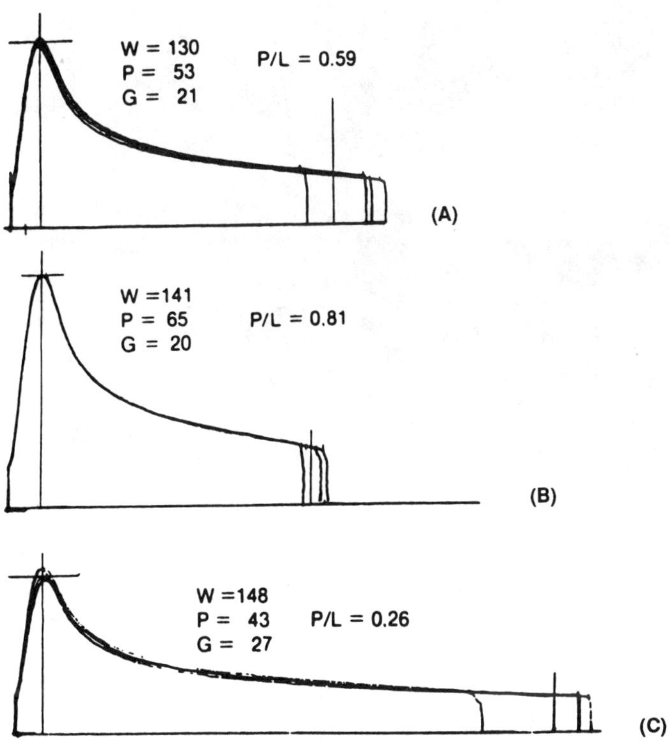

FIGURE 22 Alveograms of different flour types. (A) Normal dough, (B) short dough with little stretching properties and (C) soft dough with excess stretching properties. (Courtesy of Chopin SA, Villeneuve la Garenne, France.)

FIGURE 23 The Brabender visco/amylograph. (Courtesy of C.W. Brabender Instruments Co., South Hackensack, NJ.)

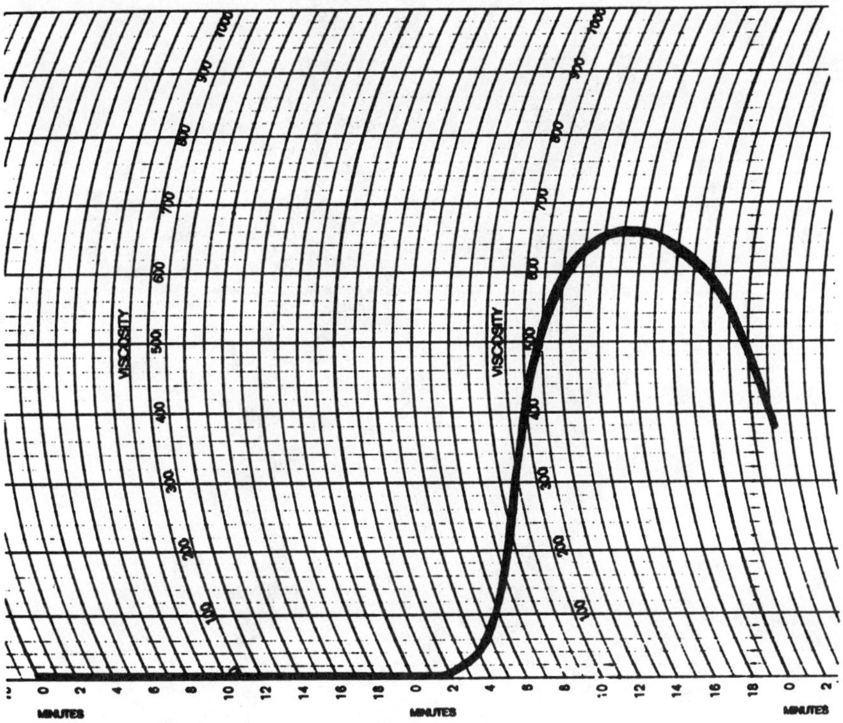

FIGURE 24 Regular amylogram of malted wheat flour by AACC method. Peak viscosity is 600 A.U. (amylograph units). (Reprinted, with permission, from Ref. 24)

Method 22–10). Swelling of the starch granules upon reaching the gelatinization temperature, together with increased concentration of the solubles in the surrounding liquid due to leaching out of starch molecules from the swollen granules causes the viscosity of the suspension to rise. Gelatinized starch becomes highly susceptible to the action of thermostable amylases which are activated by the higher temperatures. They hydrolyze and liquefy part of the total starch, thus reducing the viscosity. The recorded maximum is, therefore, a result of these two simultaneous processes (Fig. 24). Since there is a limited variability in the viscosity of gelatinizing wheat starch in the absence of amylases, the height of the viscosity curve is primarily a reflection of the amylolytic activity in the tested flour sample. The higher the activity of the starch-liquefying enzymic system, the lower the peak viscosity. Although a contribution from both α-amylase and β-amylase to this reduction in viscosity can be expected, it is predominantly the α-amylase that is responsible for the final viscosity reading. Because of a higher heat sensitivity, β-amylase is largely inactivated before the starch substrate becomes sufficiently susceptible to its action. Similarly, higher heat sensitivity of fungal α-amylases renders the test unsuitable for testing flours supplemented with such enzymes. A modified amylographic procedure had to be designed for these flours. The use of pregelatinized substrate eliminates the need for attaining temperatures above the optimum performance range of the enzyme (1) (AACC Method 22–12).

The principle of viscometry in determining amylolytic activity of wheat flour is applied in the Falling Number test (1) (AACC Method 56–81B). The test is based on measuring the time required to stir and to allow a specified viscometer-stirrer to fall a standard distance

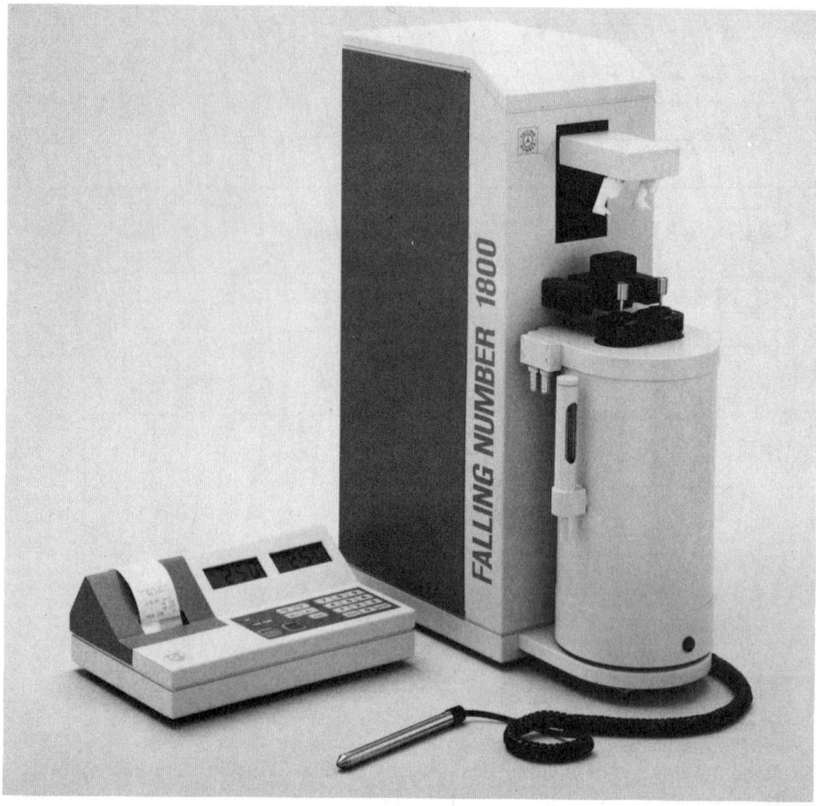

FIGURE 25 The Hagberg Falling Number apparatus. (Courtesy of Falling Number A.B., Stockholm, Sweden.)

through a gelatinized flour paste. The instrument and the testing sequence are shown in Figures 25 and 26.

IV. PHYSICOCHEMICAL TESTING

A number of tests were designed for predicting the potential bread-baking strength of wheat and wheat flour from the behavior of gluten. The *Berliner* and *Koopman method* (4) measures the swelling ability of wet gluten when immersed in an 0.1 N lactic acid solution. Higher "specific swelling factors" can be obtained with stronger glutens. A modification to this method measures the turbidity of the suspension, which is inversely related to gluten strength.

The Pelschenke test (or dough-ball test) uses a small ball of yeasted dough prepared from whole wheat meal that is immersed in water at constant temperature (1) (AACC Method 56–50). The length of time for which the ball stays in the water before it starts to disintegrate is called the "test number." It is a measure of both gluten quantity and quality. It varies from less than 30 minutes for soft wheats to more than 6 hours for very strong wheats. By dividing the "test number" by protein content in wheat, an index of gluten quality alone is obtained. The higher the index, the higher the gluten strength.

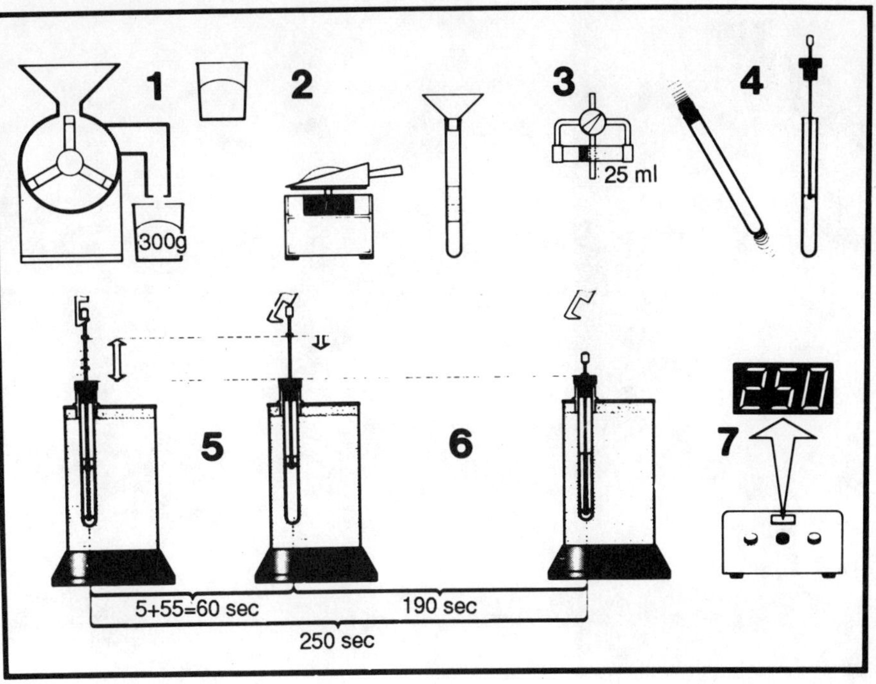

FIGURE 26 The Falling Number procedure: 1) sample preparation; 2) weighing; 3) dispensing; 4) shaking; 5) stirring; 6) measuring; 7) result. (Courtesy of Falling Number A.B., Stockholm, Sweden.)

The sedimentation test (1) (AACC Methods 56–60, 56–61A, 56–63) originally developed by Zeleny (44), measures the volume of sediment (predominantly swollen protein and occluded starch) of crude white flour suspended in dilute acetic acid. Like "test number," the "sedimentation value" is a reflection of both gluten quantity and quality but can be turned into an index of gluten quality alone if divided by protein content in the sample. It is then referred to as "specific sedimentation value."

The alkaline water retention test (38) has been useful in predicting performance of wheat flours in cookie manufacture. Another widely used test in soft wheat flour quality evaluation is the *MacMichael viscosity test.* The MacMichael viscometer measures the increase in viscosity of acidulated soft flour-water suspensions due to swelling of gluten proteins and, to some extent, of starch (1) (AACC Method 56–80). The swelling of starch becomes more pronounced if the granules are damaged. Thus, if the range in protein content is relatively narrow, the measurements reflect primarily the condition of the starch. On the other hand, if the changes in viscosity due to starch are more or less constant, the increase in viscosity is in direct relation to the swelling properties and quantity of gluten present. (The Brookfield viscometer is currently the alternate choice because MacMichael viscometer parts are no longer available.)

Physicochemical testing also includes tests evaluating gas production and gas retention capacity of dough. The AACC pressuremeter method (1) (AACC Method 22–11) measures the pressure of gas produced by a yeasted suspension of flour in an airtight and pressure gauge-equipped container (Fig. 27) after 5-hour fermentation at 30°C. Other methods for measuring gas production and/or gas retention use special instruments such as the

FIGURE 27 Pressuremeter for AACC Method 22–11 (Courtesy of National Manufacturing Co., Lincoln, NB.)

gasograph (30) or the Chopin rheofermenter (Fig. 28). The latter, apart from measuring the total and retained gas, also monitors the changes in dough volume during fermentation (Fig. 29). The Brabender maturograph (Fig. 30) measures the net results of gas production and gas loss by recording the changes in height of fermenting dough subjected to periodic punching at 2-minute intervals (Fig. 31). From the shape of the curve, optimum proofing conditions and fermentation tolerance can be established. The difference between top and bottom envelopes of the curve band reflects the changes in the height of the dough due to periodic punching and recovery and is often referred to as elasticity (35).

FIGURE 28 The Chopin rheofermentometer. (Courtesy of Chopin SA, Villeneuve la Garenne, France.)

V. FUNCTIONALITY TESTING

The flour and ingredient quality are generally determined in industrial practice by experimental baking tests. The standard procedures developed by the American of Cereal Chemists are given in Table 2. They are used for evaluation of flour quality and may be adopted for determination of baking quality of other bread ingredients. The same approach is used in evaluation of soft wheat flours for cookies, cakes, pies, etc. The standard methods do not always produce results that correlate with industrial practice. Therefore, many modifications in formulations and procedures have been made in various laboratories. The baking methods for bread and cake testing used at the American Institute of Baking are given as examples in Tables 6 and 7. The evaluation of the baked products is done by scoring for various quality parameters as specified by the respective methods (see Table 2).

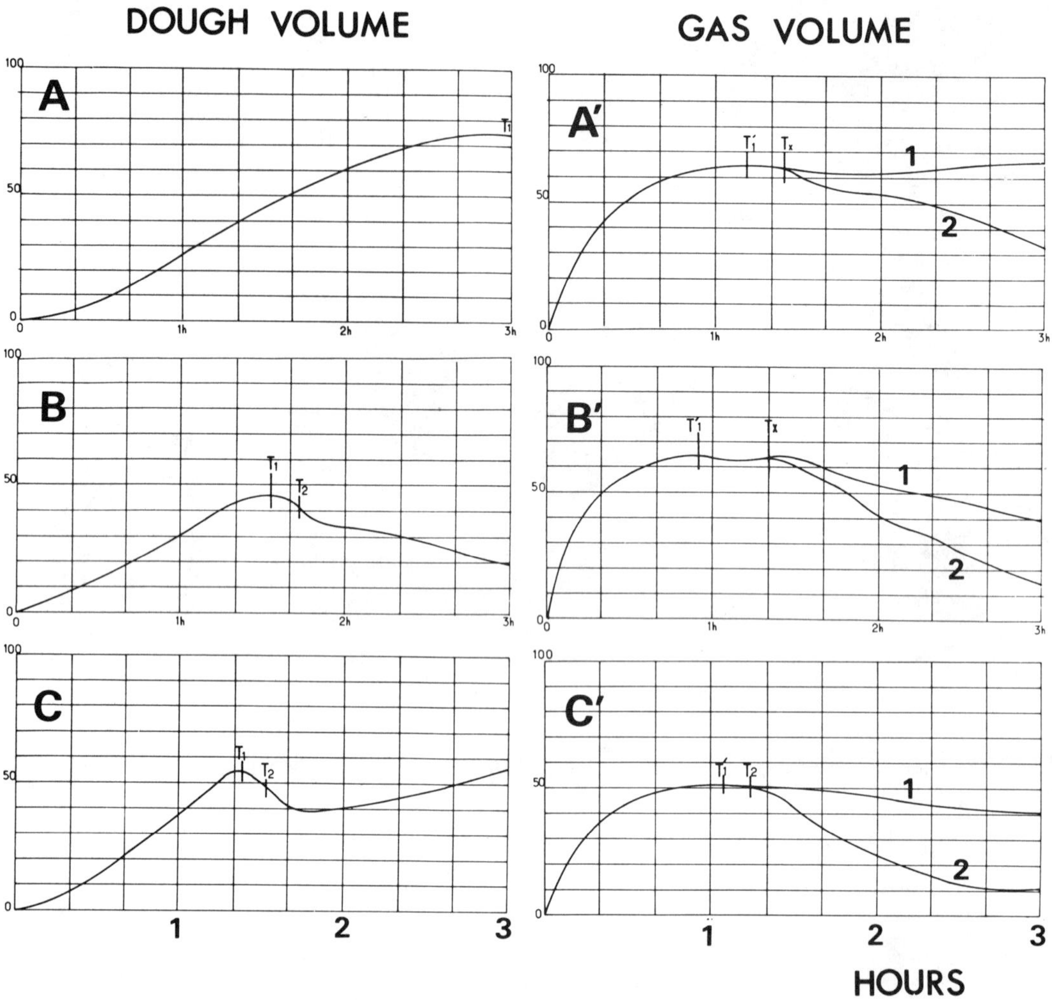

FIGURE 29 Standard rheofermentograms indicating the changes in dough volume and gas production (line 1) and retention (line 2). (A and A) Dough with good fermentation power and good tolerance. (B and B) Dough with good fermentation power but poor tolerance. (C and C) Dough with poor fermentation power and poor tolerance. (Courtesy of Chopin SA, Villeneuve la Garenne, France.)

FIGURE 30 The Brabender maturograph. (Courtesy of C.W. Brabender Instruments Inc., South Hackensack, NJ.)

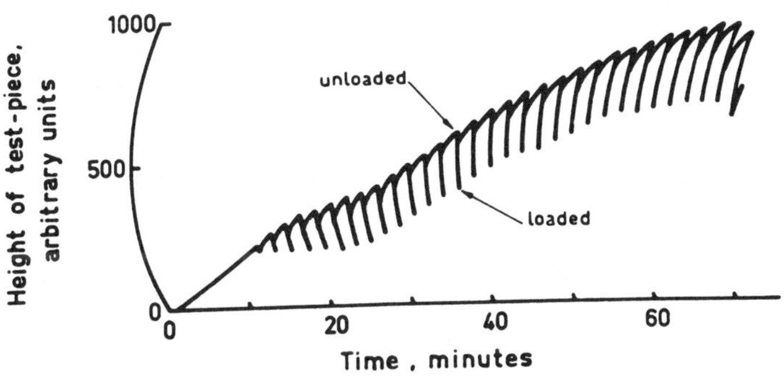

FIGURE 31 A maturogram. (Reprinted, with permission, from Ref. 3.)

TABLE 1 Some of the Most Commonly Used AACC Approved Methods and ICC Official Standards for Testing Grains and Flours

Tested property	Principle	AACC Methods[a]	ICC Standard[b]
Chemical constituent:			
Moisture	Oven-drying	44–15A	109/1
		44–16	110/1
		44–18	
		44–19	
		44–20	
	Dielectric meter	44–11	
	Vacuum drying	44–40	
	Destillation with toluene	44–51	
Protein	Automated Kjel-Foos method	46–08	
	Automated colorimetric method	46–09	
	Improved Kjeldahl method	46–10	105/1
		46–11	
		46–12	
	Micro-Kjeldahl method	46–13	
	Udy-dye method	46–14A	
	Biuret method	46–15	
Gluten	Hand washing	38–10	
	Machine washing	38–11	137
Mineral matter (ash)	Incineration	08–01	104
		08–02	
		08–03	
Crude fiber	Chemical digestion	32–10	105/1
Dietary fiber (insoluble)	Enzymatic and chemical digestion	32–20	
Crude fat	Solvent extraction	30–10	136
		30–25	
Fat acidity	Titration	02–01A	
		02–02A	
		02–03A	
Starch	Enzymatic	76–11	128
	Polarimetric	76–20	122
Starch damage	Enzymatic	76–30A	
Enzymatic activity:			
α-Amylase	Colorimetric	22–01	108
		22–06	
	Nephelometric	22–07	

TABLE 1 (Continued)

Tested property	Principle	AACC Methods[a]	ICC Standard[b]
Viscometric			
	Amylography	22–10	126
	Falling number	567–81B	107
	Pressuremetric	22–11	
	Volumetric	22–14	
	Chemical (incubation in situ)	22–15	
		22–16	
Proteases	Chemical	22–60	
		22–61	
	Spectrophotometric	22–62	
	Colorimetric	22–63	
Physical properties:			
Farinography	Dough mixing	54–21	115
Mixography	Dough mixing	54–40	
Extensigraphy	Dough stretching	54–10	114
Alveography	Dough stretching	54–30	121
Particle size	Centrifugation	50–10	
	Sedimentation		127
Physicochemical properties:			
Sedimentation value	Swelling and sedimentation	56–60	116
Oxidizing, bleaching, and maturing agents:			
Oxidizing agents	Chemical	48–02	
Acetone peroxides	Chemical	48–05	
Ascorbic acid	Chemical	86–10	
Benzoyl peroxide	Chemical	46–06B	
	Chemical	48–07	
Chlorine dioxide	Chemical	48–30	
Azodicarbonamide	Chemical	48–42	
Potassium bromate	Chemical	50–10	

[a]Approved Methods of the American Association of Cereal Chemists, 8th ed., St. Paul, MN, 1983.
[b]Standard Methods of the International Association for Cereal Science and Technology, Moritz Schäfer, Detmold, Federal Republic of Germany.

TABLE 2 Procedures for Testing Baking Quality of Wheat Flour and Ingredients

Tested property	Principle	Source
Standard procedures:		
Bread flour quality	Straight-dough baking test with long fermentation; size: pup loaf	10–09[a]
Bread flour quality	Straight-dough baking test with optimized conditions; also suitable for evaluation of effects of environment, variety, dough ingredients, protein content, flour components; used for 10–100 g flour and on 1–lb loaves; also used for blends of wheat and nonwheat flours	10–10B[a]
Bread flour quality	Sponge and dough baking tests; 800 g flour, 1–lb loaves	10–11[a]
Baking test of flour quality of sweet yeast product	Preparation of coffee cakes using 1000 g of mix	10–20[a]
Baking property of self-rising biscuit flour	Preparation of biscuits	10–31[a]
Baking quality of cookie flour	Baking sugar snap cookies and measurement of cookie spread; 225 g flour	10–50D[a]
Baking quality of cookie flour	Same as above; 40 g flour	10–52[a]
Baking quality of cake flour	Preparation of white layer cakes; 200 g flour	10–90[a]
Baking quality of angel food cake flour	Preparation of angel food cakes	10–15[a]
Baking quality of pie flour	Preparation of pie shells; 1000 g flour	10–60[a]
Baking quality of rye flour	Straight-dough process for rye bread; 100 g wheat (clear) and 100 g rye flour	10–70[a]
Baking quality of nonfat dry milk (NFDM)	Straight-dough bread process with 6% NFDM	10–85[a]
Nonstandardized procedures:		
Bread flour quality	Sponge dough method under optimized conditions, see Table 6	AIB[b]
Cake flour quality	White layer cake, see Table 7	AIB[b]

[a]Approved Methods of the American Association of Cereal Chemists, 8th Edition, St. Paul, MN, 1983.

[b]American Institute of Baking, Manhattan, KS.

TABLE 3 Generally Used Criteria for Measuring Quality of Flour for Yeast Breads

Quality criterion	Pan bread: Wholesale, conventional	Pan bread: Continuous dough	Hearth breads	Variety breads, cracked wheat Bread, rye bread	Soft rolls	Sweet goods
Wheat class[a]	HS, HW or blend	HS, HW or blend	HS	HS	HS, HW or blend	HS, HW or blend
Flour absorption	Medium to high	Medium	High	High	High	Medium to high
(% by farinograph)	60–64	59–64	63–68	65–75	59–64	60–64
Ash (%)	0.44–0.50	0.44–0.50	0.44–0.55	0.45–0.55	0.44–0.5	0.45–0.50
Alk. water retention	NS[b]	NS	NS	NS	NS	NS
Baking test: Bread; dough	Pan; straight or sponge	Pan; sponge	Hearth; straight	Pan; straight	Pan; sponge	Pan; straight
Color	Creamy white	Creamy white	Creamy white	NS	Creamy white to white	Creamy white
Enzyme activity						
Amylase						
Amylograph (BU)	475–625	525–600	400–600	350–550	475–625	475–625
Maltose (mg)	290–350	280–330	300–360	300–375	290–350	290–350
pressuremeter						
5th–hour (mm)	500–550	475–550	500–550	500–550	500–550	500–550
Falling number	200–300	200–300	175–275	200–300	200–300	200–300
Amylase analyzer	300–600	300–600	350–600	300–600	300–600	300–600
Lipase	NS	NS	NS	NS	NS	Low for prepared mixes
Protease	NS	NS	NS	NS	NS	NS

TABLE 3 (Continued)

Quality criterion	Pan bread: Wholesale, conventional	Pan bread: Continuous dough	Hearth breads	Variety breads, cracked wheat Bread, rye bread	Soft rolls	Sweet goods
Particle size						
Range (SED) = (μm)	0–150	0–150	0–150	0–150	0–150	0–150
Protein						
Quantity (%)	11.0–13.0	11.0–13.0	13.5–14.5	13.5–16.0	11.5–12.5	11.5–13.0
Quality (recording mixer values)						
Hydration	Medium	Short to medium	Medium to long	Medium to long	Medium	Short to medium
Peak development (min)	6–8	5–7	7–9	7–9	6–8	5–8
Stability (min)	7.5 minimum	8 minimum	10 minimum	10 minimum	8 minimum	8 minimum
Extensibility[d]	Medium	Medium	Medium to long	Medium to long	Medium to long	Long
Resistance to extension	Medium	Medium	Maximum	Maximum	Medium to low	Medium to low
Starch damage (%)	5.5–7.7	5.5–7.8	7.0–8.5	Low as possible	5.5–7.8	5.5–7.8

[a]HS = hard spring, HW = Hard winter.
[b]NS = not significant at present, or significance not known.
[c]Stokes equivalent diameter.
[d]Descriptive terms in each instance refer to relative values for wheat variety involved.
Source: Reprinted in a reduced version, with permission, from Ref. 10.

TABLE 4 Generally Used Criteria for Measuring Quality of Flour for Yeast Breads

			Crackers		
Quality criterion	Home baking	Biscuits	Sponge	Dough	Pastry
Wheat class[a]	HW or HW–SW blend	HW–SW blend or SW	SW or SW–HW blend	SW	SW
Flour absorption	Medium low	Low	Low	Low	Low
(% by farinograph)	52–58	50–54	48–52	48–52	48–52
Ash (%)	0.44–0.48	0.44–0.40	0.40–0.48	0.40–0.48	0.40–0.48
Alk. water retention	NS[b]	NS	58–68	58–68	50–56
Baking test: bread; dough	Biscuits; kitchen layer cakes;	Biscuits	AACC cookie; W/T 6.0–7.5	AACC cookie; W/T 7.5–8.5	AACC cookie; W/T 7.5–8.5
straight-dough bread					
Color	Creamy white	White	Creamy	Creamy	Creamy
Enzyme activity					
Anylase					
Amylograph (BU)	450–600	NS	700–800	NS	NS
Maltose (mg)	290–320	NS	200–250	NS	NS
Pressuremeter 5th–hr (mm)	400–450	NS	300–350	NS	NS
Falling number	200–300	250 minimum	250 minimum	250 minimum	250 minimum
Lipase	NS	NS	NS	NS	NS
Protease	NS	NS	NS	NS	NS

TABLE 4 (Continued)

Quality criterion	Home baking	Biscuits	Crackers		
			Sponge	Dough	Pastry
Particle size					
Range (SED) = (μm)	0–125	0–90	0–125	0–125	0–125
Protein					
Quantity (%)	9.5–11.0	9.0–10.0	9.0–10.0	8.0–9.0	8.0–9.0
Quality (recording mixer values)					
Hydration	Short	Short	Short	Short	Short
Peak development (min)	3–5	2–3	1–3	1–3	1–3
Stability	3–6	1–3	1–3	1–2	1–2
Extensibility[d]	Medium	Medium short	Medium	Medium	Medium
Resistance to extension[d]	Medium to low	Low	Low	Low	Low
Starch damage	Low as possible	Low as possible	Moderate	Low as possible	Low as possible
Viscosity					
MacMichael	90–120°	60–90°	65–90°	45–60°	45–60°

[a]HW = hard winter, SW = soft winter.
[b]NS = not significant at present, or significance not known.
[c]Stokes equivalent diameter.
[d]Descriptive terms in each instance refer to relative values for wheat variety involved.
Source: Reprinted in a reduced version, with permission, from Ref. 100.

TABLE 5 Criteria for Evaluating Flours for Cookies, Cakes, Soups, and Gravies

Quality criterion	Cookies	Cakes		Soups and Gravies (Thickeners)
		Layer	Foam	
Wheat class[a]	SW	SW	SW	SW
Flour absorption (% by farinograph)	Low; 48–52	Low; 48–52	Very low; 44–48	NS[b]
Ash (%)	0.42–0.50	0.34–0.40	0.29–0.33	0.40–0.58
Alk. water retention	50–54	54–62	52–56	48–55
Baking test: dough; batter	AACC cookie; W/T 8.0–9.5	White layer cake	Angel food cake	None
Color	Creamy	White White	NS	
Enzyme activity				
Amylase				
Amylograph (BU)	NS[b]	NS	NS	700+, high as possible
Maltose (mg)	NS	NS	NS	200, low as possible
Pressuremeter 5th–hr (mm)	NS	NS	NS	NS
Falling number	250 minimum	250 minimum	250 minimum	250 minimum
Lipase	NS	Low for prepared mixes	Low for prepared mixes	NS
				NS
Protease	NS	NS	NS	NS
Particle size Range (SED) = (μm)	0–125	0–125	20–60	0–150

TABLE 5 (Continued)

Quality criterion	Cookies	Cakes: Layer	Cakes: Foam	Soups and Gravies (Thickeners)
Protein				
Quantity (%)	7.0–9.5	7.0–9.5	5.5–7.5	9.0–10.5
Quality (recording mixer values)				
Hydration	Short	Short	Short	NS
Peak development (min)	1–3	1–2	1–1.5	NS
Stability (min)	1–2	1–1.5	1–1.5	NS
Extensibility[d]	Medium	Short	Short	NS
Starch damage[d]	Low as possible	High	Low	Low as possible
Viscosity				
MacMichael	40–65°	35–65°	30–45°	NS

[a]SW = soft winter.
[b]NS = not significant at present, or significance not known.
cStokes equivalent diameter.
[d]Descriptive terms in each instance refer to relative values for wheat variety involved.
Source: Reprinted in a reduced version, with permission, from Ref. 10

TABLE 6 White Pan Bread Control Formula[a]

Ingredients		Weight (g)
Sponge:	Bread flour	700
	Ammonium chloride	0.06
	Compressed yeast	25
	Potassium bromate (variable)	0.001
	Water	420
Dough:	Bread flour	300
	Granulated sugar	60
	Nonfat milk solids	20
	Salt	20
	Bread shortening (Trico)	30
	Water or ice (variable)	190
	Total	1765.6

Procedure

Sponge—Mixing: 1 minute at low speed; 1 minute at medium speed (Hobart A–200 or A–120 mixer with McDuffee bowl and 2–prong agitator).
Temperature: 24–25°C

Fermentation: 3–4 hours at 20°C in covered container.

Dough—Mixing: 1 minute at low speed; 4–5 minutes at medium speed. (To full gluten development.)
Temperature: 26–27°C.

Fermentation: 10 minutes at ambient temperature (covered).

Scaling weight—526 g of dough per loaf.

Pan size—11.5 cm × 23 top inside dimensions; 6.8 cm depth.

Proofing—To average height of 100 ± mm at 43.0°C ± 0.5°C (85% relative humidity).

Baking—16–18 minutes at 230°C.

Cooling—One hour at ambient temperature.

Measuring—Weight and volume by rapeseed displacement.

Scoring—Package loaves in polyethylene bags for evaluation of internal and external characteristics on following morning.

[a]Procedure of the American Institute of Baking, Manhattan, KS.

TABLE 7 White Layer Cake Formula (60% Hi-Ratio Shortening)[a]

Ingredient		Weight (g)	Mixing method[b]
A.	Cake flour	400	Blend dry ingredients for
	Granulated sugar	560	1 minute at low speed.
	Egg solids, whole	4	Add liquids and incorporate.
	Egg sholds, white	40	Add shortening and break up.
	Nonfat dry milk	50	Mix 1 minute at low speed and
	Salt	12	3 minutes at medium speed.
	Baking powder	30	
	Water	260	
B.	Water	150	Add water and scrape bowl. Mix 1 minute at low speed and 2 minutes at medium speed.
C.	Water	190	Add water and scrape bowl. Mix 2 minutes at low speed.

Record—Specific gravity and temperature of batter. Desired temperature 24° ± 1°C.
Scale—400 g of batter per 8-inch round layer cake pan.
Bake—24 min at 188° ± 191°C or until fully baked.
Cool—On wire rack at ambient temperature.
Measure—Weight and volume of baked cakes.
Score—Internal and external characteristics of cakes.

[a]Procedure of the American Institute of Baking, Manhattan, KS.
[b]Hobart N-50 mixer with 5-qt bowl and paddle beater.

REFERENCES

1. AACC, *Approved Methods of the American Association of Cereal Chemists*, 8th ed. The Association, St. Paul, MN, 1983.
2. Anonymous, Faster, better size distribution analysis of dry flour particles achieved in Microtrac instrument. *Food Process., 10*:88 (1978).
3. Bloksma, A. H., and Bushuk, W., Rheology and chemistry of dough. In *Wheat: Chemistry and Technology*, Y. Pomeranz (Ed.) Am. Assoc. Cereal Chem., St. Paul, MN, 1988, pp. 131–217.
4. Berliner, I., and Koopman, J., Determination of gluten in wheat flours. *Z. Ges. Mühlenwesen, 6*:57 (1929).
5. Black, H. C., Preston, K. R., and Dexter, J. E., Modifications to the Bühler laboratory mill to produce a flour comparable in yield and quality to the Allis-Chalmers mill. *Can. Inst. Food. Sci. Technol. J., 16*:191 (1983).
6. D'Appolonia, B. L., and Kunerth, W. H. (Eds.), *The Farinograph Handbook*, Am. Assoc. Cereal Chem., St. Paul, MN, 1984.
7. Faridi, H., and Rasper, V. F., *The Alveograph Handbook*, Am. Assoc. Cereal Chem., St. Paul, MN, 1987.
8. Farrand, E. A., Controlled levels of starch damage in a commercial U.K. bread flour and effects on absorption, sedimentation value, and loaf quality. *Cereal Chem., 49*:479 (1972).
9. Gillis, J. A., The Agtron: Photoelectric method of determining flour color. *Cereal Sci. Today, 8*: 40 (1963).

10. Halverson, J., and Zeleny, L., Criteria of wheat quality. In *Wheat: Chemistry and Technology*, Vol. I., (3rd ed.), Y. Pomeranz, (Ed.). Am. Assoc. Cereal Chem., St Paul, MN, 1988, pp. 15–45.
11. Halton, P., Significance of load-extension tests in assessing the baking quality of wheat flour dough. *Cereal Chem., 26*:24 (1949).
12. Hampel, G., The accurate determination of starch damage and overgrinding by means of the amylose number. *Getreide u. Mehl, 2*:16 (1952).
13. ICC, Method for the determination of ash in cereals and cereal products. ICC-Standard No. Standard Methods of the Int. Assoc. Cereal Sci. Technol., Moritz Schäfer, Detmold, Federal Republic of Germany, 1969.
14. ICC, Colorimetric method for the determination of alpha-amylase activity. ICC-Standard No. 108. Standard Methods of the Int. Assoc. Cereal Sci. Technol., Moritz Schäfer, Detmold, Federal Republic of Germany, 1968.
15. ICC, Determination of Besatz of wheat, ICC-Standard No. 102/1, and Determination of Besatz of rye. ICC-Standard No. 103/1. Standard Methods of the Int. Assoc. Cereal Sci. Technol., Moritz Schäfer, Detmold Federal Republic of Germany, 1972.
16. ICC, Method for using the Brabender Farinograph (under revision). ICC-Standard No. 115, Standard Methods of the Int. Assoc. Cereal Sci. Technol. Moritz Schäfer, Detmold, Federal Republic of Germany, 1972.
17. ICC, Determination of moisture content of cereals and cereal products (Basic reference method). ICC-Standard No. 109/1, Standard Method of the Int. Assoc. Cereal Sci. Technol., Moritz Schäfer, Detmold, Federal Republic of Germany, 1976.
18. Determination for the particle size distribution in flour by the Andreasen pipette method. Icc-Standard No. 127, Standard Methods of the Int. Assoc. Cereal Sci. Technol., Moritz Schäfer, Detmold, Federal Republic of Germany, 1976.
19. ISO, Wheat Flour—Physical characteristics of dough—Part 4. Determination of rheological properties using an alveograph. International Standard 5530/4. International Organ. for Standardization, Geneva, Switzerland, 1983.
20. Kent-Jones, D. W., Amos, A. J., and Martin, W., Experiments in the photo-electric recording of flour grade by measurements of reflecting power. *Analyst, 75*:133 (1950).
21. Kunerth, W. H., and D'Appolonia, B. L., Use of the mixograph and farinograph in wheat quality evaluation. In *Rheology of Wheat Products*, H. Faridi, (Ed.), Am. Assoc. Cereal Chem., St. Paul, MN, 1985, pp. 27–50.
22. Lillard, D. W., Jr., and Herbsgaard, D. M., Computer analysis and plotting of milling data: HRS wheat cumulative ash curves. *Cereal Chem., 60*:42 (1983).
23. Medcalf, D. G., and Gilles, K. A., Determination of starch damage by rate of iodine absorption. *Cereal Chem., 42*:546 (1965).
24. Nagao, S., The Do-Corder and its application in dough rheology. *Cereal Foods World, 31*:231 (1986).
25. Norris, K. H., Near infrared reflectance spectroscopy—The present and the future. In *Cereals '78: Better Nutrition for the World's Millions*, Y. Pomeranz (Ed.). Am. Assoc. Cereal Chem., St. Paul, MN, 1978, pp. 245–251.
26. Pomeranz, Y., Martin, C. R., Traylor, D. D., and Lai, F. G., Corn hardness determination. *Cereal Chem., 61*:147 (1984).
27. Pomeranz, Y., Afework, S., and Lai, F. S., Determination of hardness in mixtures of hard red winter and soft red winter wheats. I. Bulk samples. *Cereal Chem., 62*:41 (1985).
28. Preston, K. R., and Kilborn, R. H., Dough rheology and the farinograph. *The Farinograph Handbook*, W. C. Shuey and K. H. Tipples (Eds.). Am. Assoc. Cereal Chem., St. Paul, MN, 1982, pp. 38–42.
29. Prosky, L., Asp, N.–G., Schweizer, T. F., DeVries, J. W., and Furda, I., Determination of insoluble, soluble and total dietary fiber in foods and food products. Interlaboratory study. *J. Assoc. Off. Anal. Chem., 71*:1017 (1988).
30. Rubenthaler, G. L., Finney, P. L., Demeray, D. E., and Finney, K. F., Gasograph: Design, construction and reproducibility of a sensitive 12-channel gas recording instrument. *Cereal Chem., 57*:212 (1980).

31. Schellenberger, J. A., and Ward, A. B., Experimental Milling. In *Wheat and Wheat Improvement*, K. S. Quisenberry (Ed.). Am. Soc. Agron. Inc., Madison, WI, 1967, pp. 445–469.
32. Shuey, W. C., A wheat sizing technique for predicting flour milling yield. *Cereal Sci. Today, 5*:71 (1960).
33. Shuey, W. C., Flour color as a measurement of flour quality. *Bakers' Dig., 49*(5):18 (1975).
34. Shuey, W. C., and Tipples, K. H. (Eds.), *The Amylograph Handbook*, Am. Assoc. Cereal Chem., St. Paul, MN, 1982.
35. Seibel, W., Experiences with the maturograph and oven-rise recorder. *Bakers. Dig., 42*(1):44.1 (1968).
36. Swanson, C. O., and Working, E. B., Testing quality of flour by recording dough mixer. *Cereal Chem., 10*:1 (1933).
37. Udy, D. C., Estimation of protein in wheat and flour by ion-binding. *Cereal Chem., 33*:190 (1956).
38. Yamazaki, W. T., An alkaline water retention capacity test for the evaluation of cookie baking potentialities of soft winter wheat flours. *Cereal Chem., 30*:242 (1953).
39. Yamazaki, W. T., and Donelson, D. H., Evaluating soft wheat quality of early generation progenies. *Crop Sci., 13*:374 (1973).
40. Williams, P. C., Application of near-infrared reflectance spectroscopy to analysis of cereal grains and oilseeds. *Cereal Chem., 52*:561 (1975).
41. Williams, P. C., Screening wheat for protein and hardness by near infrared reflectance spectroscopy. *Cereal Chem., 56*:169 (1979).
42. Williams, P. C., and Fegol, K. S. W., Colorimetric determination of damaged starch in flour. *Cereal Chem., 46*:56 (1969).
43. Williams, P. C., and LeSeelleur, G. C., Determination of damaged starch in flour. *Cereal Foods World, 15*(4), (1970).
44. Zeleny, L., A simple sedimentation test for estimating the bread-baking and gluten qualities of wheat flour. *Cereal Chem., 24*:465 (1947).

16

Breads and Yeast-Leavened Bakery Foods

Karel Kulp
American Institute of Baking, Manhattan, Kansas

I. BREAD–BAKING INDUSTRY

A. Profile of Baking Industry

A major portion of wheat flour in the United States is utilized in commercial bakeries for the manufacture of breads and similar yeast–raised bakery foods. In 1982 this sector of the industry consumed almost 14 billion lb of flour.

B. Industry Trends

The traditional white pan bread still remains the predominant type of the industry, but other types have gained a major portion of the market. This is evident from Tables 1 and 2. The distribution of yeast–leavened products and their growth trend from 1981 to 1986 is shown in Table 3. The important component of that group are hamburger rolls and hot dog buns, which show steady use in fast food restaurants.

C. Manufacturing of Bakery Foods

The consolidation of the wholesale industry into large corporations continues. Although fully baked wholesale bakery products remain the major components of the industry (Table 1), there is a significant trend in splitting the production between the central plant (generally a wholesale bakery) and a point of distribution (a supermarket). In the central plants, the doughs for breads and yeast–leavened products are mixed, usually by a no–time dough process, as illustrated in Table 4, then frozen, and transported in refrigerated trucks to the point of distribution (called in–store bakery) where the doughs may be kept frozen and processed immediately or within 2 weeks (Figs. 1 and 2). The processing, called bake–off, includes defrosting of doughs, proofing, and baking. These doughs may also be marketed in a frozen state directly to the consumer. When manufactured for that purpose, the dough formulas contain higher than normal amounts of yeast (6%, flour basis), high levels of oxidants, and dough strengtheners to enhance the stability of doughs frozen storage, which is required to be 12–18 weeks. The advantage of baking of yeast–leavened products in in-store bakeries is the flavor and high quality texture of fully baked products. Although these

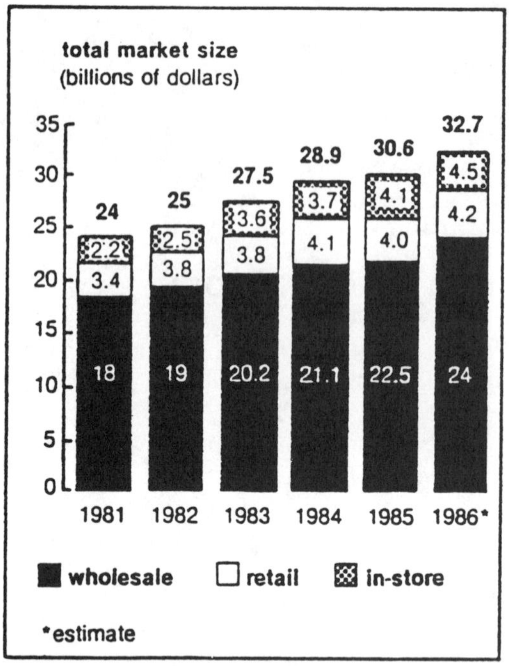

FIGURE 1 Distribution of sales volumes by wholesale, retail, and in-store bakeries during 1981–1986. From Refs. 1, 2.)

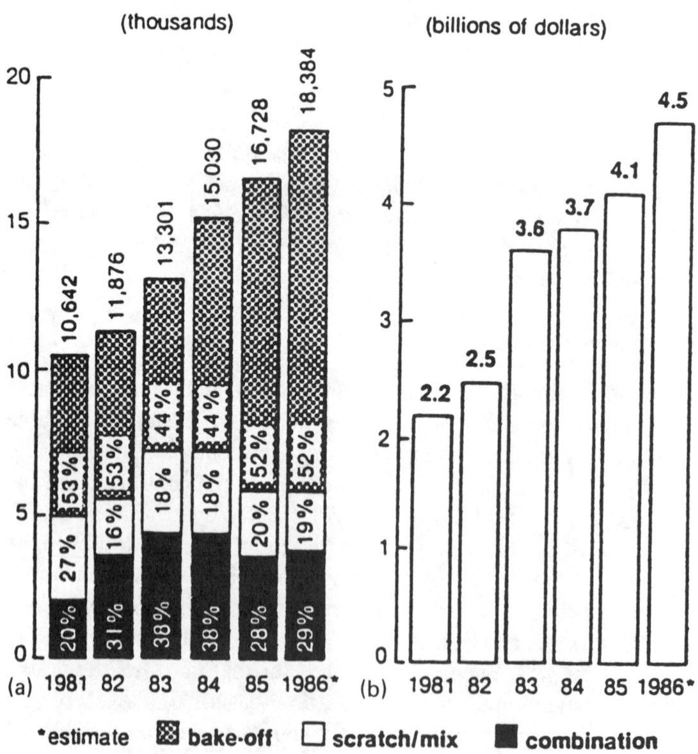

FIGURE 2 Number of in-store bakeries operating during 1981–1986 using bake-off; scratch/mix, and combination procedures (a); annual sales by volume (a) and by dollar values (b). (From Ref. 3.)

attributes are attractive to the consumer, products prepared by this type of operation require higher prices than breads from wholesale production. Consequently, the products from in–store bakeries are generally used for premium items while the wholesale fully baked products for general consumption.

II. COMMERCIAL PRODUCTION METHODS

The flow charts of the main bread processes in use are shown in Figures 3 to 5. Figures 3 and 4 depict the major bread manufacturing processes: sponge and dough and liquid ferment methods. Sponge and dough bread production predominates in the United States, accounting for about 61% of total bread production, followed by the liquid ferment process. Other processes used are the straight dough, continuous, no–time, and Chorleywood procedures. The straight dough method is mainly used in retail operations and in production of specialty breads; the continuous process lost its popularity; the no–time process is a minor production method, except for manufacturing of some specialty breads; the Chorleywood processes is common in England and many countries throughout the world. A comparison of production steps in these methods is given in Figure 5, showing the typical steps and processing times.

A. Other Yeast–Leavened Bakery Products

Per capita consumption of bakery products is shown in Table 3. Second to breads are hamburger and hot dog buns, followed by sweet yeast goods.

B. Production of Buns and Rolls

Production of buns and rolls involves mixing, dividing, rounding, proofing, molding, and panning operations, which are generally performed by integrated units that incorporate the various equipment sections needed to produce the panned dough pieces. A typical automatic unit for hamburger buns in shown in Figure 6.

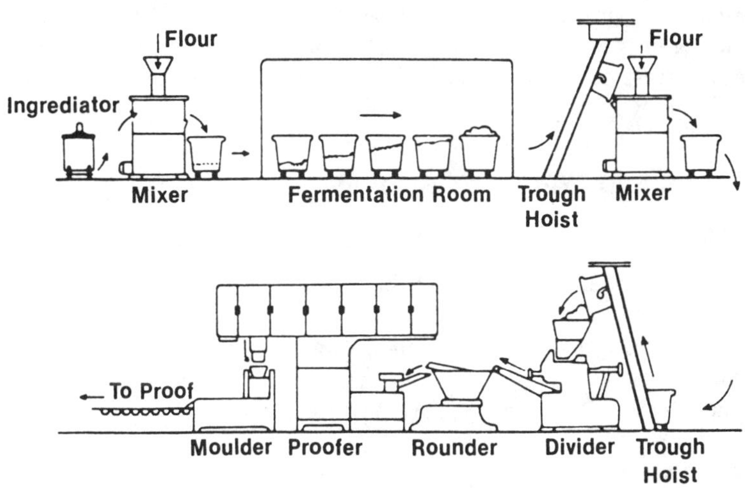

FIGURE 3 Sponge and dough bread production line. (From Ref. 4.)

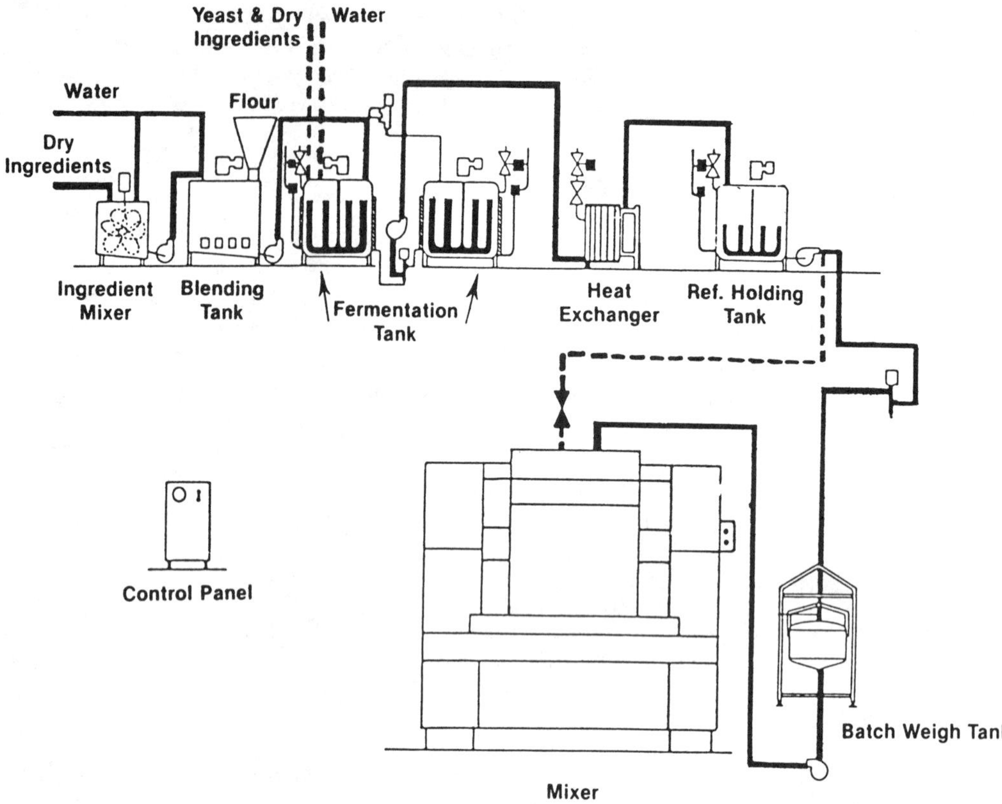

FIGURE 4 Flow chart for liquid ferment process. (From Ref. 37.)

C. Sweet Dough Equipment

The production of sweet dough products, e.g., Danish pastry, sweet rolls, coffee cakes, etc., is done either manually or by mechanical devices which perform most of the basic operations. Arrangement of process devices in an automatic sweet goods production line are given in Figure 7.

III. TYPE OF BREADS AND TYPICAL FORMULAS

A. White Pan Bread

The major type of bread manufactured in the United States is white pan bread. This type of bread is a standardized product and is regulated in respect to formulation, moisture (38% is the maximum) and enrichment (see chapter by Ranum). A typical formula for white pan bread for production of breads by major processes is shown in Table 4 along with the basic procedures. The formula for continuous bread is given in Table 5.

B. White Specialty Breads

Breads of this group are called in the market homestyle, farm, country, or range breads. They generally differ from the conventional white pan bread in internal grain and texture

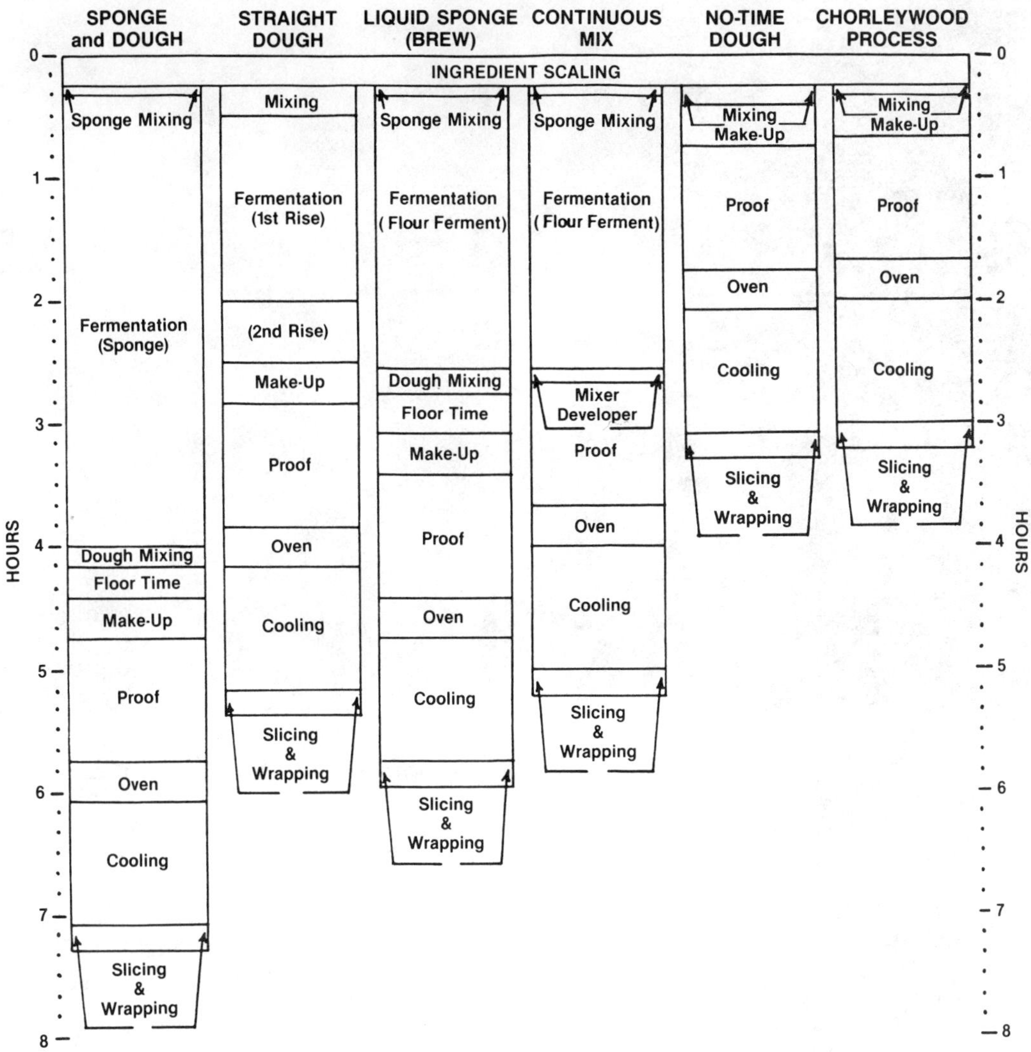

FIGURE 5 Comparison of bread-processing systems. (From Ref. 9.)

characteristics, the grain being open and their texture more coarse. These properties, reminiscent of home–made breads, are achieved mostly by not fully developing the doughs by undermixing. A typical formula is shown in Table 6. Flour used in these breads is unbleached to obtain an off–white natural crumb color. Also relatively low sugar in the formula is common for these breads.

Premium white bread (Table 7) is a dense bread usually produced by a straight dough process, described in Table 4.

White hearth breads are produced with or without lactic acid ferment, called sour by the trade. (Table 8). The main difference between pan and hearth breads is in the baking step—hearth breads are baked directly on an open hearth, whereas the others are baked in pans. The type of heat transfer during baking leads to formation of solid, crisp, flavorful crust, and other attributes associated with the type of breads. Sour hearth breads are a variation of hearth breads. Sour that is used in that type of breads is flour dough (generally rye) fer-

FIGURE 6 Model K unit for production of Hamburger buns. (Courtesy of AMF, Inc., Richmond, VA.)

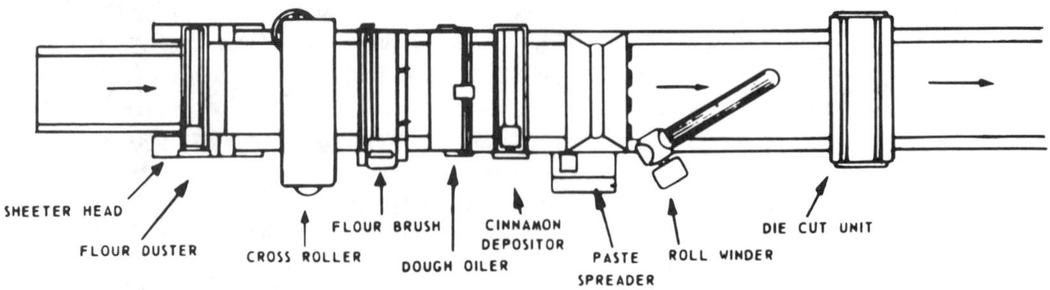

FIGURE 7 Arrangement of processing devices on an automatic sweet goods machine. (Courtesy of American Institute of Baking, Manhattan, KS.)

mented by lactic and acidic acid bacteria prepared in bakeries by natural fermentation or purchased in dry form. San Francisco hearth bread is a special sour dough bread in which the sour consists of a mixture of a special strain of yeast (*Torulopsis holmii*) and lactobacillus (*Lactobacillus sanfranciscus*).

C. Wheat breads

Typical formulas of wheat breads are given in Table 9. The most common variety of this group is "wheat" bread, which is made from a flour blend of 20–30% whole wheat flour and 80–70% patent flour. Whole wheat bread is a standardized product and has to be produced from whole wheat flour only.

D. Rye and Pumpernickel Bread

Rye breads are not popular in the United States. They are produced by different formulas (Table 10) responding to local ethnic preferences, e.g., American, Jewish, German, Swedish, Black Forest, Bohemian. The basic flour is a blend of rye flour (white and/or medium) and strong wheat patent or clear wheat flour. Addition of gluten is generally necessary to strengthen the dough.

E. Mixed Grain Breads

These types of breads are made with wheat flour and other grains, as well as vegetable materials. The grains and vegetables that have been used include corn, flax, millet, triticale, buckwheat, barley, oats, alfalfa, soy, potato, rye, rice, and sauerkraut; some of these materials are used as flours, grits, or as whole grains. Two examples of this type of bread are given in Table 11. A special method is used for production of mixed grain breads, which is also discussed under Processes.

F. Fiber Breads

For ingredients used to increase the fiber content of bread see Table 22.

G. Fruit Breads

Raisin breads (Table 12) are standardized products. Bread standards require an addition of 50% raisins (flour basis). Other fruit breads contain currants, dates, and bananas.

H. Hamburger Buns

Formula and procedure are shown in Table 13; a typical mechanical unit for their production is illustrated in Figure 6.

I. Sweet Dough (Cinnamon Rolls)

Formulas for sweet doughs and Danish pastry are given in Tables 14 and 15, along with procedures using manual preparation methods. For illustration of automatic methods, see Figure 7.

J. Flour Evaluation

In evaluation of flour and bakery products certain quality characteristics are determined. As an example, an evaluation chart used by the American Institute of Baking for determination of functional parameters in studies of flour is shown in Figure 8. A two–stage bread baking process (sponge and dough method) is used. Dough characteristics are judged at various stages: at the sponge stage after mixing, the dough–mixing time and the dough tolerance are judged. The machining character of doughs is followed and judged during dividing and molding of doughs. These judgments are made by experienced bakers. The breads produced from these doughs are scored after cooling for external and internal characteristics. These evaluations are done again by trained bakers and are subjective. External parameters include loaf volume, crust color, break and shred of the loaf. Internal parameters of importance are crumb grain, crumb color, textural character of crumb (crumb strength and texture), and sensory characteristics (flavor, aroma). Both dough and bread scores are added for evaluation when flour baking quality is evaluated.

K. Ingredient Limits of Breads and Rolls

Certain ingredients used in formulation of these products are regulated. The summary of these regulations is given in Table 16.

IV. FUNCTIONS OF BREAD INGREDIENTS

Table 17 summarizes the functions of individual bread ingredients in the bread–making process (see also Table 23 for dough strengtheners and softners).

V. INDIVIDUAL INGREDIENTS

A. Flour

The flour characteristics generally used in flour specifications are protein, ash, and amylolytic activity (indicated by Falling Number or Amylogram values). The typical U.S. values are shown in Table 18. The flour used in white breads and buns is generally a blend of spring and winter wheat flours; clear flours are the basic flours in darker bread and bun varieties; whole wheat flour is used for production of wheat and whole wheat breads.

B. Water

Analytical and hardness data are given in Chapter 17.

C. Sweeteners and Sugars

The most commonly used sugar used in bread and bun baking is high fructose corn syrup. Various parameters are found, as indicated in Chapter 17.

D. Relative Sweetness of Sugars

This property is given in Chapter 17. In estimating sweetness, one should consider that the disaccharides (sucrose and maltose) are hydrolyzed into monosaccharides by the action of yeast invertase and maltase. The sweetness should be predicted accordingly.

E. Dairy Ingredients

Stahel (11) classified dairy ingredients into two groups: dairy blends and milk replacers. Since dairy ingredients are not regulated, a great number of formulations are marketed and a

Project ______________________________

Date Mixed ______________ Date Scored ______________

SAMPLE:	DOUGH		Penalized For:
Dough Characteristics:			
Sponge (2)			A. Bold B. Flat C. Sticky
Dough Mixing Characteristics Mix Time (5)			A. Too High B. Too low
Mixing Tolerance (10)			A. Critical B. Two Minutes C. One Minute
Machinability (10)			A. Torn B. Sticky C. Bucky D. Slack E. Elastic F. Does not Seal G. Equipment Prob.
Subtotal 27			
External:	BREAD		
Volume Score (10)			A. Small B. Large
Avg. Volume, cc			
Sp. Volume, cc/g			
Uniformity of Shape (5)			A. Lack of Boldness B. Uneven Top C. Shrunken Sides D. Low Side E. Low Middle F. Flat Top G. Small End
Crust Characteristic (2)			A. Light B. Dark C. Uneven D. Dull E. Thick F. Tough G. Brittle
Break and Shred (3)			A. Wild B. None C. Shelled D. Insufficient
Subtotal 20			
Internal:			
Grain (20)			A. Open Coarse B. Thick Cell Walls C. Holes D. Non-uniform
Crumb Color (10)			A. Dull Gray B. Creamy
Crumb Strength (10)			A. Tough B. Weak
Texture (13)			A. Rough B. Core C. Crumbly D. Firm E. Gummy
Flavor and Aroma (0)			A. Satisfactory B. Unsatisfactory
Subtotal 53			
Total Score 100			

Comments: ______________________________

Evaluated By: ______________________________

FIGURE 8 Flour quality evaluation program score sheet. (Courtesy of American Institute of Baking, Manhattan, KS.)

clear distinction between these two groups is not made. Generally, on the bread label, these ingredients are referred to as dairy blends. (Table 19).

F. Yeast

Compressed yeast and other forms are given in Table 20.

G. Mineral Yeast Food

Composition of mineral yeast foods is given in Table 21. (For functionality, see Table 17).

H. Fiber

The list of high fiber ingredients for food manufacturing is growing (Table 22). The most common fiber ingredients presently used are corn bran, soy fiber, pea fiber, cottonseed fiber, and α–cellulose.

I. Dough Improvers

Dough improvers are classified as dough strengtheners and crumb softners (Table 23). The strengthening effect is generally attributed to interaction of gluten proteins with the surfactant; the crumb–softening action is explained by complexation of amylose with a surfactant. Most commonly used are lactylates and mono– and diglycerides, which are marketed singly or in blends.

J. Enzymes

Enzymes used are amylases and proteases (Table 24). Amylases added in mills are generally supplied in the form of barley malt or as fungal amylases; additionally fungal amylases are used in bakeries. Proteases for reduction of dough mixing time are added in bakeries when long mixing strong flours are encountered.

VI. STABILITY OF BAKERY FOODS

Bakery foods are perishable—they undergo physicochemical, sensory, and microbial changes as indicated in Table 25. The generic term "staling" is generally used to define changes during storage other than those caused by microbial contamination. The measurement methods of staling are summarized in Table 26. Of the methods given the determination of compressibility using Baker's compressimeter or Instron are mostly used in industrial quality control and research practice. It is advisable to follow the staling process with sensory tests, since flavor changes unrelated to firming often occur during storage. The mechanism of staling is not completely understood.

The main theories of this process are shown in Table 27, that of Schoch and French (27) being generally most accepted; a schematic representation of this mechanism is shown in Figure 9. The intact starch granules (I) swell and gelatinize during baking, which transforms the starch polymers (amylose and amylopectin) into an amorphous state (II). Upon cooling, the amylose crystallizes (retrogrades) and contributes to the initial crumb firmness of bakery foods (III). During the storage, amylopectin retrogrates (IV), which causes firming of breads. The retrogradation of amylopectin is reversible. This explains whey bread crumb can be refreshened (softened) by heat.

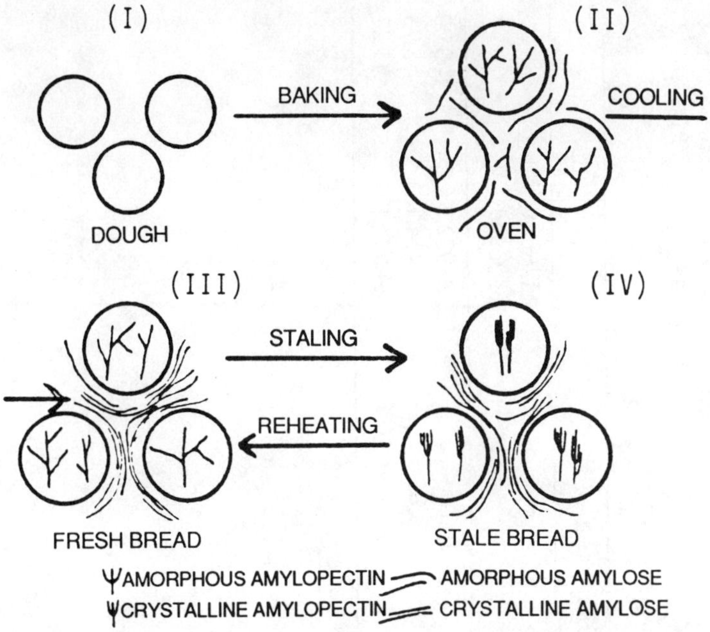

FIGURE 9 Schoch's mechanism of bread staling. (From Ref. 27.)

Retardation of microbial growth is affected by various chemical agents, as given in Tables 28 and 29; the effect of these agents on the rope–forming organism *(Bacillus subtilis)* is given in Table 30.

TABLE 1 U.S. Production of White Pan and Variety Breads in 1972, 1982, and 1986

Products category	1972		1982		1986	
	Millions of pounds	(%)[a]	Millions of pounds	(%)[a]	Millions of pounds	(%)[a]
All breads	11,418.7	100.00	10,709.6	100.00	10,922.2	100.00
White pan bread	8,657.3	75.82	6,410.2	59.85	6,414.1	58.73
Variety breads	2,761.4	24.18	4,299.4	40.15	4,508.1	41.27

[a]% expressed on basis of "all breads."
Source: Refs. 1,2,3.

TABLE 2 U.S. Production of Bread Varieties in 1972, 1982, and 1986

Product category	1972		1982		1986	
	Millions of pounds	(%)[a]	Millions of pounds	(%)[a]	Millions of pounds	(%)[a]
All bread[b] varieties	2,761.4	100.00	4,299.4	100.00	4,508.1	100.00
White, hearth, including French, Italian, etc.	549.7	19.90	918.8	21.37	976.1	21.65
Whole wheat, cracked wheat, and other dark wheat breads	777.7	28.16	1,956.1	45.50	2,196.7	48.73
Rye, including pumpernickel	426.0	15.43	383.0	8.91	398.8	8.85
Other specialty (raisin, potato, salt-free, salt-rising, etc.)	339.0	12.28	455.0	10.58	445.9	9.89
Bread, white, wheat, and rye, n.d.k.[b]	669.0	24.23	586.5	13.64	490.6	10.88

[a]% expressed on basis of "all bread varieties."
[b]White in this group refers to special white breads; n.s.k. = not specified kind.
Source: Refs. 1,2,3.

TABLE 3 Per Capita Consumption of Breads and Yeast-Leavened Products (lb per person)

Product	1981	1982	1983	1984	1985	1986
All breads	47.76	44.46	44.86	46.07	46.38	47.08
White pan	29.36	26.61	26.93	27.25	27.32	27.65
Unspecified	2.64	2.44	2.33	2.27	2.18	2.11
Various types	15.76	15.41	15.60	16.55	16.88	17.32
Hearth	4.06	3.81	3.86	4.11	4.15	4.21
Whole wheat, etc.	8.00	8.12	8.22	8.79	9.05	9.47
Rye and like	1.71	1.59	1.61	1.68	1.69	1.72
All other	1.99	1.89	1.91	1.97	1.99	1.92
Rolls	14.36	13.59	13.61	13.95	13.73	13.34
Hamburger and hotdog	12.07	11.29	11.31	11.57	11.33	11.02
Brown and serve	1.23	1.16	1.16	1.20	1.21	1.16
Hearth	1.06	1.14	1.14	1.18	1.19	1.16
Sweet yeast goods	3.20	2.88	3.12	3.32	3.27	3.47
Doughnuts	0.69	0.58	0.63	0.62	0.59	0.60
All other (coffee cakes, etc.)	2.51	2.30	2.49	2.70	2.68	2.87

Source: Ref. 4.

TABLE 4 White Pan Bread

Ingredient	Sponge/Dough: Sponge (%)	Sponge/Dough: Dough (%)	Straight dough (%)	Straight no–time dough (%)	Brew bread (%)
Flour[a]	70.0	30.0	100.0	100.0	100.0
Brew	—	—	—	—	35.0
Water	40.0	24.0	64.0	65.0	32.0
Yeast, compressed	3.0	—	2.5	3.5	—
Salt	—	2.0	2.0	2.0	1.4
Sugar or sweetener solids	—	8.5	8.0	6.0	7.0
Shortening	—	3.0	2.75	2.75	2.75
Yeast food	0.5	—	0.5	0.6	0.5
Nonfat dry milk or milk replacer	—	2.0	2.0	2.0	2.0
Fungal protease	0.5	—	0.25	0.5	—
L–Cysteine, ppm	—	—	—	40.0	—
Potassium Bromate, ppm	—	—	15.0	30.0	30.0
Ascorbic Acid, ppm	—	—	—	60.0	—
Vinegar (100 grain)	—	—	—	0.5	—
Monocalcium phosphate	—	—	—	0.25	—
Mono– and diglycerides, hydrate	—	0.5	0.5	0.75	0.5
Dough strengtheners	—	—	—	—	0.2
Calcium propionate	—	0.2	0.2	0.2	0.2

[a]About 11.2–11.7% protein, 0.44–0.46 ash, entriched according to U.S. standards (14% m.b.). All ingredients are given in bakers % (flour = 100%).

Procedures

(1) Sponge and Dough: Sponge: Sponge/dough ratio: 70/30; mixing: 1 min low, 3 min high speeds (77°F/25°C); fermentation: 3–5 hr (86°F/30°C). Dough: Mixing: 1 min low, 10–12 min high (80°F/27°C); rest period: 15–20 min. Divide into 18oz (515 g) pieces for 1 lb (454 g) loaves. Round and give 7 min rest period, sheet, shape, and pan; proof: 55 min. [(107°F/42°C)/85% rel. humidity]; bake: 18 min at 445°F (230°C).

(2) Straight Dough: One-step process. Mixing: 1 min at low and 15–20 min at high speeds (78–82°F/26.5–27.5°C). Fermentation time 2 hr at 86–95°F (30–35°C)/85% relative humidity. This procedure is used by retailers or for specialty breads.

(3) Straight Dough—No-time: Mix 1 min at low and 10–15 min (80–84°F/27–29°C). Proof for 55 min [(107°F/42°C)/85% rel. humidity].

(4) Brew Procedure: Brew: Disperse ingredients by high speed agitation (5 min, 80°F/28°C). Ferment (low agitation) for 1.5 hr (88–92°F/31–33°C). Add to dough ingredients and proceed as in sponge and dough process. Brew consists of 80.95% water, 7.75% sweetener solids, 8.0% compressed yeast, 3.8% salt, and 0.2–0.5% buffer (calcium carbonate plus ammonium sulfate); 35% of this brew is added to the dough.

Source: Ref. 5.

TABLE 5 Continuous Bread-Making Formula—Nonflour Brew

Brew ingredients	(%)[a]	Dough composition	
Water	60.32	Flour, %	100
Sugar	7.61	Brew, %	74
Yeast	2.67	Shortening blend, %	3
Yeast food	0.50	Water, %	2
Salt	2.10	$KBrO_3$, ppm	60
Milk powder	0.33	KIO_3, ppm	12
Mold inhibitor[b]	0.10		

[a]Total flour basis.
[b]Calcium propionate.
Source: Ref. 6.

TABLE 6 Home-Style Bread Formula[a]

Ingredients	Total (%)	Sponge (%)	Dough (%)
Flour, unbleached	100	65	35
Water (variable)	65	40	25
Corn syrup	3	—	3
Salt	2	—	2
Shortening	3	—	3
Yeast	2	2	—
Yeast food	0.375	0.375	—
Rolled wheat	2	2	—
Mold inhibitor	as needed	—	as needed

[a]Ingredients based on 100 parts flour. Sponge temperature, 76°F. Sponge fermentation, 4 hr. Dough temperature, 82°F. Floor time, 15–20 min. Give medium proof. Bake 420– 430°F, 30 min.
Source: Ref. 7.

TABLE 7 White Premium Bread Formula[a]

Ingredients	%
Flour	100
Sugar	4
Salt	2
Shortening	4
Butter	4
Honey	3
Milk, condensed or evaporated	28
Water (variable)	30
Yeast	3
Yeast food	0.375
Mold inhibitor	as needed

[a]Ingredients based on 100 parts flour. Dough temperature, 80°F. Ferment 1.5 hr; punch. Rest approximately 20 min. Divide, round, rest 15 min. mold. Give full proof, bake 400°F.
Source: Ref. 7.

TABLE 8 White Hearth Bread Formulas[a]

	White Hearth[b]			Sour French[c]		
Ingredients	Total (%)	Sponge (%)	Dough (%)	Total (%)	Sponge (%)	Dough (%)
Flour	100	60	40	100	60	40
Water (variable)	55	33	22	59	35	24
White sour	—	—	—	4	4	—
Yeast	2	2	—	1.75	1.75	—
Yeast food	0.375	0.375	—	0.5	0.5	—
Salt	2	—	2	1.5	—	1.5
Sugar	2	—	2	2	—	2
Shortening	2	—	2	1.5	—	1.5
Malt (nondiastatic)	1	—	1	—	—	—
Vital wheat gluten	—	—	—	2	2	—

[a]Ingredients based on 100 parts flour.
[b]Sponge temperature, 76°F, ferment 4.5 hr. Dough temperature, 80°F. Floor time, 15–20 min. Bake 400–430°F.
[c]Sponge temperature, 76°F, ferment 4 hr. Dough temperature, 80°F. Floor time, 20 min. Bake 400–430°F.
Source: Ref. 7.

TABLE 9 Wheat Bread Formulas[a]

Ingredients	Wheat bread[b]			100% Whole Wheat[c]			Honey Bran Loaf[d]		
	Total (%)	Sponge (%)	Dough (%)	Total (%)	Sponge (%)	Dough (%)	Total (%)	Sponge (%)	Dough (%)
Flour	60	60	—	—	—	—	60	60	—
Whole wheat flour	40	15	25	100	65	35	40	—	40
Wheat bran	—	—	—	—	—	—	10	—	10
Water (variable)	59	45	14	68	43	25	63	35	28
Yeast	2.5	2.5	—	3	3	—	3	3	—
Yeast food	0.5	0.5	—	0.25	0.25	—	0.5	0.5	—
Salt	2.25	—	2.25	2.25	—	2.25	2.25	—	2.25
Shortening	3	—	3	3	—	3	3	—	3
Dextrose	5	—	5	—	—	—	—	—	—
Sugar	—	—	—	6	—	6	—	—	—
Honey	—	—	—	2	—	2	—	—	—
Molasses	10	—	10	—	—	—	8	—	8
Nonfat dry milk	—	—	—	3	—	3	3	—	3
Vital wheat gluten	2	—	2	2	—	2	2	—	2
							2	2	—

[a]Ingredients based on 100 parts flour.
[b]Sponge temperature, 76°F, ferment 4 hr. Dough temperature, 80°F. Floor time, 15 min. Bake approximately 420°F.
[c]Sponge temperature, 78°F, ferment 3 hr. Dough temperature, 80°F. Floor time, 15 min. Bake approximately 450°F.
[d]Sponge temperature, 76°F, ferment 4 hr. Dough temperature, 78°F. Floor time, 15 min. Bake approximately 420°F.
Source: Ref. 7.

TABLE 10 Rye and Pumpernickel Bread Formulas[a]

Ingredients	Standard Rye Bread[b] Total (%)	Sponge (%)	Dough (%)	Pumpernickel[c] Total (%)	Sponge (%)	Dough (%)
Flour, clear	80–60	80–60	—	24	24	—
Patent white rye flour	—	—	—	—	—	—
Medium rye flour	20–40	—	20–40	—	—	—
Dark rye flour	—	—	—	38	38	—
Rye meal	—	—	—	38	—	38
Water (variable)	64	36	30	71	32	39
Yeast	2	2	—	2	2	—
Salt	2	—	2	2.5	—	2.5
Rye sour, commercial	0–5	—	0–5	—	—	—
Sour dough	—	—	—	10	—	10
Malt, nondiastatic or other sweetener	0–4	—	0–4	1	—	1
Shortening	2	—	2	1	—	1

[a]Ingredients based on 100 parts flour.
[b]Sponge temperature, 76°F, ferment 3.5–4 hr. Dough temperature, 78–80°F. Floor time, 20 min. Bake approximately 430°F with steam.
[c]Sponge temperature, 76°F, ferment 3 hr. Dough temperature, 80°F, scale, round, rest 15 min, mold. Bake approximately 430°F with steam.
Source: Ref. 7.

TABLE 11 Mixed Grain Bread Formulas[a]

Mixed Grain Bread 1[b]					
Base	%	Sponge	%	Dough	%
Crushed wheat	70	Flour	60	Flour	40
Soy grits	10	Base	15	Water	20
Rolled oats	10	Yeast	3	Salt	2
Corn meal	10	Water	45	Brown sugar	8
				Shortening	4

Mixed Grain Bread 2[c]			
Soaker Stage	%	Dough Stage	%
Whole wheat kernels	14	Mash (from soaker stage)	
Coarse rye meal	14	Water	13
Buckwheat flour	12	Clear flour (16% protein)	36
Ground barley	3	Salt	2.5
Ground millet	7	Soy four, low fat	4
Ground oats	7	Blackstrap molasses	4
Ground alfalfa	1	Buckwheat honey	4
Toasted wheat germ	3	Potato flour	3
Wheat gluten	3	Margarine	3
Water	55	Yeast	3
		Lecithin	0.375

[a]Ingredients based on 100 parts flour.
[b]Sponge temperature, 74–76°F, ferment 4.5 hr. Dough temperature, 80°F (9).
[c]Mix soaker stage 5 min at low speed, soak 8 hr. Dough temperature, 80°F. Floor time, 30 min. Bake 400°F, 30 min
Source: Ref. 7.

TABLE 12 Raisin Bread Formulas[a]

	Raisin bread[b]			Raisin oatmeal bread[c]	Raisin yogurt bread[d]
Ingredients	Total (%)	Sponge (%)	Dough (%)	Total (%)	Total (%)
Flour	100	65	35	100	100
Oatmeal	—	—	—	15	—
Wheat bran	—	—	—	5	—
Water (variable)	64	40	24	62	64
Yeast	4	2	2	3	4
Yeast food	0.5	0.5	—	as needed	as needed
Salt	2.25	2.25	—	2	2
Shortening	3	—	3	2	2
Honey or brown sugar	—	—	—	6	6
Sugar	8	—	8	—	—
Vital wheat gluten	4	4	—	3	3
Cinnamon	0.4	—	0.4	—	—
Nonfat dried milk	4	—	4	2	—
Yogurt solids	—	—	—	—	5
Raisins	63	—	63	50	50

[a]Ingredients based on 100 parts flour.

[b]Sponge temperature, 78°F, ferment 3.5 hr. Dough temperature, 80°F. Floor time, 30 min. Bake approximately 400°F.

[c]Dough temperature, 80°F. Floor time, 30 min.

[d]Dough temperature, 80°F. Floor time, 30 min.

Source: Ref. 7.

TABLE 13 Hamburger (Soft) Buns

Ingredient	Sponge (%)	Dough (%)
Flour, enriched	70.0	30.0
Yeast	3.5	—
Water	40.0	22.0
Sugar	—	12.0
Salt	—	2.0
Shortening (oil)	—	5.0
Mineral yeast food, bromate type	0.5	—
Protease	0.5	—
Dough strengtheners	—	0.5
Emulsifier, hydrated mono-diglycerides	—	0.5
Ca-propionate	—	0.2

Procedures:

1. Sponge/dough process (hamburger buns):
   Sponge: Mixing— 1 min at low and 3 min at high speeds (75°F/24°C). Fermentation time— 3 hr (86°F/30°C).
   Dough: Mixing— develop dough to very extensible stage (10 to 14 min at high speed); divide, round, and relax (2 min); flatten and place on bun pans; proof 50 min at 104–107°F (40–42°C)/80–85% relative humidity; bake at 445–465°F (230–240°C).
2. Other uses of the formula: same formula is used for production of hot dog buns. Soft rolls of various shapes and sizes are prepared similarly. The formulas for latter products may be lower in sweetness and/or shortening.
3. Also, brew process is used, similar to that for white pan bread (cf., Table 4).

Source: Ref. 5.

TABLE 14 Sweet Dough (Cinnamon Rolls)

Ingredient	%
Bread flour, enriched	100.0
Sucrose, granulated	15.0
Dextrose (corn sugar)	2.0
Defatted soy flour	5.0
Shortening	12.0
Mono– and diglycerides, hydrated	2.0
Nonfat dry milk or replacer	2.0
Salt	1.5
Mineral yeast food, bromate type	0.5
Yeast, compressed	8.0
Dough strengthener	0.25
Flavor	Variable
Water	54.0

Procedure: Mix (horizontal mixer) all ingredients 14–20 min (80°F/27°C); ferment 30 min, sheet, sprinkle with cinnamon/sugar blend, roll, cut into units; pan; proof 30–45 min (104°F/40°C)/85% relative humidity; bake 12–15 min (400°F/205°C).

Source: Ref. 5.

TABLE 15 Danish Sweet Goods

Ingredient	%
I	
Hydrated shortening	15.0
Hydrated mono– and diglycerides	3.0
Salt	2.0
Sucrose, granulated	20.0
Nonfat dry milk	2.0
High fat soy flour	3.0
II	
Water	35.0
Yeast, compressed	8.0
Whole eggs, liquid	20.0
III	
Flour: 40% pastry, 60% bread enriched	100.0

Procedure: Cream (I) in upright mixer equipped with dough hook; add (II) to (I) and blend 1 min at low and then 2 min at high speeds; add (III) and mix at low speed until incorporated; divide and place on floured trays; retard 3–4 hr (36–39°F/2–4°C); sheet and spot roll in margarine on 2/3 of sheet (use 1/3 wt. margarine per 1 wt. of dough). Fold, and retard (3 hr). Sheet dough, give 3-fold and retard overnight. Sheet, add cinnamon, and/or other fillings, roll, cut, pan, proof, and bake. Both sweet dough and Danish pastry products are prepared in various sizes and shapes. They are filled with fruit, almond, nuts, etc., fillings, iced, glazed, and crunch or streusel topped.

Source: Ref. 5.

TABLE 16 Ingredient Limits for Bread and Rolls

Ingredient	Standardized product (% flour weight)	Nonstandardized product (% flour weight)
Flour	No limit	No limit
Water	38 maximum	No limit
Shortening	No limit	No limit
Salt	No limit	No limit
Milk	No limit	No limit
Sweetener	No limit	No limit
Yeast nutrients[a]	0.25 maximum	No limit
Nonwheat flours	3 maximum	No limit
Oxidants		
Azodicarbonamide	45 ppm	45 ppm
Potassium bromate, Potassium iodate, Calcium iodate, Calcium peroxide	75 ppm, total	75 ppm, total
Ascorbic acid	No limit	No limit
Monocalcium phospahate	0.75 maximum	No limit
Enzyme active soy	0.5 maximum	No limit
Calcium propionate	No limit	No limit
Spices	No coloring spice[b]	No limit
Color	Not permitted	No limit
Emulsifers and dough conditioners		
Calcium stearoyl-2-lactylate	0.5 maximum[c]	0.5 maximum[c]
Sodium stearoyl-2-lactylate	0.5 maximum[c]	No limit
Mono- and diglycerides	No limit	No limit
Succinylated monoglyceride	0.5 maximum[c]	0.5 maximum[c]
Polysorbate	0.5 maximum[c]	0.5 maximum[c]
Ethoxylated monoglyceride	0.5 maximum[c]	0.5 maximum[c]
Diacetyl tartaric acid esters	No limit	No limit
Other safe and suitable ingredients		

[a]Does not include monocalcium phosphate.
[b]Only spices that do not impart egglike color.
[c]Total not to exceed 0.5%.
Source: Ref. 8.

TABLE 17 Functions of Bread Ingredients

Ingredient	Function
	Structure
Flour	1. Protein (gliadin and glutenin) and water form viscoelastic material, called gluten. Gluten retains gas formed by sugar fermentation and contributes to structure of dough and bread.
	2. Starch + water + heat forms a viscous paste that sets to a gel after baking. During bread storage the starch crystallizes (retrogrades and contributes to firming (major part of staling) of breads.
	3. Protein content for bread flour: 11–13%, 14% moisture basis.
	Hydration
Water	1. Combines (hydrates) protein to form gluten.
	2. Hydrates flour gums (pentosans) and mill-damaged starch granules.
	3. Solvent, dispersing agent, and medium for chemical and biochemical reactions.
	4. Aids dough mobility.
	Leavening
Yeast	1. Produces carbon dioxide, ethanol by fermentation of fermentable sugars.
	2. Conditions dough biochemically.
	3. Forms flavor precursors (by-products of alcoholic fermentation).
	4. Rate of fermentation is controlled by temperature, nutrient supply, water level, pH, sugar concentration, salt, and level and type of yeast.
	Flavor Enhancer
Salt	1. Helps control fermentation.
	2. Toughens dough by interaction with gluten.
	3. Extends required dough development (delayed addition in dough mixing decreases mixing time by 10–20%).
	Controls Fermentation
Mineral yeast food	1. Water condition—calcium salts.
	2. Yeast conditioners—ammonium salts.
	3. Dough conditioners—oxidizing agents.

TABLE 17 (Continued)

Ingredient	Function
	Energy Source for Yeast
Sugar	1. Fermentable carbohydrates (fermentation). 2. Flavor—residual sugars (sweeteners), fermentation by-products, Maillard-type compounds during baking. 3. Crust color—results of caramelization (sugars and heat) and nonenzymatic browning (reducing sugar plus amino group of proteins, amino acids, etc.). 4. Extends shelf life by increasing hygroscopicity due to presence of residual sugars and tenderizing the crumb.
	Lubrication
Shortening	1. Ease of gas cell expansion in doughs. 2. Lubricates slicing blades during bread slicing. 3. Extends shelf life. 4. Tenderizes crust.
	Nutrition and Crust Color Enhancement
Dairy products	1. Protein (high in lysine) and calcium. 2. Flavor enhancement. 3. Crust color (browning reaction and caramelization). 4. Buffering effect in doughs and liquid ferments.
	Nutrition
Enrichment	1. Standards of Identity require the following levels per lb of bread: Fe = 12.5 mg Niacin = 15.0 mg Ca (optional ingredient) = 600 mg Riboflavin – 1.1 mg Thiamine = 1.8 mg
	Retardation of Microbial Spoilage
Mold inhibitors	1. Retard mold growth and extend spoilage free life. 2. Also retard formation of "rope" by *B. subtilis*.

TABLE 17 (Continued)

Ingredient	Function
	Enhancement of Flour Strength
Wheat gluten	1. Increases dough strength (1% gluten increases protein content by 0.6%). 2. Increases water absorption [1% added gluten (flour basis) enhances absorption by 1.5% (flour basis)]. 3. Improves dough mixing of fermentation tolerance. 4. Increases bread loaf volume. 5. Especially used in formulation of specialty breads.
Diastatic malt	1. Contributes fermentable sugar maltose. 2. Contains amylases, which convert starches to sugars. 3. Enhances flavor. 4. Improves crust color. 5. Improves dough handling. 6. Extends shelf life.
Nondiastatic malt	1. Contributes sugar (maltose). 2. Enhances flavor. 3. Improves crust color.
Fungal enzymes	A. Amylases: 1. Convert starches to sugars. 2. Aid crust color. 3. Improve dough handling. 4. Extend shelf life. 5. Less heat stable than cereal amylases (they are inactivated before starch gelatinizes in the oven). B. Proteases: 1. Weaken doughs due to cleavage of peptide bonds in wheat proteins. 2. Reduce dough-mixing time. 3. Increase pan flow.
	Enzyme active soy flour
	1. Contains lipoxygenase enzyme, which acts as: a. Bread crumb whitener. b. Shelf-life improver. c. Dough strength and mixing dough tolerance enhancer.
	Soy flour
	1. Improves nutrition of product by increasing lysine. 2. Extends shelf life. 3. Enhances dough strength and dough-mixing tolerance.

TABLE 18 Analytical Indices of Bread Flours[a]

Flour type	Protein (N × 5.7) (%)	Ash (%)	Falling no. (sec)	Amylograph value (B.U.)
Hard Red Spring Wheat (HRS)				
Patent	12.0–13.5	0.42–0.48	215–240	400–500
Clear	12.5–14.0	0.65–0.80	215–240	400–500
Whole wheat	13.5–15.5	1.70–1.80	215–240	400–500
Hard Red Winter Wheat (HRW)				
Patent	10.5–11.8	0.42–0.48	220–245	400–500
Clear	11.8–12.6	0.65–0.75	220–245	400–500
Whole wheat	11.6–12.5	1.70–1.80	220–245	400–500

[a]All values reported on 14% moisture basis.
Source: Ref. 10.

TABLE 19 Protein Content (dry basis) of Dairy and Nondairy Ingredients of Dairy Blends

Ingredient	%
Nonfat	36.5
Sweet cream buttermilk	34.0
Sweet dairy whey	12.0
Caseinates and whey protein	34–90
Soy flour	50.0
Soy (further processed)	50–70
Corn flour	10

Source: Ref. 11.

TABLE 20 Yeast

Form	Storage temperature (°F/°C)	Storage time	Moisture (%)	Method of addition	Conversion factor[a]
Fresh, compressed					
Cake	34–45/2–7	3–4 weeks	67–72	Weigh and add with other ingredients or disperse in water before mixing.	1
Crumbled	34–45/2–7	3–4 weeks	67–72		1
Fresh, cream (liquid)	35–39.1–4	10–14 days	80–84	Meter and add with other ingredients.	1.5–1.8
Dry					
Active dry yeast (ADY)	Room temperature	1–12 months	6–8	Must hydrate at 105–110°F (40–43°C) for 10–15 min, then use.	0.4–0.5
Instant dry	Room temperature	1 year plus (depending on packaging)	4–6	Blend dry with other dry ingredients or delay addition in mixer.	0.33–0.4

[a]Conversion factor relates the amount of other forms of yeast to compressed yeast.

TABLE 21 Composition of Representative Yeast Foods and a Dough Conditioner

Type I	
Calcium acid phosphate	50.0%
Sodium chloride	19.35
Ammonium sulfate	7.0
Potassium bromate	0.12
Potassium iodate	0.10
Starch (filler)	23.43
Type II	
Calcium sulfate	25.0%
Ammonium chloride	9.7
Potassium bromate	0.3
Sodium chloride	25.0
Starch	40.0
Type III	
Calcium peroxide	0.65%
Ammonium phosphate	9.0
Dicalcium phosphate	9.0
Starch or flour	90.35

Source: Ref. 12.

TABLE 22 Commercially Available Fiber Ingredients

Product	Total dietary fiber (TDF) (%)	Supplier[a]
Cereal sources		
Barley's best high protein flour	35	23
Barley bran flour	70	12,23
Barley fiber	70	4
Corn bran fiber	55–95	4,8,15,19,21,30,32,38
Defatted wheat germ	20	21,37
Oat bran	15–22	4,21,23,24,38
Oat fiber	80–90	4,21,24,38,39
Oat—Snow White	95	4
Malted grain fiber (barley, rice, cornstarch)	31	7
Rice bran, defatted	35–40	12
Rice bran, stabilized	30–40	21,31
Rice fiber	70–80	4,21,38
Rice fiber/bran	70–80	21
Rice germ/bran	35–40	5,28
Sofgrain, whole kernels/kibbled (wheat, rye, oasts, triticale, barley)	35	38
Wheat bran, red and/or white	40–50	3,4,5,11,18,19,20, 21,23,27,38
Wheat germ, stabilized, full-fat	10	4
Legume and Fruit Sources		
Apple fiber	43	4,12,34,36,38
Apple rice granules	40	25
Blackcurrant fruit fiber	43	12
Citrus pulp SACs (orange or grapefruit)	62	21,26
Cocoa fiber/bran	55–75	4,21
Cottonseed fiber	95	6,16
Date bran	44	25
Dried cranberries	6–8	26
Dried pears	14	33
Fig powder	64	33
Lemon, orange, grapefruit fiber	25–70	2
Pea fiber	50–95	10,21,38
Peanut fiber	50	4,21,38
Soy fiber/bran/flakes	45–75	2,13,19,21,29
Sugar beet fiber	73–80	1,4,9,14,21
Sugar cane fiber	72–86	9,21
Sweet lupin bran flour	85	23
Tomato fiber	45	4,21,38

TABLE 22 (Continued)

Natural and man-made gums		
Carrageenan gum	75	35
Cellulose powder	90–95	17
Guar gum	90	35
Gum agar	75	35
Gum arabic	90	35
Locust bean gum	90	35
Sodium carboxymethyl cellulose	75	35
Xanthan gum, Ticaxan	75	35

Suppliers of fiber ingredients:

1. American Crystal Sugar Co.
Moorhead, MN 56560
(218) 236–4426

2. Archer Daniels Midland Co.
Decatur, IL 62525
(800) 637–5850

3. Bay State Milling Co.
Winona, MN 55987
(507) 452–1770

4. Canadian Harvest Process (1986) Ltd.
St. Thomas, Ontario, CANADA
(519) 633–5030

Canadian Harvest USA
Cambridge, MN 55008
(612) 689–5800

5. Capitol Milling Company
Los Angeles, CA 90051
(213) 628–8235

6. ConAgra, Inc
Omaha, NE
(402) 978–4050

7. Coors Food Products Co.
Golden, CO 80401
(303) 277–5760

8. Corn Products
Argo, Il 60501
(312) 563–5378.

9. Delta Fiber Foods
Minneapolis, MN 55426
(612) 541–0925

10. Dupro Division, Dumas Corp.
Golden Valley, MN 55427
(612) 541–0202

11. Fisher Mills, Inc
Seattle, WA 98134
(206) 622–4430

12. Freeman Industries, Inc.
Tuckahoe, NY 10707
(914) 961–2100

13. Grain Processing Corp.
Muscatine, IA 52761
(319) 264–4265

14. Holly Sugar Corp.
Colorado Springs, CO 80901
(303) 471–0123

15. Illinois Cereal Mills
Paris, IL 61944
(217) 465–5331

16. International Filler
North Tonawanda, NY 14120
(716) 693–4040

17. James River Corporation
Hackensack, NJ 07601
(201) 487–6400

18. King Milling Company
Lowell, MI 49331
(616) 897–9264

19. Lauhoff Grain Company
Danville, IL 61834
(217) 442–1800

TABLE 22 (Continued)

20. Mennel Milling Co.
Fostoria, OH 44830
(419) 435–8151

21. Mid American Food Sales, Ltd.
Northbrook, Il 60062
(312) 480–0720

22. Minnesota Grain Pearling Co.
Cannon Falls, MN 55009
(507) 263–3325

23. National Grain Products Co., Inc.
Minnetonka, MN 55345
(612) 476–2215

24. National Oats Co.,Inc.
Cedar Rapids, IA 52402
(319) 364–9161

25. Northbrook Granary
Northbrook, Il 60062
(312) 998–8188

26. Ocean Spray Cranberries, Inc.
Plymouth, MA 02360
(617) 747–1000

27. Ogilvie Mills, Inc.
Minnetonka, MN 55345
(800) 328–6299; (612) 936–3600

28. Pacific Rice Products, Inc.
Woodland, CA 95695
(916) 662–5056

29. Protein Technologies International
St. Louis, MO 63164
(314) 982–3175

30. Quaker Oats Co.
Chicago, IL 60654
(312) 222–7101

31. Riviana Foods, Inc.
Houston, TX 77252
(713) 529–3251

32. A. E. Staley Co.
Decatur, IL 62525
(217) 423–4411

33. Sun-Diamond Growers of CA
Pleasanton, CA 94566
(800) USA Sun–D

34. Tastee Apple Inc.
Newcomerstown, OH 43832
(614) 498–8316

35. TIC Gums, Inc.
New York, NY 10017
(212) 697–3216

36. Tree Top, Inc.
Selah, WA 98942
(509) 697–7251

37. Vitamins, Inc.
Chicago, IL 60601
(312) 861–0700

38. Wilke International, Inc.
Overland Park, KS 66212
(913) 341–8140

39. D. D. Williamson & Co., Inc.
Louisville, KY 40206
(502) 895–2438

Note: This list was compiled form responses to a survey conducted by AIB. Therefore, this list may not be complete and is not meant to be exclusive. AIB neither recommends nor endorses one supplier over others offering similar ingredients.
Source: Ref. 13.

TABLE 23 FDA Regulations Dough Strengtheners/Crumb Softeners Standardized Products

Product	Limitation	Reference[c]	Function
Calcium stearoyl–2–lactylate	0.5% max.[a]	CFR21, 136.110	Dough strengthener (excellent) Crumb softener (good)
Sodium stearoyl–2–lactylate	0.5% max.	CFR21, 136.110	Dough strengthener (excellent) Crumb softener (very good)
DATA Esters	No limit	CFR21, 136.110	Dough strengthener (excellent) Crumb softener (fair)
Mono– and diglycerides	No limit	CFR21, 136.110	Dough strengthener (none) Crumb softener (excellent)
Succinylated monoglycerides	0.5% max.[a]	CFR21, 136.110	Dough strengthener (good) Crumb softener (good)
Polysorbate 60	0.5% max.[a]	CFR21, 136.110	Dough strengthener (fair) Crumb softener (good)
Ethoxylated monoglycerides	0.5% max.[a]	CFR21, 136.110	Dough strengthener (very good) Crumb softener (poor)
Sucrose esters	GMP[b]	CFR 170.3	Dough strengthener (very good) Crumb softener (good)

[a]Total alone or in combination cannot exceed 0.5% based on flour.
[b]Use in accordance with good manufacturing practices.
[c]Code of Federal Regulations.
Source: Ref. 32.

TABLE 24 Enzymes in Baking

Enzyme	Source	Commercial form
Cereal amylases	Barley malt	Barley malt flour.
	Barley malt	Malt extract and syrup.
	Barley malt/fungal amylase	Malt syrups with non-cereal amylase.
Bacterial amylase	Bacterial (*B. subtilis*)	Powder, tablets, for retardation of staling.
Fungal amylases	Fungi (*Aspergilli*)	Powder, tablets, pouches.
Proteases	Fungi (*Aspergilli*)	Tablets, powders, malt syrup with fungal protease.
	Bacterial (*B. subtilis*)	Tablets of powder.
Bromelain (protease)	Fruits	Tablets or powders.
Lipoxygenase	Soy	Active soy flour used for blending and dough oxidation.

Source: Ref. 15.

TABLE 25 Definitions of Changes Affecting Freshness of Bread During Storage

Term describing loss of freshness	Definition	Characteristics	Cause
Staling	A series of changes that cause a decrease in consumer acceptance other than that resulting from the action of spoilage organisms.	Crust staling: loss of crispness.	Moisture migration from crumb to crust.
		Crumb staling: firming, development of crumbli– ness, loss of flavor, and emergence of stale flavor.	Retrogradation, complexation of glavorants with amylose and oxidative changes.
Microbial spoilage	Growth of microorganisms in crust or crumb during storage.	Molds (various Asper– gilli, Penicillia) wild yeasts, spore formers.	Contaminated ingredients, air, and equipment during processing.
Flavor deterioration	Part of staling (see Staling above).	Loss of fresh bread flavor and aroma.	Complexation of flavorants with amylose.
		Development of stale bread flavor.	
			Oxidative changes in flavorants, migration of flavorants from crust to crumb.

Source: Ref. 16.

TABLE 26 Methods of Determination of Freshness of Breads

Method	Principle	Reference/Use
	I. Physical Methods	
Compressibility	A uniform square of crumb is compressed to constant deformation using Baker's Compressimeter or Instron. The required compression force increases with bread age.	AACC Method 74–10 (18), most common research and control method.
Capacitance/Conductance	Both properties increase with the age of bread.	Kay and Willhoft, research method (19).
Cell-wall firmness measurements	Determines compressibility of a bread crumb pellet; eliminates effect of loaf volume.	Guy and Wren research method (20).
Differential thermal analysis	Emergence of an endothermic peak indicative of starch crystallization	Axford and Collwell (21), Russell (22).
X-ray diffraction	Change of X-ray diffraction pattern from A to B/V.	Zobel (23), research. NOTE: not usuable in bread with bacterial α-amylase.
Nuclear magnetic resonance	Measures decrease of water mobility which decreases with age of bread.	Leung et al. (24).
Rate of starch crystallization	Avrami Equation [$\theta = E_L - E_t)/(E_L - E_O) = \exp(-kt^n)$]is used to estimate the rate constant k; 1/k = time constant of crystallization of the system is generally reported (the higher the value, the slower the rate); θ is noncrystalline portion, E_L is limiting modulus and E_t, E_o moduli at times t and O, respectively	Comford et al. (25), widely used in research.
	II. Chemical Methods	
Iodine absorption	This value decreases with bread aging; it may be determined colorimetrically, iodometrically, or potentiometrically.	Pelshenke and Hampel (26), research.
	III. Sensory Methods	
Panel test evaluation of staling	Bread samples rating by panel for degree of staleness.	AACC 77–30 (18), widely used in control and research work.

Source: Ref. 17.

TABLE 27 Theories of Bread Staling

Theory according to:	Basis
Schoch and French (27)	Retrogradation of starch polymers. Amylose crystallizes rapidly and produces initial firmness. Firiming during storage is attributed to crystallization of amylopectin.
Lineback (28)	Essentially same as Schoch's except it emphasizes intergranular interaction.
Knyaginichev (29)	Formation of structured gel, consisting of starch, protein, and water.
Erlander and Erlander (30)	Interaction of gliadin and glutenin with starch chains.
Willhoft (31)	Implicated gluten in addition to starch.

Source: Ref. 17.

TABLE 28 Effect of Sorbates and Propionates on Mold-Free Shelf Life of Baking Foods

Additive	% Flour basis in dough	% Product weight on surface	Average days without mold
Variety bread			
None	0	0	3
Ca propionate	0.16	0	5
K sorbate	0	0.016	9
K sorbate	0.02	0.016	11+
Hamburger buns			
None	0	0	7
Ca propionate	0.3	0	16
K sorbate	0	0.02	23
K sorbate	0.02	0.02	28+
English muffins			
None	0	0	5
Ca propionate	1.0	0.1[a]	9
K sorbate	0.12	0.1	12
K sorbate	0.12	0.2	26+
Brown and serve rolls			
None	0	0	4
Ca propionate	0.4	0	9
K sorbate	0.05	0.06	14
K sorbate	0.05	0.12	27+
Flour tortillas			
None	0	0	2
Ca propionate	0.25	0	2
K sorbate	0.05	0.1	9
K sorbate	0.05	0.2	36+

[a]In dusting flour.
Source: Ref. 33.

TABLE 29 Effect of Antimicrobial Agents on Mold-Free Shelf Life of Bread

Antimicrobial agent	Usage (oz./CWT)	Mold-free shelf life (days)[a] Sponge/dough	No-time dough
Single agents			
Ca propionate	6	7	5
	5	6	4
	4	6	4
	3	4	4
	0	3	3
Sodium diacetate	6	7	5
	5	6	4
	4	6	4
	3	4	4
	0	3	3
200 Grain vinegar	16	5	5
	8	3	3
	6	3	3
	4	3	3
	0	3	3
Potassium sorbate	2.5	3	4
	1.5	3	3
	0	3	3
Combination of agents			
Ca propionate:Na diacetate	6:0	6	5
	6:1	5	4
	4.5:1.5	5	4
	3:3	4	3
	0:0	3	3
Ca propionate:200 grain vinegar	6:4	8	5
	5:4	6	5
	4:4	6	5
	3:4	6	4
	0:4	3	3
Ca propionate/mono–calcium phosphate	6:4	8	5
	5:4	8	5
	4:4	7	5
	4:5	6	5
	0:4	4	4

[a]Slices of bread inoculated with mold spores.
Source: Ref. 34.

TABLE 30 Effect of Antimicrobial Agents on Rope-Free Shelf Life in Bread

Antimicrobial agent	%, Flour basis	Days of rope-free shelf life
Acetic acid	0.1	3–6
	0.2	> 21
Calcium acetate	0.3	3–6
	0.4	6–10
	0.5	> 21
Sodium diacetate	0.2	3–6
	0.3	17–21
Propionic acid	0.1	3–6
	0.2	> 21
Calcium propionate	0.18	3–6
	0.23	6–10
	0.35	> 21
Sodium propionate	0.2	3–6
None—control	0.0	0–3

Source: Ref. 35.

REFERENCES

1. Anonymous, Bakery products, *Census of Manufactures*, SIC 2051 and 2052, Bureau of the Census, U.S. Department of Commerce, Washington, DC, 1977.
2. Anonymous, Bakery products, *Census of Manufactures*, SIC 2051 and 2052, Bureau of the Census, U.S. Department of Commerce, Washington DC, 1985.
3. Anonymous, Trends '86, *Bakery Production and Marketing*, 21 (5): 80–84, 86–87, 90, 92, 94–96, 98–99 (1986).
4. Semling, H. B., Jr., Continued growth in 1988, *Baking Industry*, 101(1): 21–23 (1988).
5. Kulp, K., and Dubois, D.K., Breads and sweet good in the United States, *American Institute of Baking Research Department Technical Bulletin, 4* (6) (1982).
6. Redfern, S., Brachfeld, B.A., and Maselli, J.A., Continuous mix temperatures, *Cereal Sci. Today, 9*: 190–191 (1964).
7. Ponte, J. G., Jr., Production technology of variety breads In *Variety Breads in United States*, Miller, B. S. (Ed). American Association of Cereal Chemists, St. Paul, MN, 1981, pp. 9–26.
8. Dubois, D. K.,Labeling aspects of variety breads In *Variety Breads in United States*, Miller, B. S. (Ed)., American Association of Cereal Chemists, St. Paul, MN, 1981, pp 27–35.
9. Zelch, R., American Institute of Baking, Manhattan, KS, personal communication, 1988.
10. Bennett, R., American Institute of Baking, Manhattan, KS personal communication, 1988.
11. Stahel, N. G., Milk replacers, *Proceedings of the 58th Annual Meeting of Bakery Engineers*, Chicago, IL, 1982, pp. 68–73.
12. Pyler, E. J., *Baking Science and Technology*, Volume 1, Siebel Publishing Company, Chicago, IL, 1983, pp. 578–584.
13. Vetter, J. L., Commercially available fiber ingredients and bulking agents, *American Institute of Baking Research Department Technical Bulletin, 10* (5) (1988).
14. Dubois, D. K., American Institute of Baking, Manhattan, KS, personal communication, 1988.
15. Dubois, D. K., Enzymes in baking: Classification, *American Institute of Baking Research Department Technical Bulletin*, (10) (1980).
16. Kulp, K., and Ponte, J. G., Jr. Staling of white pan bread: Fundamental causes, *CRS Crit. Rev., Food Sci. Nutr., 15*: 1–48 (1981).
17. Kulp, K., and Vetter, J., Effect of aging on freshness of white pan bread, pages 1–31 in : *Handbook of Food and Beverage Stability*, G. Charalambous, ed., Academic Press, Inc., publishers, New York (1986).
18. American Association of Cereal Chemists (AACC), *Approved Methods of the AACC*, AACC, St. Paul, MN, 1983.
19. Kay, M., and Willhoft, E. M. A., Bread staling IV. Electrical properties of the crumb during staling, *J. Sci. Food Agric., 23*: 321–331 (1972).
20. Guy, R. C. E., and Wren, J. J., A method for measuring the firmness of the cell–wall material of bread, *Chem. Ind.* (London): 1727–1728 (1968).
21. Axford, D. M. E., and Colwell, K. H., Thermal investigation of bread staling, *Chem. Ind.* (London): 467–468 (1967).
22. Russell, P. L., A kinetic study of bread staling by differential scanning calorimetry and compressibility measurements. The effect of added monoglyceride. *J. Cereal Sci., 1*: 297–303 (1983).
23. Zobel, H. F., A review of bread staling. *Bakers Dig., 47* (4): 52–53, 56–61 (1973).
24. Leung, H. K., Magnuson, J. A., and Bruinsma, B. L., Water binding of wheat flour doughs and breads as studied by deuteron relaxation. *J. Food Sci., 41*: 95–99 (1983).
25. Comford, S. J., Axford, D. W. E., and Elton, G. A. H., The elastic modulus of bread crumb in linear compression in relation to staling, *Cereal Chem., 41*: 216–229 (1964).
26. Pelshenke, P.F., and Hampel, G., Starch retrogradation, *Bakers Dig., 36* (3): 48–50, 52–54, 56–67 (1962).
27. Schoch, T. J., *Cereal Chem., 24*: 231–249 (1947).
28. Lineback, D. R., The role of starch in bread staling In *"International Symposium on Advances in Baking Science and Technology,"* Department of Grain Science, Kansas State University, Manhattan, KS, 1984, pp. 51–59.

29. Knyginichev, M. I., The nature of staleness of bread and preservation of freshness, *2h. Vses, Khim, O–va.,10:* 277 (1965).
30. Erlander, S. R., and Erlander, L. G., Explanation of ionic sequences in various phenomena, X. Protein–carbohydrate interactions on mechanism for staling of breads, *Staerke, 21:* 305–315 (1968).
31. Willhoft, E. M. A., Breadstaling. I. Experimental study. II. Theoretical study, *J. Sci. Food Agric., 22:* 176–180 (1971).
32. Dubois, D. K., Dough strengtheners and crumb softeners. I. Definition and classification, *American Institute of Baking Research Department Technical Bulletin, 1* (4): (1979).
33. Monsanto company, Potassium sorbate surface treatment for Yeast–raised products, Monsanto, Co., St. Louis, MO 1977, p. 3.
34. Brisco, R. Mould control in baked goods, *Bakers J., 39* (20):12–13, 31–32 (1978).
35. Ingram, M., Ottaway, F. L. M., and Coppock, J. B. M., The preservative action of acid substance in food, *Chem, Ind. (London)*: 1154–1163 (1956).
36. Seiling, S., Equipment demands on changing production requirments, *Bakers Dig., 43* (5): 54–56, 58–59 (1969).
37. Kulp, K., Technology of brew systems in bread production *Bakers Dig., 57* (3): 20, 22–23 (1983).

17

Soft Wheat Products

Hamed Faridi
Nabisco Brands, Inc., East Hanover, New Jersey

I. INTRODUCTION

The terms "hard" and "soft" as applied to wheats are descriptions of the texture of the wheat kernel. The flour obtained from a hard wheat kernel has a coarser particle size than does flour from a soft wheat kernel. Low protein (7–10%) flour milled from soft wheats are most suitable for making cakes and biscuits (1). The number and types of products made from soft wheat is large. A partial list is found in Table 1. All of those products have better appearance and eating quality when made from soft rather than hard wheat flour. The per capita consumption of a number of soft wheat products is shown in Table 2.

II. INGREDIENTS

A. Flour

Soft wheats differ from hard wheats in kernel hardness, a basic genetic characteristics directly inherited. When ground or milled, soft wheat fractures into significantly smaller particles than hard wheat, which is reflected in the greater "break flour yield" upon milling. Break flour is the portion of the kernel endosperm of flour fineness obtained without crushing or reduction in the milling operation. Soft wheat cultivars exhibit a wide range in kernel texture, but historically there has been a distinct difference in average flour particle size between the hardest soft wheat and softest hard wheats such that a relatively simple grinding test can classify wheat properly.

To achieve greater uniformity and consistency, specifications are adopted by millers and bakers, which vary between companies as well as from one baking plant to another. However, they all have some common features. Modernization and automation of the bakeries necessitated exact and demanding specifications on incoming flour. The following are overall specifications for soft wheat and its flour. However, these specifications change depending upon the particular end product (3).

Wheat:

High test wheat or 1000 kernel weight; uniformity of kernel size
Ease of milling and high flour yield
Low to medium protein content (8–11% on 14% moisture basis)
Moisture content not exceeding 13%
Kernel softness (PSI > 55s)
Low amylase activity; no sprout damage

Flour:

Bright and creamy color and relatively low ash content
Low to medium flour protein content (7–10 on 14% moisture basis)
Low water absorption
Low damaged starch content
Fine flour granulation
Medium mixing requirement and satisfactory dough-handling properties
High volume in cakes with satisfactory texture
Tender-bite cookies and crackers

Key factors pertinent in characterizing soft wheat flours are significant measurable parameters and operational experience. There are so many different types of baked product involved that no one criterion can meet all requirements of the numerous soft wheat products.

Flour constitutes the primary raw material to which all soft wheat products formulations are related. It provides a matrix around which other ingredients in varying proportions are mixed to form dough or batter systems.

Cookie and cracker flours (Table 3) receive no special treatment or additives, i.e., they are normally bleached or chemically matured in any way and have no chemical leavening additives such as phosphates or self-rising ingredients. Cookie flours at times are bleached very lightly as a means of reducing or controlling the "spread" of the product.

Cake flour is treated with chlorine to pH values 4.5–5.2. Table 4 shows levels of chlorine gas required to lower the pH of flour (an indication of the level of chlorination). A distinguishing characteristic of cake flour is that it is relatively fine compared to cookie flour. Characteristics of flours suitable for production or cakes are shown on Table 5.

B. Water

Water is the least known ingredient in baked products (5). The quality of water used as an ingredient can have greater effects on bakery products than is generally recognized. The amount and type of dissolved minerals and organic substances present in the water can affect the flavor, color, and physical attributes of the finished baked goods as well as the machining of doughs, marshmallows, icings, etc.(6,7).

For example, some trace metals, particularly copper, may have catalytic effects on the development of rancidity in fats and oils (8). Table 6 shows some characteristics of water related to baking. Table 7 shows the baking industry's water hardness classification. Water acts as a plasticizer and its amount added is adjusted so as to produce a dough of acceptable consistency for processing.

C. Chemical Leavening

Most soft wheat products are chemically leavened and yeast is rarely used. Use of chemical leavening will give a more tender bite to the finished product. There are four major leavening gases used in leavening of soft wheat products (Table 8). Air and water are incorporated in the dough or batter during mixing. Carbon dioxidc is mainly generated by the added sodium bicarbonate. The popularity of sodium bicarbonate as a gas source is based on its low cost, lack of toxicity, ease of handling, relatively tasteless end product, and the high purity of commercial supplies. Sodium bicarbonate has some disadvantages as well. Among them is its rather rapid rate of solution at room temperature, a feature which reduces the amount of control that can be exercised over its leavening action (9).

Specifications for purchasing baking soda should include a purity requirement and particle size limitations. The latter will be determined by the use to which the soda will be put. Powdered grade is mostly used for biscuits, cookies, and most cakes. It is used when substantial leavening is needed before baking. It dissolves rapidly and completely to assure complete availability for reaction with the acid ingredients (10).

Fine powdered is recommended for angel food cakes where long baking time dictates a requirement for rather long gas release periods, and in refrigerated and frozen doughs which need only minimal leavening before packaging and during storage. It is also used in pancake mixes and baking powders when uniformity and stability of the dry blend is essential. The particle size also facilitates rapid blending. Granular is useful in the refrigerated doughs and some doughnut mixes. the sieve analyses of these three sodas are given in Table 9. Table 10 shows the most common leavening acids that are used in combination with sodium bicarbonate to make-up baking powder. the function of a leavening acid is to promote a controlled and nearly complete evolution of carbon dioxide gas from a dough in which the latter compound exists in its dissolved or bound form. Obviously, the acid must be edible and yield edible and nearly tasteless products. It should exist in a solid form at normal storage temperatures; it should be economical and easy to handle; and it should not have any deleterious effects on other dough constitutes (e.g., it should not weaken the gluten) (8). By law a baking powder must produce at least 12% carbon dioxide upon complete reaction. That regulation establishes the level of soda in the baking powder (1). The choice of what leavening acid to use is based on a number of factors, the most important of which is the rate of reaction. Kickline and Conn (9) have published the relative reaction rates of the common leavening acids (Fig. 1). Since the soda dissolves almost immediately, it is the leavener's rate of dissolution that determines the rate of carbon dioxide release. Leaveners are, therefore, characterized, and sometimes specified by a leavening rate (Table 10). This characteristic is useful in selecting the best leaveners for a specific use. The desired rate of release will vary widely depending upon the particular use.

The baking powder offered to bakers are generally classified as "fast acting," "slow acting," and "double acting" (Table 11). Fast acting powders release most of their gas at room temperature. Slow acting powders release a portion of the available carbon dioxide during mixing, but generate most of it by reactions occurring at elevated temperatures. Double acting powders are really a version of the slow acting type, which have somewhat more gas-producing potential during mixing. Most bulk powder for bakery use is also of the double acting type. Other important factors affecting the choice of leavening acids are the taste of the product as well as their effect on dough rheology.

The reactivity of baking powders is determined by the concept of neutralization value, which is defined as the grams of soda that 100 g of acidic salt will neutralize (10). The neutralizing values of major chemical leaveners are also shown in Table 10. the leavener and soda are generally used in such proportions that after baking there remains little, if any, unchanged soda or acidic salt in the finished product. There are some exceptions to this balancing of leavener and soda when particular effects related to the pH of the final product are desired (i.e., color, taste), a notable example of this being the high pH chocolate of devils food cake. In average cases, however, the amount of soda has been fixed:

$$\text{Amount of leavener} = 100/\text{leavener N.V.} \times \text{amount of soda}$$

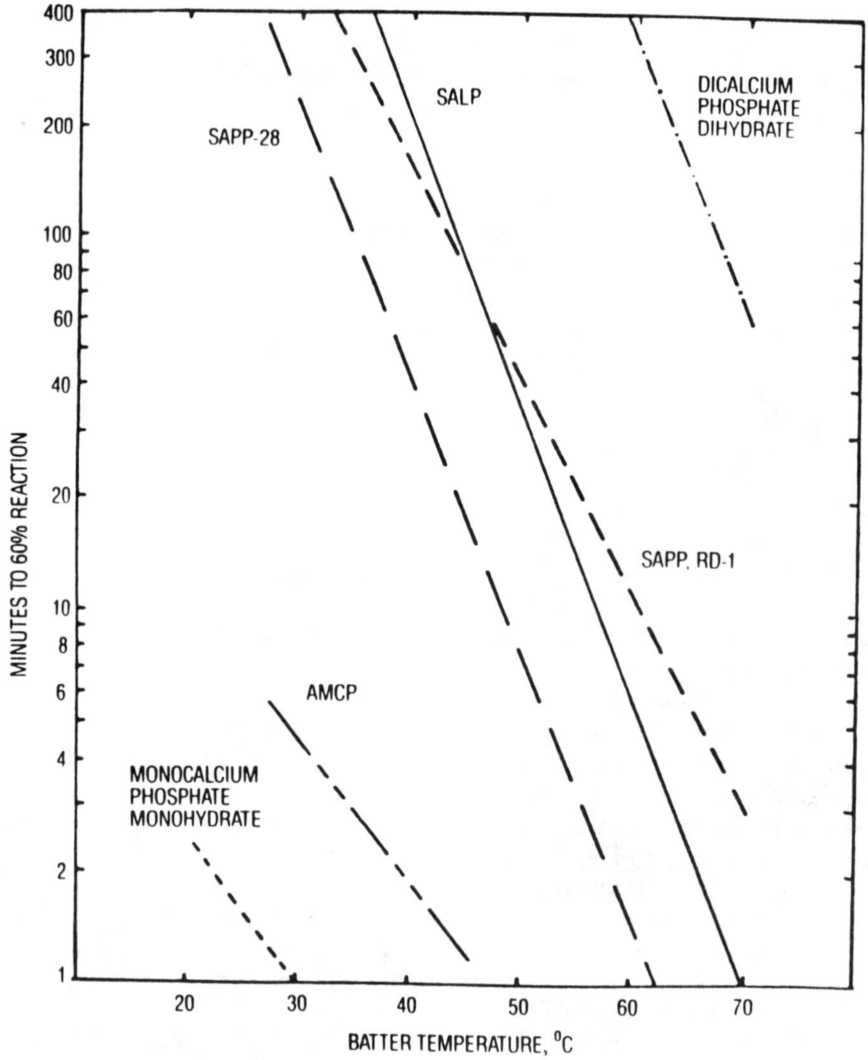

FIGURE 1 Reaction rate vs. batter temperature for various leavening acids. SAPP, sodium acid pyrophosphate; SALP, sodium aluminum phosphate; AMCP, anhydrous monocalcium phosphate; RD-1 is a grade of SAPP. From (Ref. 9.)

If two leaveners of different N.V.s are desired in a given application, two calculations are necessary after deciding what fraction of the total soda shall be neutralized by each leavener (10,11).

Figure 2 shows the typical rate of reaction curves for several leavening acids in a cookie dough medium at 25°C. The increase in temperature of baked product in the oven (Fig. 3) profoundly affects the reactivity of leavening agents. Additionally, leaveners can contribute to the internal structure of baked goods by means other than gas production. This is due to the leaveners' cationic or anionic effect on the other ingredients (9).

Of the cations present in phosphate leaveners, sodium, calcium, and aluminum ions, the latter two can contribute markedly to structure in terms of fine grain and generally thin cell

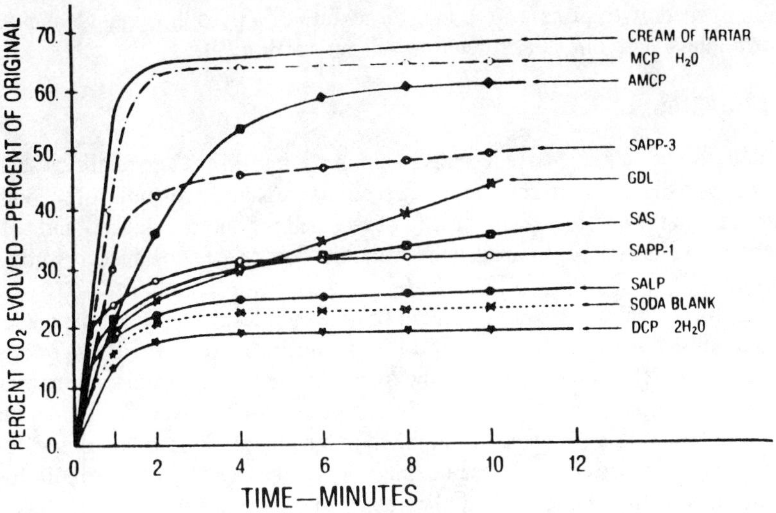

FIGURE 2 Typical rate of reaction curves for several leavening acids in a biscuit dough medium at 25°C. (From Ref. 11.)

wall size. This is particularly noticeable in white batter products. Additionally, calcium and aluminum types of leavener impart much more resiliency to baked goods than does the sodium ion. Apparently, the structure in the liquid state is also greatly strengthened by calcium and aluminum ions, thus retaining a fine bubble structure rather than coalescing to larger structures. This strengthening of the gluten is used extensively in the bread industry by means of dough conditioners. The sulfate ion in significant amounts has a deleterious effect on gluten strength. Of the anions present in phosphate leaveners, ortho and pyro ions, the

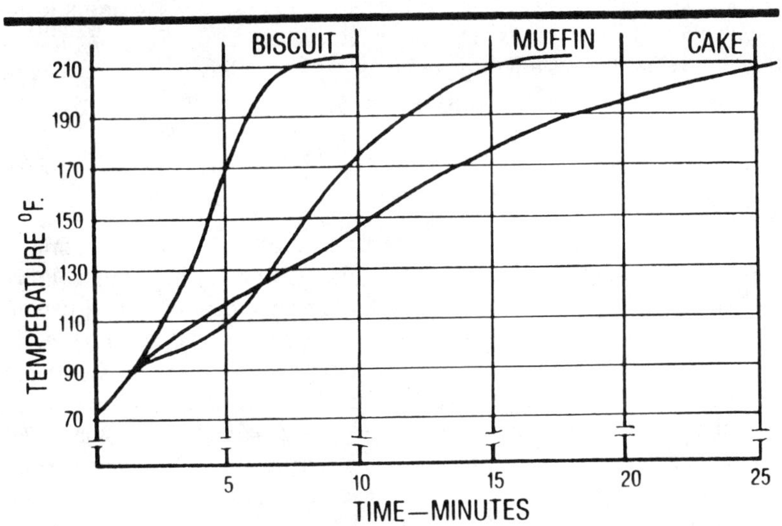

FIGURE 3 Internal temperature of baked products during baking. (From Ref. 12.)

latter will sometimes react with proteins. the moister texture of a pyrophosphate leavened product vs. an orthophosphate one is attributed to this interaction (9).

D. Fats and Emulsifiers

Fats are used in doughs and batters, surface sprays, and in cream fillings and coatings such as chocolates. The type and amount of shortenings and emulsifiers in the formula affect both the machining of the doughs and the eating quality of finished products. Coatings and fillings are dependent upon fats and oils to furnish the structural part of the substance as well as good flavor release.

In cake making, the fat functions to contain air as minute bubbles, which form the nuclei for expansion and texture during baking. In puff doughs, fat is used to create distinct horizontal layers of discontinuity which separate and expand during baking. In biscuit creams and chocolate, the physical properties of the fat must give a firm consistency at ambient temperatures but rapid melting characteristics in the mouth so that the sugar and other flavors are released quickly. Fats used in surface coatings, applied as a spray of warm oil, for savory crackers are best if they have limited absorption into the biscuit and remain as a glossy film (8).

In evaluating the stability of a fat for baking use, two properties which generally receive primary consideration are the fats organoleptic characteristics and its plastic range. Flavor and odor are of great important. The plastic range, or the fat's consistency at different working temperatures, is of major importance for baking applications. Fats usually consist of a mass of tiny individual crystals enmeshing a considerable proportion of liquid oil. The firmness of a plastic fat or shortening is influenced by a number of factors, such as temperature, the size of solid particles, the shape of the crystalline material, crystal rigidity, and the solid-liquid ratio within which the fat is plastic (13). Table 12 describes the factors affecting the quality of fat used in the formula of soft wheat products. Table 13 shows most of the fats available today for production of various soft wheat products. The fats used in the formula must be sufficiently plastic to enable good distribution during mixing and must have little if any solid fat present at temperatures above body temperature (37°C), otherwise the product will leave a greasy or waxy feel in the mouth. The solid fat indices of a selected number of bakery shortenings are shown in Table 14.

Emulsifiers are a large group of surface active agents whose function is to promote the formation and stabilization of water emulsions such as fat (oil) and water. They are used at low (,1%) levels and are classified as minor ingredients or food additives. The most common emulsifier used in soft wheat product is lecithin, derived principally from soya bean (8).

E. Sweeteners

After flour, sweeteners are the most important ingredient in the soft wheat product formulation. In addition to providing sweetness, most sweeteners provide one or more of the following functions: tenderizing, texture, yeast nutrient and fermentation control, stability, bulk agent-body, humectancy, flavor, crust color, and shelf-life extension. The selection of the proper sweetener is determined by the functions desired and the properties of the sweetener that provides these functions. The principal sweetener used in soft wheat products is sucrose. High fructose corn syrup, invert sugar, honey, glucose syrups, and molasses are used to a lesser extent. Granulated sugar is highly refined and is the most common sugar used in the baking industry. Many grades, as defined by particle size ranging from coarse sanding sugar to very fine granulated sugar, are available (16).

For example, sanding sugars are used for topping on such products as cookies where particles of sugar on the surface of the finished cookies are desirable. Fine granulated sugar is used for fillings where sugar is not completely dissolved and large particle sugar would produce a "gritty" taste. A medium particle size "bakers special" granulated sugar is the most common type used in bakery food applications. The particle size of granulated sugar is one method for controlling cookie spread during baking. Powdered sugars are divided into general categories ranging from 2X to 12X, with the higher numbers indicating increasing fine particle size. The most common powdered sugars are the 6X and 10X types, the 6X sugar being used for dusting of cake doughnuts, and 6X or 10X sugar is used in fillings and icings where a coarser particle would produce a gritty mouthfeel (Table 15). Fondant is prepared by heating a concentrated sucrose solution to boiling then cooking gradually with controlled mixing. During controlled cooling, the sugar crystallizes as very fine particles. The primary application for this product is in icings and fillings where a very smooth mouthfeel is desired.

Invert sugar is obtained from sucrose when a solution of sucrose is treated with acid and/or enzymes, this treatment splitting the sucrose molecule into its two components, glucose and fructose. Several types of invert sugar are available, the types determined by the degree of inversion or conversion to the monosaccharides. Invert sugar has moisture retention (humectant) properties which makes it useful in prolonging shelf life of many bakery foods. It is particularly effective in preserving softness in soft cookies and brownies. When used at the rate of 3–5% based on powdered sugar weight, in icings and glazes, invert sugar improves the gloss and "shine" on the surface of the icing and also improves pliability (16).

Brown sugar is a blend of sucrose and molasses and is made of fine crystals covered with a film of highly refined molasses-flavored syrup. As many as 13 grades of brown sugar are available, ranging from light to dark in color.

Corn starch is natural polymer whose molecular structure consists of glucose polymers in linear or in branched chains. When hydrolyzed, either by enzymes, by acid, or by a combination of both, the long chains are split into smaller segments. The higher saccharide are not fermentable and do not contribute sweetness to the finished bakery food. The term commonly used to indicate the degree of conversion is "dextrose equivalent," or D.E. In cake products, a 42 D.E. syrup may be used; higher D.E. syrups may cause dark crumb color in some varieties. In cookies, a 42 D.E. corn syrup is recommended and is also recommended for icing. The carbohydrate composition of various corn syrups is shown in Table 16. High fructose corn syrup is replacing the invert syrups in many applications since it is basically the same type product—a blend of fructose and dextrose. The use of high fructose corn syrup in cakes and cookies is not widespread since the sweetener provides functions other than fermentable sugars and sweetness to these products. In chemically leavened products, sucrose is not converted to glucose and fructose in the batter and remains as sucrose in the finished bakery food. Therefore, adding high corn syrup as a replacement requires formula adjustments for production of satisfactory products (16,17).

One of the primary functions of sweeteners in bakery foods is to provide sweetness to the product. The relative sweetness of various sweeteners is shown in Table 17. Sweeteners vary in solubility. Table 18 demonstrates the effect of temperature on the extent of solubility (saturation) of some sugars. various sugars also have a tendency to absorb moisture from the air depending on the air's relative humidity (Table 19) which may cause manufacturing problems for the bakeries due to caking. Table 20 compares the properties of the widely used sweeteners.

Molasses is a concentrated, brown to black, highly flavored, relatively impure syrup obtained during sugar processing and is used in some soft wheat products for its characteristic flavor, sweetness, and coloring properties (7). Molasses contains a substantial amount of invert sugar, which functions as a humectant, improving moisture retention in such products such as soft cookies. Table 21 shows characteristics of different types of molasses.

F. Chocolate

Chocolates and cocoa powder in the formula provide the distinct flavor that is very popular with the consumer. the common source of both cocoa and chocolate is the cacao bean that consists of a shell (14%) and the nib or interior kernel (86%). The shell is removed by roasting and drying, The roasted nib is then ground into a homogeneous mass called chocolate liquor, and also bitter or baking chocolate. During the grinding operation the temperature is increased to 104°F (40°C) and, because the nibs have a fat content of over 50%, the ground mass becomes fluid and can be run into molds. It is this chocolate that is principally used by bakers for the production of chocolate cakes and cookies. The composition of the chocolate liquor is given in Table 22 (13).

Cocoa results when the chocolate liquor is passed through hydraulic presses where most of the fat, or cacao butter, is expelled. The press cake remaining after the extraction of the cacao butter will have a fat content of 8–25%, depending upon the duration and degree of pressure applied. The press cake is then ground to a fine powder in micropulverizers to yield the powdered cocoa of commerce. Since cacao butter is relatively neutral in flavor, its extraction results in an intensification of flavor in the cocoa which, therefore, normally contains from 1.5 to 1.7 times as much flavor as does chocolate. Cocoas are differentiated into natural and "Dutch" process, or alkalized cocoas (Table 23). The process of "Dutching" darkens the cocoa color. Because of there lighter color and milder flavor, natural process cocoas are used primarily in the production of icings, fudges, sauces, and drinks, whereas the Dutched cocoas are more suitable as ingredients of cakes and cookies. Cocoa products may vary in their pH values from 5.2 to as high as 8.8 in heavily alkalized cocoa. The pH of the cocoa is of importance in determining the color, flavor, density, and textures of cakes. It also governs the amount and types of various chocolate used in production of soft wheat products (10).

Many types of cakes and cookies are enrobed with chocolate and other compound coatings. The advantages of enrobing are shown in Table 25. The compound coatings are formulated with stable fats and maintain their flavor, mouthfeel and richness for many months. There are four main types of coatings (Table 26). To enrobe a cookie or cake, the baked item should be at a temperature of 75–85°F. The coating is usually tempered by a controlled cooling and heating process to improve its stability. The enrobing process of baked products is demonstrated in Figure 4.

G. Salt

Two types of salts are commonly used in baking, namely, granulated or common salt and fine flake salt. Salt is used in many of the soft wheat product formulations at the 1–1.5% (on flour basis) concentrated for its flavor and flavor-enhancing properties (24).

Salt used in the production of cookies should be a fine-screened product with a rapid solubility rate. For cracker doughs the salt can be somewhat coarser in screening, but it may be an advantage to a cracker and cookie baker to inventory only one type of salt. Topping salt for crackers is characteristically a flake-type salt of some coarseness. Topping salt should be uniform in size, with a minimum of "fines," which may cause uneven feeding and poor appearance of the finished cracker.

Icings need about 1% salt to reduce the cloying sweetness of the high sugar content. A fast solubility rate should be the outstanding physical characteristic of salt for icings (24).

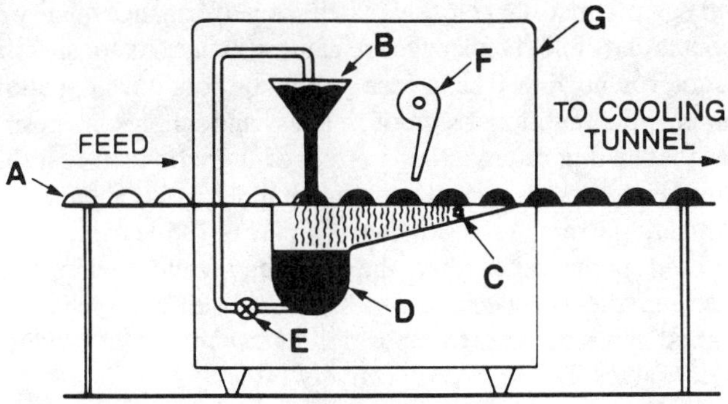

FIGURE 4 Schematic of an enrober used for coating cakes and cookies. (A) Baked items at 75–85°F temperature proceed from a conveyor belt onto a wire mesh belt. (B) The tank that discharges a curtain of liquid coating. (C) A roller under the belt for coating the bottom of items. (D) Holding tank for coating. (E) Circulating pump. (F) Air blower to intensify the gloss and leave some ripples on the coating. (G) The enrober. (From Ref. 23.)

H. Dairy Products

In baked goods, dairy products are used for color improvement, water absorption and spread-control properties, and flavor (17). It is impractical for bakers to use fresh milk for several reasons such as storage, sanitation, and shipping cost. Furthermore, it is more economical to substitute other fats for butterfat. Therefore, the form of choice is nonfat dry milk solids, with use of minor quantities of evaporated milk (from whole or skim milk) and sweetened condensed milk.

Nonfat dry milk may be spray dried or roller dried, the former being much more popular. Where reconstitution is anticipated, dried milk is usually supplied in instant form. The amount of milk in a cake batter is usually based on all the liquid in the formula being in the milk and eggs. When nonfat dry milk and dried eggs are used, cake formulas tend to maintain about these same levels of milk and eggs solids, substituting in each case solids and water for liquid milk and eggs. Formula balance is more directly influenced by the percentage of eggs, shortening, sugar, flour, and liquid; hence there is somewhat more tolerance to the amount of milk solids. Under these conditions changes may be made according to the cost and availability. Nonfat milk solids function as a structure builder and a dryer, contribute significantly to nutrition, and help develop crust color, through the presence of lactose, which caramelizes in the oven. The typical composition of dairy-based ingredients used in soft wheat products is shown in Table 27.

Whey can be regarded as skim milk from which the casein has been removed. It is the liquid residue remaining from the cheese-making process. Examples of the utilization of whey in soft wheat products is shown in Table 28.

I. Egg Products

Eggs affect the texture of bakery products as a result of their emulsifying, leavening, tenderizing, and binding actions. They add color and nutritional value and, in many cases, desirable flavor. they are essential for obtaining the characteristic organoleptic qualities of most cakes, some cookies, and many other soft wheat products (7). Eggs are available in various forms including fresh shell eggs in home baking, refrigerated bulk eggs, frozen whole eggs, egg whites and yolks, and dried egg products that closely match the performance quality of the liquid forms, but dried products and chilled bulk eggs are gaining in use. As for specific contributions to baked foods, color is notable. The golden yellow of pound cake, yellow layers, sponge cakes, and donuts is the traditional sign of richness and wholesomeness.

In the classifications of the toughening or tenderizing effect of ingredients (see Table 38), eggs play a dual role, whites being a toughener and structure builder, yolks a tenderizer through high fat content. Outstanding examples of function as part of the leavening and structural system are angel food and sponge cakes where the egg or egg white is whipped to considerable volume, with other ingredients added to the form before baking in such a way as to maintain lightness and good cell structure. Composition and type of egg products available to bakers are shown in Tables 29 and 30, respectively (10).

J. Spices, Flavors, and Colors

Fresh high quality spices and flavors offer innumerable ways for a baker to introduce variety in the baked products. Flavoring materials that have been added to bakery products include but are not restricted to natural products such as spices, chocolate, vanilla, and essential oils, artificial flavors such as vanillin, and modified flavors such as cultured butter, starter distillates, and hydrolyzed vegetable protein. Characteristics of different spices and flavors used in production of soft wheat products are shown in Table 31. Both artificial and natural coloring agents are used in the bakery industry to enhance the visual appeal of products. All food colors must be selected from the compounds approved by government agencies, principally the FDA. There are two categories of acceptable color additives: certified colors and uncertified colors. Certified colors include chemically synthesized dyes and their lakes. Lakes are made by absorbing or chemically combining water-soluble dyes with certain specified insoluble materials (usually aluminum compounds). The chemical compounds making up the list of certified color additives have been subjected to a very thorough program of toxicity tests.

The uncertified color additives include pigments extracted from fruits, vegetables, etc.; inorganic substances that are not expected to be absorbed from the body (e.g., titanium dioxide); and some synthesized materials that are more or less similar to certain natural products. Among the last group, carotenoids are particularly useful as bakery product colorants because they simulate the color of butter and eggs. Table 32 includes some of the currently permitted uncertified additives, and Table 33 lists the certified pigments.

K. Icings

Icings may be defined as "glazing or coatings of sugar an various other ingredients blended together and pleasingly flavored to suit individual tastes" (13,27). Lachmann and Voll (28) have summarized the quality features of a good icing as follows:

1. It should spread easily and handle satisfactorily at regular working temperatures when applied either by machine or by hand.
2. It should adhere to the baked product without thinning out and running off.
3. It should set firmly within the desired time limit.
4. It should not dry out too rapidly or crack after prolonged periods of storage; neither should it pick up too much moisture, become sticky or melt.
5. It should maintain its glossy appearance and its true color.
6. It should not become too gritty, or separate during storage.
7. An aerated icing should maintain its shape and not bleed or dry out.

A good icing will enhance the quality and appearance of the finished product; it will neither dry out its crust nor make it soggy. In general, icings are divided into three major categories: (1) the flat icings, which are not subject to whipping and, therefore, do not contain any air cells; (2) the highly aerated and fluffy icings (often called foam-type icings) and imitation whipped cream frostings, which are extremely light; and (3) the partially aerated products, which comprise the popular butter-cream icings and their modifications.

The formula for production of some popular icings is shown in Table 34.

III. PRODUCTS

A. Crackers

Crackers contain little or no sugar and moderate levels of fat (1,7,8,29). The dough generally contains low levels of water. There are three major types of crackers: saltine, chemically leavened, and savory. Crackers, particularly saltines, are deceptively simple food systems. In fact, the process required to produce acceptable saltines is both lengthy and complex—so complex that the significant changes occurring in the physical properties of the dough are just now being understood in detail (1,29). A brief description of the process follows. Saltine crackers are made by a sponge dough process that requires about 24 hours, much of it for fermentation (Fig. 5). Since cracker sponges are mixed just long enough to wet the flour, gluten development occurs only to a limited extent at this stage. During the next 19 hours of sponge fermentation, the consistency of the sponge changes drastically and the sponge becomes less elastic and more acidic. Figure 6 demonstrates the relationship of dough pH and cracker fermentation time.

After fermentation, the sponge is mixed with other dough ingredients and the dough-up flour. the dough is allowed to relax for 4–6 hours. After the relaxation/resting period is over, the dough is taken to the hopper of the sheeter.

Cracker doughs are laminated after exiting the sheeting rolls. the purpose of laminating the sheeted dough (into 6–8 layers) is fourfold. First, it constitutes a method of repairing a dough sheet with holes or tears. Second, by turning the folded dough 90°, stresses in the dough are made more uniform in two directions. Third, the repetitive rolling and folding of the dough accomplishes a significant amount of work on the gluten. this process makes the dough more suitable for baking a delicate and layered structure characteristic of saltine crackers. Fourth, by introducing another material, like fat or dusting flour, between layers of dough, a characteristic flaky structure could be produced after baking (8).

Multiple pairs of heavy steel rolls (called gauge rolls) gradually reduce the dough sheet thickness to that desired for cutting. Typically, there are two or three pairs, although only one pair may be used for short doughs and more than three may be used where gentle reductions in dough thickness are necessary. As a rule of thumb, the reduction in thickness should be about 2:1 for each pass through a roll pair, although ratios of up to 4:1 are used. Obviously, the greater the ratio, the more work and stress is put into the dough and the most its physical properties change (7,8).

Chemically leavened crackers are generally called snack crackers and have a final pH of about 6.5. After mixing and a relatively short rest time (2–4 hrs), the dough is sheeted to form a continuous ribbon, which is then laminated, with a light application of dusting flour between the layers. Graham crackers are chemically leavened and semi-sweet crackers in which part of the white flour (10–40%) is replaced by whole wheat flour. Figure 7 outlines the process for production of chemically leavened crackers.

Flavored or savory crackers are well accepted in the U.S. market and account for much of the recent growth in cracker sales. The intense savory flavors are produced by adding the

```
Ingredient Scaling
        |
        V
Sponge Mixing - 25 C
        |
        V
Sponge Fermentation - 19 hr.
        |
        V
Dough Mixing - Spindle Mixer [100 Rev.]
        |
        V
Dough Proof - 4 hr.
        |
        V
Dough Laminating
        |
        V
Dough Sheeting
        |
        V
Dough Cutting [12 pc = 48.6 g]
        |
        V
Salt Application
        |
        V
Baking
        |
        V
Oil Spray and Conditioning
        |
        V
Packaging [1.5 - 2.8% Moisture]
```

FIGURE 5 Production flow chart—saltine crackers. (From Ref. 29.)

appropriate flavoring agents directly to the dough or to the surface of the crackers after baking. Savory or cheese crackers are generally produced from fermented doughs. The yeast fermentation and the lower pH improve the cheese flavor. the formulation and process is basically similar to that of soda crackers, with adjustments to compensate for the fat and moisture content of the cheese (29). Typical formulas for various types of crackers (saltine, snack, and grahams) are shown in Table 35.

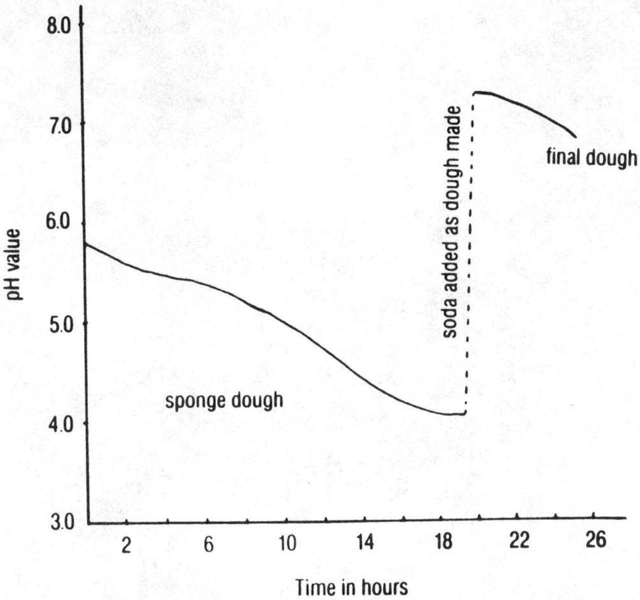

FIGURE 6 Relationship of dough pH with time during fermentation of saltine cracker sponge and dough. (From Ref. 8.)

B. Cookies

In general, cookies are products made from soft and weak wheat flour. They are characterized by a formula high in sugar and shortening and low in water. Cookie doughs are cohesive but to a large degree lack the extensibility and elasticity characteristic of bread doughs.

Relatively high quantities of fat and sugar in the dough allow plasticity and cohesiveness without the formation of a gluten network (1). In addition, and again depending of the formulation, cookie dough tends to become larger and wider as it bakes rather than to shrink as does cracker dough. This increase in size, or "spread," is the greatest single problem in process control (7,8).

The most tender eating cookies can be obtained from a given short dough only if the amount of mixing is kept to a minimum. Minimal gluten development is achieved by carrying out the mixing process in two or even three stages. Obtaining a dough of correct consistency at the end of mixing is important, and the mixing step can therefore can be considered critical.

Perhaps the best way to classify cookies is by the way the dough is placed on the baking band.

1. Rotary molded cookies

For rotary molded cookies, the dough is forced into molds on a rotating roll. As the roll completes a half turn, the dough is extracted from the cavity and falls on the baking oven band (Fig. 8A). the consistency of the dough must be such that it feeds uniformly and readily fills all of the crevices of the roll cavity under the pressure existing in the feeding hopper. At the same time, it must be possible to extract the dough pieces cleanly from the cavity without their undergoing distortion or forming "tails." Rotary molded dough must also be suffi-

```
Ingredient Scaling
        |
        v
Mixing - Upright Mixer  [30 C]
        |
        v
Dough Proofing  [2 - 3 hr]
        |
        v
Dough Laminating
        |
        v
Dough Sheeting
        |
        v
Dough Cutting
        |
        v
Salt Application
        |
        v
Baking - 4 min.
        |
        v
Spray or Seasoned Oil Application
        |
        v
Post Conditioning
        |
        v
Packing [1.0 - 3.0% moisture]
```

FIGURE 7 Outline of process for production of chemically leavened crackers. (From Ref. 29.)

ciently cohesive to hold together during baking. Dough spread and rise should usually be minimized. Doughs formulated to meet these requirements are usually fairly high in sugar and shortening but low in moisture. The development of gluten is definitely to be avoided (29). The typical dough is crumbly, lumpy, and stiff, with virtually no elasticity. Much of the cohesiveness of this type of dough comes from the plastic shortening used (1).

2. Wire-cut cookies

In the production of wire-cut cookies, a relatively soft dough is extruded through an orifice and cut to size, usually by a reciprocating wire (Fig. 8B). The dough must be cohesive enough to hold together on the band, yet short enough to separate cleanly when cut by the wire. The rate of extrusion is related to the dough consistency. Here, dough spread during baking is desirable but must be closely controlled. Another type of extruded doughs (fruit bars) is similar to wire-cut except that the extrusion of dough is continuous without wires

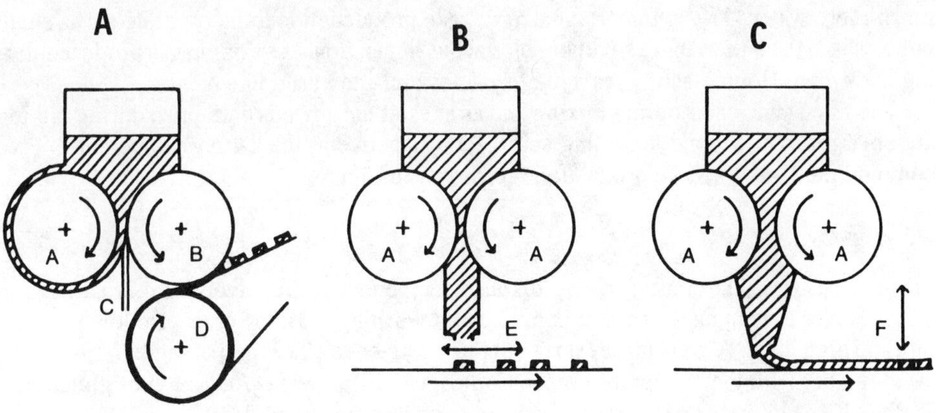

FIGURE 8 Three common methods for cookie production. (A) Rotary method: A dry and crumbly dough is compressed by a grooved roller (A) into biscuit-shaped dies engraved on a forming roller (B). The surplus dough is trimmed from the surface of the (B) roll by means of an oscillating knife (C). The shaped dough pieces are extracted onto a canvas web, which is pressed against the (B) roll by the roller (D). Very little changes in dimension occurs during baking. Therefore, complex patterns are carried through to the finished product. (B) Wire-cut method: A soft dough is extruded through dies by two contra-rotating grooved rollers. The extruded dough is cut by a reciprocating wire (A) and usually falls directly onto the oven band. (C) Bar-press method: A soft dough is extruded or co-extruded (fruit bars) and them deposited on a conveyor as a series of parallel ribbons, which are cut to suitable lengths by a cutter (F), either before or after baking. (From Ref. 29.)

and the orifices are usually designed to produce strips rather than round shapes. Fig bars are made by extruding a fig paste within a tube of dough of similar consistency (Fig. 8C).

3. Cutting machine cookies

Another method for cookie production, which is being slowly discontinued, is the cutting machine. The process and formulation of cutting machine cookies produces the familiar Christmas cookie. A dough with slightly less fat and sugar but more water than rotary mold dough is sent through multiple sheeting rolls and made into a continuous sheet. The sheeting and cutting operations for the short doughs are essentially the same as for cracker doughs. The dough shows very little elasticity, so shrinkage before cutting is not a problem. Cutting can be similar to the production of crackers in that docking and imprinting of a name or similar pattern is performed before the outline is cut. Typical formulas for the production of various types of cookies are shown in Tables 36 and 37.

C. Cakes

Probably the most characteristic and unique product made from soft wheats is cake, traditionally the centerpiece of festive and joyous celebrations (30). Pyler (13) identified three major quality factors in cake baking: (1) suitability of the individual ingredients for the specific type of cake being made, (2) the proportion in which the ingredients are combined in the cake formula, and (3) the procedure followed in the mixing and baking operations. Coughlin (31) has categorized cake ingredients (Table 38) as follows: tenderizer, moistener, drier, and flavorer. Cakes, like cookies, are quite high in both sugar and shortening. The major difference between the two is that a cake formula (Table 39) tends to contain much more water (1). An important step in cake production is to incorporate air as small bubbles into the batter. these small bubbles act as nucleation sites for the gas produced during leavening. Double-acting leavening systems are often used in cakes.

The first stage reacts during mixing and as a result the air cells entrapped during mixing are enlarged. The second leavening acid becomes active at the baking temperature. Five common methods for cake production are shown in Figure 9.

D. Pie Crusts

Pies are pastries that consist of two distinct components, a relatively thin crust portion which serves to contain the second component, the filling (13). Pie crusts are low in moisture and high in fat. A good pie crust is both flaky and tender. The ratio of ingredients (Table 40), together with the method of preparation (figure 10) prevents formation of gluten network and results in baked crusts that are more or less friable or flaky.

E. Hard Pretzels

Hard pretzel is a baked food unique both in shape and in having a hard and shiny outer surface (1). Pretzels are made from a straight or sponge dough system (Table 41), which is quite similar to a cracker formula. Pretzel dough, however, is somewhat drier and, therefore, stiffer. After the doughs are formed, they are first cooked in a lye bath, coarse salt is applied, and then they are baked. This unusual treatment, plus drying to a low level (2–4%), gives a color, sheen, flavor, and texture that is greatly appreciated as a snack and party food widely served in America. Packaging must hold moisture at this low level; well-protected pretzels retain their flavor and crispness for a year or more (1,13,29). The production scheme of hard pretzels is shown in Figure 11.

F. Puff Pastries

Puff pastries are laminated, flaky products with an open structure and a typical flavor of fried goods such as donuts (13). A puff pastry dough consists of many thin layers of dough separated by fine laminations of fat. A piece of properly made puff pastry, 1/4 inch thick, is likely to expand on baking to a height of over 2 inches, and eightfold increase in height. This is far greater expansion than is obtained with a cake batter. Because of the light flaky structure which is developed, the baked product has good eating properties. The rise achieved during baking can be attributed to two main principles: (1) flour strength, and (2) the layered structure of the unbaked material. The gas responsible for the volume increase in puff pastry is the water vapor produced from the dough water. The fat layers in the structure are impervious to water vapor and act as barriers against which the water vapor formed by the underlying dough layer can exert a pressure. In this way, the dough layers expand and separate, giving the lift required. During baking, the fat in the layers eventually melts and becomes absorbed by the dough. The "lift" and the delicate structure of baked puff pastry is a result of lamination by lapping a sheet of puff pastry upon self a number of times. As the number of laps given a dough increases, the number of individual layers increases at a much higher rate. Up to a point, the greater the number of laps (and therefore, number of layers), the greater the lift (32). After that point, additional lamination may decrease the "lift" and lower the eating quality of the pastry (Fig. 12). The ingredients for production of puff pastry dough are shown in Table 42. The production scheme for puffed pastry is shown in Figure 13.

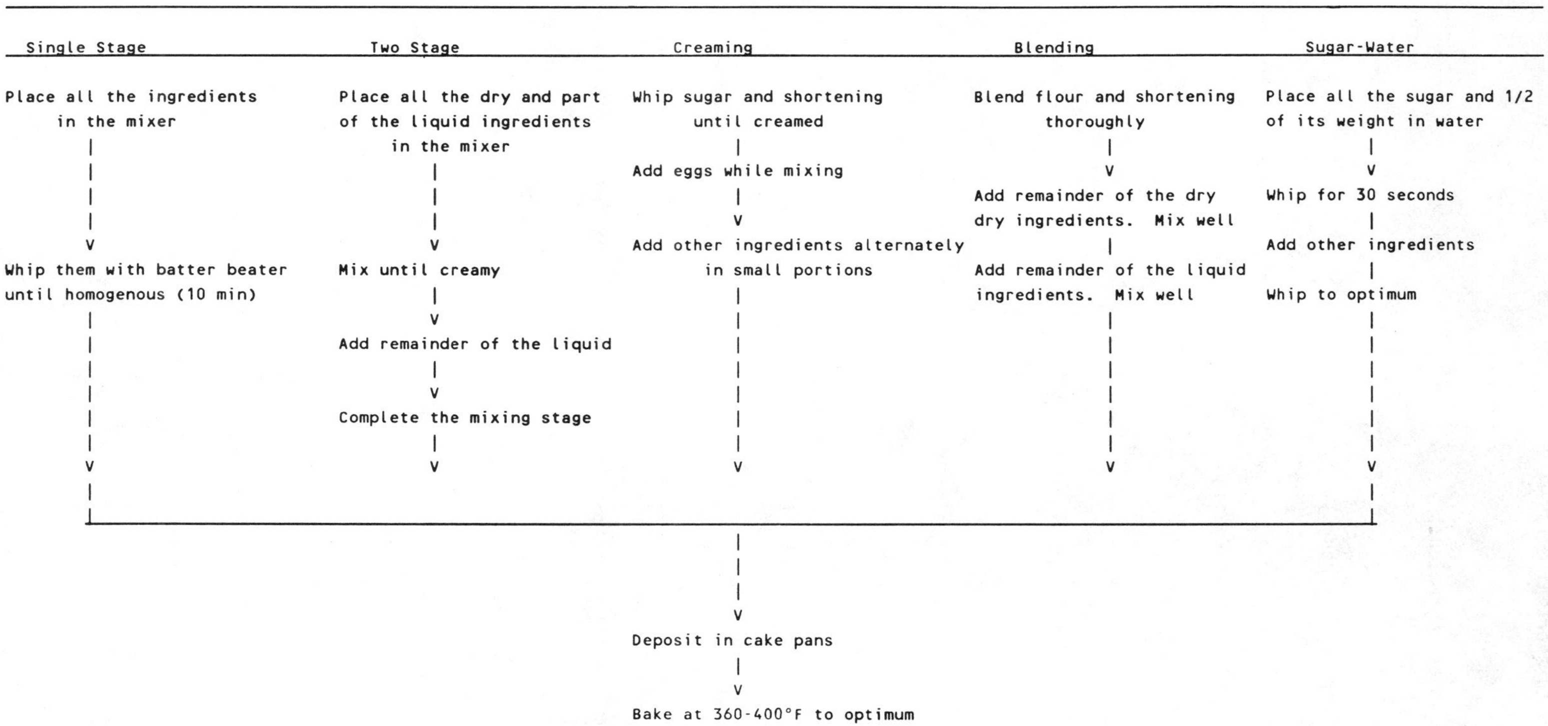

FIGURE 9 Production flow charts—cake baking.

```
Mix flour with shortening (35-40°F)
              |
              V
     Add other ingredients
              |
              V
    Refrigerate (2-3 hours)
              |
              V
   Roll to desired thickness
              |
              V
         Cut to size
```

FIGURE 10 Production flow chart—pie crust.

```
                               Mix the sponge
                                     |
                                     V
                             Ferment 6-10 hours
                                     |
                                     V
                                 Dough-up
                                     |
                                     V
          ____________________________________________________
          |                                                  |
          V                                                  V
Extrude into a desired shape                     Extrude into strands
          |                                                  |
          |                                                  V
          |                                      Clip and twist the strands
          |                                                  |
          |                                                  V
          |                                      Make a knot
          |__________________________________________________|
                                     |
                                     V
                   Pass quickly (25 sec) through a caustic
              bath (1.25% sodium hydroxide solution at 85-90°C)
                                     |
                                     V
                        Sprinkle large crystal salts
                                     |
                                     V
                   Bake at 246°C for 5 min. to 15% moisture
                                     |
                                     V
                      Dry 20 to 90 min. at 105-135°C
                         to final moisture of 2-4%
```

FIGURE 11 Production flow chart—hard pretzels.

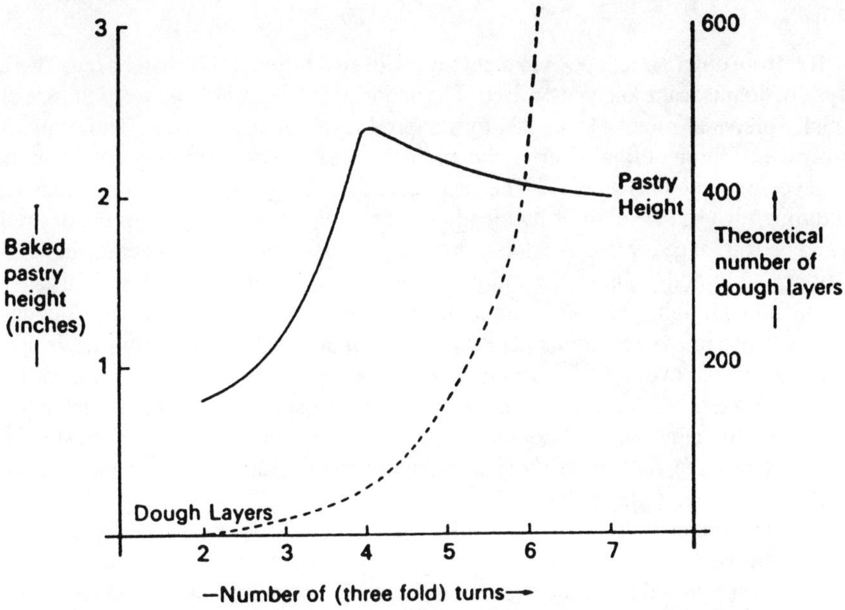

FIGURE 12 Effect of number of turns on the baked pastry height, using a low melting point fat. (From Ref. 32.)

```
  Blend ingredients (60°F)
            |
            V
        Roll the dough------------------
            |                          |
            V                          |
Spread additional shortening           |
            |                          \   Three
            V                          /   Times
         Fold 1 X 4                    |
            |                          |
            V                          |
         Refrigerate--------------------
            |
            V
     Cut to desired shapes
            |
            V
Bake at 350°F for 30-40 min.
```

FIGURE 13 Production flow chart—puff pastry.

G. Donuts

Donuts differ from other sweet goods in that they are deep fat fried rather than baked. There are two type of donuts: cake and yeast raised. The majority of bakers have adopted the use of commercially prepared mixes (Table 43) for the production of cake donuts. Those mixes require only the addition of liquid during the mixing process to transform them into batters ready for depositing into the frying fat. The basic flour for making cake donut mixes may be obtained through wheat selection or by blending hard with soft wheat flour to the desired strength and performance level. Cake donut flour for mixes is often dried to aerate and oxidize the flour and to help achieve desired performance and stability. The flour is usually unbleached, although light treatment with a maturing agent, such as chlorine dioxide, has been used by some mills. Selections of leavening is important; a slow-acting leavening agent is used so that gas evolution is not completed before the second side has expanded and been cooked or fried to its final shape. Sodium acid pyrophosphate with a controlled reaction rate is usually employed, although other such as sodium aluminum phosphate, monocalcium phosphate, dicalcium phosphate, and glucono-delta-lactone are also used in cake donut formulations (30,33).

The use of commercially prepared mixes has become as firmly established in the production of yeast-raised donuts as in the case of cake donut production. A representative formulation of a yeast-raised donut mix is given in Table 44. The extrusion process for yeast-raised donuts requires a dough that will extrude readily through the die under air pressure or a vacuum and yet will retain its shape during proofing. This type of dough is most readily obtained by the sponge dough method in which the sponge is fermented at 80°F (27°C) with 6% yeast for 1–2 hours (13).

Proofing conditions should be relatively dry and warm, with a temperature range of 118–122°F and a relatively humidity of 35–45% to accelerate skin formation on the proofing dough pieces. Proof time will normally run from 20 to 25 minutes. Frying temperatures should be held to within 400°F, with frying times ranging from 45 to 60 seconds per side, depending on donut size. A recent innovation in donut proofing is the application of microwave, or ultra-high frequency energy to heat the product. Microwave energy causes an instant, uniform rise in temperature throughout the dough piece instead of a temperature gradient from the external surface into the dough interior as is the case in conventional proofing. As a result, normal proofing times of 20–25 minutes are reduced to about 4 minutes (13). The three methods for production of yeast raised donuts are shown in Figure 14.

H. Refrigerated/Frozen Doughs

By definition, refrigerated dough is unbaked, flour-based product which is stored between 32–45°F (34). the original refrigerated dough was chemically leavened biscuit with an appropriate shelf life of 3 weeks. Today, the refrigerated dough industry encompasses a wide range of products. Such products are packaged in cans or in plastic films and require full baking before consumption (35).

There are two types of refrigerated doughs available. One is cookie dough packaged in plastic film. It requires only a small amount of leavening action and has a high solute content ($A_w = 0.8$). Thus, it requires little preservation besides refrigeration. The other type is packaged in cans and represents the majority of refrigerated dough products. They are leavened dough products, which include a variety of biscuits, cookies, sweet rolls, and dinner rolls. That type of refrigerated dough is unique because preservation of dough against both microbial spoilage and loss of leavening action is required. All refrigerated dough products are leavened chemically. Baking soda (sodium bicarbonate) is used as the source of leavening gas CO_2. The leavening acid most commonly used to liberate CO_2 from soda in dough is

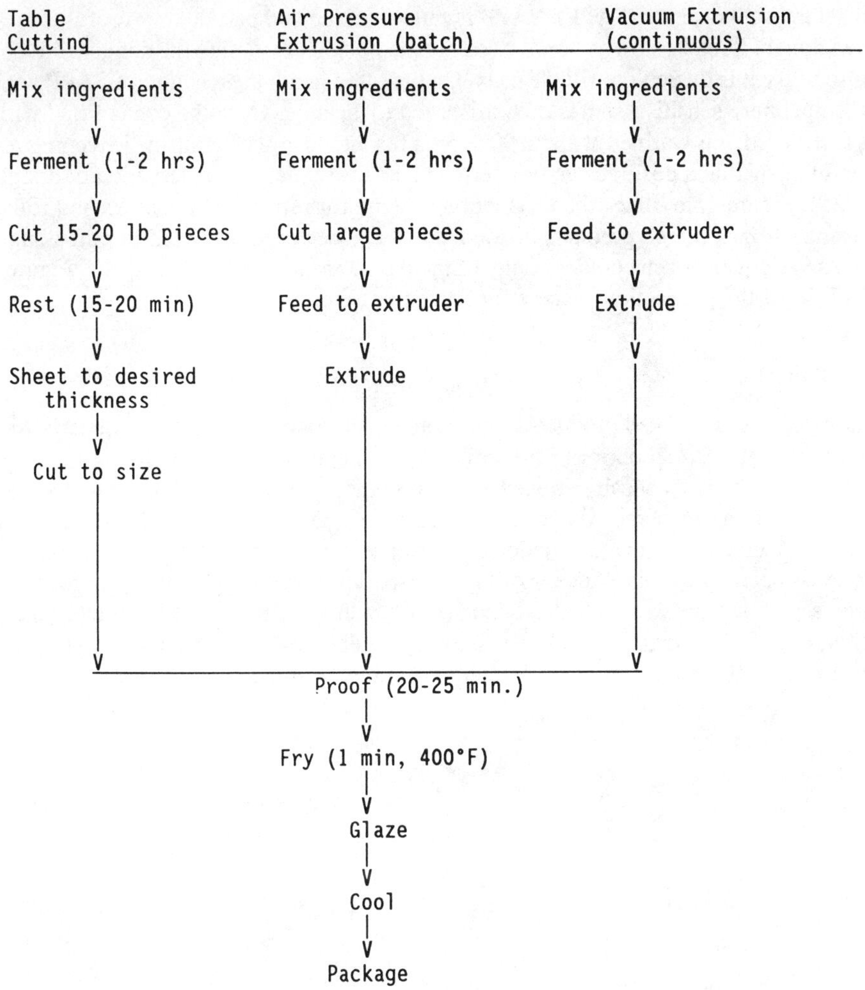

FIGURE 14 Three common flow charts for production of yeast-raised donuts. (From Ref. 33.)

sodium acid pyrophosphate (SAPP). SAPP is rather unique and particularly useful for refrigerated dough. It reacts slowly with the soda in the dough, preventing the dough from becoming so puffy it is difficult to fill the cans. On the other hand, the reaction of SAPP with soda is temperature sensitive so it can be increased by heating. Once the can is filled with doughs and closed, it is warmed at about 90°F for a few hours to accelerate the leavening reaction and to generate a positive can pressure which is essential for preservation of dough quality (35). Figure 15 outlines the production scheme for refrigerated cookies and rolls.

Consumer frozen dough products, made from soft wheat, are restricted to puff dough items and a variety of frozen cookies, many of them made as institutional products. Figures 16 and 17 detail the production process for frozen unbaked puff pastry.

IV. SCORING

The making of uniform baked products requires adequate control over the raw materials in their production and strict control of the finished product quality.

Most bakeries have set specifications for various attributes of their baked products. For example, loaves of cakes obtained in experimental baking tests or produced in commercial bakeries are subjected to regular evaluation or scoring to determine their physical characteristics on a comparative basis. Since scoring is based largely upon personal judgment in which the scorer visualizes an ideal product with which he compares the product under evaluation, it is obvious that no absolute values can be obtained (13). A typical cake score data sheet and a cake score report are shown in Figures 18 and 19, respectively.

```
         Mix the ingredients
                  |
                  V
                 Rest
                  |
                  V
                Sheet
                  |
                  V
                 Cut
                  |
                  V
        Wrap in aluminum foil
                  |
                  V
          Stack in a "can"
                  |
                  V
            Seal the can
                  |
                  V
   Proof at a raised temperature (90°F)
  to generate internal pressure of 15 psi
                  |
                  V
              Refrigerate
```

FIGURE 15 Production flow chart for refrigerated biscuits.

```
                    Mix the ingredients
                             |
                             v
              Store in refrigerator (4-6 hrs)
                             |
                             v
                           Remix
                             |
                             v
     Extrude laminated dough with plastic shortening
                             |
see Fig. /                   |
  17     \                   |
                          Sheeter
  (two layers of dough with a layer of shortening in between)
                             |
                             v
                          Folder
                             |
                             v
                          Sheeter
                             |
                             v
The final dough has over 100 alternating layers of dough and shortening
                             |
                             v
                       Cutter, filler
                             |
                             v
                  Freezing tunnel (-20°F)
                             |
                             v
                         Packaging
```

FIGURE 16 Production flow chart—frozen unbaked puff pastry.

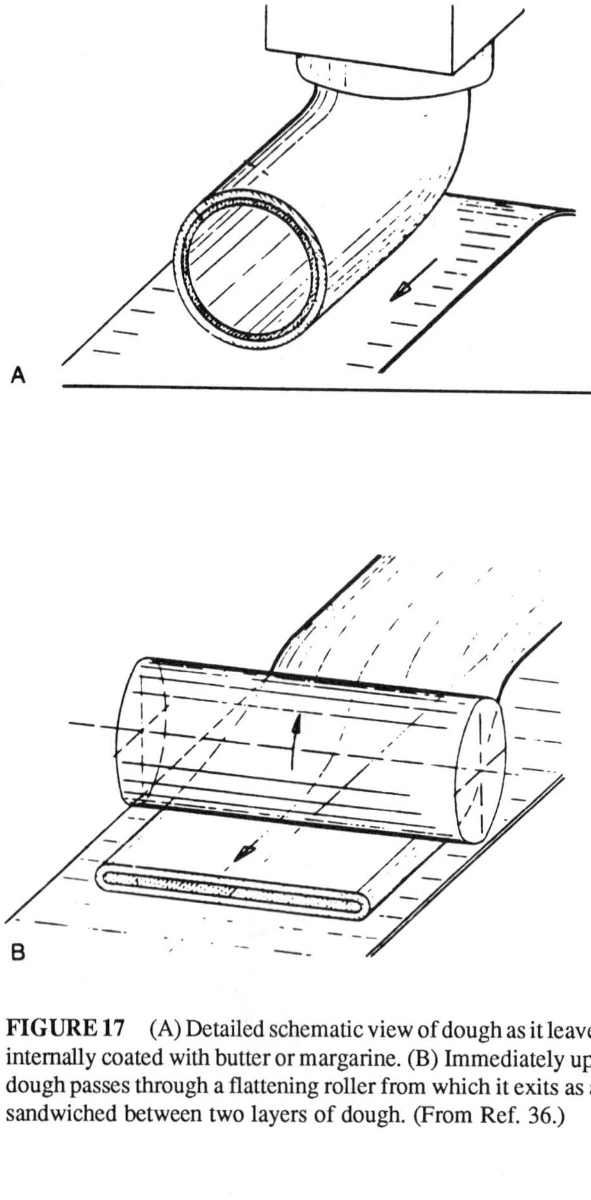

FIGURE 17 (A) Detailed schematic view of dough as it leaves extruder. The cylinder of dough is internally coated with butter or margarine. (B) Immediately upon leaving extruder, the cylinder of dough passes through a flattening roller from which it exits as a single layer of butter or margarine sandwiched between two layers of dough. (From Ref. 36.)

	Perfect score	Sample score	Penalized for: (check faults)
EXTERNAL			
Volume	15		☐Too small ☐Too large
Color of crust	5		☐Not uniform ☐Light spotty ☐Dark ☐Dull ☐Sugar ring ☐Grease ring
Symmetry of form	10		☐High center ☐Low center ☐High side ☐Low side ☐Peaked ☐Uneven ☐Burst
Character of crust	5		☐Thick blistery ☐Tough ☐Rubbery ☐Flaky ☐Moist ☐Dry
EXTERNAL SUBTOTAL	35		

	Perfect score	Sample score	Penalized for: (check faults)
INTERNAL			
Grain	15		☐Open coarse ☐Non-uniform ☐Thick cell walls ☐Holes
Color of crum	10		☐Gray ☐Dark ☐Streaky ☐Dull ☐Non-uniform ☐Light
Aroma	10		☐Strong ☐Lack of ☐Foreign ☐Musty ☐Sharp
Taste	20		☐Flat ☐Foreign ☐Salty ☐Soda ☐Sour ☐Unpleasant aftertaste
Texture	10		☐Rough-harsh ☐Lumpy ☐Too compact ☐Too loose ☐Crumbly
INTERNAL SUBTOTAL	65		
Total Score	100		

FIGURE 18 Cake score data sheet. (courtesy of American Institute of Baking.)

Cake type

Date/time baked

Sample number

Date/time received

Bakery name

Date/time scored

Plant location (address)

Weight as received (unwrapped)

Checker

Diameter (inches)

Examiner

EXPLANATION OF SCORE

EXTERNAL

Volume

The desired volume varies in different sections of the country, and according to different types of cakes. Nevertheless, in order to render an unbiased value for volume on the enclosed score, standards for yellow layer have been established.

Color of Crust

A good live color is desired. It should be uniform and free from spots or streaks.

Symmetry of Form

The ideal layer should be symmetrical without low edges, high or low centers.

Character of Crust

A good crust is thin and tender. It should not be thick or rubbery; neither should it be too tender so that it breaks easily.

INTERNAL

Grain

The grain is the structure formed by the strands of gluten, including the area they surround. The cell structure varies considerably with the different types of cake. Uniformity of size with thin walled cells is desirable. Coarseness, thick cell walls, uneven cell size, and large holes are indicative of poor grain.

Color of Crumb

No definite tint can be established for the color of crumb. However, it should be bright with some luster. The surface should present a uniform shade without streaks or dark patches.

Aroma

Aroma as here used is recognized by the organs of smell. The aroma may be noted as sweet, rich, fresh, musty or flat. The ideal layer has a pleasant, rich, fresh, sweet, natural aroma.

Taste

The most important attribute of good cake is that it has a pleasant and satisfying sweet taste.

Texture

Texture is determined by the sense of touch. It depends on the physical condition of the crumb and is influenced by the grain. It is an expression of the pliability and smoothness of the crumb. The ideal texture is soft and velvety, without weakness, and should not crumble.

FIGURE 19 Cake score report. (Courtesy of American Institute of Baking.)

TABLE 1 Products Made from Soft Wheat

Biscuits	Pancakes
Cookies	Doughnuts
Crackers	Oriental noodles
Wafers	Thickening agent for soups and soup mixes
Pretzels	Crumbs for coating fish and meat products
Cakes	Breakfast cereals
Pastry products from pie crust to sweet Danish pastry	Flat breads
Waffles	Ice cream cones

TABLE 2 Per Capita Consumption of Soft Wheat Bakery Products in the United States (lb/person)

Product	1982	1983	1984	1985	1986	1987[a]
Sweet yeast goods	3.39	3.54	3.67	3.52	3.53	3.54
Doughnuts	0.60	0.64	0.62	0.58	0.57	0.58
All others (coffee cakes, etc.)	2.79	2.90	3.05	2.94	2.96	2.96
Soft cakes	5.54	5.24	5.46	5.34	5.31	5.28
Snack cakes	3.84	3.66	3.94	3.90	3.90	3.88
All other	1.70	1.58	1.52	1.44	1.41	1.40
Pies	2.13	1.98	2.01	1.98	1.95	1.94
Snack pies	1.21	1.13	1.14	1.13	1.11	1.11
All other	0.92	0.85	0.87	0.85	0.84	0.83
Cake–type doughnuts	1.92	1.78	1.80	1.79	1.76	1.75
Cookies	9.57	9.48	10.05	10.17	10.64	10.90
Sandwich	2.57	2.54	2.69	2.72	2.85	2.93
Marshmallow	0.37	0.37	0.39	0.40	0.41	0.43
Wafers for ice cream sandwiches	0.24	0.24	0.258	0.26	0.27	0.27
All other	6.39	6.33	6.72	6.79	7.11	7.27
Crackers	7.91	7.86	7.87	7.76	7.86	7.86
Graham	0.78	0.77	0.77	0.76	0.77	0.79
Saltines	2.55	2.55	2.54	2.50	2.54	2.61
Cracker sandwiches	0.48	0.50	0.49	0.50	0.49	0.51
Cracker meal and crumbs	0.17	0.17	0.17	0.17	0.17	0.16
Pretzels	0.80	0.80	0.79	0.78	0.77	0.75
Other crackers	3.11	3.08	3.10	3.06	3.11	3.04

[a]Estimated by the International Trade Administration (ITA).
Source: Ref. 2.

TABLE 3 Characteristics of Flour Used for Cookies/Crackers Production

Product type	Wheat blend[a]	Typical Ranges: Protein (%)	Ash (%)	Viscosity (°M)[b]	Spread factor (W/T)[c]
Cookies					
Wire cut	SWW-SRW	7.0–8.0	.38–.42	20–40	8.5–10.0
Rotary	SWW-SRW	8.0–9.0	.38–.42	30–50	7.5–9.0
Deposit	SWW-SRW	7.5–8.5	.38–.42	25–45	8.0–9.5
Crackers					
Soda-sponge	SRW-HRW	9.0–10.0	.41–.45	55–70	6.5–8.0
Soda-dough	SRW	8.5–9.5	.39–.43	40–55	7.5–9.0
Snack	SRW	8.5–9.5	.40–.44	40–55	7.5–9.0

[a]SRW = Soft Red Winter; SWW = Soft White Winter; HRW = Hard Red Winter.
[b]MacMichael value.
[c]Width/thickness.
Source: Ref. 4.

TABLE 4 Chlorination Level For Production of Bleached Flour[a]

Oz. Chlorine/ 100 lb Flour		pH
None		5.87
1.00		5.28
1.25		5.15
1.50		5.03
1.75	1.70[b]	4.94
2.00		4.86

[a]Values are valid for flours with 0.36% ash content. The higher the ash content of a flour, the greater the resistance to pH change.
[b]Recommended treatment for optimum baking results.
Source: Courtesy of M. C. Harris.

TABLE 5 Attributes of a Good Cake Flour

Ability to carry the high amounts of sugar in the formula.
Ability to develop a strong network structure without making the product tough.
Fine particle size.
Chlorination (1100–239 ppm), pH range 4.5–5.2.
Low protein content (7–8%) soft red or white flour.

TABLE 6 Characteristics of Water as an Ingredient for Production of Soft Wheat Products

Normal pH range 6.5–6.8
Variable composition dissolved solids range 150–500 ppm
Water hardness[a]
 Soft water
  Dough stickiness
  Lower absorption
  Lower gas retention
 Hard water
  Dough buckiness
  Increase in the mixing time

[a]Water hardness = quantity of calcium carbonate or calcium sulfate in a given sample
Source: Ref. 5.

TABLE 7 Baking Industry's Water Hardness Classification

Classification	Hardness (ppm of $CaCO_3$ or $CaSO_4$)[a]
Very soft	0–15
Soft	15–50
Medium hard	50–100
Hard	100–200
Very hard	Over 200

[a]50–100 ppm considered best for baking.
Source: Ref. 6.

TABLE 8 Major Leavening Gases Used in Leavening of Soft Wheat Products

Gas	Source	Comments
Air	Mixing process	Nitrogen solubility in water is low. Forms nucleation sites during mixing.
Water vapor	Water as an ingredient	Because of high boiling point has a limited effect.
Ammonia	Ammonium bicarbonate	Decomposes completely. No residual salts. Used in products that are baked to near dryness such as cookies and crackers.
Carbon dioxide	Bicarbonates, mainly sodium	Needs an acid for appropriate leavening activity (see Table 10).

Source: Ref. 1.

TABLE 9 Physical Characteristics of Commercial Sodium Bicarbonate

Cumulative percentage retained by	Powdered No. 1 (%)	Fine powdered No. 3 (%)	Fine powdered No. 3DF (%)	Granular No. 4 (%)	Granular No. 5 (%)	Extra coarse granular No. 6 (%)
29 mesh[a]	—	—	—	—	—	—
42 mesh	trace	—	—	—	trace	13.00
65 mesh	NA[b]	—	—	—	27.00	91.00
100 mesh	0.50	0.01	—	1.50	92.50	99.32
170 mesh	17.50	3.50	trace	80.50	99.00	99.60
200 mesh	30.00	9.00	0.5	91.00	99.70	99.65
325 mesh	63.00	30.00	15.0	97.50	99.80	99.76
400 mesh	76.00	36.00	45.00	98.50	99.90	99.86
Apparent density (lb/cu ft)	55.6	47.0	52.5	56.25	45.2	42.8

[a]The various screens are given by mesh number. In cases where mesh does not correspond with screen number, the comparison is as follows: 29 mesh is equal to U.S. Standard No. 30; 42 is equal to U.S. No. 45; and 65 to U.S. to No. 70. In all others the mesh designation and the U.S. Standard numbers are identical.
NA = not available.
Source: Ref. 10.

TABLE 10 Some Common Leavening Acids

Chemical name	Chemical formula	Common name or abbreviation	Rate at room temperature	Neutralizing value	Other characteristics
Monocalcium phosphate monohydrate	$CaH_4(PO_4)_2H_2O$	$MCP.H_2O$	Very fast	80	Used in combination with slow acids to aid nucleation
Monocalcium phosphate anhydrous	$CaH_4(PO_4)_2$	MCP	Slow	83	Slow reacting at room temperature, reacts in oven
Dicalcium phosphate dihydrate	$CaHPO_42H_2O$	DCP	None	33	Slightly alkaline at room temperature. Trigger too late for most products. Used in frozen cake batters.
Sodium acid pyrophosphate	$Na_2H_2P_2O_7$	SAPP	Slow to fast	72	Wide range of reactivity, release CO_2 in the oven, used in production of cake doughnuts
Sodium aluminum sulfate	$Na_2SO_4{\cdot}Al_2(SO_4)_3$	SAS	Slow	100	Little reaction until heated, too slow to be used alone. Used in double acting household bak ing powder. May accelerate fat rancidity.
Sodium aluminum phosphate hydrate	$NaH_{14}A_{13}(PO_4)_8$ $_4H_2O$	SALP	Slow	100	Temperature triggered
Sodium aluminum phosphate anhydrous	$NaH_{14}A_{13}(PO_4)_8$	SALP	Medium	110	Primarily temperature triggered
Potassium acid tartrate	$KHC_4H_4O_6$	Cream of Tartar	Medium fast	50	First leavening acid developed, relatively expensive
Glucono–delta–lactone	$C_6H_{10}O_6$	GDL	Slow	55	Very expensive. Some uses in cake doughnuts.

[a]Parts by weight of sodium bicarbonate which will be neutralized by 100 parts of the baking acid under standardized conditions. Neutralizing values may vary greatly as dough composition is changed.
Source: Refs. 7,9.

TABLE 11 Types of Baking Powder

Type	Ingredient	%
Fast acting		
Formula No. 1		
	Tartaric acid	5.97
	Cream of tartar	44.90
	Sodium bicarbonate	26.73
	Starch	22.40
Formula No. 2		
	Monocalcium phosphate	33.43
	Sodium bicarbonate	26.73
	Starch	39.84
Slow acting		
Formula No. 1		
	Sodium acid pyrophosphate	40.38
	Sodium bicarbonate	30.59
	Starch	29.03
Double acting		
Formula No. 1		
	Monocalcium phosphate	13.28
	Sodium aluminum sulfate	19.92
	Sodium bicarbonate	26.73
	Starch	40.07
Formula No. 2		
	Monocalcium phosphate	6.68
	Sodium aluminum sulfate	21.38
	Sodium bicarbonate	26.73
	Starch	45.21

Source: Ref. 10.

TABLE 12 Factors Affecting Quality of Shortening

Keeping quality (resistance to the development of rancidity)
Flavor (readily absorbs foreign odor)
Shortening value
Creaming properties (ability to incorporate air during mixing)
Range of plasticity (fat should be soft and workable, but not liquid, over the range of temperature at which is likely to be used)
Texture
Color
Sensitivity to light (causes off-flavor and rancidity)

Source: Ref. 14.

TABLE 13 Classification of Shortening Used in Production of Soft Wheat Products

Type	Source	Advantage	Disadvantage
I. Animal Fats			
Butter	Contains > 80% buffer fat	Distinct and intense flavor – image of quality	High cost, relatively short shelf life
Neutral lard	Leaf and kidney fat from hog	Light color, bland flavor short shelf life	High cost, relatively short shelf life
Open kettle rendered lard	Back and leaf fat from hog	Distinct flavor	
Dry rendered lard	Back and leaf fat from hog	Mild flavor, good	
Prime steam lard	Back and leaf fat from hog	Resistance to rancidity, grainy texture	
Hydrogentated lard	Hydrogenated and deodorization of hog fat	Firm, white fat, odorless and tasteless	
Modified lard	Ester intercharge of hog fat	Good creaming property	
Hydrogenated modified lard	Hydrogenation of modified lard	Wide temperature range of plasticity, improved stability	
Stabilized lard	Addition of stabilizers such as BHT to hog fat	Excellent stability	
Oleo oil	Fresh beef and/or mutton fat	Mild flavor	
Rendered pork fat	Intestinal and trimmed hog fat	Good stability due to natural antioxidants	
II. Vegetable Fats and Oils			
Coconut oil,	Copra or dried coconut kernel	Light color, bland, very stable	
Hydrogenated coconut oil	Hydrogenation of whole coconut oil, with melting points of 84°, 92° and 110°F		
Plasticized coconut oil	Chilling and whipping of hydrogenated coconut oil (10% air incorporated)	Improved creaming properties	Melting destroys creaminess

TABLE 13 (Continued)

Type	Source	Advantage	Disadvantage
Palm kernel oil	Oil from kernel of palm fruit	Very similar to native coconut oil	
Refined corn oil	Germ portion of common corn		
Refined cottonseed oil	Cotton seed	Light color, mild flavor	Low storage stability
Refined soya oil	Soybean seeds	Pale color, pleasant flavor	Light sensitive with time, develops beany odor
Refined peanut oil	Peanuts	Bland flavor, very stable against rancidity	
III. Hydrogenated Shortening			
All vegetable hydro–geneated shortenings	A blend of refined vegetable oils especially cottonseed and soya oils	Bland flavor, excellent stability	
All animal hydro–genated shortenings	A blend of animal fats, hydrogenated and modified by inter esterification	Bland flavor, white, smooth texture, good creaming quality and stability	
Mixed vegetable and animal hydrogenated shortenings	Mixtures of animal fats and vegetable oils	Bland flavor, white, smooth texture, good creaming quality and stability	
IV. Compound Shortenings			
All vegetable compounds	A mixture of oils and hard fats, hydrogenated		Low stability
Mixed animal and vegetable compounds	A mixture of animal fats and refined vegetable oils		Low stability
V. Other Fats			
Margarine	Oleo or coconut oil, hydrogenated soya or cottonseed oil	Butter like flavor, good stability	
Mixed fish oil	A mixture of refined, deodorized and hydrogenated fish oil with lard or vegetable oil	Bland flavor	Low stability

Source: Ref. 14.

TABLE 14 Solid Fat Indices of Some Bakery Shortenings at Different Temperature

Type of shortening	S.F.I. 50°F (%)	S.F.I. 70°F (%)	S.F.I. 80°F (%)	S.F.I. 92°F (%)	S.F.I. 100°F (%)	S.F.I. 110°F (%)
Lard	24–26	18–20	12–14	4–5	2.5–3	None
Plastic animal and vegetable shortening	31–35	22–26	20–25	15–19	13–15	8–10
Plastic emulsified animal and vegetable shortening	30–34	22–26	21–25	17–19	13–15	7–9
Plastic vegetable shortening	26–32	17–23	15–21	10–16	9–11	6–8
Plastic emulsified vegetable shortening	26–32	17–23	16–21	11–16	9–11	5–7
Fluid aerating shortening	4–8	3–5	2–4	0.5–2.5	None	None

Source: Ref. 15.

TABLE 15 Typical Screen Analysis of Granulated Sucrose

Tyler screen mesh (% on)	U.S. screen mesh (% on)	Confectioners AA or medium granulated	Sanding	Fine granulated	Extra–fine granulated	Bakers special or fruit granulated	Standard powdered or 6X	Extra–fine powdered or 10X	Fondant and icing	Dehydrated fondant drivert or dri–fond
8	8									
10	10	5.6								
14	16	59.0								
20	20	72.4	9.3	trace						
28	30	7.4	49.2	4.3	0.1	trace				
35	40	0.4	37.6	74.5	13.8	0.4				
48	50		3.3	18.6	40.2	1.7				
80	80		0.3	2.3	40.6	24.8				
100	100					32.2	0.3	0.1		
150	140					31.6	1.8	1.4		
200	200						6.6	2.4		
270	270						8.2	3.0		
325	325						10.8	7.0		
Through last sieve		0.2	0.3	0.3	5.0	9.2	72.3	86.1	99.0[a]	99.0[a]

[a] The extremely fine grain size of fondant and icing sugar makes the regular screen analysis impractical. However, paractically all particles will pass through a 325 mesh standard sieve having 0.0043 cm openings. The average particle size of fondant and icing sugar is abour 20 μm, i.e., 0.0020 cm.

Source: Ref. 17.

TABLE 16 Carbohydrate Composition (glucose and glucose polymers) of Commercially Available Corn Syrups

Type of conversion	DE	Mono–	Di–	Tri–	Tetra–	Penta–	Hexa–	Hepta–	Higher
Acid	30	10.4	9.3	8.6	8.2	7.2	6.0	5.2	45.1
Acid	42	18.5	13.9	11.6	9.9	8.4	6.6	5.7	25.4
Acid–enzymes[a]	43	5.5	46.2	12.3	3.2	1.8	1.5	—	29.5[b]
Acid	54	29.7	17.8	13.2	9.6	7.3	5.3	4.3	12.8
Acid	60	36.2	19.5	13.2	8.7	6.3	4.4	3.2	8.5
Acid–enzymes[a]	63	38.8	28.1	13.7	4.1	4.5	2.6	—	8.2
Acid–enzymes[a]	71	43.7	36.7	3.7	3.2	0.8	4.3	—	7.6[b]

[a]Carbohydrate composition of acid–enzyme syrups will vary as a result of different processes used. The values given here are to be considered only as examples of ranges of values available commercially.
[b]Includes heptasaccharides.
Source: Ref. 18.

TABLE 17 Relative Sweetness of Various Sweeteners[a]

Sweetener	Relative sweetness
Sucrose	100
Dextrose	75
Fructose	130
Maltose	30
Lactose	15
Galactose	60
Honey	100
High fructose corn syrup	
42%	100
55%	100–110
90%	120–160
Molasses	70–90
Invert	110
Brown sugar	85–90
Malt syrup	30
Corn syrup	
42 DE[b]	50–55
62 DE	70
70 DE	75
95 DE	80

[a]Relationship may vary depending on concentration and formulation.
[b]DE = Dextrose equivalent.
Source: Refs. 13,18.

TABLE 18 Effect of Temperature on Sweetener Solubility

Temp. (°F)	Invert (%)	Sucrose (%)	Glucose (%)
68	63	67	48
77	66	68	51
86	75	69	55
104	79	71	62
120	82	73	71
140	87	74	75

Source: Courtesy of American Institute of Baking.

TABLE 19 Water Absorbed by Sweeteners at Various Relative Humidities

% RH	Invert	Glucose	Corn syrup 64DE	Sucrose
91	56	57	45	44
80	34	33	27	27
70	26	24	19	19
10	3	5	3	0

Source: Courtesy of American Institute of Baking.

TABLE 20 Sweeteners and Sugars

	Comparison of sweeteners							
	Sucrose	High fructose corn syrup	Liquid invert (medium)	42 DE[a] corn syrup	62 DE corn syrup	70 DE corn syrup	95 DE corn syrup	Glucose
Solids (%)	100	71	77	80.3	82.0	82.5	71	91
Moisture (%)	—	29	23	19.7	18	17.5	29	8.5
Composition (dry basis, %)								
Glucose	50	50	50	19	39	50	93	100
Fructose	50	42	50	—	—	—	—	—
Maltose	—	3	—	14	28	29	3	—
Higher saccharides	—	5	—	67	33	21	4	—
DE	—	—	—	42	62	70	95	
Fermentables (%)	100	95	100	33	67	82	97	100
Minimum storage temperature (°F)	—	29	32	35–38	35–38	35–38	57–66	60

[a]DE = Dextrose equivalent.
Source: Ref. 19.

TABLE 21 Characteristics of Different Types of Molasses

Type	Total sugars (%)	Sucrose (%)	Invert sugar (%)	Ash (%)	Color
Open–kettle	68–70	43–45	24–26	1–2	Reddish–yellow
First centrifugal	60–66	48–50	13–15	14–15	Light yellow
Second centrifugal	56–60	36–38	20–22	20–22	Reddish
Blackstrap	52–55	25–27	27–29	27–29	Dark brown

Source: Ref. 7.

TABLE 22 Composition of Chocolate Liquor

Composition	Percent
Moisture	1.70
Fat	54.00
Theobromine	1.08
Caffeine	0.42
Other nitrogen substances	0.42
Starch	8.21
Crude fiber	2.65
Other carbohydrates	17.32
Total ash	3.04
Water-soluble ash	0.72
Water-insoluble ash	2.32
Acid-insoluble ash	0.02

Source: Ref. 20.

TABLE 23 Composition of Cocoa Powder (moisture and fat-free basis)

Composition	Natural process (%)	Dutch process (%)
Ash	6.3	10.3
Theobromine	1.9	2.8
Caffeine	0.5	0.5
Tannins	14.6	14.0
Protein	28.1	27.0
Sugar	2.4	2.3
Starch	14.6	14.0
Cellulose	22.0	21.2
Pentosans	3.7	3.4
Acids	3.7	3.4
Other substances	1.2	1.1

Source: Ref. 21.

TABLE 24 Various Types of Chocolate Used in Production of Soft Wheat Products

Type	Characteristics	Application
Natural cocoa powder	Powder ground from the press cake remaining after chocolate liquor is removed. Fat content varies 8–25%. More intense flavor than chocolate.	Icings, compound coatings
Dutched (alkalized) cocoa powder	Dutching darkens the color and improves solubility of chocolate. The process involves addition of alkaline solutions to partially roasted beans. The roasting continues until desired color level is obtained.	
Lightly alkalized		Cake mixes, bakery ingredients
Moderately alkalized		Icings, coatings
Heavily alkalized		
Chocolate, chocolate liquor chocolate paste	Obtained by grinding cocoa nibs. Contains not less than 50% cocoa butter.	Baking chocolate
Sweet chocolate	Mixture of chocolate with sugar and/or dextrose with or without addition of cocoa butter.	
Milk chocolate	Mixture of chocolate, sugar, and milk solids (no less than 12%).	

Source: Ref. 22.

TABLE 25 Advantages of Enrobing of Soft Wheat Products

Improves appearance
Enhances flavor
Improves eating quality
Improves shelf life
Good moisture barrier
Prevents drying out
Maintains crispness
Nonsticky surface
Structural strength

Source: Ref. 23.

TABLE 26 Formulation of Various Coatings Used on Soft Wheat Products

Ingredients	Chocolate coating (%)	Chocolate type compound coating (%)	Cocoa compound coating (%)	White and pastel color coating (%)
Chocolate liquor (50% cocoa butter)	15	15	—	—
Cocoa powder (11% cocoa butter)	—	—	8	—
Cocoa butter (90°F melting point)	29	—	—	—
Domestic hard butter (derived from soybean or cottonseed oil, melting point of 100°F)	—	29	—	—
Lauric hard butter (derived from palm kernel and coconut oil, melting point range 92–109°F)	—	—	36	35
Sugar	44	44	44	53
Nonfat dry milk	12	12	12	12
Lecithin	>1	>1	>1	>1
Color and flavor	—	—	—	Some

Source: Ref. 23.

TABLE 27 Typical Composition of Dairy-Based Ingredients

Ingredient	Moisture (%)	Protein (%)	Fat (%)	Lactose (%)	Ash (%)	Calcium (%)	Phosphorus (%)
Nonfat dry milk (NFDM)	3.0	35.9	0.8	52.3	8.0	1.31	1.02
Dried buttermilk	2.8	34.3	5.3	50.0	7.6	1.25	0.97
Casein	7	88	1	—	4	—	—
Caseinate	3–5	90–94	0.7–1.0	—	6–7	—	—
Dried whey	4.5	12.9	1.1	73.5	8.0	0.65	0.59
Whey protein concentrate	2	20–60	2–9	18–60	3–18	—	—
Lactose	0.5	0.1	0.1	99.0	0.2	—	—

Source: Ref. 25.

TABLE 28 Utilization of Whey in Foods

Food	Quantity of whey solids (%)	Contributions of whey components
Baked goods–sweet goods crackers (percent of flour weight)	3.0	Flavor, texture, shorter dough time, improved keeping quality
Dry mixes	10.0	Tenderizing, color, flavor
Confections	10.0	Flavor, body, moisture retention
Icings, frostings	6.0	
Batter mix (for frying)	5.0	Color, flavor
Whey–soy blends for food manufacture (use fat–free soy flour)	40.0	Masks soy flavor, high protein

Source: Ref. 26.

TABLE 29 Composition of Egg Products

Component	Liquid whole egg (%)	Dried whole egg (%)	Liquid egg white (%)	Dried egg white (%)	Liquid egg yolk (%)	Dried egg yolk (%)
Solids	27	96	14	95	51	96
Protein	13.3	47.2	11.6	78.6	16.7	31.4
Fat	11.5	40.8	—	—	31.6	59.4
Nitrogen–free extract	1.10	3.90	0.80	5.42	1.2	2.25
Glucose	0.32	1.07	0.40	2.71	0.21	0.40
Ash	1.00	3.55	0.80	5.42	1.50	2.82
Lecithin	1.52	5.41	—	—	4.18	7.84

Source: Ref. 7.

TABLE 30 Types of Egg Products available to Bakers

Type	Container	Conversion factor[a]
Shell eggs, fresh or storage	30–doz case	0.84
Bulk liquid eggs	Tank truck	1.00
Whites		1.00
Yolks		1.00
Whole eggs		1.00
Frozen egg products	Friction–lid metal cans	
Whites		1.00
Whole eggs		1.00
Whole eggs plus extra yolk		1.00
Yolks		1.00
Whole eggs plus corn syrup[b]		1.11
Egg yolks plus salt[b]		1.11
Egg yolks plus sugar[b]		1.11
Whole eggs plus salt[b]		1.11
Dried egg products[c]	Bags, boxes, and drums	
Whites, spray dried		7.24
Whites, pan dried (flake albumen)		7.00
Whole eggs, standard		3.70
Whole eggs, glucose–free		3.70
Yolk solids, standard		2.72
Yolk solids, glucose–free		2.72
Whole egg solids, plus sugar[d]		2.33
Whole egg solids, plus corn syrup solids[d]		2.33
Yolks plus sugar[d]		1.49
Yolks plus corn syrup[d]		1.49

[a]Multiply kg of product by this factor to obtain equivalent kg of bulk liquid egg, white, or yolk.
[b]Based on frozen product containing 10% (dried weight) of carbohydrate, this will vary depending on supplier.
[c]In addition, dried whole eggs and dried yolks (any style) can be procured in free– flowing form.
[d]Based on dried products containing 33% carbohydrate. May vary among different suppliers.
Source: Ref. 7.

TABLE 31 Natural Spices and Flavorings Most Commonly Used in Soft Wheat Products

Type	Plant source	Origin	Compatibility	Application
Cinnamon	Cinamomum cassia	China, Indochina, Indonesia	Sugar, chocolate, coffee, apple, peach, pear	Cookies and cakes, rolls, danish pastry
Nutmeg	Seed of a peachlike fruit tree	Tropics, Indonesia, East and West Indies	Sugar, apple, cherry, peach, pear, coffee	Cookies, cakes, pies, doughnuts
Vanilla	Fruit of an orchid	Tropic and semi–tropics	Sugar, fruity flavors	Cakes, cookies
Mace	Skin of nutmeg	As nutmeg	Sugar, custard, cherry	Cookies, cakes, pies, fillings, eclairs
Cloves	Unopened bud of clove tree	Indonesia, Zanzibar, Madagascar	Sugar, various stewed fruits	Cakes, fillings
Ginger	Root of a tuberous perennial plant	Tropics, Jamaica, India	Sugar pies	Fillings, cookies, cakes, wafers,
Allspice	Pea–size fruits	Jamaica, Mexico, Guatemala	Sugar, blueberry, pineapple	Cookies, cakes, pies, tarts, fruit cakes, doughnuts
Cardomom	Fruit of a plant from the ginger family	India, Sri Lanka, Guatemala	Sugar, blueberry, grape, apple, pumpkin	Danish pastry, fillings, pies
Poppy	Tiny seeds of a plant from the poppy family	Netherlands, Poland, Argentina		Toppings for rolls, crackers, pastries, cakes, doughnuts
Caraway seed	Fruit of a plant from the parsley family	Netherlands, Poland	Sugar, apple, peach, apricot, rye	Corn muffins, cheese rolls, cakes, cookies
Aniseed	Fruit of a plant from the parsley family	Egypt, Spain, Mexico, Turkey	Sugar, most stewed flavor sweet rolls	Icings and fillings, coffee cakes

Source: Ref. 7.

TABLE 32 Uncertified Color Additives[a]

Colorant	Restrictions
Annatto extract	—
β–apo–8'–Carotenal	15 mg per lb or pt
β–Carotene	—
Beet powder	—
Canthaxanthin	30 mg per lb or pt
Caramel	—
Carrot oil	—
Cochineal extract (carmine)	—
Cottonseed flour (toasted, etc.)	—
Fruit juice	—
Grape color extract	—
Paprika and its oleoresin	—
Riboflavin	—
Saffron	—
Titanium dioxide	Maximum of 1%
Turmeric and its oleoresin	—
Vegetable juice	—

[a]Uncertified colors are derived from plant or mineral raw materials by various processing techniques and have been approved for food use in many cases because of a long history of satisfactory prior usage.
Source: Ref. 10.

TABLE 33 Certified Color Additives[a]

Permanent listing	Provisional listing
FD&C Red No. 3	FD&C Yellow No. 6 Lake
FD&C BLue No. 2	FD&C Red No. 3 Lake
FD&C Yellow No. 5	FD&C Blue No. 1 Lake
FD&C Green No. 3	FD&C Blue No. 2 Lake
FD&C Blue No. 1	FD&C Green No. 3 Lake
FD&C Red No. 40	FD&C Yellow No. 5 Lake
FD&C Red NO. 40 Lake	
FD&C Yellow No. 6	

[a]May be used for coloring soft wheat products generally in amounts consistent with good manufacturing practices.
Source: Ref. 10.

TABLE 34 Formula for Production of Icings

	Flat type	Chocolate fudge	Vanilla butter creme	Vanilla fudge	Chocolate flavored butter creme	Topping marshmallow
Powdered sugar	82	63	59	73	50	51
Invert sugar	4	4	5	3	5	25
Corn syrup	4	4	—	3	—	—
Water	10	10[a]	8	8[a]	8	22
Flavoring	as desired	—	vanilla	vanilla	vanilla	as desired
Chocolate fudge base	—	15	—	—	—	—
Shortening (emulsified type)	—	4	20	9	22	—
Salt	—	as desired	—	1	—	—
Butter	—	—	8	3	8	—
Nonfat dry milk	—	—	—	3	2	—
Dutched cocoa	—	—	—	—	5	—
Gelatin	—	—	—	—	—	2

[a]Boiling water.
Source: Ref. 28.

TABLE 35 Typical Formulas for Production of Various Types of Crackers

				Saltine		Cheese snacks	
Ingredients	Graham	Cheese cracker straight dough	Lunch biscuit	Sponge	Dough	Sponge	Dough
Flour	80	100	100	65	35	75	25
Graham flour	20	—	—	—	—	—	—
Shortening	12	12	15	—	—	12	—
Lard	—	—	—	—	11	—	—
Sugar	25	—	3	—	—	—	—
Molasses	5	—	—	—	—	—	—
Invert syrup	5	—	—	—	—	—	—
Cheese	—	25	—	—	—	—	25
Salt	1	1	1	—	2	—	1
Soda	variable	0.5	—	—	variable	—	0.5
Calcium phosphate	—	—	2.5	—	0.5	—	0.25
Ammonium bicarbonate	0.5	—	—	—	—	—	0.25
Water	20	30	30	29	—	25	5
Sponge	—	10	—	10	—	—	—
Milk powder	—	—	4	—	—	—	—
Yeast	—	—	—	0.25	—	0.25	—
Paprika	—	—	—	—	—	1	—

Source: Ref. 14.

TABLE 36 Typical Formulas for Various Types of Cookies

Ingredients	Vanilla wafer	Chocolate wafer	Vanilla sugar cookies	Oatmeal cookie	Lemon crisp	Lemon snap	Chocolate snap	Short bread	Chocolate chip cookie	Chocolate nut cookie	Ginger snap	Semi-hard sweet	Social tea	Petit boure
Flour	100	100	100	100	100	100	100	100	100	100	100	100	100	100
Powdered sugar	70	70	45	70	60	45	35	60	45	55	19	25	20	28
Shortening	30	30	35	35	25	15	—	18	50	50	22	15	15	—
Lard	—	—	—	—	—	—	20	—	—	—	—	—	—	—
Butter	5	—	—	—	10	—	—	12	—	—	—	—	—	20
Invert syrup	10	8	7	—	—	—	3	—	0.8	—	—	10	15	—
Corn syrup	—	—	—	—	—	—	6	—	—	—	—	—	—	—
Molasses	—	—	—	—	—	—	—	—	—	—	50	—	—	—
Honey	—	—	—	—	—	—	—	—	—	—	—	4	—	—
Milk powder	5	6	3	5	2	—	6	1	2.5	2.5	—	—	1.5	—
Eggs (frozen or powdered)	0.5	2	5	7	4	1.5	—	6	—	10	—	—	1	—
Water	45	—	30	30	35	45	25	15	15	25	5	16	10	23
Soda	1	—	1	0.8	0.8	1.2	1.5	2	—	0.5	2	0.75	0.8	0.5
Calcium Phosphate	0.5	—	0.5	0.25	0.5	—	—	—	0.25	0.25	—	—	—	0.2
Ammonium Bicarbonate	0.5	—	1	0.5	1	0.2	—	0.5	—	0.25	—	0.6	0.2	0.2
Lecithin	—	—	—	—	—	—	—	—	—	—	—	—	0.2	—
Bisulfite	—	—	—	—	—	—	—	—	—	—	—	0.05	0.1	—
Corn starch	—	7	—	—	—	—	—	—	—	—	—	5	2.5	—
Potato flour	—	—	—	—	—	—	—	—	—	—	—	—	—	14
Salt	2	1.5	2.5	1.5	0.8	1	1	1.2	1	1.25	2	1.5	1	0.6
Cocoa powder	—	10	—	—	—	—	10	—	—	10	—	—	—	—
Chocolate chip	—	—	—	—	—	—	—	—	50	—	—	—	—	—
Chopped nuts	—	—	—	—	—	—	—	—	12.5	10	—	—	—	—
Raisin	—	—	—	25	—	—	—	—	—	—	—	—	—	—
Rolled oats	—	—	—	45	—	—	—	—	—	—	—	—	—	—
Ginger	—	—	—	—	—	—	—	—	—	—	1	—	—	—
Flavor	Butter Vanilla	—	—	—	Lemon Oil	Lemon Oil	Vanilla	Vanilla	Butter Vanilla	Butter Vanilla	Butter Lemon	—	Butter Vanilla	—
Color	—	—	—	—	Yellow	—	—	—	—	—	—	—	Butter	—
Spice	—	—	—	Cinnamon Nutmeg	—	—	—	—	—	—	—	—	—	—

Source: Ref. 14.

TABLE 37 Typical Formula for Production of a Co-extruded Fig Bar

Ingredient	Dough	Jam
Flour	100	—
Fig	—	100
Sugar	40	50
Invert syrup	5	20
Shortening	20	—
Eggs	8	—
Milk powder	3	—
Water	20	—
Soda	0.5	—
Calcium phosphate	0.5	0.25
Ammonium bicarbonate	0.5	—
Salt	0.1	0.75
Glucose	—	30

Source: Ref. 14.

TABLE 38 Functional Categories of Ingredients Used in Cake Production

Ingredient	Toughener	Tenderizer	Moistener	Dryer	Flavorer
Flour	X			X	
Crystalline sugar				X	X
Liquid sugar or syrup			X		X
Shortenings		X	X		
Milk solids	X			X	X
Egg white	X		X		
Egg yolk		X	X		
Whole egg	X	X	X		
Dried egg	X			X	
Cocoa powder				X	X
Chocolate		X		X	X
Leavening agents		X			
Salt					X

Source: Ref. 31.

TABLE 39 Formula for Production of Four Popular Cakes[a]

	White layer cake	Yellow layer cake	Chocolate cake	Devil's food cake
Flour (low protein and chlorinated)	100	100	100	100
Sugar	100	140	150	155
Shortening (hydrogenated)	45	50	60	55
Egg white	60	—	—	15
Whole egg	—	60	70	60
Evaporated milk	40	—	—	—
Whole milk	—	110	—	—
Non fat dry milk	—	—	15	22
Baking powder	6	6	6	4
Salt	2.5	4	4	5
Chocolate liquor	—	—	20	—
Cocoa powder	—	—	—	20
Water	—	—	100	130

[a] On flour basis.
Source: Ref. 13.

TABLE 40 Typical Formula for Production of Pie Crust

Ingredient	%[a]
Flour (low protein)	100
Shortening	65
Salt	2–4 (dissolve in water)
Sugar	3–8
NFMS	2–6
Water (chilled)	25–35

[a] On flour basis.

TABLE 41 Typical Hard Pretzel Formula

Ingredient	%[a]
Sponge	
Flour	20
Water	10
Yeast	0.63–1.25
Dough	
Flour	80
Water	25
Salt	1.2
Shortening	3

[a]On flour basis.

TABLE 42 A Typical Formula for Production of Puff Pastry

Ingredient	%[a]
Flour	100
Salt	2
Shortening (including that added during rolling)	100.
Water	20

[a]On flour basis.

TABLE 43 Donut Mix Composition

Ingredient	%
Bread flour	35.00
Cake flour	65.00
Fine granulated sugar	35.00
Dextrose	3.5
Salt	1.5
Mace	0.7
Egg yolk solids	3–5
Nonfat dry milk	5–8
Shortening	6
Sodium bicarbonate	1.5
Sodium acid pyrophosphate	2
Defatted soy flour	0–6
Vanilla	to suit

Source: Ref. 13.

TABLE 44 Yeast-Raised Doughnut Mix Formula

Ingredient	%[a]
Flour	100
Sugar	8–12
Shortening	8–14
Nonfat dry milk	2–6
Egg solids	2–4
Salt	2–3

[a]On flour basis.
Source: Ref. 13.

V. REFERENCES

1. Hoseney, R. C., *Principles of Cereal Science and Technology*, American Association of Cereal Chemists, St. Paul, MN, 1986, pp. 245–276.
2. Semling Jr., H.B., Continued growth in 1988, *Baking Industry*, (January):21(1988).
3. Faridi, H., and Finley, J. W., Improved wheat for baking, *CRC Critical Review in Foods and Nutrition*, 28: 175–209(1989).
4. Tanilli, V. H., Characteristics of wheat and flour for cookie and cracker production, *Cereal Foods World*, 21: 624–628, 644(1976).
5. Collins, J. J., Water as ingredient, *Proceedings of the American Society of Bakery Engineers*, 1978, pp. 36–41.
6. McFaul, R., Quality of water for the baking industry, *Proceedings of the American Society of Bakery Engineers*: 70–75(1981).
7. Matz, S. A., and Matz, T. D., *Cookie and Cracker Technology*, AVI, Westport, CT, 1978.
8. Manley, D. J. R., *Technology of Biscuits, Crackers and Cookies*, Ellis Horwood Ltd. Publ. Co., Chichester, UK, 1983.
9. Kickline, T. P., and Conn, J. F., Some functional aspects of leavening agents, *Bakers Digest*, *44*(4): 36–40(1970).
10. Matz, S. A., *Ingredients of Bakers*, Pan. Tech International, McAllen, Texas, 1987.
11. Labaw, G. D., Chemical leavening agents defined, *Proceedings of the American Society of Bakery Engineers*, 1977, pp. 77–81.
12. Reimann, H. M., Chemical leavening, *Proceedings of the American Society of Bakery Engineers*, 1982, pp. 83–89.
13. Pyler, E. J., *Baking Science and Technology*, Siebel Publ. Co., Chicago, IL, 1988.
14. *Biscuit and Cracker Handbook*, Biscuit and Cracker Manufactures Association, Washington, D.C., 1981.
15. Ellinger, R. H., The development and uses of fluid shortenings, *Bakers Digest, 36*(6): 65–69(1962).
16. Dubois, D. K., Sweeteners: Classification, Properties, and Functions in Bakery Foods, *American Institute of Baking Technical Bulletin*, Vol. VI, Issue 6, Manhattan, KS, 1984.
17. Bohn, R. T., and Junk, W. R., Sugars, in *Bakery Technology and Engineering*, S. A. Matz (Ed.), AVI Publ. Co., Westport, CT, 1960, pp. 170–187.
18. Nesetril, D. M., Corn sweeteners, their type and uses, in *Baking, Bakers Digest*, 41(3): 28–32(1967).
19. Labaw, G. D., High fructose corn syrup for bread products, *Bakers Digest*, 51(5): 104–106(1977).

20. Schoen, M., in *Chemistry and Technology of Food and Food Products*, 2nd ed., M. B. Morris (Ed.), Interscience, New York, 1951.
21. Fincke, H., *Handbuch der Kakaoerzeugnisse*, 2nd ed., Springer, Berlin and New York, 1965.
22. *Bakers Digest, Source '85*, Sosland Pub. Co., Shawnee Mission, KS, 1985, p.17.
23. Wing, D. H., Enrobing of baking products, *Proceedings of the American Society of Bakery Engineers*, 1975, pp. 136–141.
24. Strong, L. R., The functional properties of salt in bakery products, *Bakers Digest, 43*(1): 55–57(1969).
25. Dairy-based ingredients for food products, DRINK, UDIA, *Source '85*, Sosland Pub. Co., 1985, p.39.
26. Gillies, M. T., *Whey Processing and Utilization*, Noycs Data Corporation, Park Ridge, NJ, 1985.
27. McGee, O. L., Bakers helper, In *Baking Science and Technology*, E. J. Pyler, Siebl Publ. Co. Chicago, IL 1988.
28. Lachman, A., and Voll, H., Structure and behavior of icings, *Bakers Digest, 43*(2): 40–45,73(1969).
29. Hoseney, R. C., Wade, P. and Finley, J. W., Soft wheat products, in *Wheat Chemistry and Technology*, 3rd ed., Y. Pomeranz (Ed.), American Association of Cereal Chemists, St. Paul, MN, 1988, pp. 407–456.
30. Loving, H. J. and Brennis, L. J., Soft wheat uses in the United States, in *Soft Wheat: Production, Breeding, Milling, and Uses*, W. T. Yamazaki and C. T. Greenwood (Eds.), American Association of Cereal Chemists, St. Paul, MN, 1981, pp. 169–207.
31. Coughlin, F. J., *The Evolution of a Cake Formula and Formula Balance*, Procter and Gamble, Cincinnati, OH, 1947.
32. McGill, E. A., Puff pastry Production, bakers Digest, 49(1), 28–38, 1975.
33. Flick, O. B., Production of yeast-raised doughnuts, *Proceedings of the American Association of Bakery Engineers*: 174–183(1971).
34. Allenson, A., Refrigerated doughs, *Bakers Digest*, 56(5): 22–24(1982).
35. Chen, R. W., Refrigerated doughs, *Cereal Foods World*, 24(2): 46–47(1979).
36. Cleven, F., and Weber, L., A new method for the continuous production of puff and Danish pastry dough, *Bakers Digest*, 51(5): 138 (1977).

18

Breakfast Cereals

Ronald E. Tribelhorn
Colorado State University, Fort Collins, Colorado

I. INTRODUCTION

In this century, cereal grains have found significant uses as breakfast foods. Breakfast cereal technology has evolved from the simple procedure of milling grains for cereal products that require cooking to the manufacturing of highly sophisticated ready-to-eat products that are convenient and quickly prepared.

Some of the first cereal products to be sold were milled grains of wheat and oats that required further cooking in the home prior to consumption. Efforts have been made to reduce the amount of in-home preparation required for these products through methods of precooking or special additives. Also, the advent of the microwave oven has made home use of these cereals more convenient.

Published information on the various processes and formulations of products that can be classified as breakfast cereals is limited. The proprietary nature of the breakfast cereal industry limits the information base to patent records and publications by individuals not directly associated with the industry. This chapter summarizes to the largest extent possible state-of-the-art breakfast cereal technology, explores current processing techniques, formulations, and products, and provides a simplified guide to processing cereal grains into breakfast cereal food products.

II. CEREAL CATEGORIES

Breakfast cereal products can be categorized into types based on use or physical nature of the product. Breakfast cereals will be discussed according to the following categories and respective definitions.

1. Traditional cereals that require cooking—cereals which are sold in the market as processed raw grains. Wheat or oat cereals typically consumed hot would be examples of this type of cereal.
2. Instant traditional hot cereal—cereals which are sold in the market as cooked grains and requiring only boiled water for preparation. Wheat and oat cereals are again examples.
3. Ready-to-eat cereals—a group of cereals manufactured from grain products having been cooked and modified such that they may be subdivided into flaked, puffed, or shredded products.
4. Ready-to-eat cereal mixes—cereals combined with other grains, legumes, or oil seeds and dried fruit products. Granola cereal mixes best describe this type of breakfast cereal.
5. Miscellaneous cereal products—cereal products which cannot be included in any of the above types because of specialized process or end use. Cereal nuggets and baby foods are products which may be considered in this category.

Familiarity with the U.S. cereal industry and products available permits one to identify additional products falling into the above categories.

III. INGREDIENTS AND ADDITIVES

Most breakfast cereal products contain large amounts of cereal grains and have very few additives. Additives are mostly used to improve the texture of the cereal or to change functional characteristics of the final product. A list of additives and common uses in breakfast cereals is given in Table 1.

Since cereal products are often the only food consumed for the morning meal, they are supplemented with vitamins and minerals to improve the nutritional quality. Addition of these ingredients is based on a specific percentage of the daily required intake for adults. The reader is referred to applicable literature on the addition of vitamins and minerals to cereal products for specific information.

IV. CEREALS REQUIRING COOKING

Breakfast cereal products which require cooking are mostly made form wheat and oats. Corn has been used extensively in the southern United States as a cooked grit. Few hot breakfast cereals are made from rice or rice products.

A. Wheat-Based Products

Wheat-based cereal products are made from wheat middlings (farina) obtained from the milling process. They are mostly endosperm pieces which are free from bran and germ. Careful attention has been given to the particle size of these middlings with the most acceptable products having the following minimum particle size specifications.

Through U.S. #20—100%
Through U.S. #45—<10%
Through U.S. #100—<3%

"Quick cook" cereals are prepared by steaming grain particles at elevated temperatures and pressure followed by flaking. Generally, this method of processing reduces the preparation time of the cereal to about one-third the time of the traditional raw cereals. Particle size during steam cooking is critical to an acceptable product since the extent of gelatinization during processing and the mouthfeel are affected by particle size (1). The smaller the particle, the more surface area available for heat transfer and, thus, the better degree of cook attainable.

Hard wheats have been found acceptable for making farina-based cereals since the end product does not get pasty (1). One disadvantage to hard wheat is the longer process time required to achieve a good cook and to gelatinize the starch in the wheat (2). Generally, no other cereal or cereal fractions are used as ingredients in these formulations.

Disodium phosphate has been used as a suitable additive to reduce the in-home preparation time to cook wheat cereals (3). Disodium phosphate modifies the pH of the mixture and promotes hydrolysis of the starch during heating. Soaking farina in a solution made of sodium hydroxide and ammonium salt during processing has also been found effective in reducing the time required to cook wheat products (4).

Gums incorporated into farina cereals are said to reduce cook time (5). The gum creates a viscous liquid which surrounds the particles and increases the heat transferred during cooking and adds to the overall texture of the cooked cereal. Special thickeners such as carboxymethylcellulose have also been used for improving texture and cookability of cereals (6). The thickeners are especially effective for instant cereal products since their presence results in prepared cereal consistencies similar to those made using traditional methods.

Special hydration mediums which are mixtures of sodium chloride, sodium tripolyphosphate, sodium bicarbonate, sodium carbonate, tetrasodium EDTA, and water have been effective in reducing cooking time of whole cereal grain products (7). These hydration mediums are claimed to tenderize cereal tissues at lower temperatures and pressure than the traditional steam cook process and to reduce the amount of in-home preparation time in half. The medium is removed by washing at the end of the process followed by drying.

Proteolytic enzymes have been used to prepare an instant type farina product (1). Proteolytic enzyme treatment of farina permits water access to starch granules within the middlings particles, thus reducing the time for water absorption and gelatinization.

Fruit-flavored farina products are gaining consumer acceptance. Techniques have been identified to make acceptable fruit flavored farina products through the addition of thickeners, preservatives, and special processing techniques [8.9]. General process information for farina based cereal products is listed in Table 2.

B. Oat-Based Products

Breakfast cereals are easily made from oat cereal (1). They have acceptable flavors and good nutritional qualities. To make these cereals, oats are first cleaned to remove foreign materials, then roasted. Roasting serves to soften the endosperm of the cereal and make the hull brittle, which permits decortication and easy separation of the groats from the hulls. The groats are then crushed between two steel rolls to form flakes. The flaked oats are cooled, mixed with any additives, and packaged (Fig. 1). "Quick cook oats" are made in a fashion similar to the traditional method outlined above. A cutter machine is employed between the dehulling and rolling steps, which slices the groats into pieces measuring from one-third to one-half the thickness of the whole groat. This step enables more surface area to be exposed to water and heat during the in-home preparation, reducing the cooking time to approximately 5 minutes. Traditional in-home preparation of oats requires 10– 15 minutes (1).

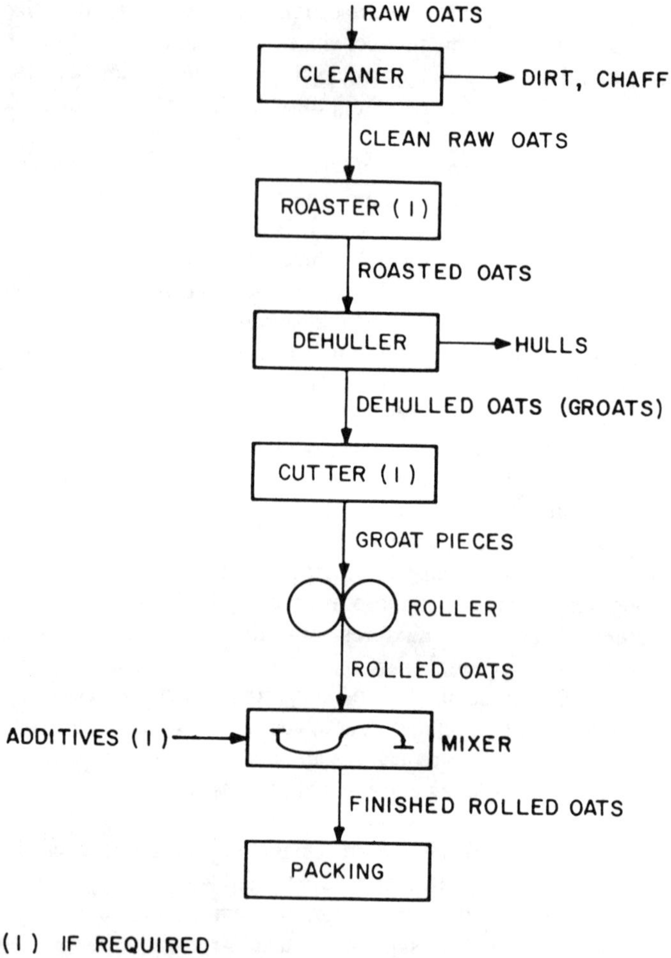

FIGURE 1 Process line to make oat cereals.

Oat-based hot cereals, are manufactured from a single cereal component. Dehulled oats or "oat groats" is the most acceptable raw material for manufacture of this product.

Additives to oat-based cereals are primarily used to improve the texture of the cereal when prepared. Pregelatinized starch (10), gums (11), and cereal hydrolysate (12) have been used effectively to improve the texture of oat cereals. The hydrolysate and gums both add consistency and thickness to the product, making it smooth and creamy in texture.

Proteinaceous fractions of oat groats have been used to prepare a quick-cooking oat cereal (13). These fractions when combined with processed oats at a level between 3.5 and 5.5% by weight results in a mixture that requires only the addition of boiling water to prepare.

A critical problem with oat cereals is the tendency toward thick consistency and excessive pastiness after extended heating. This is a problem for restaurants and institutional facilities where oatmeal is served from steam trays over several hours. Milk and cream addition (14) and monoglycerides (15) have been used effectively to reduce the sticky nature of the hydrated oat product. Specific process information for oat cereals is given in Table 3.

V. READY-TO-EAT (RTE) CEREAL PRODUCTS

The largest market for breakfast cereal products is for the ready-to-eat varieties. Consumer popularity of RTE cereal types is due to the minimal time required to prepare these products. RTE cereals are generally manufactured from cereal grains as flakes, shreds, or shaped materials.

A. RTE Cereal Grains—Flakes

Flaked cereal grains were the first form of ready-to-eat cereal products available to the consumer. Processes to make the flakes are simple and result in products that are well cooked and have acceptable flavor.

The traditional method (1.16–18) of manufacturing a flaked cereal product begins by cleaning the cereal grain, followed by decortication and milling to break the whole grain into pieces that are one-third to one-half the size of the original whole grain. These pieces are mixed with other ingredients as required and steamed for 2 hours or longer under pressure. The steamed mass is then broken into small segments and dried under carefully controlled conditions. The grit pieces are then either tempered for 24 hours (17) or directly flaked between steel rolls. The resulting flakes are then dried and toasted under high temperature to give a suitable flavor and color. A traditional process flow is shown in Figure 2.

Variations of this process have been used to improve and introduce new flaked cereal products (17–21). the primary difference between these processes and the traditional flaking process is the replacement of the steam-cooking step with extrusion processing. In additional to cooking the grains, extrusion allows formation of uniform pellets and thus flakes, which can be made from single or multicomponent formulations. A typical process flow to make a flaked product from the output of an extruder is also shown in Figure 2.

Flaked cereals are manufactured mostly from corn and wheat. A typical flaked cereal might have as a formulation approximately 90% cereal component plus 8% sugar, 1% salt, and 1% malt. Recently developed flaked breakfast cereals have included multiple cereal formulations and cereal formulations in combination with other grain seeds such as soybean.

Efforts have been made to use oats in flaked products (18.19). In one instance, a mixture of oat flour with other cereal grains was used to make a flaked cereal (19). Ranges for ingredients are as follows:

Oat flour: 60–70%
Rice flour: 7–12%
Soy flour: 5–10%
Sucrose: 5–15%
Lecithin: 0.5–.15%
Salt: 2–4%
Casein: 1.5–3.5%

Formulations for flaked cereals made directly from the output of an extruder have also been considered (22). In this case, oat type flakes were being manufactured from a formulation consisting of 57% oat flour, 33% water, and 10% sugar.

Flaked cereal products contain few additives since the process does not require specialized treatment of raw materials. Sugar and malt are frequently added to flaked cereal formulations in order to improve the flavor. The presence of sugar and malt in the formulation also promotes the Maillard reaction, which adds color and flavor to the flakes.

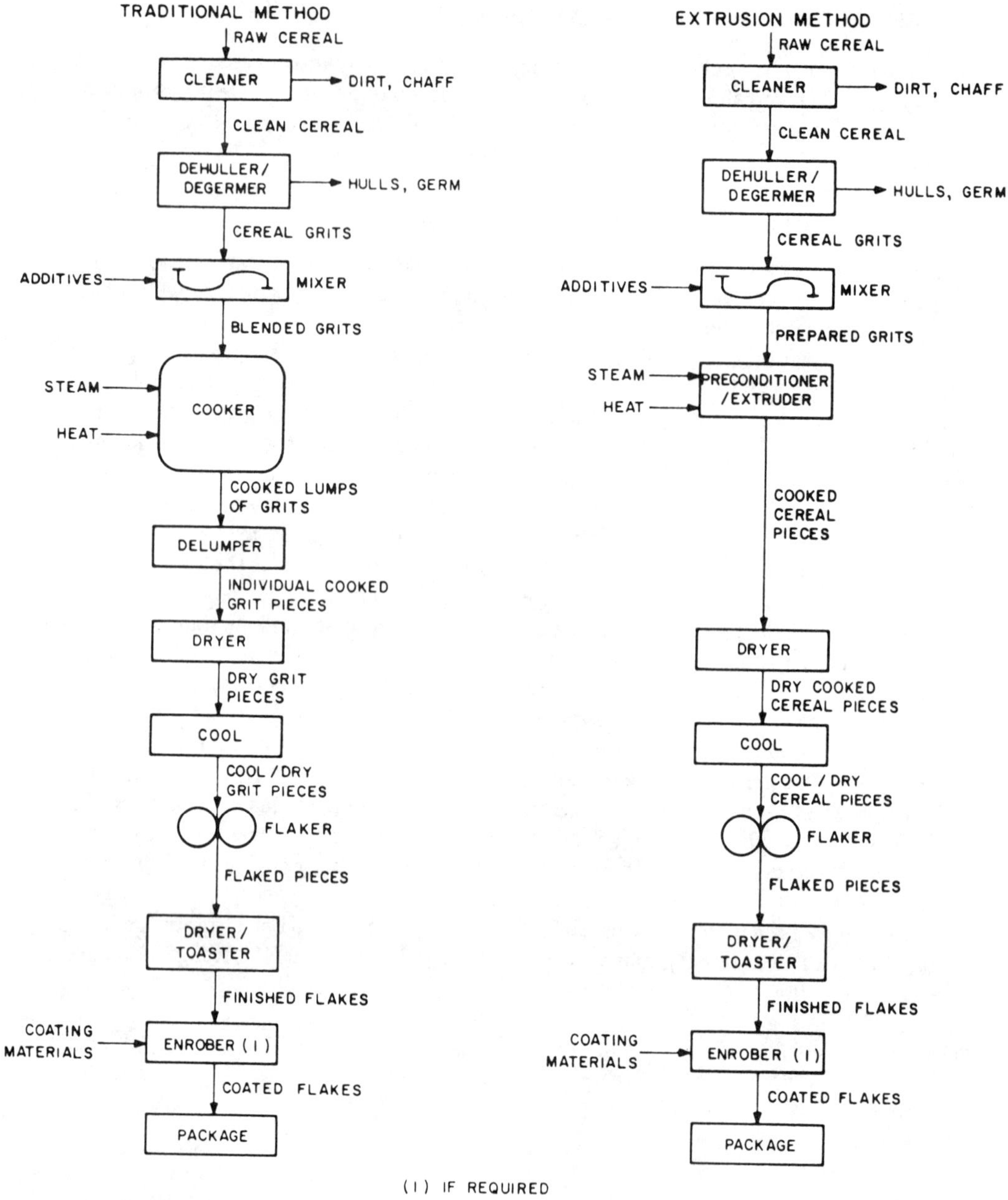

FIGURE 2 Process lines for cereal flake production.

An organic silicone-based additive which reduces the degree of liquid absorption when milk is added to a flaked cereal has been proposed (23). While results showed promise, use of this additive has been limited.

Papain has been used in formulations which incorporate soybean to enzymatically treat the bean and improve processing capability (24). sodium bicarbonate has been used to reduce unacceptable flavors in formulas containing soybean (25). Processing descriptions, additives, and ingredients found in flaked products are shown in Table 4.

B. RTE Cereal Grains—Puffed Products

Cereal grains are well suited to expansion type processes for the manufacture of breakfast cereals. The starch contained in these grains can be modified easily by forcing water into the starch molecules at high pressures and temperatures. Rapid release of this high pressure and water results in rupture of the starch granules forming a cellular structure piece which is uniformly delicate and brittle in texture. Puffed breakfast cereal products are popular because processes used permit shaping of the final product, making the product more appealing to the consumer.

Puffing of cereals was originally done using puffing guns (1,16). These devices were designed to heat and pressurize cereal products with steam, hold the cereal at these conditions over a specific period of time, and then release the steam pressure quickly from the chamber. The rapid release of pressure coupled with the flashing of the moisture results in an expanded or puffed product. Several variations to the traditional puffing apparatus have been proposed to improve (28) or change the traditional batch process puffing gun into a continuous process (26,27,29) and to improve operational conditions and controlling parameters for these processes (30).

Extrusion has also been found useful in making puffed or expanded cereal pieces. The extruder permits shaping of cereal grains, as well as cooking and mixing of cereal formulations. Extrusion has been used to process grains prior to puffing gun expansion and to make uniformly shaped spherical breakfast cereals (18,31). Direct expansion from the extruder has also been proposed to eliminate the puffing gun step (32–34). Dry raw materials are introduced into the extruder or, in some instances, into a preconditioner before extrusion. The raw cereal grains are allowed to pass through the extruder, which is normally operating at a high temperature and pressure. After passing the extruder screw, the mixed ingredients undergo a rapid pressure reduction upon exiting the die and expansion of the starch occurs. By carefully designing the die, shaped products with various textures may be achieved. The product is then dried and coatings are applied.

Improvements in extruders and extrusion techniques now enable new special products such as multiple cross-section products to be made. One product discussed (25) was made from two extruders combining flow into one outlet die. Using this design, one extruder might be used to process the cereal grain which would form the outer shell of the processed cereal piece. A second extruder would process materials which might fill the inner sections of the cereal piece with a soft filling or colored cereal product. These extrudates can be flaked or puffed to produce a product with varying colors or cross-sectional appearance and textures.

Another convenient way to manufacture puffed products is to follow cooking, which could be done in an extruder, with high-temperature drying in an oven or dryer. Extrusion, in this instance, is used to cook and gelatinize the cereal before puffing. Typically ovens are set for very high temperatures (274–288°C). Materials are held for very short times at this temperature to avoid excess heat, which can cause off-flavors. The high oven temperature drives off moisture, rapidly resulting in expansion or puffing of the individual pieces.

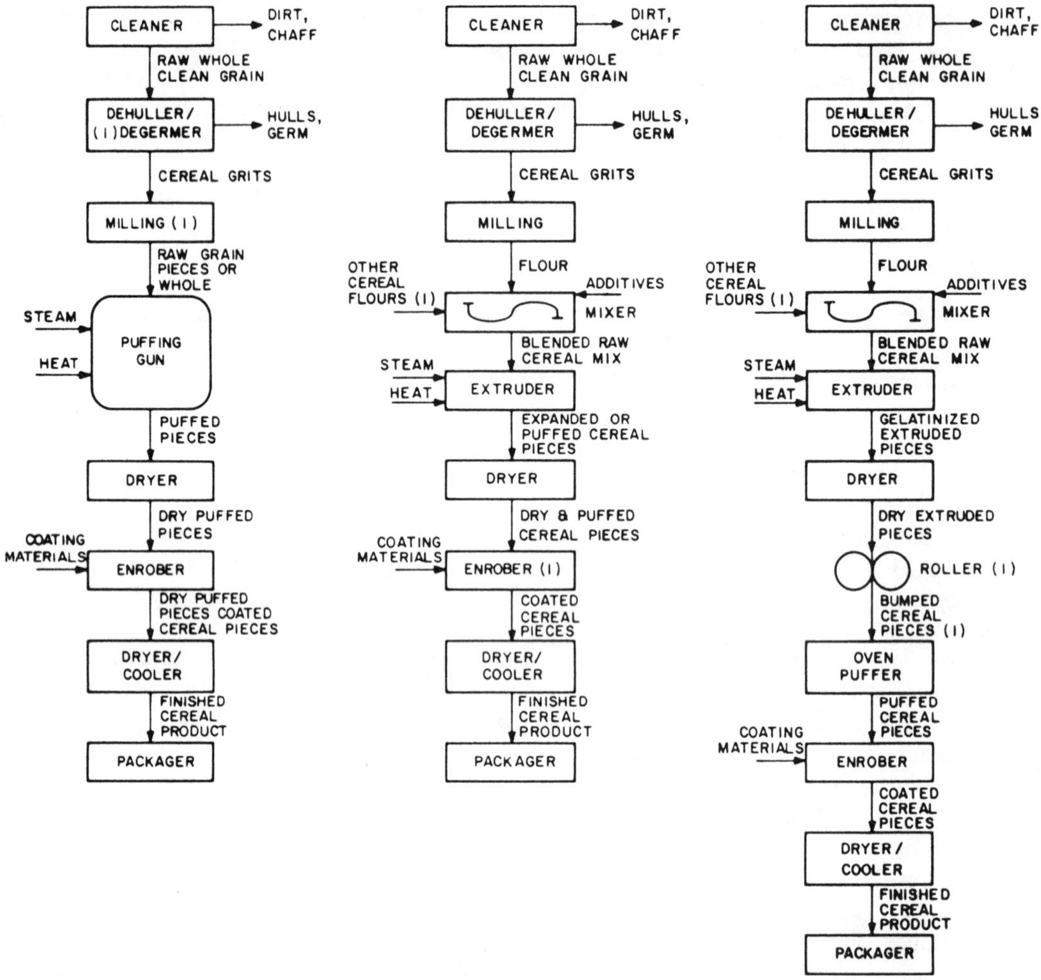

FIGURE 3 Process lines used to make puffed or expanded cereal pieces.

Fluidized bed driers (31) and microwaves (36) have been used to replace drier ovens with successful results. Processing schemes typically used for the manufacture of expanded breakfast cereals are shown in Figure 3.

Whole cereal grains which have been cleaned and sometimes decorticated are used as the raw materials for puffing guns. Such products as puffed wheat and rice are made in this fashion. A typical formulation when using extrusion processing will include one or two major cereal components such as corn and wheat. Additives, if necessary to achieve a special product characteristic, are mixed into the cereal components prior to cooking. For example, addition of special starch fractions to the cereal components have been used and shown to improve the expanded cereal piece by making it more delicate and uniform in texture and appearance (37). The addition of starch permits control of expansion and texture in these products and may be added at levels between 5 and 50%, depending on the type of starch used. Sugar adds flavor to cereal products but can also provide a certain degree of control on expansion (17).

Sodium bicarbonate has been used in the puffing process in the presence of soybean to reduce the unacceptable bean flavor in the final product (25). Sodium bicarbonate has also been used as an ingredient with wheat and honey to make a honey-graham flavored cereal (38). Other additives, when required, are typically added during the coating process. Typical and published formulations and processes dealing with both extrusion and puffing gun technology are given in Table 5.

C. RTE Cereal Grains—Shredded

Shredded breakfast cereals are made from whole cereal grains and primarily wheat. New shredded products are being developed which use alternative cereals, such as rice or corn and some are now offered with soft center, fruit-flavored confections.

Typical processes have been described for making shredded products (1,16). Cereal grains are cleaned and boiled in water until tender. The moisture in these cooked grains is allowed to equilibrate in a tempering process for several hours before processing the grains into shreds with steel rollers. For this step, one roller is smooth and the other roller is corrugated to form the small strands. The strands exiting the rollers are cut into the bite-sized pillow-shaped products and dried or baked. The finished shaped products are then packaged or coated as required.

Extrusion is also being used to make shredded products (16). This process has been used more extensively for production of shreds from flinty corn and rice cereal is more versatile and permits incorporation of other cereal products, ingredients, and additives. Minor ingredients such as salt and sugar are included in some of these formulations. A few applications are shown in Table 5.

A large portion of the shredded cereal market is for bite-size, shredded biscuits. For this product, strands of the formed dough are passed between two rollers, one smooth and one with knives placed radially, which pinch and cut the strands into bite-size pieces (39).

D. Miscellaneous Cereal Products

A number of cereal products available in the market cannot be placed in any of the above categories because of their unique qualities. This section is devoted to these processes.

1. Granola

Granolas have become popular because they are an alternative to the processed cereal products discussed above. Granolas are typically a composite of several kinds of grains and additives, which are clustered together. A typical process for this type of product utilizes flaked products such as rolled oats, barley, or other cereal grains and combines these with other noncereal products, such as nuts, coconut, oil, water, and spices. These mixes are then dried, toasted, and broken into pieces for consumption (16).

2. Cereal Granules

Cereal granules are a cereal product made by a unique process (1). A stiff dough is made from cereal grain flours, such as wheat and barley, salt, dry yeast, and water. The dough is stored under controlled conditions for several hours, shaped into loaves, and then transferred to an oven to bake. The oven is set for 400°F, and the dough is baked for 2 hours. The baked loaves are cooled, then broken into pieces with shredding knives and returned to an oven set for 250°F for 2 more hours of baking. The pieces are removed from the oven and broken up into small granules.

3. Baby Food Cereals

Baby food cereals are typically fed not only for breakfast, but also for other meals as supplements to breast milk or to introduce infants to solids. The majority of baby foods sold are canned. However, there are some dry cereal products available that are used as RTE cereals. These are usually made from rice, barley, and sometimes wheat. The traditional method for making these cereal products is to cook the decorticated and degermed cereals in water followed by drum drying to form the flaked cereal products. Extrusion has been used in the manufacture of high protein cereal flakes for baby foods (40). Few additives and preservatives are used.

VI. COATINGS FOR BREAKFAST CEREALS

Breakfast cereals that have been processed into a flake or puffed product have very little flavor. The high temperatures used to process the cereals can distill flavors from the foods, making flavor addition or enhancement necessary for product acceptability. Sweet flavors are generally the most acceptable for breakfast cereals, and, therefore, the majority of coating materials have a sugar base. The largest problem with sugar coatings and their application is control of the quantity metered onto each cereal piece and the resulting appearance of the coating after application. Careful selection of the application process and additives helps overcome these problems.

Application of flavor coatings on cereals is difficult due to the hygroscopic nature of cereal products after they have been processed. Sugars required water as the solvent, which is absorbed rapidly by processed cereal products. The rapid absorption causes sticky pieces which are difficult to dry, and in some instances the coated cereals adhere to each other. To minimize this problem, the amount of water in the coating step is reduced or a coating technique is used.

An acceptable coating has been achieved by heating sugar with water to a hard candy condition and spraying it on cereal products under constant pressure to keep the cereal pieces fluidized (17, 41). The enrobed material is then heated to fuse the sugar to the cereal product.

An acceptable and novel method of providing sweetened cereals is through the use of a spun sugar coating. The sugar is spun into a blanket, processed cereal pieces are placed on top of the blanket, and then a second blanket is placed on top of the cereal pieces. The spun sugar is compressed around the cereal pieces and dried. The coating is said to dissolve in milk upon contact (42).

An alternative to coating is to provide the sweetener through sugar bodies. Sugar bodies are made by combining sugars with other ingredients, cooking these ingredients together to form a hard candy, cooling carefully, adding flavors and colors, followed by a process of kneading and vacuum expanding the material to shape the material into individual pieces (43).

Most cereal coatings use sucrose as the main ingredient. Variations in formulation can be achieved by changing the mix of sugar types and other additives to achieve a certain hardness or appearance to the coating. Variations and acceptable ranges for sugar components have been outlined (44) and are as follows:

Sucrose: 10–70%
Glucose: 1–30%
Invert sugar: 5–25%
Water: 15–40%

Application of sucrose and sugar syrups generally results in cereal pieces that stick together and a coating that crystallizes into an opaque appearance. To overcome these problems, nonsucrose sugar materials such as glucose or invert sugar are incorporated (44,45). Honey is one form of nonsucrose sugar which is used extensively since the flavor characteristic is acceptable and a transparent coating can be achieved.

Additives to coating formulations are generally limited to flavor enhancement and to improving the adhesion of the coating to the cereal. Acetic acid in combination with sodium acetate has been added to cereal coatings (44) and, the proper proportions, provides some flavor enhancement. To reduce problems with using water as a solvent, emulsifiers have been added to create emulsions of oil, water, and sugar (46). This reduces the amount of water that will penetrate the cereal piece and reduces the amount of drying that is required.

Thickening agents are sometimes employed to change the consistency of coatings. For the sugar body method of sweetening, several thickeners such a methylcellulose, gum alginate, dextrin, pregelatinized starch, and other hydrophilic colloids are considered acceptable (43). Another method of coating cereals has been devised and involves the addition of an encapsulating colloid material with a solvent, which is used as a carrier for the sweetener (47). The solvent is evaporated, and, in the process, the colloid and sweetener merge with one another to coat the cereal, giving a frosted appearance.

While sugar materials are considered to have limited nutrition, they are suitable as energy sources for breakfast cereal foods. There are also concerns about the effects on dental caries. The addition of monosodium dihydrogen phosphate at levels between 0.5 and 1.5% proved effective in reducing the occurrence of caries through consumer testing of the product (48).

VII. CONCLUSION

Breakfast cereal processors continue to develop new formulations and processes to improve and increase the selection of consumer products. A substantial number of the new products being introduced are tailored to compete with health foods. By-products of the milling industry continue to find application in RTE type breakfast cereals. Bran products such as wheat and oat bran are found in cereal products and are gaining popularity because of medical evidence supporting a connection between intake and reduction of cholesterol levels in the blood.

Processing is becoming more flexible due to the application of twin-screw extruders. These machines can be set up to perform multiple operations and provide a consistent and specific history to the material being processed. With the ability to modify and control the shear environment in these extruders, it should be possible in the future to use a wider range of specialized cereals, grains, additives, and sweeteners to result in a number of new and unusual products.

TABLE 1 Additives Commonly Used in Breakfast Cereals

Additive	Common usage in cereals
1. Annatto	Add butter or peach color to cereal products
2. Barley malt flour	Typically added to raw materials Source for α– and β– amylase (49) Enhances starch hydrolysis (50) Flavor enhancer
3. Butylated hydroxyanisole	Prevention of oxidative rancidity of oils (49) Added to food directly and/or to packaging materials. Sum of two methods no more than 50 ppm
4. Butylated hydroxytoluene	Prevention of oxidative rancidity of oils (49) Added to food directly and/or to packaging materials. Sum of two methods no more than 50 ppm
5. Brown sugar	Flavor enhancer added at raw material stage or a part of coating for cereal Effective in control of gelatinization of starch (49) Control of puff and cook (49)
6. Calcium carbonate	Mineral supplement (51) Dough conditioner (51) Dough–firming agent (51) Viscosity modification (17) pH modification (acidic) (17)
7. Calcium caseinate	Mineral supplement (52)
8. Carrageenan	Emulsifier (51) Thickener or gelling agent (51) Stabilizer (51) Used in cereal coatings
9. Carboxymethyl cellulose	Used for binding and thickening cereal–coating formulations Emulsifier (51)
10. Cereal malt syrup	Used as sweetener and flavor additive for raw materials and sugar coatings (49) Inhibits crystallization of sucrose during coating Imparts sheen to product surface (49) Color (Maillard reaction)
11. Corn syrup	Used as sweetener and flavor additive for raw materials and sugar coatings (49) Inhibits crystallization of sucrose during coating Imparts sheen to product surface (49) Color (Maillard reaction)

TABLE 1 (Continued)

Additive	Common usage in cereals
12. Dextrose	Sweet flavor enhancer (49) Color (Maillard reaction)
13. Dicalcium phosphate	Acidification of dough (52) Dough modifier (52,53)
14. Disodium phosphate	Promotes acid hydrolysis of starch in cereals (52)
15. Ferric phosphate	Iron supplement (52) Cooking aid for hot cereals at levels of 0.2–2.0% (52)
16. Fructose	Flavor enhancer added at raw material stage or a part of coating for cereal Effective in control of gelatinization of starch (49) Control of puff (49)
17. Gelatin	Thickener for sugar coatings Dough modifier (17,54)
18. Glycerine	Crystalline modifier—prevents improper crystalline formation in cereal coatings (49)
19. Guar gum	Adhesive used for coatings and glazes on cereal products
20. Gum arabic	Adhesive for icings, glazes, and sugar coatings (49,54)
21. Invert sugar	Used to control and modify crystalline structure (49) Used in sugar coatings Cereal flavor enhancer
22. Lecithin	Emulsifier (17)
23. Malic acid	Additive to cereal products to impart acidic flavors Used in Fruit–flavored products
24. Modified corn starch	Control of density of puffed cereal products (17,55,56) Alter strength characteristics of puffed cereal products (17,55,56) Used as additive to improve shelf life of cereals (17,55,56) Humectant (17,55,56) Flavor enhancer (17,55,56) Binder for oils and fats in cereal products (17,55,56)

TABLE 1 (Continued)

Additive	Common usage in cereals
25. Molasses	Flavor enhancer added at raw material stage or a part of coating for cereal Effective in control of gelatinization of starch (49) Control of puff and cook (49)
26. Maltodextrins	Carrier for water-insoluble flavors (49) Thickener and body improver in cereal flours (50) Reduce volatilization of flavor from mixes of cereals (49)
27. Monosodium dihydrogen phosphate	Additive to cereal coatings to prevent dental caries (48)
28. Oils or fats	Control of dough strength and viscosity during processing (17) Control of puffed cereal textures (17) Improvement or modification of mouth feel and product functional characteristics Food energy additive
29. Papain	Proteolytic enzyme used primarily in cereal foods containing soybean products Improves or alters cooking characteristic of cereals (1)
30. Salt	Flavor improvement Buffer
31. Sodium alginate	Gelling or thickening agent for fruit filled cereals
32. Sodium bicarbonate	pH modifier in doughs (51) Flavor modifier cereal products containing soybean (25)
33. Sodium phosphate (monobasic)	pH modifier in doughs (52) Control and enhancement of starch hydrolysis Antioxidant (52) Flour modifier (52)
34. Sodium phosphate (dibasic)	Buffer (51) Mineral supplement (51)
35. Sorbitol	Used as a sweetener in cereal flour products at acceptable levels of 1–3% (49) Crystalline modifier and control additive for sugar coatings (49) Humectant (49)

TABLE 1 (Continued)

Additive	Common usage in cereals
36. Sugar (sucrose)	Flavor enhancer added at raw material stage or a part of coating for cereal Effective in control of gelatinization of starch (49) Control on degrees of puff and cook (49)
37. Tricalcium phosphate	Anticaking agent (49,51) Buffer (49,51) Mineral supplement (49,51)
38. Water	Control of expansion in puffed cereal products Control of degree of cook in expanded cereal products Solvent for sugar and other cereal coatings
39. Wheat starch	Control of product texture in cereal food products (49) Control of protein levels in cereal flours through dilution Control of cellular structure of puffed cereals (17,55,56) Control of puffed cereal texture during extrusion and puffing (17,55,56) Adjustment of rate of hydration in water (17,55,56) Retention of oil in cereal flour mixtures (17,55,56)
40. Yeast	Flavor modification and enhancement Dough modifier Leavening agent (49)
41. FD&C colors (yellow, red)	Color modification to cereals Added before extrusion at 300 ppm (17) Soluble in water and glycerine

TABLE 2 Processing Information for Farina Type Hot Cereals

Cereals used	Process description	Process objective	Additives	Additive level	Additive function	U.S. Patent granted
Wheat (middlings)	Milling	Uniform grit for cooking	Gum	0.5–2.5%	Thickener to suspend particles and to improve cooking capability	No. 2,890,115
Wheat (middlings)	Steam/drum dry	Instant wheat cereal	Sodium carboxy–methylcellulose	0.1–10%	Viscosity modifier for drum drying	No. 3,526,513
Wheat (middlings)	Steam/drum dry	Fruit–Flavored instant wheat cereal	Sodium carboxy–methylcellulose	0.1–10%	Viscosity modifier for drum drying	No. 3,526,514
			Fruit preserves	4.5–40%	Flavor enhancer	
Wheat (whole)	Tempering in salts and acid with drying	Quick cook whole kernel wheat product	Sodium hydroxide	5–15%	pH modifier	No. 3,190,754
			Ammonium hydroxide	0.5–3%	pH modifier	
			Citric acid	1%	pH modifier to promote hydrolysis of starch	
Wheat	Tempering in pH modified solution followed by drying	Quick cook wheat cereal	Salts of sodium and ammonia	1–3%	pH modifier and protein dissociating solubi–lizing agent	No. 3,526,511 No. 3,190,754
Wheat (middlings)	Temper at 16% MC, 220°F, flake, and dry	Instant wheat cereal without additives	None	—	—	No. 3,620,761

TABLE 3 Processing Information for Oat Type Hot Cereals

Cereals used	Process description	Process objective	Additives	Additive level	Additive function	U.S. Patent granted
Oats	Thin flaking and dry heat	Quick cook oat cereal	None	—	—	No. 3,490,915
Oats	Flaking followed by addition of milk, cream or cream substitute followed by drying	Quick cook nonsticky oat cereal	Milk Sugar Cream	20–25% 3–4% 3–4%	Flavor, reduce stickiness Flavor, Flavor, texture modifier	No. 3,494,769
Oats	Mixing in additives to quick cook oat flakes	Quick cook oat cereal	Polysaccharide Gum	0.1–1.0%	Encapsulate oat flakes with rapidly hydrating gum material to form gelatinous films on flakes	No. 2,999,018
Oats	Oat flakes are tempered, steam cooked, reflaked, dried, and blended with hydrolysate	Instant cook oat cereal	Cereal hydrolysate	2.25–2.5%	Add texture to the hydrated cereal mixture	No. 3,640,729
Oats	Oat fractions added to oat flakes tempered /steamed/cooked/reflake dried and blended with hydrolysate	High protein instant oat cereal	Oat fraction	4.25–5.0%	Improved nutrition without cooking	No. 3,640,728
Oats	Pregelatinized starch added to oat flakes, then tempered, steam cooked, reflaked, dried	Instant oat cereal	Pregelatinized starch material	2–5%	Improved texture and cookability during preparation	No. 3,704,134 No. 3,520,695
Oats	Quick cook oat flakes cooked and mixed with monoglycerides	Cooked oat cereal	Monoglyceride	0.3–3.0%	Reduce pastiness upon prolonged heating	No. 3,113,868

TABLE 4 Processing Information for Cereal Flakes

Cereals used	Process description	Process objective	Additives	Additive level	Additive function	U.S. Patent granted
Oats, Rice	Steam cooking extrusion/flake/dry, and toast	Crisp light nutritious flake	Soy flour	5–10%	Protein source	No. 3,318,705
			Sucrose	5–15%	Expansion control	
			Lecithin	0.5–0.15%	Emulsifier	
			Salt	2–4%	Flavor	
			Milk casein	1.5–10%	Supplement	
Oats, Rice	Pelleting grains/steam cook/dry/flake/toast	Crisp light nutritious flake	Soy flour	5–10%	Protein source	No. 3,121,637
			Sucrose	5–15%	Expansion control	
			Lecithin	0.05–0.15%	Emulsifier	
			Salt	2–4%	Flavor	
			Milk casein	1.5–10%	Supplement	
Oats	Cook/extrude/flake/dry/toast	Oat flakes	None	—	—	No. 3,345,183
Oats, rice	Cook/extrude/dry/flake/toast	Crisp flaked cereal	Soy flour	5–10%	Protein source	No. 2,998,317
			Sucrose	5–15%	Expansion control	
			Lecithin	0.5–0.15%	Emulsifier	
			Salt	2–4%	Flavor	
All cereals	Cook/extrude/cool/flake/toast	flaked cereal	Sugar	8%	Flavor, expansion control	No. 3,062,657
			Salt	1%	Flavor	
			Malt	1%	Flavor	
Oats, wheat	Cook/extrude/dry/toast	Flaked cereal by die design	Sugar	10%	Flavor, expansion control	No. 3,732,109
All cereals	Enzyme presoak for soy blend with cereal/extruded flake	Soy fortified flake product	Soy isolate	As required	Protein supplement	No. 3,664,848
			Proteolytic enzyme	445 ppm of soy isolate	Remove bitter flavor	
					Improve functionality	
All cereals	All	Soy, fortified flakes	Sodium bicarbonate	0.2–2%	Suppress soy flavor	No. 3,814,824
Rye	Puff gun/grind extrude/flake/toast	Better flake	None	—	—	No. 3,556,802

TABLE 5 Processing Information on Puffed and Expanded Cereals

Cereals used	Process description	Process objective	Additives	Additive level	Additive function	U.S. Patent granted
Rice, wheat, corn	Puffing gun	To expand cereal grains into formed precooked product	Pregelatinized starch	5–50%	Expansion or puffing aid	No. 2,954,295 No. 2,954,296
Rice	Puffing gun	To expand cereal grains into formed precooked product	None	—	—	No. 3,128,690
Mixtures of cereals	Puffing gun	To expand cereal grains into formed precooked product	Vegetable oil	1.5–5%	Reduce sticking of puffed units to each other and to equipment	No. 3,246,690
			Sucrose	5%	Expansion control flavor	
			Salt	2%	Flavor	
			Coloring	0.03%	Food enhancement	
Corn wheat	Co–extrusion/dry/ puffing gun	To combine two dissimilar products into a distinctive product	None	—	—	No. 3,499,766
All types of cereals	Extrusion cooking	One–step process expansion and cooking equipment	None	—	—	No. 3,117,006 No. 3,458,322 No. 3,458,321 No. 3,482,992 No. 3,054,677

TABLE 5 (Continued)

Cereals used	Process description	Process objective	Additives	Additive level	Additive function	U.S. Patent granted
Corn, wheat, oats	Extrusion cooking/drying/enrobing	Puffed peanut butter–flavored RTE cereal	Sugar	10%	Flavor, puff control	No. 3,723,131
			Peanut butter	13.5%	Flavor	
			Oil	1.5%	Carrier base for flavor	
Wheat flour	Preprocessing/extrusion	Honey graham flavored cereal	Sugar	20%	Flavor	No. 3,554,763
			Molasses	4%	Flavor	
			Honey	4%	Flavor, crystalline control	
			Salt	0.8%	Flavor	
			Sodium bicarbonate	0.4%	Buffer, pH adjustment	
Oats	Separation of oats into frac–tions for removal of β-glucans	Highly expanded cereal product	None	—	—	No. 4,620,981
Wheat or corn or rice	Extrusion/drying/puffing either hot air or oil fry	Puffing by double processing	Oil	2–6%	Puff control	No. 3,703,379
Any cereal	Extrusion/case harden/bump/dry air expansion	Precise control of puff	None	—	—	No. 3,453,155

REFERENCES

1. Matz, S. A., *The chemistry and Technology of Cereals as Food and Feed*, AVI Publishing Company, Inc., Westport, 1959, pp. 547–566.
2. Daniels, R., *Breakfast Cereal Technology*, Noyes Data Corporation, Park Ridge, NJ, 1974.
3. Ellinger, R. H., *Phosphates as Food Ingredients*, CRC Press, Cleveland, OH, 1972.
4. Matthews, R. D., *Quick Cooking Wheat Process*, U.S. Patent Number 3,190,754, 1965.
5. Cantor, S. M., Faller, L., and Wolnak, B., *Quick-Cooking Cereal Product and Process of Producing Same*, U.S. Patent Number 2,890,118, 1959.
6. Hyldon, R. G., *Instant Wheat Farina*, U.S. Patent Number 3,526,513, 1970.
7. Rockland, L. B., *Production of Quick-Cooking Cereal Grains*, U.S. Patent Number 3,526,514, 1970.
8. Gralak, B. G., and Hyldon, R. G., *Flavored Instant Grain Products*, U.S. Patent Number 3,526,514, 1970.
9. Spring Jr., F. E., *Cook-in-the-Bowl Cereal*, U.S. Patent Number 3,620,761, 1971.
10. Ronai, K. S., and Spanier, H. C., *Process for Preparing an Instant Oat Cereal*, U.S. Patent Number 3,704,134, 1972.
11. Huffman, G. W., and Moore, J. W., *Instant Oatmeal*, U.S. Patent Number 2.999,018, 1961.
12. Ronai, K. S., and Spanier, H. C., *Instant Oatmeal Product* U.S. Patent Number 3,640,729, 1972.
13. Ronai, K. S., and Spanier, H. C., *Instant Oat Cereal Product*, U.S. Patent Number 3,640,728, 1972.
14. Tressler, D. K., *Process for Preparing Oats Cereal*, U.S. Patent Number 3,6796,769, 1970.
15. Lee, L. J., *Process for Preparing a Cooked Cereal and the Resulting Product*, U.S. Patent Number 3,113,868, 1963.
16. Fast, R. B., Breakfast cereals: Processed grains for human consumption, *Cereal Foods World*, *32*, (3):241–244 (1987).
17. Harper, J. M., *Extrusion of Foods*, Volume II, CRC Press, Inc., Boca Raton, FL, 1981, pp. 78–86.
18. Clausi, A. S., Vollink, W. L., and Michael, E. W., *Breakfast Cereal Process*, U.S. Patent Number 3,218,705, 1967.
19. Clausi, A. S., Vollink, W. L., and Michael, E. W., *Breakfast Cereal Process*, U.S. Patent Number 3,121,637, 1964.
20. Gulstad, C. O., *Method for Making Rye Flakes from Exploded Kernels*, U.S. Patent Number 3,556,802, 1971.
21. Vollink, W. L., *Process for Making Breakfast Cereal Flakes*, U.S. Patent Number 3,062,657.
22. Post, J. G., and Smith N. F., *Ready-to-Eat Oat Cereal Biscuit*, U.S. Patent Number 3,732,109.
23. Maloney Jr., J. F., and Craig, T. W., *Process for Altering the Moisture Absorptive Characteristics of Ready-to-Eat Breakfast Cereal*, U.S Patent Number 3,595,670, 1971.
24. Bedenk, W. T., *High Protein Ready-to-Eat Breakfast Cereals Containing Soy Isolate*, U.S. Patent Number 3,689,279, 1972.
25. Bedenk, W. T., *Breakfast Cereals Containing Soy Material and Sodium Bicarbonate*, U.S. Patent Number 2,814,824, 1974.
26. Mogi, S., Toei, R., and Yamaguchi, M., *Apparatus for Making Puffed Foods*, U.S. Patent Number 3,456,565, 1969.
27. Tsuchiya, T., Long, T., and Hreha, K., *Method and Apparatus for Continuous Puffing*, U.S. Patent Number 3,231,387, 1966.
28. Meiland, W., *Puffing Gun*, U.S. Patent Number 3,299,799, 1967.
29. Paugh, G. W., *Continuous Puffing Method*, U.S. Patent Number 3,908,034, 1975.
30. Holtz Jr., W. E., and Reinhart, R. R., *Method for Producing and Improved Puffed Cereal*, U.S. Patent Number 3,660,110, 1972.
31. Reesman, S. H., *Process for Producing Puffed Dough Pieces*, U.S. Patent Number 3,506,454, 1970.
32. Bixby, D. G., and Helmke, K. L., *Preparation of Ready-to-Eat Peanut Butter Containing Cereal*, U.S. Patent Number 3.723,131, 1973.

33. Gordon, W. A., Hempenius, W. L., and Kirkwood, R. R., *Process for Preparing a Highly Expanded Oat Cereal Product,* U.S. Patent Number 3,620,981, 1986.
34. Bowman, J. W., *Method of Producing a Cereal Product,* U.S. Patent Number 3,104,975, 1963.
35. Vollink, W. L., and Scharschmidt, R. K., *Process for Making a Puffed Multi-Phased Cereal Product,* U.S. Patent Number 3,499,766, 1970.
36. McAlister, R. E., *Microwave Puffing of Cereal Grain and Products Made Therefrom,* U.S. Patent Number 3,682,651, 1972.
37. Murray, D. G., Marrota, N. G., and Boettger, R. M., *Manufacture of Farinaceous-Based Amylose-Containing Food Products By Means of an Extrusion Type Process,* u.S. Patent Number 3,554,763, 1971.
38. Fast, R. B., Hershak, B. O., and Spotts, C. E., *Preparation of Ready-to-Eat Cereal Characterized by a Honey-Graham Flavor,* U.S. Patent Number 3,554,763, 1971.
39. Graham Jr., W. R., and Grogg, B., *Method of Making Shredded Cereal Product,* U.S. Patent Number 3,054,677, 1962.
40. Lutz, R. J., *Extruded Cereal product and Its Preparation,* U.S. Patent Number 4,044,159, 1977.
41. Massmann, W. F., Michael, E. W., and Vollink W. L., *Process for Hard Candy Coating Food Particles,* U.S. Patent Number 2,689.796, 1954.
42. Shoaf, M. D., Groesbeck, C. W., and Cowart, D. G., *Dry Food Products in Spun Filaments and Method of Making Same,* U.S. Patent Number 3,615,671, 1971.
43. Christianson, G., *Flavored Ready-to-Eat Breakfast Cereal,* U.S. Patent Number 3,686,001 1972.
44. Fast, R. B., *Process for Preparing a Coating Ready-to-Eat Cereal Product,* U.S. Patent Number 3,318,706, 1967.
45. Vollink, W. L., *Process for Producing a Candy Coated Cereal,* U.S. Patent Number 2,868,647, 1959.
46. Lyall, A. A., and Johnston, R. J., *Emulsified Oil and Sugar Cereal Coating and Incorporating Same,* U.S. Patent Number 3,959,498, 1976.
47. Schade, H. R., Baggerly, P. A., and Woods, D. R., *Frosted Coating for Sweetened Foods,,* U.S. Patent Number 4,079,151, 1978.
48. Muhler, J. G., *Process for Producing a Stabilized Ready-to-Eat Breakfast Cereal Product,* U.S. Patent 3,467,529, 1969.
49. Furia, T. E., *Handbook of Food Additives,* The Chemical Rubber Company, Cleveland, OH 1968.
50. Whistler, R. L., Bemiller, J. N., and Paschall, E. F., *Starch Chemistry and Technology,* 2nd edition, Academic Press, Inc., Orlando, FL, pp. 87–123, 615–616.
51. National Academy of Sciences, *Food Chemicals Codex,* Prepared by the Committee on Specifications, Food Chemicals Codex, of the Committee on Food Protection National Research Council, Washington, D.C., 1972.
52. Rechcigl Jr., M., *CRC Handbook of Nutritional Supplements* Volume I, *Human Use,* CRC Press, Inc., Boca Raton, FL, 1983, p. 155.
53. Van Wazer, J. R., *Phosphorus and Its Compounds,* Volumes 1 and 2, Interscience Encyclopedia, Inc., New York, 1966.
54. Glicksman, M., Utilization of natural polysaccharide gums in food industry, *Advances in Food Research, 11*: 109 (1962).
55. Smith, P., The functions and utilization of starch in extruded foods, in *Symposium on Extrusion: Process and Product Development,* Central State Section, American Association Of Cereal Chemistry, St. Paul, MN 1971.
56. Feldberg, C., *Extruded starch-based snacks, Cereal Science Today, 14*(6): 211 (1969).

19

Pasta: Raw Materials and Processing

Brendan J. Donnelly
North Dakota State University
Fargo, North Dakota

I. INTRODUCTION

The word pasta is a generic term used in reference to the whole range of products commonly known as spaghetti, macaroni, and noodles. Italians, who are the largest consumers of pasta products in the world, call them *pasta Alimentare* (alimentary paste). Germans refer to them as *Teigwaren* (pasta goods), the French *pates alimentaires* (1). Production and consumption of the various pasta products, which number approximately 150 in the United States, include long spaghetti and macaroni; short cut products such as elbows, shells, and noodles; and such specialty products as bow ties, rigatoni, lasagna, etc. (2) (Fig. 1). More formal definitions of pasta products are incorporated in the Standards of Identity for macaroni and noodle products as outlined in the U.S. Code of Federal Regulations (CFR) (Table 1)(3). In the CFR macaroni products are defined as a class of food each of which is prepared by drying formed units of dough made from semolina, durum flour, farina, flour, or any combination of two or more of these, with water and with or without one or more of optional ingredients specified in the CFR.

The CFR standards allow the use of optional ingredients in order to enhance some of the cooking qualities. There is provision made for the use of egg white solids ranging from 0.5 to 2.0% of the weight of the finished goods—also the use of disodium phosphate for a quick-cooking pasta product. Seasoning is permitted in the form of onions, celery, garlic, salt, and bay leaf. Gum gluten and concentrated glycerol monosterate are also allowed in pasta products within specified limits. Whole milk and nonfat milk solids are also permitted under the standards, as are 12.5% soya flour; 3% vegetable solids (includes spinach, carrots, and tomato solids); whole wheat, and enriched (not less than 4 mg and not more than 5 mg thiamine, not less than 1.7 mg and not more than 2.2 mg riboflavin, not less than 27 mg and not more than 34 mg niacin or niacinamide, and not less than 13 mg and not more 16.5 mg of iron (Fe), in each pound) pasta.

II. HISTORICAL NOTES

Italy is generally regarded as the original home of pasta products. They appear to have been developed both in Italy and China or Japan. Certainly Italy is the country with which pasta

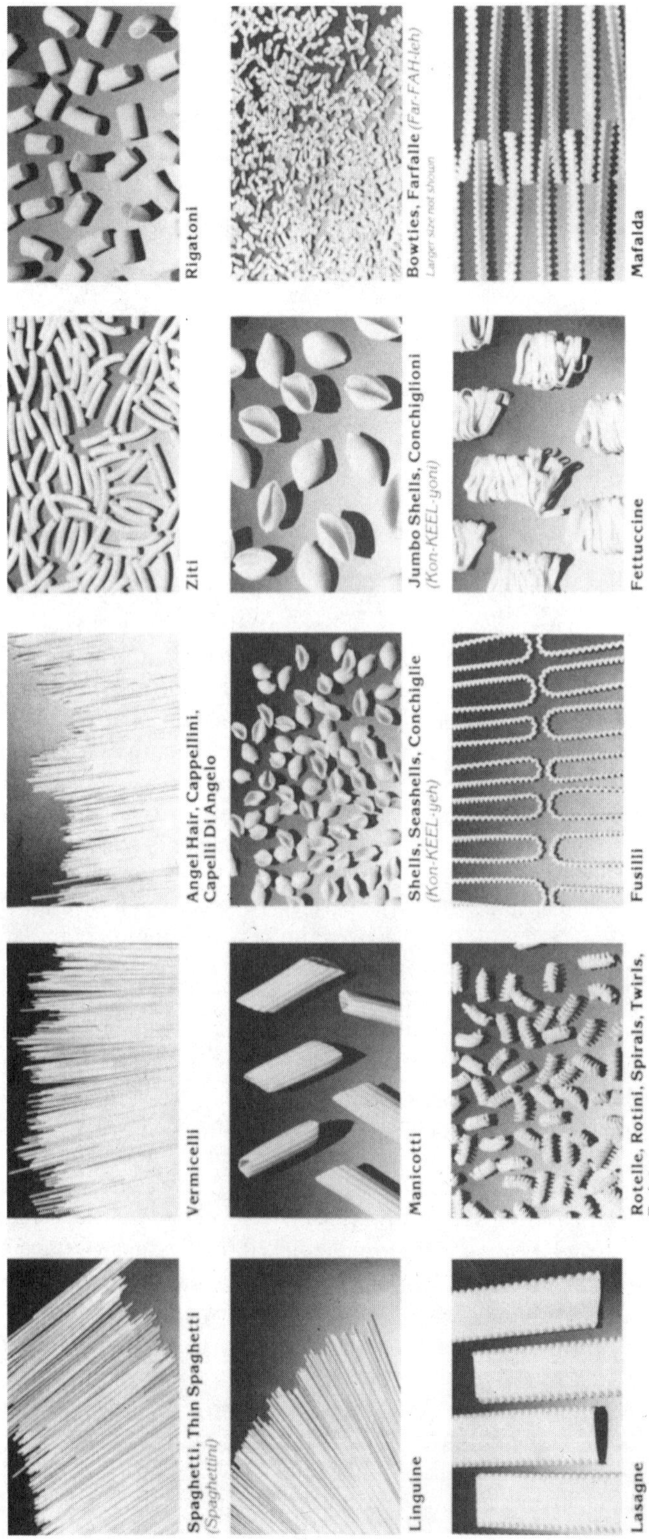

Spaghetti, Thin Spaghetti (*Spaghettini*)
Vermicelli
Angel Hair, Cappellini, Capelli Di Angelo
Ziti
Rigatoni

Linguine
Manicotti
Shells, Seashells, Conchiglie (*Kon-KEEL-yeh*)
Jumbo Shells, Conchiglioni (*Kon-KEEL-yoni*)
Bowties, Farfalle (*Far-FAH-leh*)
Larger size not shown

Lasagne
Rotelle, Rotini, Spirals, Twirls, Twists
Fusilli
Fettuccine
Mafalda

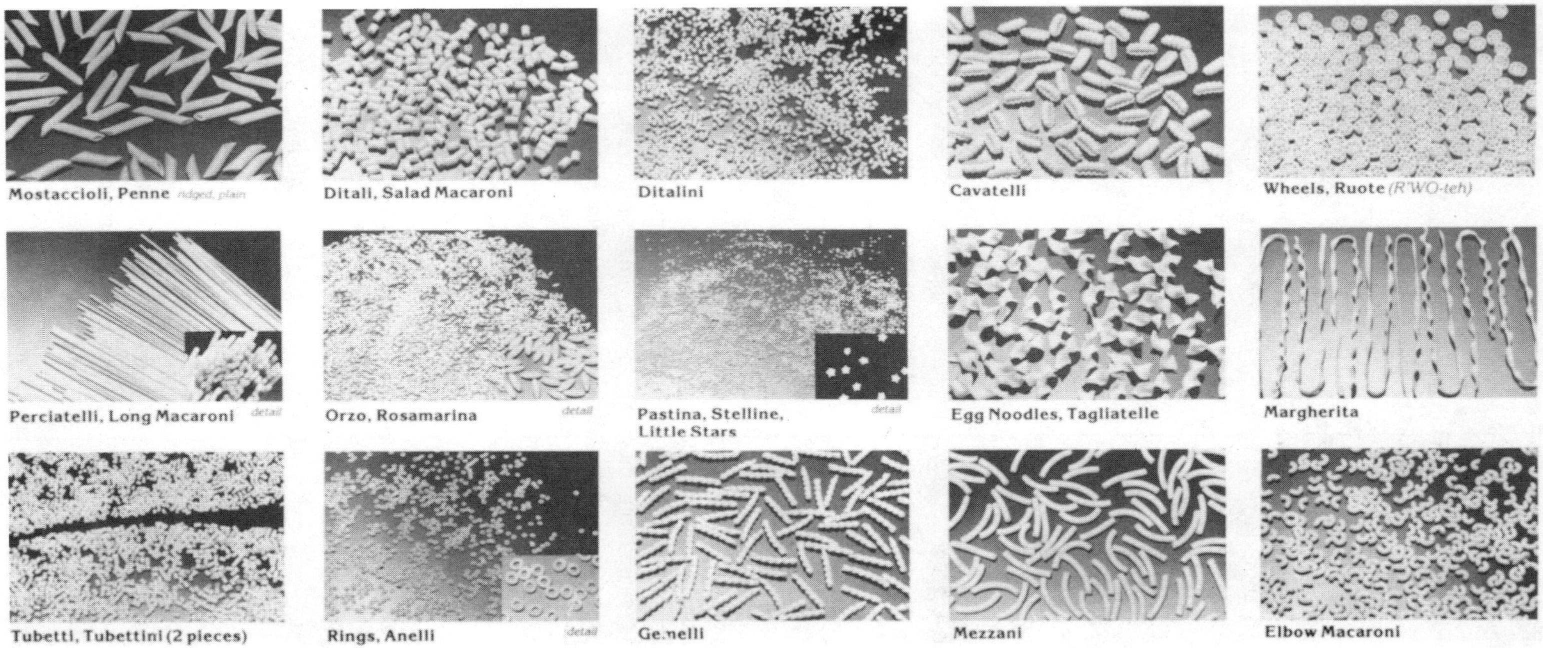

FIGURE 1 Pasta products of varying shape and size. Reproduced by kind permission of the National Pasta Association, Washington, D.C.

products are most readily identified (1). Pasta products as we know them were first made in Italy about 800 years ago. In the fifteenth century, Italians learned how to make noodles from the Germans, who had previously learned the process in their travels to Asia. The Germans called this product *Nudeln*, which is still the more popular type of pasta consumed in Germany (2). The Italian climate, especially around Naples, was particularly favorable for the drying of pasta products. After the Italian pasta products industry quickly developed, it spread rapidly to France and elsewhere in Europe.

The pasta industry in the United States was started commercially by Antoine Zerega, who in 1848 founded A. Zerega and Sons, Inc. in Brooklyn, New York(2).

Originally, all pasta products were home-made. By about 1800, the first mechanical devices for the manufacture of pasta products appeared in Italy. Around 1850 the first hand-operated mechanical presses came into existence. By 1860 more elaborate presses were made, but the increasing popularity of pasta products required that more efficient production processes be developed. These were developed in Italy and France, and subsequently in Germany (1). At the turn of the twentieth century, equipment comprising mixers, kneaders or gramolas, hydraulic extrusion presses, and drying cabinets became available. It wasn't until 1934 that the French, Swiss, and Italians developed continuously operated automatic presses replacing the batch systems (1). Today all presses are of the continuous type for all makes and shapes of pasta products. Production lines can now process 2000 kg per hour of pasta in one continuous automatic operation from mixing and extrusion, to drying and packing. The recent advent of high temperature drying, and in some cases microwave drying of short goods, has improved plant-operating efficiencies and product throughput with some indications of product quality improvement.

III. RAW MATERIALS FOR PASTA PRODUCTS

The pasta products macaroni, spaghetti, vermicelli, and noodles are manufactured primarily from semolina, durum granulars, and flour produced from the milling of durum wheat. These are the preferred raw materials for the production of superior quality pasta products. To a lesser extent, farina and flour from common wheats are also used. In addition, pasta can be processed from blends of these nondurum materials with durum semolina or flour. However, these products usually suffer a deficiency with respect to some quality attributes such as color or cooking quality. The degree to which blending is practiced is usually dependent on availability and price of the raw materials (wheat) and competitive pricing and consumer acceptance in a given market.

A. Durum Wheat

This is the raw material of choice for the production of pasta products. Durum wheats are better adapted to semiarid climates than are common wheats, and as a result, durum growing is more widespread in the semiarid zones around the world (Table 2) (4). Production of durum wheat in the United States is concentrated in an area known as the "durum triangle" in north central North Dakota. The states of Minnesota, South Dakota, and Montana produce durum wheat in areas bordering the state of North Dakota, and durum is also produced in southern California and Arizona (desert durum) (Table 3)(5). Desert durums have different wheat kernel and quality characteristics when compared with durum grown in the Upper Great Plains states (Table 4) (5). They generally have higher test weight, higher 1000 kernel weight, higher percentage of larger or plumper kernels, and lower moisture content. Protein content of the desert durums are 1.0 percentage point lower (dry matter basis) than their counterparts up north.

The first durum wheat variety of importance to be grown in the Upper Great Plains was Arnautka(6). This variety was introduced from Russia to the United States in 1856. It was not until about 1900 that this wheat came into fairly extensive production along with another variety from the U.S.S.R. called Kubanka. Over the years durum wheat variety or cultivar development and introduction has undergone considerable change, particularly since the early 1940s (6,7). Since 1943, most of the predominant cultivars grown in the United States were developed and released by North Dakota State University in cooperation with the ARS-USDA (Table 5). The changing distribution of varieties illustrates that as new varieties are developed and released, they are soon accepted by the producers and the older varieties are gradually replaced due to improved agronomic, disease resistance, and/or quality traits.

One of the primary factors establishing the economic value of durum wheat is its grade. According to the U.S. Grain Standards for wheat, durum wheat is classified into six possible grades (Table 6)(8). The basis for this classification depends upon:

Minimum test weight per bushel
Amount of damaged kernels
Maximum limits of foreign material, shrunken and broken kernels and defects
Maximum limits of mixtures of wheats of other classes

Durum wheat is further divided into three subclasses, namely:

1. Hard Amber Durum Wheat—must contain 75% or more vitreous kernels of amber color
2. Amber Durum Wheat—must contain 60% or more but less than 75% vitreous kernels of amber color
3. Durum Wheat—contains less than 60% vitreous kernels of amber color

It should be noted that the protein content of durum wheat is not a grading factor. However, a minimum of about 11.0% protein in semolina is necessary to make pasta products having good cooking quality. Durum millers generally prefer the higher numerical grades within the subclass Hard Amber Durum because of higher potential semolina milling yields.

Characteristics, other than grade, which distinguish durum wheat from the common bread wheats grown in the United States are outlined in Table 7.

B. Semolina

The word semolina is derived from the Italian word *semola* and the French equivalent *semoule.*

Pasta products are manufactured principally from the three main milled products of durum wheat, namely semolina, durum granulars, and durum flour. Farina and flour from common wheat are also used but to a lesser extent in the United States than elsewhere. For the production of good quality pasta, the particle size of the semolina should not be too coarse nor too fine. Semolina milling is unique in that the objective of the process is to prepare granular middlings with a minimum of flour production. To mill durum wheat into semolina, several steps are necessary (9). These include wheat cleaning, tempering, milling, and purifying.

Details of the milling process will not be developed here, however, it is important to know something about the process, as the main objective relates to the processing of semolina into pasta. First, the wheat is cleaned to remove foreign material and shrunken and broken kernels. Foreign material is matter other than durum wheat. An excessive amount of it can be detrimental to the appearance and eating quality of the finished pasta product (10). Tempering (conditioning) the wheat to a moisture content of approximately 16.5% toughens the seed coat so that an efficient separation of bran and endosperm can take place. If properly conditioned the endosperm will be reduced to the proper granular size distribution with minimum production of flour. Proper sifting and purifying will allow for maximum semolina yield, large bran flakes with a minimum amount of bran powder.

The process of durum milling is a complex procedure of repetitive grinding and sieving. The tempered wheat is ground on a series of corrugated break rolls, the objective being to open up and scrape the wheat kernels to release the endosperm from the bran. A second set of reduction rolls having finer corrugations is used to grind the middlings (semolina) to the proper size. Various vibrating sieves are used between grinding steps to allow for efficient reduction of the endosperm to proper granular size. Final steps of milling involves purifying to separate as much of the small bran particles and flour from the semolina. A commercial durum mill will produce 64% semolina and 9% flour from a good grade of durum wheat. A typical product diagram of a commercial durum mill is shown in Figure 2 (10).

Semolina is defined as the purified middlings of durum wheat which has been ground so that all of the products pass through a No. 20 U.S. sieve and not more than 3% shall pass through a No. 100 U.S. sieve. Particle size distribution for typical U.S. commercial semolina is shown in Table 8.

Durum flour is the purified endosperm of durum which is ground fine enough so that at least 98% passes through a U.S. No. 70 sieve.

Most U.S. pasta manufacturers prefer semolina that has uniform particle size rather than a coarse-ground semolina. With fine granulation less trouble is encountered in mixing the semolina and water to form a uniform dough for extrusion. If semolina is not uniform, but consists of fine as well as coarse particles, the fine particles will tend to absorb water faster than the larger particles, resulting in white specks in the pasta.

Durum granular is usually used in short-cut pasta such as shells or elbows.

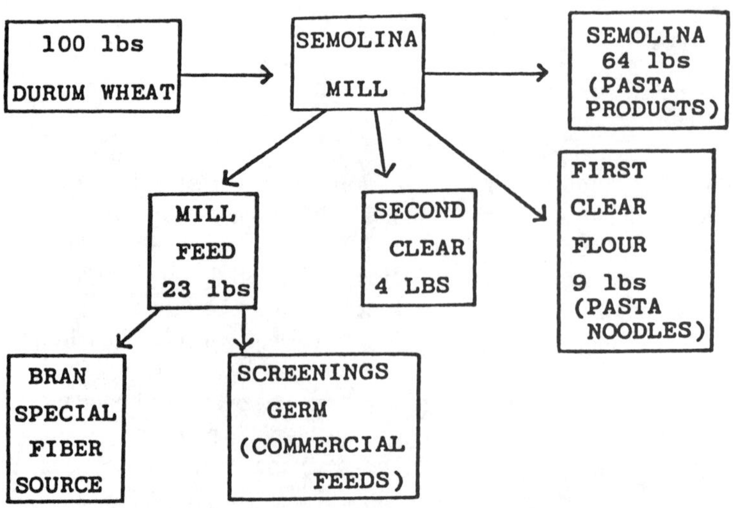

FIGURE 2 Milling of Durum Wheat Into Semolina. (From Ref. 1.)

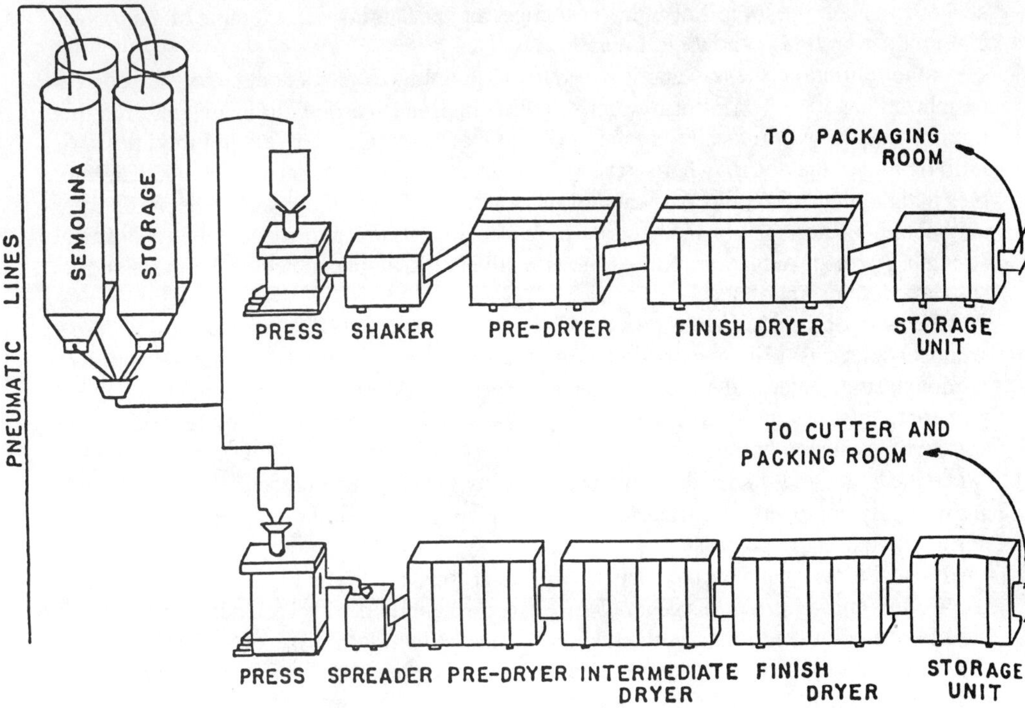

FIGURE 3 Long goods line. (From Ref. 10.)

Durum flour is used primarily in the manufacture of noodles and noodle products since it has been determined that the fine flour yields a smoother and more homogeneous mix with the egg ingredient (2). Durum flour is classified according to its ash content into various grades such as Fancy Durum Patent Flour, Durum Patent Flour, Straight, and first clear.

C. Water

The water used for pasta products should be pure, have no off-flavors, and be suitable for drinking. Since pasta is processed below pasteurization temperatures, the bacterial count of the water is directly related to the bacterial count of the finished product. Consequently, only pure water of low total plate count should be used (9). Water composition considered best suited for pasta processing is shown in Table 9 (2). However, the recent advent of high temperature drying and microwave drying of pasta has resulted in lower levels of microbial counts in pasta products than previously experienced with conventional temperature drying (11– 13).

IV. PASTA PRODUCTION

A. Extrusion

Pasta extrusion and drying has evolved to the point where continuous, high throughput presses and dryers are standard throughout the industry, with outputs known to be as high as 7000 kg/hr (15,000 lb/hr) at a Barilla short goods line in Parma, Italy. A typical flow diagram for processing either short or long goods pasta using conventional temperature (up to 55 °C) drying is shown in Figure 3 (10). The essential steps include the continuous press, shaker/spreader, predryer, finish dryer, storage, and packaging. An example of a commercial production line is shown in Figure 4.

In the continuous press, water is added to the semolina to give a dough moisture content of approximately 31%. Uniform water/semolina mixing is carried out in a counterrotating mixing chamber with vacuum applied prior to extrusion. Counter rotating mixing shafts limit balling of the dough, whereas the applied vacuum reduces formation of small air bubbles in the dough and limits oxidation of the xanthophyll/lutein pigments. The presence of air bubbles in pasta gives the product a chalky appearance and reduces its mechanical strength. Pigment oxidation reduces the attractive yellow appearance of the pasta and its subsequent consumer appeal.

The heart of the continuous press is the extrusion auger, which kneads the dough into a homogeneous mass prior to extrusion through a die. Auger speed and temperature control of the dough contributes to the quality of the pasta product. Most modern presses are equipped with sharp-edged augers having a uniform pitch over this entire length. The auger fits within a grooved extrusion barrel, which helps the dough move forward and reduces friction between the auger and the inside of the barrel during the extrusion process. Extruder barrels are normally equipped with water-cooled jackets to hold the pasta temperature near 40°C during the extrusion process.

The inside surface of the die influences product appearance and extrusion rate (10). Current manufacturing processes generally use bronze dies fitted with Teflon inserts to extend the life of the die and improve the quality of the product with respect to product surface smoothness and appearance. Stainless steel dies are frequently used in high volume presses to prevent distortion of the dies. These dies use Teflon– or non–Teflon coated bronze inserts.

B. Drying

Another critical step in pasta processing is the drying phase. Moist pasta from the extruder needs to be dried from 31% to approximately 12% moisture so that the product will be hard, retain its shape, and store without spoiling. Regardless of dryer design, and temperature–humidity–airflow control, problems can arise if the pasta is not dried carefully (1). For example, if dried too slowly, the product will tend to spoil or become moldy during drying. On the other hand, if dried too rapidly, moisture gradients will occur, which can cause the product to crack or check. Checking can occur either during the drying cycle or as long as several weeks after the product has been packaged. If large stresses are present due to improper drying, any change in relative humidity can result in a checked product. Therefore, it is essential that the pasta product is dried using a drying cycle tailored to meet that product's requirements.

Up to 1974, Conventional Temperature Drying (CTD) of pasta, that is, using drying temperatures up to 55°C (131°F), was considered to be the latest in technology. Drying times were about 16 hours for long goods and 8 hours for short goods. Many of these lines are still in commercial operation. A typical example of a CTD or low temperature (LT) drying cycle for spaghetti is shown in Figure 5.

High Temperature Drying (HTD) was introduced into commercial drying lines in 1974. HTD raised drying temperatures from 55°C to 75°C, which resulted in shorter drying times (10 hours for long goods, 4.5 hours for short goods), lower bacterial counts, and improved end-product quality (Fig. 5)(14–17).

In the last few years, the evolution of pasta-drying technology has increased drying temperatures from 75°C to 100°C and above. These drying cycles are referred to as Very High Temperature Drying (VHTD) or Ultra High Temperature Drying (UHTD) (18). The advan-

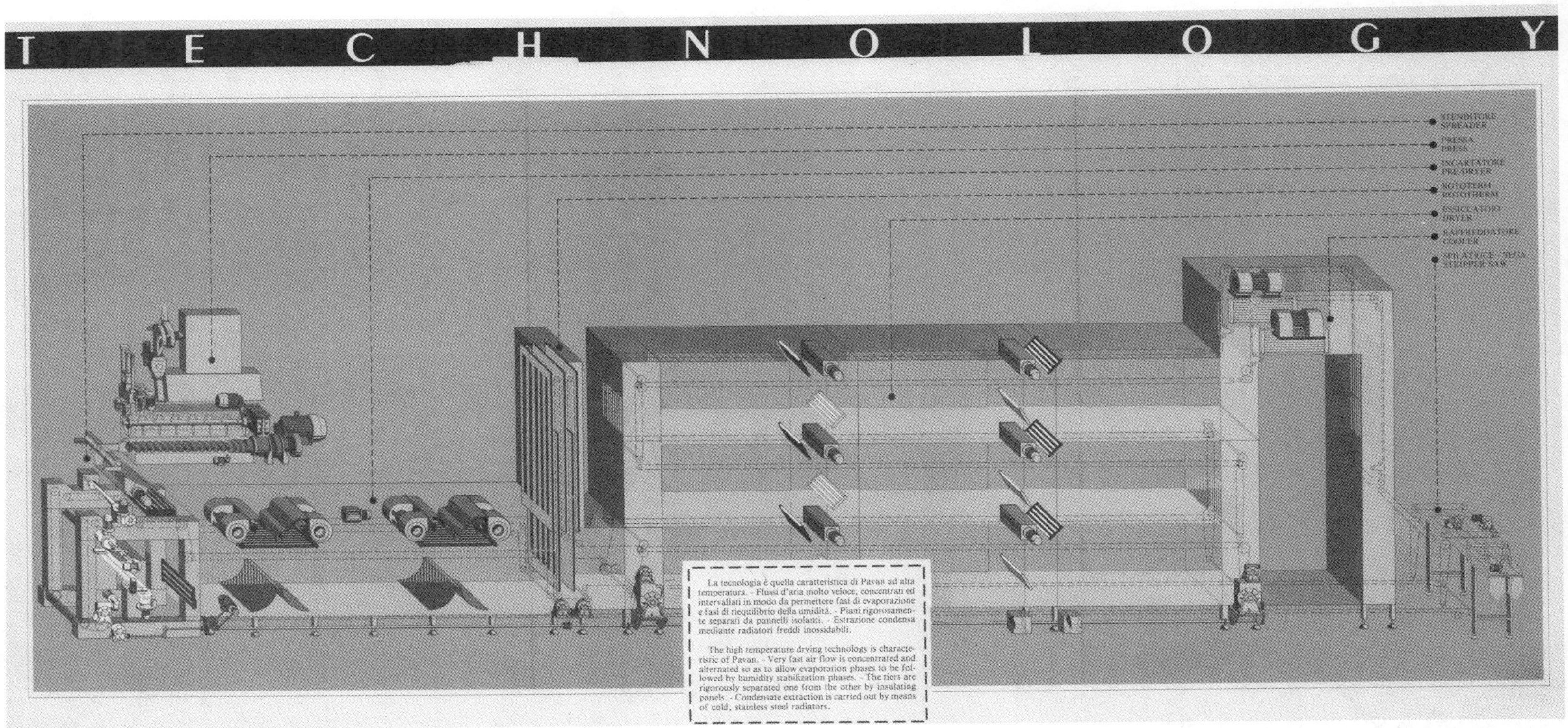

FIGURE 4 Commercial production line. (Reproduced by kind permission Pavan, Padova, Italy.)

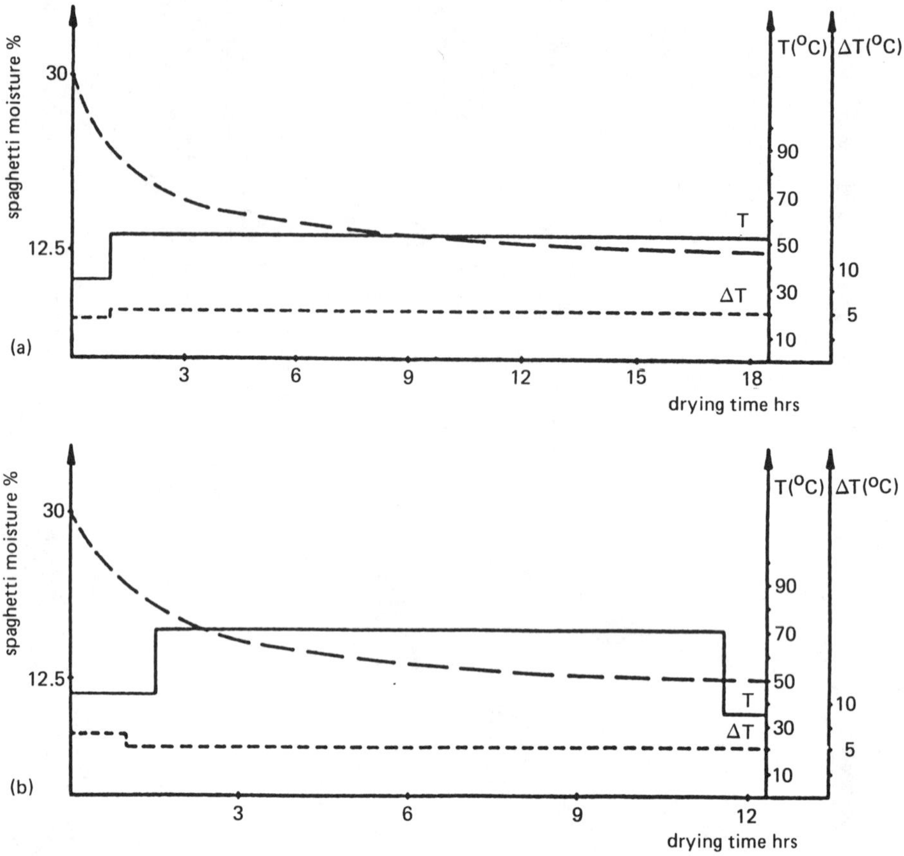

FIGURE 5 Processing diagrams for low and high temperature drying. T = Air temperature inside dryer, taken on a dry-bulb thermometer, ΔT = difference between temperatures taken inside dryer on dry-bulb and wet-bulb thermometers. T and ΔT are used in a formula to find the relative humidity inside the dryer. (From Ref. 18.)

tages of VHTD include significantly reduced drying times (5.5 hours for long goods, 2.5 hours for short goods), improved end-product quality (Table 10), and reduced investment and operating costs. A typical VHTD profile is shown in Figure 6.

Microwave technology has also been successfully applied to the drying of short goods pasta (11). Microwave drying of long goods pasta has not been commercially feasible to date. The drying system for short goods consists of three stages: a conventional hot air predryer, a microwave–hot air second stage, and an equalized third stage. Figure 7 shows a typical drying curve for elbow macaroni at a production rate of 1500 kg/hr (3300 lb/hr). As shown in the figure, drying time is less than 2 hours. Advantages claimed for microwave drying include requiring 1/3 to 1/4 floor space of conventional dryers; reduced drying times; improved product color and cooking quality; reduced plate counts; reduced sanitation costs; and reduced operating costs.

C. Packaging

As pointed out by Walsh and Gilles (9) there are thousands of different sizes, shapes, and types of packages in which pasta products may be sold. However, they all perform the same

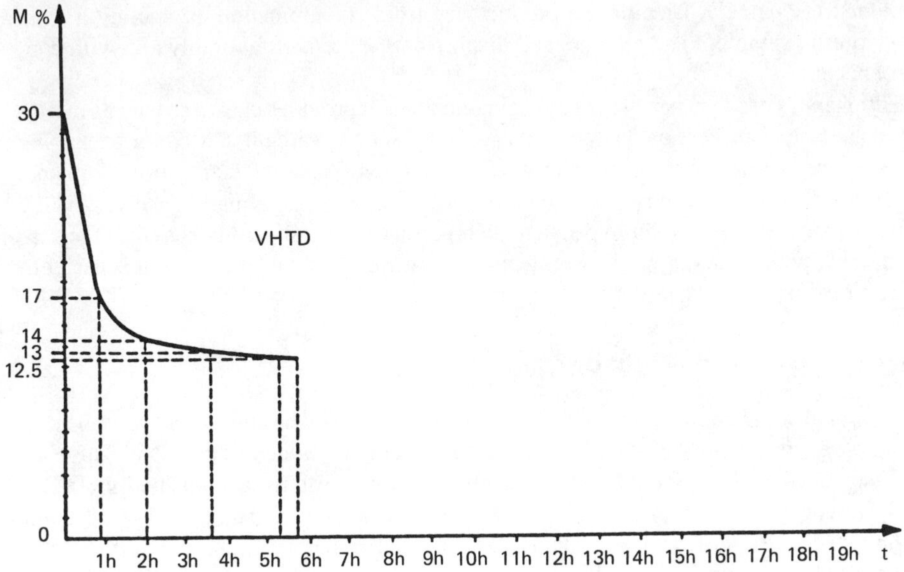

FIGURE 6 Long-cut pasta product moisture curve. M = moisture; t = drying time. (Ref. 18.)

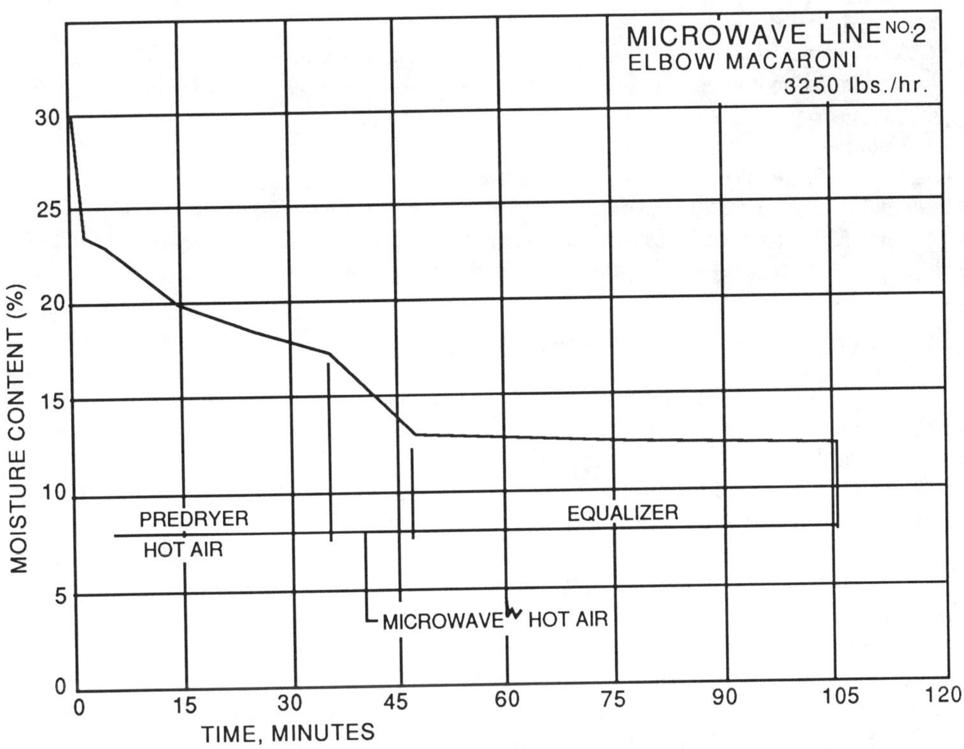

FIGURE 7 Typical drying rate curves for pasta drying with microwaves. (From Ref. 11.)

basic functions, namely, keeping the product free from contamination, protecting it from damage during shipment and storage, and displaying the product favorably and with consumer appeal.

Cellophane is used primarily for packaging noodles. It provides clarity as well as insect- and moisture-proof protection. Low density polyethylene bags are often used for packaging pasta products. These offer the same protection but do not have the clarity of cellophane. They also have the disadvantage of being difficult to stack on supermarket shelves. Many U.S. pasta manufacturers prefer to package their products in cardboard boxes since they are easy to stack, provide good physical protection for the product, and advertising is easier to print and read on boxes than plastic containers.

V. NUTRIENT PROFILE OF PASTA

The nutritional composition of pasta is shown in Table 11 (19). The amino acid composition of pasta products are presented in Table 12,and the vitamin/mineral content in Table 13 (19). Pasta products are a good source of many essential nutrients, as indicated in Tables 11–13. However, like any single food source, it does not supply all the nutrients for a complete human diet. Protein in pasta contains all 8 essential amino acids; therefore, 0.68 kg of pasta of 12.0% protein would supply the required daily amounts of protein for a normal person. Pasta is particularly low in lysine. Therefore, when assessing nutritional quality, the combined value of other foods that are consumed with pasta should be taken into consideration. For example, most often pasta is consumed with meat, meat sauce, and/or cheese. The protein present in the meat and/or cheese usually balance that in pasta, so that the combined nutritional quality of the food is very high.

As a low fat food, dry pasta contributes only small amounts of fat to the diet. Nonegg pasta contains no cholesterol and little saturated fats. Over 63% of the fat in nonegg dry pasta is made up of the essential fatty acid linoleic acid.

Pasta has recently become a popular source of complex carbohydrate for athletes, and also for those embarking on a low sodium, low fat, high complex carbohydrate diet. Carbohydrate loading is a technique that has gained popularity among athletes to increase muscle glycogen stores for prolonged exhaustive exercises such as running marathons (20). Pasta, as a source of complex carbohydrate, is a valuable part of this diet.

VI. PROTEIN-FORTIFIED PASTA PRODUCTS

Over the years numerous studies have been conducted on producing protein-fortified pasta products. Researchers have looked at incorporating soy flour (21), nonfat dried milk (22), wheat flour, oat flour, corn germ, and casein (23), corn soy milk (24), and fish protein concentrate (25) to enhance the protein quality of the pasta. However, all suffered from some undesirable consumer acceptability trait, whether it be color, taste, texture, or mouthfeel. A study conducted by Seyam et al. (26) attempted to reevaluate the merits of incorporating protein sources into pasta using semolina as a base. More than 60 different high protein materials were used in their study. Of these 35 were selected for further investigation after preliminary evaluation of the original 60 showed processing and quality problems. The sources of protein included commercially available soy products, wheat germ, defatted cotton seed meal, defatted oat, fish protein concentrate, corn protein isolate, low lactose whey, egg albumin solids, and edible bean protein. In general their results showed that again, the protein fortified products and formulations all suffered some deleterious quality trait, and low degree of consumer acceptability because of poor product color cooking quality, or taste.

VII. FACTORS INFLUENCING PASTA QUALITY

A. Pasta Processed from Semolina/Farina

Although stated earlier that durum wheat semolina is the raw material of choice for the production of pasta products, almost any type of wheat may be and is used for producing pasta products (27). Fernandez et al. (28) have shown that the mill streams of durum were more yellow than those obtained from bread wheats and that they gave lower absorptions, which is an advantage in pasta processing since less water has to be removed in drying. Subsequent reports by Mousa et al. (29–30) have shown that pasta processed from durum wheat granular mill streams (GMS) with other bread wheat GMS produces a better pasta product in appearance and cooked properties than pasta processed from the GMS of bread wheats alone. Wyland and D'Applonia (17) studied the influence of drying temperature and farina blending on pasta quality. Blends were prepared that contained the durum semolina and 0.5, 10, 20, 40, 60, and 100% of each class of hard red spring and hard red winter wheat farina. Temperatures of 40, 60, 70, and 80°C were used in drying the spaghetti after extrusion. Results showed that increasing drying temperature improved spaghetti color, increased cooked firmness, and decreased cooking loss and cooked weight values. Increasing the proportion of HRS and HRW wheat farina in the farina–semolina blends brought about a decrease in cooking loss, cooked weight, and spaghetti color. Drying at the higher temperatures improved spaghetti firmness. Wyland and D'Appolonia (17) concluded that a good quality pasta product can be obtained by incorporating a certain percentage of farina with semolina and that the quality of these products can be improved by using high temperature drying.

A report by Kim et al. (31) assessed the influence of HRW wheat farina on spaghetti cooking quality and color compared to spaghetti processed from semolina. They found that dry spaghetti made from farina of 14.3% mill extraction appeared almost the same as durum spaghetti, except its color was a faded yellow. Spaghetti cooked firmness was slightly lower and stickier than the 100% semolina product. Farina of fine, intermediate, and coarse granulation reportedly gave spaghetti of good quality. However, when granulation was at flour fineness, spaghetti had an uneven surface, brown color, and gave higher cooking losses.

A newly released report by Kim (32) reviewed the noodle-making properties of U.S. produced wheats, namely Dark Northern Spring (DNS), Hard Red Winter (HRW), and Western White (WW) Wheats. This publication provides an up-to-date evaluation of the effects of wheat flour protein on noodle quality and cooking quality. Kim notes that noodles can be considered as Chinese type and Japanese or Korean type. In Chinese type noodles, lye water is used in processing, whereas Japanese or Korean noodles are processed with salt water. Some general conclusions drawn from Kim's studies indicate that the optimum protein content for Korean (Ramyon) noodles is 9.28–9.62%, regardless of whether the product was made from a blend of HRW–WW or DNS–WW, and that there were no significant differences in cooking quality (cooked weight, cooked volume, cooking loss) between the blended products. The report provides a good review of the limited published data on noodle processing.

B. Pasta Processed from Sprout-Damaged Grain

Germination (sprouting) of grain before it is harvested can be a problem when rain and cool weather prevents or slows down normal harvesting operations. Pasta manufacturers are particularly sensitive to using semolina milled from sprouted durum wheat in their pasta-processing operations since it can affect end-product quality. Several studies have been conducted on the problems of sprouting in terms of end-product quality (33–39). Some general conclusions from those studies indicate that test weight, kernel distribution, protein content, milling performance, pasta color, and cooking quality were not adversely affected by increasing sprout damage (decreasing Falling Numbers). The only major adverse effect appeared to be higher semolina speck counts and spaghetti shelf stability (35–36). It was also noted (36) that sprout damage levels of 4.0% or higher (Falling Numbers of 120 or less) resulted in pasta products having high potential for checking and cracking in storage. Commercial manufacturers of spaghetti are concerned not only with the problems mentioned above but also with the tendency of spaghetti processed from sprout-damaged grain to stretch and fall off the rods during the drying cycle. Because of such concerns a number of U.S. pasta manufacturers will not process pasta from semolina with Falling Numbers less than 300 (40). Since research results indicate that pasta can be processed without any apparent problems with Falling Numbers of 250 commercial manufacturer's use of 350 and higher seems an unusually large margin of safety not warranted by reported data.

C. Pasta Processed from Low Test Weight (T.Wt) Wheat

Test weight (lb/bu or kg/hl) is an important quality trait of wheat as it can effect numerical grade designation and milling performance. Recently Dexter et al. (41) reported on the relationship of durum wheat test weight to spaghetti quality. It was found that T.Wt. had a highly significant effect on durum wheat quality when considered in the absence of environmental damage. As T.Wt. decreased, durum wheat milling potential declined because of the combined effects of lower semolina yield, higher semolina ash, and duller semolina color. They found that semolina ash and semolina color were affected by growing season, and that low T.Wt. resulted in improved spaghetti cooked firmness and resilience because of a strong negative relationship between T.Wt. and wheat protein. The impact of T.Wt. on wheat grading was similar to that reported previously by Watson et al. (42).

D. Lipids and Their Role in Determining Spaghetti Cooking Quality

Lipids, although a minor constituent of pasta, appear to influence the physical properties of cooked spaghetti. Fabriani et al. (43), in their studies on the changes in lipids during processing of pasta products, found that lipids were less amenable to extraction in the pasta product than semolina. Such results suggested that lipids undergo chemical changes and/or complexation as a result of the mechanical action of the screw on the semolina dough during extrusion. Subsequently Barnes et al. (44) reported that about 90% of the free lipids in semolina became bound during processing and, more particularly, during the drying process. Laignele (45) has written a review on lipids and their role in pasta and pasta products. A more recent study by Matsuo et al. (46) has shed some new light on the role of lipids in pasta products. The lipids found to exert marked influence on spaghetti cooking quality were the monoglycerides. They not only reduced surface stickiness but also improved tolerance to overcooking. The ability of monoglycerides to form water-insoluble complexes with amylose is thought to result in this reduced stickiness (47).

Unsaturated monoglycerides, for example, monolinolein and monolinolein, are more effective in forming complexes with amylose at lower temperatures (30–40°C) than at higher temperatures (60°C). Monoglycerides with saturated fatty acids, on the other hand, are very effective in complexing amylose in amylose solution at the higher temperature of 60°C. Therefore, if complex formation can be assumed to follow similar temperature dependence in doughs, the temperature at which dough development occurs during processing may be the deciding factor as to the effect of various lipids on the cooking quality of pasta.

Monoglyceride–starch interactions in suspension were studied by Kim and Robinson (48) at temperatures ranging from 30°C to 90°C. They found that at 30°C amylose bound small amounts of lipids due to its extended conformation. At gelatinization temperatures near 60°C, and the resultant conformational change of amylose from extended to helical form, the amount of monoglycerides bound to starch increased reaching a maximum at 90°C. Matsuo et al (46) proposed that if such a mechanism existed in a dough system, it would explain their results, which showed a decrease in pasta cooked stickiness for high temperature dried pasta with added monoglycerides compared with low temperature dried pasta.

E. Protein Quantity vs. Quality and Impact on Pasta Cooking Quality

The cooking characteristics of pasta products are the ultimate tests in determining its quality. In general, cooked pasta should be neither "mushy" nor "rubbery." It should retain its shape during cooking and be firm to the bite (al dente). Cooking time is important in terms of relative speed of cooking and tolerance to overcooking.

Three major components of cooking quality assessment include cooked weight, cooking loss, and cooked firmness (texture). Cooked weight (expansion volume) is a measure of the water-absorbing capacity of the pasta during cooking and should be three times the weight of the dry material. Cooking loss is the percent solids lost to the cooking water. Cooked firmness determines the chewing characteristics of pasta. Cooked weight and cooking loss are relatively easy to measure, but objective measurements of firmness and stickiness has been the subject of study over many years (49–56). Objective firmness tests are now used routinely for cooked spaghetti, since it was shown they have a high positive correlation (r = 0.812) with taste panel scores. Research has also shown there is a significant positive correlation between cooking quality and protein quantity and quality (57–64). In general, results show that higher protein and stronger gluten protein in semolina produces pasta with better overall cooking quality and tolerance to extended cooking than do lower protein, weaker gluten products.

Because of the positive correlation between stronger gluten and improved pasta cooking quality, considerable research has been directed towards the development and interpretation of prediction tests related to gluten quality. An article by Dick (65) reviewed some of the tests used to predict durum wheat and pasta quality. In his review, Dick discusses such prediction tests as the mixograph, farinograph, wet gluten, sodium dodecyl sulfate (SDS) sedimentation, electrophoresis, and chromatography and their relevance to pasta quality. This review and the references therein are a valuable source of information on the development of the various test procedures used to predict pasta quality.

F. Regrinds and Impact on Pasta Quality

Commercial pasta processing is such that anywhere from 5 to 15% of product end up as regrinds. These are reprocessed into pasta after blending back with the raw material prior to mixing and extrusion in the press. A study by Donnelly (66) has shown that blending regrinds beyond 15% can adversely reduce the color of the final product (spaghetti) and cooked firmness and also reduce the tolerance to extended cooking times when compared with a 100% semolina processed spaghetti as control.

VIII. PASTA QUALITY EVALUATION

No standard procedure exists for the determination of pasta quality in terms of appearance, color, and cooking quality. Pasta quality is such a subjective matter than what is acceptable in one country is not necessarily acceptable in another. Therefore objective/subjective evaluation of pasta in laboratories around the world evolved with the perceived needs of the indigenous consumer. Two examples provided here reflect these differences.

In Italy the evaluation of pasta cooked stickiness, firmness, and bulkiness is widely accepted and applied (67). The test is performed on spaghetti of 1.60–1.65 mm or 1.70–1.75 mm diameter cooked under standard conditions for 10–11 min according to the diameter. At least three expert tasters assess the cooked product for:

1. *Stickiness*, which is the state of surface disintegration of the cooked product, estimated by visual inspection, with or without the aid of a standard reference pasta.
2. *Firmness*, which is the resistance of the cooked pasta when chewed or flattened between the fingers or sheared between the teeth.
3. Bulkiness, which is the degree of adhesion of pasta strands after cooking and is evaluated visually and manually.

The characteristics are then reported as in Table 14. Overall pasta cooking quality is determined by summing the score obtained for each characteristic, multiplying the sum by 33.3, and dividing by 100. The final value of the cooking quality is correlated to a description.

Spaghetti with a total score equal to or below 40 is poor; above 40 but below or equal to 50 is not satisfactory; above 50 but below or equal to 70 is fair; above 70 but below or equal to 80 is good; above 80 is excellent.

By contrast, an evaluation procedure used in the United States involves objective procedures that not only assess the cooking quality of pasta but also that of the raw materials used in the processing of the pasta (68). Computer and statistical analysis of quality evaluation data provides for overall quality rating within major and minor fault parameters (Table 15). Major emphasis is placed on such quality traits as wheat protein, semolina and spaghetti color, and spaghetti cooked firmness. Faults in any of these traits change the acceptability of the sample quickly. An advantage of this type of computer scoring system is its flexibility for adjustment to meet changing quality demands.

IX. CONCLUSION

This chapter provides an overview of factors which can influence the processing and quality attributes of pasta products. Processing of pasta has evolved over many years from an art to a highly sophisticated continuous raw materials blending, mixing, extrusion, drying, and packaging technology as we know it today. In order to produce superior quality pasta, attention has to be paid to the source and quality of raw materials used; the quality of water mixed with the raw materials to form the dough prior to extrusion; the quality of other ingredients used in the dough mix (eggs, etc.); extrusion conditions; and drying conditions (LT.HT.UHT, or microwave).

Taken in total, producing high quality pasta is much more complicated than it might first appear. The roles played by plant breeders, cereal chemists, producers, millers, and the grain markets can have significant influence on what the pasta manufacturers will use and process in their plants. This in turn will ultimately affect the consumer's perception of the product as a desirable food. In the United States the American per capita consumption of pasta products has risen from 7 lbs (3 kg) in 1968 to 15 lbs (6.8 kg) in 1987 (69) and more recently is estimated to be 16.3 lbs (7.4 kg) (70). It is anticipated that per capita pasta consumption in the United States will continue to increase, which bodes well for the future of the industry. An industry study indicates that in the 10-year period between 1986 and 1995, consumption of dry and packaged pasta will double in value from $1.5 billion to just over $3.0 billion (Table 16) (70). Similar growth is projected for both canned and frozen pasta but at significantly lower dollar value.

Contributing to the increase in pasta consumption trends is the phenomenal growth of fresh pasta. Projections are that fresh pasta sales will average 17% annually over the next decade to total $720 million in 1998 or 10% of the total pasta market (71). Fresh pasta from the extrusion press has approximately 31% moisture, and since many fresh pastas are sold unpackaged, manufacturers have no labeling requirements as do the dry packaged products. The products can be processed from nondurum wheat flours or farina and many contain egg. Once fresh pasta is processed, shelf life becomes an important consideration in getting the product to the consumer. In general, fresh pasta has a shelf life of about 30 days. Some distributors freeze the product during the distribution phase and let it "slack" out to get some extra shelf life days. According to Shaffer (72), future improvements in packaging and oxygen removal will help preserve the fragile flavors of fresh pasta while extending its shelf life.

Concomitant with the increasing consumption of fresh pasta is the fact that more and more pasta companies are discovering the marketability of flavored pasta. Ostrove (73) points out, we now have a situation in the pasta market where the product has become more than simply a vehicle for sauces, but is available in nearly every conceivable flavor and color as a compliment to any meat, fowl, or fish. Dehydrated powders of vegetable flavors are generally preferred as overall quality of the products tends to be better. Powders such as spinach, carrot, tomato, corn, broccoli, and others as well as spices like saffron and flavors like vanilla, mushroom, cayenne, and curry are becoming more and more popular. Fruit-flavored pasta, as well as walnut and grand marnier–sweetened cream sauce, etc., suggests utilization of pasta for desert.

So the breadth and depth of pasta utilization is being continually explored and reflects some of the optimism being projected by the industry for increased consumption trends.

As this chapter was completed, a new book was published by the American Association of Cereal Chemists entitled *Durum: Chemistry and Technology* (67). This is an excellent source of information for anyone interested in pursuing the history, genetics, production, composition, quality, and utilization of durum wheat.

TABLE 1 Code of Federal Regulation: Pasta Products

Macaroni	Such products are tube-shaped and more than 2.8 mm but not more than 7 mm in diameter.
Spaghetti	This macaroni product is tube-shaped or cord-shaped (not tubular) and more than 1.5 mm but not more than 3 mm in diameter.
Vermicelli	This macaroni product is cord-shaped (not tubular) and not more than 1.5 mm in diameter.
Noodle Products	This class of food is prepared by drying formed units of dough made from semolina, durum flour, farina, flour, or any combination of two or more of these,with liquid eggs, frozen eggs, dried eggs, egg yolks, frozen yolks, dried yolks, or any combination of two or more of these, with or without water, and with or without one or more of the optional ingredients specified in the CFR. The finished noodle product must contain not less than 87% of total solids. The total solids of noodle products contains not less than 5.5% by weight of the solids of egg or egg yolk.

Source: Ref. 3.

TABLE 2 Principal Durum-Growing Areas in the World

Afghanistan	Iraq
Algeria	Jordan
Argentina	Libya
Australia	Italy
Canada	Spain
Ethiopia	Syria
France	Tunisia
Greece	Turkey
India	United States
Iran	U.S.S.R.

Source: Ref. 4.

TABLE 3 Durum Wheat Production For Major U.S. Producing States

	Crop Year (million metric tons)				
State	1987	1986	1985	1984	1983
Arizona	0.10	0.11	0.11	0.20	0.13
California	0.14	0.18	0.20	0.26	0.17
Minnesota	0.03	0.04	0.05	0.04	0.03
Montana	0.14	0.12	0.04	0.10	0.11
North Dakota	2.09	2.18	2.60	2.14	1.48
South Dakota	0.08	0.04	0.07	0.09	0.06
Total U.S. Production	2.58	2.66	3.06	2.82	1.99

Source: Ref. 5.

TABLE 4 Durum Wheat Harvest Data

	Upper Great Plains		Pacific Southwest	
	1987	1986	1987	1986
Test weight (kg/hl)	75.5	76.4	80.6	81.5
Vitreous kernels (%)	90	85	93	87
Moisture (%)	12.1	12.4	8.1	7.9
Protein dry basis	16.4	15.2	15.3	14.4
(@12% moisture)	14.4	13.4	13.5	12.7
1000 kernel weight (g)	38.4	41.2	49.0	53.0
Kernel size distribution (%)				
Large	43	45	69	82
Medium	54	52	30	17
Small	3	3	1	1
Ash (%)	1.69	1.75	1.65	1.73
Falling number (sec)	298	364	456	427

Source: Ref. 5.

TABLE 5 Durum Wheat Cultivars Grown in the Upper Great Plains

Cultivar	Origin	Date[a]	Cultivar	Origin	Date[a]
Arnautka	Russia	1856	Stewart-63	Canada	1964
Pelissier	Algeria	1900	Leeds	U.S.—ND	1966
Kubanka	Russia	1900	Rolette	U.S.—ND	1971
Pentad	Russia	1911	Wakooma	Canada	1972
Mindum	U.S.—MN	1917	Ward	U.S.—ND	1972
Golden Ball	South Africa	1918	Botno	U.S.—ND	1973
Carleton	U.S.—ND	1943	Crosby	U.S.—ND	1973
Stewart	U.S.—ND	1943	Macoun	Canada	1973
Vernum	U.S.—ND	1947	Rugby	U.S.—ND	1973
Nugget	U.S.—ND	1950	Cando	U.S.—ND	1975
Sentry	U.S.—ND	1954	Coulter	Canada	1977
Langdon	U.S.—ND	1956	Calvin	U.S.—ND	1978
Ramsey	U.S.—ND	1956	Edmore	U.S.—ND	1978
Towner	U.S.—ND	1956	Vic	U.S.—ND	1979
Yuma	U.S.—ND	1956	LLoyd	U.S.—ND	1983
Wells	U.S.—ND	1960	Monroe	U.S.—ND	1985
Lakota	U.S.—ND	1960	Renville	U.S.—ND	1988

[a]Approximate date released, distributed, or first grown in the United States.

TABLE 6 Grades and Grade Requirements of Durum Wheat[a]

Grade	Maximum limits of							
							Wheat of other classes[d]	
	Minimum test weight per bushel (lb)	Heat-damaged kernels (%)	Damaged kernels (total)[b] (%)	Foreign material (%)	Shrunken and broken kernels (%)	Defects (total)[c] (%)	Contrasting classes (%)	Wheat of other classes[e] (%)
U.S. No. 1	60.0	0.2	2.0	0.5	3.0	3.0	1.0	3.0
U.S. No. 2	58.0	0.2	4.0	1.0	5.0	5.0	2.0	5.0
U.S. No. 3	56.0	0.5	7.0	2.0	8.0	8.0	3.0	10.0
U.S. No. 4	54.0	1.0	10.0	3.0	12.0	12.0	10.0	10.0
U.S. No. 5	51.0	3.0	15.0	5.0	20.0	20.0	10.0	10.0
U.S. Sample grade	U.S. Sample grade shall be wheat that (1) does not meet the requirements for the grades U.S. Nos. 1,2,3,4, or 5, or (2) contains eight or more stones, two or more pieces of glass, three or more crotalaria seeds (*Crotalaria* spp.), two or more castor beans (*Ricinus communis*), four or more particles of an unknown foreign substances(s) or a commonly recognized harmful or toxic substance(s), or two or more rodent pellets, bird droppings, or equivalent quantity of other animal filth per 1,000 g of wheat; or (3) has a musty, sour or commercially objectionable foreign odor (except smut or garlic odor); or (4) is heating or otherwise of distinctly low quality.							

[a]Reprinted, with modifications, from USDA (1985). The subclass Hard Amber Durum shall be Durum Wheat with 75% or more of hard and vitreous kernels of amber color. The subclass Amber Durum Wheat shall be Durum Wheat with 60% or more but less than 75% of hard and vitreous kernels of amber color. The subclass Durum Wheat shall be Durum Wheat shall be Durum Wheat with less than 60% of hard and vitreous kernels of amber color.

[b]Includes Heat-damaged kernels.

[c]Defects (total) include damaged kernels (total), foreign material, and shrunken and broken kernels. The sum of these three factors may not exceed the limit for defects.

[d]Unclassed Wheat of any grade may contain not more than 10% of wheat of other classes.

[e]Includes contrasting classes.

TABLE 7 Characteristics of Durum Wheat Compared with Common Bread Wheats

Separate species of wheat
Tetraploid (28 chromosomes) vs. hexploid (42 chromosomes) for the common wheats
Hardest of all wheat classes
Amber in color
Higher test weight and 1000-kernel weight
Kernel sizes larger in relation to height and width
Endosperm has higher level of xanthophyll pigments, which gives semolina its bright amber appearance
Pasta products made from durum semolina have better cooking stability

Source: Ref. 7.

TABLE 8 Particle Size Distribution for Typical U.S. Commercial Semolina

U.S. sieve No.	Sieve Opening (mm)	% Total Semolina
On 20	0.86	0.0
On 40	0.38	22.8
On 60	0.23	57.6
On 80	0.18	14.6
On 100	0.14	9.3
Through 100		1.7

Source: Ref. 7.

TABLE 9 Water Composition for Pasta Production

	Maximum limits (mg/liter)
Calcium or magnesium carbonate	180–200
Sulfates	70–90
Silicates	25–30
Nitrates, nitrites	5–10
Chlorides	5–10
Organic matter	10–40
Optimum pH	6.6–6.9

Source: Ref. 2.

TABLE 10 Impact of VHTD on Quality of Pasta

Quality parameter	Result
Color	Improved
Available lysine	Unchanged
Plate count	Negligible
Enzyme activity	Absent
Cooked firmness	Improved
Cooked stickiness	Reduced

Source: Ref. 18.

TABLE 11 Nutritional Composition of Pasta (per 100 g)

Pasta	Moisture (%)	Ash (g)	Calories	Protein (g)	Fat (g)	Total carbohydrates	Dietary fiber (g)
Dry	9.8	0.7	366	12.8	1.6	75.1	1.8
Range (dry)	5.2–12.0	0.6–0.8	—	12.1–14.2	0.2–2.6	—	1.4–2.5
Cooked	71.5	0.18	117	4.3	0.6	25.2	0.5–0.68
Egg noodles							
Dry	9.2	0.9	380.6	14.0	4.2	71.7	1.7
Range (dry)	7.9–11.5	0.7–1.1	—	12.0–14.4	3.1–5.6	—	0.86
Cooked	70.4	0.7	125	4.1	1.5	23.3	1.1–2.6

Source: Ref. 19.

TABLE 12 Amino Acid Content of Pasta Products

Amino acids	Uncooked macaroni (mg/100 g food)	Cooked macaroni (mg/100 g food)
Histidine	310	98
Isoleucine	580	180
Leucine	1,060	330
Lysine	290	90
sulfur amino acids (Met & Cys)	630	195
aromatic amino acids (Phy & Tyr)	1,110	330
Threonine	410	130
Tryptophan	170	53
Valine	650	200

Source: Ref. 19.

TABLE 13 Vitamin/Mineral Content of Pasta Products

Nutrients per 100 grams dry weight		Enriched Macaroni Product	Enriched Noodle Product	U.S. RDA
Thiamine	mg	1.07	1.08	1.5
Niacin	mg	8.35	7.94	19.0
Riboflavin	mg	0.48	0.49	1.7
Vitamin B_6	mg	.084	.088	2.0
Folacin	ug	12.0	16.0	400.0
Vitamin E	mg	trace	trace	30.0 (IU)
Pantothenic Acid	mg	0.3	10.0	
Biotin	ug	1.0		300
Iron	mg	4.09	4.91	18.0
Calcium	mg	20.0	32.0	1200.0
Magnesium	mg	48.0	71.0	400.0
Phosphorus	mg	148.0	215.0	1000.0
Potassium	mg	152.0	191.0	[a]
Sodium	mg	8.0	21.0	[a]
Zinc	mg	1.0	1.76	15.0
Copper	mg	0.247	0.260	2.0
Manganese	mg	0.525	0.690	[a]
Selenium	ug	90.0	85.0	[a]

[a]No U.S. RDA has been established.
Source: Ref. 19.

TABLE 14 Evaluation of the Quality of Cooked Pasta

Score	Stickiness	Bulkiness	Firmness
0	Totally	Totally	Absent
20	Very High	Very High	Rare
40	High	High	Insufficient
60	Rare	Rare	Sufficient
80	Almost absent	Almost absent	Good
100	Absent	Absent	Excellent

Source: Ref. 67.

TABLE 15 Durum Program Faulting and Scoring Values

	Range[b]		Effect on evaluation score[c]	
Variable[a]	Minor fault	Major fault	Minor fault	Major fault
Test weight (lb/bu)	–2.2	–3.1	—	–1
1000-kernel weight (g)	–2.1	–5.1	—	–1
Small kernels (%)	+5	+10	—	–1
Wheat protein (%)	12.5	11.5	–1	–2
Total extraction (%)	–2.5	–3.5	–1	–2
Semolina extraction (%)	–3.0	–4.0	–1	–2
Semolina color	–10	–15	–2	–3
Specks per 64.5 cm^2 (10 in^2)	+10	+15	—	–1
Semolina protein (%)	11.5	11.0	–1	–2
Visual spaghetti color	–1.0	–1.5	–2	–3
Firmness (g-cm)	–1.5	–2.25	–1	–2

[a]Other variables measured and recorded that do not contribute to the final score are large and medium kernels (%), total extraction, falling number, and mixogram.
[b]Wheat and semolina protein percentages are fixed lower limits for faults. All other values represent the deviation from the average of the standards required to warrant a minor or major fault.
[c]These values are subtracted from the beginning score of four.
Source: Ref. 68.

TABLE 16 The Market for Pasta by (Type)($ millions)

Year	Dry and packaged	% of Total	Canned	% of Total	Frozen	% of Total
1974	$565.4	64.0	$264.0	29.9	$55.1	6.2
1975	586.4	62.6	284.2	30.3	66.2	7.1
1976	638.1	61.6	320.5	30.9	77.3	7.5
1977	751.7	61.8	372.6	30.6	92.3	7.6
1978	793.5	61.6	385.8	30.0	108.1	8.4
1979	814.2	59.6	418.2	30.6	134.8	9.9
1980	999.7	64.4	394.4	25.4	158.1	10.2
1981	1,136.0	66.3	408.6	23.8	169.0	9.9
1982	1,192.8	65.4	402.9	22.1	227.7	12.5
1983	1,294.4	64.8	428.7	21.5	274.6	13.7
1984	1,336.7	64.2	425.8	20.5	318.0	15.3
1985 E	1,369.8	62.2	454.8	20.7	376.5	17.1
1986 E	1,531.4	63.2	463.4	19.1	426.6	17.6
1995 P	3,272.7	66.8	762.5	15.6	863.1	17.6

E = estimate
P = projection
Source: Ref. 70.

REFERENCES

1. Hummel, Ch., *Macaroni Products: Manufacture, Processing, and Packing*, 2nd ed., Food Trade Press, Ltd., London, 1966, p.1.
2. Winston, J., *Macaroni, Noodles, Pasta Products*, In Publishing Corporation, New York, 1971.
3. Food and Drug Administration: Code of Federal Regulations; Title 21: Ch. 1, Subchapter B, Part 139, revised April, 1983, U.S. Government Printing Office, Washington, D.C.
4. *FAO Production Yearbook.* Food and Agriculture Organization, Rome, 1985.
5. *U.S. Wheat Crop Quality Report*, U.S. Wheat Assoc. Inc., Washington, D.C., 1987.
6. Sibbitt, L. D., and Gilles, K. A., Spring wheat varieties: their development, production, and utilization. *Proceedings of the Fifth National Conference on Wheat Utilization Research*, Agricultural Research Service-USDA, 1967.
7. Donnelly, B. J., and Gilles, K. A., Changes in production and utilization of durum wheat and durum products. *Proceedings of the Ninth National Conference on Wheat Utilization Research*, Agricultural Research Service-NC-40, 1976.
8. USDA Official United States Standards for Grain. Section 1.9, revised May 25, 1985, Federal Grain Inspection Service, U.S. Dep. Agric., Washington, D.C.
9. Walsh, D. E., and Gilles, K. A., Wheat: production and utilization, in *Macaroni Products*, The AVI Publishing Co., Westport, CT, 1974.
10. Banasik, O. J., Pasta processing. *Cereal Foods World, 26* (4):166 (1981).
11. Smith, F. J., Microwave-hot air drying of pasta, onions and bacon. *Microwave Newsletter, 12* (6):1979.
12. Manser, J., The influence of the bacteria count of pasta products during the drying process. *Buhler Diagram, 64*:9 (1978).
13. Pavan, G., High temperature drying improves pasta quality. *Food Engineering International* (Feb.) 37 (1980).

14. Gruennenfelder, F., personal communication 1987.
15. Manser, J., High temperature drying of pasta products. What is high temperature drying and how does it effect pasta products? *Buhler Diagram.* *69*:11 (1980).
16. Dexter, J. E., Matsuo, R. R., and Morgan, B. C., High temperature drying: effect on spaghetti properties. *J. Food Science*, *46*:1741 (1981).
17. Wyland, A. R., and D'Applonia, B. L., Influence of drying temperature and farina blending on spaghetti quality. *Cereal Chem.*, *59*(3):199 (1982).
18. Pavan, G., Pappotto, G., Ballini, N., and Didone, G., New technological developments in the production of dry pasta foods. Paper presented at Detmold, Germany, March 6–7, 1986.
19. *The Nutrient Profile of Pasta.* National Pasta Association, Arlington, VA, 1984.
20. Costill, D., Sports nutrition: The role of carbohydrates. *Nutrition News, 41*(1):1978.
21. Paulsen, T. M., A study of macaroni products containing soy flour. *Food Tech.*, *15*:118 (1960).
22. Globe, E. F, Anderson, P. W., and Goldman, P. F., Macaroni made with nonfat milk. *Cereal Sci. Today*, *12*:510 (1967).
23. Winston, M., and Winston, J., Investigation of the effect of protein additives on macaroni. *Tee. Moliteria*, *23*(14):486 (1972).
24. Fleetwood, W., Pasta and protein. *War on Hunger*, *5*:8(9) (1971).
25. Matsuo, R. R., Bradley, J. W., and Irvine, G. N., Effect of protein content on the cooking quality of spaghetti. *Cereal Chem.*, *49*:707 (1972).
26. Seyam, A. A., Breen, M. D., and Banasik, O. J., Study of the use of the unique functional characteristics of wheat in product development. Final Project Report to National Wheat Institute, Washington, D.C., June, 1976.
27. Irvine, G. N., Durum wheat and pasta products. In *Wheat Chemistry and Technology*, Pomeranz, Y. (Ed.), Ch. 15, American Association of Cereal Chemists, St. Paul, MN, 1978, pp. 777–796.
28. Fernandez, J. L. A., Shuey, W. C., and Maneval, R. D., Bread wheat granular mill streams with a potential for pasta production. 1. Physical and Analytical Properties. *Cereal Chem.*, *53*:308 (1978).
29. Mousa, E. I., Shuey, W. C., Maneval, R. D., and Banasik, O. J., Farina and semolina for pasta production. I. Influence of wheat classes and granular mill streams on pasta quality. *Association of Operative Millers* (July):4083 (1983).
30. Mousa, E. I., Shuey, W. C. Maneval, R. D., and Banasik, O. J., Farina and semolina for pasta production. II. Effect of granular mill stream blends on pasta quality. *Association of Operative Millers* (March): 4234 (1984).
31. Kim, H. I., Seib, P. A., Posner, E., Deyoe, C. W., and Young, H. C., Milling hard red winter wheat to farina. Cooking quality and color of farina spaghetti compared to semolina spaghetti. *Cereal Foods World*, *31*:810 (1986).
32. Kim, S. K., Research on noodle making properties of U.S. wheats. Project Report to U.S. Wheat Associates, Inc., Washington, D.C., March 31, 1989.
33. Bean, M. M., Keagy, P. M., Fullington, J. G., Jones, F. T., and Mecham, D. K., Dried Japanese noodles. 1. Properties of laboratory prepared noodle doughs from sound and damaged wheat flours. *Cereal Chem.*, *51*:416 (1974).
34. Bean, M. M., Nimmo, C. C. Fullington, J. G. Keagy, P. M., and Mecham, D. K. Dried Japanese noodles. II. Effect of amylase, protease, salts and pH on noodle doughs. *Cereal Chem.*,*51*:427 (1974).
35. Dick, J. W., Walsh, D. E., and Gilles, K. A., The effect of field sprouting on the quality of durum wheat. *Cereal Chem.*, *51*:180 (1974).
36. Donnelly, B. J., Effect of sprout damage on durum wheat quality. *Macaroni J.*, *62*(11):8 (1980).
37. Ibrahim, R. H., McDonald, C. E., and Donnely, B. J., Experimental high temperature drying of spaghetti. Paper presented at 66th Annual Meeting of the American Association of Cereal Chemists, Denver, CO, 1981.

38. Matsuo, R. R., and Dexter, J. E., Sprout damage in durum wheat and spaghetti. Paper presented at the 66th Annual Meeting of the American Association of Cereal Chemists, Denver, CO, 1981.
39. Dick, J. W., A review of sprouting of north american wheats as related to pasta processing and end-product quality. *Getreide Mehl und Brot, 36* (Jahrgang-Heft 7):187 (1983).
40. Kuhl, S., personal communication, 1987.
41. Dexter, J. E., Matsuo, R. R., and Martin, D. G., The relationship of durum wheat test weight to milling performance and spaghetti quality. *Cereal Foods World, 32*(10):772 (1987).
42. Watson, C. A., Banasik, O. J., and Sibbitt, L. D. Relation of grading and wheat quality factors to end-use quality characteristics for durum wheat. *Macaroni J., 58*(11):10 (1977).
43. Fabriani, G., Lintas, C., and Quaglia, G. B., Chemistry of lipids in processing and technology of pasta products. *Cereal Chem., 45*:454 (1968).
44. Barnes, P. J., Day, K. W., and Schofield, J. D. Commercial pasta manufacture: changes in lipid binding during processing of durum wheat semolina. *Z. Lebensm. Unters. Forsch., 172*:373 (1981).
45. Laignelet, B., Lipids in pasta and pasta products. In *Lipids in Cereal Technology*, P. J. Barnes (Ed.) Academic Press, London, 1983, p.269.
46. Matsuo, R. R., Dexter, J. E., Bordeau, A., and Doun, J. D., The role of lipids in determining spaghetti cooking quality. *Cereal Chem. 63*:484 (1986).
47. Eliason, A. C., and Krog, N., Physical properties of amylose-monoglyceride complexes. *J. Cereal Sci.,3*:239 (1985).
48. Kim, Y. J., and Robinson, R. J., Effects of surfactants on starch in a model system. *Staerke, 31*:293 (1979).
49. Binnington, D. S., Johannson, H., and Geddes, W. F., *Cereal Chem.*, 2:149 (1939).
50. Matsuo, R. R., and Irvine, G. N., Spaghetti tenderness testing apparatus. *Cereal Chem. 46*:1 (1969).
51. Matsuo, R. R., and Irvine, G. N., Note on an improved apparatus for measuring spaghetti tenderness. *Cereal Chem., 48*:554 (1971).
52. Walsh, D. E., Measuring spaghetti firmness. *Cereal Sci. Today, 16*(7):202 (1971).
53. Voisey, P. W., Larmond, E., and Wasik, R. J., Measuring the texture of cooked spaghetti. 1. Sensory and instrumental evaluation of firmness. *J. Inst. Can. Sci. Technol. Aliment., 11*(3):142 (1978).
54. Voisey, P. W., Wasik, R. J., and Lougheed, T. C., Measuring the texture of cooked spaghetti. 2. Exploratory work on instrumental assessment of stickiness and its relationship to microstructure. *J. Inst. Can. Sci. Technol. Aliment., 11*(4):180 (1978).
55. Dexter, J. E., Kilborn, R. H., Morgan, B. C., and Matsuo, R. R., Grain research laboratory compression tester: Instrumental measurement of cooked spaghetti stickiness. *Cereal Chem., 60*:139 (1983).
56. Dexter, J. E., Matsuo, R. R., and Morgan, B. C. Spaghetti stickiness: Some factors influencing stickiness and relationship to other cooking quality characteristics. *J. Food Sci., 48*:1545 (1983).
57. Matsuo, R. R., and Irvine, G. N. Effect of gluten on the cooking quality of spaghetti. *Cereal Chem., 47*:173 (1970).
58. Walsh, D. E., and Gilles, K. A., The influence of protein composition on spaghetti quality. *Cereal Chem., 48*:544 (1971).
59. Matsuo, R. R., Bradley, J. W., and Irvine, G. N., Effect of protein content on the cooking quality of spaghetti. *Cereal Chem. 47*:707 (1972).
60. Wasik, R. J., and Bushuk, W., Relation between molecular-weight distribution of endosperm proteins and spaghetti-making quality of wheats. *Cereal Chem., 52*:322 (1975).
61. Dexter, J. E., and Matsuo, R. R. Influence of protein content on some durum wheat quality parameters. *Can. J. Plant Sci., 57*:717 (1977).
62. Dexter, J. E., and Matsuo, R. R., The effect of gluten protein fractions on pasta dough rheology and spaghetti-making quality. *Cereal Chem., 55*:44 (1978).
63. Grzybowski, R. A., and Donnely, B. J., Cooking properties of spaghetti: factors affecting cooking quality. *J. Agric. Food Chem., 27*:380 (1979).

64. Dexter, J. E., and Matsuo, R. R. Relationship between durum wheat protein properties and pasta dough rheology and spaghetti making quality. *J. Agric. Food Chem.*, *28*:899 (1980).
65. Dick, J. W. Rheology of durum. in *Rheology of Wheat Products*, H. Faridi (Ed.), American Association of Cereal Chemists, St. Paul, MN, 1985, p. 219.
66. Donnelly, B. J., Pasta regrinds: Effect on spaghetti quality. *J. Agr. and Food Chem.* *28*:806 (1980).
67. Fabriani, G., and Lintas, C. (Eds.), *Durum Wheat: Chemistry and Technology*. American Association of Cereal Chemists, St. Paul, MN, 1988.
68. Nolte, L. L., Youngs, V. L., Crawford, R. D., and Kunerth, W. H., Computer program evaluation of durum and hard red spring wheat. *Cereal Foods World*, *30*:227 (1985).
69. Ronzoni, R. Ronzoni lauds pasta industry's record of energy and growth. *Milling and Baking News.*, *67*:1 (1988).
70. New Industry study sees solid growth for pasta over coming decade. *Pasta J.* *2*:15 (1986).
71. Doglione, N. The changing Italian pasta manufacturing industry. *Pasta J.*, *71*(51):21 (1989).
72. Shaffer, J., The fresh pasta industry makes dough. *Pasta J.*, *71*(3):17 (1989).
73. Ostrove, M. V., The flavored pasta explosion. *Pasta J.*, *17*(3):24 (1989).

20

CEREAL-BASED SNACK FOODS

Joseph A. Maga
Colorado State University, Fort Collins, Colorado

I. INTRODUCTION

Snack foods have always been an important part of American life, and they represent a diverse and continually expanding and changing group of food items whose sales in 1986 approached $25 billion. An overview as to the diversity, dollar volume, and percent market share is presented in Table 1. If one combines snack categories in Table 1 that are either totally or partially cereal based, cereals amounted to 47.3% of the total snack food market and had sales approaching $12 billion. On a worldwide basis, the United States is the leading snack food–consuming country, followed by the United Kingdom, West Germany, and France (2).

The snack food market is continually changing relative to product categories, and although most snacks are not primarily consumed for their nutrients, many snacks are formulated and marketed with nutrition in mind. Typical of this concept are many granola snacks, which were initially viewed as being wholesome and healthful but lately have been changed to sweet, chocolate-covered reward foods (3). As a result, sales in this category have fallen close to 20% in one year (1) because they lost their image as a "health food." Dried fruit snacks are also going through a similar decline due to the introduction of a wide range of fruit-based snacks that have less emphasis on nutrition. Because of the concern about salt in the American diet, manufacturers have changed formulations to lower salt or used salt substitutes. Some potato chips are now even marketed as no salt added chips. These changes, coupled with aggressive advertising, have maintained consumption trends for most snacks. In fact, one of the snacks that is usually salted, popcorn, actually staged a 26% increase in market growth primarily due to the growing popularity of microwave popcorn (1). Several companies are marketing potato chips, which normally contained 27–44% fat due to the deep-fat frying process, that now have only approximately 20% fat. Several snack food producers have recently promoted snacks that are fiber sources. One such item is formulated so that a 1.3 oz pack contains 10 grams of fiber (4).

The increased popularity of certain ethnic snacks, such as corn and tortilla chips, has significantly increased this segment of the snack industry. The popularity of Mexican foods has expanded into large eastern United States population centers, and along with Mexican drinks and main courses, Mexican snack food use has also expanded.

Packaging has had a dramatic influence on the snack food industry since currently approximately 95% of all snacks are sold in packages that weigh less than 35 grams (2).

Many snacks that traditionally have been sold in health food stores have now found their way into supermarket and convenience stores since small product lines have been bought by major food companies who have the advertising budget to heavily promote more nutritious snacks. The American snack food industry is dominated by well-established large companies, but there are many efficient regional and local processors and distributors. They exist because many snacks are low in density and thus are relatively expensive to transport long distances. Also, most snacks are rather fragile and have relatively short shelf lives, thus making their local or regional production more advantageous.

II. COOKIES AND CRACKERS

The most popular cereal-based snacks are cookies and crackers. Most cookie doughs or batters are not fermented by are just blended, formed and baked. In contrast, most cracker doughs are fermented before forming and baking.

The Biscuit and Cracker Manufacturers' Association reported that in 1986 slightly over 2 billion pounds of cookies were produced along with close to 1.5 billion pounds of crackers (1). In crackers, soft wheat flour usually represents around 90% of the product, whereas amounts of flour as low as 25% are used in cookies. On a per capita basis, as can be seen in Table 2, cookies are more popular than crackers, with sandwich-type and saltines being the most popular in each category.

Several brands have been around for generations (Oreos—75 years, Barnum's Animal Cookies). On the other hand, many others have relatively short histories. A classic example is now what is called soft cookies. They were nationally introduced in 1983 and were designed to simulate freshly baked at home cookies that have a softer center portion than the outside. Most are produced by using different sugar sources. The outside portion usually used granulated sugar, whereas liquid fructose or other liquid sugar sources are used inside. The liquid sugar portion is more hydroscopic and thus gives a moister texture. In 1985, the market for this type of cookies expanded to $450 million but fell in 1986 to $300 million (1). This clearly demonstrates the dynamics of the industry such that a relatively simple formulation modification resulted in a product that now shares 10% of the cookie market.

Other recent trends in the cookie and cracker industry include the introduction of numerous novel shapes, more chocolate-based items, and gourmet products, both domestic and imported.

A phase of the cookie industry that is also seeing a dynamic change is the Oriental or fortune cookie market. The traditional fortune cookie now has competition from such products as Wish Biscuits, which are gourmet fortune cookies with sophisticated sayings, as well as chocolate-covered fortune cookies.

Cookies and crackers have also undergone significant formulation changes with an eye towards nutrition emphasizing all natural ingredients, fruit juices as sweetening agents, lower salt, and added fiber.

A diverse range of flavor-coated crackers has also appeared where essentially the same formulation is used to make the base product but different flavors and colors are applied to the cracker surface. Popular flavors have been cheese, bacon, taco, and, currently, Cajun.

Nutritional losses are usually greater during the baking of cookies than for crackers mainly due to the interaction of protein fractions with the high levels of reactive carbohydrates that are present in most cookie formulations (5).

The small size of most cookies and crackers makes an accurate estimate of nutritional content relative to average serving size somewhat difficult. Therefore, the industry has agreed that a serving of cookies is 1 oz, which represents 3 sandwich, 8 sugar wafers, or 3 chocolate chip cookies. On the other hand, an average serving of saltines is 4, whereas an average serving of soup or oyster crackers is 18–20.

Historically, cookies and crackers have been called baked snacks and can be divided into sweet baked snacks (cookies) and salty-based snacks (crackers). Cookies, in turn, can be classified as sugar, bagged, icebox, sandwich, and bar-type. Typical formulations are shown in Table 3. Crackers, in turn, can be divided into soda crackers, which are fermented and include soda and saltine crackers; sprayed crackers, which are usually round and chemically leavened; and cheese crackers, which are fermented and contain cheese as a dough ingredient. Typical cracker formulations are shown in Table 4.

III. CORN AND TORTILLA CHIPS

After potato chips, corn and tortilla chips represent the second largest category of salted snacks. They currently are a $1.5 billion industry in the United States (1). Current new related items include a host of flavored products as well as snacks made out of native American blue corn. Acreage of blue corn is increasing, and the crop is being used exclusively for snack and blue corn flour tortillas.

Of all cereal-based snacks, the method for making corn and tortilla chips is unique, in that it has a long history, going back to the Aztecs, and as compared to soda cracker or pretzel manufacture, for example, it does not really involve sophisticated equipment and thus is adaptable to a small operation.

Authentic corn and tortilla chips start with a product called masa, which is either made fresh or bought in the dry or instant form. A general flow diagram for the production of masa and its subsequent use to produce corn or tortilla chips is outlined in Figure 1. The major difference between tortilla chips and corn chips is the extra baking step required in the manufacture of tortilla chips. Additional information on tortilla chips is provided in Chapter 2.

The key to the process is the initial nixtamalization or alkaline cooking and steeping step involving whole corn. In the past, not much attention was given to corn type in the manufacture of fresh masa that was to be immediately converted to chip products. However, corn quality is quite critical in the manufacture of dry masa. To date, corn hybrids that have a hard endosperm with a small dent, white cob, little or no tendency to form red streaks in the pericarp, easily removable pericarp, uniform kernels, and low dry matter losses during processing are highly preferred since such hybrids have been found to increase dry masa yield 10–20% (8).

In the actual process, good quality whole corn is cooked for up to 3 hours at 80–100°C with frequent stirring in 120–300% excess water containing 0.1–2.0% hydrated lime. The cooked corn is then permitted to steep, usually overnight. The cooking and steeping steps permit the endosperm to hydrate and soften, encourages partial starch gelatinization, and disrupts and partially dissolves the pericarp. Subsequent washing removes the pericarp and residual lime. the resulting cooking/steeping liquor normally contains 2–6% of dissolved and suspended solids (8), which is usually discarded. The washed material is then traditionally stone ground, resulting in fresh masa, which is then sheeted, cut, and either baked and fried to produce tortilla chips or just fried after sheeting and cutting to make corn chips. In the case of corn chips, the masa can also be directly extruded and cut into the frying oil instead of being sheeted and cut.

If reconstituted masa flour is used, a large particle size should be used for corn chips. This will provide for interruption in the dough structure so that air and water vapor can escape during frying. If small particle flour is used, excessive puffing occurs during frying,

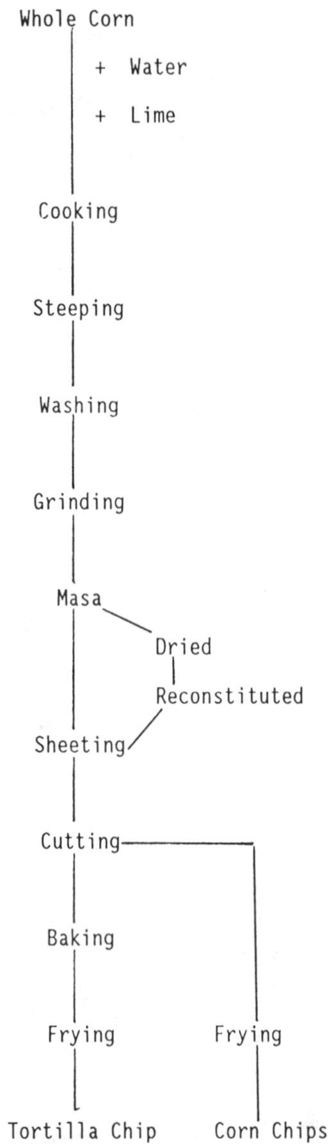

FIGURE 1 Production of masa and the resulting corn tortilla chips. (From Ref. 8.)

resulting in a chip that will absorb more oil during frying and be easily broken. Corn chips average 35% fat, whereas good quality tortilla chips have 25% (8). The preliminary baking step with tortilla chip manufacture sets the structure, thereby minimizing oil uptake during frying. Corn chips are usually made from a blend of white and yellow corn masa, which produces a chip that is light in color. Corn chips made from 100% yellow corn masa usually have an objectionable burnt flavor, which is attributed to the degradation of carotenoids during the frying step (8).

Many of the smaller manufacturers prefer to use dry masa flours that come in a wide range of colors and particle sizes, thereby permitting the optimization of masa to specific product. The need for capital equipment and labor are drastically reduced when masa flour is used. However, masa flour usually lacks some of the characteristic flavor associated with fresh masa, and usually the texture of products produced using masa flour is not as good. In addition, products made from masa flour stale faster than those made from fresh masa. They also have a shorter shelf life relative to rancidity. Since whole corn is used to produce masa, germ oil is present, which can go rancid if the flour is not properly processed and stored. The shelf life for masa flour is approximately 6 months (8). As a result of the above limitations, most large producers of corn and/or tortilla chips prefer to use fresh masa.

With either masa base, flavors and colors can either be added to the dough, which can result in significant flavor loss due to volatilization or degradation during frying, or they can be added either as a powder or in the form of an oil spray after frying.

IV. PIZZA

The toppings on a pizza or its style may vary, but the crust is still cereal based. Aside from the traditional thin crust pizza, many companies offer products such as thick crust, French bread pizza, pizza rolls, croissant pizza, and whole wheat pizzas.

The three major segments of the pizza industry are deli-style fresh pizzas that are traditionally offered in restaurants and fast food outlets but are now also available in many supermarkets, via home delivery, or as frozen pizzas. By far, the fresh baked in-store pizza still has the largest market ($15 billion), but the home delivery ($2 billion) and frozen ($1 billion-plus) markets are both rapidly expanding. The frozen market has especially benefited from the introduction of acceptable microwaveable pizzas, while the home delivery market has expanded primarily due to advertising and major expansion in delivery outlets with well over 20,000 outlets now in operation (1).

Pizza dough must function in several ways. First, it must have good extensibility so that it can either be mechanically rolled or hand stretched, and second, it must be strong enough to support sauces and toppings before and during baking. The flour of preference is milled from American Dark Northern Spring wheat or a good quality Hard Red Winter wheat having 13–14% protein.

The two major types of pizza are thin crust or Neopolitan, which is usually round in form and baked directly on the oven hearth, and thick crust or Sicilian, which is usually baked in deep pans. Both types of dough use the same basic formulation composed of yeast, salt, water, and flour. A typical base formulation is shown in Table 5. In addition, optional ingredients include sugar, malt, shortening or oil, and dough conditioners. These ingredients give the dough better extensibility, increased water absorption, improved crust color, taste and flavor, better crispness and chewiness, and better freezing properties.

Once fermented, rolled, and topped, thin crust pizza is usually baked at 500°F, whereas thick crust pizza is baked at 450°F. Baking time for both types ranges from 10–15 min. Typical amounts of dough required for various sizes of thin crust pizzas are shown in Table 6, and the general processing steps involved in making pizza are summarized in Figure 2.

V. IMPORTED AND MISCELLANEOUS

Due to the fact that many Americans view imported goods as having better quality and due to increased travel by Americans to foreign lands, a significant number of imported cereal-based snacks are finding their way into the domestic market. As seen in Table 7, most of these products come from Canada, as well as Denmark, West Germany, Japan, and the United Kingdom. Most products are in the form of biscuits, cakes, wafers, and similar snack items.

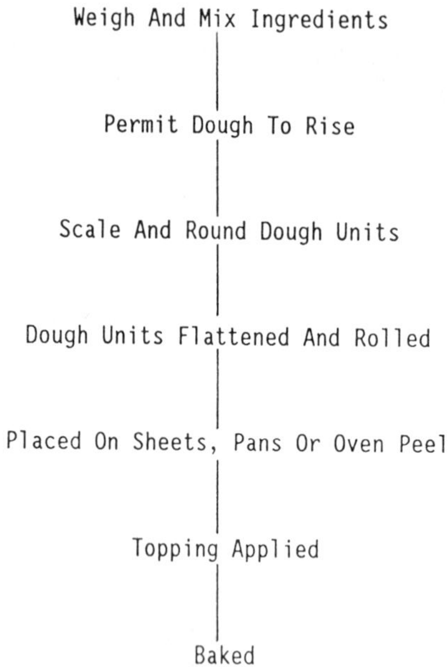

FIGURE 2 Typical steps in pizza manufacture.

Miscellaneous cereal-based snack foods are products that clearly do not fit any of the other major categories and thus cover a multitude of products produced domestically as well as being imported. Typical products in this category include rice cakes, crackers, pasta snacks, sesame sticks, bagel chips, baked whole grain rice snacks, and ethnic specialties.

Of special interest are various rice-based Oriental snacks whose popularity is increasing in the United States. Both glutinous (sticky) and nonglutinous rice is used, and, as seen in Table 8, their starch expansion properties are significantly different.

The classification of Japanese rice crackers relative to rice type is summarized in Figure 3, and a typical flow diagram for the manufacture of glutinous rice crackers is shown in Figure 4. Rice crackers made from glutinous rice are soft in texture, whereas nonglutinous rice produces rice crackers that have a hard and rough texture.

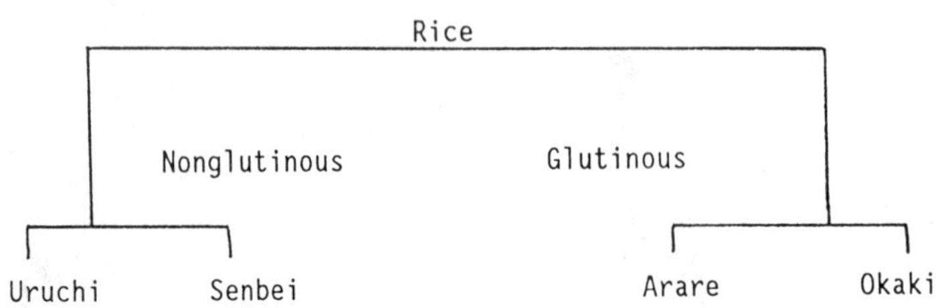

FIGURE 3 Classification of Japanese rice crackers. (From Ref. 9.)

Glutinous Rice
↓
Washed
↓
Soaked
↓
Drained
↓
Crushed
↓
Steamed
↓
Cooled
↓
Kneaded
↓
Frozen
↓
Cut
↓
Dried
↓
Flavor Coated
↓
Baked
↓
Dried
↓
Packaged

FIGURE 4 Processing of glutinous rice crackers. (From Ref. 9.)

VI. EXTRUDED SNACKS

The extruder and various modifications thereof can be used to make a wide variety of cereal-based snacks that in 1986 represented a $456 million market (1). The first commercially expanded, extruded corn snack was introduced in 1946, although the technology had been demonstrated 10 years earlier (10).

Extrusion technology offers many advantages over other traditional forms of processing (Table 9), the greatest of which is that it is a continuous process, which lends itself well to the production of both expanded and dense snacks. A general flow diagram of a typical process is shown in Figure 5. The major components of an extruder are shown in Figure 6, and various configurations for single and twin-screws are summarized in Figure 7. By changing screw configuration, a wide variety of extruded snacks can be produced.

Today three general categories of extruded snacks are recognized. The first includes highly expanded curls, ball, and related products. These snacks are called second generation since their manufacture is more complex than for the production of snacks such as potato chips, which are called first generation. They are basically made by introducing materials such as low moisture degerminated corn grits, into a extruder where the major source of energy comes from the dissipation of the mechanical energy input, thereby causing highly expanded end products, which are then usually dried and coated with flavorings, which usually contain oil, salt, and cheese powder.

The second category, called third generation, includes more technically sophisticated filled or tube-type snacks. Some require the use of two extruders, with the first cooking the

Ingredient Blending

|

Moisture Adjustment

|

Extrusion

|

Cutting

|

Drying

|

Flavor Addition

|

Packaging

FIGURE 5 Extrusion processed snacks.

dough while the second is used to form product shape. Using this technology, a wide variety of shapes and textures can be obtained.

The third category centers around pellet-based snacks that are formed as dense pellets via extrusion but are expanded before consumption using traditional frying or baking techniques. These are also considered to be third-generation snacks and have the advantages that the extruded pellets, also called half-life or intermediates, are very shelf stable and dense, thus they can be shipped long distances for economical transportation costs. Some typical flavored formulations are shown in Table 10. The manufacturing interrelationships between second-and third-generation extruded snacks are shown in Figure 8.

Almost any cereal can be extruded, but if expansion is a major objective, the number of functional cereals is commercially limited to dry-milled dehulled and degerminated corn grits or rice flour. Rice is interesting in that it gives a crisper product that is blander in flavor than corn. Thus, rice-based extruded snacks are more compatible with a wide range of added flavors, whereas in the case of corn-based extruded snacks, the added flavor should be compatible with corn flavor. Cereals that have significant amounts of lipid, such as oats, are more difficult to expand due to dough slippage within the extruder and usually require high moisture and high temperature conditions before significant puffing will occur. Another excellent puffing cereal is sorghum, which is not used much in snack foods in the United States because many of the varieties are heavily pigmented, which produces a dark extrudate. Wheat flour can also be used but does not give as much expansion as corn grits or rice flour. Noncereal flours such as potato and tapioca also give excellent expansions.

Purified or modified cereal starches can also be used. Their generally higher cost as compared to cereals is somewhat offset by the facts that they can give excellent expansion and produce very bland extrudates that can carry a wide range or added flavors. However, their amylose/amylopectin ration is somewhat critical since a high amylopectin content starch will produce fragile extrudates of rather poor density. Approximately 5–20% amylose in the starch will significantly improve expansion as well as texture. The better

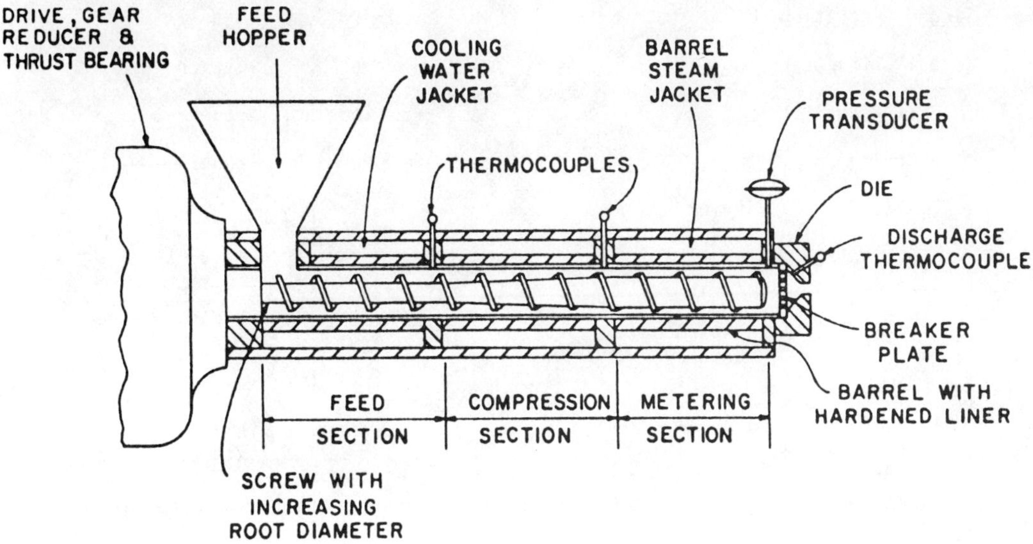

FIGURE 6 Typical extruder components. (From Ref. 11.)

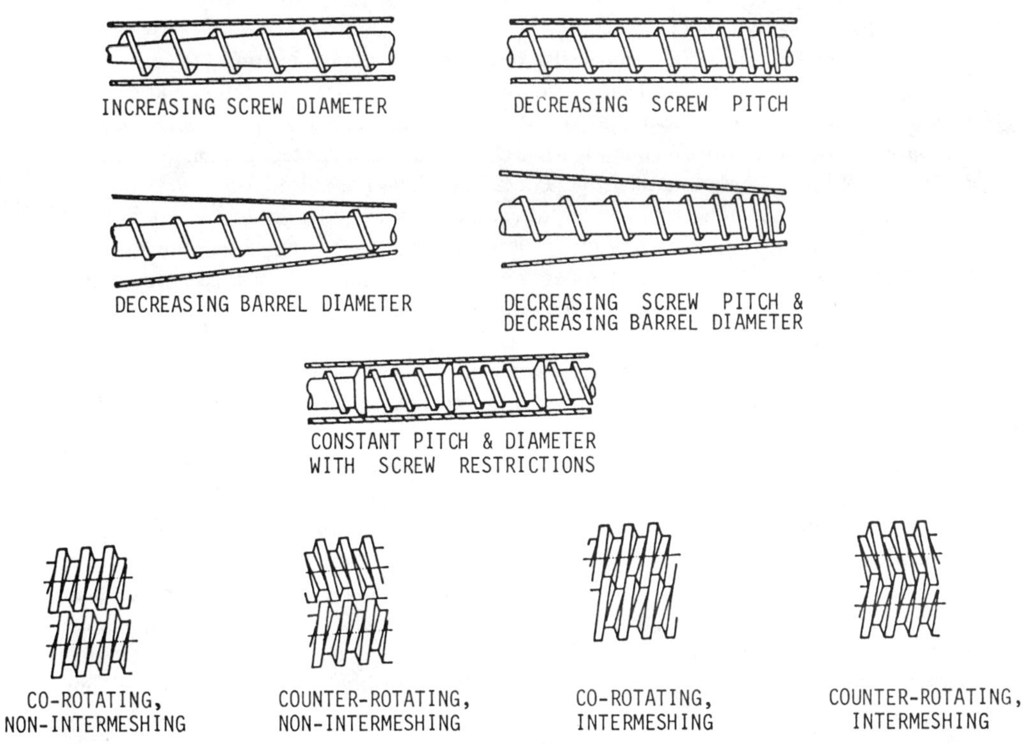

FIGURE 7 Single and twin-screw extruder configurations. (From Ref. 11)

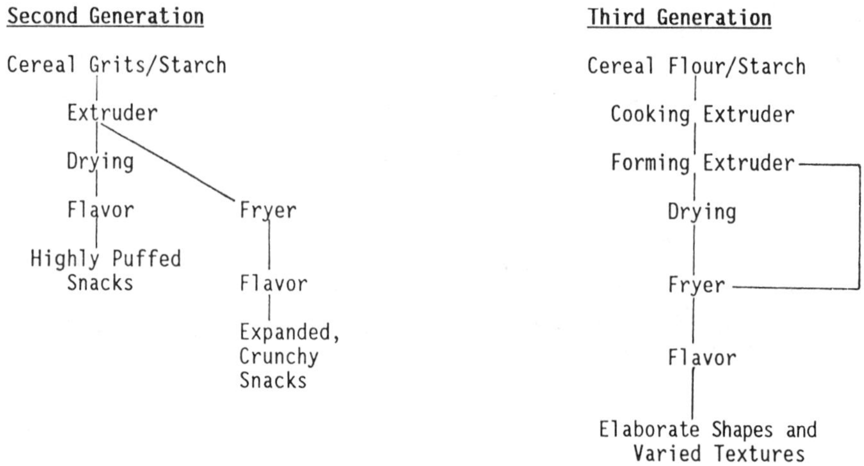

FIGURE 8 Extrusion processing of second- and third generation cereal-based snacks. (From Ref. 10)

expansion properties of amylose as compared to amylopectin is attributed to the phenomena that unbranched amylose can orientate itself easier and thus exit the extruder orifice in a more functional fashion required for expansion than branched amylopectin. Chemically modified starches, including cross-bound, pregelatinized, and phosphate, acetate, and hydroxymethyl derivatives, have been found to give good results, especially in the manufacture of half-life products (12).

Numerous other factors can influence the degree of puffing. For example, the amount of moisture in the feed material will influence the resulting dough temperature, and thus the amount of starch gelatinization that occurs during the extrusion process. an undergelatinized starch will not fully puff. If a starch becomes dextrinized, expansion also decreases. If conditions are optimum, the gelatinized starch will be able to form a very flexible dough that is easily expanded immediately after exiting the extruder due to the expansion of the associated steam when it reaches atmospheric pressure. In low moisture, high temperature doughs, a significant amount of water is lost as steam but the expanded product may still require further drying to obtain a shelf-stable product.

This loss of water vapor also reduces the retention of flavorings added to the dough prior to extrusion. Flavor compounds that are quite volatile or water soluble will be lost to the atmosphere as the product exits the extruder. Thus most manufacturers prefer to add flavorings after extrusion. Several of the major flavor houses have introduced flavorings which they claim are thermostable and can withstand extrusion conditions, but some loss of flavorants can still be expected.

Another important variable affecting puffing is the dough residence time within the extruder. Most larger units are equipped with a variable drive screw, which can be used to control the amount of starch gelatinization occurring within the extruder. Cereal particle size can dramatically influence the degree of product puffing. If particles are too small, especially with corn, they are difficult to feed and dough slippage can occur in the extruder, thereby decreasing puff. Some extruders are especially sensitive to this condition and thus require a uniform particle size feed material. The presence of added salt will also usually decrease puffing, and therefore, if extruded puffed products are to be salted, it is usually done in a subsequent step after extrusion. However, in the extrusion processing of half-life products, salt is added to the feed material so that product puffing is discouraged.

The addition of coloring agents prior to extrusion is usually not suggested since most colors tend to either thermally degrade or chemically react during extrusion. Also, with highly expanded products, physical fading of added color becomes apparent since the resulting expanded structure dilutes the amount of color per unit volume.

A major advantage of extruder-produced expanded snacks is that only one basic unit is needed since mixing, cooking, and forming are accomplished in a continuous fashion during extrusion. Also, by varying factors such as ingredients, moisture, extrusion temperature, and residence time, a wide variety of snacks ranging from dense to highly expanded can be produced. The variety of snacks that can be produced with the same unit can be expanded by having numerous dies of different shapes, thereby the same base formulation can be used to produce many interesting shapes and forms of the same product.

Historically, extruders had a single screw of relatively simple design that would produce a specific snack under a set of relatively rigid conditions; however, most new extruders are designed to give the operator a wide range of operating parameters. Typical of the new breed of extruder is the twin screw unit, which give the operator the option of producing a wide range of snacks having diverse textural properties.

Also, if an extruder is operated under conditions of high moisture and relatively low temperature, the unit can essentially be used as a dough-forming device, which, when equipped with a knife attachment, provides a continuous means of directly feeding a frying unit. This approach is used by some producers of corn chips. Due to the versatility of extruders, their use should continue to expand in the production of cereal-based snacks.

VII. POPCORN

Popcorn is a snack that is unique to America, with most Europeans having the opinion that corn was meant for animal consumption. However, in the United States, its popularity and the forms in which it is available continue to grow. The major reason for its popularity is that it is relatively inexpensive and easy to prepare even in homes. Again the microwave oven has also added greatly to its growth. Microwave popcorn accounts for 50% of popcorn sales (1). In 1986, $192 million was spent for microwave popcorn, while sales in the other two popcorn categories, candy glazed, and prepopped plain and flavored, were $106 million and $84 million, respectively. This does not include annual sales of over $400 million sold through outdoor/indoor leisure and foodservice outlets, $350 million in movie theaters, or $150 million that is sold as traditional unpopped kernels. When these various aspects of the industry are combined, total popcorn sales in 1986 approached $1.3 billion.

The commercial concept of microwave popcorn was introduced in the mid-1970s. At that time, microwave popcorn was sold in the frozen food section of grocery stores. Room temperature–stable microwave popcorn was introduced in 1984. Many brands of shelf-stable microwave popcorn are on the market today.

Several cereals including rice, wheat, and some sorghum varieties will pop to some extent when exposed to heat at atmospheric pressure, but none can match the 20–40 times expansion resulting from good quality popcorn. In the manufacture of puffed breakfast cereals, wheat or rice can be made to expand only several time in volume.

The composition and structure of raw popcorn kernels are such that popping occurs due to the high proportion of flinty or horny endosperm present in popcorn as compared to other cereals (13). Popcorn is also unique in that each kernel is covered with a tough by elastic pericarp that prevents moisture from gradually escaping the kernel. When the raw kernel is heated, moisture within the kernel is converted to steam and when the resulting pressure is sufficient, an explosion occurs.

The major measure of good quality popcorn is the amount of expanded product resulting from a specific weight or volume of unpopped corn. This in turn is influenced by numerous factors including popping method, differences in kernel maturity, kernel moisture level, and genotype. Kernel moisture is apparently quite critical, and for most popcorn varieties, 13–14% seems to be optimum. If the moisture content is lower, many kernels only partially open since not enough steam is produced to cause maximum expansion. On the other hand, if the kernels are too moist, small volume also results because too much steam pressure develops and the kernel pops prematurely. Also, the resulting popped kernels are rough and jagged. Volume can either be measured using the Official Volume Test (OVT) or the Weight Volume Test (WVT) (14). With the OVT, unpopped volume is compared to popped volume, whereas with the WVT, the pounds of raw popcorn are compared to the cubic inches of popped corn. Of the two, WVT is preferred since most raw popcorn is sold by weight and resold by popped volume.

Popped kernel tenderness is another important quality attribute and is indirectly related to degree of expansion since as expansion increases, so does apparent tenderness. Tenderness is also related to the amount of hard core at the base of each kernel. Usually the tough pericarp or hull is highly fragmented during popping and large amounts of free hulls are not desirable. Pericarp thickness also influences popped product texture with a thin pericarp being desirable, but usually the larger the kernel, the thicker its pericarp.

Most popcorn processors are not especially concerned with popcorn flavor, since usually butter and salt are added to popped products that are called plain flavored popcorn. In the case of caramel corn, the caramelized sugars contribute most of the flavor, and with flavored popcorns, the same is true. However, freshly popped corn has a very distinctive and pleasing flavor that certainly contributes to the overall acceptability of popcorn.

The overall shape of popped corn can also be of significance, especially to the processor who is making caramel corn. In this instance, varieties that produce a round or ball shape, often called mushroom popcorn, are preferred because they can be more uniformly coated and are not as fragile to required mixing and tumbling associated with the coating process. The other major shape characterization is butterfly popcorn, because the popped kernels have a highly irregular surface since the kernels actually turn inside out upon popping. Butterfly popcorn retains salt better and usually its texture is considered to be better than mushroom popcorn. Butterfly popcorn is typical of south American cultivars.

Popped kernel color can vary from chalky white to cream color, while hull color can vary from white to deep yellow. Some processors prefer lighter hull colors because residual hulls will not be as noticeable in the popped product. Most varieties that produce a mushroom puff are quite yellow.

The methods of harvesting and storage can also influence pop quality. Some farms still harvest on the cob and slowly cure the kernels on the cob for several weeks, whereas others combine at harvest when the corn is at 16–17% moisture. the kernels are permitted to naturally equilibrate in moisture. Popcorn harvested at high moisture levels needs to be dried, but the process must be done slowly or the kernels may develop cracks which then reduce popping quality. If properly harvested and stored, popping ability can continue to improve for up to a month. Properly stored raw popcorn can have a shelf life of over a decade, and thus it represents a cereal-based snack commodity where long-term storage is not a problem.

Commercial popcorn can be popped using wet (oil) or dry (air) techniques. Some of the advantages of wet popping include usually a better flavor and odor since oil is an intimate part of the process, and simplicity of operation. The process requires relatively inexpensive equipment. On the other hand, wet popping requires more oil than dry popping, irregular oil coverage results, the formation of smoke may be a problem, and equipment cleanup is quite difficult. With dry popping, more expensive equipment is required, but it better lends itself to a continuous operation and thus is more compatible with the production of caramel and flavored popcorn.

During the popping of corn, weight losses in the neighborhood of 15% can be expected. The major loss comes from water loss during popping. Good quality raw popcorn should be in the 13–14% moisture range at time of popping, and immediately after popping popcorn normally has a moisture content of approximately 2.5%. If the popped product is sifted, there is also the loss of hulls to consider. Also, some pieces of popped popcorn break during handling and thus usually represents another loss.

The packaging of popped popcorn is a special problem since freshly popped corn is very hydroscopic and will absorb moisture from the air even at 20% relative humidity. Moisture uptake dramatically influences texture and the product becomes elastic and chewy.

Traditional popcorn has historically been flavored with butter or flavored/colored oils and salt. A typical formulation would be 73% popped corn, 22% oil, and 5% salt (14). Caramel and cheese-coated popcorn are also available but have a relatively short shelf life if not properly packaged, again because of their ability to absorb moisture from the environment thus becoming sticky.

VIII. CEREAL MEAL BARS

The cereal meal bar area is in a period of dramatic change. Once products known as granola bars had a very positive image a products that were healthful snacks. However, most granola bars are now formulated akin to candy bars, although a few are still cereal, fruit, and nut based. Sales have fallen the last several years, with 1986 sales being $359 million, down from $439 million in 1985(1). Typical formulations for this type of product are shown in Table 11 and the general steps in manufacture are outlined in Figure 9.

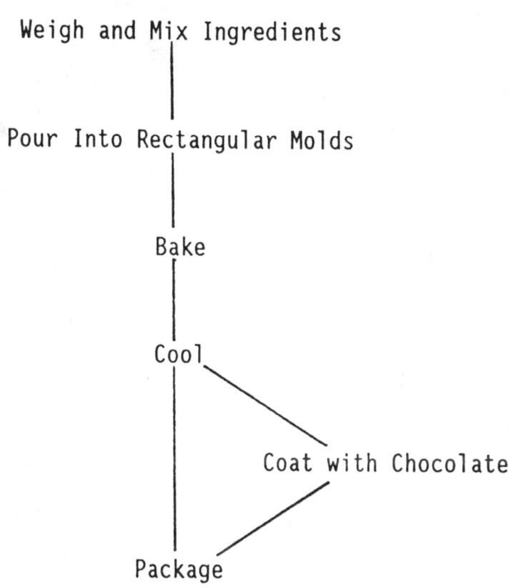

FIGURE 9 Typical steps in cereal meal bar manufacture.

IX. PRETZELS

Pretzels are well established as an American snack with total sales in 1986 reaching $300 million with close to 90% being in the form of hard pretzels. The remaining 10% are soft pretzels, especially frozen soft pretzels. These products are usually consumed warm.

There are five general market segments for pretzels in the United States. The first is the regional market, where the northeastern United States is by far the leader. Local regional producers accounted for approximately $120 million in sales in 1986 (1). Next is the distribution of pretzel brands nationally through warehouses as well as national marketers who deliver pretzels door-to-door. The fourth category is regional full-line snack manufacturers who have pretzel lines, and the last is the soft pretzel market.

Pretzels are made from fermented dough. Traditional curled pretzels are usually machine formed then passed through a lye dip for approximately 10 seconds containing either 0.5% sodium hydroxide or 2% sodium carbonate at 180–210°F. Salt at a final level of 2% is added and the product baked at 450°F for 4–5 minutes. Pretzels come from the oven at around 15% moisture and go through a drying tunnel set at around 250°F for 25–90 minutes to reduce final moisture to 2.0–2.5%. Stick pretzels are extruded, cut, and then follow the same general procedure as described above. Typical formulations for twist and stick pretzels are shown in Table 12 and their manufacturing process summarized in Figure 10.

X. TOASTER PASTRIES

Toaster pastries, initiated in 1964, are another unique American product. They are essentially individual flat pastries with a fruit-type filling layered inside. Their size and thinness permits them to be heated in a toaster or microwave to provide a warm snack product. Frozen forms with more sophisticated fillings are also available. Sales for this cereal-based snack category approached $200 million in 1986.

These are made by depositing a thin layer of fruit filling having a composition as shown in Table 13 onto a thin layer of pastry dough. A top layer of dough is applied and rectangular shapes cut. This is followed by baking, cooling, and packaging.

Weigh and Mix Dough Ingredients

Form Dough Stick

Proof

Form Twists

Caustic Tanks

Drain

Salt

Bake

Drying Oven

Package

FIGURE 10 Typical steps in the production of pretzels. (From Ref. 7.)

TABLE 1 Composition and Dollar Value (Millions) of the 1986 U.S. Snack Food Market

Category	% of Market	$ Millions
Candy	29.9	7,400
Cookies/Crackers[a]	22.8	5,672
Potato chips	12.0	2,977
Corn/Tortilla chips[a]	5.9	1,462
Snack cakes/Pies[a]	5.0	1,240
Nut meats	4.7	1,160
Frozen pizza[a]	4.3	1,063
Imported and miscellaneous	2.4	606
Dried fruit	2.1	508
Extruded snacks[a]	1.8	456
Hot snacks[a]	1.8	400
Popcorn[a]	1.5	382
Granolas[a]	1.4	359
Pretzels[a]	1.2	293
Fabricated chips[a]	0.8	203
Toaster pastries[a]	0.8	193
Total		24,769

[a]Totally or partially cereal-based
Source: Ref. 1.

TABLE 2 American Per Capita Consumption of Cookies and Crackers in 1986

Product category	Pounds
Cookies	9.79
Sandwich	2.62
Marshmallow	0.38
Wafers	0.25
All other	6.54
Crackers	7.96
Saltines	2.58
Graham	0.78
Pretzels	0.77
Cracker sandwiches	0.51
Cracker meal and crumbs	0.17
All other	3.15
Total cookies and crackers	17.75

Source: Ref. 1.

TABLE 3 Typical Cookie Formulations

	Sugar		Bagged		Icebox		Sandwich		Bar	
Ingredient	lb	oz	lb	oz	lb	oz	lb	oz	lb	oz
Sugar	2		1	8	1	8	1	8	2	
Salt		1/2		1/2		1/2		1/2		1
Milk powder		2		1		—	—	—	—	1
Shortening	1	8	1	1	2	12		12	1	4
Corn syrup		2		—		2	—	—	1	7
Eggs		8		4		10	1	8		1
Water		12		7		—	—	—	—	8
Vanilla		1		1		1		1	—	—
Cake flour	4		2		2	4	2	4	4	6
Baking powder		2		1/2	—	—	—	—	—	—
Bread flour	—	—	—	—	2	4	—	—	—	—
Butter	—	—	—	—	—	—		12	—	—
Cinnamon	—	—	—	—	—	—	—	—		1
Baking soda	—	—	—	—	—	—	—	—		1
Raisins	—	—	—	—	—	—	—	—	2	
Walnuts	—	—	—	—	—	—	—	—	1	

Source: Ref. 6.

TABLE 4 Typical Cracker Formulations

	Soda cracker	Sprayed cracker	Cheese cracker
Sponge ingredients (lb)			
	70	—	75
Flour	30	—	23.5
Shortening	4	—	7.5
Malt	0.02	—	—
Yeast	0.23	—	0.3
Cheese, cheddar	—	—	10
Dough ingredients (lb)			
Flour	30	100	25
Shortening	5.8	7.5	—
Salt	1.4	1.0	1.0
Malt	0.92	1.5	1.0
Water	0.8	28	0.8
Sodium bicarbonate	0.63	0.9	0.45
Sugar	—	5.0	—
Nonfat dry milk	—	2.5	—
Invert syrup	—	2.5	—
Monocalcium phosphate	—	0.75	—
Cheese	—	—	4.0

Source: Ref. 7.

TABLE 5 Typical Pizza Dough Formulation

Ingredient	Pound	Ounce
Yeast		8
Water	2	
Salt		4
Flour	13	8
Optional:		
Sugar		2
Malt syrup		2
Shortening		2
Dough conditioner		1/8–1/4

Source: Ref. 6.

TABLE 6 Dough Weights for Various Sizes of Thin Crust Pizzas

Pizza size	Ounce	Grams
Individual Size	2.5	75
12-in. Round	10	300
14-in. Round	13	380
16-in. Round	18	500

Source: Ref. 6.

TABLE 7 Value of Imported Cereal-Based Snacks in 1986

Country	Value in $ Million
Canada	75.2
Denmark	58.0
West Germany	19.5
Japan	14.4
United Kingdom	11.8
Belgium	8.9
Italy	8.1
Mexico	7.8
France	7.1
Netherlands	5.8
Spain	5.1
Hong Kong	3.1
Israel	2.9
Colombia	1.9
Switzerland	1.4
Greece	1.2
South Korea	1.2
Norway	1.1

Source: Ref. 1.

TABLE 8 Expansion Rates of Various Starches

Starch source	Rate of expansion
Rice (glutinous)	100
Rice (nonglutinous)	10.8
Potato	17.0
Sweet potato	15.5
Wheat	10.8
Corn	9.2

Source: Ref. 9.

TABLE 9 Advantages of Extrusion Processing

Continuous processing
High productivity
High product quality
Low labor and flour space requirements
Versatility
Minimum amount of effluent
Energy efficient

TABLE 10 Typical Extruded Pellet Snack Flavoring Formulations

Ingredient	Percent
Cheese Flavor:	
Starch	53
Water	27
Cheddar cheese powder	18
Salt	2
Potato and Onion Flavor:	
Water	34.6
Starch	29.6
Potato granules	27.8
Soy concentrate	5.5
Salt	2.3
Onion powder	0.2

Source: Ref. 12.

TABLE 11 Typical Cereal Meal Bar Formulations

Ingredient	Nut Based (lb)	Fruit and nut based (lb)
Flour	400	400
Shortening	140	140
Sugar	200	200
Cinnamon	2	2
Allspice	1	2
Ginger	—	3
Skim milk powder	20	20
Salt	5	5
Soda	4	6
Acid powder	4	2
Coconut	100	—
Nuts	160	—
Raisins	—	240
Pecans	—	60
Molasses	200	300
Liquid whole eggs	60	48
Water	32	40

Source: Ref. 15.

TABLE 12 Typical Pretzel Formulations

Ingredient	Twist (lb)	Stick (lb)
Flour	160	160
Shortening	2	4
Malt	2	4
Yeast and yeast food	[a]	[a]
Ammonium bicarbonate	1 oz	4 oz
Sodium bicarbonate	3 oz	—
Water	8 gal	7.5 gal

[a]As required.
Source: Ref. 7.

TABLE 13 Typical Filling Formulation for Toaster Pastries

Ingredient	Percent
Sugar	31.25
Corn syrup	25.00
Water	20.83
Evaporated apples	12.50
Strawberry puree	8.32
Dehydrated cirtus pulp	0.83
Carboxymethylcellulose	0.33
Citric acid anhydrous	0.33
Algin	0.21
Agar	0.16
Locust bean gum	0.13
Sodium benzoate	0.08
Imitation flavor	0.014
FD&C Red No. 2	0.008
FD&C Red No. 5	0.008

Source: Ref. 6.

REFERENCES

1. Editors, Annual snack food review, *Snack Food, 76*:(6)M2–M36 (1987).
2. Blenford, D. E., U. K. second largest snack market in world, *Snack Food, 76*(11)–51(1987).
3. Sloan, A. E., Reward foods: A new product niche, *Cereal Foods World, 32*(5)–370 (1987).
4. Editors, Nutritional snacks being mainstreamed, *Snack Food, 76*(3)–24–25 (1987).
5. Bednarcyk, N. E., Nutrient contributions of cookies and crackers, *Cereal Foods World, 32*(5)–367–369 (1987).
6. Sultan, W. J., Recipes for cookies, In *Practical Baking*, AVI Publ., Westport, CT, 1980, pp. 407–459.
7. Matz, S. A., Fermented dough products, In *Cookie and Cracker Technology*, AVI Publ., Westport, CT, 1968, pp.137–149.
8. Gomez, M. H., Rooney, L. W., Waniska, R. D., and Plfugfelder, R. L., Dry corn masa flours for tortilla and snack food production, *Cereal Foods World, 32*(5)–372–377 (1987).
9. Li, C. F., and Luh, B. S., Rice snack foods, In *Rice: Production & Utilization*, B. S. Luh (Ed.), AVI Publ., Westport, CT 1980, pp. 690–711.
10. Harper, J. M., Snack food, In *Extrusion of Foods*, Vol. II, CRC Press, Boca Raton, FL, 1981, pp. 66–78.
11. Harper, J. M., Food extrusion, In *Extrusion of Foods*, Vol I, CRC Press, Boca Raton, FL, 1981, pp. 1–19.
12. Matz, S. A., Puffed snacks, In *Snack Food Technology*, 2nd ed., AVI, Westport, CT, 1984, pp. 150–165.
13. Reeve, R. M., and Walker, H. G., The microscopic structure of popped cereals, *Cereal Chemistry, 46*:227–241 (1969).
14. Matz, S. A., Snacks based on popcorn, In *Snack Food Technology*, 2nd ed., AVI, Westport, CT, 1984, pp. 138–149.
15. Smith, W. H., Formulations, In *Biscuits, Crackers and Cookies*, Vol. 2, Applied Science Publ, London, England, 1972, pp. 153–274.

21

MALTED CEREALS: PRODUCTION AND USE

Richard E. Pyler and David A. Thomas
Coors Brewing Company, Golden, Colorado

Malting of the wild barley cultivar *Hordeum spontaneum* to provide a food source for man and accidental microbial infection of the resultant "mash" to produce an alcoholic beverage probably occurred in Neolithic times (1). Intentional malting was accomplished on tables for generations after that for domestic use in baking, brewing, or distilling. Malting was necessary to convert hard, insoluble cereal grains to softer, sweeter kernels by allowing them to germinate in thin layers with frequent manual turning. Drying of the germinated grain was frequently left up to the sun's rays or to small wood or peat-fired kilns. Later, floor maltings that handle several tons of barley per batch were built alongside small breweries and distilleries and are still used today in some locations.

Since the mid-nineteenth century, however, pneumatic malthouses have been employed by commercial maltsters in order to malt very large batches of cereal grains (up to 500 tons) in an economic and efficient manner. Pneumatic malthouses use large fans to draw or push prescribed volumes of conditioned air through the respiring cereal grain bed to remove heat, purge carbon dioxide, and provide oxygen to the growing embryo. Without these fans and the perforated or "false" bottoms installed in steep tanks, germination compartments and kiln floors, the temperature in large batches of grain would not be easy to control.

Many cereal grains can be malted but barley is the cereal most often malted. For this reason, more is known about the internal chemical changes in barley malting, but most of the information can be applied to the malting of other cereals, like wheat, rye, triticale, oats, and sorghum.

I. THE MALTING PROCESS

Simply stated, malting is a brief initiation of the normal botanical growth of a cereal grain followed by an abrupt interruption of this growth with hot, dry air in the malt kiln. This starting and stopping of botanical growth is expertly timed by the maltster to achieve minimum malting loss and maximum malt modification. The three main steps of the malting process are steeping, germination, and kilning.

Figure 1 represents a simplified schematic of the malting process. In this scheme, cereal grains and water are combined in specific ways with the transfer of gases, like oxygen and

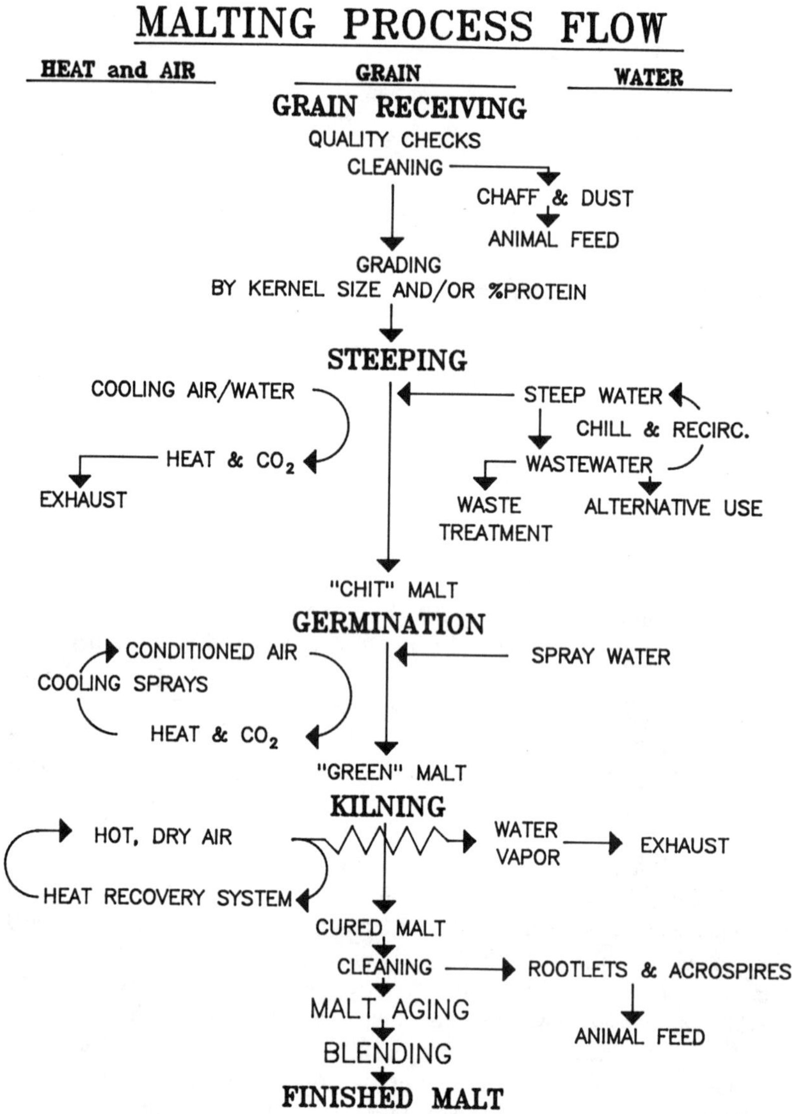

FIGURE 1 Simplified schematic of the malting process.

carbon dioxide, and the timed application of cooling and heating, to produce finished malt and by-products. Recycling of air, heat, and water is practiced to varying degrees in different malthouses. Some recovery of water and heat is usually done internally in most malt houses to improve malting economics but may even by used in such seemingly unrelated industries as fish farming, providing heat to buildings, or growing potted plants in greenhouses. It is likely that the future will hold much more in the way of recycling and recovery to continue to improve economics and lessen the environmental impact of malthouses.

After harvest, the cereal grains are usually stored for a time, typically one month to a year, to allow the grains to break normal dormancy before malting is started. The degree of dormancy is determined by several factors, including growth environment and cereal variety. Techniques have been developed to help expedite the breaking of dormancy by artificially heating the grain for a specified time (2). Laboratory tests are routinely carried out to assess germination vigor and other quality characteristics before grain is brought in for malting.

The many complex changes that occur during malting are lumped together in one term, "modification," and include degradation of cell wall materials, solubilization of the protein matrix, and production of amylolytic enzymes that will be used to advantage later in brewing or baking. Of these, hydrolysis of β-glucan, which makes up 70% of the endosperm cell walls of barley (3), is the principal rate-limiting step in the modification of barley. Other cereals, wheat, for instance, contain less β-glucan (20% total) but contain more pentosan (70% total) in endosperm cell walls (4).

The maltster must ensure that sufficient hydration of the cell wall material occurs and that β-glucan hydrolases are synthesized and released into the endosperm. These enzymes solubilize the high molecular weight β-glucans from the cell wall and ultimately degrade them to smaller molecules. If malt cell wall modification is started but not completed prior to kilning, i.e., if the malting process is too short or the barley quality is poor, then high molecular weight β-glucan materials will be solubilized but not degraded, which can cause a number of technological difficulties. In brewing, these include impeded wort separation, reduction in extract yield, slowed beer filtration, and the formation of undesirable precipitates and hazes in beer.

The food reserves of the mature barley grain can be classified into two groups, those required immediately for embryonic respiration and growth, and those that are stored in the endosperm in an insoluble form for later use by the embryo (during growth), the brewer (in the mash vessel), or the baker (bread functionality). In barley, the respiratory substrates consist of sucrose and raffinose, which are concentrated in the embryo and aleurone layer and used during the first 24 hours of growth (5). They provide the immediate energy needed by the embryo to begin respiration and are used for enzyme synthesis in the aleurone layer, which surrounds the endosperm, in response to embryo-secreted hormones like gibberellic acid. Long-term food reserves in the endosperm include starch, protein, β-glucan, and pentosan and are more technologically important to the final user of the malt. The chemical characteristics and physical relationship of β-glucans, proteins, and pentosans to starch granules in the endosperm make enzymic hydrolysis (via malting) imperative prior to the use of these cereals in most food processes. The maltster desires extensive degradation of endosperm cell walls, moderate solubilization of matrix proteins, adequate production of enzymes, and limited amylolysis of native starch granules. It is this balancing of the natural hydrolytic and cytolytic processes comprising malt modifications that account for the art and science of malting expertise and modern malting technology.

A. Steeping

Just as the growing cereal seed in the field requires a period of water uptake before seed germination can begin, the maltster must provide for this in the malthouse. The hydration of cereal grains in the malting process is facilitated by a sequence of soaking, air rests, and/or spray cycles, dictated by the rate of water uptake, germination characteristics of the grain, and type of malting equipment in place. Modern malting methods usually require 24–80 hours to increase kernel moisture to 40–48% in steeping, while almost 300 years ago, winter steeping periods of 150–174 hours were not uncommon (6) (Table 1).

In practice, the maltster must provide enough water contact to allow rapid absorption, but not too much, especially in the case of water-sensitive grains. These grains will actually show a reduction in germination vigor if exposed to an excess of water. Water sensitivity, not to be confused with normal seed dormancy, is caused by a layer of surface water which inhibits the transfer of oxygen to the respiring embryo.

The temperature of the water is also critical in steeping. The steep water must be warm enough to allow for rapid water uptake and germination, but not too warm, which would allow natural microflora on the surface of the grain to grow. This could lead to the competition with the embryo for oxygen and may produce off-flavors in the final malt product. Warmer temperatures can also exacerbate the respiration in water-sensitive grains. Normal steeping water temperatures vary from 10 to 20°C.

Replacement of depleted oxygen and removal of heat generated by grain respiration is achieved by drawing attemperated and humidified air down through the steep tank during air rest cycles, by immersing the grain in a series of fresh water soaks (which can produce a heavy demand on water treatment facilities), or, more recently, by recirculating chilled, aerated water through overhead high volume spray nozzles. Aeration of the steep water is important and must be repeated several times during a normal steeping cycle. The rate of reduction of dissolved oxygen from steep water has been recently measured at about 1 ppm/hr in the initial soak cycle and 10 ppm/hr in a third soak (7). This means that dissolved oxygen can be depleted from steep soak water in as little as one hour, therefore the need for frequent changes or aeration of steep water. Some steep vessels use a water/air "geyser" system in the center of the tank to continuously move the grain and aerate it.

During water imbibition, the cereal grain swells as much as 25% in volume. This swelling, due mainly to absorption of water by grain protein, must be handled by periodic rousing of the grain during water soak cycles with blasts of compressed air. This rousing or "tank blowing" also serves to aerate the water and loosen foreign material from the surface of the grain.

The functions of steep water, once absorbed into the grain, include initiation of cell elongation, respiration, and secretory activity of the embryo and activation of enzymes. The maltster must also ensure that the endosperm is sufficiently hydrated for these enzymes to hydrolyze the food reserves stored in the endosperm. When the cereal grain is hydrated and the embryo shows signs of rootlet growth, it is "steeped-out" to a germination vessel.

B. Germination

Germination has been defined, in a strict botanical sense, as the process which starts with water imbibition, proceeds through intermediate phases of enzyme activation and mitosis, and ends with elongation of the radicle or rootlet (8). Germination in the malting process, however, generally covers the growth phase, which occurs after steeping and before kiln drying. Paradoxically, the rootlet emergence, called "chitting," which signifies the end of botanic germination, usually occurs prior to the end of steeping and before the "germination" phase of malting has begun. Although the imbibition of water by the grain is effectively complete at the end of steeping, some malting methods, particularly pneumatic germination compartments with high air throughput and subsequent water evaporation, require additional water application during germination to maintain adequate grain hydration. This water is usually applied through fine mist sprays attached to the germination turning machine. However, excessive mechanical turning of the malt, usually with augers, in order to apply water during germination, may cause grain damage.

Germination requires the most time in the malting process and, as Table 1 shows, has been the area most improved over the past 300 years of malting technology (9). This 10-fold reduction in germination time has been realized by improved varieties through cereal grain breeding research, tighter control of cereal agronomic properties, and application of new malting technology, as well as an increased scientific understanding of exactly what occurs during malt modification.

Current germination methods and equipment are almost as plentiful as malthouses. By far the most common design is pneumatic compartment maltings, also known as Saladin boxes. These are rectangular, open-top vessels with slotted, false floors that hold the steeped grain in beds 3 to 5 feet deep. Air, which has been attemperated (10–18°C) and humidified 90+% relative humidity) by water sprays, is passed through the bed by updraft or, less commonly, downdraft fans at a fairly high rate (200–1000 m^3/hr/ton), which removes the heat of respiration from the bed. This also provides oxygen to the respiring embryo and prevents the grain from souring.

Most germination methods involve batch processes, with as many as several dozen germination compartments in large malthouses. Continuous malting can also be done in germinators that move the malt through belts, augers, or drums and aerate and apply spray water automatically. Lower labor costs and more rapid germination are claimed to offset the higher capital and maintenance costs of continuous malting plants.

Some maltsters also use modification aids like gibberellic acid, which speeds up modification, or potassium bromate, added to slow proteolysis and, thereby, reduce malting loss. This practice will vary with the quality of the harvested grain, customer requirements, and, of course, malting philosophy. Detailed descriptions of the physiology of malt modification have been reviewed elsewhere (10,11).

The maltster must balance the quality, germination characteristics, and steep-out moisture of the grain against germination temperature, water applications, and germination cycle time to produce the desired green malt characteristics prior to kilning. However, in contemporary malthouses producing at, or near, capacity, germination cycle time is often preset by schedule and is not very flexible. The maltster must then "steer" the modification of the malt by adjusting germination air-on temperature, water applications, and addition of malting aids, if used. Air temperatures are often controlled in the winter by mixing cold, fresh air with warmer, recirculated air. In the summer, however, large refrigeration units are used to cool the air, and these represent a considerable energy demand for malthouses. In traditional floor maltings, without refrigeration, hot weather malting would not be attempted due to lack of ability to cool the "pile" or "piece."

Analytical methods to evaluate the degree of modification during germination range from simple on-the-spot checks by the maltster to more involved laboratory techniques. "Rubbing out" or smearing the green malt between thumb and index finger is certainly the simplest technique for determining whether the endosperm is "friable" or soft. This technique is used by virtually all practical maltsters, who can easily be recognized by the layer of chalky, white malt starch on their "maltster's thumb."

Recent malting literature is full of more sophisticated laboratory techniques for determining green malt modification, such as warm water extraction, cell wall fluorescence staining, scanning electron microscopy, and soluble protein estimation. No doubt methods will continue to be developed that help the production maltster determine the exact degree of malt modification prior to kilning.

C. Kilning

The objectives of kilning are to arrest botanical growth and internal modification, to reduce moisture for grain storage, and to develop color and flavor compounds in the malt. By carefully selecting the cycle times, temperature, relative humidity, and volume of air drawn through the kiln bed, the maltster largely determines the characteristics of the finished malt. High volumes of warm, dry air will produce lightly colored, enzymic malt, while reduced flow of hot, wet, recirculated kiln air will "stew" the malt and produce high color and more toasted malt flavors.

The general relationship between enzyme activity level and color in flour and extract of a few representative malts is shown in Table 2. Although individual maltster's definitions may vary, diastatic malts with high amylolytic enzyme activity will have light color (pale malts), while crystal, chocolate, and black patent malts are very dark and have little or no enzyme activity due to heat denaturation of the enzyme proteins. Each malt in the list has it's own use for specific brewing and baking applications. Color and enzyme activity, as well as other malt characteristics, are tightly identified in the customer's specification.

The technological development of commercial malt kilns started with simple gravity-flow kilns with perforated floors and evolved to those with variable speed fans and multiple floors. The distinctive "Pagoda-style" roof architecture of many kilns at floor maltings, still used at many distilleries in the highlands of Scotland, were built for function, rather than aesthetics, although they certainly are classic buildings. The high domed chimney of these kilns acts to create a natural draft which draws the heat and peat smoke (used in Scotch whiskey manufacturing for the distinctive flavor) up through the drying malt bed. The germinated malt is spread in shallow layers from 4 to 12 inches deep. During kilning, workers manually turn the malt with pitchforks or wooden shovels to ensure even drying and contact with peat smoke. Fans were later added to these kilns, and any of late design, to speed up the drying process, improve uniformity, and conserve energy by recirculating dry air during the last phases of kiln drying and roasting. Today's kilns can be comprised of single or multiple floors, rotating drums, or even vertical kilns. Bed depth can be as little as 12 inches or as much as 6 feet in deep bed kilns, with kiln cycle times of 9–48 hours. Forced air volumes also vary considerably between malthouses and can be set very high to achieve rapid drying (>10,000 m^3/hr/ton) or low (<2,000 m^3/hr/ton) to slow the drying process, thereby allowing further malt modification. Kiln airflow is usually updraft, although downdraft kilns are also being used, with reported advantages for both. The initial drying stage in kilning is designed to remove most of the free and bound moisture in the kernel while protecting against enzyme denaturation with lower temperatures, usually 40–65°C. The roasting or curing stage of kilning immediately follows drying and is responsible for most of the color and flavor compounds produced in malt. Final roast temperatures of 75–100°C are common in the United States for typical lager malts, although specialty malts are routinely produced outside of this temperature range. Most maltsters monitor the on-and-off grain temperature and humidity very closely during kilning and use these to signal when the malt moisture is right for the next scheduled temperature raise, especially the critical final roasting temperature. Malt kiln technology has been covered in great detail by Hough et al. (12).

After the final roasting cycle is complete, the kiln and malt are cooled by drawing fresh air through the fans. Samples are taken immediately from the kiln bed and final malt quality or grade is determined by a variety of tests. Some maltsters simply score flavor, aroma, and friability or crushability of the malt by rapid methods to determine malt grade. Others store the malt in temporary surge bins for several days and wait for complete laboratory results before assigning a malt grade and malt storage bin. Full malt analysis might include percent warm water extract, enzyme activity levels, extract and flour color, as well as traits specific for the intended commercial use of the malt. Complete malt analyses are described in detail by the U. S. (13, 14), Great Britain (15), and European (16) malting communities.

D. Malt Aging

An important final step in producing malt for brewing is the aging of whole malt prior to grinding. Although the exact mechanisms of malt aging remain to be elucidated, the required storage apparently involves the slow diffusive balancing of moisture within the kernel. While the water taken up in steeping and germination has been absorbed over a period of several days, removal of the water during kilning is much more rapid. Fresh kilned malt, at 3–5% kernel moisture, can be thought of as having a dry outer periphery at perhaps 1–3% moisture, and a "wetter" internal core of 5–8% moisture or higher. This malt, if milled immediately after kilning, would produce a mixture of very dry finely divided flour and larger "clumps" of higher moisture. These different particles would tend to agglomerate and reduce extractability. Indeed, fresh-kilned malt has been known to cause lower extract yield, reduced enzyme activity, and poor performance in mash filtration and spent grain handling. For these reasons, most brewers specify a minimum malt age before milling, usually at least one month.

Cereal grain variety and quality, degree of modification achieved in germination, and kilning conditions largely determine the characteristics of the final malt product. Commercial maltsters will usually blend several different types of malts, based on final malt analysis, to meet customer specifications.

II. BARLEY MALTS IN BREWING

The primary use of barley malt produced in the United States is in the production of alcoholic malt beverages, primarily beer, where it represents the primary non-water ingredient. More than 4.7 billion pounds of barley malt was used by U. S. brewers in 1986 in the production of 178 million barrels of beer (17). Barley malt has been the traditional ingredient of choice for this purpose throughout the world because it is widely adapted to a variety of climates and soil conditions and thus is readily available, because its balance of amylolytic and proteolytic enzymes produces a highly fermentable wort without requiring added nutrients, and because its husk remains attached during harvest, which provides protection to the growing acrospire during malting and provides a filtration medium in the brewing process. Barley malt also provides sufficient enzymes for the production of fermentable sugars for adjuncts commonly used today.

Barley and barley grasses make up the genus *Hordeum* of the *Gramineae*. Two distinct species of barley are used for malting and are classified according to the characteristics of the grain spikelets during ripening. These spikelets are arranged radially from the nodes on the developing head of grain. When all six spikelets are fertile, six rows of grain will be formed, looking down the length of the barley head. This barley is therefore called six-rowed (*Hordeum vulgare*). In two-rowed barley (*Hordeum distichon*), only the two central spikelets are fertile, and two rows of grain are produced.

Malt from six-rowed barleys comprise the predominant ingredient in beers produced in the United States, whereas malts from two-rowed barleys are used in most other countries. The preference for six-rowed barleys in the United States is based on their adaption to growing areas near the early brewing centers (New York, Cincinnati, Milwaukee, Detroit, and St. Louis) and their historically higher level of enzyme activity which allowed the use of adjunct or nonmalt sources of starch. Two-rowed malts continue to be used in the U. S. as the sole malt source of one major western brewer and in some of the "super-premium" beers produced by several U. S. brewers. Typical analysis of two-rowed and six-rowed barley malts made for brewing are shown in Table 3.

A. The Brewing Process

Brewing may be conveniently divided into four distinct operations: mashing, fermentation, aging, and finishing. The following is a brief description of the process, concentrating in each area on the contributions of malt to the changes which occur. Fuller descriptions of the chemistry, biochemistry, and technology of the process are available from several sources (18–20).

1. Mashing

The objective of the mashing process is to convert malt components and any adjunct material used into a fermentable substrate consisting of fermentable sugars derived from malt and adjunct starch, amino acids derived from malt protein, vitamins and minerals, nucleic acid breakdown products, and other materials in the malt and adjunct which are either naturally soluble or are solubilized by the malting and/or mashing process.

A brewing adjunct is a starch-rich material which is used primarily as a source of fermentable sugars. Common adjuncts used in the United States are derived either from corn or rice. Corn is usually used in the form of grits, which represent endosperm chunks derived from the dry milling process, or in the form of highly fermentable syrups. The rice used is a by-product of the rice-milling operation and consists of those rice kernels that are broken during milling. The different adjuncts used by foreign and domestic brewers have been reviewed by Pyler and Thomas (21) and Canales (22).

Typical brewing practice in the United States utilizes a double mash system in which the malt and adjunct are mashed separately and are then combined to finish the process. In this system, the adjunct is combined with a portion of the malt, usually about 10% based on adjunct weight, and water at 45 to 55°C. The mash is carried through a series of temperature raises and holds in order to gelatinize and solubilize the adjunct starch. This requires holds of varying duration at various temperatures to provide time for malt α-amylase to liquefy and solubilize the adjunct starch. The primary goal in the cereal mash is the liquefaction of starch, not the production of fermentable sugars. The final stage in the cereal mashing schedule is typically an atmospheric or pressure boil to ensure complete gelatinization.

While the cereal mash is being prepared, malt is mashed in with water, usually between 40 and 50°C, and a similar time-temperature profile is followed. The mash is held at 40–50°C for 30–60 min in order to allow proteolytic enzymes time to produce free amino acids which represent the only source of nitrogen available for yeast metabolism. At the same time, various other malt enzymes are solubilized and begin to degrade and solubilize carbohydrates, nucleic acids, etc., to smaller molecules. At the appropriate time, the temperature of the malt mash is raised by addition of the boiling adjunct mash to a temperature of 60–67°C, where the action of the amylases and the production of fermentable sugars are very rapid. The hold at this temperature is of a duration sufficient to impart the desired degree of fermentability to the subsequent wort. The temperature is then raised again to about 75°C to complete the conversion of starch and starch-derived dextrins to short-chain oligosaccharides.

Two other mashing systems or modifications of them are in common use. One is termed the single mash infusion system, wherein malt is mashed in at a constant temperature of about 65°C and is held for 30 minutes to 2 hours. During this period, malt enzymes convert the protein and starch into fermentable materials. The second mashing system, termed decoction mashing, is similar to the double mashing system used in the United States in that the mashes are carried through a series of temperature raises and holds. In decoction mashing, a portion of the mash is removed to another vessel where it is heated to boiling or other predetermined temperature before being added back to the main mash. Temperature raises occur during the recombination of the mashes. Many modifications of these systems are in use in various parts of the world, depending upon the characteristics of the malt used and the type of beer being produced.

Separation of the sugar/amino acid–rich extract from the insoluble materials derived from the malt and adjunct is traditionally accomplished by means of a lautering vessel. The mash is introduced into the lautering vessel, whose bottom consists of a slotted plate, and the mash is allowed to settle undisturbed for a period of time to allow the insoluble material to settle and form a filter bed. Once the bed has been formed, filtration by gravity begins and the filtrate is returned to the lautering vessel until it is of suitable clarity. The extract is then run into the brew kettle. After all the extract has been filtered in this manner, the liquid entrained in the insoluble spent grains still contains considerable amounts of fermentable sugar, and these are recovered by sparging, i.e., by adding hot water to the top of the lautering vessel to wash the remaining extract from the filter bed.

Several other devices are also used for mash filtration. The mash filter is designed as a plate and frame filter, the plates and frames being separated by filter cloths. These cloths provide support for the filter bed. In this case, the mash is pumped through the filter, followed by sparge water. The "Strainmaster" is another unit used for mash separation. The rectangular vessel is traversed in its bottom half by a series of perforated tubes which function much the same as the slotted bottom of the traditional lauter tun described above.

The extract in the kettle is typically boiled for about 2 hours, during which time the hops are added in two or three portions. Worts are boiled to (a) sterilize the wort prior to fermentation, (b) form color, flavor, and aroma compounds through the Maillard reaction, (c) inactivate any remaining enzymes, (d) reduce the wort volume to allow standardization to fermentation specific gravity, (3) drive off undesirable flavor and aroma compounds, (f) coagulate soluble, high molecular weight proteins which may cause problems later in processing or in finished product, and (g) extract compounds from hops which are responsible for the characteristic bitterness and aroma of beer.

When the boil is complete, the hop residue is removed (if whole hops have been used) and the extract is transferred to the coolship or whirlpool where it is cooled slightly and the mostly proteinaceous material precipitated during the boil is allowed to settle for separation. The clarified wort is decanted off, cooled to fermentation temperatures, and aerated. This completes the brewhouse phase of the process, which typically lasts between 6 and 10 hours.

2. Fermentation

Fermentation is usually carried out at 8–10°C for 5–7 days. During this period, the fermentable sugars produced during malting and mashing are consumed, as are certain of the free amino acids which function as the primary source of assimilable nitrogen. The latter are necessary for the production of yeast mass that occurs during fermentation and which may amount to five times the amount originally added. Near the end of the primary fermentation, the yeast either flocculates and settles to the bottom of the tank or rises to the top.

3. Aging

Fermentation ends before all of the fermentable sugar has been consumed, and the beer is transferred with the yeast remaining in suspension to other tanks for the aging, lagering, or secondary fermentation step. Here, most of the remaining fermentable sugar is consumed, various flavor and/or aroma active materials either rise or decline in concentration, natural carbonation is effected, and the flavor of the beer matures. Aging may last anywhere from one week to several months. Filtration through diatomaceous earth removes residual yeast and suspended material before the beer is blended to appropriate alcohol levels. A final polish filtration leaves the beer ready for packaging. If the final filtration is designed to leave the beer free of microorganisms, the packaged product need not be pasteurized and can be sold as "draft beer." Currently, while most packaged beer in the United States is subjected to pasteurization, one major brewery pasteurizes none of its products and other brewers are also beginning to produce unpasteurized beers.

This brief description of the brewing process illustrates the various contributions of malt to the process and to the finished product. Among these contributions are (a) enzymes for the formation of fermentable sugars, free amino acids, nucleic acid degradation products, etc., (b) a source of fermentable carbohydrates and free amino nitrogen from the malt starch and protein, (c) minerals and vitamins necessary for vigorous yeast metabolism, and (d) color, flavor, and aroma compounds derived from Maillard and caramelization reactions which occur during kilning and mashing.

The development in recent years of technology for the production of microbial enzymes in production-scale quantities has led to their use in brewing to supplement the natural enzymes of malt, to allow the use of nontraditional adjuncts such as barley, and to alleviate specific problems which may occur in the brewhouse. The uses of such enzymes have been described recently by Beckerich and Denault (23).

B. Light Beers

Light or low-calorie beers represent a growing proportion of the total beer market in the United States today. These products may be produced by (a) dilution of normally produced beer to a predetermined calorie level, (b) the use of microbial enzymes (amyloglucosidase) in the mash tun or fermenter to convert all of the starch to fermentable sugar, (c) the use of highly fermentable corn syrups as adjuncts, or (d) through the addition of sterile filtered malt extract to the fermenter, which allows the natural malt enzymes to convert more, but not all, of the starch to fermentable sugar. Most of the light beer produced in the United States employs microbial enzymes.

III. BAKING APPLICATIONS

Another major user of cereal malts has been the baking industry. Malts used by the baking industry are produced in the same general manner as those used for brewing purposes with minor changes depending upon the type of grain being malted. This process has been described by Gill (24). Briefly, the grain is steeped, germinated, and kilned as described above and then further processed depending upon the type of malt product being produced.

For the production of malt flours, the kilned malt is cleaned to remove the rootlets and acrospires that form during germination and is milled to the appropriate particle size.

The production process for malt extracts and syrups is more extensive and mirrors to a large extent the mashing step of the brewing process described above. The malt is mashed in with water at a controlled temperature, typically between 35 and 45°C. The mash is then carried through a controlled time/temperature profile to ensure the proper breakdown of malt protein to soluble protein and free amino acids and of the malt starch into fermentable sugars, primarily glucose and maltose.

At the end of the mashing process, the mash is filtered through the bed composed of malt husk and other insoluble material and the extract is boiled to sterilize it, to inactivate the enzymes, and to impart, through Maillard-type reactions, desirable flavor and aroma to the extract. After boiling, the extract is cooled and precipitated protein is allowed to settle. After filtration, the extract is concentrated to about 80% solids in a multiple effect evaporator.

For the production of malt syrups comprising mixtures of barley malt and corn or wheat as adjuncts, the adjunct is usually cooked separately in the presence of a small amount of the malt to gelatinize and liquefy its starch and then is added to the remainder of the malt mash to convert its starch into fermentable sugars.

Malt is also available as dried malt extract, and, in this case, the malt syrups produced as described above are subjected to spray drying.

The temperatures used during the various phases of production must be carefully controlled as they will affect both the color of the resulting extract and its enzyme activity. Before the widespread availability of microbial enzymes, diastatic malts were produced without the boiling step mentioned above, and the concentration steps were carried out at lower, carefully controlled temperatures. Currently, many diastatic malts are produced by the addition of enzymes of microbial origin to the finished, nondiastatic syrups (24).

The composition of malt products varies depending upon the conditions employed in the malting and concentration processes, the cereal grain malted, and the type and level of adjunct used (Table 4). It also varies depending upon whether a diastatic or a nondiastatic malt is considered.

Table 5 shows the composition and properties of two malt syrup products. Solids levels in syrups are typically about 75–80% for reasons of economy in shipping and storage and because high solids levels prevent microbial growth. Reducing sugars, expressed as maltose, range from 55 to 75% and protein levels may be quite variable depending on the type of adjunct used and its level and on the extent of amylolysis and proteolysis which occur during mashing. The color of the syrups and dissolved dried extracts is generally between 60 and 500 on the Lovibond scale because of the occurrence of Maillard-type reactions during the kilning, mashing, and boiling processes.

Malt flours, syrups, and extracts contribute a number of functional components to the bakery products to which they are added. The malting process results in the production of fermentable sugars, low molecular weight nitrogenous materials, enzymes (especially amylolytic and proteolytic enzymes), and various flavor and color compounds. In addition, the cereal which is malted generally brings with it its complement of vitamins and minerals (Table 6). Diastatic and nondiastatic malts differ generally only in that the latter are processed so as to inactivate their enzymes.

Many of the components of malt products have been demonstrated to have desirable effects on baked products. Fermentable sugars, including glucose, maltose, and maltotriose, contribute to gas production in doughs both immediately in the case of glucose, and later in the process in the case of the higher sugars. Sound wheat flours used for baking purposes generally contain some β-amylase activity, but no α-amylase or proteolytic activity. They also contain low levels of starch which was damaged during the milling process and which is susceptible to amylolytic activity. This damaged starch is available to β-amylase action in doughs but the extent of breakdown is limited by the inability of β-amylase to hydrolyze or bypass α-1, 6 linkages in starch. The addition of diastatic malt products provides both α-and β-amylases to the dough and the α-amylase is able to make available for maltose production a significantly greater proportion of the damaged starch for subsequent fermentation by the yeast. Among the advantages derived from the action of the amylolytic enzymes are increased gas production in the dough, better moisture retention in the crumb, and slower staling due to the hygroscopic nature of the dextrins produced by α-amylase and the more extensive breakdown of starch. In addition, the preformed sugars present in malt products and their mineral and vitamin contents ensure rapid and vigorous yeast activity. The presence of reducing sugars and low molecular weight protein breakdown products leads to desirable crust color and flavor development due to the formation of Maillard-type reaction products.

Finney et al. (28) have shown that cereal malts were acceptable replacements for the sucrose that is normally added to bread dough formulations. With optimum levels of malt (as measured by loaf volume), adding sucrose at levels of 0 and 6% had no effect, indicating that all of the sugar necessary for fermentation had been supplied preformed in the malt or was produced from damaged starch by the malt enzymes. They also demonstrated that triticale malts were most effective in increasing loaf volumes at a given addition level because of their higher diastatic power and higher levels of soluble protein. Malt syrups were also more effective than corresponding flours, probably because they were added at higher levels and, because of the manner in which they are produced, they contributed more fermentable sugars. Malt use levels vary widely in baking applications depending to some extent upon whether its primary contribution is to be its enzyme activity or its flavor and/or color. Table 7 shows typical levels range from 0.25 to 33% based on flour. A more detailed discussion of the use of cereal malts in the baking industry has been provided by Pyler (29).

IV. MALTS IN DISTILLED SPIRITS

Another common use for cereal, especially barley, malt is in the production of distilled spirits, where it functions primarily as a source of fermentable sugars.

The mash bill for distilled spirit production may contain only barley malt, as in the case of Scotch malt whiskey, or may contain barley malt along with other cereals, as in the case of Scotch grain whiskey, Bourbon, and Canadian and rye whiskeys.

The malt used in the production of Scotch malt whiskey is steeped and germinated in the normal manner and the green malt is taken to the kiln. During the kilning process, peat is burned directly under the kiln floor for a specified period. The absorption by the green malt of certain constituents from the peat smoke impart desirable organoleptic properties to the malt and subsequently to the distilled spirit. This is followed by low temperature drying in order to minimize enzyme denaturation.

The malt so produced is then mashed at 60–65°C, a temperature range within which both α-and β-amylases are very active, in order to maximize conversion of starch into fermentable sugars. After filtration of the insoluble material, the extract is pitched with yeast without prior boiling to allow amylolysis to continue during the fermentation process.

Most other distilled spirits are produced using barley malt as only a fraction of the total mash bill. For bourbon whiskey, for example, the mash bill must contain at least 51% corn but may contain up to 70%. Malt may constitute only 10–15% of the total cereal. The use of unmalted cereals requires their prior gelatinization and liquefaction, and this is accomplished using a portion of the malt as described for the brewing process. After gelatinization has been effected, the mash is cooled to about 63°C, and the remaining malt is added to complete the conversion of starch to fermentable sugars.

Among the cereals utilized as adjuncts in various products are wheat, rye, oats, and unmalted barley.

A more complete discussion of the production of distilled spirits has been prepared by Nagodawithana (30).

V. NONALCOHOLIC BEVERAGES

A variety of nonfermented malt-based beverages is prepared and consumed throughout the world. Among these are hot or cold water infusions of ground malt and malted milk. The latter is typically prepared from a malt extract, which may be combined with milk products and dried. If milk is present, the drink is prepared by dissolving the powder in warm water; if not, warm milk is used.

A fuller discussion of nonfermented malt-based beverages is given by Briggs (31).

VI. OTHER CEREAL MALTS

This discussion has been limited to the production and uses of barley malt as it is by far the most widely used in the United States. Other cereals, primarily rye, wheat, triticale, and sorghum, are also malted in other parts of the world and find use in applications similar to those described above.

TABLE 1 Approximate Cycle Times During the Last 300 Years of Malting

	1686	1886	1986
Steeping (days)	6–8	3–7	1–3
Germination (days)	10–35	8–28	3–6
Kilning (days)	4–6	2–4	1/2–2

TABLE 2 Different Malt Types Showing General Relationship Between Final Enzyme Activity and Malt Color

Malt type	Enzyme activity	Color
Diastatic	High	Light
Brumalt		
Lager		
Pilsener		
Pale ale		
Mild ale		
Munich		
Diamber		
Cara-Pils		
Amber		
Dextrine		
Caramel		
Crystal		
Chocolate		
Black patent	Low	Dark

TABLE 3 Typical Analysis of Six– and Two–Rowed Barley Malts Produced for Brewing

Characteristic	Six–rowed malt	Two–rowed malt
Moisture, %	4.0	3.9
Wort color, SRM[a]	1.85	1.75
Wort clarity	Clear	Clear
Soluble protein, %	5.41	5.22
Total protein, %	12.59	11.90
S/T ratio[b]	43.0	43.9
Diastatic power, db, °L[c]	155	166
Extract, fine, as is, %	76.4	77.7
Extract, fine, db, %	79.6	80.9
Extract, coarse, as is, %	75.0	76.2
Extract, coarse, db, %	78.1	79.3
F–C difference, %	1.5	1.6
α–Amylase, D.U.[d]	45.1	49.0
Viscosity, cps	1.47	1.5
pH	5.68	5.70
Conversion,[e] min	5–7	5–7
Growth[f] %		
0–1/4	0	0
1/4–1/2	2	2
1/2–3/4	5	6
3/4–1	84	80
Over 1	9	12
Assortment, %		
On 7/64 sieve	51.4	80.9
On 6/64 sieve	40.1	15.6
On 5/64 sieve	7.7	3.1
Thru 5/64	.8	.4
Bushel weight, lb	39.0	40.1

[a]Standard Reference Method, American Society of Brewing Chemists.
[b]Soluble to total protein ratio.
[c]Degrees Lintner.
[d]Dextrinizing Units.
[e]Time taken at 70°C to reach point where starch is degraded by malt α–amylase to the point that it will no longer react with iodine.
[f]Acrospire growth as a percentage of kernel length.
Source: Courtesy of Great Western Malting Co.

TABLE 4 Proximate Analysis of Barley Malt Flour (%, dry matter)

Component	Percent	
Protein (%)	11.5–14.5[a]	8–11[b]
Lipid (%)	1.5–2.5	2–3
Ash (%)	2–3	2.2
Fiber	2–6	—
Nitrogen–free extract (%)	68–74	—
Starch (%)	—	58–60
Nucleic acids (%)	—	0.2–0.3
Cellulose (%)	—	5
Amino acids (%)	—	1–2

[a]Values this column from Ref. 25.
[b]Values this column from Ref. 26.

TABLE 5 Analysis of Malt Products

	Nondiastatic malt syrup	Malted barley and corn syrup
Solids (%)	78.5–80.0	78.5–80.0
Acidity (%)	0.9–1.4	0.7–1.1
Reducing sugars (%)	54–62	57–67
Protein (%)	5.0–6.3	3.0–4.0
pH	5.1–5.8	5.1–5.8
Ash (%)	0.8–1.4	0.6–1.0
Color (Lovibond)	200–370	100–250

Source: Adapted from Ref. 24.

TABLE 6 Vitamin Content of Malted Barley

Vitamin	Amount (ppm)	
Biotin	0.22[a]	0.2[b]
Niacin	90.2–149.6	90–150
Pantothenic acid	4.4–17.2	4.4–5.6
Pyridoxine	5.9–7.9	6–8
Riboflavin	1.98–3.96	2–5
Thiamine	2.86–3.52	1.2–6.8
Choline	Trace	—
Inositol	Trace	—
Ascorbic acid	—	Absent[c]
Folic acid	—	0.4

[a]Values this column from Ref. 25.
[b]Values this column from Ref. 27.
[c]Values in unmalted barley.

TABLE 7 Typical Use Levels of Malt Syrups in Baked Goods

Product	Malt type	Use level (%) (flour basis)
White bread	LD	0.25–0.50
Soft rolls	LD	1.0–2.0
Hard rolls	MD	2.0–2.5
White hearth bread	LD	0.25–3.0
Hamburger buns	LD	0.25–0.50
Rye breads	DB	3.0–6.0
Sweet doughs	LD	2.0
English muffins	LD	0.5–1.0
Pretzels	ND	2.0–6.0
Soda crackers	SP	1.0
Cookies	ND	18.0

LD = low diastatic; MD = medium diastatic; DB = dark bread type; SP = special; ND = nondiastatic.
Source: Adapted from Ref. 29.

REFERENCES

1. Corran, H. S., *A History of Brewing*, Newton Abbot: David and Charles, 1975, p. 11.
2. Taylor, J. A., and Palmer, G. H., Drying and Dormancy in Malting Barleys, Progress Reports on Research and Development, Home-Grown Cereals Authority, 1982, pp. 11–16.
3. Fincher, G. B., Morphology and chemical composition of barley endosperm cell walls. *J. Inst. Brew., 81*:116–112 (1975).
4. Bacic, A., and Stone, B. A., A (1-3) and (1-4) linked β-glucan in the endosperm cell walls of wheat. *Carbohydr. Res., 82*:372–377 (1980).
5. MacLeod, A. M., Studies on the free sugars of the barley grain. *J. Inst. Brew., 58*:270–276 (1952).
6. Tryon, T., *The New Art of Brewing Beer, Ale and Other Sorts of Liquors*, London: Salisbury, 1691, pp. 45–57.
7. Maule, A. P., Aerated Conical and Flat-Bottomed Steeping Systems in Theory and Practice. In *Proceedings of the Second Aviemore Conference on Malting, Brewing and Distilling*, Campbell, I., and Priest, F. G. (Eds.), The Institute of Brewing, London, 1987, pp. 251–254.
8. Wellington, P. S., Germination and seedling emergence. In *The Growth of Cereals and Grasses*, F. L. Milthorpe, J. D. Ivins (Eds.), Butterworths, London, 1965, pp. 3–19.
9. Thomas, D. A., The magic of malt. In *Beer and Brewing*, Vol. 6, V. Thomas (Ed.), Brewers Publications, Boulder, CO, 1987.
10. Palmer, G. H., and Bathgate, G. M., Malting and brewing. In *Advances in Cereal Science and Technology*, Y. Pomeranz (Ed.), AACC, St. Paul, MN, 1976, pp. 237–324.
11. Macleod, A. M., The physiology of malting. In *Brewing Science*, J. R. A. Pollock (Ed.), Academic Press, London, 1979, pp. 146–224
12. Hough, J. S., Briggs, D. E., and Stevens, R., The technology of malting and kilning. In *Malting and Brewing Science*, Chapman and Hall, London, 1971, pp. 108–148.
13. American Society of Brewing Chemists, *Methods of Analysis of the ASBC*, 7th ed., St. Paul, MN, 1976.
14. American Association of Cereal Chemists, *Approved Methods of the AACC*, St. Paul, MN, 1984.
15. The Institute of Brewing, *Recommended Methods of Analysis*, London, England, 1982.
16. European Brewery Convention, *Analytica-EBC*, 4th ed., Brauerei-und Getranke-Rundschau, CH8047, Zurich, Switzerland, 1987.
17. *Brewers Digest, 62*(5):16 (1987).
18. Briggs, D. E., Hough, J. S., Stevens, R., and Young, T. W., *Malting and Brewing Science* 2nd ed., Chapman and Hall, New York, 1981.
19. Pollock, J. R. A., (Ed.), *Brewing Science*, Academic Press, New York, 1979.
20. Broderick, H. M., (Ed.), *The Practical Brewer*. Master Brewers Association of the Americas, Madison, WI, 1977.
21. Pyler, R. E., and Thomas, D. A., Cereal research in brewing: Cereals as brewers adjuncts. *Cereal Foods World, 31*:681 (1986).
22. Canales, A. M., Unmalted grains in brewing. In *Brewing Science*, Vol 1, J. R. A. Pollock, (Ed.), Academic Press, New York, 1979, pp. 223–278.
23. Beckerich, R. P., and Denault, L. J., Enzymes in the preparation of beer and fuel alcohol. In *Enzymes and Their role in Cereal Technology*, J. E. Kruger, D. Lineback, and C. E. Stauffer, (Eds.), American Association of Cereal Chemists, St. Paul, MN, 1987, pp. 335–355.
24. Gill, R. J., Malts used in baking. *Proc. Am. Soc. Bakery Engrs.*, 1982, 61.
25. Matz, S. A., and Matz, T. D., *Cookie and Cracker Technology*. AVI Publishing Co., Westport, CT, 1978, P. 125.
26. Harris, G., The structural chemistry of barley and malt. In *Barley and Malt*, A. H. Cook, (Ed.), Academic Press, London, 1962, p. 435.
27. Scriban, R., In *Brewing Science*, Vol. 1, Pollock, J. R. A. (Ed.), Academic Press, New York, 1979, p.52.
28. Finney, K. F., Shogren, M. D., Pomeranz, Y., and Bolte, L. C., Cereal malts in breadmaking. *Bakers Digest*, 46(1):36 (1972).

29. Pyler, E. J. *Baking Science and Technology*, 2nd ed. Siebel Publishing Co., Chicago, 1973, pp. 423–430.
30. Nagodawithana, T. W., Yeasts: Their role in modified cereal fermentations. In *Advances in Cereal Science and Technology*, Vol, VIII, Y. Pomeranz (Ed.), American Association of Cereal Chemists, St. Paul, MN, 1986, pp. 15–104.
31. Briggs, D. E., *Barley*. Chapman and Hall, London, 1978, pp. 560–586.

22

Cereal Enrichment

Peter Ranum
Atochem North America, Buffalo, New York

I. INTRODUCTION

The United States has had a cereal enrichment program since 1941. Following are some of the discoveries and events that led to its adoption (1).

1. The discovery in the early 1900s that certain chemical compounds (vitamins) and minerals in foods are essential for good health.
2. The synthesis of thiamin by R. R. Williams in 1936. The reduced cost of synthetic vitamins made enrichment economically feasible.
3. The finding that a large proportion of the vitamins and minerals naturally occurring in wheat were being removed by the milling process (2).
4. The alarming incidence of nutritional deficiency diseases, especially pellagra (niacin deficiency), among the poorer population groups of the United States.
5. The early advocation by James F. Bell of General Mills Inc. and Dr. Russell Wilder of the American Medical Association (AMA) of cereal enrichment.

In 1939 a few bakers started enriching bread voluntarily after the AMA Council on Foods published a resolution encouraging that it be done. The Food and Nutrition Board (FNB) of the National Research Council (NRC) was formed in 1940. One of its first actions was to propose guidelines for cereal enrichment. In 1941 the U.S. Food and Drug Administration (FDA) issued standards of identity for enriched flour and farina. Enrichment of white flour and bread became mandatory during World War II. This order was recalled after the war but by then 19 states had passed their own mandatory enrichment laws. Since the war the U.S. enrichment program has been expanded to cover other cereal staples: cornmeal, rice, and pasta.

II. NATURE OF ENRICHMENT

Fortification can be defined as simply the addition of nutrients (vitamins, minerals, protein, or amino acids) to food. The term enrichment was coined to describe a special type of fortification practiced under the following circumstances:

1. The food being enriched is an expensive dietary staple, consumed by the general population on a daily basis. This applies to many of the cereal grain products like wheat flour, bread, cornmeal, and rice. These foods make excellent vehicles for getting nutrients to the people who need them the most.

2. The nutrient being added is present in the unprocessed food at significant levels, and the amount added is to make up for what was lost in normal processing, such as milling.

Whether a cereal product should be enriched with a particular nutrient for this reason is readily evident from its Index of Nutritional Quality (INQ), shown in Table 1 for a number of nutrients in different cereals products. The INQ is the ratio of the percent nutrient requirement to the percent of the recommended energy intake (2000 kcal/day) provided by the food (3). An INQ near 1.0 indicates that the food has a good nutrient to caloric balance for that particular nutrient. INQs far below 1.0 show that the food is low in the nutrient compared to its caloric content, suggesting that fortification would be in order.

As an example, the INQ for niacin is shown to have dropped from 1.4 in wheat to 0.2 in white flour. Enrichment brings it back up to 1.5. Other nutrients showing similar reduction are iron, zinc, magnesium, copper, thiamin, and pyridoxine. Riboflavin is low in wheat; enrichment brings it to higher than natural levels. All cereals are shown to be poor sources of calcium. The optional calcium enrichment standards make then fair sources of calcium with INQs near 1.0

3. In order for a particular nutrient to be included in an enrichment program, there must be good evidence that it is in short supply within the target population. There must be evidence of *need* for that nutrient. This can be very difficult to establish, which probably accounts for the fact that only four nutrients are presently required under the standards.

Sometimes there is such a great need for a nutrient that is added at amounts far above natural levels or to foods that do not naturally contain it. Such addition is more properly called fortification. This is the case with vitamin A. There is so great a need for it in developing countries that is required in all the Title II PL-480 (The U.S. Food for Peace Program) commodities.

4. In order for enrichment to work, it has to be invisible. It must not cause any change in the appearance, flavor, or even the price of the food. This prevents the addition of riboflavin to rice (it gives an unnatural yellow color) and the restoration of magnesium in wheat flour (flavor problems). Vitamin E is quite expensive; its addition to a basic food staple would result in a noticeable increase in cost.

III. GOVERNMENT REGULATIONS AND GUIDELINES

An enrichment program is normally set up, regulated, and enforced by the government. As such, it is a public health program designed to ensure that the general population receives adequate amounts of essential nutrients.

While the addition of nutrients to food is recognized to be an effective way of maintaining and improving the overall nutritional quality of the food supply, unregulated fortification can cause problems. Some foods could get superfortified, creating nutrient imbalances. To regulate this the FDA suggests the following circumstances under which it would be appropriate to add nutrients to food (4):

1. To correct a recognized dietary deficiency.
2. To restore nutrients lost in processing.
3. To balance the nutrient content in proportion to the caloric content.
4. To avoid nutritional inferiority in new products replacing traditional foods.
5. To comply with other programs and regulations.

The FDA further suggests (5) that the following conditions be met when adding nutrients to foods:

1. The intake of the nutrient is below a desirable level in diets of a significant number of people.
2. Food used to supply the nutrients is likely to be consumed in quantities that make a significant contribution to the diet of the population in need.
3. The addition of the nutrient is not likely to create an imbalance of essential nutrients.
4. The nutrient added is stable under proper conditions of storage and use.
5. The nutrient is physiologically available from the food.
6. There is reasonable assurance against excessive intake reaching a level of toxicity.

A number of countries have cereal enrichment programs, including the United States, Canada, and the United Kingdom. Sadly, most of the developing countries, which stand to gain the most from such programs, do not have them.

IV. ENRICHMENT STANDARDS

The U.S. and Canadian cereal enrichment standards are shown in Table 2. These standards specify the types and levels of nutrients required to be in enriched cereal products. In the United States, enrichment can be achieved either at the mill or the bakery. Canada requires that all the enriching be done at the mill. You cannot sell unenriched flour in Canada, but you can in the United States.

The U.S. FDA does not require enrichment; it only specifies what the product must contain if it is labeled or advertized as being enriched. Enrichment is required for most cereal products purchased by the USDA Agriculture Stabilization and Conservation Service (ASCS) (Table 3) and for products sold in states with mandatory enrichment laws. Since there are 37 states with such laws, enrichment is a virtual requirement for anyone wishing to ship interstate. Enrichment is required in all of Canada.

Revisions in the U.S. enrichment standards for flour and bread were made in 1974 for vitamins and in 1981 for iron. These revisions put the standards as single, minimum levels instead of a range and adjusted the bread standards so they were 63% of the flour standards, since bread generally contains around 63% flour. Previously there were separate flour and bread standards that did not match. It used to be possible for the baker to enrich to a higher level by adding nutrients rather than by using enriched flour.

An expanded cereal enrichment/fortification program was proposed for cereal grain products by the National Academy of Sciences, Food Nutrition Board (6) in 1975 (see Table 4 for levels in bread). This was never adopted in the United States, largely because of lack of support from the FDA. A few bakers tried them out voluntarily but it never met with much commercial success. A similar proposal in Canada did result in expanded optional standards, but little use has been made of them.

Some baking companies have marketed white breads claimed to be nutritionally equivalent to whole wheat. To do this they add all the nutrients, including fiber, needed to make up the difference between those in white bread and those in whole wheat bread. An example of one such product is shown in Table 4.

V. SOURCES OF NUTRIENTS

The source of different nutrients used in enriching cereals is given in Table 5. The amount of a nutrient source to add to meet a particular standard has to be adjusted on the basis of the concentration of the nutrient in the chemical compound.

In the case of vitamins, molecular weight (MW) adjustments must be made if the chemical source of the nutrient used differs from the compound specified by the FDA (4). For example, the U.S. RDA for thiamin is 1.5 mg/day of thiamin hydrochloride, whose MW is 337.3. Thiamin mononitrate, with a MW of 327.4, would have an effective vitamin concentration of 100 (337.3/327.4) = 103%. Fat-soluble vitamins have U.S. RDAs in International Units (IU). Commercial sources of these vitamins are not pure compounds. Their concentrations are given in IU/g or IU/mg, as shown in Table 5. In order to facilitate the addition of vitamin B_{12}, which is needed in very low amounts, vitamin producers offer it in diluted form.

The stability of added vitamins, shown in Table 6, should be considered when fortifying foods (7). Niacin is quite stable under all conditions. Thiamin can show significant losses at high pH, such as present in chocolate cakes and tortillas. Thiamin mononitrate has been found to be the most stable form for cereal enrichment (8). Riboflavin can be lost when exposed to sunlight. This can be a problem when enriched pasta is packaged with clear plastic window, as is often done. Vitamin A can show heavy loses if not in a stabilized form. The ASCS specifies that vitamin A be encapsulated and, when present in the cereal, show no more than 20% loss of activity after 21 days storage at 45°C (9). There are no major problems with the stability of folic acid, pyridoxine, or pantothenic acid. The normal application of bleaching, maturing, and oxidizing treatments at the mill was found to have no effect on added vitamins (10).

Iron is often a problem in cereal enrichment. The characteristics of the different iron sources commonly used in enrichment are given in Table 7. As a general rule, ferrous sulfate should be used whenever possible because of its higher bioavailability. When not possible, because of long shelf-life requirements, reduced iron is recommended.

Cereal enrichment has proven to be a cost-effective way to improve a nation's nutritional health. The current cost of adding the three B vitamins and iron required under the U.S. standards is about 5 cents per 100 lb of cereal product. Since the average per capita consumption of enriched milled cereals runs around 100 lb per year, enrichment costs each person a nickel per year. For that everyone receives, on the average, 25% of their U.S. RDA for iron, 30% for niacin and riboflavin, and 50% of their thiamin requirement.

It is not very likely that any change will occur in the U.S. cereal enrichment program. There is no good evidence of any particular nutrient being deficient or excessive enough to justify a change. Fortification of nonstandard or newly developed foods, however, will continue to be a hotly debated issue. Food producers will want to add nutrients whenever it helps them sell their product. The Food and Drug agencies will try to keep this from getting out of control. On a worldwide basis, iron and vitamin A are the two micronutrients in greatest demand. There is a lot of work being done to increase the amount of these two nutrients in the food supply.

TABLE 1 Index of Nutritional Quality for Selected Cereal Grain Products

Product	Pro	Ca	P	Fe	Zn	Mg	Mn	Cu	B_1	B_2	Ni	PA	B_6	Folic acid
Barley	0.7	0.1	1.1	0.6	0.8	1.1		1.6	0.5	0.2	0.9		0.8	
Cornmeal	0.7	0.0	0.4	0.3	0.3	0.7	0.2	0.5	0.5	0.1	0.3	0.3	0.7	
Millet	0.9	0.1	1.9	2.3	0.7	1.7	2.5	2.3	3.0	1.4	0.7	0.5	1.2	1.3
Oats	1.1	0.3	2.1	1.3	1.2	2.0		1.5	2.1	0.4	0.3			
Rice, brown	0.6	0.2	1.2	0.5	0.7	1.8	5.2	0.7	1.3	0.2	1.3	0.8	1.4	0.9
Rice, white	0.6	0.1	0.5	0.2	0.4	0.5	1.6	0.4	0.3	0.1	0.4	0.5	0.4	
Rye	1.1		2.3	0.4	1.5	1.8	4.0	1.3	1.7	0.8	0.5		0.9	
Wheat flour	0.8	0.1	0.1	0.2	0.2	0.3	0.8	0.4	0.2	0.1	0.2	0.3	0.1	0.2
Whole wheat flour	1.2	0.2	2.2	1.1	1.0	2.0	6.1	1.0	2.2	0.4	1.3	0.6	1.0	0.7
Wheat	1.1	0.2	2.3	1.1	1.2	1.8	6.0	1.3	2.1	0.4	1.4	0.6	1.1	0.6
Enriched cornmeal	0.7	0.7	0.4	0.9	0.3	0.7	0.2	0.5	1.6	0.8	1.3	0.3	0.7	
Enriched rice	0.6	0.6	0.5	0.9	0.4	0.5	1.6	0.4	1.6	0.8	1.0	0.5	0.4	
Enriched flour	0.8	1.1	0.1	1.3	0.2	0.3	0.8	0.4	2.3	1.3	1.5	0.3	0.1	0.7

Pro = protein; Ni = niacin; PA = pantothenic acid.
Source: Based on data of chemical composition of whole grains plus additional data from Ref. 12 using the formula: INQ = [2000/caloric content] × [nutrient content/nutrient RDA].

TABLE 2 U.S. and Canadian Cereal Enrichment Standards

Product	Thiamin	Ribo–flavin	Niacin	Iron	Calcium
		United States (mg/lb):			
Wheat flour	2.9	1.8	24	20	(960)
Self-rising flour	2.9	1.8	24	20	960
Corn meal and grits	2.0–3.0	1.2–1.8	16–24	13–26	(500–750)
Self-rising corn meal	2.0–3.0	1.2–1.8	16–24	13–26	(500–1750)
Rice	2.0–4.0	1.2–2.4	16–32	13–26	(500–1000)
Pasta					
(macaroni and noodles)	4.0–5.0	1.7–2.2	27–34	13–16.5	(500–625)
Bread	1.8	1.1	15	12.5	(600)
		Canada (mg/100 g):			
Wheat flour	0.44–0.77	0.27–0.48	3.5–6.4	2.9–4.3	(110–140)
Bread	0.24	0.18	2.2	1.76	(66)

Figures in parentheses are optional, the rest are mandatory. Single level values indicate minimums. Canada has optional flour enrichment standards for vitamin B_6 (0.25–0.31 mg/100 g), folic acid (0.04–0.05), pantothenic acid (1.0–1.3), and magnesium (150–190). Calcium is required in Newfoundland.

TABLE 3 Enrichment Standards for Government-Purchased ASCS[a] Commodities as of 1990

Product	Thiamin (mg/lb)	Riboflavin (mg/lb)	Niacin (mg/lb)	Iron (mg/lb)	Calcium (mg/lb)	Vitamin A (IU/lb)
Wheat flour—U.S.	2.9	1.8	24	20	1	
Wheat flour—export	2.9	1.8	24	20	500–625	10,000–12,000
Soy-fortified flour	2.9	1.8	24	20	500–625	10,000–12,000
Corn meal—U.S.	2.0–3.0	1.2–1.8	16–24	13–26		
Corn grits—U.S.	2.0–3.0	1.2–1.8	16–24	21–26		
Corn masa flour	2.0	1.2	16	13–26		
Corn meal and soy fortified corn meal (export), bulgur, soy-fortified bulgur, and soy-fortified sorghum grits	2.0–3.0	1.2–1.8	16–24	13–26	500–750	10,000–12,000

[a]ASCS is the Agriculture Stabilization and Conservation Service of U.S. Department of Agriculture, Kansas City, MO.

TABLE 4 Expanded Enrichment/Fortification

Nutrient	Typical level (mg/kg) in bread		
	Canada[a]	NRC/NAS[b]	Commercial whole white[c]
Thiamin	2.4	4.0	4.0
Riboflavin	1.8	2.3	2.3
Niacin	22.0	33.0	33.0
Pyridoxine	1.4	2.8	1.9
Folic acid	0.24 +	0.4 +	0.56
Pantothenic acid	6.0	—	4.6
Vitamin A (IU/kg)	—	6000	—
Iron	18	28	28
Calcium	660	1240	830
Magnesium	900	—	630
Zinc	—	14	16
Manganese	—	—	26
Copper	—	—	2.3

[a]The levels shown for Canada are based on the required and optional Canadian flour enrichment standards in Table 2.
[b]The NRC/NAS are based on the expanded cereal enrichment standards proposed by the National Research Council (6).
[c]The last column shows a commercial bread product produced by the W.E. Long Independent Bakeries Co. (11).

TABLE 5 Nutrient Sources

Nutrient[a]	Code	Source:[b] chemical compound	Nutrient activity (%)[c]
Vitamin A	A	Vitamin A Acetate	500 IU/mg
		Vitamin A Palmitate	250 IU/mg
Thiamin	B_1	Thiamin Mononitrate	103
		Thiamin Hydrochloride	100
Riboflavin	B_2	Riboflavin	100
Niacin	B_3	Nicotinic Acid (Niacin)	100
		Niacinamide	100
Folacin		Folic Acid	100
Pantothenic acid	B_5	D–Calcium Pantothenate	82
Pyridoxine	B_6	Pyridoxine Hydrochloride	83
Cyanocobalamin	B_{12}	1% Vitamin B_{12}	1
Ascorbic acid	C	Ascorbic Acid	100
		Sodium Ascorbate	89
Vitamin D	D_3	Vitamin D_3	100 IU/mg
Tocopheryl	E	Alpha Tocopheryl Acetate	.05 IU/mg
Calcium	Ca	Calcium Sulfate, anhyd.	29
		Calcium Sulfate, dihydrate	23
		Calcium Carbonate	40
Copper	Cu	Cupric Carbonate Basic	53
		Cupric Sulfate, anhyd.	40
Iodine	I	Calcium Iodate	62
		Potassium Iodate	59
Iron	Fe	Hydrogen Reduced Iron	98
		Electrolytic Iron	99
		Ferrous Sulfate	32
		Ferric Orthophosphate	29
		Ferrous Fumarate	33
Magnesium	Mg	Magnesium Carbonate	25
		Magnesium Oxide, heavy	60
		Magnesium Sulfate	14
Maganese	Mn	Manganese Sulfate	32
Phosphorus	P	Dicalcium Phosphate, anhyd.	29
		Tricalcium Phosphate	39
Zinc	Zn	Zinc Sulfate	36
		Zinc Oxide	80

[a]The nutrient is the form of the nutrient specified by the U.S. FDA (4).
[b]The source is the form of the nutrient actually used.
[c]The nutrient activity shows the % nutrient in the source.

TABLE 6 Stability of Vitamins Under Different Conditions

Vitamin	pH 7	Acid	Alka-line	Air	Light	Heat	Cooking loss (%)
Vitamin A	S	U	S	U	U	U	0–40
Vitamin C	U	S	U	U	U	U	0–100
Vitamin B_{12}	S	S	S	U	U	S	0–30
Vitamin D	S	S	U	U	U	U	0–40
Niacin	S	S	S	S	S	S	0–75
Vitamin B_6	S	S	S	S	U	U	0–40
Vitamin B_1	U	S	U	U	S	U	0–80
Vitamin B_2	S	S	U	S	U	U	0–75
Vitamin E	S	S	S	U	U	U	0–75

S = stable; U = unstable.
Source: Ref. 7.

TABLE 7 Characteristics of Different Iron Sources Used in Cereal Enrichment

Characteristic	Reduced Iron	Electrolytic Iron	Ferrous Sulfate	Ferric Orthophos–phate
Stability in flour	Good	Good	Poor	Excellent
Bioavailability	Fair	Good	Excellent	Poor
Color	Black	Black	Tan	White
Magnetic	Yes	Yes	No	No
Labeling required	"iron"	"iron"	"ferrous sulfate"	"ferric orthophos-phate"
Cost	low	higher	low	high

REFERENCES

1. Lavallee, D. F., A history of the policy of cereal-grain enrichment in the United States. Master's thesis, Syracuse University. Syracuse, New York, Dec., 1984.
2. Schultz, A. S., Atkin, L., and Frey, C. N., The vitamin B1 content of wheat, flour and bread. *Cereal Chem.*, *16*:643 (1939).
3. Hansen, R. G. An index of food quality. *Nutr. Rev.*, *31*:1 (1973).
4. *U.S. Code of Federal Regulations*, Title 21, Part 104 B, Fortification Policy.
5. General principles governing the addition of nutrients to food. U. S. Federal Register, *39*(116):20900 (1974).
6. NAS/NRC. Proposed fortification policy for cereal-grain products. National Academy of Sciences, Washington, D.C., 1975.
7. Birdsall, J. J., Stability and availability of nutrients. In *Technology of Fortification of Foods*, National Academy of Sciences, Washington, D.C., 1975.
8. Borenstein, B., Vitamin A fortification of flour and cornmeal. *Northwest. Miller*, *276*(2):18 (1969).
9. Houser, N. D. Amendment 1 to wheat flour/soy-fortified bread wheat flour export announcement WFSF-2. *USDA ASCS*, Jan 26, 1988.
10. Ranum, P. M., Loewe, R. J., and Gordon, H. T., Effect of bleaching, maturing, and oxidizing agents on vitamins added to wheat flour. *Cereal Chem.*, *58*:32 (1981).
11. Dwyer, R., personal communication, 1982.
12. Drake, D. L., personal communication, 1988.

23

Nutritional Quality of Cereals and Cereal-Based Foods

Gur S. Ranhotra
American Institute of Baking, Manhattan, Kansas

I. CEREAL-BASED FOODS IN OUR DIET

For a vast majority of the world population, cereal-based foods are the major source of energy and nutrients. Even in the North American diet, they provided, at the turn of the century, about 40% of the total energy. Today, however, they provide only about 20– 25%. Cereals do not contain vitamin B_{12}, and they are low in certain important nutrients. However, they are the major source of several of the approximately 40 different nutrients we need for good health.

Although collectively all cereal-based foods provide about 20–25% of the total calories in our diet, wheat-based foods alone provide about three-fourths of this amount. The remaining one-fourth of the energy is provided by a variety of foods based on other principal (corn, rice, rye, oats) and minor (barley, triticale, sorghum, millets) grains.

II. CHEMICAL COMPOSITION OF WHOLE GRAINS

When a series of samples of wheat or any other grain is chemically analyzed, a wide range of values for each analyzed constituent is encountered. Thus, the values shown in Table 1 (1–3) are only representative values. Moisture, protein ($N \times 5.7$ or other appropriate factor), fat, ash (mineral matter), and fiber are the major (proximate) components routinely analyzed. Available carbohydrate values are determined through calculation and equal the difference between 100 and the sum of the percentages of the analyzed components.

Fiber, until recently reported as crude fiber (only the cellulose and lignin components), is now increasingly being determined and reported as total dietary fiber. Since total fiber values are severalfold greater than crude fiber values, calculated carbohydrate values are invariably lower when fiber is considered as total, rather than crude, fiber.

Cellulose, noncellulose polysaccharides (hemicelluloses, pectic substances, gums, and mucilages), and lignin (not a carbohydrate) are the major components of total dietary fiber. Some of these are soluble in water, while others are not. From a health point of view, this distinction is becoming important in that insoluble fiber may be beneficial in diseases of the gastrointestinal tract, while soluble fiber components may help lower elevated blood cholesterol and sugar levels. Table 2 lists the total and soluble fiber content of several cereal-based foods (4).

III. NUTRIENTS IN WHOLE GRAINS

In contrast to most animal-derived foods and leguminous seeds, the protein content of cereal grain (particularly corn, rice, and barley) is low (Table 1). The nutritional quality of protein in grains is also low because of the deficiency of lysine—the most limiting amino acid—and certain other essential amino acids. This impairment of protein quality has no nutritional significance in the North American diet, however, since other foods (meat, milk, cheese, eggs) usually included in the diet are high in lysine. In addition, most Americans consume protein well in excess of the body's need.

Whole grains are low in fat, and the fat that is present is high in polyunsaturated fatty acids. Grains contain no cholesterol, and grain-based foods are the major source of complex carbohydrates—starches and fiber—in our diet. Whole grains are also a significant source of several vitamins and minerals. This, however, does not include calcium, nor phosphorus when it is present as phytic acid phosphorus. Not only is phytic acid phosphorus poorly absorbable, it may also interfere with the absorption of other mineral elements.

IV. PROCESSING OF CEREAL GRAINS

Preparation of cereal-based foods usually requires some degree of processing of whole grains. Milling is a major such process. Milling often alters proximate composition in several ways. For example, the milling of wheat to obtain white flour (less than 80% extraction) eliminates the bran and germ fractions, both of which are rich in nutrients. Some processes may cause changes in the nutrients themselves. This may include hydrolysis of phytic acid and polysaccharides. Losses of nutrients can occur during heat treatment, e.g., baking and cooking. Besides heat, other factors which influence nutrient stability include: time, pH, moisture (water activity), light, oxygen, metals, oxidants, enzymes, and possibly certain additives. Nutrient losses are not the only consequence of heat treatment; heat treatment sometimes improves the overall nutrient profile of food products.

The extent to which nutrient makeup can be altered by milling of whole grains is evident when compositional data (5) on wheat flours of different extractions are compared (Table 3). Comparing whole wheat and white breads (Table 4) provides further evidence of this.

V. NUTRIENTS IN BREAD AND RELATED PRODUCTS

North Americans consume only about half (about three 1-oz slices) the amount of bread currently consumed in most European countries. Bread products differ appreciably in their nutrient profile as the compositional values listed for white and whole wheat bread suggest (Table 4) (6). Because of the high fiber content (about 2 g total dietary fiber per slice), whole wheat bread is low in energy as compared to white bread. In contrast, it is high in various vitamins and minerals except those which are added to white flour when flour is enriched. This enrichment adds thiamin, riboflavin, niacin, iron, and the optional nutrient calcium.

Over 90% of the white bread commercially produced in the United States is enriched white bread. Enrichment standards for white bread and various other products are discussed in a separate chapter. Although the addition of calcium to white flour is optional, newer knowledge on the health benefits of calcium (prevention and management of osteoporosis, colon cancer, and high blood pressure) has prompted millers and bakers to now include calcium. An expansion of the current enrichment program to include five additional nutrients (zinc, vitamin A, vitamin B_6, folic acid, and magnesium) was proposed by the National Academy of Sciences in 1974.

Enriched white bread (as is true for all good products which contain added nutrients or make nutritional claims) is required to carry a nutritional label. Dietary standards, called U.S. Recommended Daily Allowances (U.S. RDAs), used for nutrition label purposes are also discussed in a separate chapter.

While enriched white bread must conform to standards of identity, no such standards exist for variety breads. Use of enriched white flour, however, results in several of the variety breads being enriched.

Variety breads (breads other than the standard white pan bread) represent about one-third of the total bread market in the United States today. Several of these products, obtained from four major U.S. cities, were recently analyzed (7–9) for various nutrients (Table 5). It is apparent that these products are good sources of protein and fiber in our diet (fiber values listed are insoluble fiber values). They are also a good source of several vitamins and minerals, yet they are low in calories. The sodium content of these products, however, tends to be high. Some reduction of this sodium may be in order in view of the relationship between sodium and high blood pressure in sodium-sensitive individuals.

VI. NUTRIENTS IN PASTA PRODUCTS

Pasta products are usually made with semolina, which is milled from durum wheat. Several types of pasta products are marketed, but spaghetti, egg noodles, and macaroni are the most popular. These are usually made with semolina enriched according to the standards discussed in a separate chapter. Several brands of these products, obtained from major U.S. cities, were recently analyzed (10). Table 6 summarizes this information (cooked products only). Since the products were cooked without fat or salt added, they were low in fat and virtually free of sodium. Considered collectively, one 5.8 oz serving of the cooked products would provide 20% or more of the U.S. RDA (or safe intake) of thiamin and manganese, 15% or more of iron and niacin, 10% or more of protein, copper, phosphorus, and riboflavin, and 5% or more of zinc and magnesium. These contributions consider nutrient losses (mainly potassium and water-soluble vitamins) that occurred during the cooking process.

VII. NUTRIENTS IN FLOUR AND CORN TORTILLAS

Tortillas made with wheat flour or corn are an important food item in the diet of many Americans. As values listed in Table 7 show (11), they can be a significant source of various nutrients in our diet and they are low in calories. Unlike flour tortillas, salt is rarely added in the making of corn tortillas. Because the corn used in tortilla production is lime treated, the calcium content of corn tortillas is quite substantial. The process of tortilla making also renders niacin in the corn (which is mostly bound and, thus, poorly available) more absorbable. As a result pellagra (niacin deficiency disease) is rarely seen in populations where tortillas are consumed in several forms—as bread, fried, filled, and topped.

VIII. NUTRIENTS IN SELECTED SWEET GOODS, SNACK FOODS, AND FAST FOODS

While bread, pastas, tortillas, and other staple food items such as rice, oats, sorghum, and millets are nutrient-dense foods, such is not the case for sweet goods and several of the snack and fast food items where flour is a major ingredient. This is because these products tend to be high in fat and/or sugar and, as such, are high in calories in proportion to the nutrients which they contain. Many of us enjoy these products, and, as a result, the U.S. Department of Agriculture has grouped these foods under a new food category called the "Fats/Sweets/Alcohol" group. Table 8 lists the nutrient composition of some selected sweet goods, snack food, and fast food items (6,12,13).

IX. NUTRIENTS IN BREAKFAST CEREALS

Breakfast cereals are traditionally divided into two major categories; ready-to-eat and to-be-cooked cereals. Table 9 presents nutritional information on selected ready-to-eat cereals. These values are based on information gathered from personal communications (14). Values shown are sample means. The uniformity of some values is the consequence of label declaration (not determined values) for added nutrients. Over 90% of the ready-to-eat breakfast cereals produced in the United States today are nutrient fortified. Several breakfast cereals tend to be good sources of fiber and a number of micronutrients. However, they also seem to be high in sodium, which results from ingredients used during processing.

X. NONWHEAT CEREAL FOODS

Besides wheat, rice and corn are the other major cereals widely consumed worldwide. In most Asian countries, the consumption of cooked rice exceeds 1/2 lb a day. Although white rice is universally preferred, it is low in several nutrients (1) as compared to brown rice (Table 10). In the United States, white rice is routinely enriched according to the standards discussed in a separate chapter, with the exception that riboflavin (imparts color) is omitted. As discussed earlier, corn-based foods (tortillas, for example) are the dietary staple for many in Latin America and other countries. Corn, however, has a lower nutritive value than most other grains, particularly because of the lower content of protein and niacin. The nutritional quality of oats is superior to other cereal grains, but the consumption of oats in the United States other than as a breakfast cereal, is modest. Other than beer, the use of barley in food products is also modest. Although rye is the main bread grain of many European countries, its consumption in North America is also modest. Triticale (cross-breeding of rye and wheat) flours are now commercially available in the United States but have not found wide acceptance in food products. Sorghum and the millets are the staple foods in many countries of Asia, Central America, and Africa. In the United States, however, only about 1% of the crop produced is used for food purposes.

XI. ROLE OF CEREAL-BASED FOODS IN HEALTH AND DISEASE

While infectious and deficiency diseases have virtually disappeared in the North American population, we are now plagued with the high incidence of chronic degenerative diseases such as heart disease, colon and breast cancer, stroke, diabetes, osteoporosis, and hypertension. These are multifactorial diseases, but diet may play a significant role in their prevention and management. Toward this end, several dietary goal and guideline reports have been published in recent years. In 1985, the U.S. Department of Agriculture and the Department of Health and Human Services published a joint report (revised version of a 1980 report) titled "Dietary Guidelines for Americans" (15). Table 11 lists these guidelines and briefly summarizes the role that cereal-based foods can play in achieving these guidelines.

XII. CONCLUSIONS

Cereal-based foods provide about one-fourth of the total energy in the American diet. They are also an important source of the many of the 40 or so nutrients we need for good health. The quality of protein in cereal-based foods is low as compared to animal-derived foods, but this has no nutritional consequence in the American diet, which traditionally includes many animal-derived foods.

Processing of grains to produce food products may result in nutrient losses, but processing can also improve the nutritional profile of foods through inactivation of undesirable microorganisms and certain antinutrients. Nutritional profile also improves when cereal-based foods are nutrified with selected nutrients or processing aids, which are also nutrients. Several dietary recommendations made in recent years suggest that we include more complex carbohydrates (starches and fiber) in our diet; grain-based foods are a major source of complex carbohydrates.

TABLE 1 Chemical Composition of Whole Grains

Grain	Proximate components (g/100 g)						Micronutrients (mg/100 g)							
	Water	Protein	Fat	Ash	Fiber[a] (crude)	Carbo-hydrates (total)	Calcium	Phos-phorus	Iron	Sodium	Potassium	Thiamin	Ribo-flavin	Niacin
Wheat (hard red)	13.0	14.0	2.2	1.7	2.3	69.1	36	383	3.1	3	370	0.57	0.12	4.3
Wheat (soft red)	14.0	10.2	2.0	1.7	2.3	72.1	42	400	3.5	3	376	0.43	0.11	3.6
Wheat (durum)	13.0	12.7	2.5	1.7	1.8	70.1	37	386	4.3	3	435	0.66	0.12	4.4
Corn (field)	13.8	8.9	3.9	1.2	2.0	72.2	22	268	2.1	1	284	0.37	0.12	2.2
Rice (brown)	12.0	7.5	1.9	1.2	0.9	77.4	32	221	1.6	9	214	0.34	0.05	4.7
Oats	8.3	14.2	7.4	1.9	1.2	68.2	53	405	4.5	2	352	0.60	0.14	1.0
Barley (pearled)	11.1	8.2	1.0	0.9	0.5	78.8	16	189	2.0	3	160	0.12	0.05	3.1
Rye	11.0	12.1	1.7	1.8	2.0	73.4	38	376	3.7	1	467	0.43	0.22	1.6
Sorghum	11.0	11.0	3.3	1.7	1.7	73.0	28	287	4.4	—	350	0.38	0.15	3.9
Millet (Proso)	11.8	9.9	2.9	2.5	3.2	72.9	20	311	6.8	—	430	0.73	0.38	2.3
Triticale	12.0	12.8	1.9	1.7	2.2	71.6	29	429	4.5	4	385	0.81	0.27	1.4

[a]Total dietary fiber values are several-fold higher than crude fiber values.
Sources: Refs. 1–3.

TABLE 2 Total Dietary Fiber and Soluble Fiber Content of Selected Cereal-Based Foods

Product	Fiber (g/100 g)	
	Total	Soluble
Rye bread	3.5	1.2
White bread	1.8	0.9
Whole wheat bread	8.5	1.9
Graham crackers	10.1	1.8
Saltine crackers	3.9	1.6
Barley (pearled)	10.8	2.8
Macaroni (cooked)	0.9	0.2
White rice (cooked)	0.8	0.1
Brown rice (cooked)	2.4	0.2
All bran cereal	30.8	5.9
Corn bran cereal	21.0	5.2
Corn flakes	1.6	0.6
Oat bran (dry)	14.9	7.2
Quick oats (cooked)	1.6	0.8
Puffed wheat	7.2	3.4
Shredded wheat	11.8	1.6
Wheat germ (plain)	19.6	4.7

Source: Ref. 4.

TABLE 3 Composition of Flours of Various Extractions in Series Milled from Manitoba Wheat (14% Moisture Basis)

	Extraction of flour (%)			
	70	80	85	100
Proximate Components (%)				
Protein (N × 5.7)	12.9	13.4	13.7	13.8
Fat	1.17	1.45	1.72	2.52
Ash	0.41	0.60	0.76	1.55
Fiber (crude)	Trace	0.13	0.33	2.17
Carbohydrates (as starches)	70.9	69.6	68.0	63.7
Micronutrients (mg/100 g)				
Thiamin	0.07	0.21	0.29	0.37
Riboflavin	0.07	0.08	0.10	0.17
Niacin	0.85	1.11	1.35	5.56
Iron	1.42	1.67	2.24	3.08
Sodium	2.1	2.1	2.9	3.4
Potassium	83	113	148	316
Calcium	13	16	19	28
Magnesium	27	45	63	143
Copper	0.22	0.27	0.36	0.65
Zinc	1.17	1.65	2.18	3.77
Phosphorus	98	141	190	350
Phytate phosphorus	30	64	97	345
Chlorine	48	49	45	39

Source: Ref. 5.

TABLE 4 Nutrients in a 25-g Slice of Commercially Produced Bread

	White bread (enriched)	Whole wheat bread
Energy (kcal)	67	61
Proximate Components (g)		
Water	9.2	9.6
Protein	2.1	2.4
Fat	1.0	1.1
Ash	0.5	0.6
Fiber (total dietary)	0.6	2.0
Carbohydrates	12.2	11.4
Micronutrients (mg)		
Calcium	31.5	18.0
Iron	0.7	0.9
Magnesium	5.2	23.2
Phosphorus	27	65
Potassium	28	44
Sodium	128	159
Copper	0.03	0.08
Zinc	0.15	0.42
Thiamin	0.12	0.09
Riboflavin	0.08	0.05
Niacin	0.94	0.96
Pantothenic acid	0.11	0.18
Vitamin B_6	0.01	0.05
Folic acid	0.009	0.014

Source: Ref. 6.

TABLE 5 Nutrient Profile of Selected Variety Breads Commercially Produced in the United States

	Bread Type								
	Pumpernickel	Raisin	Oatmeal	Whole wheat (100%)	Cracked wheat	Mixed grain	Italian	French	Pita
Proximate Components (%)									
Moisture	39.1	33.7	37.1	39.4	36.2	38.5	36.8	35.9	35.7
Protein	8.4	7.8	8.4	9.9	8.8	9.3	9.5	9.5	9.1
Fat	2.8	4.4	4.4	4.1	4.0	3.9	3.4	2.7	1.8
Ash	2.6	1.8	2.1	2.4	2.1	2.3	2.1	2.0	1.7
Dietary fiber (insoluble)	4.2	2.2	2.2	5.7	3.5	4.3	1.2	1.1	1.0
Carbohydrates (available)	43.0	50.1	45.9	38.5	45.5	41.8	47.0	48.9	50.8
Energy (kcal/100 g)	231	272	257	230	253	239	257	258	256
Micronutrients (mg/100 g)									
Calcium	56	68	62	70	52	84	84	75	82
Phosphorus	164	108	123	222	153	166	108	101	90
Magnesium	52	27	37	88	51	58	29	27	24
Iron	3.8	3.2	3.1	3.3	3.4	3.5	2.9	3.2	2.4
Zinc	1.6	0.8	1.0	2.0	1.3	1.4	1.0	0.9	0.9
Copper	0.26	0.20	0.21	0.30	0.21	0.26	0.20	0.22	0.16
Manganese	1.3	0.5	1.0	2.5	1.4	1.4	0.5	0.5	0.5
Sodium	695	421	601	553	544	588	616	599	477
Potassium	204	255	125	260	174	198	103	115	113
Thiamin	0.29	0.29	0.35	0.22	0.38	0.27	0.34	0.39	0.47
Riboflavin	0.33	0.23	0.23	0.15	0.23	0.24	0.23	0.24	0.25
Niacin	3.6	2.4	3.1	4.4	4.2	4.3	4.2	4.4	4.1
Folic acid	0.04	0.04	0.03	0.05	0.04	0.05	0.03	0.03	0.02
Vitamin B_6	0.18	0.05	0.05	0.18	0.10	0.13	0.05	0.05	0.27
Pantothenic acid	0.37	0.34	0.36	0.65	0.56	0.53	0.39	0.42	0.43

Source: Refs. 7–9.

TABLE 6 Nutrient Profile of Cooked Pasta Products

	Spaghetti	Egg noodles	Macaroni
Proximate Components (%)			
Moisture	64.0	70.6	66.1
Protein	5.4	4.9	4.9
Fat	0.8	1.4	0.7
Ash	0.3	0.3	0.2
Dietary fiber (insoluble)	0.7	0.6	0.6
Carbohydrates	28.8	22.2	27.5
Energy (kcal/100 g)	147	123	138
Micronutrients (mg/100 g)			
Calcium	7	10	7
Phosphorous	61	69	55
Magnesium	18	19	16
Iron	1.5	1.7	1.7
Zinc	0.5	0.6	0.5
Copper	0.1	0.1	0.1
Manganese	0.3	0.3	0.3
Sodium	0	3	0
Potassium	34	26	31
Thiamin	0.27	0.24	0.22
Riboflavin	0.10	0.09	0.11
Niacin	2.13	1.93	2.24
Pantothenic acid	0.16	0.16	0.11
Vitamin B6	0.03	0.03	0.03
Folic acid	0.009	0.01	0.008

Source: Ref. 10.

TABLE 7 Nutrients in Flour and Corn Tortillas

	Flour tortillas	Corn tortillas
Proximate Components (%)		
Moisture	25.2–29.3	45.2
Protein	7–8.1	5.2
Fat	3.6–10	3.1
Ash	1.7–2.9	1.4
Dietary fiber (insoluble)	NA	4.1
Carbohydrates	51.3–59.0	41.1
Energy Content (kcal/100 g)	282–313	206
Micronutrients (mg/100 g)		
Phosphorus	61–98	269
Calcium	27–103	205
Magnesium	NA	63
Iron	1–1.9	1.5
Zinc	NA	1.1
Copper	NA	0.13
Manganese	NA	0.40
Sodium	371–781	9
Potassium	89–111	185
Thiamin	0.18–0.42	0.08
Riboflavin	0.07–0.32	0.05
Niacin	1.6–4.3	2.11
Folic acid	NA	0.015
Pantothenic acid	NA	0.16
Vitamin B_6	NA	0.27

NA = not available.
Source: Ref. 11.

TABLE 8 Nutrients in Selected Sweet Goods, Snack Foods, and Fast Foods

	White cake (chocolate icing)	Pie (apple)	Doughnut (cake)	Cookies (chocolate chips)	Crackers (saltines)	Bagels (plain)	Pizza (sausage)	Biscuit (milk added)
Proximate Components (%)								
Moisture	20.2	50.9	21.2	4.7	4.0	33.6	44.2	28.7
Protein	4.3	2.0	5.1	4.8	9.3	10.3	10.0	7.4
Fat	15.5	10.1	23.1	23.0	9.9	1.4	10.2	11.9
Ash	1.3	0.7	1.7	1.3	3.1	1.7	NA	3.6
Fiber (crude)	0.2	0.5	0.2	0.1	0.4	0.2	NA	0.2
Fiber (insoluble)	NA	NA	NA	NA	NA	0.8	NA	NA
Carbohydrates (total)	58.7	36.4	49.0	66.2	73.7	53.0	33.1	48.4
Energy (kcal/100 g)	382	239	419	473	428	263	264	333
Micronutrients (mg/100 g)								
Calcium	79	9	44	28	59	39	130	208
Phosphorus	74	23	220	71	94	100	149	456
Iron	0.96	0.90	1.46	2.17	2.62	3.33	1.00	2.05
Magnesium	19	8	23	24	28	30	NA	24
Zinc	0.37	0.17	0.51	0.50	0.62	0.89	NA	0.63
Copper	0.11	0.05	0.13	0.11	0.17	0.19	NA	0.11
Manganese	NA	NA	NA	NA	NA	0.58	NA	NA
Sodium	281	153	557	360	1335	494	663	935
Potassium	108	85	109	129	136	94	125	200
Thiamin	0.10	0.11	0.24	0.13	0.40	0.39	0.12	0.43
Riboflavin	0.15	0.08	0.21	0.21	0.56	0.25	0.17	0.38
Niacin	0.95	0.92	1.75	1.86	6.54	4.59	1.26	3.00
Folic acid	0.005	0.005	0.008	0.009	0.018	0.020	NA	0.006
Pantothenic acid	0.20	0.15	0.39	0.15	0.27	0.39	NA	0.36
Vitamin B_6	0.024	0.026	0.035	0.018	0.043	NA	NA	0.045

NA = not available.
Sources: Refs. 6,12,13.

TABLE 9 Nutrients in Selected Ready-to-Eat Breakfast Cereals

	All-Bran	Corn Flakes	Grape Nuts	Shredded Wheat	Rice Crispies
Proximate Component (%)					
Moisture	3.0	2.6	3.2	6.4	2.3
Protein	14.3	8.1	11.7	10.9	6.8
Fat	1.8	0.3	0.4	1.3	0.7
Ash	6.5	2.9	2.7	1.5	2.9
Fiber (crude)	7.0	0.4	1.6	2.3	0.3
Fiber (insoluble dietary)	29.9	1.1	4.8	9.3	0.2
Carbohydrates (total)	74.4	86.1	82.0	79.8	87.4
Energy (kcal/100 g)	249	389	357	352	395
Micronutrients (mg/100 g)					
Calcium	81	3	38	41	14
Iron	15.9	6.3	4.3	3.2	6.3
Magnesium	373	12	67	170	36
Phosphorus	932	63	251	363	121
Potassium	1234	92	334	327	104
Sodium	1128	1238	695	2	1200
Zinc	13.2	0.3	2.2	2.5	1.7
Copper	1.14	0.07	0.33	0.50	0.25
Manganese	NA	0.08	NA	3.07	0.99
Ascorbic acid	53	53	NA	NA	53
Thiamin	1.3	1.3	1.3	0.3	1.3
Riboflavin	1.5	1.5	1.5	0.3	1.5
Niacin	17.6	17.6	17.6	4.6	17.6
Pantothenic acid	1.71	0.18	0.95	0.81	0.70
Vitamin B_6	1.8	1.8	1.8	0.3	1.8
Folacin	0.35	0.35	0.35	0.05	0.35
Vitamin B_{12}	NA	NA	5.3	NA	NA
Vitamin A (IU)	4409[a]	4409	4409	NA	4409

NA = not available
[a]Expressed in International Units (IU).
Source: Ref. 14.

TABLE 10 Nutrients in Brown and White Rice

	Brown rice	White rice
Proximate Components (%)		
Moisture	12.0	12.0
Protein	7.5	6.7
Fat	1.9	0.4
Ash	1.2	0.5
Fiber (crude)	0.9	0.3
Carbohydrates (total)	77.4	80.4
Energy (kcal/100 g)	360	363
Micronutrients (mg/100 g)		
Calcium	32	24
Phosphorus	221	94
Iron	1.6	0.8
Sodium	9	5
Potassium	214	92
Thiamin	0.34	0.07
Riboflavin	0.05	0.03
Niacin	4.7	1.6

Source: Ref. 1.

TABLE 11 Dietary Guidelines for Americans and the Role of Cereal-Based Foods in Achieving These

Guidelines	Cereal-based foods
Eat a variety of foods	Constitute one of the four basic food groups.
Maintain desirable weight	Most are nutrient, and not calorie dense. Thus, they provide satiety and prevent overconsumption of calories.
Avoid too much fat, saturated fat, and cholesterol	Whole grains are low in fat and saturated fat and contain no cholesterol. Most cereal-based foods (except sweet goods and snack foods) also conform to this pattern.
Eat foods with adequate starch and fiber	Are the major source of starches and fiber (complex carbohydrates) in our diet.
Avoid too much sugar	Are low in sugar with the exception of sweet goods.
Avoid too much sodium	Whole grains are virtually free of sodium. While most processed foods are not, some reduction in their sodium content can be achieved through reformulation.
If you drink alcoholic beverages, do so in moderation	Do not contain alcohol.

Source: Ref. 15.

REFERENCES

1. Watt, B. K., and Merrill, A. L., *Composition of Foods: Raw, Processed, Prepared.* Ag. Handbook No. 8, U.S. Department of Agriculture, Washington, D.C., 1975.
2. Lorenz, K., Reuter, F. W., and Sizer, C., The mineral composition of triticales and triticale milling fractions by X-ray fluorescence and atomic absorption. *Cereal Chem.*, *51*:534 (1974).
3. Michela, P., and Lorenz, K., The vitamins of triticale, wheat and rye. *Cereal Chem.*, *53*:853 (1976).
4. Anderson, J. W., *Plant Fiber in Foods*. HCF Diabetes Research Foundation, Inc., Lexington, KY, 1986.
5. McCance, R. A., Widdowson, E. M., Moran, T., Pringle, W. J. S., and Macrae, T. F., The chemical composition of wheat and rye and of flours derived therefrom. *Biochem. J.*, *39*:213 (1945).

6. McQuilkin, C., and Matthews, R. H., *Provisional Tables on the Nutrient Content of Bakery Foods and Related Items*. Nutrient Data Research Group, U.S. Dept. Agriculture, Washington, D.C., 1981.
7. Ranhotra, G., Gelroth, J., Novak, F., Bohannon, F., and Matthews, R., Proximate components in selected variety breads commercially produced in major U.S. cities. *J. Food Sci., 49*:642 (1984).
8. Ranhotra, G., Gelroth, J., Novak, F., and Matthews, R., Minerals in selected variety breads commercially produced in four major U.S. cities. *J. Food Sci., 50*:365 (1985).
9. Ranhotra, G., Gelroth, J., Novak, F., and Matthews, R., B vitamins in selected variety breads commercially produced in major U.S. cities. *J. Food Sci., 50*:1174 (1985).
10. Ranhotra, G. S., Gelroth, J. A., and Novak, F. A., Nutrient profile of pasta products commercially produced in the United States. *Nutr. Rep. International, 33*:761 (1986).
11. Ranhotra, G. S., Nutritional profile of corn and flour tortillas. *Cereal Foods World, 29*:729 (1985).
12. Ranhotra, G., Pizza products: Nutritional and dietary aspects. *Cereal Foods World, 30*:704 (1984).
13. Ranhotra, G. S., Gelroth, J. A., and Novak, F. A., Nutritional composition of bagels commercially produced in the United States. *Cereal Foods World, 30*:209 (1985).
14. Douglass, J. S., Matthews, R. H., and Hepburn, F. N., *Composition of Foods: (Breakfast Cereals) Raw, Processed, Prepared.* Agri. Handbook No. 8-8. Human Nutrition Information Service, U.S. Department of Agriculture, Washington, DC, 1982.
15. USDA/USDHHS. *Nutrition and Your Health: Dietary Guidelines for Americans*. U.S. Department of Agriculture and U.S. Department of Health and Human Services, Washington, D.C., 1985.

24

Ethanol Production from Cereal Grains

Bruce E. Dale
Texas A&M University, College Station, Texas

I. INTRODUCTION

Ethanol has been used as a fuel for internal combustion engines since their invention. Reports on the use of alcohol as a motor fuel were published in 1907 and detailed research was conducted in the 1920s and 1930s. Interest in ethanol as a motor fuel has followed cycles of fuel shortages and/or low feed grain prices. Any material that contains sugar can potentially be fermented to produce several kinds of alcohols. This chapter discusses only the use of cereal grains to make ethanol, with an emphasis on ethanol from maize (corn).

II. BASIC FLOWSHEET FOR ETHANOL PRODUCTION FROM CEREAL GRAINS

The conversion of sugar into ethanol is represented by the following equation:

$$C_6H_{12}O_6 \rightarrow 2C_2H_5OH + 2CO_2 + \text{heat}$$
$$\text{1 glucose} \rightarrow \text{2 ethanol} + \text{2 carbon dioxide}$$

The maximum possible yield is about 51 pounds of ethanol from every 100 pounds of glucose. The attainable yield is somewhat less, about 47–49 pounds of ethanol per 100 pounds of glucose. Approximately equal weights of (a) ethanol, (b) a protein-enriched fermentation residue, and (c) carbon dioxide are produced per unit dry weight of raw material when maize, sorghum, and wheat are the feedstocks (raw materials). The technology to make ethanol from grain is well established, and details are available in numerous handbooks. An outline of the process technology is given in Figure 1.

III. ETHANOL YIELDS FROM VARIOUS CEREAL GRAINS

While maize is the most widely used raw material for ethanol production, as outlined in Figure 1, many other grains can and have been used. The probable ethanol yields (1,2) from

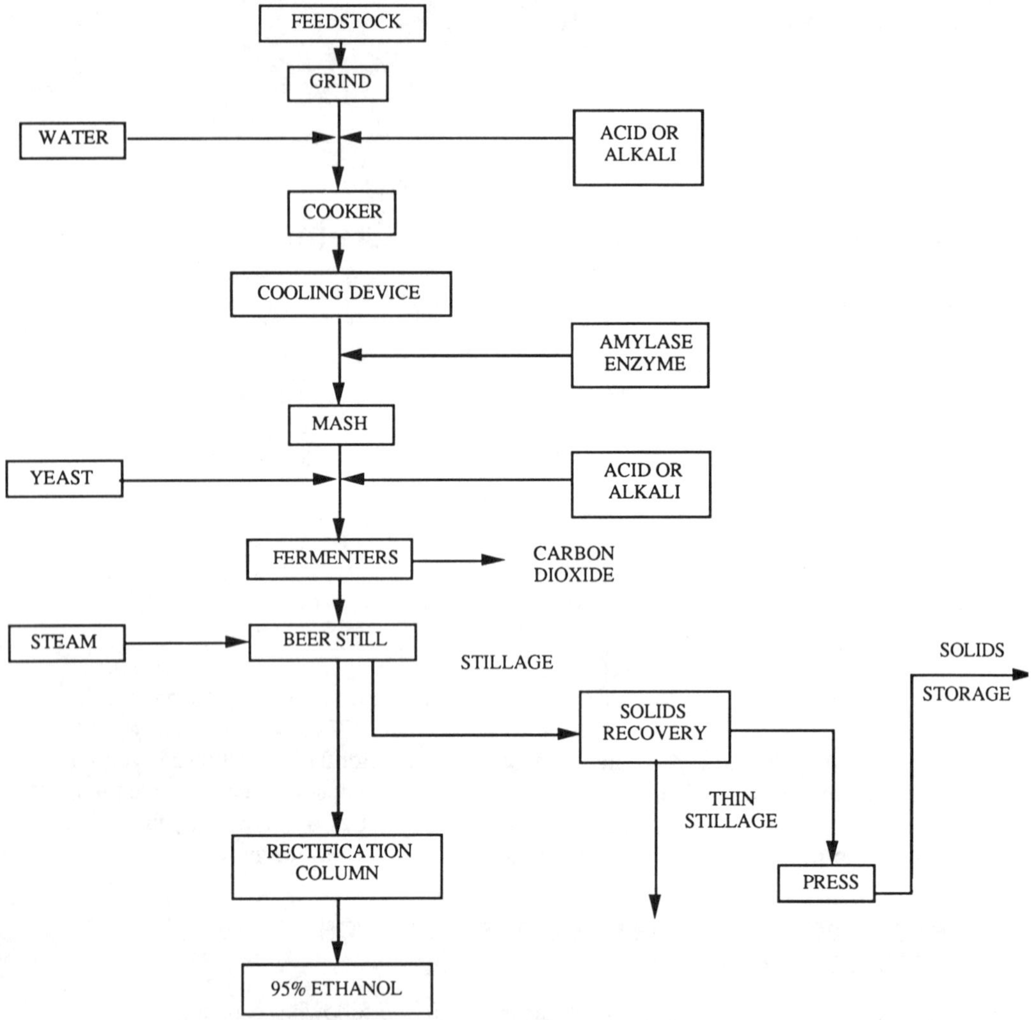

FIGURE 1 Simplified flowsheet for ethanol production from cereal grains.

the various cereal grains are summarized in Table 1. While these yields vary, they tend to be in the range of 1000 liters/ha/yr.

IV. ECONOMICS OF ETHANOL PRODUCTION

The cost of ethanol production are composed of capital (investment) costs and operating (production) costs. All costs given here are in U.S. dollars. The capital costs were approximately \$1.50 to \$2.00 per annual gallon of capacity in 1983 (1). The capital costs of smaller plants are proportionally more than those for larger plants per gallon of annual capacity. The cost of raw materials makes up 50–70% of the operating costs of grain ethanol plants. Therefore, such plants are extremely sensitive to changes in raw material prices. The cost of raw material ranges from about \$1.10 per gallon of ethanol from sorghum and maize to about \$2.90 per gallon from rice (2). Feedstock costs per gallon of ethanol from various cereal grains are given in Table 2 along with the average conversion rates and prices paid for the grain upon which these feedstock costs are based (2). Other cost data are summarized in Table 3 for various sizes of ethanol production facilities (2).

V. COMPARISON OF ETHANOL WITH OTHER FUELS

As mentioned above, ethanol has been used as a motor fuel for many years. Brazil in particular operates a large fraction of its automobile fleet on ethanol or gasoline-ethanol blends ("gasohol"). There are a number of advantages of ethanol compared with other fuels, particularly gasoline. For instance, ethanol has a very high octane number compared to gasoline, thereby increasing engine power and reducing noise. In addition, ethanol burns more cleanly than gasoline and produces lower levels of carbon monoxide, oxides of nitrogen, and total hydrocarbon emissions than does gasoline. The lower vapor pressure of ethanol also reduces direct emissions to the air and therefore minimizes air pollution in this important way as well. When ethanol is blended with gasoline, the overall improvement in the octane rating and reduction in air pollution are roughly proportional to the fraction of ethanol in the blended fuel. These and other reasons have caused cities such as Denver to require the use of oxygenated fuels such as gasohol during times of peak air pollution. Ethanol therefore commands a price premium compared to gasoline and is marketed for these characteristics in the United States. Over 800 million gallons of ethanol were blended with U.S. gasoline supplies in 1988.

There are also some significant disadvantages of ethanol when used in engines designed to burn gasoline or in gasoline-ethanol blends. Among these are vapor lock as well as difficulties in starting and extended periods of engine sluggishness (particularly in colder weather). The most serious disadvantage in blending gasoline and ethanol is that under some conditions the alcohol may separate from the gasoline. The tendency to separate is increased at lower temperatures, as the quantity of water absorbed increases, and also as the quality of the gasoline decreases. Research and experience have shown that most of the problems with gasoline-ethanol blends can be overcome by proper adjustment and maintenance of the engine. Ethanol used alone is an excellent fuel with many advantages over gasoline. However, a larger volume of ethanol will always be required to travel the same distance. Some of the key fuel properties of ethanol, gasoline, and gasoline-ethanol blends are found in Table 4.

VI. OTHER MARKETS FOR ETHANOL

In addition to its fuel uses mentioned above, ethanol is an important industrial chemical in its own right. Tax-paid fermentation alcohol has medicinal, pharmaceutical, and flavoring applications. To prevent illegal consumption as a beverage, much alcohol is denatured (treated with other chemicals such as benzene, methyl isobutylketone, etc.) and used industrially. Denatured alcohol is found in antifreeze and as a general purpose solvent. In addition, ethanol is a raw material for hundreds of chemicals. Among the most important of these are acetaldehyde, ethyl acetate, acetic acid, and glycols. These chemicals can be manufactured by both glucose fermentation and synthetic (from hydrocarbons) routes. For example, most industrial ethanol is currently manufactured from ethylene. In some circumstances, it has been found desirable to reverse this process and make ethylene, a very high volume chemical (U.S. production of ethylene was over 16 million tons in 1988) from ethanol. If sufficiently inexpensive ethanol is available, this is a potential product of great importance. Another interesting option is to produce synthetic gasoline from alcohols such as methanol and ethanol by Mobil Corporation's MTG process (4). For instance, New Zealand now produces a large quantity of synthetic gasoline in this way from methanol derived from large offshore natural gas fields.

VII. OTHER PRODUCTS FROM GLUCOSE

It is important to realize that the sugars generated from starch hydrolysis can be fermented to a wide variety of products, taking advantage of the inherent flexibility of fermentation. Of the many possible products of glucose fermentation, those which conserve as much of the mass of the sugar feedstock molecule as possible will have an inherent economic advantage. Therefore, oxychemicals rather than the more reduced hydrocarbon-type chemicals are preferred. Table 5 summarizes some of the more important oxychemicals potentially available by fermentation, their recent production history and cost, and also the major uses or derivatives of each oxychemical. These materials cost in the range of $0.50 to $1.00 (U.S.) per pound. At current corn prices (approximately $3.00 per bushel) the cost of preparing a fermentable sugar syrup from corn is in the neighborhood of $0.10 to $0.15 per pound of sugar when byproduct credits are taken. These price differentials between the sugar feedstock and the oxychemical product require very efficient conversion and separation technologies. A dominant economic factor is the yield of oxychemical product per unit weight of the sugar feedstock. For well-developed fermentation processes, the cost of the sugar feedstock generally represents approximately 30–60% of the total production costs (5). In addition to the oxychemicals listed in Table 5, some of the other candidates for production from glucose are lactic acid, polyhydroxybutyric acid, and single cell protein.

VIII. VALUE AND MARKETS FOR ETHANOL INDUSTRY BY-PRODUCTS

Fermentation alcohol plants produce two principal byproducts: the solid residue that remains after the alcohol has been extracted and carbon dioxide. This solid residue, called distillers' grain (DDG) or spent grains, is produced at the rate of about 18 dry pounds per bushel of corn fermented. It is a valuable material and is commonly used as a livestock feed supplement. The economics of ethanol fuel production are strongly dependent on marketing this valuable by product which is worth in the range of $110 to $150 per ton depending on the price of competing feeds. An additional premium in value may be justified due to the fact that much of the DDG protein will resist breakdown by rumen microorganisms and will pass down to the intestinal tract where it can be directly absorbed by the animal. Sales of distillers' grains also help insulate a fermentation ethanol facility from the adverse effects of rising costs of grain.

For every gallon of ethanol produced, approximately 6.3 pounds of carbon dioxide (CO_2) will also be formed. Carbon dioxide is used in many oil fields to recover additional oil. It can also be combined with hydrogen gas to produce methanol. Carbon dioxide is frequently used to preserve the quality of many agricultural products and as a refrigerant. It has obvious uses in the manufacture of carbonated beverages. The possible uses (and hence the possible value) of CO_2 will be very site-specific because costs of CO_2 recovery are about equal to the value of the product (approximately $30 per ton). However, some ethanol plants are located such that they can profitably recover and sell their carbon dioxide.

IX. CONCLUSIONS

Ethanol use as a motor fuel, either by itself or blended with other fuels, is well established. Ethanol has some characteristics that make it particularly valuable as a fuel or in a blend.

Likewise, the technology of ethanol production from cereal grains is well established and can be practiced at various scales. There are a variety of other possible uses of ethanol in addition to the fuel uses. Many other oxychemicals can potentially be produced from the glucose derived from grains. The cost of ethanol and other oxychemicals from glucose fermentation is strongly dependent on the cost of the grain.

TABLE 1 Ethanol Yields from Various Cereal Grains

Crop	Grain yield (ton/ha/yr)	Ethanol yield from grain (liters/ton)	Ethanol yield (liters/ha/yr)
Maize	1.7–5.4	360	600–1,940
Wheat	1.5–2.1	340	510–710
Barley	1.2–2.5	250	300–625
Rice	2.5–5.0	430	1,100–2,200
Sorghum	1.10–3.7	350	350–1,300
Oats	2.4–4.8	240	580–1,160

TABLE 2 Approximate Raw Material Costs for Ethanol Production from Cereal Grains

Feedstock	Average feedstock cost U.S. dollars/unit (1979)	Dollars (U.S.) per gallon of ethanol
Maize	2.69/bushel	1.14
Sorghum	2.40/bushel	1.08
Wheat	3.46/bushel	1.36
Rye	2.36/bushel	1.07
Oats	1.46/bushel	1.43
Barley	2.35/bushel	1.24
Rice	228.80/bushel	2.88

TABLE 3 Approximate Costs of Ethanol Production in Various Facilities (U.S. Dollars, 1979)

Cost item	Small on-farm[a]	Large on-farm[a]	Large community with DDGS[b]
Feedstock	1.00	1.00	1.10
Labor and utilities	0.33	0.24	0.26
Supplies	0.09	0.09	0.09
Indirect costs	0.16	0.11	0.11
Byproduct credit	(0.54)	(0.48)	(0.46)
Capitol costs	0.30	0.17	0.19
Total cost	1.34	1.13	1.29

[a]Ethanol from on-farm units is assumed to be 95% ethanol, 5% water.
[b]Ethanol is assumed anhydrous, the byproduct protein feed is dried.

TABLE 4 Properties of Ethanol and Other Fuels

Property	Gasoline	Ethanol	Gasohol[a]
Density (lb/gal)	6.2	6.5	6.2
Energy content (Btu/gal)	121,000	84,600	117,400
Pump octane	87–98	98–102	90–91
Intake air/fuel ratio	14.9	9.0	14.6
Heat of vaporization (Btu/lb)	176	396	180
Vapor pressure (Psia at 100°F)	7.2	2.5	8.2

[a]Defined as 10% anhydrous ethanol and 90% gasoline (volume/volume).

TABLE 5 Oxychemicals from Glucose

Chemical	1981 U.S. production (million pounds)	Current price ($ U.S./ pound)	1981 commercial value (million dollars)	Major use or derivative
Ethylene glycol	4,055	0.275 to 0.285	1,281	Polyethylene terephthalate, antifreeze
Acetic acid	2,706	0.265	511	Vinyl acetate, cellulose acetate
Acetone	2,167	0.31	483	Solvents, methyl, and other methacrylates
Isopropyl alcohol	1,644	0.31	507	Acetone,solvents
Adipic acid	1,210	0.57	653	Nylon 66
Butanol	823	0.335	251	Solvents, butyl acrylate
Acrylic acid	691	0.58	276	Polymers
Methyl ethyl ketone	626	0.37	260	Solvents
Propylene glycol	480	0.44	208	Unsaturated polyester resin
Glycerol	370	0.805	259	Drugs,cosmetics
Citric acid	235	0.71 to .775	192	Food,drugs

Source: Adapted from Ref.7.

REFERENCES

1. National Research Council, *Alcohol Fuels Options for Developing Countries*, National Academy Press, Washington, D.C., 1983, pp. 5,7.
2. United States Department of Agriculture, *Small Scale Fuel Alcohol Production*, U.S. Government Printing Office, Washington, D.C.., March 1980, p.V–13.
3. Parsons, R. A. (Ed.), *Feed and Fuel from Ethanol Production*, Northeast Regional Agricultural Engineering Service (NRAES), Riley Robb Hall, Cornell University, Ithaca, New York, 14853, March 1982, p. II–6.
4. Gasoline or olefins from an alcohol feed, *Chem. Eng.* (April 21):86 (1980).
5. Dale, B. E., and Linden, J. C., Fermentation substrates and economics, Chap. 6 in *Annual Reports on Fermentation Processes, Vol. 7*, Academic Press, New York, 1984, pp. 107–134.
6. Ng, T. K., Busche, R. M., McDonald, C. C., and Hardy, R. W. F., Production of feedstock chemicals, *Science*, *219*:733–740 (1983).

Index